Judy Lessler

D1531883

Business Survey Methods

WILEY SERIES IN PROBABILITY AND MATHEMATICAL STATISTICS

Established by WALTER A. SHEWHART and SAMUEL S. WILKS

Editors: *Vic Barnett, Ralph A. Bradley, Nicholas I. Fisher, J. Stuart Hunter, J. B. Kadane, David G. Kendall, David W. Scott, Adrian F. M. Smith, Jozef L. Teugels, Geoffrey S. Watson*

A complete list of the titles in this series appears at the end of this volume

Business Survey Methods

Edited by

BRENDA G. COX
Mathematica Policy Research, Inc.

DAVID A. BINDER
Statistics Canada

B. NANJAMMA CHINNAPPA
Statistics Canada

ANDERS CHRISTIANSON
Statistics Sweden

MICHAEL J. COLLEDGE
Australian Bureau of Statistics

PHILLIP S. KOTT
U.S. National Agricultural Statistics Service

A Wiley-Interscience Publication
JOHN WILEY & SONS, INC.
New York • Chichester • Brisbane • Toronto • Singapore

This text is printed on acid-free paper.

Copyright © 1995 by John Wiley & Sons, Inc.

All rights reserved. Published simultaneously in Canada.

Reproduction or translation of any part of this work beyond
that permitted by Section 107 or 108 of the 1976 United
States Copyright Act without the permission of the copyright
owner is unlawful. Requests for permission or further
information should be addressed to the Permissions Department,
John Wiley & Sons, Inc., 605 Third Avenue, New York, NY
10158-0012.

Library of Congress Cataloging in Publication Data:
Business Survey Methods / edited by
  Brenda G. Cox ... [et al.].
      p. cm. — (Wiley series in probability and mathematical
  statistics. Applied probability and statistics)
    "A Wiley-Interscience publication."
    Includes index.
    ISBN 0-471-59852-6
    1. Economic surveys—Congresses. 2. Social sciences—Statistical
  methods—Congresses. I. Cox, Brenda G.   II. Series.
  HA31.2.S874 1995
  330′.0723—dc20   94-20611
                                                               CIP

Printed in the United States of America

10 9 8 7 6 5 4 3 2

Contents

PART B SAMPLE DESIGN AND SELECTION

PART C DATA COLLECTION AND RESPONSE QUALITY

PART F PAST, PRESENT, AND FUTURE DIRECTIONS

Contributors

Rich Allen, U.S. National Agricultural Statistics Service, Washington, DC

Carole A. Ambler, U.S. Bureau of the Census, Suitland, Maryland

David Archer, Statistics New Zealand, Wellington, New Zealand

François Ben-Zur, Bank of Israel, Jerusalem, Israel

Paul P. Biemer, Research Triangle Institute, Research Triangle Park, North Carolina

David A. Binder, Statistics Canada, Ottawa, Canada

K. R. W. Brewer, Australian National University, Canberra, Australia

Ronald M. Carpenter, Statistics Canada, Ottawa, Canada

B. Nanjamma Chinnappa, Statistics Canada, Ottawa, Canada

G. H. Choudhry, Statistics Canada, Ottawa, Canada

Anders Christianson, Statistics Sweden, Stockholm, Sweden

Young I. Chun, U.S. Bureau of Labor Statistics, Washington, DC

Richard L. Clayton, U.S. Bureau of Labor Statistics, Washington, DC

Michael J. Colledge, Australian Bureau of Statistics, Belconnen, Australia

Brenda G. Cox, Mathematica Policy Research, Inc., Princeton, New Jersey

Lawrence H. Cox, U.S. Environmental Protection Agency, Research Triangle Park, North Carolina

Don A. Dillman, Washington State University, Pullman, Washington and U.S. Bureau of the Census, Suitland, Maryland

Cathryn S. Dippo, U.S. Bureau of Labor Statistics, Washington, DC

Ronald S. Fecso, U.S. National Agricultural Statistics Service, Washington, DC

Joseph K. Garrett, Nielsen Marketing Research, Northbrook, Illinois

Leopold Granquist, Statistics Sweden, Stockholm, Sweden

Greg Griffiths, Australian Bureau of Statistics, Belconnen, Australia

Rachel M. Harter, Nielsen Marketing Research, Northbrook, Illinois

Michael A. Hidiroglou, Statistics Canada, Ottawa, Canada

Vicki J. Higgins, U.S. Bureau of the Census, Suitland, Maryland

Stanley M. Hyman, U.S. Bureau of the Census, Suitland, Maryland

Ayah E. Johnson, Henry M. Jackson Foundation for the Advancement of Military Medicine, Rockville, Maryland

Ruth Ann Killion, U.S. Bureau of the Census, Suitland, Maryland

Phillip S. Kott, U.S. National Agricultural Statistics Service, Fairfax, Virginia

John G. Kovar, Statistics Canada, Ottawa, Canada

Sylvia Leaver, U.S. Bureau of Labor Statistics, Washington, DC

Hyunshik Lee, Statistics Canada, Ottawa, Canada

Rodney J. Lewington, Statistics New Zealand, Wellington, New Zealand

Susan Linacre, Australian Bureau of Statistics, Belconnen, Australia

Tzen-Ping Liu, Statistics Canada, Ottawa, Canada

Brian MacDonald, U.S. Bureau of Labor Statistics, Washington, DC

Thomas L. Mesenbourg, U.S. Bureau of the Census, Suitland, Maryland

Nash J. Monsour, U.S. Bureau of the Census, Suitland, Maryland

Shaila Nijhowne, Statistics Canada, Ottawa, Canada

Esbjörn Ohlsson, Stockholm University and Statistics Sweden, Stockholm, Sweden

M. Chris Paxson, Washington State University, Pullman, Washington

Danny Pfeffermann, Hebrew University, Jerusalem, Israel

Mark Pierzchala, U.S. National Agricultural Statistics Service, Fairfax, Virginia

Leon Pietsch, Australian Bureau of Statistics, Belconnen, Australia

Richard Platek, Statistics Canada (retired), Ottawa, Canada

J. N. K. Rao, Carleton University, Ottawa, Canada

Jacob Ryten, Statistics Canada, Ottawa, Canada

Joan Sander, U.S. Bureau of Labor Statistics, Washington, DC

Carl-Erik Särndal, University of Montreal, Montreal, Canada

Richard S. Sigman, U.S. Bureau of the Census, Suitland, Maryland

K. P. Srinath, Statistics Canada, Ottawa, Canada

Peter Struijs, Statistics Netherlands, Heerlen, Netherlands

John Tarnai, Washington State University, Pullman, Washington

Robert D. Tortora, U.S. Bureau of the Census, Suitland, Maryland

Richard Valliant, U.S. Bureau of Labor Statistics, Washington, DC

Frederic A. Vogel, U.S. National Agricultural Statistics Service, Washington, DC

George S. Werking, U.S. Bureau of Labor Statistics, Washington, DC

Patricia J. Whitridge, Statistics Canada, Ottawa, Canada

Ad Willeboordse, Statistics Netherlands, Voorburg, Netherlands

William E. Winkler, U.S. Bureau of the Census, Suitland, Maryland

David A. Worton, Statistics Canada (retired), Prescott, Canada

Preface

A nation's official statistics are directly affected by the quality of the data derived from its surveys of businesses, farms, and institutions. Yet methodology and standards for these surveys vary tremendously across countries and statistical agencies—unlike the situation for surveys of persons and households. Reasons for this disparity are diverse, but most relate to the difficult design and execution problems such surveys encounter for which solutions are not readily available in the research literature. The International Conference on Establishment Surveys (ICES) was organized to address this problem. This monograph is a product of that conference.

The idea of a conference on this topic grew out of a December 1990 telephone conversation between Brenda G. Cox, then a visiting research fellow at the U.S. National Agricultural Statistics Service, and B. Nanjamma Chinnappa, Director of Statistics Canada's Business Survey Methods Division. As associate editor of the *Journal of Business & Economic Statistics*, Cox was recruiting papers for a special session on business survey methods for the 1991 Joint Statistical Meetings. Having identified speakers from Statistics Canada, Cox and Chinnappa discussed how lack of published methods and communication among researchers was a stumbling block for progress in solving business surveys' unique problems. They decided that an international conference was needed to (1) provide a forum to describe methods in current use, (2) present new or improved technologies, and (3) promote international interchange of ideas.

Obtaining sponsors for the conference was the first step. An unusual feature of the ICES is its dual identity as the fourth conference in the series of survey research methods conferences organized by the American Statistical Association (ASA) and as the tenth symposium in the series of Statistics Canada international symposia.

Each year since 1984, Statistics Canada has organized an international symposium focusing on a specified topic in survey methods and systems. Past symposia have covered topics such as analysis of survey data (1984), small-area statistics (1985), missing data in surveys (1986), statistical uses of administrative data (1987), impact of high technology on survey taking (1988), analysis of data in time (1989), measurement and improvement of data quality (1990), spatial issues in statistics (1991), and the analysis of longitudinal surveys (1992). Chinnappa approached Assistant Chief Statistician Gordon J. Brackstone, who quickly obtained permission for the topic to be the subject of the 1993 symposium.

In January 1991, Cox approached Judith M. Tanur, chair of ASA's Survey Research Methods Section (SRM), to request that SRM cosponsor the conference with Statistics Canada. SRM has adopted a program of international conferences as a vehicle for encouraging discussion and publication of current research on survey methods and applications. Previous survey methods conferences have dealt with panel surveys (1986), telephone survey methodology (1987), and measurement errors (1990). In October 1991, SRM approved the topic for the next conference in its series. ASA's Business & Economic Statistics Section joined SRM as a sponsor in early 1992. Other professional societies sponsoring the conference include the International Association of Survey Statisticians (who cosponsored the three previous ASA survey methods conferences), the American Agricultural Economics Association, the National Association of Business Economists, and the Statistical Society of Canada.

Throughout 1991, members were recruited for the conference's organizing committee. Organizing committee members and their institutional affiliation are as follows:

Brenda G. Cox, Mathematica Policy Research, Inc., chair and monograph editor for Part D: Data Processing;

B. Nanjamma Chinnappa, Statistics Canada, monograph editor for Part F: Past, Present, and Future Directions;

Michael J. Colledge, Australian Bureau of Statistics, monograph editor for Part A: Frames and Business Registers;

Phillip S. Kott, U.S. National Agricultural Statistics Service, monograph editor for Part B: Sample Design and Selection;

Anders Christianson, Statistics Sweden, European coordinator and monograph editor for Part C: Data Collection and Response Quality;

David A. Binder, Statistics Canada, monograph editor for Part E: Weighting and Estimation;

David Archer, Statistics New Zealand, coordinator for invited papers;

Daniel Kasprzyk, National Center for Education Statistics, coordinator for contributed papers; and

Bernard Wong, Australian Bureau of Statistics, Asian and Pacific Rim Coordinator.

Researchers conducting business surveys and analyzing their data products tend to be scattered across government agencies and private organizations. An innovation for this conference was the recruitment and use of contact persons who generously contributed time and resources, consulted with the organizing committee in setting up the conference program, and publicized the conference within their organizations.

In October 1991, the committee met to plan the content of the conference. Naming the conference presented the first problem because no term encompassed the diverse types of organizational entities of interest. The committee decided to use "establishment survey" as a general term to cover all surveys of organizational entities, and it decided to refer to the conference as the International Conference on Establishment Surveys: Survey Methods for Businesses, Farms, and Institutions. In this monograph, however, "establishment" had to be more narrowly defined. Instead, this book uses "business" to describe the diverse array of organizational entities surveyed by government agencies and the private sector.

Every stage of a business survey poses problems quite different from those encountered in surveying persons and households. To ensure that all unique features were identified, the committee developed an outline of key issues for business surveys and distributed it to contact persons for comment. The revised version of this outline furnished the topics to be addressed in the conference.

As for previous ASA survey methods conferences, papers were solicited for publication as a separate monograph by John Wiley & Sons. The organizing committee decided that the monograph should be a textbook describing the theory and practice of business surveys. Stages of the survey process are addressed in the first five parts, and crosscutting topics are covered in the last part. Each part includes papers that describe standard techniques to deal with business surveys' unique problems as well as new technologies and innovative research. For the monograph, topics that cut across application areas were preferred over those that focused on unique problems for a particular type of business.

In January 1992, the organizing committee issued the call for monograph papers, asking researchers to submit abstracts for proposed articles. In June 1992, the organizing committee used the submitted abstracts to choose 35 papers for inclusion in the monograph. Monograph authors presented their papers at the conference. Before and after the conference, monograph editors worked with the authors to improve the readability and coherence of the articles, to impose uniformity in format and writing style, and to highlight interrelationship between papers.

Having chosen the monograph articles, the organizing committee turned its attention to identifying topics and speakers for the conference's invited paper sessions. Invited papers differed from monograph papers because of their focus on a particular application area or their specialized nature, which led to keen interest for researchers engaged in similar research. Thirty invited papers were solicited for presentation at the conference.

Finally, a call for contributed papers was issued in June 1992. To widen the conference and encourage attendance, the committee only required that contributed papers relate to an aspect of the design, implementation, or analysis of business surveys. More than 100 contributed papers were presented at the conference.

In setting up the conference, the committee became aware of the diversity of software packages being used in business surveys for matching and record linkage, data capture, computer-assisted interviewing, automated coding, and data analysis. To take advantage of this abundance, we asked Alan R. Tupek of the National Science Foundation to set up a software demonstration laboratory where attendees could talk one-to-one with software specialists about the capabilities of a software package, as well as experiment with the software.

The joint Canadian/American sponsorship led us to choose a location convenient to both Washington, DC and Ottawa. Statistics Canada symposia are traditionally held in Ottawa, whereas ASA's survey research methods conferences are typically in the United States. Buffalo, New York was chosen for its border location, and the conference took place on June 27–30, 1993. Over 400 researchers attended the conference, representing more than 30 countries besides the United States and Canada. Unlike previous ASA survey methods conferences, ICES produced a proceedings volume of the invited and contributed papers, which together with this monograph provide a permanent record of the papers presented at the conference.

Without financial support, conferences such as this one could not be organized nor could edited monographs be produced. The following research organizations generously contributed funds to support the ICES:

Australian Bureau of Statistics
The Dun & Bradstreet Corporation
Finland Central Statistics Office
National Science Foundation
Nielson Marketing Research
Research Triangle Institute
Statistics Canada
Statistics Denmark
Statistics New Zealand
Statistics Sweden
United Kingdom Central Statistical Office
U.S. Bureau of the Census
U.S. Bureau of Economic Analysis
U.S. Bureau of Labor Statistics
U.S. Energy Information Administration
U.S. National Agricultural Statistics Service
U.S. National Center for Education Statistics
U.S. National Center for Health Statistics
U.S. Small Business Administration
University of Michigan, Survey Research Center
Westat, Inc.

In addition, ASA and IASS contributed funds to support the project.

Readers must decide for themselves how well the committee achieved its goal of a textbook describing business survey methods. Our intended audience includes newcomers to business surveys as well as experienced researchers actively engaged in conducting business surveys. The book's focus is on the unique aspects of business surveys; activities common to all surveys are not addressed. A general background in survey methodology will aid the reader in appreciating the techniques described in the various chapters. In addition, Part B (Sample Design and Selection) and Part E (Weighting and Estimation) require knowledge of survey sampling theory at the level of William G. Cochran's *Sampling Techniques*, 3rd edition, New York: John Wiley & Sons, 1977.

The editorial committee would like to take this opportunity to thank the many conference supporters who provided needed advice and encouragement and reviewed draft manuscripts. Our sincere thanks go to Lee Decker and Claudine Donovan of the American Statistical Association for ICES meeting arrangements and logical planning. Elizabeth Finnerty and Denise Dunn of Mathematica Policy Research, Inc. and Brenda K. Porter and Mary Ann Rowland of Research Triangle Institute provided clerical and secretarial support in producing the monograph. Denise Lockett and Paula Ray of the Kelton Group supplied editorial support. Paul Biemer of Research Triangle Institute gave invaluable advice on the logistical details associated with organizing the conference and producing the monograph. Finally, our involvement as editors was made possible by employers who generously supported the time and resources we needed for the project: U.S. National Agricultural Statistics Service, Research Triangle Institute, and Mathematica Policy Research, Inc. (Cox); Statistics Canada (Binder, Chinnappa); Statistics Canada and the Australian Bureau of Statistics (Colledge); Statistics Sweden (Christianson); and U.S. Bureau of the Census and U.S. National Agricultural Statistics Service (Kott).

BRENDA G. COX
DAVID A. BINDER
B. NANJAMMA CHINNAPPA
ANDERS CHRISTIANSON
MICHAEL J. COLLEDGE
PHILIP S. KOTT

Business Survey Methods

CHAPTER ONE

Unique Features of Business Surveys

Brenda G. Cox
Mathematica Policy Research, Inc.

B. Nanjamma Chinnappa
Statistics Canada

A simple but useful view of a *nation* is to consider it as a society or group of people who live within politically determined geographic boundaries and whose social and economic activities are regulated by a common government. To further the nation's progress, the nation's government and its citizens need information to measure how the nation is performing so that policies can be made and their implementation monitored. The primary purpose of a nation's statistical agencies is to meet this need through the collection of social and economic statistics. Because they have no generally accepted definitions, we define *social statistics* as those statistics that relate to people and their activities as individuals and *economic statistics* as those statistics that relate to the organizational entities that conduct the economic activities of a nation. Social statistics are usually collected through surveys of persons or households, while economic statistics are collected through surveys of organizational entities. In this monograph, we refer to these two survey types as *social surveys* and *economic surveys*, respectively.

Unlike the situation for social surveys, economic surveys have too few commonly accepted, practiced, and published methodologies. The Subcommittee on Measurement of Quality in Establishment Surveys drew special attention to this phenomenon in describing economic surveys conducted by the U.S. government (Federal Committee on Statistical Methodology 1988, p. 1) noting that:

> The collection of data from establishments is not new. Some establishment-based data series have been continuous since the early part of this century, and many

Business Survey Methods, Edited by Cox, Binder, Chinnappa, Christianson, Colledge, Kott.
ISBN 0-471-59852-6 © 1995 John Wiley & Sons, Inc.

predate household surveys. Nonetheless, in contrast with household surveys, for which a rich literature has emerged over the past 5 decades, very little in the way of theoretical or evaluative work on survey quality has been published for establishment surveys.

The comparative shortage of literature and the government's approach to establishment surveys have resulted in a situation unique to establishment surveys. Today, there are few commonly accepted approaches to the design, collection, estimation, analysis, and publication of establishment surveys. Establishment surveys abound in rich variety, with little standardization of design, practice, and procedures.

Implicit in this comment is the notion that surveys of businesses, farms, and institutions need a separate treatment in the literature because their survey design and application problems are quite different from those of social surveys. In spite of their diverse application areas, there is also a remarkable commonality in the types of problems encountered in designing and implementing economic surveys and censuses.

Documentation of methods to tackle these problems and exchange of ideas across nations can serve as an avenue for improving the methodology for economic surveys. An important first step in this regard has been the International Round Table on Business Frames, which has provided since 1986 ". . . a forum for statistical agencies with mutual interest in the development, maintenance, and use of business registers to exchange ideas and discuss possible solutions to a wide range of similar problems" (Castles and Sarossy 1991, p. i). Another milestone in this regard has been the work of the Federal Committee on Statistical Methodology (1988) in documenting quality issues for economic surveys and censuses in the United States.

The International Conference on Establishment Surveys (ICES) was organized to continue this effort to document methods for economic surveys. ICES expanded the scope of previous work on economic surveys to include all types of surveys of organizational entities, both public and private, and for all types of purposes, not just economic reporting. In this monograph, we refer to these surveys as *business surveys.*[1] This monograph contains selected papers from that conference. Other papers from the conference may be found in *Proceedings of the International Conference on Establishment Surveys* (American Statistical Association 1993).

This monograph presents methodology that is being used by many agencies and nations to address the variety of problems encountered in designing and implementing business surveys. This chapter focuses on the unique features of business surveys, describing these differences for each functional survey activity.

[1]The conference used "establishment" as a generic term for surveys of businesses, farms, and institutions. In this monograph, "establishment" is used in the more conventional sense of an economic unit at a single physical location.

1.1 DEFINITIONS AND GENERAL ISSUES

The monograph adopted its expanded definition of ''business'' and ''business survey'' to fill a void. The English language does not contain a term that describes the diverse organizational entities that we wish to cover, so we invented one. Note that our definition does not conform to common usage, which regards a business as ''a commercial or sometimes an industrial enterprise'' (Merriam Webster, Inc. 1993, p. 154). Statistical agencies in the United States and Canada have no official definition for the term ''business'' but informally use it as a generic term to describe the units surveyed in economic surveys.[2] Statistics Canada does define a *business entity* for its business register as:

> An economic transactor having the responsibility and authority to allocate resources in the production of goods and services, thereby directing and managing the receipt and disposition of income, the accumulation of property, and borrowing and lending, and maintaining complete financial statements accounting for these responsibilities (Statistics Canada 1986).

This monograph uses ''business'' and ''business survey'' in an even broader sense to describe organizational entities and the associated surveys that study the characteristics or attributes of these organizational entities.

As an example of the diversity of the application areas for business surveys under our definition, consider the following types of businesses:

- *Businesses* (in the dictionary sense): retail and wholesale stores, manufacturers, construction companies, mining operations, financial institutions, transportation companies, public and private utilities, service providers, and so on.
- *Farms*: crop and livestock operations, agribusinesses, vineyards, family farms and ranches, plant nurseries, cooperatives, and so on.
- *Institutions*: schools, prisons, courts, hospitals, local governments, professional and trade associations, and so forth.

The unifying attribute of these business surveys is that they are designed to study the organization; individuals within these organizations are surveyed only as spokespersons for the organization.

Many business surveys are conducted to provide data needed for calculating key economic indicators that monitor the economy over time and for constructing official statistics such as national accounts. These economic surveys tend to be similar in terms of the statistical and survey methodologies they require.

---

[2]Personal communications with Michael J. Colledge, then at Statistics Canada, and Nash Monsour and Richard Sigman of the U.S. Bureau of the Census.

Economic surveys have these dominating characteristics that are not evident to the same extent in social surveys:

- Businesses tend to have very skewed distributions with many small businesses and very few large businesses. Therefore, size greatly affects the precision of survey estimates.
- The rapid rate of change in economic data creates a demand for quick estimators/indicators of these changes, ideally "as they are happening."
- A wealth of alternative data sources are available for businesses (such as from administrative records and remotely sensed data on land use), allowing these data to substitute for survey data or be used for editing and imputing.
- Most economic statistics have an integrating framework, such as the system of national accounts, within which they must fit.
- Reliable aggregated estimates are available later from other sources (e.g., marketing data on crops) that allow evaluation and validation of the survey estimates.

These characteristics pose interesting and difficult challenges to survey designers and lead to the need for special techniques for frame development, sample design, data collection and processing, and estimation and reporting.

1.2 FRAMES AND BUSINESS REGISTERS

Social and demographic surveys typically have readily available frames for use in sampling. In the United States for instance, face-to-face interview surveys of the general population use public use files from the decennial Census to construct the area frame and obtain size measures for use in sampling. Telephone surveys are another feasible alternative because more than 93 percent of the household population have telephones (Thornberry and Massey 1988, p. 29). Again, sampling needs can be met using publicly available databases, such as telephone exchange code listings.

For businesses, there are no such sources of publicly available data for building sampling frames and constructing size measures. Typically, government agencies create business frames through expensive and error prone matching and linkage operations, combining data from multiple administrative databases such as tax and employment files. The resulting database may then be augmented through direct data collection to delineate the structure of large operations and to collect auxiliary data needed for sampling.

As with surveys of the general population, the resultant frame information is regarded as confidential and cannot be released to the public. In countries (such as the United States) with a decentralized system of statistics, legal safeguards often prevent frame data from being shared across national statistical agencies. For its surveys of agricultural production, for instance, the U.S. National Agricultural Statistics Service cannot use results from the Census of Agriculture (conducted by the U.S. Bureau of the Census) and must build its own frame independently (Clark and Vacca 1993). Not only does this result in duplication of effort, it also leads to discrepancies between the two data series that are unrelated to sampling.

Data sharing is less of an issue for countries (such as Canada) with centralized statistical offices because operations are all in house. A recent innovation for such agencies is the development and maintenance of integrated business registers that are intended to serve as the sole basis for frame development and sample selection for business surveys, regardless of the application area. In Chapter 2, Colledge describes the steps involved in creating and maintaining frames and business registers.

Confidentiality considerations and legal restrictions prevent the private sector from using these government-maintained list frames for market research and other purposes. For these private sector surveys, area frame sampling approaches tend to be precluded because businesses are difficult to enumerate in the field and size measures (number of businesses and their characteristics) are available only for large geographic units that are too expensive to enumerate. Random digit dialing techniques are also infeasible due to the low incidence of business phones among telephone numbers and the multiplicity of phones attached to some businesses. Most private surveys select their business samples from privately operated lists constructed from telephone listings, credit rating services, and so forth. These frames can have serious coverage problems particularly for new operations, sole proprietorships, zero-employee firms, and businesses in service industries (Cox et al. 1989, pp. 16–18).

In constructing the frame and later in interviewing, complex definitional issues frequently arise. Many problems relate to the issue of "What is a business?" For instance, it is not uncommon for salaried professionals such as accountants and statisticians to consult in their spare time. Should free-lancers be considered as another form of sole proprietorship and included in the target population? If included, how will data be obtained to construct the frame? Similar questions for agricultural surveys relate to, "What is a farm?" and "When should those who farm as a hobby be included in the target population?"

Other questions arise around the issue, "When is a business considered to be in operation?" A corporation that has ceased operations may continue paying registration fees as court cases are pursued and property settlements are arrived at. An owner may consider his/her business to be "in operation" even if no income was generated during the past year. Is a farm not a farm when no crops are being planted and no livestock are being raised?

Related questions revolve around the issues of "When should one consider a business to have been born?" and "What events should lead to a determination that a business has died?" Definitional problems such as these arise because the concept being studied—"being in business"—is more like a continuum than a discrete event. Frame developers must even reckon with the fact that, unlike the human population, dead businesses *can* come back to life. In Chapter 4, Struijs and Willeboordse discuss the treatment of births, deaths, and other changes in business registers.

Additional problems arise because of the complexity associated with defining the unit(s) to be recorded on the frame. The sampling unit for a business survey is not a natural unit and is often defined in terms of the data being collected for the survey. Thus, a retail business may be defined in terms of its products, services, or location; a farm in terms of the commodities produced; and an institution in terms of the clientele it services or the authority that administers it. In contrast, the sampling unit for a social survey is a natural and commonly understood unit such as a person or household and is independent of the data being collected.

Similarly, the hierarchical structure of the units in social surveys is reasonably straightforward to delineate, ranging from the person to the family, community, and so forth. As Nijhowne describes in Chapter 3, the hierarchy of units of business surveys can be quite complex, ranging from the business location to administrative and legal structures, with the appropriate unit for a survey depending on the data items required. Businesses are rarely static with splits, mergers, and growth contributing to the complexity of the population being studied.

Large operations are often characterized by complex legal and organizational structures that are difficult to relate to the units used for sampling and reporting. Clarifying these relationships for large businesses has led many business registers to include a separate data collection step to profile the legal, operational, and administrative structures of large, complex organizations. Pietsch describes such operations in Chapter 6.

Small businesses present quite different problems for frame building and maintenance. Their extreme volatility makes it difficult to obtain accurate and timely frame information on their births, deaths, and other changes. About 1 percent of all small businesses cease operations *each month* in the United States. The large number of these small units presents particular problems in terms of maintaining coverage while controlling costs. In the United States, for example, 49 percent of employers (i.e., businesses with employees) have fewer than five employees. Small businesses with no employees (i.e., the self-employed) account for 45 percent of all business tax returns (Phillips 1993). Because of the skewness of the underlying quantities being estimated, the total contribution of these small businesses to estimates tends to be relatively small. This anomaly leads to interesting approaches for their treatment in business surveys. In some applications, data from administrative records are substituted for direct data collection from small businesses.

As Archer describes in Chapter 5, the volatility of the business universe makes maintaining a business register a complex, labor-intensive operation. Advance planning and extensive coordination are needed to appropriately handle the voluminous information being fed into the register from diverse sources such as administrative databases, survey feedback, media announcements, and so forth. Care has to be taken to avoid biasing future survey samples by inappropriate use of input data such as survey feedback, for instance, and to preserve time series information needed for later analyses.

An important component of developing and maintaining business registers is the assignment of standard industrial classification (SIC) codes (Monk and Farrar 1993). Because SIC codes are used to create the frames for particular application areas, classification errors can lead to undercoverage of targeted industries and reductions in sampling efficiency associated with inclusion of out-of-scope units. In addition, many analysis domains are defined based upon SIC codes. Not only must the SIC codes be defined correctly for individual businesses, the codes themselves must be sufficiently detailed and current to describe the nation's economy. Technological advances and other changes in a nation's economy require periodic revisions of the industrial classifications. In Chapter 7, MacDonald describes how a SIC system revision can be gracefully incorporated into the business register while simplifying estimation of the analytical effects of the revision on historical time series.

1.3 SAMPLE DESIGN AND SELECTION

Until recently, business surveys tended to use longstanding designs based upon nonprobability, judgment, or voluntary samples. Because the primary purpose of many business surveys is the estimation of change, some study planners have argued that change can be measured well enough by comparing responses from common respondents over time, irrespective of how respondents are chosen. The argument is as follows: "Biased samples serve the purpose well enough, as long as their bias remains the same over time." In addition, the availability of periodic census benchmarks often allows these change ratio estimates to be tethered to a known level. Increasingly, however, these nonprobability designs are being replaced by probability sampling methods which are more robust, produce estimates that are defensible, and have sampling errors that are measurable.

Business surveys tend to estimate population totals and other statistics for which operation size greatly affects precision. The businesses themselves tend to be highly skewed with a small number of large businesses accounting for a substantial percentage of total production. At the opposite end of the size distribution, business populations have many very small operations, which change quite rapidly over time with high birth and death rates. As Sigman and Monsour describe in Chapter 8, standard sampling practice in this situation is to stratify by size and oversample large operations, often surveying all of the very large operations. Hence, efficient sample designs require the use of list frames that incorporate known or projected measures of size for each business.

The rapidly changing nature of businesses causes any list frame to become outdated very quickly and therefore incomplete in terms of coverage of the target population. One way to ensure coverage of the population may be to use an area frame instead of a list frame. Whereas a list frame is a list of businesses, an *area frame* is a collection of nonoverlapping geographic units or area segments that, taken together, span the country or population of interest. These area segments become the sampling units, which are selected using stratified, multistage designs. Rules are established that link the businesses in the target population to the area segments. All businesses linked to the selected area segments are enumerated. A business that spans several area segments is linked uniquely to one of these segments or linked to all of them with its data apportioned among them using multiplicity adjustment factors. Although area frames provide complete coverage, they generally lead to clustered and therefore inefficient sampling designs. The size measures available for these area segments are not highly correlated with the businesses' sizes, resulting in a loss of control over the sample's size distribution and therefore larger than desirable variances for sample estimates.

To take advantage of the list frame's information on business sizes while retaining the complete coverage feature of the area frame, some organizations use multiple frame sample designs. Essentially, these designs select independent samples from both list and area frames with rules formulated to account for multiple probabilities of selection. For instance, the U.S. National Agricultural Statistics Service (NASS) uses a multiple frame design with two frames for sample selection for its Quarterly Agricultural Survey:

- a list frame of agricultural operations derived from various agricultural databases and
- an area frame constructed by dividing the total land area of the United States into sampling units or "segments," with acres used as the size measure in selection.

For its multiple frame estimates, NASS removes the multiplicity associated with selecting from the full-coverage area frame and the overlapping list frame by using the area frame to represent only those (nonoverlap) operations that are not included in the list frame. Kott and Vogel describe this approach in Chapter 11.

Because "time is money," business owners generally resent spending time responding to surveys. Because business populations are not that large and in addition larger operations are sampled at high rates, response burden quickly becomes another issue to be addressed in sample designs for business surveys. Consider, for example, a stratum sampled at a rate of 25 percent each year— a not uncommon circumstance. Over a 4-year period, one might expect each business to be sampled once. If simple random samples are selected independently each year, however, on average:

- 31.6 percent will not be sampled at all.
- 42.2 percent will be sampled once.

- 21.1 percent will be sampled twice.
- 4.7 percent will be sampled three times.
- 0.4 percent will be sampled four times.

The response burden problem is compounded by the fact that large units are selected with high probabilities (often 1) both *over time* and *across surveys* at a point in time. In addition, economic analyses usually require measures of change over time, which are measured with greater precision when there is a planned overlap between year-to-year samples. In Chapter 9, Ohlsson tells how permanent random numbers can be used to control the frequency with which frame units enter successive samples over time, while Srinath and Carpenter discuss other methods in Chapter 10.

Institutional surveys often use two-stage sampling designs and two distinct data collection efforts, one targeted at the institution and the other at the institution's "clients" (e.g., schools and their pupils). The institutional survey collects data about the institution for the institutional-level analyses. Some of these institution-level data may be used in the client-level analyses as well. The sampled institutions are also asked to provide client information needed to create the second-stage frame for sampling clients. Typically, the sampled clients are interviewed on site at the institution (see, for example, McLemore and Bacon 1993, McMillen et al. 1993, and Swain et al. 1993).

Sampling business populations often requires the development of innovative statistical procedures. In Chapter 12, Garrett and Harter discuss the use of Peano key sequencing in market research surveys. Peano keys are used to order a list of businesses geographically; the sample can then be automatically updated for business births and deaths. In Chapter 13, Johnson describes another unusual approach—using a household survey in a network sampling approach to identify businesses for interview. Such a technique could prove useful, for instance, in surveying zero-employee businesses which tend to be missing or inadequately described in business frames and registers.

1.4 DATA COLLECTION AND RESPONSE QUALITY

Collecting data on business organizations is very different from collecting data on individuals. Examples of the questions that survey designers must answer include the following:

- What level of the business organization is best able to answer survey questions—the establishment, the enterprise, or something in between?
- In terms of job title or position, who is the person within the business organization most likely to know or be able to find the answers to survey questions?
- Will permission have to be obtained from the owner or chief executive officer prior to completing the questionnaire?

- What techniques are effective in getting past "gate keepers" who limit access to upper management?
- What records are available to the firm for use in answering survey questions?
- Can survey questions be structured to conform to the business' record-keeping practices, including its fiscal year?
- Are there particular times of the year when data are more readily available?
- What is the best way to collect data that may be viewed as confidential business information?

As is clear from the above, minimizing nonresponse and measurement error in business surveys requires advance planning and creative data collection approaches.

In the past, business surveys have lagged behind social surveys in their adoption of new statistical methodologies such as cognitive techniques for questionnaire development. Yet these new techniques have shown great promise in improving the quality and efficiency of survey estimates. As Biemer and Fecso point out in Chapter 15, attention must focus on the total error in survey estimates. In particular, business surveys need to devote much more attention to identifying sources of measurement error and quantifying the effect of these errors on survey quality.

Household surveys have long grappled with the issues associated with designing survey questionnaires that work for the wide variety of human respondents encountered, from the cooperative to the uncooperative, from the affluent to the needy, from the college-educated individual to the high-school dropout, and so forth. Business surveys have human respondents as well and hence face these same problems. In addition, they must also deal with the wide variation in organizational structures, management practices, products produced and services rendered, and so forth. In Chapter 16, Dippo et al. describe strategies for developing and evaluating the data collection process, including the use of cognitive methodology in designing the survey questionnaire.

A key feature of business questionnaires is that the bulk (or at least a substantial fraction) of the collected data tends to be quantitative and continuous rather than categorical and discrete. In contrast, most household survey questionnaires (except, for example, surveys of income and expenditure) contain categorical questions having yes/no or multiple choice responses, with few questions requiring a quantitative response. Quantitative "how many" or "how much" questions pose more difficult response problems for cooperating individuals and naturally tend to elicit greater nonresponse and inaccurate responses. More effort is required from the respondent, and some answers may be unknown or unknowable. In the U.S. Farm Costs and Returns Survey, for instance, farmers are asked to report their annual fuel expenses and then asked the percentage of these expenses associated with each commodity they produce! Hard-to-report information such as this are often required by data users for incorporation into national accounts and other economic analyses.

Besides this difficulty in answering quantitative questions, issues of sensitivity and confidentiality arise for most business surveys. The information that business surveys request often constitutes "confidential business information" for which the business restricts outside access to prevent its becoming known to its competitors. Because the respondent is reporting for an organization rather than for himself/herself, consultation with and/or permission from one or more management staff may be needed before releasing such data.

Obtaining acceptable levels of response for the total questionnaire and for individual items, then, is a continuing concern for business surveys (see, for instance, Mesenbourg and Ambler 1993, Milton and Kleweno 1993, and Wallace 1993). Compounding the problem is the widespread use of mail data collection, for which lower response rates are common. As Christianson and Tortora report in Chapter 14, an international survey of statistical agencies suggests that mail data collection is used as the exclusive data collection mode for more than half of all business surveys; for the remaining surveys, mail is frequently one of several modes used in combination. The predominance of mail collection can be attributed to (1) the ready availability of business names and addresses on business frames and registers, (2) the need for advance letters and questionnaires to obtain the owner/manager's permission to respond, and (3) the time required by the respondent to consult accounting and other business records. As Paxson et al. describe in Chapter 17, however, business mail surveys can achieve reasonable levels of response by adapting Dillman's (1978) total design method, which was originally developed for social surveys.

More recently, attention has focused on the use of new technologies to reduce data collection costs while maintaining or even improving data quality. In Chapter 18, Werking and Clayton discuss a 4-year study in which reluctant mail respondents were converted using computer-assisted telephone interviewing and then transferred to less expensive touchtone and voice recognition telephone collection for future waves of the survey. In Chapter 19, Ambler et al. discuss the steps needed to collect economic data via electronic data interchange.

The problems are quite different when data from administrative sources are used in lieu of survey data. Nonresponse tends to be less of a concern, especially if the administrative data are mandated for regulatory reasons. Problems relate primarily to concepts used in the administrative system and its procedures for editing and imputing data. The survey organization has little control over these factors, and, worse still, the factors can change with little notice or consultation with the survey organization. In the extreme, the source itself can disappear (e.g., as happens with customs forms when free trade is introduced). Statistical agencies try to influence the administrative data concepts and formats and maintain close contact with the administrative data agencies to resolve these problems.

1.5 DATA PROCESSING

A distinguishing characteristic of business surveys is the large amount of information available for editing and imputing. The frame itself can be an abundant source of information. Information may also be available from administrative databases and past business surveys. This richness of information can prove to be an effective basis for editing and for replacing missing data using imputation techniques such as those described by Kovar and Whitridge in Chapter 22.

In business surveys, there is much more extensive use of longitudinal and other logical edits than is customary or acceptable for social surveys. There is also much more scope for macroediting and data conflict resolution for business surveys because of checks and balances in economic data, especially from integrating frameworks such as the system of national accounts. It is not uncommon for logical imputation to be used to create entire data records for total nonrespondents in business surveys. Very large operators have responses so different from the norm that expert imputation may be considered preferable to statistical imputation. Questions tend to have related responses so that logical inconsistencies are easily detected and corrected in the recorded data. One consequence of this vast scope for editing, as Granquist notes in Chapter 21, is that there is a growing belief that business survey data are overedited, thereby wasting survey resources that could be better used elsewhere. He argues that selective editing should be used, targeting resources to those errors having most effect on survey results.

Matching and record linkage is also widely used in business surveys. Frames and business registers are often created by combining data records from two or more administrative data sources. In multiple-frame approaches, matching is used to identify businesses with multiple selection opportunities through their inclusion on more than one frame. Errors in matching and linkage lead to undetected duplicates (overcoverage) and inappropriate elimination of records due to mismatching (undercoverage). The extent of error in the matched data is dependent on the quality of the input data and the accuracy of the matching operation. In Chapter 20, Winkler profiles (1) automated approaches for matching and record linkage and (2) methods for estimating the extent of matching error.

Another feature that distinguishes business surveys from social surveys is that public-use data tapes are not typically released for business surveys, due to the small size of the business universe and the associated difficulty in preserving confidentiality. In addition, entries in tables and results from analysis often have to be suppressed to prevent breaching the confidentiality of individual responses. In Chapter 24, Cox provides a conceptualization of the problem from a mathematical perspective and then profiles methodologies in use to address the issue of confidentiality protection.

Increasingly, business surveys are adopting automated approaches for editing, imputation, and other data processing operations. In Chapter 23, Pierzchala describes automated approaches appropriate for business survey data, how available software should be evaluated, and the results of documented applications. General features are described for many editing and imputation software packages. Other application areas for automated procedures include coding (Miller 1993), record linkage (Nuyens 1993), confidentiality protection in tabular releases (Sande 1984), and mapping survey results (Cowling et al. 1993).

1.6 WEIGHTING AND ESTIMATION

Business surveys tend to use quite simple sampling approaches, such as stratified simple random sampling from a list frame. The list frame design strata tend to form good weighting classes for nonresponse adjustment so that constructing nonresponse-adjusted sampling weights is straightforward. Explicit models for nonresponse can then easily be incorporated into the estimation approach.

However, as Hidiroglou et al. discuss in Chapter 25, weighting and estimation for business surveys are not as simple as they might initially appear. Auxiliary data are often used in ratio, poststratification, regression, and raking ratio approaches to improve the efficiency of point estimates. These auxiliary data are obtained from administrative databases and previous survey data. The availability of such auxiliary data is another unique feature of business surveys.

Usually, the parameters being estimated are totals, often for rather fine subdivisions of the population such as states or provinces. Because businesses have skewed populations, outliers pose difficult problems in obtaining efficient domain estimates. As Lee describes in Chapter 26, it is usually not the observation or its weight that is excessive, but rather the expanded value (the weight times the observation). Outliers can also take the form of very small expanded values when large values are the norm. These outliers tend to be associated with misclassified list frame records (e.g., large operations assigned to a small size stratum) and (in multiple frame surveys) with large area frame operations not found on the list fame.

Business statistics are often demanded for very small geographic areas— much smaller than can be supported by the survey's sample size. For instance, the U.S. National Agricultural Statistics Service allocates survey samples to facilitate efficient state-level estimation, yet must deal with the need for county-level estimates—a very fine geographic subdivision indeed (Flores-Cervantes et al. 1993). As Rao and Choudhry discuss in Chapter 27, the sample sizes for such small areas make direct estimation impractical; instead model-based approaches are needed that "borrow strength" from related areas to create indirect estimators that increase the effective sample size and thus decrease the variance. Auxiliary data from administrative records and recent censuses are used to build such models.

Other forms of complex estimation also occur in business surveys. Much of this estimation is driven by the need to monitor the nation's economy. In Chapter 28, Leaver and Valliant discuss price indexes, perhaps the most complex quantity that business surveys estimate. The concept being measured—change in the cost of living from one time period to another—is an economic construct that is both difficult to define and difficult to measure. Developing adequate variance estimates for such complex statistics presents special challenges. The survey designer must also deal with the conceptual problems associated with inability to obtain price data over time for seasonal items, new product lines, phased-out products, and so forth.

Because of the rapid and dynamic changes in economic data, timeliness of estimates is crucial for business surveys. This places considerable strain on survey schedules. Business surveys usually produce quick preliminary estimates to satisfy the need for early indicators; these preliminary estimates are then subject to revision as additional survey data or market intelligence become available. Large discrepancies between preliminary and final estimates can lead to errors in decision-making and loss of confidence in the statistical series. Research is needed to identify methods that produce preliminary estimates that are good predictors of the final revised estimates.

Time-series techniques are emerging as an important research tool in improving estimates. In Chapter 29, Pfeffermann et al. present methods to detect or predict turning points in economic time series. Their chapter relates to another unique feature of business surveys—namely, use of business surveys to create leading economic indicators whose movements signal changes in the nation's economy. These leading indicator series may consist of preliminary estimates or related statistics.

The complexity of the concepts being measured in business surveys together with the difficulty of measuring them may make model-based methods an attractive alternative to traditional design-based methods for some applications. In Chapter 30, Brewer describes model-based versus design-based approaches, using two Australian surveys to illustrate their objectives and methods. Brewer argues for a combined estimation strategy with a design approach that accommodates both methods.

1.7 PAST, PRESENT, AND FUTURE DIRECTIONS

Knowledge of the history of a nation's economic surveys is needed to understand the origins of current methodology and changes needed in the future. In Chapter 31, Allen et al. describe the effect that changing economic conditions and advances in statistical technology has had on the agricultural program of the United States. In Chapter 32, Worton and Platek describe the origins and evolution of Statistic Canada's business surveys. Quite different in populations studied, these two chapters illustrate that much of our current programs were shaped by dedicated personnel who aggressively fought for procedures to correct deficiencies in quality, content, and coverage of their nation's business surveys.

A continuing theme for today's business surveys is the need to standardize concepts and approaches across surveys to expedite integration of survey results. Estimates derived from business surveys are a critical source of information for the construction of the nation's system of national accounts, which are used to establish economic policy, monitor economic trends, and develop long-range plans for the nation's economy. Lewington describes the interrelationship between a nation's economic surveys and its national accounts in Chapter 33.

To construct national accounts, data are combined across surveys of diverse types of businesses. Generation of reliable national accounts is, therefore, dependent on the quality of the data derived from the individual surveys. Equally important for national accounts is the ability to integrate data derived from these surveys. That is, do the combined data sources provide complete coverage of all types of businesses without overlaps of their target populations? Are definitions and classifications consistently applied across surveys? Are economic components of the national accounts being collected in a uniform manner across surveys and across time? In Chapter 34, Griffiths and Linacre describe these and other facets of data quality of business surveys.

Finally, Ryten looks to the future in Chapter 35, reminding us that the business surveys of the future must reflect changing economic conditions as well as technological advances. The way businesses organize themselves in the future may make the establishment no longer appropriate as a data collection unit, for instance. The rising costs of traditional government surveys may require tradeoffs of survey frequency and detail to free up funds to meet the new information demands of the next century.

A crucial objective of business surveys has always been the measurement of change over time. For many business surveys, the need to avoid confounding such change estimates had led to reluctance to adopt new technologies. Quality improvements come at the price of a ''break'' or discontinuity in the time series. Because of the deep interest and attention paid to economic time series, survey managers are criticized when their use of new survey advances improves survey estimates but impedes interpretation of the underlying time series. Another challenge for the future is to accelerate the adoption of innovative techniques that promote quality improvement while protecting the historical validity of the underlying time series.

1.8 CONCLUDING REMARKS

The unique features of business surveys relate primarily to the conceptual and practical problems in defining the basic units being surveyed and the characteristics of the variables for which data are to be collected. The composition and complexity of these units, their distribution in the target population, and their volatility over time create difficult problems for all steps of survey design and implementation. The use of survey outputs to build integrated economic models such as national accounts creates a need for standardizing concepts across business surveys and allows evaluation and validation that can improve the quality of their outputs.

This chapter illustrates only some of the many unique features of business surveys. The other chapters of this monograph present these topics in detail. We hope that the statistical concepts that motivate these chapters stimulate further discussion and challenge survey practitioners to extend these methodologies to meet their particular needs.

REFERENCES

American Statistical Association (1993), *Proceedings of the International Conference on Establishment Surveys*, Alexandria, VA.

Castles, I., and G. Sarossy (1991), ''Introduction,'' *Sixth International Round Table on Business Frames*, Belconnen: Australian Bureau of Statistics, p. i.

Clark, C. Z. F., and E. A. Vacca (1993), ''Ensuring Quality in U.S. Agricultural List Frames,'' *Proceedings of the International Conference on Establishment Surveys*, Alexandria, VA: American Statistical Association, pp. 352–361.

Cowling, A., R. Chambers, R. Lindsay, and B. Parameswaran (1993), ''Mapping Survey Data,'' *Proceedings of the International Conference on Establishment Surveys*, Alexandria, VA: American Statistical Association, pp. 150–157.

Cox, B. G., G. E. Elliehausen, and J. D. Wolken (1989), *The National Survey of Small Business Finances: Final Methodology Report*, RTI/4131-00F, Washington, DC: Board of Governors of the Federal Reserve System.

Dillman, D. A. (1978), *Mail and Telephone Surveys: The Total Design Method*, New York: Wiley.

Federal Committee on Statistical Methodology (1988), *Quality in Establishment Surveys*, Statistical Policy Working Paper 15, PB88-232921, Washington, DC: U.S. Office of Management and Budget.

Flores-Cervantes, I., W. C. Iwig, and R. R. Bosecker (1993), ''Composite Estimation for Multiple Stratified Designs from a Single List Frame,'' *Proceedings of the International Conference on Establishment Surveys*, Alexandria, VA: American Statistical Association, pp. 380–384.

McLemore, T., and W. E. Bacon (1993), ''Establishment Surveys at the National Center for Education Statistics,'' *Proceedings of the International Conference on Establishment Surveys*, Alexandria, VA: American Statistical Association, pp. 93–98.

McMillen, M., D. Kasprzyk, and P. Planchon (1993), ''Sampling Frames at the National Center for Education Statistics,'' *Proceedings of the International Conference on Establishment Surveys*, Alexandria, VA: American Statistical Association, pp. 237–243.

Merriam-Webster, Inc. (1993), *Merriam-Webster's New Collegiate Dictionary*, 10 ed., Springfield, MA.

Mesenbourg, T. L., and C. A. Ambler (1993), "Response Improvement Initiatives in the 1993 Economic Censuses," *Proceedings of the International Conference on Establishment Surveys*, Alexandria, VA: American Statistical Association, pp. 473–477.

Miller, D. (1993), "Automated Coding at Statistics Canada," *Proceedings of the International Conference on Establishment Surveys*, Alexandria, VA: American Statistical Association, pp. 931–934.

Milton, B., and D. Kleweno (1993), "USDA's Annual Farm Costs and Return Survey: Improving Data Quality," *Proceedings of the International Conference on Establishment Surveys*, Alexandria, VA: American Statistical Association, pp. 945–949.

Monk, C. H., and C. H. Farrar (1993), "Standard Industrial Classification (SIC) Codes," paper presented in the International Conference on Establishment Surveys, Buffalo, NY.

Nuyens, C. (1993), "Generalized Record Linkage at Statistics Canada," *Proceedings of the International Conference on Establishment Surveys*, Alexandria, VA: American Statistical Association, pp. 926–930.

Phillips, B. D. (1993), "Perspectives on Small Business Sampling Frames," *Proceedings of the International Conference on Establishment Surveys*, Alexandria, VA: American Statistical Association, pp. 177–184.

Sande, G. (1984), "Automated Cell Suppression to Preserve Confidentiality of Business Statistics," *Statistical Journal of the United Nations: ECE*, **2,** pp. 33–41.

Statistics Canada (1986), "Business Entity and Standard Statistical Unit: Definitions and Procedures for Implementation," Internal working document, Ottawa.

Swain, L., M. Brodeur, S. Giroux, K. McClean, J. Mulvihill, and J. Smith (1993), "Frame Creation for Institutional Surveys at Statistics Canada," *Proceedings of the International Conference on Establishment Surveys*, Alexandria, VA: American Statistical Association, pp. 771–774.

Thornberry, O. T., Jr., and J. T. Massey (1988), "Trends in United States Telephone Coverage Across Time and Subgroups," in R. M. Groves, P. P. Biemer, L. E. Lyberg, J. T. Massey, W. L. Nicholls II, and J. Waksberg (eds.), *Telephone Survey Methodology*, New York: Wiley, pp. 25–49.

Wallace, M. E. (1993), "Response Improvement Initiatives for Voluntary Surveys," *Proceedings of the International Conference on Establishment Surveys*, Alexandria, VA: American Statistical Association, pp. 478–483.

Frames and Business Registers

CHAPTER TWO

Frames and Business Registers: An Overview

Michael J. Colledge[1]
Australian Bureau of Statistics

The survey frame is a vital component in the survey process. It establishes a survey's starting point by defining the population to be surveyed and by providing the information needed for stratifying, sampling, and contacting businesses. Provision of a survey frame accounts for a significant proportion of the total expenditures on a business survey. For a program of business surveys conducted by a national statistical agency, providing the frames and correcting frame errors can exceed 20 percent of the total program budget.

Frame inadequacies can lead to operational nightmares, massive nonsampling errors, and misleading interpretation of survey results. For example, an employment survey based on a frame that includes only large businesses can give a misleading impression of overall trend when small businesses have different growth rates. On the other hand, suppose the frame provides complete coverage of all businesses, large and small, but in doing so includes a substantial proportion of units that are out of business. Efforts to contact defunct units selected for the sample and distinguish them from nonrespondents can significantly increase survey costs. Sampling variances can also be inflated as the result of the zero values associated with null units.

Frames play a significant role in the integration of survey data. Surveys are not often conducted in complete isolation. Usually they are repeated or carried out within the framework of a broad statistical program, whether that of a national statistical agency or a commercial enterprise. It is widely recognized

[1]This chapter was written while the author was at Statistics Canada. I thank Brenda G. Cox for helpful comments and Pierre Piché for assistance in preparing the document. The views expressed are the author's and do not necessarily reflect the official position of Statistics Canada or the Australian Bureau of Statistics.

Business Survey Methods, Edited by Cox, Binder, Chinnappa, Christianson, Colledge, Kott.
ISBN 0-471-59852-6 © 1995 John Wiley & Sons, Inc.

that integration of data across surveys is highly desirable (Australian Bureau of Statistics 1969, Bonnen 1981). Provided the data can be meaningfully related, there will be a synergic gain; combined data sets contain more information than the sum of the separate parts (Colledge 1990). In this context, the choice of frames is important because it can facilitate or inhibit integration. For example, suppose the impact of labor costs and capital investment upon productivity is to be studied based on data from a manufacturing survey combined with data from employment and capital expenditure surveys. The lower the level of aggregation at which survey output can be related, the greater the information content. However, combining data at a low aggregate level requires that the surveys have relatable units, stratified in similar ways. This, in turn, depends upon harmonization of the survey frames.

Given that many surveys are conducted within a program framework, an efficient procedure for providing frames for individual surveys it to maintain a multipurpose frame database that meets each survey's needs. For a program of business surveys, this database is referred to as a *business register*. Two advantages accrue from using a business register as opposed to separate systems for each survey. First, a register simplifies data integration by ensuring that the survey frames are based on relatable sets of units. Second, the consolidation of frame maintenance activities produces economies of scale.

Figure 2.1 illustrates a statistical program with eight regularly conducted

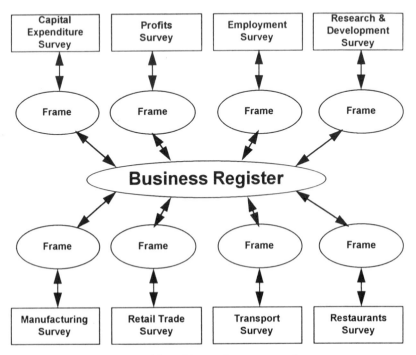

Figure 2.1 Provision of frames for a program of surveys.

business surveys. (In practice, a national statistical agency may conduct 100 or more business surveys.) In the absence of a register, each survey operation must maintain its own frame. Without a doubt, these operations will duplicate each other in accessing and processing information for frame updating. If a decision were made to share the updating information between each pair of frames, this would imply 56 frame data flows, each requiring its own protocol. Introduction of a register simplifies data sharing by reducing the number of flows and protocols to 16.

The next five chapters focus on the construction, maintenance, and use of registers designed to provide frames for programs of business surveys. This chapter summarizes the features of frames and business registers and thereby introduces the following chapters. Underlying concepts, including the definition and classification of statistical units, are described, followed by their application in the creation, maintenance, and use of a business register. Quality considerations and current research topics are also outlined. Frames for one-time-only surveys and specialized techniques for institutional and agricultural surveys are not discussed. Wright and Tsao (1982) describe frames in a general context, emphasizing methods for handling frame imperfections. Their paper includes an annotated bibliography through 1982. Sands (1993) outlines the frame creation process for two ad hoc business surveys.

2.1 BUSINESS SURVEY FRAMES: GENERAL CONCEPTS

To harmonize the terminology used, this and subsequent chapters on survey frames give explicit definitions for commonly used terms. However, the reader is cautioned that national practices vary substantially (Struijs and Willeboordse 1992). Even definitions provided by the U.N. Statistical Office (1990) in the International Standard Industrial Classification (ISIC Revision 3) are not universally applied.

The term "business" is an example; it has no unique, commonly accepted definition. In this monograph, *business* is used in a general sense to refer to an economic unit engaged in the production of goods or services. Commercial enterprises, departments of government, farms, institutions, and nonprofit organizations are all businesses under this definition. This definition is broader than colloquial usage where "business" implies commercial or commercial and industrial activities only.

A *survey* is an operation involving the collection of data to be used for statistical purposes, and a *survey program* is a set of surveys conducted by a single statistical organization or commercial organization. Surveys are designed to satisfy data requirements, expressed in terms of a population of interest and a set of data items describing its members. A *business survey* is a specific type of survey in which the units of interest are businesses or parts of businesses. The set of units about which data are sought is referred to as the *target population*. For operational reasons, this set of units often cannot be assembled exactly, which leads to the concept of the *sampled population*, the set of units from which the survey sample is actually selected. Ideally, the sampled and target populations should coincide. In practice, the sample population is as close to the target population as is operationally feasible. For example, the target population for a monthly survey of retail trade might be all businesses engaged in retail activity at any time during the month. The sampled population, however, might be based on an administrative list that includes only those businesses registered for employee payroll deductions.

A *statistical unit* is a unit defined for statistical purposes. A *target statistical unit* is the unit about which data are required and the basis for aggregation of individual data. A *collection unit* is the unit from which data are collected, not necessarily the same as the target statistical unit.

2.1.1 Characteristics and Types of Survey Frames

The *survey frame* is defined as the set of units comprising the sampled population with identification, classification, contact, maintenance, and linkage data for each unit. *Identification data* are items that uniquely identify each unit such as name, address, and alphanumeric identifiers. *Classification data* are items required to stratify the population and select the sample, usually variables such as size, industrial and regional classification, and (depending upon the units) institutional sector and legal form. *Contact data* are items required to locate units in the sample, including the contact person, mailing address, telephone number, and previous survey response history. *Maintenance and linkage data* are items needed if the survey is to be repeated or is part of a program of surveys. They include dates of additions and changes to the frame and linkages between frame units. Collectively, the identification, classification, contact, maintenance, and linkage data items are referred to as *frame data*.

There are two distinctive types of frames. As the name suggests, a *list frame* is a list of units and the associated frame data. In business surveys, the starting point for constructing a list frame is typically an administrative list such as the set of businesses that are registered for value-added taxes or for employee payroll deductions. An *area frame* is a set of geographic areas from which areas are selected and the associated units enumerated.

A survey may have more than one frame. In a multistage design, there is a frame for each stage. For example, the first-stage frame for the Canadian trucking origin and destination survey comprises transport businesses, while the second-stage frame is the shipping documents for each business selected in the first stage (Statistics Canada 1990).

In a multiphase design, the frame for the second (or subsequent) phase of sampling is the subset of units comprising the sample for the previous phase supplemented with additional classification data for these units, usually obtained by survey. Colledge et al. (1987) discuss a two-phase design used by Statistics Canada to sample income tax records for extraction of financial data not captured during taxation processing. Records selected in the first-phase sample form the second-phase frame. They are assigned a detailed industrial code which is used to stratify and select the second-phase sample (Armstrong and St-Jean 1993).

As described in Chapter 11, a single-stage, single-phase survey design may select units from two or more frames, with appropriate multiple-frame procedures for ensuring that duplication in coverage is removed or adjusted for in estimation. Isaki et al. (1984) describe the dual-frame methodology used for the U.S. Retail Trade Survey. The list frame is constructed from businesses registered for payroll deductions. Businesses that are nonemployers or newly registered are not included in this primary frame. To improve coverage of the target population, the list frame is supplemented by an area frame of land segments covering the entire United States. Segments are selected and all businesses within these segments are enumerated. The resulting businesses are matched to the list frame; those businesses found in the list frame are excluded from further consideration. The unmatched units are added to the sample to enable estimation for businesses not in the list frame. Choudhry et al. (1994) describe a more sophisticated variant of this approach, incorporating a two-stage area frame design.

Repeated surveys have the same frame requirements for each repetition. Thus, substantial overlap is likely between the frame data from one survey occasion to the next. Furthermore, although frames for different surveys within a program are not generally the same, an overlap or planned exclusivity usually exists between them. To integrate data across surveys, the relationships between each survey's population units have to be understood. These factors lead to the use of a general-purpose business register.

2.1.2 Distinctive Features of Business Frames

The concepts presented thus far are applicable to any survey frame, whether based on populations of persons, households, or businesses. In common with social surveys, a business survey program must consider demands for regional breakdowns and for international comparability. In contrast to social surveys, however, business surveys have distinctive characteristics that influence frame development and maintenance methods (Colledge 1989).

First, businesses tend to be heterogeneous in terms of size. Typically, the largest 1 percent of businesses in a national population accounts for over 50 percent of economic activity. Table 2.1 illustrates the assets, revenue, and profits coverage of large enterprises in Canada. Because of their impact on survey estimates, very large businesses are usually sampled with certainty in business surveys.

Second, it is often difficult to decide the units within a large business about which and from which to obtain data. Organizational structures, bookkeeping practices, and accounting systems affect the decision. To illustrate this point, Figure 2.2A indicates the structure of a business operating a chain of retail stores. The business is organized into three divisions with responsibilities for

Table 2.1   Cumulative Percentage of Assets, Revenues, and Profits of Largest Canadian Enterprise Groups, 1988

Cumulative % Enterprises Ranked by Size	Cumulative Assets (%)	Cumulative Revenue (%)	Cumulative Profits (%)
0.01	56.3	30.4	36.3
0.1	74.9	51.9	61.2
1.0	85.4	68.4	75.7
5.0	91.3	81.4	83.8
10.0	93.8	87.1	87.7
25.0	97.1	94.1	94.2
50.0	99.1	98.2	98.2
75.0	99.8	99.6	99.7
100.0	100.0	100.0	100.0

Source: Catalogue 61-210, Statistics Canada, Ottawa, Canada.

wholesale, transport, and retail activities. Two divisional offices are in different parts of the country from the head office. The retail division operates 11 sales outlets, which are grouped into branches for management purposes. Given this organization, the outlets may be capable of reporting employment and sales data. But can they report operating surplus or capital expenditure intentions? If not, from which organization units within the business are these data likely to be available?

Third, because economic performance varies by industry, few business statistics have utility without an industrial breakdown. Classification by industrial activity is vital, for instance, in analyzing financial operating ratios or research and development budgets.

Fourth, in countries with market economies, the system of national accounts (SNA) requires a wide range of business data. As Lewington notes in Chapter 33, a national program of business surveys must address SNA concepts and requirements.

These distinctive characteristics determine desirable features of business survey frames, as summarized below and elaborated in the remainder of this chapter:

- The use of standardized statistical units facilitates better specified units with improved coverage for individual surveys and increases the scope for integration of the resulting statistical products.
- The use of standardized industrial and geographic classification schemes leads to more accurate classification of business units and survey products and facilitates data integration.
- List frames based on administrative data sources are preferable to area frames because they involve lower costs and more efficient sampling.
- A business register and standardized procedures for defining frames and

Organizational Units

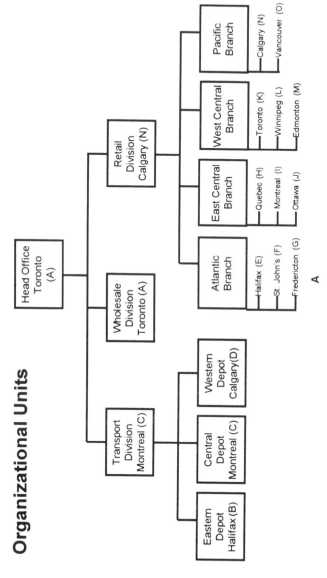

Figure 2.2 (A) Organizational structure of a hypothetical large retail business. (Statistics Canada Model.)

28

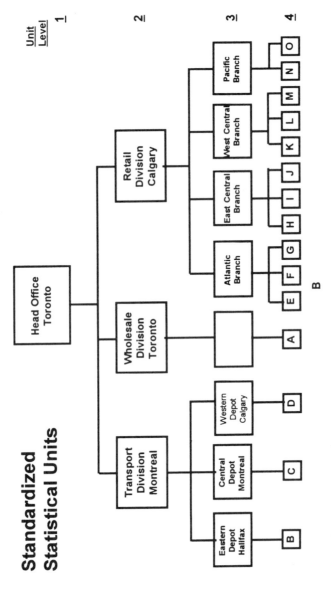

Figure 2.2 (*Continued.*) (B) Standard statistical structure of a hypothetical large retail business (Statistics Canada Model).

drawing samples lead to better coordination of surveys within a program, more control of respondent burden, and data integration.

2.2 TYPES OF UNITS AND STANDARDIZED STATISTICAL UNITS

In creating business frames, many types of units have to be considered. A large business invariably divides itself into several overlapping sets of units for legal, administrative, and operational purposes. At least two distinct sets of units can usually be identified:

- a legal structure containing one or more legal units which provide the basis for ownership, enter into contracts, employ labor, and so forth; and
- an operating structure, reflecting the way the business makes and enacts decisions about its use of resources and production of goods and services.

Businesses maintain accounting records to satisfy external administrative requirements and to support management decisions and control. Within a business accounting system, the data items reflect decision-making and reporting needs at each organizational level. Thus, the data-reporting capability of units within a business depends upon their level within the organizational hierarchy.

A program of business surveys can be viewed as a process by which input data from business accounting records are transformed into statistical outputs designed to meet users' needs. Because such needs may not be aligned with data supply, an essential aspect of this transformation is a set of standard concepts and definitions to relate inputs to outputs. Key to this process is the definition of standard statistical units.

Users of survey data have an insatiable demand for detail. However, cost and response burden impose limits. Requests for survey data must be matched to bookkeeping practices. It is unreasonable to request more detail than can be extracted from business accounts. Within this general constraint, satisfying users' demands implies that a survey's conceptual framework should enable data collection at the maximum level of detail that businesses maintain. In the context of survey frames, detail refers to the industrial and geographical breakdowns for which a business can provide economic data.

2.2.1 Defining Standard Statistical Units

Small businesses tend to engage in a single type of activity at a single location, making collection of geographical and industrial detail easy. For such businesses, economic production can be assumed to belong to a single industry at a single location. In other words, a small business can be regarded as a single statistical unit for data collection purposes.

For large businesses, the situation is not so simple. A large business may engage in a range of activities, vertically or horizontally integrated, in many locations around the country. At various sites, for example a business may have activities ranging from cutting lumber to manufacture and sale of wooden furniture (vertical integration). Another business may be involved in transport, electronic communications, and hotels (horizontal integration) across the country.

Industrial and geographical breakdowns are required in all cases. However, it is impractical to expect a large business to define its own industrial and geographical breakdown consistently over time and with respect to other businesses. Instead, in collaboration with the large business, the survey taker should determine (1) the appropriate "standard" statistical units about which data are to be obtained and (2) the organizational units from which to collect these data based on its organizational structure and bookkeeping practices. The process of setting up statistical reporting arrangements for a business is referred to as *profiling*. The use of standard statistical units ensures consistency across businesses and over time (see Chapter 6).

The statistical units about which data are obtained may differ from the collection units from which data are collected. For example, a business may decide that the chief accountant at the head office will report sales for the production plants which are the target statistical units. Figure 2.3 illustrates the various different types of units and their interrelationships.

There are many types of business data: production, commodity, employment, financial, capital expenditure, research and development, training, waste management, use of new technology, and so forth. Different collection arrangements may be appropriate for different types of data. For example, sales, shipments, revenues, employment, and wages and salary data may be available for each physical location of a business, whereas operating surpluses may be known only for divisions or branches that encompass several locations. Financial data may be obtainable only for the entire business.

The variety of survey objectives and data collected requires several types of statistical units within a business, and hence more than one set of standard statistical units. However, the underlying requirement that data across surveys be integrable implies that: (1) as few as possible standard unit types should be defined consistent with the goal of maximizing data detail within the bounds

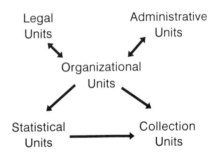

Figure 2.3 Relationships between types of units.

of data availability; and (2) the standard unit types should relate to one another at as low an organization level as possible. The latter feature ensures that data from different types of standard units within the same business can be brought together at the finest level of industrial and geographic detail. A hierarchy of standard statistical units is ideal.

There is no international uniformity regarding the most appropriate *statistical units model*—that is, the required number of different standard statistical unit types and their definitions. It is generally acknowledged, however, that a least two distinct standard unit types are necessary: one for the collection of production data and the other for financial data. Most statistical agencies define more. For example, Colledge and Armstrong (1995) describe the four-level hierarchy of statistical units used by Statistics Canada. Figure 2.2B indicates a result of applying this particular standard unit model to the organizational structure of the retail business shown in Figure 2.2A. The units model adopted by the Australian Bureau of Statistics (1989) provides an example of a five-level hierarchy. The ISIC Revision 3 model contains four standard units, not in a hierarchy. Whatever the model used, the objective is the same: to derive from the organizational structure of a business, however complicated, a standardized statistical structure for collecting survey data.

Chapter 3 discusses definitions of standard statistical units, including those contained in ISIC, Revision 3. It also outlines the relationships of these units to business bookkeeping practices, to required statistical outputs, and to national accounts concepts.

2.2.2 Defining Changes in Standard Statistical Units

For repeated surveys, frame information needs to be carried forward from one survey occasion to the next, thus enabling continuity (or controlled rotation) of the sample, which in turn leads to more reliable estimates of change and to more efficient data collection and handling of nonresponse. This consideration gives rise to the requirements to define criteria for the creation (*birth*) and the disappearance (*death*) of each type of standard statistical unit. Because the alternative to a unit's death as its continued existence, the criteria for unit death implicitly determine the criteria for unit continuity.

Births and deaths for businesses are considerably more difficult to establish than those for human populations. They depend upon the type of unit being considered, for instance. Birth and death criteria for a business' legal and administrative units are determined by legal and administrative rules and can be quite different from the criteria for birth and death of the business' organizational units. Furthermore, none of these criteria may be appropriate for defining the births and deaths of statistical units.

Consider an incorporated business comprising one legal unit with two divisions, one manufacturing and one wholesale (Figure 2.4A). Suppose each division has its own payroll account with the taxation office so that each can provide a complete set of production data, but financial data are maintained

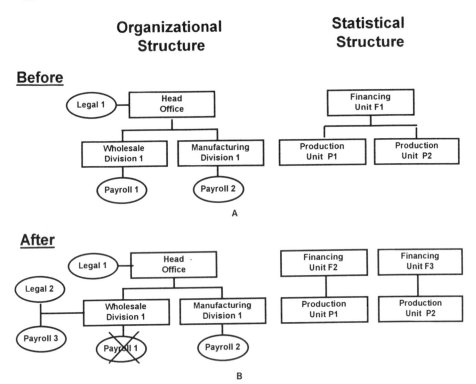

Figure 2.4 (A) Organizational and statistical structures before change (Hypothetical Case). (B) Organizational and statistical structures after change (Hypothetical Case).

only for the business as a whole. For a production survey, this business can be represented as two statistical units which happen to coincide with the two administrative units defined for payroll purposes. For financial statistics, one statistical unit is defined encompassing the entire business.

Now suppose that taxation reasons lead the business to split off one division to form a separate legal company which it wholly owns and which it operates just as before. A new legal unit has been born. The taxation office will likely require that the payroll account associated with the split-off division be closed, and that a new account be opened for the new company. Thus, one administrative unit dies, another continues, and another is born. Because operations have not changed, production statistics should continue to be based on the same two (production) statistical units as before, although one unit is now a new legal company. If all finances are still controlled from the business' central office, financial statistics should be collected from the existing financial unit, though it now spans two legal entities. However, if the new company assumes control of its finances, the existing financial unit should die and two units should be born, one for the parent company and one for the new company (see Figure 2.4B). This relatively simple example indicates the complexity in defining births and deaths.

For survey frames, births and deaths are the most significant events, but other changes occur. Units merge or split with or without losing their identities. They start new activities or cease existing ones. They expand or contract in size. They change their location. In Chapter 4, Struijs and Willeboordse provide a comprehensive framework for classifying such changes.

2.3 INDUSTRIAL CLASSIFICATION SYSTEMS

Most economic statistics, particularly those for production, are meaningful only with a breakdown by industry. For example, although gross domestic product is published for all industries combined, an industrial breakdown is needed to analyze changes in its value. Because they ensure consistent classification across surveys and over time, standard industrial classification (SIC) systems are an integral part of the conceptual framework for frames and business registers.

The basic objective of a standard classification by industry is to allocate the economic activities of each statistical unit to an industry (class) chosen from a set of mutually exclusive industries that account for all activities in the economy. Underlying the definition of these industries is the goal of realistically reflecting the way economic activities are organized, in as much detail as possible given the units being classified, business bookkeeping practices, and data availability.

Within a classification scheme, the industries are usually organized into a hierarchy, having three or four levels, which facilitates successive degrees of aggregation. For example, in Statistics Canada's 1980 Standard Industrial Classification, some 850 industries are organized into about 300 groups within 75 major groups, which in turn are contained within 17 divisions.

2.3.1 Units to Be Classified

The statistical unit being classified must be considered in defining industrial classification systems. In particular, the unit's size affects the precision of the classification. Lower-level units with their narrow range of activities can be more precisely classified than higher-level ones that tend to engage in a wider range of activities. In principle, then, different types of statistical units require different industrial classification systems. In practice, though, most national statistical agencies (Statistics Canada being an exception) use a single system for all types of units.

2.3.2 Definition of Industries

Given the type of standard unit for which the classification is to be designed, the traditional starting point for defining industries is to form a matrix within which every statistical unit within the economy can be placed. The rows of the matrix are production activities, and the columns are goods and services produced. Each industry is defined as a cell or group of cells within the matrix. The objective is to ensure that each industry is composed of units primarily engaged in the same kind of activity and producing similar goods and services. This approach was initially developed for manufacturing industries but has been broadened to other sectors. The general principles invoked when carrying it out are the following:

- The definition of most industries is based on activities deemed to be *primary* to that industry—that is, activities that characterize the industry. An activity is usually considered as primary for one industry only.
- Primary activities should account for a large proportion of the total output of the units classified to an industry. In other words, the industry should have a large *specialization ratio*, defined as the ratio of the industry's output of primary activities to its total output.
- A large proportion of the output of activities defined as primary to an industry should be produced by units within that industry. That is, the industry's coverage ratio should be high, where *coverage ratio* refers to the ratio of the output from primary activities for that industry as compared to the whole economy's total output from these activities.
- Each industry should be sufficiently large that it is economically significant, of demonstrated interest to users, and not subject to confidentiality considerations in publishing results.
- The classification should facilitate international comparison, by being as close as possible to international or multinational standards.

In practice, there are many variations on this approach, for example, basing the classification of some industries on inputs or on class of customers. Classification of government institutions requires special consideration. Possible approaches are summarized in documentation emerging from design of the 1997 U.S. Standard Industrial Classification (Triplett 1993).

2.3.3 Assignment of Industry Codes

Given a classification system, the assignment of industry codes to statistical units requires the formulation of coding conventions. For example, codes are generally based on the unit's activities over a 1-year period, not its day-to-day activities. To prevent units with two major activities or outputs of nearly equal importance from flip-flopping between industries, the coding procedures usually contain *resistance rules* which limit the frequency of changes.

2.3.4 Revision of Industrial Classifications

Any industrial classification has to be revised over time to account for changing economic activities. New technology creates new industries and causes old ones to disappear. Changes in business organization and bookkeeping practices must also be taken into account. The desire to update an industrial classification is balanced by resource implications and the loss of data continuity associated with revisions. A change of classification has a major effect on data collection and outputs and is never lightly undertaken. The average lifetime of a nationally defined standard industrial classification before the next revision is 10–15 years.

2.4 GEOGRAPHICAL CLASSIFICATION SYSTEMS

Geographical breakdowns of statistical data are required for purposes such as allocating public funds for regional development or private investments in resorts, hotels, or shopping centers. Often such decisions require statistics from several business surveys—for instance, retail trade, employment, and capital investment surveys. So these data can be readily combined, *standard geographical classification* systems should be used. Definition of a standard geographical classification is more straightforward than for industrial classification because the requirements are easier to specify and international comparability is not a major factor.

Two basic starting points exist for developing geographical classifications for business statistics: (1) administrative systems with a regional dimension such as postal codes, electoral districts, local municipalities, provinces, states, and regions and (2) the land-based statistical system used for the population census.

Virtually all geographical systems are hierarchical in that sets of smaller areas roll up into larger ones. In Canada, for example, the postal code comprises six alphanumeric characters; the first three refer to forward sortation districts within which the remaining three characters determine specific areas.

In general, geographical systems cannot be easily related to one another except at high levels of aggregation such as province, state, or metropolitan area. Nevertheless, classification at this high level of commonality is usually sufficient for stratification purposes and for many business statistics.

As Nijhowne indicates in Chapter 3, the major problem associated with geographical classification is not the lack of a system, but that the activities of large businesses are dispersed and cannot easily be allocated to specific areas.

2.5 BUSINESS REGISTER CREATION AND MAINTENANCE

The preceding sections outline the conceptual framework for a program of business surveys in terms of frame requirements and the need for standard statistical units, standard industrial and geographical classification systems, and rules for handling changes. This section describes how such concepts can be operationalized through the creation, maintenance, and use of a register capable of providing frames for most, if not all, business surveys within a program. In Chapter 5, Archer discusses register operations in more detail.

2.5.1 Creation and Maintenance Principles

Although a business register may have secondary objectives, its primary role is to provide good quality frames for a program of surveys. This implies that the register should: (1) contain standard statistical units from which a frame can be generated for each survey, providing complete, current, and non-duplicated coverage of the survey's target population and facilitating data integration over survey repetitions and across surveys; (2) contain accurate classification and contact data for the statistical units to enable efficient survey operations; (3) generate frames at minimum total cost for given quality over survey repetitions and across the program of surveys; and (4) be readily accessible to, usable by, and updatable by survey operations.

In principle, a register can be built from scratch by enumerating the businesses within the required area, which means the whole country for a national register. In practice, enumeration is an expensive process, used only occasionally for registers covering small areas. The more cost-effective approach is to base the register on a list that is maintained for an administrative or commercial purpose. Examples of administrative lists are businesses registered for value-added taxes, employers making payroll deductions or unemployment insurance payments for employees, and corporations filing income tax returns. Possible commercial sources are businesses registered by chambers of commerce or by telephone or electricity utilities.

The benefit of an administrative or commercial source is that it provides not only a relatively inexpensive starting point for constructing the register but also ongoing information for register maintenance. The legislation supporting the administrative process may set limits on the use of the administrative data that results (for example, see Jabine and Scheuren 1985).

However, a single administrative or commercial list usually cannot satisfy all of the register's data requirements. Its coverage may be inadequate or its data items may be insufficient for classification and contact. Invariably, the register will use supplementary information from other sources, in particular (1) other administrative or commercial lists, (2) direct surveys conducted by the register itself, and (3) surveys for which the register provides frames (see Figure 2.5). In addition, a survey frame drawn from a list-based register may be supplemented by an area frame.

Units of different sizes create different problems for register development and maintenance. For small businesses, the main problems are the large numbers of units to be maintained, the high birth and death rates, and the difficulty in obtaining accurate classification data. For large businesses, the problems are characterized by difficulties in defining appropriate statistical units and reporting arrangements for complex organizations and in keeping up to date with organization changes. In view of these differences, register creation and maintenance procedures are usually structured according to unit size, with reliance on administrative data for smaller units and on direct data collection for larger ones.

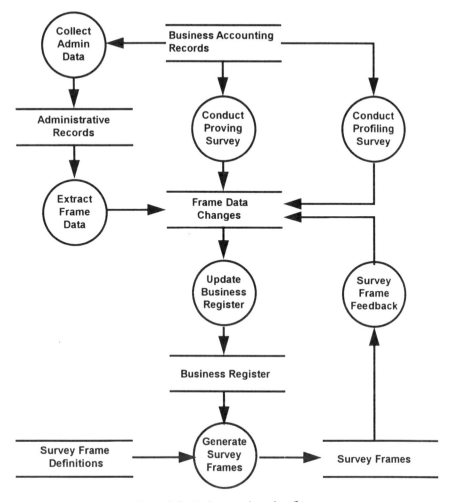

Figure 2.5 Business register data flows.

2.5.2 Data from Administrative and Commercial Sources

The use of administrative or commercial data sources is appropriate for the
creation and maintenance of register records for small businesses because administrative and statistical units usually coincide. When no single source provides complete coverage of all statistical units, the primary source can be supplemented with data from other administrative or commercial systems that
include additional businesses. The problem with this approach is that the data
from multiple sources must be matched and the duplicated units identified and
removed to prevent upward bias in survey estimates. Frequently, the sources
do not share a common identification numbering system, and thus matching

involves linkage of records based on name, address, and other characteristics. In view of the large number of units to be linked, automated record linkage software is a vital component of register systems. As described by Winkler in Chapter 20, a variety of linkage algorithms are in use around the world. One characteristic they have in common is that as the first step, they cast the business's name and address data into a standard format.

Although automated record linkage techniques enable fast and inexpensive matching, doubtful links have to be examined manually, which is a costly operation. In addition, links are missed, leading to unknown amounts of duplication. These effects leads most statistical agencies to rely on one, or at most two, administrative source(s) for register creation and maintenance. The choice of sources varies according to the available data sources. Statistics New Zealand's register is based on value-added tax data. The U.S. Bureau of the Census and the Australian Bureau of Statistics use payroll deduction data. The United Kingdom Central Statistical Office uses value-added tax and payroll data. Statistics Netherlands takes data from Chambers of Commerce.

In using administrative or commercial sources for register maintenance, register staff lose control over the volume, frequency, and quality of incoming data. Coordination, cooperation, and contracts with the corresponding administrative or commercial agencies are essential to ensure data quality and to prevent unexpected changes or disappearance of data when administrative processes change. In France and some Scandinavian countries, administrative and statistical registration functions are integrated within a single organization. The Institute National de la Statistique et des Études Economiques (INSEE) in France, for instance, registers businesses for administrative and statistical purposes simultaneously. This reduces duplication in data collection, but at the expense of having a system which is more complex and difficult to change.

Classification and contact data from administrative or commercial sources invariably have to be recoded in a suitable format for the register. In particular, business activity descriptions from administrative forms must be converted to standard industrial classification codes. Thus, automated and computer-assisted industrial and geographical coding systems are common features of the software for register operations. Miller (1993) describes automated coding at Statistics Canada. Frequently, an administrative description does not contain sufficient information to assign a unique and accurate code, and additional manual investigation is required. For example, Estevao and Tremblay (1986) report that industry codes could be assigned for only 70 percent of the activity descriptions obtained from the payroll deduction source used by Statistics Canada's business register.

2.5.3 Data from Register Surveys

Data from administrative and commercial sources alone are insufficient for creation and maintenance of a business register. Administrative units may not be appropriate as statistical units for larger businesses. Information about births and deaths may be untimely. Industrial, geographical, and size classification data may be incomplete or inaccurate. To make up for deficiencies, registers collect data from businesses directly. These *register surveys* vary in function and format according to the data required and the size of the business.

For small businesses, the primary purpose of a register survey is to obtain or validate unit activity status and classification data. The survey instrument is typically a mail questionnaire or a brief telephone interview. Such surveys are referred to as *proving surveys*.

For large businesses, register staff focus on determining the business organization and bookkeeping practices, defining standard statistical units and classifications, and establishing reporting arrangements. This process, referred to as *profiling*, involves a preparatory investigation of the business, a face-to-face interview with senior representatives, and a subsequent review of the information collected with survey staff. In Chapter 6, Pietsch describes the profiling process and its relationship to survey procedures.

2.5.4 Data Feedback from Surveys

The surveys for which the register provides frames are another source of information for register maintenance. Business surveys often include questions designed to update frame data. Frame data acquisition by this means can become an integral part of a register's maintenance program.

For efficient operation of repeated surveys, the register must record unit, classification, or contact information obtained during survey operations so that it is available for subsequent occasions. For example, evidence that a business is no longer operational or that the contact address has changed should be recorded. Indications from a survey response that the business has reorganized, suggesting possible changes in its statistical structure, should also be fed back to the register for further investigation.

Survey feedback channeled through the register can benefit other surveys. For example, information that a business is bankrupt will remove the need for nonresponse follow-up by a subsequent survey. However, as a frame data source, survey feedback has its limitations. In general, survey processes will not detect new businesses or shifts in industrial activity. Thus, survey feedback must be regarded as supplementary to other data sources for register maintenance, not as the complete updating mechanism.

Furthermore, there is a technical limitation to the use of survey feedback in the case of repeated sample surveys in which there is controlled overlap of the sample from one repetition to the next. The procedures for updating the register with survey feedback have to be carefully devised; otherwise bias may be introduced into samples for subsequent survey occasions. For example, suppose that survey units identified as dead were simply deleted from the register. Then a future sample, drawn with deliberate (usually maximum) overlap with the previous one, would have fewer dead units and would under represent the number of dead units in the population. The same limitation applies to surveys within a program for which there is controlled overlap between samples.

2.5.5 Other Aspects of Register Functionality

One component of a survey frame specification is the reference period. Different surveys may have different reference periods: month, quarter, year, and so forth. To generate a frame with the appropriate reference period for each survey, the register must be able to provide more than today's view of its data holdings. The register must possess *temporal functionality*, that is, the capacity of providing a snapshot of frame data for any specified reference period. For example, as of January, year Y, the register should be capable of generating lists of units (1) in scope for the January, year Y monthly retail survey and (2) in scope for the year Y-1 annual retail survey. The first list requires a current snapshot, the second a historical one. As described by Cuthill (1989), procedures for providing this functionality vary in complexity, from a *rollback database* in which snapshots are recorded and can be retrieved but never updated, to a *full temporal database* that enables snapshots for any reference period to be recorded, updated, and retrieved as they were at any point in time.

Given the high volume and diversity of updating information received by the register, maintenance processes must be automated. However, many situations require manual investigation. Thus, automated scheduling of manual maintenance operations is also a useful facility. Systems for measurements and control of quality and performance are vital not only to ensure optimal allocation of register resources but also to inform survey managers of the likely impact on survey estimates of register updating processes and errors.

From time to time, an element of the conceptual framework underlying the register changes. For example, the definitions of statistical units are modified to reflect changes in business organization structures, or a revised industrial classification system is introduced. Such changes cause discontinuities in frame data and hence in survey estimates, which can be accommodated only by means of parallel runs for a period of time. Register functionality enabling production of frames under the old and new concepts is thus useful. In Chapter 7, MacDonald describes how revisions in the industrial classification can be handled.

2.6 OTHER USES OF BUSINESS REGISTERS

In addition to providing frame data, a register may have other functions. For registers maintained by national statistical agencies, these secondary functions are becoming increasingly important and must be considered in register design and operation (see Chapter 5).

2.6.1 Measurement and Control of Respondent Burden

As the source of survey frames and samples, the register is ideally placed to record which units are currently in sample for which surveys. In addition, the register's functionality can be extended to control overlap between successive samples of a repeated survey and across surveys. In Chapter 9, Ohlsson reviews burden control techniques that begin with the assignment of a permanent random number to each statistical unit in the register. These numbers form the basis for random selection and controlled overlap of samples across surveys and over time.

2.6.2 Provision of Lists for External Clients

Registers maintained by national statistical agencies may be used to provide frames and lists for organizations outside the agency, usually on a commercial basis. Except for registers specifically set up for joint administrative and statistical purposes, such activities are invariably curtailed by privacy considerations and by confidentiality provisions of the statistical act under which the agency operates. Commercially run business registers are not limited by such constraints.

2.6.3 Production of Small Area Economic Statistics

The industrial, geographical, and size classification data for the lowest level of statistical units provide a basis for the register to produce rudimentary small area statistics. If this function is considered important, the classification detail in the register may be expanded beyond that needed for survey frames. Typically, three specific types of data enhancement are required to supplement data obtained from administrative sources.

By definition, the geographical component of small-area data is of particular importance; accuracy is crucial because users can easily detect errors in geographical coding. In many cases, however, geographical codes for small statistical units are derived from an address field in the administrative record. These addresses may not correspond to the business' physical location, but to the organization or individual providing the administrative data such as the business' payroll agency, taxation accountant, or some other office. Therefore, the geographical code must be validated to ensure that they refer to the business' actual location.

The information available from administrative sources may not be sufficiently detailed, precise, or current for small-area statistics. For example, counts of businesses by sales range may be wanted, whereas the register draws its basic data from payroll deduction records having only numbers of employees and remittances. For distribution industries, statistics on floor area are commonly requested. Such information is unlikely to be available from any administrative source.

2.6.4 Production of Business Demographic Data

Demand is growing for information on *business demography*—the business equivalent of births, marriages, deaths, health, and prosperity (Haworth and Kerr 1993). Governments and businesses need such information in deciding how to achieve higher employment, increased capital formation, profit enhancement, and so on. Generally, these decisions involve when and how to invest resources, to make organizational changes, to introduce new policies regarding business conduct, and so forth. Such decisions can be translated into specific questions concerning business demography.

For small businesses the questions pivot on the incidence of "infant mortality." How many births occur in a year? How is life expectancy distributed? When does an infant business reach "maturity" in the sense of being indistinguishable from the average business in the same sector? What are the characteristics of "infant mortality?" What are the attributes of healthy "infants?" What are the typical sequences of events in the passage from birth to maturity? What changes of direction occur? Which forms of assistance promote infant health and progress to maturity? What are the attributes of "old age?" What determines whether a small business becomes large, stays small, or reaches old age?

For large businesses the questions concern the business counterparts of marriage, divorce, and reproduction. How many mergers, amalgamations, and takeovers occur? How many split-ups or spin-offs? What are the attributes likely to lead to these events? What are the most typical sequences of events? What forms of assistance or constraint are effective?

These questions underlie the drive elaborated in Chapter 4 for precise definitions and data concerning business formation, cessation, births, deaths, amalgamations, mergers, takeovers, spin-offs, and changes of organization, activity, and industry. The capacity to produce such data goes well beyond the basic requirements to generate frames. It can only be achieved through longitudinal tracking of a cohort of businesses over time; cross-sectional snapshots are insufficient.

2.7 QUALITY AND PERFORMANCE MEASUREMENTS

Quality and performance measurements guide decisions regarding the development and operation of a business register in the achievement of client satisfaction. The primary objective of a national business register is to provide frames for the survey program conducted by the agency. Thus, the register's most important clients are the survey operations it supports. In serving these clients, however, the register is indirectly meeting the needs of external clients—that is, the end users of the agency's survey products. Secondary objectives such as the provision of basic, small-area, and demographic statistics directly involve external clients. The focus of quality and performance measurements depends upon the priorities assigned to the various clients.

2.7.1 Definition of Register Quality

Frame quality is not simply a function of data accuracy, although it is important. The quality of any statistical product, including a frame, can be thought of as having four dimensions: relevance, accuracy, timeliness, and cost (Fecso

1989). Measuring register quality requires an evaluation along these dimensions of the register's principal outputs—frames for surveys.

The relevance of a register may be measured in terms of (1) the extent to which the register provides frames matched to survey target populations, (2) the accessibility and ease of use the generated frames, (3) the extent to which register data are actually used, and (4) the degree to which use of the register facilitates integration of data outputs across the survey program.

Sometimes, a register's capability is not fully exploited because one or more surveys use an alternative frame generation function instead of the register. This situation is not uncommon in cases where survey operations predate the register. It limits register quality.

Register errors may be caused by the inherent limitations of input data, or by delays and errors in data acquisition and processing. Register accuracy may be measured (for each survey, and overall) in terms of (1) coverage errors: (missing units, duplication, and extraneous units), (2) classification errors (units not classified or misclassified by industry, geography, or size), (3) contact errors (units with incomplete or incorrect contact data), (4) the impact of these errors on survey collection and processing operations, and (5) the impact of these errors on survey outputs.

Timeliness can be measured in terms of (1) time required to generate frames relative to the rate at which frame data change over time and (2) the currency of frame data.

Costs can be measured in terms of: (1) the costs incurred in providing frames, and the ratios of frame costs to overall costs for each survey; (2) the frame related costs incurred by survey operations in supplementing or duplicating register operations; and (3) the total register processing costs as a proportion of total survey program budget.

Quality measurements must be analyzed within the context of register unit counts, processing volumes, and changes in counts and volumes. For example, the proportion of units erroneously indicated as active when actually dead must be seen within the context of the annual unit death rate experienced by the register. A large annual death rate combined with a high cost of obtaining timely information about deaths limits the resources that are worth expending to improve accuracy in this respect. The effects on survey outputs must also be considered in assessing the significance of errors. For example, survey estimates can be adjusted to account for dead units.

Register products are a function of inputs and processing. To complement an evaluation of products, quality measurements of register input data and processes are required. For administrative and commercial data sources, the basic issues are which databases to access and how best to use them. Thus, measurement requirements center on content deficiencies, definitional discrepancies, item completion and error rates, time lags, and acquisition costs. For processes, the main concerns are efficiency and minimizing the impact of processing errors and delays on output data quality. Thus, measurement needs include processing volumes, costs, error rates, backlogs, and computer processing and systems development audit trails.

2.7.2 Users and Uses of Quality Measurements

Register managers use input data volumes, acquisition costs, and error rates in negotiating arrangements with suppliers and assessing alternative inputs. They analyze processing volumes, costs, backlogs, and error rates to determine changes in processing priorities, procedures, and training to reduce costs and error rates. In ongoing discussions with clients, register staff use unit volumes, changes in volume and error rates, and the associated impact on the frames which the register generates.

As the register's clients, survey managers consider unit volume changes, error rates, and associated impacts on survey outputs as the basis for (1) discussing possible improvements in frame data supply with register staff; (2) compensating for artifacts in survey estimates due to changes, delays, or other irregularities in register operations; and (3) providing users of survey outputs with explanations of frame-related artifactual changes remaining in estimates.

The survey program's senior managers review data acquisition and processing costs and statistical unit volume changes, error rates, and associated survey output impacts, in conjunction with cost and impact measurements for other survey functions. Their goal is to allocate developmental and operational resources to the register and the other survey functions on a scientific basis— for example, through use of total error model as described by Linacre and Trewin (1989).

2.8 RESEARCH AND DEVELOPMENT

Many statistical agencies engage in ongoing research and development of techniques and systems for business frames and registers. Current areas of research include the following topics:

- the use of multiple databases in creating and maintaining a register—in particular, monitoring the quality of incoming administrative data and developing multiple frame techniques;
- coordination of statistical and administrative functions including (1) cooperative arrangements between statistical and administrative agencies, (2) combining administrative and statistical unit registration in an all-purpose register, and (3) dealing with confidentiality constraints and with legally active but operationally inactive units.
- area frame development and maintenance, particularly definition of sampling units and size measures, computer-generated maps, replacement of area frames by list frames, and use of area frames for measuring undercoverage and other errors;
- frame maintenance strategies such as procedures for use of survey feedback without biasing the selection of future samples and determination of the appropriate frequencies for validating, reclassifying, and reprofiling units;

- industrial classification, including use of different classification schemes for different types of units, limitations in precision of classification code assignments based on activity descriptions, and automated and computer-assisted coding;
- the use of postal codes and grid reference systems for geographical classification;
- developments in automated record linkage systems;
- quality measurement and control of frame data, such as assessing and compensating for the impact of frame errors on survey estimates and using measurements in allocating developmental or operational resources;
- performance and cost measurement and control—in particular, use of performance data in allocating resources—and assessment of, and compensation for, artifacts in survey estimates due to changes, delays, or other irregularities in register operations;
- measurement and control of respondent burden, including systems for recording and summarizing contacts with individual units and for control of overlap across surveys; and
- use of a business register for small-area and business demographic data.

Meeting annually since 1986, the International Round Table on Business Survey Frames provides a forum for discussion of these topics and a source of technical reports. Selected papers from the first five Round Tables are being brought together as a single volume (Statistics Canada 1994). Papers from the sixth and seventh Round Tables are available from the host countries, the Australian Bureau of Statistics (1991) and Denmark's Bureau of Statistics, or from the individual authors.

REFERENCES

Armstrong, J., and H. St-Jean (1993), "Generalized Regression Estimation for a Two-Phase Sample of Tax Records," *Proceedings of the International Conference on Establishment Surveys*, Alexandria, VA: American Statistical Association, pp. 402–407.

Australian Bureau of Statistics (1969), "Australian Year Book," Belconnen, Australia.

Australian Bureau of Statistics (1989), *Statistical Units Definitions and Rules*, technical report, Belconnen, Australia: Industry Division.

Australian Bureau of Statistics (1991), *Sixth International Round Table on Business Survey Frames*, Belconnen, Australia: Industry Division.

Bonnen, J. (1981), "Improving the Federal Statistical System: Issues and Options," Washington, DC: The President's Reorganization for the Federal Statistical System.

Choudhry, H., M. J. Colledge, and J. Mayda (1994), *Redesign of the Hungarian Monthly Retail Trade and Catering Survey*, working paper, Ottawa: Statistics Canada.

Colledge, M. J. (1989), "Coverage and Classification Maintenance Issues in Economic Surveys," in D. Kasprzyk, G. Duncan, G. Kalton, and M. P. Singh (eds.). *Panel Surveys*, New York: Wiley, pp. 80–107.

Colledge, M. (1990), "Integration of Economic Data: Benefits and Problems," *Proceedings of Statistics Canada Symposium: 90 Measurements and Improvement of Data Quality*, Ottawa, Canada: Statistics Canada, pp. 51–63.

Colledge, M., and G. Armstrong (1995), "Statistical Units, Births and Deaths at Statistics Canada after the Business Survey Redesign," *Selected Papers from the First Five International Round Tables on Business Survey Frames*, Ottawa: Statistics Canada, in press.

Colledge, M., V. Estavao, and P. Foy (1987), "Experience in Coding and Sampling Administrative Data," *Proceedings of the Survey Research Methods Section, American Statistical Association*, pp. 529–534.

Cuthill, I. (1989), "The Statistics Canada Business Register," *Proceedings of the Annual Research Conference*, Washington, DC: U.S. Bureau of the Census, pp. 69–86.

Estevao, V., and J. Tremblay (1986), *An Evaluation of the Assignment of Standard Industrial Codes from PD-20 Data*, technical report, Ottawa: Statistics Canada.

Fecso, R. (1989), "What Is Survey Quality: Back to the Future," *Proceedings of the Survey Research Methods Section, American Statistical Association*, pp. 88–96.

Haworth, M. F., and J. R. Kerr (1993), "The Demography of the Business Sector and the Use of Statistics Derived from Business Register and Insolvency Records," *Proceedings of the International Conference on Establishment Surveys*, Alexandria, VA: American Statistical Association, pp. 8–17.

Isaki, C., C. Konschnik, and N. Monsour (1984), *Reselection of an Area Sample for the Retail and Service Surveys*, technical report, Washington, DC: U.S. Bureau of the Census.

Jabine, T. B., and F. Scheuren (1985), "Goals for Statistical Uses of Administrative Records: The Next 10 Years," (including discussion), *Journal of Business and Economic Statistics*, **3**, pp. 380–403.

Linacre, S. J., and D. J. Trewin (1989), "Evaluation of Errors and Appropriate Resource Allocation in Economic Collections," *Proceedings of the Annual Research Conference*, Washington, DC: U.S. Bureau of the Census, pp. 197–209.

Miller, D. (1993), "Automated Coding at Statistics Canada," *Proceedings of the International Conference on Establishment Surveys*, Alexandria, VA: American Statistical Association, pp. 931–934.

Sands, M. S. (1993), "Frame Creation for the Survey of Minority Owned Business Enterprises and the Survey of Women Owned Businesses," *Proceedings of the International Conference on Establishment Surveys*, Alexandria, VA: American Statistical Association, pp. 775–780.

Statistics Canada (1980), *Standard Industrial Classification*, Catalogue 12-501E, Ottawa, Canada: Minister of Supply and Services.

Statistics Canada (1990), *Trucking in Canada*, Catalogue 53-222, Ottawa, Canada: Minister of Supply and Services.

Statistics Canada (1994), *Selected Papers from the First Five International Round Tables on Business Survey Frames*, Ottawa: Statistics Canada.

Struijs, P., and A. Willeboordse (1992), "Terminology Definitions and Use of Statistical Units," presented at the Seventh International Round Table on Business Survey Frames, Copenhagen, Denmark.

Triplett, J. (1993), "Economic Concepts for Economic Classifications," Survey of Current Business, **73,** No. 11, U.S. Department of Commerce, Bureau of Economic Analysis, Washington, DC.

U.N. Statistical Office (1990), *International Standard Industrial Classification of All Economic Activities—Revision 3*, Sales No. E.90.XVII.11, New York: United Nations.

Wright, T., and H. Tsao (1983), "A Frame on Frames: An Annotated Bibliography," in T. Wright (ed.), *Statistical Methods and the Improvements of Data Quality*, New York: Academic Press, pp. 25–72.

CHAPTER THREE

Defining and Classifying Statistical Units

Shaila Nijhowne[1]
Statistics Canada

Economic statistics describe the behavior and activities of businesses and the transactions that take place between them. National statistical agencies conduct business surveys to compile a variety of economic statistics. To carry out these surveys, a frame is required of target statistical units, the units of observation to be surveyed or sampled. This chapter discusses the definition and classification of statistical units required for a program of economic statistics.

In the context of this chapter, a *business* is an economic transactor with the autonomy, authority, and responsibility for allocating resources for the production of goods and services. It may consist of one or more legal transactors. The term "business" encompasses farms, incorporated and unincorporated businesses, and government enterprises engaged in the production of goods and services. It covers government institutions and agencies engaged in the production of noncommercial or nonmarketed services, as well as organizations such as unions, professional associations, and charitable or nonprofit organizations providing services to their members or to the general public.

The universe of businesses includes small businesses engaged in one or a very few activities, as well as large and complex businesses engaged in many different activities, horizontally or vertically integrated. The activities of busi-

[1]The assistance of Statistics Canada's Standards Division staff is gratefully acknowledged. This chapter draws on ideas developed by Gérard Côté. Frank Pope and Daniel April offered comments and suggestions. Moreno Da Pont, Cathy Connors, and Sunanda Palekar assisted with final preparation of the chapter and index. Comments from Peter Struijs, Michael J. Colledge, Leon Pietsch, and Ad Willeboordse were very helpful. The views expressed in this chapter are those of the author and do not necessarily reflect those of Statistics Canada.

Business Survey Methods, Edited by Cox, Binder, Chinnappa, Christianson, Colledge, Kott.
ISBN 0-471-59852-6 © 1995 John Wiley & Sons, Inc.

nesses can be undertaken at, or from, one or more geographical locations or areas.

An *activity* takes place when resources, such as labor, capital, raw materials, and other intermediate inputs, are combined to produce goods or services associated with a production process. A production process may produce alternative products, joint products, or a principal product and a by-product. An activity may encompass one simple process, such as fabric weaving resulting in the production of woven fabric. It may also consist of a series of separated or integrated production processes, such as casting, forging, welding, assembling, and painting, to produce an automobile (U.N. Statistical Office 1990, p. 9). Some activities carried out within the business result in the production of goods and services for internal consumption. Examples of such activities, referred to as *ancillary activities*, are support services provided to the rest of the business by head offices and accounting and computer departments.

Businesses have units *at which* or *from which* they undertake the economic activity of producing goods and services. Production usually takes place *at* a particular location—for example, at a mine, a factory, or a farm. On the other hand, the activity of producing services may take place *from* a certain location. For example, the activity of a construction business is often treated, for statistical purposes, as being delivered from the office of the construction company, where accounts are maintained and work is planned and organized. Construction crews and machinery are directed to building sites. These sites are not regarded as the geographical location of the construction activity, even though they are the location of the physical output of the activity. The same situation exists for services, such as those of engineering consultants, where the activity is regarded as delivered from the office of the business, by persons who deliver the service to the client at another location.

In large and complex businesses, the units at which or from which production takes place are grouped for management, administrative, and decision-making purposes into hierarchical structures. Higher-level *organizational units* own, control, or manage the lower-level *production units* at which production decisions are made or production takes place. A business may be structured along geographical, legal, functional, or operational lines. Businesses may have one structure or several structures to carry out different functions or to serve different purposes.

In these businesses, management of the business' financial affairs usually occurs at a higher organizational level than does management of production operations. The accounting systems of businesses usually reflect this management structure by mirroring the hierarchy of management responsibility for the business' operations. The accounts required to support the management and decision-making functions, whether financial or production, are usually maintained for the corresponding level of management responsibility.

Businesses also have a *legal structure*. They define and register themselves in terms of legal units for the ownership of assets. These legally constituted units or groups of units form the legal base of the business. A business derives its autonomy from the common ownership and control of its resources irrespective of the number of legal units under which it registers them. Businesses usually use their legal structure and its associated legal units to submit corporate tax returns to government revenue authorities. They may use the same or different units for other administrative purposes such as remitting payroll or value-added taxes to government authorities.

In small businesses, the operational and legal structures often coincide and may even be embodied in a single unit. For large businesses, the operational structure may be different from the legal structure, coinciding with it only at the highest level of the business. In such cases, the organizational and production units of the business's operational structure may differ from the units of their legal structure.

For economic analysis, two main types of data are required to describe the economic activities of businesses: (1) financial statistics organized by institutional or other sectors and (2) production statistics classified by industry and (in some countries) by geographical area. Usually the data are required for activities carried out within, or from within, domestic boundaries. The two types of data are required separately, as well as integrated into the system of national accounts (see Chapter 33).

Given the varied sizes of businesses, their unique operational structures and accounting systems, the multiplicity of their activities, and the many geographical locations at or from which they operate, it is necessary to define *standard statistical units*. These units facilitate the collection of integrated, consistent, comparable, and unduplicated financial and production data across all businesses, large and small.

Economic statistics draw upon the accounting records of businesses. The records that are maintained in support of financial decision-making, management, and control provide the data required for financial statistics. Such records include consolidated profit and loss accounts and balance sheets of assets and liabilities. The source of information for production statistics and labor income statistics are cost accounts. These cost accounts record operating revenues earned from the sale of goods and services and the associated costs, wages and salaries, depreciation, and operating profits.

3.1 DEFINITION OF STANDARD STATISTICAL UNITS

The target statistical unit required for consistent and unduplicated coverage of the universe of businesses depends upon the measurement objective of the survey—that is, the data to be collected and the desired industrial homogeneity and geographical precision.

A survey of the practices of national statistical agencies has shown that there is no international standardization of statistical units on business registers (Struijs and Willeboordse 1992). Instead, across countries many differences are found in definitions and in the terminology used to describe the units. The same types of units are sometimes given different names, and the same names may denote different concepts. For this reason, this chapter adopts a set of names and definitions, which do not necessarily correspond to any country's practices. For a specific country's methodology and definitions, interested readers should consult papers presented at the International Round Tables on Business Survey Frames (Statistics Canada 1995) or documentation from national and international statistical offices.

Most national agencies distinguish units at two levels: (1) the level at which major financial decisions are made and (2) the level at which production decisions are made or at which production takes place. Because large and complex businesses maintain financial and production data for different levels of the business in their accounting systems, a minimum of two types of statistical units must be defined which generally bear a hierarchical relationship to one another. In this chapter they are referred to as the enterprise (for financial statistics) and the kind-of-activity unit (KAU) or the establishment (for production statistics).

Whether the KAU or the establishment is used for production statistics depends upon the level of management responsibility, the completeness of accounting records, and the geographical precision desired for the survey. In practice, both types of units may engage in principal and secondary activities, but the accounting records of the KAU tend to be more complete than those of the establishment. In addition, activities of a KAU may be carried out at or from one or more locations or geographical areas, whereas the activities of an establishment are undertaken at or from one geographical location or area.

The International Standard Industrial Classification of all economic activities (ISIC Rev. 3) uses the terms ''local unit'' for a unit that engages in more than one activity at one location, and it uses the term ''establishment'' for the unit that engages in only one activity at one location (U.N. Statistical Office 1990, p. 19). In this chapter, *establishment* is used to cover both situations because statistical agencies commonly label their production units as ''establishments'' even though they engage in principal and secondary activities at a single location or area.

In terms of industrial activity, large and complex enterprises own and control a number of production units undertaking different activities and will inevitably be less homogeneous than their component units. The more narrowly the KAU or establishment is defined in terms of economic activities, the more homogeneous the resulting industrial statistics will be. However, the actual organization of businesses' production and accounting records constrains the definition of production units, unless the statistician is willing to artificially allocate direct costs and overhead expenses. For industrial statistics, many agencies consider it desirable to minimize such allocation.

3.1.1 Types of Standard Statistical Units

For financial statistics, the enterprise is the standard statistical unit of interest. The *enterprise* is the organizational unit of the business that has autonomy with respect to financial and investment decision-making, as well as authority to allocate resources for the production of goods and services. The enterprise is the level at which financial and balance sheet accounts are maintained from which international transactions, an international investment position (when applicable), and the consolidated financial position can be derived. The enterprise has KAUs or establishments under its ownership and control. The legal basis of the enterprise can consist of one or more legal units, and it can own and control one or more KAUs or establishments. For compiling domestic accounts for the system of national accounts, the enterprise is defined to include only those KAUs or establishments whose activities are conducted within or from within domestic boundaries.

For production statistics, the KAU or the establishment is the unit used. This unit has management responsibility for production operations. Both KAUs and establishments engage in a principal economic activity, but they may also engage in secondary activities belonging to other classes of the industrial classification. The choice of KAU versus establishment depends upon the relative level of management responsibility and geographical precision desired. In large countries there is often a need to compile production data for small geographical areas and to know the geographical area at which or from which economic activity is conducted. Where there is limited need for geographical precision and emphasis is placed on autonomy (the level of managerial responsibly with respect to production decisions) and the direct availability of operating revenues, operating costs, and detailed commodity outputs, inputs, and value added, the choice of unit is the *kind of activity unit*, the production unit whose activities are conducted at or from more than one geographical location or area. When the location of economic activity is of concern, the unit of choice will be the more narrowly defined *establishment*, the production unit whose activities are conducted at or from one geographical area, and for which operating revenues by commodity and associated costs can be obtained and value added can be derived. To derive value added for production units belonging to large and complex businesses, countries using the establishment as defined here may need to supplement data directly available for the establishment with information from a higher-level unit on purchased services, operating profits, and other needed data.

To cover all the productive activities of an enterprise, it is necessary to measure the costs associated with *ancillary activities*, the production of goods and services for consumption within the enterprise. Two issues arise, one relates to whether ancillary activities need to be identified as separate units and the other (which is the subject of a later section) concerns their industrial and geographical classification.

Generally, ancillary activities are not treated as the activities of separate producing units in their own right; their costs are simply added to the principal activity of the producing unit to form the economic activity of a KAU or establishment. However, when an ancillary activity is conducted at a different geographical location from the main activities of the enterprise, it is useful to identify it as a separate unit for collecting costs, labor, and capital-employed data, even though the activity does not generate revenues. The ancillary unit can then be assigned its own geographical location, whatever the choice of industrial classification. (For a more complete definition and discussion of principal, secondary, and ancillary activities, see U.N. Statistical Office 1990, pp. 9–12.)

The need for special information about every geographical location at which activity takes place may also require another statistical unit, a *local-kind-of-activity unit* (LKAU). Every geographical location of a KAU is an LKAU. If a country uses an establishment that covers a geographical area, it may also need to identify every geographical location at which activity takes place. This lower-level unit may be required to compile specific information about all industries or about particular industries. Some countries using the KAU or establishment for production statistics collect information about investment or employment for every geographic location. Others use this unit to compile information for particular industries, such as the number of rooms and occupancy rates of tourist accommodations, hotels, motels, and campgrounds.

There is yet another unit useful for particular statistical purposes. The *global enterprise* is the unit of common ownership and control whose boundaries, in terms of economic activities, can transcend the domestic boundaries of a country. At a level higher than the enterprise, it is used to identify the country in which ownership and control of domestic enterprises resides. It refers both to commercial enterprises and to labor unions and organizations, where membership or ownership may transcend domestic boundaries. The relationship of the units is diagrammatically shown in Figure 3.1.

Whether or not a particular type of unit needs to be recorded and maintained on a business register depends upon whether the statistical program needs to survey it directly. As a rule, it is necessary to record only two statistical units on the business register, namely, the enterprise and the KAU or establishment. For some purposes, the ancillary unit is useful. Though specific information

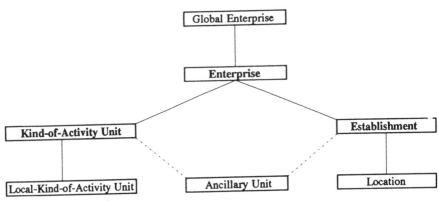

Figure 3.1 A diagram depicting the logical hierarchical relationships among the types of units of the statistical structure.

about geographical locations below the KAU and the establishment may be useful, it may not be necessary to maintain the LKAU on the business register if the required information can be obtained from the KAU or establishment. Similarly, the global enterprise is usually not needed as a separate unit on the business register. As long as the global enterprise that owns each statistical enterprise is identified, data for the global enterprise can be compiled from that of the enterprise.

3.1.2 Identification of Standard Statistical Units

The statistical units of large and complex businesses are delineated in a process referred to as *profiling*. Profiling identifies the business, its legal structure, its operating structure, and the production and organizational units used to derive the statistical units for the business (see Chapter 6). Once identified, the enterprise and its constituent KAUs or establishments constitute the statistical units of the business' *statistical structure*. In delineating the statistical structure, functional or other groups in the organizational structure may be ignored and the constituent units regrouped to form the units of the statistical structure. For multi-establishment or multi-KAU enterprises, the statistical structure may not coincide with the legal structure in which ownership of assets are registered.

3.1.3 Collection Arrangements

The first step in data collection is the identification of the statistical unit about which data are to be collected. However, the identification of the statistical unit does not mean that data must necessarily be collected from that unit. An appropriate collection strategy has to be devised by consulting businesses and examining their record-keeping practices.

The unit *from* which data are collected is the *reporting unit*. If records are maintained at a higher level of the business, the reporting unit may differ from the statistical unit *about* which data are required. A business' adoption of centralized computer-based accounting systems may mean that the desired data about each KAU or establishment can be obtained from a central source such as the head office of the enterprise.

Similarly, if data about LKAUs or locations are available from the KAU or establishment, it may not be necessary to survey them directly or to maintain detailed information about them on the business register. If some surveys use different reporting units from others, the data have to be aggregated or disaggregated to the same level of statistical unit for comparison.

3.2 INDUSTRIAL AND GEOGRAPHICAL CLASSIFICATION

A program of economic statistics must do more than assemble all the desired economic data at the level of each statistical unit. The economic phenomena to be described as statistics must be grouped by the characteristics that are being analyzed. The two main characteristics are industry and geographical location.

3.2.1 Industrial Classification

Because economic statistics are collected to study the behavior of economic units engaged in producing goods and services, an *industrial classification*, consisting of homogeneous groups of activities, is required to classify economic information about statistical units.

Industrial classifications are a means of grouping units by similarity of characteristics. The more homogeneous the classes of the classification, the more analytically useful the data. Generally, the two main characteristics considered relevant for an industrial classification are (1) similarity of goods and services produced with reference to demand and markets served and (2) supply considerations such as similarity of inputs, production functions, processes, and technologies used. In principle, classifications could also be designed to group units with similar debt equity rations, labor force structures, capital–labor intensities, or other business characteristics, but these are not usually regarded as necessary criteria for the design of general-purpose standard industrial classifications (see Ryten 1992).

The International Standard Industrial Classification of all economic activities (U.N. Statistical Office 1990, p. 12) states that:

> The main criteria employed in delineating divisions and groups (the two- and three-digit categories respectively) of ISIC, concern the characteristics of the activities of producing units which are strategic in determining the degree of similarity in the structure of the units and certain relationships in an economy. The major aspects of the activities considered were (a) the character of the goods and services produced; (b) the uses to which the goods and services are put; and (c) the inputs, the process and the technology of production.

Grouping units by similarity of supply-side characteristics (e.g., production processes inputs, and production functions) does not always lead to the same groupings that are derived from similarity of demand-side characteristics (e.g., goods and services produced and markets served). Sometimes units can be combined into industries in such a way that they are homogeneous both in terms of the production of goods and services that serve a particular market and in terms of inputs and processes used. This is the ideal situation. In all other cases, choices have to be made. Supply-side or demand-side criteria have to be chosen, or mixed criteria have to be used. Using supply-side criteria for an industrial classification provides a framework of statistics useful for the study of industrial performance and productivity. The needs of demand analysis of markets and market share can largely be met by data on the output of goods and services in terms of products. However, it is often useful to study the structure of the industry (or industries) of origin of demand-side product groupings. (For further discussion of conceptual issues underlying industrial classification, see U.S. Office of Management and Budget 1993).

Industrial classifications are designed for a particular type of statistical unit. There are two approaches with respect to defining the interdependent characteristics of the economic activity of the statistical unit and the industries of the industrial classifications. One approach considers it important to reflect the actual organization of production. This approach defines the economic activity of an industry, taking into account the actual combinations of physical activities undertaken in the production units of businesses, for which the businesses can themselves provide the required outputs and associated inputs. Where representing the actual organization of production is considered important, narrow homogeneity requiring artificial allocation is not the goal for defining the statistical unit for industrial statistics. The combinations of physical activities found in the statistical units are reflected in the industry classes of the classification. However, the lower the degree of specialization and the greater the combinations of goods and services produced by individual statistical units, the more likely it will be that the classes of the industrial classification will not encompass all the unit's physical activities. When units have a large range of products, the industries to which they have been classified by their principal activities will have secondary activities contained in the industrial statistics produced for that activity. The advantage of this approach is that it reflects the actual way in which production is organized. To achieve homogeneity, the classification clusters goods and services produced together, so that businesses can themselves provide the required data and the need for artificial allocation is minimized. The other approach prefers to articulate narrowly defined economic activities in terms of activities or commodities produced, without reference to the activities or groups of products actually produced together, for which businesses keep records and accounts.

The extent of estimation undertaken by different countries varies. Because of the importance of an industry and the need to delineate it separately, some countries divide vertically integrated production units producing different products into distinct statistical units. To do this, the business itself must be able to provide output data, a transfer price, and the majority of associated inputs. Other countries create separate classes in the industrial classification for integrated production. Some countries define statistical units based on actual production units and compile production statistics for them, and then subdivide them into more narrowly defined analytical units for national accounts. [As country practices vary, readers should refer for further information to ISIC Rev. 3, the industrial classification of the U.N. Statistical Office (1993); to NACE, the industrial classification of the Statistical Office of the European Communities (1993); and to the industrial classifications of national statistical offices.]

3.2.2 The Relationship between Industrial and Commodity Classifications

A *commodity classification* is required to collect information about the statistical unit's outputs and inputs of goods and services. It is useful to have an independent commodity classification which groups commodities in an aggregation structure that reflects market or commodity rather than industry characteristics. However, it is also useful to be able to group commodities into principal and secondary outputs of industries and to define industries in terms of commodity outputs as well as activities. When industries are defined in terms of a particular technology and they produce a range of products, it may not be possible to define them in terms of commodity outputs. Examples of such industries are machine shops, iron foundries, and plastic extruders. (For a discussion of the flexibility of businesses to change and adapt their output of goods and services to market conditions, see Chapter 35.) However, in many cases, industries are composed of production processes that create quite specific products, and therefore the industry can also be defined in terms of products produced. In such cases, commodity outputs can be used as proxies for activities.

An industrial classification reflects the structure of the economy and the specialization of production. Industries are defined in terms of activities. In principle, the necessary homogeneity criteria for a supply-based industrial classification can be based on a dispersion index that measures the extent of similarity of the input structures of the statistical units to be classified. However, even supply-side industries can sometimes be defined in terms of commodity outputs, provided that the criteria used to define commodities at the lowest level of the commodity classification are the same as those adopted for defining industries. If, for example, it is proposed to define a wooden cabinet industry or an industry of cast and forged products and suitably defined products exist in the commodity classification, then (given that data with respect to the value of their output are collected) statistical units can be clustered into industries with reference to commodity output and industries can be required to meet predetermined thresholds of specialization, coverage, and size.

Grouping of units into industries based on the similarity of their activities can therefore be refined by using a commodity classification to collect data on the value of goods and services produced by KAUs or establishments. The homogeneity criteria of specialization and coverage and the criterion of economic significance or size can be used to confirm the validity of the goods-producing industries of an industrial classification. An industry's *specialization ratio* is the ratio of its principal commodity output to its total output. Its *coverage ratio* is the ratio of its principal commodity to the total output of that commodity by all industries.

Minimum thresholds of specialization, coverage, and size can be set for the creation of industry classes in an industrial classification. For services, most countries are only now beginning to develop a classification and gather statistics in the commodity dimension.

3.2.3 Classification Systems for Production and Financial Statistics

The classification of statistical units determines the classification of data collected about them. Generally, industrial classifications are designed for the classification of the production unit or lower-level statistical unit. Often, for large and complex businesses, activities are more specialized at the level of the KAU and the establishment than at the level of the enterprise. Industries can be more narrowly defined for these lower-level units than for large and complex enterprises. Because large enterprises may be vertically and horizontally integrated and may cover a much wider range of activities than KAUs or establishments, some national statistical agencies find it useful to have an industrial classification for enterprises separate from that for KAUs or establishments.

Traditionally, industrial classifications designed for production statistics subdivide the economy into primary, secondary, and tertiary industries at the higher levels of the classification. Frequently, the higher levels of the classification are used for coding enterprises. However, the activities of many large, vertically integrated enterprises cut across the traditional distinctions of extractive, manufacturing, and services divisions, making it necessary to classify them by principal activity. Enterprise classifications can be designed to have analytically meaningful sectors that bring together the vertically integrated activities of enterprises, even though such sectors do not address the issue of horizontally integrated enterprises.

There is now a great deal of interest in integrating financial and production statistics. For comprehensive analysis of business behavior, it is useful to consider the option of an "integrated system" of industrial classifications. For this integrated system, the four-digit level of the traditional industrial classification, being designed to classify the production units of businesses, would take into account the activity combinations of KAUs or establishments, after which it should be possible to group these classes into two hierarchies. The first hierarchy would enable the activities of KAUs or establishments to be represented at the three-digit and two-digit levels of the industrial classification in the traditional way. The second hierarchy would regroup the four-digit classes in a manner that reflects the vertically integrated activity combinations of large and complex enterprises to provide the higher levels of an enterprise classification. The design of the four-digit classes would have to accommodate both hierarchies. The enterprise classification would have an independent four-digit level for the classification of enterprises. By being able to map the four-digit level of the traditional classification into the highest level of the enterprise classification, this approach to integrating the activities of enterprises with those of production units would have the added advantage of providing two analytical frameworks for production statistics. (For a description of the Canadian establishment and enterprise classifications, see Nijhowne and Côté 1991, Statistics Canada 1980, and Statistics Canada 1986.)

Standard industrial classifications with standard aggregations are required to enhance data comparability. However, one or even two standard aggregations, whether for production or financial statistics, cannot serve all analytical purposes. Special aggregations will always be required for particular analyses. For instance, for calculating financial ratios, the choice may be to separate the universe of single-unit enterprises from that of multi-unit enterprises and to handle them through special aggregations.

3.2.4 Geographical Classification

For regional analysis, the geographical location of economic activity is also of interest. Geographical classifications define areas useful for economic analysis. Geographical classifications may consist of different sets of geographical areas that have hierarchical relationships (see Statistics Canada 1992 for an example). An important set of areas is the hierarchy of local, regional, provincial, or state administrative areas that have jurisdiction over the collection and disposition of funds available for economic development and support and over the development of economic infrastructure and social programs. One set of geographical areas can therefore be built up from areas defined by administrative and political boundaries.

A different set of geographical areas is based on economic concepts such as the urban core and the labor market. Sometimes called *metropolitan areas*, these areas are delineated by first identifying the *urban core*, a group of the lowest-level administrative areas of the geographical classification that have a specified minimum population and population density. Adjacent areas are then analyzed in terms of the location of the labor force's dwellings and the proportion of residents who commute to work in the urban core. A *labor market area*, so delineated, reflects the commuting patterns of the labor force. The required data usually come from a population census. Both social and economic data can be classified to geographical areas for integrated analysis. (For a discussion of the application of this concept, see Dubuisson 1983 and Nadwodny et al. 1990.) The concept of urban core and labor market area can be widened to create wider economic areas encompassing the hinterlands that support the economies of the urban areas, or distinctions can be made between urban and rural areas. Quite separately, ecological zones can be delineated for the classification of pollutant emission by businesses for environmental statistics.

The production of data classified by geographical location requires that the statistical units on the register be coded to the areas of the geographical classification. Just as large enterprises are likely to encompass a wider range of activities than their individual production units, so also large enterprises often control production units located in different geographical areas. Whereas production units can be coded to particular geographical areas and regions, it is quite likely that enterprises cannot be geographically coded. As a consequence, production statistics are produced for small areas, but financial statistics are often produced only for the country as a whole. To compile production statistics by industry for small areas, a ruling must be made as to the geographical location from which activities such as construction, transportation, and tele-communications are carried out.

3.2.5 The Classification of Statistical Units

Classifications usually have hierarchical structures. They are generally de-signed for use with a particular statistical unit. It is customary to code the statistical unit to the lowest level of the classification and record the code on the business register.

Survey programs take two basic forms. They may collect a few variables from the whole population of businesses large and small. On the other hand, they may collect a large array of data from the units belonging to specific classes of the industrial classification or specific geographical areas. Classi-fying the statistical unit facilitates the design of questionnaires targeted to col-lect data from a particular industry group or for a particular geographical area. In combination with size variables such as sales, assets, and employment, the industrial classification enables the design of effective sampling strategies.

The hierarchical structures of classifications are useful in themselves. Each level of a classification is designed to serve an analytical purpose, but the hierarchy also serves another purpose. It enables the integration of statistics produced from different surveys by providing a structure within which data can be produced at higher levels, when the samples are inadequate to produce data at the most detailed classification level.

Depending upon the definition of the classes of the industrial classification, KAUs or establishments are assigned standard industrial classification (SIC) codes based upon their activities, input structures, goods and services pro-duced, or a combination of these attributes.

Ancillary units can be assigned their own geographical location, whatever the choice of industrial classification. In particular, some countries treat the head offices of large and complex businesses as separate units and assign them to the industrial class of their main activity. Others assign them to one partic-ular class of the industrial classification or create a "head office" class, and yet others assign them to the main industry of the units they support.

The universe of units does not remain the same over time. Units are clas-sified to an industry on the basis of their predominant activity. Units go out of business, new units come into being, and the predominant activity of units change. Because the activity of an enterprise depends upon the activity of its constituent production (and ancillary) units, mergers and acquisitions may not affect the classification of the production units, but they affect the classification of the enterprise (see Chapter 4).

A single KAU or establishment may undertake activities that are classified in different industries. Unless the classification contains combined activity classes, a unit with two economic activities is classified by its predominant activity. In a subsequent time period, the relative composition of the unit's activities may change, causing a change in its predominant activity, which in turn requires a shift in its industrial classification. In these cases, it becomes necessary to carry the date of the change on the register and to build in resistance rules for data compilation. This is accomplished by setting up systems through which the unit's classification for compiling data can be kept the same over the year during which subannual surveys occur, so that the classification of the unit remains the same for subannual and annual surveys covering the same variables. Resistance rules may also be required to reduce spurious fluctuations over time in annual reference period statistics. *Resistance rules* specify the conditions under which the industrial classification of a unit is unchanged, even though its predominant activity changes.

3.3 CLASSIFICATION REVISION

The characteristics on which industrial and geographical systems are based change over time. Classifications have to be revised at periodic intervals to remain relevant and analytically useful. After a revision, intertemporal statistical comparison is facilitated if the codes of both the old and the new classification are carried on the business register, making it possible to link data for units coded to the old classification with data for units classified to the new classification. (This subject is discussed in Chapter 7.)

3.3.1 Industrial Classifications

Industrial classifications are designed to reflect the structure of the economy, the organization of production, and the analytical requirements of users. All three change over time. New activities come into being, and existing activities grow and decline. For industries for which commodity output is a good proxy for activities, homogeneity and size can be used for creation or for confirmation of the classification's validity. When homogeneity and size thresholds are derived by an analysis of the output of establishments in a particular time period, degradation of the homogeneity ratios and of economic significance creates a need to revise the classification.

To reflect the structure of the economy and remain relevant, classifications have to be revised at periodic intervals, with all the attendant problems of maintaining the historical comparability of data series. It is customary to revise industrial classifications every 10 years. New or emerging industries are recognized in the interim by subdividing data at the industry level.

A program of economic statistics requires the production of time-series data. Classification revisions create the need to link data series based on different classifications. Usually, even the lowest-level statistical units have principal and secondary activities. Data comparisons, made after microdata for individual statistical units are aggregated into industry classes, are distorted by the secondary activities buried with the principal activity in the industrial aggregates.

3.3.2 Geographical Classification

Changes in administrative boundaries make it necessary to update geographical classifications periodically. The conceptual definition of types of economic areas may remain unchanged, but economic growth and changes in the location of industry may cause changes in the boundaries of geographical areas. When a geographical classification is revised and the areas of which it is composed change their boundaries, the production units of businesses must be recoded.

The periodicity of revision depends upon the frequency with which new data become available. Information about changes in administrative boundaries may be available annually, whereas population censuses may be the only source of the place-of-work data needed to define labor market areas, making the redefinition of metropolitan areas dependent upon decennial censuses.

3.4 INTERNATIONAL COMPARABILITY OF INDUSTRIAL STATISTICS

International comparability of industrial statistics requires use of the same statistical units and classifications. Provided that the same commodity classification is used, the total output of two economies can be compared in the commodity dimension. However, a comparison of industrial structure, performance, and productivity needs the full range of output, input, and value-added data by industry. If two countries are not using a common industrial classification, one country's data must be converted to the other's classification. Even if the two countries use the same statistical unit and industrial classification, industrial comparability will be affected by the extent to which the production units in these different countries have different principal and secondary activities reflected in their commodity mix of outputs.

To convert data to another industrial classification, one of two approaches can be adopted: (1) output data aggregated by industry can be converted by assigning commodity outputs to their principal producing industries and splitting industry inputs in the same proportions for assignment, or (2) each individual statistical unit belonging to each industry can be classified to the other classification. The second approach is preferable. However, after assignment, the nature and composition of the outputs and inputs of the industries being compared should be examined to understand the nature of the commodity mix of outputs and the extent of secondary output in each country's industrial statistics (see Ryten 1992). Use of an international commodity classification aids in comparing commodity statistics and in analyzing the activity/commodity composition of industrial statistics.

3.5 CONCLUDING REMARKS

An economic statistics program needs to define standard statistical units to produce consistent and integrated statistics about the financial and production activities of businesses, large and small. To produce analytically useful industrial data, the statistical units need to be classified to suitable industrial and geographical classifications. Because classifications have to be revised from time to time, the codes of the different vintages of the classifications need to be carried for the register's statistical units to facilitate intertemporal comparability.

Common commodity classifications across countries aid in comparing commodity statistics. Until the same industrial classification is used across countries, carrying the codes of the international classification or of other national classifications against the register units is desirable to facilitate the regrouping or conversion of data and therefore their analysis.

REFERENCES

Dubuisson, R. (1983), *Metropolitan Area Concepts in Canada and Selected Foreign Countries*, Geography Series Working Paper 4, Ottawa: Statistics Canada, Geography Division.

Nadwodny, R., H. Puderer, and R. Forstall (1990), ''Metropolitan Area Delineation: A Canadian–U.S. Comparison,'' paper presented at the Annual Meeting, American Population Association, Toronto, Canada.

Nijhowne, S., and G. Côté (1991), ''Industrial Classifications: Widening the Framework,'' paper presented at the International Conference on the Classification of Economic Activities, Williamsburg, VA.

Ryten, J. (1992), ''Inter-Country Comparisons of Industry Statistics,'' paper presented at the International Conference on the Classification of Economic Activities, Williamsburg, VA.

Statistical Office of the European Communities (1993), ''General Industrial Classification of Economic Activities within the European Communities'' (NACE), Rev. 1, unpublished document, Luxembourg: EUROSTAT.

Statistics Canada (1980), *Standard Industrial Classification 1980*, Ottawa: Minister of Supply and Services, Canada.

Statistics Canada (1986), *Canadian Standard Industrial Classification for Companies and Enterprises 1980*, Ottawa: Minister of Supply and Services, Canada.

Statistics Canada (1992), *Standard Geographical Classification SGC 1991*, Ottawa: Minister of Industry, Science and Technology, Canada.

Statistics Canada (1995) *International Round Table on Business Survey Frames, 1–5: Selected Papers*, Ottawa, Canada.

Struijs, P., and A. Willeboordse (1992), ''Terminology, Definitions and Use of Statistical Units,'' paper presented at the 7th International Round Table on Business Survey Frames, Copenhagen, Denmark.

U.N. Statistical Office (1990), *International Standard Industrial Classification of all Economic Activities—Revision 3* (ISIC Rev. 3), Series M, No. 4, Rev. 3 (Sales No. E.90.XVII.11), New York: United Nations.

U.S. Office of Management and Budget (1993), ''Issues Paper 1,'' *Federal Register*, Vol. 58, No. 60, Part III, pp. 16991–17000.

CHAPTER FOUR

Changes in Populations of Statistical Units

Peter Struijs and Ad Willeboordse[1]
Statistics Netherlands

When and how to record changes of statistical units is the source of many intriguing problems for business registers. For example: How should births be recognized? When should identification numbers be changed for statistical units? How should the register track units? The way changes are recorded has a large impact on statistical aggregates. This impact needs to be considered before deciding how the register should deal with change.

In this chapter, we discuss the effect of changes of statistical units on aggregates and the practical implications for register maintenance. The chapter has the following objectives: (1) to define the types of changes of statistical units in a systematic way, (2) to discuss business demography, and (3) to outline the impact on time series of changes in the business population. A basic classification of changes is introduced which can be used for all types of statistical units. The classification is demonstrated for the kind-of-activity unit (KAU). Establishment and enterprise changes are also briefly treated.

4.1 BUSINESS CHANGES AND STATISTICAL AGGREGATES: AN ILLUSTRATION

The following example is used throughout the chapter to illustrate the varied effects on estimation of a register's treatment of business changes.

[1]The authors thank Johan Lock of Statistics Netherlands for useful suggestions and comments.

Business Survey Methods, Edited by Cox, Binder, Chinnappa, Christianson, Colledge, Kott.
ISBN 0-471-59852-6 © 1995 John Wiley & Sons, Inc.

Suppose the theory Nijhowne presented in Chapter 3 is applied to create annual statistics on the printing industry: Code 2221 of the international standard industrial classification (ISIC) of the U.N. Statistical Office (1990). The statistics are based on KAUs with at least five employees. As Table 4.1 illustrates, the evolution of the industry can be roughly summarized as growth followed by stabilization. This is the aggregate result of what has happened to the individual units.

Consider the hypothetical case of Clear Printing Co. This independent book printing company was started in 1989 by a former employee of another printing company. After 2 years of strong growth, Clear Printing created a subsidiary company for its transport activities. In 1992, Clear Printing and its transport company merged with Freedom Publishing Co., the principal client of Clear Printing, to form Freedom Printing and Publishing Co. (FPPC). All activities were integrated, including record keeping. No separate accounting records were maintained after the merger.

The history of Clear Printing is an example of what underlies the figures in Table 4.1. The histories of the individual KAUs account for the net evolution of the industry. Consider how Clear Printing might have been represented in Table 4.1. In particular,

- What is the first year in which Clear Printing was included? Inclusion of a business depends on when (or if) the register identifies that the business exists and that it is large enough to include. The time of inclusion also depends on whether businesses are counted immediately after they are known to be large enough, or in the next year.

- Were the transport activities included in the table after the subsidiary was split off into a separate company? This depends on how the KAU definition deals with economically integrated legal units. A principal task of the register is to determine what (if any) statistical consequences should be drawn from changes recorded in administrative data.

- What happened after the merger? Depending on its classification rules, the register may classify FPPC in either the printing or the publishing industry. FPPC may even be split into two statistical units. The register may make the change immediately or in the next year.

Table 4.1 Printing Industry[a]: Fictitious Example

Characteristic	1990	1991	1992	1993
Number of KAUs	485	495	500	495
Number of Employees	5300	5400	5450	5500
Turnover (Sales in Millions)	$530	$545	$560	$560

[a]Includes only KAUs with five or more employees.

While accounting for the effect of business changes on estimated totals appears clear-cut (i.e., a mere addition of individual data), this examination suggests that the answer is not that simple.

The apparent pattern of growth followed by stabilization in the aggregate data is also not as unambiguous as it may appear. Individual changes are generally small compared to the aggregates, but can be large compared to *aggregate change*, which is the *difference* between statistical aggregates (totals). Aggregate change is sensitive to the register's procedures for dealing with changes at the unit level. Clear Printing is only one example of the many types of changes taking place that can result in inclusion or exclusion of KAUs from tabulations depending on decisions made by register staff.

Aggregate change statistics are somewhat subjective, because categorizing business changes requires statistical decisions that may not be clear-cut. For example, the printing industry may be in decline, or it could be doing better than Table 4.1 shows. This much variation in the real-world changes underlying Table 4.1 may sound farfetched, but consider the following scenario. If manufacturers tend to split off ancillary service activities, service industries will appear to increase over time and the manufacturing industry will appear to decline. Ironically, the core activities of manufacturing may be unchanged or even increase.

Clearly, registers must establish a consistent way to deal with business changes. To do so, register staff must identify the types of changes that occur in businesses and then use this information to construct a standard classification for business changes. Factors underlying changes in aggregates can then be discerned. For instance, the contribution of each change class can be specified, allowing conclusions such as; ''Newborn KAUs were responsible for 10 percent of turnover in 1991, while KAUs that were reclassified from other industries into the printing industry account for 5 percent.'' In addition, businesses that have changed in structure can be isolated, enabling study of the evolution of units unaffected by change (other than growth or shrinkage).

The need for information on change is not restricted to understanding the evolution of industries. Public attention also focuses on the number of births and deaths, the employment generated by new businesses, the profitability of mergers, the life expectancy of businesses, and related issues (Ryten 1992). The demand for such data is not restricted to the KAU or the establishment. Changes in ownership and composition of enterprises are well worth monitoring. All these matters can be addressed only if the types of changes are well-defined.

4.2 CHANGES CONCERNING INDIVIDUAL STATISTICAL UNITS

What types of changes should be identified? We are concerned with changes in *statistical units* that should be recorded in the register. To do this, we must know which administrative changes are statistically relevant and which are not. This distinction is mainly derived from the definition of the statistical unit (see Chapter 3). Changes identified at the level of the statistical unit clarify how populations, as provided by the register, are related in time. Relevant in their own right for business demography, these changes also specify the contribution of each change class to differences in statistical aggregates over time.

The comparison of the population of statistical units at different moments is the first step in deriving change categories (Struijs and Willeboordse 1995). For example, if we compare populations of statistical units on January 1 of 1992 and 1993, we may find that a number of old units are missing and that new units have appeared. Have the old units really disappeared, and are the new ones truly new? Did the other units really remain the same? Determining which units the two populations have in common is important. Moreover, a unit from one population may have links with several units of the other, as when statistical units merge.

When comparing units from different populations, two criteria, then, must be taken into account: (1) the number of related units in the old and new populations and (2) the continuation of the identity of the units involved. Other criteria may be statistically relevant, depending on the type of statistical unit being compared (e.g., establishment or enterprise). However, these two factors apply to all types of statistical units and to all unit definitions. In combination, the two criteria define the mutually exclusive change classes shown in Table 4.2.

The first category of Table 4.2 is the simplest; one unit in 1992 is related to one unit in 1993, and it is considered to have the same identity. This does not mean that it is the same in all possible respects, but that it is predominantly the same. For instance, the industrial code could have changed into a related one, or the size class may have changed. An important reason to identify these *changes of characteristics* is that such changes can result in the inclusion or

Table 4.2 Basic Classification of Changes: Classes and Defining Criteria

Change Class	Number of Units Involved[a]	Identity Continued
1. Change of Characteristic	1:1	Yes
2. Change of Existence		
1. Birth	0:1	No
2. Death	1:0	No
3. Change of Structure		
1. Concentration		
1. Merger	x:1	No
2. Takeover	x:1	Yes
2. Deconcentration		
1. Break-Up	1:y	No
2. Split-Off	1:y	Yes
3. Restructuring	x:y	Yes or No

[a]Number of units before and after the change: $x > 1$, $y > 1$.

exclusion of the unit from specific statistics. In the printing industry example, which is based on KAUs with at least five employees, the growth of Clear Printing leads to its inclusion in the tabulations from a specific date.

Changes of existence involve units that are not related to any unit of the population to which they are being compared. Consequently, there is no continuity of identity. This class includes two subclasses: (1) units of the 1993 population that are new in all respects and (2) units of the 1992 population whose operations have ceased altogether and are not continued in other units. These subclasses are referred to throughout this paper as *births* and *deaths*, respectively.

In other context, births as defined here might be referred to as "pure births" or "Greenfield births" to distinguish them from other uses of the term. For example, a register manager is likely to refer to any unit with a new identification number as a "birth." A survey statistician may use the term "birth" for all units that are new with respect to a particular sampling stratum or statistical population. In this chapter, units that are new to a particular statistical population are referred to as the *inflow*. Similar comments apply to use of the term "death." The departure of units from the population of a statistic is referred to here as *outflow*.

Changes of structure involve more than one unit from the population in 1992 or 1993. Three variations can occur: (1) One unit changes into more than one unit, (2) two or more units change into one unit, or (3) two or more units change into two or more units. (An example of the second type of change in structure is the emergence of Freedom Printing and Publishing.) The three situations can be referred to as *deconcentration*, *concentration*, and *restructuring*, respectively.

Deconcentration and concentration are subdivided to reflect the effect of the change on the unit's identity. During deconcentration, a unit either breaks up without any one unit retaining the identity of the original unit, or one or more units split off from a unit, which is generally larger and retains its identity. These two subclasses are referred to as *break-ups* and *split-offs*, respectively. The situation is reversed in the case of concentration. The unit emerging from concentration may or may not be essentially the same as one of the units before the change. Units either merge and lose identity or one unit takes over one or more units, which are generally smaller. These two subclasses are referred to as *mergers* and *takeovers*, respectively. Restructuring is the most complex change that takes place. It serves as a "not elsewhere classified" (nec) class.

Each combination of the two classifying criteria (number of units involved and continuation of identity) leads to one specific class, and therefore the classes are mutually exclusive. For instance, if the size class of a unit changes because of a takeover, the change is *not* classified as a change of characteristic, as this is still a many-to-one situation. In the same vein, a merger is not considered to be the sum of a number of deaths plus one birth. As previously noted, the terms "birth" and "death" are reserved for essentially new or vanished units.

The change classes are not only mutually exclusive, but they cover all possible cases. Thus, they meet the two basic requirements of classifications. One possible category has been omitted—namely, the one-to-one situation with loss of identity. For example, if a business ceases all its operations and sells its office building to an entirely new business, the old and the new business can be linked. However, the identity of the old business is not continued in the new business. For most purposes, this situation need not be distinguished from the event of a death followed by a birth. As a rule, one can do without the one-to-one, with-loss-of-identity category when explaining the relationship between subsequent populations. If needed, a subdivision of the categories of birth and death (birth related to death, and vice versa) can serve as an alternative to this class.

When applying the basic classification of changes to specific types of statistical units, such as the KAU or the establishment, the two criteria have to be worked out. That is, register management must specify (1) how the number of units involved in a change is to be determined and (2) how to decide whether two units are predominantly the same. The latter is far from trivial and involves a value judgment about what factors contribute to the identity of a unit and to what extent. Is the name of a business essential? Or is ownership, machinery, or location more important in establishing identity? The answers depend on the statistical unit in question, which in turn depends on the economic process being described.

4.2.1 Changes in Kind-of-Activity Units

The basic classification of changes can be applied to all types of statistical units. We illustrate its application for the KAU in this section. Again, changes are based on comparing the population of units at different moments in time. The two criteria of the basic classification are established first. Then the classification of changes is extended with subcategories relevant to KAUs. Finally, two important aspects of applying the classification of changes to KAUs are discussed: (1) the availability of information on changes and (2) the determination of the date at which a change took place.

The application of the basic classification to changes of KAUs depends on the definition of the KAU and the role it plays in a country's statistical system. As explained in Chapter 3, the KAU is used for statistics on the production process. In its role as the unit where goods and services are produced, the KAU is a combination of production factors. We use this fact in examining the two criteria for classifying changes.

Units Involved in a Change

The first question is, "Which KAUs are involved in a change?" That is, "Which units of the old population (again, say on January 1, 1992) are considered to be related to which units of the new population of January 1, 1993?" This information can be derived from what happened to the production factors.

If a KAU of the 1992 population has one or more production factors partly or entirely in common with a KAU of the 1993 population, then these units are considered to be related. However, to avoid unnecessary complexity, unimportant links are ignored. For a link to be disregarded, it must be insubstantial compared to the production factors of the KAU before the change and the KAU after the change.

As an example, the emergence of FPPC has been charted in Figure 4.1. Here, we assume that Clear Printing includes the transport activities. The blocks show how the legal units are assumed to relate to statistical units. (Identification of the statistical units always takes place before changes can be classified, of course. Because statistical units are the subject of Chapter 3, the unit structure in this example has been kept simple.) FPPC is a KAU, because its operations are integrated and accounting records are only available for FPPC as a whole. The figure also shows the production factors of the KAUs, and it includes all KAUs existing on January 1 of 1992 or 1993 that are related through production factors partly or entirely shared. Lines that do not connect KAUs (i.e., that start or end in the void) denote production factors that are new in 1993 or that are not used by any KAU after 1992.

Extra Printing Co. has been included in the figure because it has a connection with FPPC. One of its employees joined FPPC, so there is a link between Extra Printing and FPPC in terms of production factors. Obviously, this link is insubstantial. From the point of view of Extra Printing and FPPC, it involves only part of one of the production factors, and this part is small. Thus, this link should be ignored. In terms of the classification of changes, only Clear Printing, Freedom Publishing, and FPPC should be considered as related, resulting in a two-to-one situation. What happened to Extra Printing is thereby reduced to a one-to-one situation.

Identity

Continuation of identity, the second criterion of the classification of changes, can also be defined using production factors. However, the comparison is not between the individual production factors, but the factors in combination. Because it is the combined factors that make a KAU unique, the identity of a KAU depends on the degree to which the same combination of production factors occurs before and after the change. Any situation is possible, from sameness in all respects to no similarity at all. In between, the production factors that are retained have to be evaluated to decide if identity has continued in spite of the change. Identity is maintained if the continued production factors are important, relative to the old and new KAU.

Applied to Figure 4.1, this means that the production factors of Clear Printing and Freedom Publishing are compared to those of FPPC. The production factors of the two units existing in January 1992 are almost entirely embodied in FPPC. Even the management of Clear Printing and Freedom Publishing merged. Evidently, FPPC is essentially different from each of the earlier two KAUs. The situation can be characterized as two-to-one without continuation of identity, and the change is a merger.

January 1, 1992   January 1, 1993

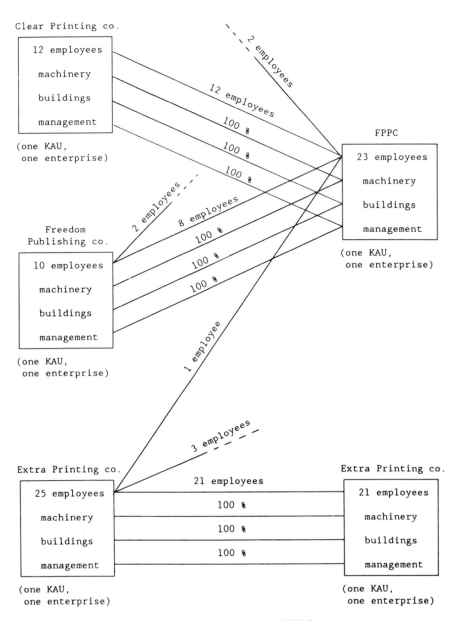

Figure 4.1 The emergence of FPPC.

The example is clear-cut, but one can imagine a situation in which Clear Printing is much bigger than Freedom Publishing, the management of Clear Printing becomes the management of FPPC, and the production factors of FPPC are largely the same as those of Clear Printing. Then, a takeover would have taken place rather than a merger. As to Extra Printing, this KAU has retained its identity. It either has experienced no change, in terms of the classification of changes, or has experienced a change of characteristic. If the link between Extra Printing and FPPC had been more substantial, all three KAUs would have been regarded as part of a restructuring (Category 3.3 in Table 4.2).

To decide on the degree of continuation of the combined production factors of a KAU, the factors need to be quantified and weighed. This is necessary, for instance, when the choice between merger and takeover is not as obvious as in the example. The simplest approach is to select one factor that can be quantified easily, such as number of employees, and to assume continuity of the KAU if that factor is continued for, say, more than 60 percent. Thus, if more than 60 percent of the employees of Clear Printing went to work for FPPC and if more than 60 percent of the employees of FPPC had previously worked for Clear Printing, the change would be considered a takeover of Freedom Publishing by Clear Printing. If less than 60 percent of the employees of FPPC had worked for Clear Printing, it would be considered a merger.

Other production factors also need to be quantified and weighed. Ideally, one should let the relative importance of production factors depend on the economic activity and size of the KAU. For agriculture, land is obviously important for unit continuity. As another example, a change of management in a small service KAU has more impact on identity than does a management change in a large manufacturing KAU. Because the economic activity of a KAU is closely related to its production factors, comparing economic activities of KAUs assists in judging unit continuity.

Some changes of characteristics are not conceptually possible because they imply loss of identity. An activity change from printing (ISIC 2221) to bookbinding (ISIC 2222) does not imply discontinuity, but a hypothetical change from printing to manufacturing coke oven products (ISIC 2310) does. In the latter case, a death and a birth must have occurred. The practical rules for deciding the continuation of identity can include a list of ''allowed'' activity changes. For example, only activity changes within an industry (say, ISIC two-digit level) and changes from one industry to a vertically related one (e.g., from wholesale of food to retail sale of food) could be allowed.

So far it has been assumed that only production factors are relevant to the criterion of identity. However, the identity of a unit depends not only on its constituent parts, but also on its position relative to others, in particular its clients. A change of customers, for instance, is an indication for discontinuity. For a KAU with a single location, a change of location may affect its identity, as the location is important in relation to the KAU's market. A restaurant moving over a long distance can hardly be considered the same restaurant; it must start from scratch to win customers in the new environment. A KAU's relationship to its environment is also indicated by trade name (whether or not it presents itself as the same as before) and ownership.

Let us now return to the first criterion, the number of units involved. We have not been explicit about when the links between KAUs are sufficiently insubstantial that they can be ignored. This involves the same kind of evaluation of significance of production factors as was needed to determine continuation of identity. A threshold in terms of production factors has to be established for taking links into account. Such a threshold would prevent linking Extra Printing and FPPC based upon the movement of one employee. The threshold would be relative to the KAUs concerned rather than absolute. We do not derive such a threshold here, as in practice the need for it is limited. In most cases the less substantial links are not known. In any case, the reasoning would follow the same lines as for the continuation of identity.

Subcategories of Change

An obvious subdivision of the basic classification of changes is one based upon the characteristics that can change. The main characteristics of a KAU are size class and kind of economic activity. These are the characteristics used to define statistical populations, such as the hypothetical printing population shown in Table 4.1. To link subsequent populations, changes in these defining characteristics should be distinguished in classifying changes.

Apart from changes in the characteristics used in defining populations, changes in other characteristics such as ownership or trade name can be distinguished. These distinctions are useful, because such changes indicate that, although the identity of a KAU is essentially preserved, the KAU is no longer the same in all respects. Information about such characteristics may help alleviate the arbitrariness inherent in the choice between continuity and loss of identity.

Other useful subdivisions depend on the criteria used in defining the KAU. As explained in Chapter 3, several elements play a role in defining the KAU, of which independence, homogeneity of activity, and availability of accounts are the most frequently applied. When independence (i.e., autonomy of decision-making with respect to the production process) is the only criterion, changes in independence will correspond to changes in the basic classification, and no subdivisions are needed. For instance, in the basic classification, the amalgamation of two similar autonomous units into one autonomous unit is a merger. A subdivision entitled "merger because of loss of autonomy" is unnecessary because that occurs by definition when independence is the only criterion.

However, when there are other defining criteria, changes resulting from the different criteria should be tracked. For instance, if homogeneity is a defining criterion, a KAU can be split if a secondary activity surpasses a certain threshold. Similarly, if availability of accounts is a defining factor, the mere introduction of a new accounting system in a business can lead to splitting KAUs.

Of course, reverse cases also exist. Such changes should be distinguished from instances in which an economic decision-making unit is split into two or more such units (and the other way round). Subcategories in the classification of changes like "split-off for reasons of homogeneity" would allow this distinction.

Information Requirements

In practice, statistical changes can be established only after receiving signals that something has changed. Apart from feedback from statistical surveys based on the register, such signals typically come from the administrative sources feeding the register. As mentioned earlier, administrative information cannot be interpreted *a priori* as relevant statistical information. After establishing the statistical units involved, the kind of change and the date of occurrence (the date on which the change took place in reality) have to be determined.

The usefulness of administrative information depends on the statistical information required and the contents and quality of the administrative sources. In principle, complete information on production factors is needed to establish the kind of change; in practice, operational procedures dictate the amount of information needed. Continuity can be established in a simple or in a sophisticated manner, using rules of thumb or complicated weighing schemes. Much information is needed in any case to identify the statistical units.

The example of the transport company split off from Clear Printing illustrates this process. Upon receiving a signal that the transport company has been created, the register first checks the KAUs involved. It may then decide that Clear Printing and the transport company together form a single KAU, because together they are still a single economic unit fulfilling the definition of the statistical unit. Thus, the administrative registration of the transport company does not have any statistical effect. The example also shows that the identification of units and classification of changes go together as they are usually based on the same information.

Apart from the kind of change, determination must be made of the date the change occurred, or at least the first period in which the change is to be reflected in statistical output. This date may differ from the date of the administrative change, as administrative registration can be premature or delayed. In theory, the moment of change is derived from the classification rules (for changes of characteristics) or from the definition of the KAU itself (for other changes). For instance, the date at which a KAU was born is equivalent to the date that the KAU criteria were met.

However, changes often stretch over periods of time rather than occurring in an instant. A birth, for instance, may take some time. Between the initial idea to start a business and the actual delivery of products, the business may develop a business plan, secure financial backing, rent offices, obtain machinery, hire staff, and so forth. If the KAU's definition requires being engaged in market transactions, the birth will only be recognized after the startup period. If the definition merely states that production factors must be engaged (building rented, employees contracted, etc.), the birth will take place earlier. In practice, the register may decide to use the moment at which major financial commitments are made, the moment of no return, as the birth date. Then statistical coverage of initial investments is secured.

4.2.2 Changes in Establishments and Enterprises

The basic classification of changes can also be applied to the two other types of statistical units mentioned in Chapter 3—that is, the establishment and the enterprise. Experience in applying change classifications for these units is limited (worldwide). We address the main aspects.

Establishment Changes
As with KAUs, the application of the classification of changes to other types of statistical units depends on the unit's definition and the role the unit plays in a country's statistical system. The establishment is used for the statistical description of (certain aspects of) the production process, as is the KAU. A classification of establishment changes is therefore similar to the classification discussed for KAUs. However, there are notable differences.

Geographical location, as explained in Chapter 3, is an important defining criterion for the establishment. Location also affects the way to apply the criterion of continuation of identity in classifying changes and may even alter the judgment of which units are involved in a change. Hence, a comparison of the locations of establishments plays a major role in classifying change. In addition, the region where an establishment is located is an important characteristic as it is used in defining statistical populations.

In some statistical systems, the establishment is used as a proxy for economic decision-making units such as the KAU. However, when an establishment involved in a change is not autonomous, does not have market sales, or does not have complete accounting records, it fails as a proxy for a KAU, and in this instance application of the classification of changes would also fail as a proxy for KAU changes. For example, if two establishments that equate to two separate KAUs merge into one KAU consisting of two establishments, no change will have occurred to the establishments. If FPPC consisted of two establishments, corresponding to the former Clear Printing and Freedom Publishing, the merger would not have been reflected in establishment statistics.

Enterprise Changes
As explained in Chapter 3, the enterprise is the unit used for statistics on financial processes; it consists of one or more KAUs. Control over the enterprise's operations is crucial to the financial role of an enterprise. Therefore, decisions regarding which enterprises are involved in changes, and the determination of unit continuation must take control into account.

Generally, control is linked with ownership and is exercised through the board of directors of the business. Consequently, continuity of control depends on continuity of shareholding of the apex (the parent or holding company) of the enterprise. If an enterprise gets a new parent company from outside the enterprise, continuity of control is lost. In other situations, it may be hard to trace the controlling interest especially if it is dispersed. One solution is to identify changes of controlling interest by means of a characteristic that indicates when there is continuation of control. Then control itself would cease to be an explicit criterion for continuity.

In deciding about continuity of enterprises involved in a change, what happens to the enterprise's KAUs is important, because the KAUs generate the return on assets. Thus KAU continuation may be needed to guarantee the continuation of the financial role of the owner enterprise. Another reason to consider continuity of KAUs is that decisions about allocation of production capacity (i.e., decisions on which KAUs exist) are made at the enterprise level. Using continuity of KAUs as a criterion for continuity of enterprises implies that the size of continued KAUs and parts of KAUs have to be measured and weighed. Collins (1991) developed a method for measuring the continuity of enterprises.

The classification categories for enterprise changes can be subdivided. The most important subdivision concerns changes of characteristics, particularly those characteristics that define populations of enterprises for statistics. Examples are changes in activity or institutional sector and size (in terms of employment or a financial measure such as total net assets). Detecting changes in ties with foreign units is also helpful, as the enterprise may be part of an international group with the apex inside the country or abroad.

4.3 AGGREGATE CHANGES: BUSINESS DEMOGRAPHY

The economic structure of a country and the changes therein can be described at the microlevel in terms of individual statistical units, characteristics of these units, and changes in these statistical units. Information about the structure and evolution of *populations* of units is obtained by aggregation. For instance, annual statistics might be compiled in which units are tabulated according to characteristics such as establishments by region or KAUs by kind of economic activity and size class. Another example is to aggregate for a specific change class and produce statistics on the number of births of KAUs by kind of economic activity.

These examples belong to the domain of *business demography*, the statistical description of a population of units and changes therein. Obviously, the classification of changes is important for business demography. It allows systematic reports on changes, and, more importantly, tables can be connected that refer to different time periods. This function is more demanding than merely reporting on categories of change.

Populations of statistical units can be defined by specifying the type of unit, the selection values for characteristics, and the time period in question. An example is the number of KAUs with at least five persons employed on January 1, 1992 in the printing industry (ISIC 2221) to which Clear Printing belonged. Using the classification of changes, this population can be linked to the one on January 1, 1993. Table 4.3 shows how this can be done.

The only changes of characteristics included in the table are those affecting the number of population units. For instance, changes in economic activity within the printing industry have been ignored. No change affected both characteristics; otherwise a line would have been inserted to indicate the combined change. Because the population has a lower size cutoff, crossing the threshold is a major cause of inflow and outflow. The number of births (and, to a lesser degree, deaths) is generally low for cutoff statistics. A large business size from the start is uncommon, except when it results from a change of structure.

A possible effect of changes of structure is illustrated by the FPPC case. Suppose FPPC is a KAU to be classified in ISIC 221 (publishing whether or not connected with printing). Neither Clear Printing nor its successor belongs to the 1993 population of the printing industry, which is reflected in Table 4.3 in the outflow column. In general, as a consequence of new values of characteristics, changes of structure may result in an inflow, an outflow, or a combination of inflow and outflow. The units remaining from a break-up, for instance, might stay in the industry but leave the population because of their smaller sizes. But even changes of structure that occur entirely "inside" the population affect the number of units in most cases.

Table 4.3 Business Demographic Change in the Printing Industry[a] in 1992–1993: Fictitious Example

Category	Inflow	Outflow
Number of KAUs, January 1, 1992	500	—
Change of Characteristic		
Change of Economic Activity	2	2
Change of Size (Employment)	33	28
Change of Existence		
Birth	0	—
Death	—	5
Change of Structure		
Concentration	1	10
Merger	1	8
Takeover	0	2
Deconcentration	4	0
Break-Up	0	0
Split-Off	4	0
Restructuring	0	0
Gross Demographic Change	40	45
Number of KAUs, January 1, 1993	—	495

[a]Includes only KAUs with five or more employees.

If a change causes both inflow and outflow, the table contains either the net or the gross change. *Gross change* records the total number of population units involved in a particular type of change. *Net change* balances inflow and outflow for each category of change. Gross figures are more informative; the total number of changes of structure (mergers for example) cannot be derived from the net figures. If a unit classified in the printing industry merges with a unit from outside this industry, and the emerging unit is classified in the printing industry, net figures for the printing industry would not reflect any change at all. As we shall see later, gross figures also facilitate relating statistical aggregates based on different populations.

Thus, relating populations that refer to different time periods involves more than classifying changes and counting them. For instance, a merger implies (1) an outflow of at least two units from one or more populations and (2) an inflow of one unit into a later population. To relate populations, one has to specify the populations to which the units belong.

Comparing Table 4.3 to other business demographic tables, two situations are worth examining:

- The time periods compared are the same for the tables.
- The time periods compared are different, but apart from the dimension of time, the populations of the tables are the same.

An example of the first situation is a set of business demographic tables on different industries, such as the printing and the publishing industries; each table compares the situation on January 1, 1992 to January 1, 1993. If the tables are constructed like Table 4.3 and the numbers are gross, the tables are additive. However, when adding the figures, we might adjust for activity changes from printing to publishing, and vice versa. This amounts to compiling a consolidated table. Although the numbers involved may be small, the classes of origin of incoming units and destination of outgoing units may be specified for activity changes.

In fact, relating populations requires that a decision be made for each characteristic about the level of the classification where changes are to be identified. For example, changes of activity at the four-digit level of ISIC may be recognized. This could be the same level used for stratification. By choosing a standard level, confusion can be avoided.

An example of the second situation is an annual statistic as compared to a monthly statistic for the same industry. Relating such statistics is complicated because short-term changes do not always add up to the long-term changes. If only annual populations are compared, a birth is not measured if the unit dies within the same year. The same remark applies to subsequent structural changes within a year. One approach might be to define accumulated changes (e.g., "split-off followed by death"), but this may not be practical. Solutions to the problem of relating short-term and long-term statistics depend heavily on how changes are detected and recorded in the register.

Apart from specifying the relationship between populations as in Table 4.3, the categories of change can be used for statistics on the categories themselves. For instance, there is substantial demand for information on births and deaths. The number of births, as well as most other categories of change, can be established by checking signals received from administrative sources feeding the register. However, the signals received are generally insufficient to derive the number of deaths. Many administrative registers do not accurately record deaths, because the units do not report properly. In such situations, statistics on deaths can only be made by means of register samples or censuses.

4.4 AGGREGATE CHANGES: TIME SERIES

Business demographic information is very useful in itself, but its value is increased considerably if related to information on economic variables, such as output of production, number of employees, or value added. Conversely, data on variables can gain substantially from being linked to business demographic information. It may be interesting to know, for instance, that during 1992 employment increased from 5450 to 5500 in the printing industry (KAUs with at least five persons employed in ISIC 2221), but it would deepen the insight to know the demographic background, as illustrated in Table 4.4.

What does Table 4.4 show? First of all, it reveals that those KAUs that have not changed in terms of the classification of changes during 1992 experienced

Table 4.4 Employment in the Printing Industry[a] by Category of Business Demographic Change in 1992–1993: Fictitious Example

		Employees	
Category	KAUs	January 1, 1992	January 1, 1993
Situation on January 1, 1992	500	5450	—
Outflow			
Change of Economic Activity	2	15	—
Change of Size	28	180	—
Death	5	30	—
Change of Structure	10	125	—
No Demographic Change	455	5100	5050
Inflow			
Change of Economic Activity	2	—	215
Change of Size	33	—	195
Birth	0	—	0
Change of Structure	5	—	40
Situation on January 1, 1993	495	—	5500

[a]Includes only KAUs with five or more employees.

a decline in employment. This is quite different from the overall change in the industry. Another remarkable fact is that the effect of activity changes on employment overshadows the overall change. The same goes for changes of structure. Such facts cast doubt on the significance of information on net change alone. They demonstrate that business demographic information can provide a perspective that is essential for balanced judgment of overall change.

However, even if business demographic information is supplied, the interpretation of a time series like that of Table 4.4 is not straightforward. It is not always obvious how statistical data relate to economic reality. For instance, the formation of FPPC had a remarkable statistical consequence; the flourishing Clear Printing left the population of the printing industry. Perhaps the fact that its printing business went so well induced the merger, causing the departure from the printing industry.

The example of FPPC also illustrates the consequences for time series of the conceptual difference between the printing industry (i.e., all KAUs classified in ISIC 2221) and the printing activity in the economy as a whole. The production capacity of Clear Printing is not counted in the 1993 printing industry, because the industry in which FPPC is classified includes publishing connected with printing. Moreover, there are other industries that contain printing activities as secondary (and of course ancillary) activities, and the printing industry also includes secondary activities. Such aspects of classification systems have to be taken into account when interpreting time series.

The interpretation complexities inherent to classification systems are magnified when dealing with change. Heterogeneity may be intrinsic to a classification system, but changes occurring to statistical units that affect the heterogeneity of industries complicate the interpretation of time series.

Also complicating the interpretation of time series is the fact that most changes stretch over a period of time. The changes of economic activity that had such an impact on total employment in the printing industry example could have been the result of steady, relative growth as secondary activities became principal activities. Moreover, units can be on a balance point between industries for quite some time. In such cases the changes of activity are not instantaneous, and the timing of reclassification is somewhat arbitrary.

Finally, the interpretation of time series depends on the quality of the register information, in the sense that changes in aggregates can be the result of delayed updates or corrections. Business registers are never complete and never entirely up-to-date, and they inevitably contain errors. This is due to differences between administrative data and economic reality, incompleteness of administrative sources, and time lags in data processing. Our knowledge of reality is volatile; what we know or think we know about the situation today may need to be corrected or supplemented tomorrow.

Fortunately, there are several options for enhancing the interpretation of time series. The statistician can:

- provide the business demographic background of statistical aggregates;

- provide information on changes in *homogeneity ratios*, which are measures of the specialization and coverage of industries;
- apply classification resistance rules; and
- provide information on corrections and delayed changes.

We have demonstrated the importance of the first measure, providing information on the contribution of change classes to aggregate change. If the described population has an artificial boundary, such as the lower size cutoff in the example, it is particularly important to provide business demographic information, to give an idea of the consequences of the boundary. However, the provision of such information is subject to practical limitations. For instance, when aggregates are based on samples, infrequent changes cannot be estimated accurately.

If not hampered by accuracy problems, specification of changes of relatively rare occurrence can be worthwhile, because such changes can have a decisive influence on net figures. This was the case for the changes of structure shown in Table 4.4. However, publication of such changes may not be possible because of the risk of disclosure of individual business information. Nevertheless, statisticians need this information for their own assessment of changes in statistical outcomes. The staff compiling statistics about the population to which FPPC belongs have to know that FPPC is the successor of both Clear Printing and Freedom Publishing so that the microlevel data can be compared.

The second measure, the provision of information on homogeneity ratios, sheds some light on the composition of production capacity and the changes therein. For instance, if a unit obtains a secondary activity because of a takeover, this affects the homogeneity ratios. However, homogeneity ratios can also change without business demographic changes taking place. For a proper understanding of time series, it is necessary to know how the homogeneity of industries depends on the classification system and the definition of statistical units (Struijs 1991).

The third measure, the application of classification resistance rules, reduces the arbitrariness of reclassification for units that are balanced between classes. For industrial classifications, these rules stipulate that reclassification is only carried out if the change of principal activity is not likely to reverse. This may only become evident after some time. It implies a deviation from the correct, instantaneous classification, but can be justified by the inevitable uncertainty of timing.

The fourth measure, the provision of information on corrections and delayed changes, can take several forms. In terms of data accuracy, the most satisfying approach is to regard all statistical information as provisional and to keep publishing revised data forever. This is not practical, of course, although one or two data revisions may be feasible. Regardless, changes in statistical aggregates that do not correspond to ''real'' economic changes should be specified in text or notes. This poses a severe but essential requirement that registers track whether recorded changes are corrections or delayed changes.

Finally, the measures are not complete without a classification revision policy and a general strategy for changes in methods. Statistical measurement instruments—classifications and definitions—need periodic revision due to structural economic changes, new information needs, and new methodological insights. However, such revisions may severely hamper the interpretation of time series as discussed in Chapter 7.

4.5 CONCLUDING REMARKS

The application of the methods described in this chapter depends upon the specifics of the statistical system involved, including the definitions of statistical units used, the classification systems applied, the statistical program and its objectives, available resources, and business register practices. Moreover, these different aspects of a statistical system cannot be dealt with in isolation. Unit definitions affect the contents of classifications; both influence the method to deal with changes. Homogeneity ratios are relevant in the context of classification design as well as change, and so forth. In summary, all the statistical information published by a statistical agency should be subject to coordinated definitions and rules of application.

Uniform rules for the definition of statistical populations also assist in coordinating aggregate data. In particular, the methods of dealing with changes in statistical aggregates should be coordinated to obtain a consistent and realistic picture of economic changes. For example, if a reclassification takes place, statistical outputs should assimilate that change in a coordinated fashion, especially if they apply to the same reference period but are based on populations drawn from the register at different dates.

The methods described in this chapter have important implications for business register practice. Application of these methods is only possible if the register provides the necessary facilities. In particular, the register has to link the units involved in each change, determine the category of each change, and identify the date (or at least period) of occurrence. Specification of corrections and delayed changes is also quite involved. The next chapter indicates the extent to which these statistical objectives may be realized in practice.

REFERENCES

Collins, R. (1991), "Proposed Definitions and Classifications of Longitudinal Continuity and Change for Global Enterprises and Enterprises," *Proceedings of the Sixth International Round Table on Business Survey Frames*, Coolangatta, Australia: Australian Bureau of Statistics, pp. 190–196.

Ryten, J. (1992), "The Demography of Small Businesses: Describing the Missing Evidence," paper presented at the International Conference on Small Business, Montreal, Canada.

Struijs, P. (1991), ''The Concept of Industry in Transactor-Based Industrial Classifi-cations,'' *Proceedings of the 1991 International Conference on the Classification of Economic Activity*, Williamsburg, VA, pp. 364–383.

Struijs, P., and A. J. Willeboordse (1995), ''Towards a Classification of Changes,'' *Selected Papers from the First Five International Round Tables on Business Survey Frames*, Ottawa: Statistics Canada, in press.

U.N. Statistical Office (1990), *International Standard Industrial Classification of All Economic Activities (ISIC)*, Statistical Papers, Series M, No. 4, Revision 3, New York: United Nations.

CHAPTER FIVE

Maintenance of Business Registers

David Archer
Statistics New Zealand

The uses of a statistical agency makes of its business register shape its maintenance priorities. The register's core function is to provide survey frames efficiently and with measurable reliability. Additional uses include producing register-based statistics, monitoring and managing respondent burden, and serving as a source of commercial revenue. Because register maintenance is costly in person time and money, it is essential that register objectives be clearly understood and that maintenance activities focus on these objectives. Maintenance sources need to be carefully selected, emphasizing those that (1) provide information on all register units and (2) allow the register manager to influence the quality of received information. Use of too many maintenance sources can lead to disparate levels of quality and duplication.

Maintenance strategies involve merging information obtained from administrative records, membership lists, survey feedback, and the news media. Priorities are assigned that balance the often competing demands placed on the register by survey staff, producers of register-based statistics, and respondent burden managers. These demands include the need for (1) timely, accurate, and comprehensive information on births; (2) consistent updating of contact and classification data for existing units; (3) timely and complete identification of deaths; and (4) maintenance of structural information linking units. The vital role of a business register within a statistical agency means that it must also include quality control mechanisms to meet users' needs. The register's source data, input and classification processes, and output must be continually evaluated and matched against the objectives.

Business Survey Methods, Edited by Cox, Binder, Chinnappa, Christianson, Colledge, Kott.
ISBN 0-471-59852-6 © 1995 John Wiley & Sons, Inc.

5.1 REGISTER USES AND REQUIREMENTS

A business register is a multiple-use statistical tool. It forms the hub of a statistical agency's business statistics program and has a wide variety of uses, including (1) supporting survey design, operation, and management; (2) monitoring respondent burden; (3) producing statistics and customized analyses; (4) generating lists for sale outside the agency; and (5) storing historical information.

The register's core use is for sample design and survey operations. In this respect its basic objectives are to identify (1) new units (enterprises and establishments); (2) contact data (name, address, etc.) and changes in these data; (3) deaths of units; (4) ownership links between units; and (5) changes in characteristics such as ownership, industrial activity, geographical location, and size. Just as a survey has a defined target population, a register has a population of units that should be included. This population covers, and often exceeds, the union of the target populations of the individual surveys to which the register provides frames.

5.1.1 Survey Operations

In support of survey operations, the register's function is to identify births and deaths of units and to ensure that name and address information is complete and up to date. It must also establish and maintain ownership links between units to ensure that confidential data about individual businesses are not released. In most statistical agencies, the register provides frame data for a wide range of surveys, including subannual surveys of production and retail sales, annual surveys collecting detailed financial information, censuses of businesses (e.g., manufacturing), employment surveys, price index surveys, surveys of overseas transactions and activities, and also household surveys and population censuses that require classification of employers by industry and location.

Because a register is continuously being updated but survey populations and maintenance tasks often relate to past time periods, the register must have some form of historical capability. For example, the sample for an annual financial survey with a reference period of December 31, 1993 was selected in February 1994. Because the survey processing cycle may continue until 1995, at some later date the survey manager may want to recreate the units' ownership or activities as they were on December 31, 1993. On the other hand, monthly surveys require survey frames updated monthly. If the register were to store only the most recent information about units, storage requirements would be minimized, but the register's usefulness would be limited and there would be no audit trail.

An option for storing historical information that satisfies most needs is to archive *snapshots* of the register at key points in time (e.g., once a year for annual statistics, or when a survey sample is selected). This approach minimizes storage requirements while preserving key historical information. It allows a unit's ownership structure to be recreated at a point in time, but it does not provide a complete audit trail.

A more complex option is to *timestamp* some or all of the data on the register. Timestamping involves attaching effective dates to register data items. Its application can vary in extent from timestamping only changes in unit ownership to timestamping all changes (i.e., legal name, address, industry, geographical classification, etc.). While timestamping provides a complete audit trail for register maintenance, it also leads to a more complex register design, significantly increasing the number of operating rules and data storage requirements.

5.1.2 Survey Design

In an ideal situation, the register provides frames for all business surveys conducted by the agency and is the sole source of classification information and sample design variables. The register's ability to function effectively in this role depends upon how well it covers the surveys' target populations. Undercoverage can be a significant source of error in survey design, so the register's coverage should be as complete as possible, and the degree and type of its undercoverage should be quantified. In addition, the register should link enterprises to the establishments they own. Providing data about statistical units at both these levels gives survey statisticians a wider range of design options.

To determine whether units are members of a survey's target population, the register must access and store a wide range of classification data items, including industrial activity, geographical location, sector of the economy, type of business, size, and overseas activities (e.g., importing/exporting, overseas ownership). To enable the design to meet survey objectives, these register data and the associated rules governing changes to these data must be kept current and accurate.

The annual Enterprise Survey conducted by Statistics New Zealand provides an example of using a register for survey design. The survey objectives are to publish economic data such as value added for 61 industry groups covering most nonagricultural economic activity. A stratified random sample of enterprises is selected from the register using three variables: industry, sector of ownership, and number of employees. Industry groupings are formed using the New Zealand Standard Industrial Classification. Businesses with multiple establishments and those classified to central and local government sectors are selected with certainty. The remaining businesses are stratified by number of employees and industry groupings and random samples selected the resulting 187 strata.

5.1.3 Statistical Data

The business register can produce statistics based on snapshot tabulations of statistical units, classified by activity. A time series based on repeated register

snapshots can be used for analyses of (1) changes in job numbers within specified business groupings, (2) origins of specific types of business, (3) growth
of businesses by industry, and (4) survival rates and operating lifetimes of
small businesses. If the individual units in the snapshots can be linked across
time, the resulting longitudinal data are an even richer information source, as
demonstrated in Chapter 4.

5.1.4 Sales of Lists and Other Commercial Uses

National statistical agencies often face externally imposed revenue targets designed to reduce the required level of government funding. A register created
primarily to service the needs of business surveys has several revenue generating capabilities. It can (1) supply business name and address lists together
with industrial activity and geographical classification; (2) carry out contract
mailing operations; (3) assign industrial and geographical classification codes
to businesses in a customer's database; and (4) supply unit information, excluding names and addresses, for use by customers in analysis. The extent to
which a register can be used for revenue generation is limited by the agency's
legal framework and its own policy decisions.

Such uses tend to place stricter accuracy requirements on the register than
are normally justified for statistical purposes, and thus can produce spinoff
benefits for the agency's internal users. Commercial customers examine register products in detail. For instance, they may request lists of businesses by
location, industry, and size. Completeness is expected. If the register includes
an excessive number of businesses that are dead, duplicated, or incorrectly
classified, the customer loses faith in register products.

5.1.5 Respondent Burden Monitoring and Management

The business register can assist in monitoring and managing respondent burden. Options include using the register to (1) implement sample overlap-con-
trol techniques, (2) store information on incidence and level of burden, and
(3) coordinate a respondent management system. The most common role a
business register plays in respondent burden management is its use for *sample
overlap control*, a respondent burden control technique applied at the sample
selection stage. Statistics New Zealand uses this method to reduce the burden
on small respondents, selecting them for only one sample survey. Every unit
on the frame is assigned a random number, and the sample selected is a portion
of the random number range. Other statistical agencies, including the Australian Bureau of Statistics and Statistics Sweden, use variations of the technique
(see Chapter 9).

Statistics Netherlands has introduced another method for managing respondent burden (Kloek 1992). All business surveys draw their frames from the
register. A standard stratification for economic activity by business size is used.
A burden indicator is assigned to each questionnaire, based on its estimated

completion time. When a business is included in a sample, its burden indicator is increased by the burden value associated with its survey questionnaire. The next unit sampled within a given stratum is selected from among those having the lowest overall burden indicator.

Another approach to respondent burden is to use the business register as an information source for a system that actively manages respondents with significant levels of burden. This system can, for example, include special collection arrangements, face-to-face visits, and survey results supplied to respondents. Such an approach is valuable for larger register units which are likely to be in several surveys (see Chapter 6).

A topical respondent burden issue is the overlap between agriculture, business, and household surveys. A self-employed individual can be included in agricultural, business, and household surveys. Ideally, the statistical agency should maintain links between the business, agricultural, and dwelling registers to control the inclusion of individuals across the range of surveys. At Statistics New Zealand, the agricultural and business registers are fully integrated. However, no register of dwellings is currently maintained, although it is under consideration.

5.1.6 Survey Management

A business register may also be part of an integrated survey management facility, which combines frame functions with survey processing and respondent relations, thereby providing an effective operating environment. Using standard tools, survey managers can select samples, initiate mailouts, capture and code data, monitor response, follow up nonrespondents, estimate nonresponse, and edit data. Figure 5.1 illustrates the survey management framework within the Singapore Department of Statistics. Here the functions of the survey control system are (1) capturing, validating, and retrieving on-line data; (2) generating periodic and ad hoc reports for monitoring survey progress, tracking field work, and analysis; and (3) generating mail lists and reminders. This illustrates the growing use of registers, facilitated by modern database technology.

5.2 PRIMARY MAINTENANCE ACTIVITIES

The key register maintenance activities are (1) introducing births; (2) updating names, addresses, and supporting information; (3) identifying deaths; (4) removing duplicates; and (5) updating links between units. These activities involve a variety of information sources, including register proving surveys, survey feedback, comparisons with administrative systems, nonresponse followups, and news media searches.

Maintenance activities depend on the statistical units model and functionality of the register. For instance, if the register has a historical dimension

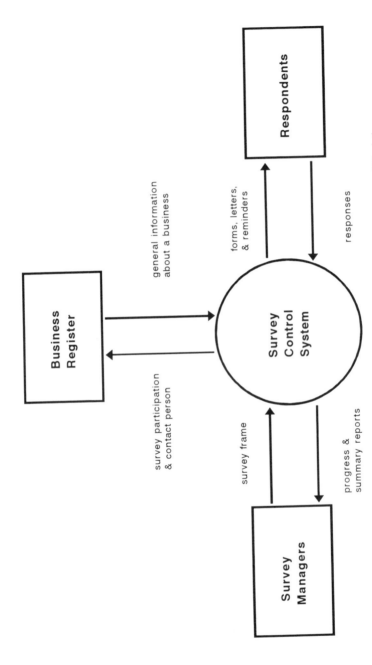

Figure 5.1 Survey management system model at the Singapore Department of Statistics.

allowing units to be linked in time, the maintenance function must include such links. Timestamping changes requires maintenance effort, particularly if registration dates and actual occurrence dates are both recorded.

A central issue in register maintenance is the assignment of identification numbers to statistical units. If assignment is based on economic considerations (e.g., the emergence of a new economic unit) rather than on administrative signals (e.g., legal name changes), then economic judgments are required that cannot be based on information from administrative sources alone.

5.2.1 Identifying Births

The birth of a statistical unit implies that it is entirely new, and that a new identification number should be assigned to it. In deciding whether a unit is new, operational rules such as these are needed:

- An activity change within the same industrial classification group does not imply loss of identity of an establishment, but a change from one manufacturing industry to a quite different one does.
- The transfer of a business to another region means loss of identity and a new market for the business, implying that a new establishment unit should be created.
- For a small service-industry business, continuation of management is a condition for continuation of identity.

Identifying births typically involves a quarter of the total resources needed for register operation. For example, the Statistics New Zealand register identifies approximately 50,000 probable births each year from registrations for value-added taxes. Of these, 32,500 are actual births and the remainder are changes in ownership of existing units, legal name changes, or refer to businesses that never actually operated. At any given point in time, births occurring within the previous 12 months account for about 11 percent of live register units.

Priorities can be placed on processing birth information according to the significance of the unit. In descending order, the priority depends on whether the unit is:

- automatically in the sample for one or more surveys using the register (i.e., in a certainty stratum);
- in the frame population for one or more surveys but not necessarily in a sample (e.g., in a retail trade survey a large number of small retailers are on the frame, but not selected into the sample);
- used to produce register-based statistics;
- used to produce name and address lists for external list brokering services; and

- included on the register but not within any survey's target population and not used to produce register-based statistics (e.g., businesses that are registered to operate but are dormant).

For some data, timely and comprehensive identification of births is critical. For example, a capital expenditure survey requires that new businesses be identified before they start investment in capital purchases. If births are late being added to the register or if they cannot be identified, survey results are biased (see Chapter 34).

5.2.2 Sources of Birth Information

The need to have comprehensive and timely birth information usually means that no single source of information is adequate. Typically, several sources are used, including administrative records (e.g., tax records), membership lists (e.g., accountants society, trade associations), and news media searches.

Administrative records are a good source of birth information. In most countries, the tax system is the main source because of its almost complete coverage of all industrial activity and its relatively low registration threshold. For example, in New Zealand, which has a value-added taxation system, the tax office provides comprehensive birth information for any registered legal unit, including self-employed individuals, businesses with and without employees, trusts, and private and nonprofit organizations.

Media such as newspapers and business magazines are a valuable, complementary source of birth data because they often provide timely identification of key births, such as those in certainty strata composed of large businesses. Also administrative sources—in particular, tax systems—do not usually give comprehensive information on the births of establishments or local units belonging to legal units with multiple locations. For example, the most timely source of information on a large supermarket opening is likely to be a newspaper advertisement. Such a unit would be in scope for a retail trade survey, and, ideally, the survey manager will have that information before the unit begins trading.

Membership lists are generally the least useful source of births. They are often less well maintained than administrative lists, and the criteria for inclusion in the lists are generally broader than is required for a business register. For example, a trade association membership list is likely to include (1) retired or inactive members and (2) businesses that provide only ancillary services to the trade. In short, membership lists are likely to be useful only in the absence of a comprehensive administrative source.

5.2.3 Birth Processing Activities

Births of statistical units suggested by administrative records or membership lists should be confirmed before being recorded in the register. Births of enterprises must be distinguished from births of establishments. Changes of ownership or structure (e.g., an establishment that changes legal owner) must be recognized. Businesses that register for an administrative tax but never start operations must be identified. Source information must be validated and additional details collected that the register needs. The most cost-effective way of confirming births is by a mail proving survey that collects or validates unit name, address, and activity information. These survey data provide the basis for identifying genuine births and adding them to the register.

The precise treatment of birth information depends upon the quality of the source, the significance of the units to which the data refer, and the available resources. For example, the United Kingdom Central Statistical Office (CSO) confirms only those businesses (legal units) whose turnover suggests they have more than 10 employees. CSO then surveys such units to obtain a list of local kind-of-activity units. Smaller businesses are assumed to operate at only one location. By contrast, Statistics New Zealand confirms every new GST registration, mainly because the New Zealand tax agency supplies name, address, and industrial activity information, but no measure of size or information needed to identify changes of ownership of existing register units.

5.2.4 Updating Contact and Classification Information

Updating name, address, and unit activity information is normally the most demanding register maintenance activity. Accurate data must be ensured for contact information such as legal and trading names and addresses, industrial and geographical classification codes, and size indicators. The main sources used to maintain such information are survey feedback, proving surveys, administrative systems, and the news media.

Survey Feedback

Feedback of data from subannual surveys is particularly important because the surveys themselves are often the most up-to-date information source. Feedback data usually relate to name and address changes and to changes in activity. They must be added to the register as soon as received to be available for the next occasion the corresponding units are surveyed.

Survey feedback of changes in classification data items cannot be used for stratification and sample selection for future occasions of the same survey. Bias would result because register units in the target population but not in the sample cannot be similarly updated. This problem can be avoided by using indicators that show the source of change data, enabling these data to be used for purposes that do not bias sample selections.

To preserve register integrity, feedback of information from surveys should be coordinated. Through contractual arrangements with the survey managers, the register manager can agree on the nature and timing of changes. Such arrangements are in place at Statistics New Zealand, where in a typical month the business register receives and acts on 2000 pieces of information from 10 different surveys.

Proving Surveys

Periodic updates based on proving surveys ensure that the register is consistent at a point in time, and that the survey design variables, particularly size indicators and classification codes, are accurate. In addition, proving surveys can identify ownership and legal structure changes. Statistics New Zealand, for example, undertakes a proving survey for non-farm businesses in February each year. (Farms are updated as part of the June agriculture survey.)

Administrative Sources

Administrative records are also a source of updating information such as addresses and geographical and industrial classification codes. Data from tax returns can be a very useful source of size information. A significant limitation of most tax systems, however, is that they provide information on changes for legal units and not local units.

5.2.5 Identifying Deaths

A regular and comprehensive program for detecting and eliminating dead units reduces (1) the numbers of post office returns ("gone, no address") and (2) nonresponse problems encountered by surveys. It also provides a higher quality product for sales of lists to external customers. Dead units cost customers money and diminish the register's sales potential.

From a statistical perspective, it is important that deaths be dealt with on a regular basis. Ad hoc purging of accumulated deaths distorts register-based statistics and time series. There should be routine monitoring of units to identify deaths. Because the identification of deaths is not an exact process, it is advisable to retain dead units in the register, flagged with a death indicator. Checks of potential births against the register sometimes indicate that a "dead" unit has undergone change in legal structure and is no longer dead but has reappeared as a birth.

5.2.6 Death Processing Activities

Deaths in samples are usually picked up by survey staff and passed to the register for updating. For these portions of the business register, deaths are thus recorded on a timely basis, although this survey feedback should not be used in future sample selection if bias will result. In other instances, the news media may provide the most timely information on deaths—for example, when a large multiestablishment retailer closes a shop.

For the remaining units on the register, proving surveys or comparisons with administrative data are required to provide information on deaths. An administrative system is often appropriate for dealing with the large number of smaller, individually insignificant units. The suitability of a particular administrative system depends on its coverage and its rules governing deaths. Many tax authorities do not declare units as dead until all taxes have been paid, often years after the business ceased operations. This is too slow to meet register objectives.

Another category of deaths is those occurring to units that are survey non-respondents, and for which there is no relevant administrative source. Survey follow-up procedures need to distinguish between nonrespondents that have ceased operations—and hence are ineligible—and eligible nonrespondents that are operating but simply not responding. If nonresponding businesses are not routinely checked, dead units will gradually accumulate in the register (and the sample). This will eventually cause unacceptable biases in surveys and statistics based on counts of register units, and it will inhibit the statistical agency's ability to sell lists.

5.2.7 Updating Links Between Units

A business register aims to provide lists of economic units upon which statistics can be based. To do this, it must maintain information on the relationships between units. In this context, businesses may be classified into three basic groups, each with its own priorities and requirements.

The first group comprises businesses that can each be represented as a single enterprise owning and operating a single establishment (or, depending on the particular units model, as a single legal unit having a single local unit). This group accounts for the majority of businesses. Maintaining structural information for these units involves identifying when the ownership of an establishment (local unit) has been transferred from one enterprise (legal unit) to another. This can be achieved via a proving survey, as previously described.

The second group comprises businesses that can be represented as a single enterprise with multiple establishments. Three percent of the enterprise-level units in the Statistics New Zealand register fall into this category. Structural information for the group is also relatively easy to maintain with a proving survey targeting the enterprise.

The third group comprises businesses with more complex structures—in particular, large corporations with a diverse range of industrial activities. Such businesses have large turnovers and employment, and therefore they are usually represented in several survey populations and selected with certainty. Thus, they are a maintenance priority. Their structures must be accurately recorded in the register by identifying the groups of enterprises (legal units) with common ownership and the establishments (local units) they collectively own. Because this sort of structural information tends to change constantly, and because the corresponding survey collection units also change frequently, dedicated specialist resources are often needed to maintain these businesses, as further elaborated in Chapter 6.

5.2.8 Removing Duplicates

When several updating sources are used, a potential exists for duplication of businesses within the register. For instance, this can happen when one source identifies businesses by their legal names and another source uses trading names. Duplicate units cause problems when respondents are approached multiple times by the same survey. Unidentified duplication can also be a significant source of bias in samples and result in substantial errors in published statistics, as noted in Chapter 34. The process for dealing with duplicates in the register differs from that for deaths, because it involves the complete removal of the duplicate unit or structure from the register, not just the setting of a flag.

To minimize duplication, some form of name and address matching should be incorporated into register operations (see Chapter 20). Depending upon its sophistication, the system may range from support of manual matching to a fully automated search for matched units. To use computerized matching effectively, it is important to have strict adherence to standard formatting rules for legal and trading names, physical addresses, and other unit details. Before a potential birth is added to the register, it should also be matched against the register to check whether a unit with a similar name and/or address is already present. Checking for duplicates should be regularly applied to the whole register. Geographical checks using street address are useful.

5.3 REGISTER DATA SOURCES

When maintenance data are being drawn from several sources, and in particular when automated matching procedures are involved, care must be taken in dealing with conflicting information to avoid the introduction of duplicates. Two forms of controls are advisable. First, updating of the register should be performed only by register staff. Second, there should be an active duplicate identification process. The preceding section discussed the suitability of various data sources in relation to specific maintenance activities. This section focuses on the factors that influence the choice of sources to use.

5.3.1 Survey Feedback

When feedback from surveys using the register is obtained, immediate action must be taken to update the register. There is no choice. Survey managers and respondents usually become angry when information they provided is not used.

5.3.2 Administrative Data Sources

Administrative systems are often the most efficient source of register maintenance data, and they involve no respondent burden directly imposed by the statistical agency. For effective use of an administrative source, the administrative unit identifiers must be stored in the register. Factors that must be taken into account by the statistical agency in deciding which administrative sources(s) to use are outlined in the following paragraphs.

A stable supply of information is important. Administrative sources are affected by changes in government policy. The tax system is particularly susceptible. The government can change tax policy, altering thresholds and criteria for furnishing returns, and so forth, without considering the effect on data supplied to the business register. To have security of access and to make effective use of the data, the register manager must develop a relationship with the administrative source, thereby influencing when and how changes are made to the administrative data. Usually, legal and formal relationships in addition to personal business relationships must be established between the administrative source and the statistical agency. The administrative source must also be one with which it is appropriate for the statistical agency to work. The independence and reputation of the statistical agency must be maintained.

The characteristics of an administrative system determine the use that can be made of its data—factors such as the population of businesses covered by the system, the frequency with which the information is collected, and the time taken to process information before it is made available to the statistical agency. The definitions used by the administrative source are also important. For example, the rules for births and deaths may be quite different. In New Zealand, where value-added tax data are used, the information received on deaths of units is untimely because of administrative rules. The tax office does not regard a unit as inactive until all assets are sold and outstanding taxes are paid, whereas the register would like to flag the unit as inactive the moment it stops business activity. Units may remain alive on the tax system for years after they would be considered dead by the register.

The quality of administrative data also effects their utility. Relevant factors include: (1) the quality of the administrative data collection vehicle; (2) the likelihood of accurate responses, bearing in mind the administrative purpose (e.g., legal and illegal misreporting to a tax system); (3) the degree of data item compliance; and (4) the quality of data capture and edit methods.

5.3.3 Register Proving Surveys

Although more expensive than administrative sources on a per-unit basis, proving surveys are an essential tool. They give the register manager greater control over data collection (e.g., questionnaire design and content) and processing (e.g., use of statistical edit standards) than do administrative sources. And they enable the manager to update units consistently.

5.3.4 Membership Lists

If suitable administrative sources exist, membership lists are not frequently used because they usually are limited in coverage, are poorly maintained, and contain unwanted units.

5.3.5   News Media Sources

The news media are a valuable source of supplementary data, of particular value in tracking the more visible businesses in the economy. Trade magazines and newspapers tend to be the most useful of the various types of news media.

5.4   QUALITY-PERFORMANCE MEASUREMENTS AND CONTROL

Measurement of register quality is the precursor to implementing quality control. Measures of coverage, coding accuracy and stability, range and type of errors, numbers of duplicates, timeliness of births, timeliness of unit information, and analyses of edit failures all reflect register quality and performance. Feedback from survey managers, sample designers, and commercial customers are also valuable in assessing quality.

5.4.1   Coverage

How well the register covers the universe of businesses is an important measure of quality. Knowledge of coverage characteristics enables assessment of the quality of survey design information and the representativeness of survey and register-based statistics. Several measurement methods are available. First, a field enumeration can be taken over a sample of geographical areas. This involves walking around the selected areas and enumerating the businesses, determining their industrial classification, and estimating their size. Comparison of these results with register units for the same areas provides estimates of register undercoverage and of incorrectly addressed or classified units. The limitation of this method is that it only covers businesses that are readily visible.

A second method involves comparison of the register with an administrative system for which coverage is known. For example, because of the high proportion of businesses they cover and the well defined areas of undercoverage, tax systems can be used to assess the number and characteristics of register units and to provide estimates of coverage. Statistics New Zealand compares all enterprise-level units on its business register against tax records. While this exercise was started to expand the coverage of the register to additional industry groups, it presented the opportunity to check undercoverage for industry groups included in the register. Computerized matching techniques in conjunction with proving surveys are used to collect information about units paying taxes but not represented on the register.

5.4.2   Accuracy of Coding

One method of measuring the accuracy of register coding (e.g., of industry, legal character, and institutional sector, etc.) is to select and retain a sample of source documents. After a defined period, register staff recode the units based on the original source documents. This sort of recoding exercise provides data on coding error rates and the amount of coding based on inadequate source data.

5.4.3   Other Measures

Quality measures can also be derived by monitoring the number of duplicates identified, the timeliness of births, and the accuracy of data from administrative sources, as well as by analyzing edit failures, error records, and complaints from survey respondents. Quality measurements can be obtained by maintaining a record of all errors for a defined period, which identifies the types of errors and their incidence. Register staff and users must record information on errors as they are detected. Although they are time consuming to produce, these data make register management aware of errors and their magnitudes, and enable quality measurement and control work to be prioritized.

5.4.4   Quality Assurance Techniques

Once quality measures have been established, controls can be designed and put in place over coding, administrative source data, and register input and output.

Quality Control of Coding
To monitor the quality of coding processes on a regular basis and to check on individual coders, a control system can be built into register operations. Simmons (1972) describes procedures based on a three-way independent sample verification. Booth and Mudryk (1985) outline the approach used at Statistics Canada.

Quality Control of Administrative Data
When using information from an administrative system, the register must have some form of quality control mechanism. This might involve checking every incoming record or a selected sample. For example, register staff at the United Kingdom Central Statistical Office check each incoming value-added tax record to validate filing keys, classification, turnover, and address (Perry 1990). Errors thus identified are manually corrected. The U.S. Bureau of Census applies statistical quality control to incoming tax data (Monk and Hanczaryk 1991). The results are fed back to the tax office to help it monitor and improve the quality of data supplied.

Quality Control of Data Capture
Edit checks are necessary to control the quality of data received from register proving surveys, survey feedback, and other sources. The aim of such edits is to reduce data input errors (e.g., by rejecting values that are out of range), to

guard against incomplete or incorrect maintenance procedures (e.g., verifying that the geographical code of a business has been updated following a change of business location address), and to minimize the possibility of invalid changes to classification codes.

Three control systems can be built around a data capture process: (1) editing messages to alert the operator when a potential error has occurred; (2) tracking the type and frequency of edit changes made, and by whom; and (3) systematic re-editing of a sample of each staff member's work. Reports on the number and type of editing changes and errors made, and by whom, can be used to target problem areas. Procedures that seem to be causing problems can be reviewed and modified. Additional systematic checks can be conducted on the work of staff who are performing below target accuracy levels, and training and documentation can be provided to improve performance.

5.5 CONCLUDING REMARKS

The use of business registers by national statistical agencies is a relatively recent practice. Most development work to date has focused on conceptual and classification issues. Only recently have statisticians started to realize the full potential of such registers. At the same time, privacy issues and the resources allocated to national statistical agencies are under close scrutiny. Thus, register managers are being forced to examine register objectives, organization, and resource allocation. The search to define and achieve register quality will continue.

REFERENCES

Booth, J., and W. Mudryk (1985), "Quality Control at Statistics Canada," *Communications in Statistics Theory and Methods*, **14**, No. 11, pp. 2589–2593.

Kloek, W. (1992), "Policies to Improve the Image of the CBS with Respondents for Economic Statistics," paper presented at the Seventh International Round Table on Business Survey Frames, Copenhagen, Denmark.

Monk, C. H., and P. S. Hanczaryk (1991), "Quality Assurance of Administrative Record Data," technical report, Washington, DC: U.S. Bureau of the Census.

Perry, J. (1990), "Business Registers—The Search for Quality," *Statistical Journal of the United Nations*, **ECE 7**, 55–61.

Simmons, W. R. (1972), "Operational Control of Sample Surveys," *Laboratories for Population Studies*, Manual Series No. 2, New York.

Singapore Department of Statistics (1982), "System Requirement Specification Report."

CHAPTER SIX

Profiling Large Businesses to Define Frame Units

Leon Pietsch[1]

Australian Bureau of Statistics

Information about units on the business register must be kept up to date to facilitate consistent and accurate reporting in statistical surveys. As Archer describes in Chapter 5, updating information can be obtained as feedback from surveys that use the register as a frame, from surveys specially conducted to update the business register, and from administrative sources. For the largest businesses in the economy, however, there are practical problems with all three sources of updating information.

Several national statistical agencies have responded to the difficulty of determining the statistical structures of complex large businesses by introducing profiling—that is, direct contacts with large businesses to define their structures. This chapter describes the concepts underlying the profiling of large and complex businesses, discusses the organizational aspects of profiling operations, and examines the impact that profiling has had for some agencies. When a profiling operation is established, other tasks and responsibilities may be seen as a natural extension to the profiling responsibility; these tasks are also discussed.

[1]The author wishes to thank colleagues at the Australian Bureau of Statistics (ABS) who made extensive and valuable comments during the drafting of this chapter, especially John Struik and John Billing. Valuable contributions were also received from staff of Statistics Canada, Statistics Netherlands, and Statistics New Zealand. Nevertheless, the views expressed are the author's and do not necessarily reflect those of ABS or other statistical agencies.

Business Survey Methods, Edited by Cox, Binder, Chinnappa, Christianson, Colledge, Kott.
ISBN 0-471-59852-6 © 1995 John Wiley & Sons, Inc.

6.1 THE PROBLEM OF DEFINING STATISTICAL UNITS FOR LARGE BUSINESSES

A large business often comprises many legal units linked by common ownership or control. Internally, it is likely to have a complex structure including multiple legal units unified under several operating divisions. Divisions often have multiple establishments and physical locations. In Chapter 3, Nijhowne describes how the statistical structure of a business relates to its legal and organizational structures.

Survey feedback is unlikely to contain sufficient information for updating the statistical structure of a large business. For example, most surveys do not request information about the various levels of units within a business, which is needed for comprehensive updating of legal units, organizational units, physical location, and so on. Similarly, administrative records are unlikely to record the structural information required to define and maintain the corresponding sets of statistical units.

Proving surveys directly address the problem of obtaining information about all units. However, the complex and unique internal arrangements characterizing large businesses make it difficult to collect accurate information about the business' structure.

The extent to which standard statistical units correspond to legal units varies between countries, depending on the requirements of data users, the nature of available updating sources, legal structures adopted by businesses in response to corporate and taxation legislation, data availability for various units, and so on. The more closely statistical units are defined to reflect the business' ability to report data, especially on a subannual basis, the more likely it is that the units will not correspond with legal units or easily defined combinations of legal units.

Furthermore, a business may swap organizational units between legal units, combine them within a legal unit, or subdivide them into separate legal units. Businesses change their treatment of overhead expenses and transfer pricing. Takeovers and mergers can occur overnight, changing the business' structure. All this can happen without the physical capital used in an organizational unit being changed or rearranged. The only visible changes are those in the legal and accounting framework.

6.1.1 Impact of Largest Businesses on Published Statistics

These factors result in a high risk that the register's representation of the economy's largest businesses is incorrect in some respects. Any survey that uses the register as a frame will reflect the undercoverage, overcoverage, misclassification, and other inconsistencies in these representations. Because they reflect the economy's largest businesses, errors have a significant impact on the accuracy of statistics produced from the survey, as well as on the consistency between statistics produced by surveys within a program.

6.1.2 Respondent Burden

The largest businesses in the economy are likely to be selected as respondents for most surveys for which they are in scope. Their respondent burden is minimized only if the statistical agency ensures that the statistical units included in survey frames correctly represent the business' organization and are aligned with structural units for which data can be reported with least effort.

6.2 PROFILING AS A RESPONSE TO THE PROBLEM

While each national statistical agency has developed a unique approach to profiling, there are common elements. At a minimum, *profiling* involves personal contact with a large business to gain insight into its legal and organizational structures and to agree upon a way of recording those structures on the business register. For such businesses, a clearly preferable way to define the statistical structure may not exist; negotiation may be required, especially regarding the definition of establishment level units.

Some agencies view profiling with its personal contact with a business as a one-time job, after which maintenance is handled by normal proving survey techniques. In other countries, profiling is seen as the first step of an ongoing program of maintaining a special relationship with the economy's largest businesses.

Profiling can be viewed as part of a broader, coordinated strategy for improving an agency's business statistics program that recognizes and responds to a variety of needs.

6.2.1 Consistency of Data Reporting

In recent decades, many resources have been applied to develop integrated statistical concepts and systems to (1) ensure that economic activity in a country is measured without gaps or duplication, (2) integrate outputs from different surveys, and (3) allow comparisons of different parts of the economy and different economies. A centralized business register is a major component of such an integrated system. Other components include a standard industrial classification, a standard institutional sector classification, and standard data items to be included in the various surveys that use the business register as a frame.

However, integration of statistical concepts and systems does not guarantee that aggregate statistics produced from the systems are consistent or complete. While the definitions within the system are integrated, data collection may be quite fragmented, especially for large, multifaceted businesses.

A large business does not usually provide the statistical agency with a single, integrated set of data describing its operations. Rather, it responds to a number of separate surveys which vary in focus. Some surveys target the enterprise, while others target establishments. Some deal with only that part of the business classified to a specific industry. Some require data on a subannual basis, while others collect annual information. Some focus on goods and services produced, others on factors of production, and still others on finance.

In addition, different people in different parts of the organization may complete the statistical questionnaires. One person may interpret requests somewhat differently from another, and no one is likely to ensure that the data are consistent across survey requests. Thus, if the data collected in this way are aggregated (e.g., within the national accounts), it is highly likely that there will be internal inconsistencies between the various data items collected from any individual large business.

Profiling is a partial means of addressing this problem. Its major focus is ensuring that the register's statistical units correctly reflect the business' organizational structure, without gaps or overlap. This is a prerequisite for consistently accounting for any particular activity within the business across the full range of statistical surveys.

There is a second consistency-related benefit from profiling. Because a senior representative is usually contacted and the need for consistency is highlighted, the business itself is more likely to monitor the consistency of the data they provide. Sometimes, as a result of profiling, the business even reviews the standards of consistency and comparability in its internal management information reports.

6.2.2 Reduction of Respondent Burden

While the demand for statistical information is increasing, businesses are trying to limit their respondent burden. They are becoming increasingly conscious of the load imposed on them by government regulations and requirements. Profiling provides statistical agencies with an opportunity to promote the importance of the survey data. Profiling normally reduces respondent burden because the statistical agency and the business together decide the best way for the business to report statistical information. When the business' accounting structure is the criterion used in defining statistical units, use of readily available data is maximized.

In routine survey operations, the agency's approaches to a business are likely to be fragmented. Often, relatively junior staff pursue information for a specific survey. The contact by profiling staff, on the other hand, is a broad-based approach. The statistical agency thus presents itself to the business "with a single face." This gives the business an opportunity to provide general feedback about difficulties it has with the agency's surveys, such as inconsistencies, duplication, and confusion resulting from multiple approaches. When profiling staff address these problems, goodwill and increased cooperation are generated. While some problems may be difficult to handle, many are easily correctable misunderstandings.

Sometimes profiling highlights previously incorrect reporting by the business. In these circumstances, the revised reporting procedures may increase respondent burden.

6.3 ORGANIZATION OF PROFILING

To justify the cost of profiling a business, there must be the expectation that the results will have an impact on the agency's statistical output or reduce respondent burden or processing costs. Factors to be considered in the organization of profiling are outlined in the following sections.

6.3.1 Criteria for Choosing Businesses to Be Profiled

There are six criteria for choosing businesses to be profiled based on size, complexity, propensity to change, respondent relations, alternative sources of data, and frequency of use.

Size
The business chosen for profiling should be significant in terms of its contribution to the total economy or to some critical aspect of the economy. A relatively small number of businesses account for a significant portion of economic activity in most countries.

In practical terms, the cost of profiling a given number of the economy's largest businesses can be weighed against the coverage they represent. Sizes of businesses and their coverage can be measured in terms of employment, turnover, value added, or similar variables. Often, employment is the only available indicator. If there is a choice, the various potential indicators can be expected to give similar results, although there may be exceptions because some industries have a higher-than-average turnover or value added per employee.

If the largest 1 percent of businesses are profiled, it is likely that coverage will be obtained for over 50 percent of the economy. Profiling a much smaller percentage of the largest businesses will probably still give 30 percent coverage. If available resources are very limited, profiling only the 10 or 20 largest businesses can still yield significant benefits.

Either before or after initial profiling, it may be judged that some large businesses have such stable or simple internal structures that they can be adequately tracked using proving surveys. Such businesses need not be kept in the profiling portfolio. This frees resources to focus on smaller but more complex or volatile units, which present a greater likelihood of significant impacts on statistical results.

Complexity
Profiling is worthwhile for businesses with complex structures because it is difficult to get the required frame information from proving surveys. The com-

plexity may reflect myriad legal relationships within the business, joint ventures with other businesses, or a diversity of activities across several industries or institutional sectors.

Propensity to Change or Volatility in Structure

A business that is continually changing its structure requires more frequent profiling. Such changes may reflect internal reorganization, or the acquisition or disposal of component units. Profiling often reveals imminent changes in structure. In this way, important updates to the business register can be made and respondent contacts revised in a timely fashion.

Respondent Relations

If a business has been a poor or difficult respondent, profiling is likely to improve its relationship with the statistical agency. This is of particular importance for the largest businesses, in view of the impact they have upon data outputs.

Reliability of Alternative Updating Data

When alternative, reliable sources are available for updating certain types of businesses, it is less important to include them in the profiling portfolio. For example, central bank records for financial institutions may be a superior source to the taxation records used to update other businesses and thus reduce the need for profiling.

Frequency of Use

Some segments of the economy are rarely incorporated into surveys, such as religious organizations and the nontrading government sector. Such businesses may be considered unimportant in profiling terms. On the other hand, these organizations have a significant economic impact and, in some countries, are in scope for business statistics such as employment. If data need to be collected from them profiling can be effective in determining how best to do so, because they often do not have standard accounts and lines of authority.

6.3.2   Choice of Staff to Conduct Profiling

Profiling has an important public relations role. Thus, profilers need to be good ambassadors for the statistical agency. Because the profiler is seeking an overview of the business, initial contact is often with a senior manager such as the company secretary or chief accountant. Profilers need to quickly understand the business complexities that such senior officers explain, and they need to recognize the implication for the agency's surveys. In addition, to discuss the problems a business may be having with existing data collections or related topics, the profilers need a broad understanding of the statistical agency's operations.

Staff meeting the above criteria are likely to be found in the agency's middle management, rather than among the junior staff used for regular surveys. In choosing staff, there may be a tradeoff between those with a good understanding of commercial business structures and practices and those familiar with the conceptual and operational frameworks of the statistical agency. Some agencies recruit business accounting specialists on a contract basis.

Organizational Structure

Normally, profiling is handled by a special unit within the statistical agency. To establish a personal relationship between the large business and the statistical agency, profilers should be given a portfolio of businesses for which they have continuing responsibility. A pool of support staff is also needed, including those who implement changes to the business register. In geographically large countries, the profiling unit should have staff located in the major business centers to facilitate face-to-face visits to business sites. The number of large businesses identified for profiling and the breadth of the responsibilities that the profilers conduct for those businesses determine the number of staff required.

As a first step in developing procedures and expertise, Statistics Canada used survey staff from middle and upper middle management levels on a part-time basis. This was a temporary measure. Because a large part of the profiling process involves research, assembly of materials, and the subsequent updating of the register after profile completion, profiling is now carried out by specialized staff. Using profiling specialists facilities communication between the persons responsible for the various steps in the process and ensures a smooth flow of work.

When profiling commences, an intensified effort is required to establish a relationship with the large business and potentially to correct for a backlog of errors in the statistical agency's records. Over time, the profiling work settles into a maintenance operation, and, on average, fewer resources need to be devoted to the task.

The primary focus of profilers is to ensure correct and expedient updating of register data for these large businesses. Inevitably, however, profilers will serve as liaison points between the profiled businesses and other parts of the statistical agency and become involved in additional tasks. A judgment has to be made regarding the level of effort that should be devoted to such efforts.

Secondary tasks have influenced the profiling approach adopted by Statistics New Zealand. Profiling was undertaken for about 6 years, and then updating by mail questionnaire was used. Both approaches were reasonably successful, but some problems did occur. As a result, Statistics New Zealand adopted a new approach that recognizes that the data required to define different levels of the statistical structure of a large business often reside in different sites across the business. Their profiling unit, closely associated with regular register operations, decides which legal units and enterprises are associated under common ownership. Data sources include journals, newspapers, and annual reports. A separate unit known as the complex business unit (CBU) profiles complex businesses and collects all business data for them. The CBU ensures that there are no gaps, omissions, or differences in approach used by different parts of the same business. The CBU compares data supplied for subannual surveys with that for annual surveys. Primarily all face-to-face contact and negotiation with complex businesses is done by the CBU.

6.3.3 Frequency of Profiling

A regular (perhaps annual) review of the organizational and statistical structures of each business is important. The need exists to tap into market intelligence on businesses to identify when structural changes have occurred. This can be done through business columns of newspapers and trade magazines and company reports (published by businesses listed on the stock exchange).

6.3.4 Rules and Procedures

Because profilers establish a close relationship with businesses, they become acutely aware of respondent burden. At times, they may be tempted to allow considerations of respondent burden to outweigh the statistical agency's requirements for detail. To resolve such tensions, precise rules and procedures must be established governing the definition of units on the register and the latitude allowed in individual cases. The procedures should include a formal mechanism whereby the agency's subject matter specialists are given the opportunity to comment on the proposed definitions of statistical units for specific businesses.

6.3.5 Resources and Costs

As the experience of various agencies indicates, profiling requires significant resources in addition to those normally used to maintain the business register.

Australian Bureau of Statistics
In the first year of operations, the Australian Bureau of Statistics (ABS) devoted about 30 person-years to profiling and updating register information. Even with this level of staffing, 2 years were needed to finalize the initial profiles of the 500 largest businesses.

As the profiling operation moved into a maintenance phase, staff levels were scaled down to less than half that at commencement. This level is expected to be sufficient to reprofile 500 large businesses at least once per year. As profilers gain experience, ways of streamlining profiling are being explored. There is a growing confidence at ABS that more use can be made of proving survey questionnaires to replace face-to-face visits. However, face-to-face visits continue to be essential when there are significant structural changes in a business.

Statistics Canada

Statistics Canada's profiling portfolio includes about 270 businesses, each of which is operated and managed as a single entity and prepares consolidated financial statements on an annual basis. When initially profiled, most statistical structures for these business units were out of date and required significant effort to update.

At approximately $3000 Canadian each, the initial profiles were costly. Early indications are that once the initial profile is completed and updated on the register, the cost of an annual review of those profiles will be considerably less. Experimentation with turn-around forms, business unit portraits, more extensive use of facsimile communication, and so on, may reduce costs further.

Statistics Netherlands

Statistics Netherlands has six staff members in its large-business unit, with responsibility for about 170 businesses. Two staff members have been involved in profiling by face-to-face interviewing. With this staffing level, not all businesses can be reprofiled every year. Those that are not visited are contacted by letter. The current capacity has not been sufficient to keep units up to date; planned extensions will increase staffing levels to 10.

Statistics New Zealand

Within Statistics New Zealand, the organizational structure comprises a profiling unit and a separate complex business unit. The profiling unit has one staff member responsible for about 200 large businesses. The complex business unit has seven members, each of whom is responsible for about 30 global enterprises. CBU staffing may be revised with the added responsibility of collecting balance of payments statistics.

If profilers become involved in additional coordination tasks, additional resources will be required. The liaison work requested by both businesses and agency survey takers can be unending and expensive. When an agency introduces profiling, resources must be allocated to liaison activities that did not occur before profiling. On the other hand, profilers must prioritize their work to ensure that their activities are cost-effective.

6.4 IMPACT OF PROFILING

It is difficult to measure the full impact that profiling has on register quality and on the agency's statistical output, especially if no measurement procedures and tools were in place when profiling began. Anecdotal evidence illustrates the failures of conventional procedures and the benefits of profiling. This section summarizes the profiling experience in Australia and Canada.

6.4.1 Australian Bureau of Statistics

For the Australian Bureau of Statistics, initial profiling made many corrections to the register. Now that regular profiling has been established, changes to business structures are reflected in a more timely way than before.

The major problems identified by profilers were related to takeovers and changes of ownership of businesses or segments of businesses. These problems may have predominated because profiling started at the end of a period of rapid takeover activity and other changes in the corporate sector.

Most of the affected business units were already recorded on the register but were shown under previous ownership. Therefore, frames taken from the register were broadly correct in terms of identifying business activity and survey benchmarks, but were inaccurate in identifying the names and addresses of businesses to whom survey forms should be sent. These errors resulted in lost time spent in correcting frames and in poor relationships with respondents.

In addition to updating the register's ownership structures, profiling also resulted in updates to the industry and employment data used for stratification. Most industry code changes were within broad industry sectors and did not impact the unit's eligibility for inclusion in a survey. Some changes might have gone undetected even if the unit had been sampled.

There were also instances where incorrect statistical units were recorded on the register, resulting in surveys collecting data from the wrong units. The resultant errors in survey estimates, amounting to millions of dollars, were detected and corrected by profiling.

Improved ABS Relationships

ABS' introduction of profiling has had a significant impact beyond the creation of more accurate and up-to-date frames for users of the register. Profiling has improved its standing in the business community; ABS has gained significant ground in public relations.

Businesses responded positively to having a single ABS contact point (in the profiling unit), who coordinates the despatch of annual survey questionnaires and with whom businesses can discuss problems, including some unrelated to these surveys.

Response rates in surveys have often improved significantly for these businesses. Consequently, fewer resources are spent following up late returns. The amount of imputation required for missing returns has decreased, though independent subject matter specialist contacts are still made.

Introduction of New Statistical Units

Profiling in Australia was instituted at the same time that the "management unit" was introduced as a new statistical unit at the enterprise level. The availability of appropriate data is a criterion in defining the unit, so that management units closely mirror the operating divisions of large businesses. Profiling was essential in implementing the management unit concept because significant changes were required for businesses for which the former legal-unit-based enterprise bore little relationship to the business' actual operating arrangements.

Introducing a statistical unit that reflected the availability of information within the business was important to improved relationships between ABS and surveyed businesses. It did, however, make it difficult to quantify the impact of profiling as the effects of profiling and new units were confounded.

6.4.2 Statistics Canada

For Statistics Canada, profiling resulted in significant changes to the register, in terms of changing business units and in adding new units. Also, standard industrial codes were revised for existing units, covering about 2.5 percent of the total revenue of the profiled units.

Improvements to Surveys
The impact on Statistics Canada's annual surveys will not be evident until a survey cycle underway at the time of writing is completed. However, the monthly surveys experienced an immediate impact. The Survey of Employment Payrolls and Hours, for example, evidenced an increase in reported employment for those businesses for which recent profiles had been made. A systematic method of measuring the impact of profiling on outputs of Statistics Canada is currently under development.

Improved Relationships
The development of the profiling unit, along with the development of associated tools such as the "Portrait System" (which generates a diagram of the business unit structure), has made the register more user friendly and has had a positive impact on the business community. Through the Portrait System, it is now feasible to exchange frame information with the businesses without having to meet face-to-face. The result is cost savings for both businesses and Statistics Canada.

6.5 EXTENSIONS OF THE PROFILING TASKS

Earlier sections of this chapter refer to the extension of profiling beyond the responsibility for updating the statistical structures on the business register. There are two main roles that can develop: (1) a general liaison role and (2) a data integration role.

6.5.1 General Liaison

Profiling involves close personal contact between the profiling staff and the business' senior representatives. In establishing such contact, the profiler stresses that the statistical agency is attempting to understand the business' structure and its accounting records so that respondent burden can be lessened. The business is encouraged to notify the profiler of any future structural changes.

Naturally, the business will take the opportunity to raise concerns in its relationship with the agency. Some concerns will be satisfied by a simple explanation. Others will require further action by the profiler or by others in the agency. The issues raised by a business can be wide-ranging and may relate to surveys that do not use the register as a frame.

To assist in coordinating contacts with the business, profilers should be aware of and perhaps coordinate any face-to-face visits being planned to the business by the agency's other subject-matter areas.

Coordination of Data Collection

If the business is included in several surveys (as most large businesses are), coordination issues may be raised. Some can be satisfied only if the profiler (or someone else in the agency) has an ongoing coordination role. This role may be little more than ensuring that individual inquiries are addressed to the appropriate contact person in the business, and that the business knows which survey forms it is likely to receive.

To take coordination a step further, the profiler may take on the role of a "postbox," whereby questionnaires are channeled through the profiler. Questionnaires handled in this way may be limited to annual surveys. The approach is likely to slow questionnaire movement and therefore may be inappropriate for subannual surveys. In any case, subannual surveys are more likely to settle into a regular pattern and to require less liaison support.

If this approach is adopted for at least some questionnaires, the profiler has a good opportunity to check that the business' structure as recorded on the register is workable in practice. Also, using the profiler as a postbox precludes unnecessary multiple approaches by subject-matter areas to the business.

Electronic Data Interchange

Another emerging coordination issue relates to the electronic capture of data. As businesses develop automated accounting systems and "the paperless office," statistical agencies will face increasing pressure to accept data through electronic media (see Chapter 19). If this occurs, there will be a strong incentive to (1) avoid having separate electronic capture for each survey and (2) establish ways of integrating data collection for the surveys. Furthermore, there will be additional pressure on the statistical agency to accept data that are already recorded in the business' accounting system without further manipulation by the business. Such developments are best handled collectively rather than on a survey-by-survey basis. Profilers are likely to play a major role in coordinating or even developing the appropriate facilities.

Statistical Units Model

Because of their experience in considering how businesses operate and record their operations, profilers can play a valuable role in developing statistical concepts, definitions, and operating procedures that reflect the "real world" that business statistics should describe.

6.5.2 Data Integration

Even if profiling ensures that all components of a large business are recorded without gaps or duplication on the register, inconsistent responses to the separate surveys sent by the statistical agency may nonetheless be provided by different parts of the business. Such problems can be addressed in several ways.

Coordinated Data Collection and Editing

Profilers can improve liaison with the businesses in their portfolios by coordinating the dispatch and receipt of the survey forms. This approach can be taken a step further as the profiler's role extends to include not only the collection of the forms but also editing and even compiling the data received from large businesses. This establishes a mechanism that can check the internal consistency of various data sets. Because the staff of a coordinating unit will have less detailed knowledge than will subject-matter specialists, relative responsibilities have to be carefully defined.

Data Confrontation

A complementary approach is to establish a database for the largest businesses into which key data from separate collections are placed. Similar items from different collections are compared to highlight inconsistencies. Inconsistencies between data reported in different collections can result from misunderstanding of data item definitions, problems of consolidation of data items when aggregating lower-level units to higher-level units, and so on. When inconsistencies are found, they need to be resolved with the aid of subject-matter specialists whose collections require the data, with the respondent's help if necessary.

Statistics Netherlands has established a "TOP100 database" which records employment, input, output, gross value added, operating surplus, fixed capital formation, depreciation, earnings, and unusual credits and charges. These data are particularly useful in checking consistency and can link financial and production activity.

The most important variable for checking the proper delineation of units covered by various surveys is the number of employees. If the number of employees reported for a unit in one survey does not equal the corresponding item(s) in other surveys, it is immediately apparent. In addition, employment is a strictly additive variable when aggregated across units. An alternative measure such as revenue is not always additive because intracorporate transfers will be netted out on consolidation. However, checking based on employment may miss problems relating to head office charges, transfer charging, and other accounting concepts. Profits are likely to be most sensitive to such problems.

Consolidation Questionnaire

An approach being tested by Statistics Canada assigns responsibility to a large business to show that the statistics it provides are internally consistent. Businesses are asked to complete a consolidation questionnaire, which requires them to report items such as operating revenue, capital expenditures, and operating profit for each level of the business. Where intracorporate transactions have been consolidated out in compiling the group's business accounts, the data must be shown on both a consolidated and unconsolidated basis on the consolidation questionnaire.

In this way, it should be possible to relate variables such as balance sheet items and indirect revenue or expense items, which are often recorded only for higher-level units in business accounts, to lower-level units. Conversely, it should be possible to reconcile revenue and purchase details that are only recorded for lower-level units with aggregates for higher-level accounts. At the time of writing, there were insufficient results available to judge the success of the consolidation questionnaires, but they are a development to watch with interest.

6.6 CONCLUDING REMARKS

An effective way to improve the accuracy of the business register is to profile large and complex businesses. The development of large business profiling should be considered in the context of a broader strategy for improving the quality of statistics produced and the relations with survey respondents.

CHAPTER SEVEN

Implementing a Standard Industrial Classification (SIC) System Revision

Brian MacDonald[1]

U.S. Bureau of Labor Statistics

The primary purpose of a business register is to provide reliable survey frames for a wide array of surveys measuring the economic activities of various target populations. For frame building and data analysis, an industrial classification system is crucial in classifying businesses according to the primary goods and services they produce. A national standard industrial classification (SIC) system reflects the industrial structure of a country's economy while accounting for the diversity of economic activities (see Chapter 3). In the United States, the SIC assigns four-digit numerical codes to establishment-level units in the business register.

It is not uncommon for businesses to change their activities or otherwise restructure themselves. To maintain data quality, the industrial codes for individual businesses recorded on the register are periodically updated using data from register proving surveys. For example, as part of register maintenance, the U.S. Bureau of Labor Statistics (BLS) conducts the Annual Refiling Survey, which verifies classifications for approximately one-third of the units in the register. By surveying each business every 3 years, accurate and current industrial codes are ensured. Besides verifying industry, the survey also updates geographical and ownership codes.

A nation's SIC system must be periodically updated to reflect underlying

[1] I would like to thank Edmund Glad and Kenneth LeVasseur for their assistance in researching and editing this chapter, as well as Fred Conrad, Shail Butani, Shaila Nijhowne, and Michael Colledge for their useful comments. The views expressed are mine and do not necessarily reflect those of the U.S. Bureau of Labor Statistics.

Business Survey Methods, Edited by Cox, Binder, Chinnappa, Christianson, Colledge, Kott.
ISBN 0-471-59852-6 © 1995 John Wiley & Sons, Inc.

changes in the economy. Such revisions create an enormous, additional task of reclassifying all businesses in the register while routine frame building and maintenance activities continue. This chapter describes the implementation of an SIC system revision, as well as its effect on a business register. Topics covered include planning the revision, establishing a concordance between new and old classification systems, preserving the historical continuity of time series, and funding. The process is illustrated by using BLS's implementation of the 1987 U.S. revision of the SIC system (MacDonald and LeVasseur 1992).

7.1 DEFINING THE NEED FOR AN SIC SYSTEM REVISION

Rapidly changing national and world economics cause a nation's SIC descriptions and structures to become outdated. New technologies such as personal computers and video equipment span new industries which do not fit into existing activity classifications. Other industries die out or become unimportant to the economy. Hence, the ability of a nation's SIC system to accurately represent business activities diminishes with time. To maintain its usefulness, the SIC system must be periodically revised by redefining industry descriptions and updating the underlying structure. Industries whose scope and outputs have shifted or waned must be restructured, and emerging or expanding industries must be included. Up-to-date industry codes greatly increase a register's utility.

Generally, industrial classification systems are revised every 10–15 years to account for new economic activities and to adjust definitions and classifications in response to technological changes, institutional reorganization, and other events. The United States and Canada are both planning an SIC revision in 1997. Several European countries have revised their SIC systems recently.

International competition increasingly influences the economic structure of industries. Thus, a revision must also address the effect of dynamic world markets. International comparability of industry definitions is another factor to take into account. Recently, interest has focused on designing revisions that closely match an international or multinational standard such as the International Standard Industrial Classification of All Economic Activities (ISIC) or the General Industrial Classification of Economic Activities within the European Communities (see Beekman 1992a, 1992b).

7.2 IMPLEMENTATION GOALS FOR AN SIC REVISION

A revised classification causes additional complications for register staff who are already routinely updating SIC codes for individual businesses based upon information received from administrative sources, survey feedback, and proving surveys (see Chapter 5). In the initial stages of implementing an SIC revision, the primary goals should be established. Typical goals are: (1) to update the register's SIC codes to the new classification system at a single point in time, (2) to produce economic and statistical data under both old and new classification systems for at least one point in time, and (3) to update register industrial classification codes as soon as possible after establishment of the new SIC system. Such objectives provide a central focus for the revision and ensure that the resulting business register is the culmination of a successful and timely application of the new SIC system.

7.2.1 Importance of One Overlapping Reference Period

The first two goals—updating of SIC codes at one time and simultaneously producing data under both systems—are imperative and interrelated. Maintaining some register units under the old system and others under the new system is undesirable, even for a short period. Thus, the sizable task of conversion to the new system is complicated by the need to accomplish it all at once. Also, to measure the impact of the SIC change on economic variables, the units must be classified under both systems for at least one reference period (e.g., see Beekman 1992a concerning the Netherlands' revision). This overlap allows the production of quantitative measures of concordance between the two classification systems, it enables the creation of statistical tools such as conversion tables and concordance coefficients for improving historical continuity.

7.2.2 Timeliness of Revision

Updating the register's industrial codes soon after establishing a new SIC system is highly desirable. The revision's purpose is to introduce an up-to-date classification system for measuring economic activity. The longer the time between development and implementation of the new SIC system, the less up-to-date the register's classifications and the less relevant the resulting statistical estimates.

It is important to note that the register must convert to the new classification system before the surveys for which it supplies frames. Thus, a delay in survey implementation is magnified by register conversion because data users cannot implement the new classification system themselves until after the register has done so. A delay of, say, 1 year in converting the register is of less consequence than the lack of consistency across series which would otherwise result, and it is also of less consequence than the corresponding inability to measure according to a single yardstick or to compare yardsticks (Popkin 1991).

7.3 PLANNING FOR AN SIC REVISION

Implementing an SIC revision within appropriate time and budget constraints requires comprehensive planning, systematic examination of problems encountered in prior revisions, evaluation of the SIC code reassignment process, and adequate funding and cost containment practices.

Part of planning for an SIC revision is ensuring adequate funding for the duration of the revision. Multiyear funding commitments are critical. The multiyear funding approach used for the 1987 U.S. SIC revision proved to be a substantial improvement over the annual funding approach used for the 1972 revision.

When developing a strategy for implementing an SIC revision, certain tasks and procedures require substantial lead time. When a wide variety of computer systems and administrative procedures require modification, the complexity of the revision becomes greater. The impact on historical time-series data must also be evaluated, although continuity problems should not be allowed to unduly inhibit the revision (e.g., see Monk 1992).

7.3.1 Reclassification

Reclassification is the process by which businesses are reassigned SIC codes based upon the revised system definitions. The extent of the reclassification process directly affects the need for system enhancements and the planning lead time.

The revised codes may be established through mechanical or manual methods. Where a business's SIC code can be reclassified directly to a new SIC code, electronic data processing can be used for *mechanical reassignment* of the unit. A *manual reassignment* of a business's SIC code occurs when the business must be surveyed and have its new SIC code assigned by knowledgeable staff.

A systematic approach that ensures accurate manual coding of a business involves use of a special register proving questionnaire. The two-step process (1) corrects or verifies the previously assigned code based on the old SIC system and then (2) reclassifies the unit using the new SIC system.

7.3.2 Lead Time and Complexity

Sufficient lead time is required to analyze the existing structure of the economy, evaluate technological and institutional changes, and develop a plan for carrying out the revision. The greater the complexity of the revision, the more calendar time and funding is required for all phases of its planning and implementation. The planning process must include adequate allowance for creating a new SIC Manual to serve as the cornerstone of the new system. Tasks requiring sufficient lead time include: (1) redesigning, programming, testing, and installing computer systems; (2) redesigning, printing, and distributing the register proving questionnaires used to determine new SIC codes; (3) developing, printing, and distributing training and reference material; (4) hiring additional staff; and (5) training staff.

7.3.3 Systems Enhancements

An SIC system revision provides the opportunity to improve the accuracy of register data and to enhance data collection and processing systems. These improvements can be developed in advance during the planning phase and introduced at the time the revision is implemented. Possible system enhancements include: (1) revised proving questionnaires, redesigned to increase response rates and simplify processing, based on cognitive testing, use of bar codes, and optical scanning techniques; (2) improved editing procedures for input data; (3) quality control procedures; and (4) computer-assisted coding software.

System and data enhancements for surveys can also be coordinated with the SIC system revision. Such enhancements further improve the overall value of the register. For example, when BLS carried out its 1987 SIC revision, expanded, more universal, establishment-level details were added to the register (MacDonald 1987).

7.3.4 Cost Containment Practices

In implementing a revision of register SIC codes, costs can be minimized by adapting existing materials and procedures rather than developing totally new ones. This applies to computer systems, surveys forms, training documents, and other items. For example, the register's regular proving survey can be suspended for a year and replaced by a special proving survey to collect data from industries where a new SIC code cannot be mechanically reassigned. The resources normally devoted to the regular proving survey may need augmenting. Existing SIC codes can thus be verified and reassigned, and new codes can be determined. Also, when the SIC computer system is being modified, a conversion routine can be developed to assign new SIC codes mechanically from the old ones, and a validity check for manually reassigned codes can be included. These cost-containment practices together with a multiyear funding approach will ensure that register revision expenditures remain within projected targets.

7.3.5 Total Elapsed Time

When the extent of SIC changes is known, implementation planning can take place concurrently with completion of the new SIC system and manual. As a result, the total elapsed time from the initial stages of system redesign to its full implementation can be less than the sum of its two main components— planning and implementation. For instance, the elapsed time for the 1987 U.S. SIC system revision was just over 5 years, while planning and implementation each required approximately 3 years.

7.4 ESTABLISHING CONCORDANCE BETWEEN NEW AND OLD CLASSIFICATION SYSTEMS

The relationship between old SIC codes and new ones can be one to one, one to many, or many to one, or even many to many. Clearly defined changes in the classification structure (one-to-one and many-to-one) allow mechanical reassignment of the corresponding SIC codes on the register. However, where old industries have been split into two or more new ones (the one-to-many and many-to-many situations) the exact relationship between old and new SIC classifications is not uniquely defined and codes for the corresponding businesses have to be manually reassigned.

Major SIC redefinitions and industry splits can disrupt the continuity of time-series data for the affected industries. Concordance coefficients and conversion tables overcome this disruption by establishing the relationship between the old and new systems. Based on data produced under both old and new classifications, they create the linkages needed to bridge gaps in time series.

7.4.1 Concordance Coefficients and Conversion Tables

Concordance coefficients are conversion factors based on a measured reallocation of data at aggregate industry levels which reflect the changes between the old and new SIC systems. They are an important tool for measuring the impact of an SIC revision, and for revising time-series and sample estimates. They quantify the reduction or expansion of relevant SIC codes and are of great benefit to data users.

Based on factors such as employment and number of statistical units, concordance coefficients are calculated for each industry. They show (1) how much each industry has changed, (2) where the movements took place, and (3) between which industries the movements occurred and in which direction.

The first step in establishing concordance is to create conversion tables relating the old and new SIC systems. These tables aid register users in making the transition to the new system, in understanding the relationship between the old and new codes, in discerning the industrial scope of changes, and in understanding how the revision affects the historical continuity of estimates.

Major conversion tables provide a comparison of SIC codes in the old and new systems. For example, the first major table in BLS's 1987 SIC Manual Conversion Tables shows all 1972 and 1987 industries by type (if any) of change, as illustrated in Table 7.1 (U.S. Department of Labor 1987). On the vertical axis are the old SIC codes, with newly created SIC codes indented to the right. On the horizontal axis, the column descriptions indicate what has happened to the old codes.

The expanded column descriptions are:

- same SIC code and industry scope;

Table 7.1 Major Conversion Table

SIC Code 1972	SIC Code 1987	Same SIC Code and Same Industry Scope	Different SIC Code but Same Industry Scope	Whole 1972 SIC Becomes Part of 1987 SIC	Splits	Expanded Industry Scope	Reduced Industry Scope	Expanded and Reduced Industry Scope	New 1987 SIC Code	Deleted 1972 SIC Code	SIC Code Reused Industry Scope Different
2257						X					
2258		X		X							
2259											
2261		X									
2262		X									
2269		X									
2271				X						X	
2272				X						X	
	2273								X		
2279				X						X	
2281				X		X					
2282				X		X					
2283					X						
2284				X		X				X	
2291				X						X	

Source: U.S. Department of Labor, Bureau of Labor Statistics.

- different SIC code but same industry scope;
- entire old industry now part of a new industry;
- industry split into two or more industries;
- expanded industry scope;
- reduced industry scope;
- expanded and reduced industry scope;
- new 1987 SIC code;
- deleted 1972 SIC code; and
- SIC code reused but with different industry scope.

7.4.2 Subsidiary Conversion Tables

Subsidiary conversion tables show the movements of economic activities and/or the changes in industrial scope. They include (1) lists of all industries within each category—for example, all SICs with the same code and industrial scope under both systems—and (2) changes in industrial scope (additions and subtractions) from old to new system and from new to old. In BLS's revision these changes are shown at all four levels of the classification system (division, major group, group, and industry). Relationships from old to new and from new to old are exhibited, and crosswalks between the two systems are provided. These tables provide the basis for comparing the two classification systems and for converting data between them. By summarizing the revision, the tables assist in validating SIC changes in the split industries and in converting SIC codes mechanically for other industries.

Table 7.2 illustrates one of BLS's subsidiary conversion tables. Changes in industrial scope between the 1972 and 1987 classification systems are displayed at the 1972 three-digit level. For instance, in the column entitled "Subtractions," the SIC code for fluid power equipment has changed from 3561 to 3594. In the column entitled "Additions," four 1972 classified industries have been changed to the 1987 SIC code 3594.

7.4.3 Split Industries

A modified register proving questionnaire can be used to collect information for units in old industries that are being split into two or more new industries. For units in all other industries (including those with no change), the codes can be converted mechanically. During BLS's 1987 revision, a file was created that contained the SIC codes under both classification systems for the 800,000 establishments in 112 split industries. This file, combined with a crosswalk for all other industries, allows data users to convert any and all units to the new system and serves as the basis for converting historical micro-data to the new classification system.

Table 7.2 Subsidiary Conversion Table

SIC Code[a] 1972	SIC Code[a] 1987	Subtractions	Three-Digit Level	SIC Code[a] 1987	SIC Code[a] 1972	Additions
3561P	3594P	Fluid Power Equipment	356	3565P	3551P	Packaging Machinery
3565	3543	Industrial Patterns		3567P	3433P	Incinerators, Metal: Domestic and Commercial
3566P	3594P	Hydrostatic Drives (Transmissions)				
3569P	3594P	Fluid Power Pumps and Motors				
3573P	3695P	Magnetic Disks	357	3575P	3661P	Teletypewriters
3576	3596	Scales and Balances, Except Laboratory				
			358	3585P	3699P	Electric Comfort Heating Equipment
			359	3593P	3728P	Fluid Power Cylinders and Actuators
				3594P	3561P	Fluid Power Equipment
				3594P	3566P	Hydrostatic Drives (Transmissions)
				3594P	3569P	Fluid Power Pumps and Motors
				3594P	3728P	Fluid Power Pumps and Motors
				3596P	3776P	Scales and Balances, Except Laboratory

[a]In the SIC Code columns, P indicates that only a transfer of activity has occurred.

Source: U.S. Department of Labor, Bureau of Labor Statistics 1987.

7.5 PRESERVING HISTORICAL CONTINUITY OF TIME SERIES

Historical time-series data are used to analyze current economic events and to forecast future ones. The most significant impact of an SIC revision is its effect on the continuity and use of time series. For instance, real differences in employment trends by industry group are confounded with changes in the definitions of industry groups. Preservation of time-series data while introducing a more modern classification system involves a tradeoff at the planning stage, with the ultimate goal being the capacity to measure economic conditions more accurately.

7.5.1 Historical Series

Data for newly combined SIC codes can be retabulated without damaging historical continuity because reclassification of the corresponding units is not required. However, when industries have been redefined, the cost of time-series reconstruction is considerable. To preserve time-series, linkages must be created where the break occurs due to introduction of the new classification system. A concordance must be established, and the changes must be quantified through the creation of concordance coefficients. To determine these coefficients, data for the same time period must be produced under both systems.

7.5.2 Caveats of Concordance Coefficients

Full conversion of a series of historical data to the new system is possible but must be carried out very carefully, because the concordance is not always exact and a precise relationship cannot always be established. In a *backcast prediction*, prior years' data are modified to match the revised SIC codes. The backcast data cannot precisely reflect economic reality because the revised definitions apply to the industries of the present and not necessarily to those of the past. Also, assignment of some new SIC codes may require additional data that are unobtainable for older data sets.

It is often difficult to determine what portion of a historical series can be successfully converted with a concordance, and it is also difficult to decide when the conversion coefficients cease to be accurate. Application of macro-data coefficients (or even unit-level data coded to both old and new systems) work well for short periods of time. However, the assumptions underlying the coefficients become invalid over longer periods where the new system's economic structure differs substantially from that of the old. For example, disaggregating the manufacture of plastics products into several industries, including one for plastics plumbing fixtures, is meaningless for a period before such fixtures existed. Disaggregation is more likely to be valid at later stages in the industry's emergence, but, even then, coefficients developed in later years will tend to overstate the new industry's size in the more distant past. Thus, retrospective creation of a lengthy time series for a new classification system is not always possible or appropriate.

7.5.3 BLS's Conversion

Conversion from the old to the new 1987 SIC classification system occurred during the first quarter of 1988. This was the only period for which aggregate employment and wage data were produced under both systems. From the aggregates, coefficients were developed to show the relationship between employment data under the two classification systems. These coefficients can be applied to historical aggregates to approximate the same data under the new classification system. The BLS register staff provided concordance coefficients to data users and survey managers but did not convert historical register data to the new classification system.

As an example, the bituminous coal-mining industry was split into two industries in the 1987 U.S. revision: surface mining and underground mining. The coefficients for the 1972-based industry were calculated as 42.74 and 57.26, showing that in the first quarter of 1988, 42.74 percent of bituminous-coal-mining employment was reclassified to the new surface-coal-mining industry while 57.26 percent was reclassified to the new underground coal-mining industry. Application of these percentages to historical employment data for bituminous coal mining generates approximate historical data for the two new industries.

Impact on Current Employment Statistics (CES) Survey
The Current Employment Statistics (CES) survey, an important user of the BLS register, collects data from a sample of 375,000 businesses and publishes monthly estimates of employment, hours, and earnings by industry within 3 weeks of the survey reference period. The register provides the survey frame. As expected, the impact of the 1987 U.S. SIC revision on CES data was greatest at the detailed industry level, with approximately one-third of all published CES series being affected. Several problems were encountered in industries where sampled businesses were reclassified to different SIC codes.

All CES data for 1988 and later were converted to the new classification system as of mid-1990. During the period from 1988 until mid-1990, CES staff developed 1989 estimates and benchmarks, converted the sample to the new SIC system, and ensured adequate coverage for the survey. To keep conversion costs to a minimum, data for years before 1988 were adjusted only for those industries where the change in scope was considered small. The adjustment was performed using the concordance coefficients developed from the one quarter of register overlap data. For these same industries, data on hours and earnings were also adjusted, based on the same new employment figures. Once the data were revised, seasonal factors were recomputed and applied using the X-11 ARIMA seasonal adjustment procedure (Dagum 1983).

A description of CES's survey frame and data adjustment process was published with the release of the first data available under the new SIC system (U.S. Department of Labor, 1989). Because two-thirds of the CES published series were unaffected by the revision, data for those industries were identical for both classification systems. Series for which the change in scope was large were not reconstructed, and therefore they begin in 1988 (Getz 1990).

Impact on Occupational Employment Statistics (OES) Survey

The OES survey is conducted in nonagricultural industries over a 3-year cycle to produce industry-specific, occupational employment estimates. The data are tabulated in an occupation/industry matrix to show total employment and to project future employment. As with the other employment programs, the impact of the revision on OES data was negligible for the broadest aggregates but important for some detailed industries.

Because 1988 was the first year of the new classification, the OES staff retabulated the entire 1987 data matrix under the new system. OES respondents for 1987 were reclassified based on the microconversion file for the split industries. Industry employment totals for 1987 were then estimated using the aggregate data coefficients described above. As a final step, occupational employment estimates were recalculated based on the new industry levels (McElroy and Walz 1990).

7.5.4 Summary

Though the effects of an SIC revision on time series are detrimental for some industries, at least these effects can be quantified through the calculation of concordance coefficients. Conversion tables display the relationship between the old and new SIC systems. Overall, a revision will improve the utility of time-series data in the future, particularly for industries that were previously inadequately structured or defined (Bishop and Werbos 1989).

7.6 FUNDING AN SIC REVISION

System revisions are costly. Updating BLS's register for the 1987 SIC revision cost about $9.8 million spread over a 5-year period. Thus funding considerations are critical. No matter how carefully developed, a plan cannot succeed if the resources required to implement it are inadequate. The challenge is to design a revision and implementation strategy that is a balance between the necessary changes to the SIC system and the need to fit within budgetary constraints. Also, funding must be sufficient to support the workload for the entire period from planning to implementation, typically several years. The funding formula must ensure adequate funding for the complete process from start to finish.

7.6.1 Funding Factors

Several factors influence the funding required to implement an SIC revision: (1) the size, complexity, and scope of the revision, in particular the number of split industries; (2) the size and complexity of the business register; (3) the time table, and how compressed it is; and (4) the incidental costs of converting data sets within the same organization to the new SIC codes. The most important factor is the size, complexity, and scope of the revision and the number of split industries.

The cost will be relatively modest when the structure of the classification system remains unchanged and the revision simply reflects institutional and technological changes along with emerging and declining industries. However, expenditures can increase severalfold when the revision involves fundamental changes to the basic SIC structure and numerous redefinitions requiring the manual reassignment of new codes, and when implementation includes wide-ranging system enhancements. A radical reorganization of an SIC system may not only increase the number of unit codes to be manually reassigned but may also require new methods of collecting the data needed for code assignment.

7.6.2 Quantitative Impact of Revision

The quantitative impact of a revision—how many units actually change SIC codes—is more significant than the size of the register. If the revision is limited to a few industries or to industries with relatively few units (e.g., the consolidation of declining industries), or can be done mechanically, the cost will be lower than when changes are widespread.

For example, in BLS's 1987 SIC revision, 4.9 million units in 893 of 1005 industries were mechanically converted to the new system. This included the following situations: (1) industries with no change, (2) industries that changed code but not scope, and (3) industries that merged to form larger ones. Old industries that were split into two or more new industries were the more difficult and costly to convert. About 800,000 units in 112 split industries required labor-intensive analysis to assign the proper code, with a much higher per-unit cost of processing.

7.6.3 Size and Complexity of Business Register

Depending on the scope of the revision, the register's size or complexity may change dramatically. Register maintenance costs may rise due to an increased number of validity checks and other related processing costs. For example, before the 1987 revision, BLS's business register contained approximately 5.7 million units. It has since expanded to 7.0 million units, reflecting some growth in the economy, but, more substantially, the wider application of unit-level reporting accompanying the revision.

7.6.4 Sufficient Time for Planning

Investing sufficient time and thought in the planning phase can reap substantial benefits later. Good planning can speed up implementation and help in cost control. An unduly compressed time schedule increases costs through overtime expenses, greater chance of error, and increased incidence of bottlenecks.

7.6.5 Effect of Revision on Data Users

Besides the costs of implementing the revision borne by the statistical agency, the effect of the revision on data users must also be considered. Concordance coefficients and conversion tables do not completely eliminate the cost to users of converting an existing business survey database to the new SIC codes.

7.7 CONCLUDING REMARKS

An SIC revision involves long-range, comprehensive planning to address the scope, effects on time series, and funding. When planning and implementing an SIC revision of a business register, care is needed to ensure a result that meets or exceeds the primary goal of improving the analysis and estimation of economic data, while simplifying the conversion process for data users (MacDonald 1991).

REFERENCES

Beekman, M. M. (1992a), "Development and Implementation of a New Standard Industrial Classification," *Netherlands Official Statistics*, **7**, pp. 18–26.

Beekman, M. M. (1992b), "International and National Standard Classifications," paper presented at the Third Independent Conference, International Association for Official Statistics.

Bishop, Y. M., and P. J. Werbos (1989), "An Interagency Review of Time-Series Revision Policies," Report of the Subcommittee on Guidelines for Making and Publishing Revisions and Corrections to Time Series, Washington, DC: U.S. Office of Management and Budget.

Dagum, E. B. (1983), *X-11 ARIMA Seasonal Adjustment Method*, Catalog No. 12-564E, Ottawa: Statistics Canada.

Getz, P. M. (1990), "Establishment Estimates Revised to March 1989 Benchmarks and 1987 SIC Codes," *Employment and Earnings*, **37**, Washington, DC: U.S. Bureau of Labor Statistics, pp. 6–34.

MacDonald, B. (1987), "The New BLS Universe Data Base System," paper presented at the Second International Round Table on Business Survey Frames, Washington, D.C.

MacDonald, B. (1991), "Administering and Implementing a Dynamic Classification

System,'' paper presented at the International Conference on the Classification of Economic Activity, Williamsburg, VA.

MacDonald, B., and K. LeVasseur (1992), ''Implementing a Revision to an Industrial Classification System,'' paper presented at the Seventh International Round Table on Business Survey Frames, Copenhagen, Denmark.

McElroy, M., and A. Walz (1990), ''Occupational Employment Based on 1972 and 1987 SIC,'' *Monthly Labor Review*, **113,** Washington, DC.: U.S. Bureau of Labor Statistics, pp. 49–51.

Monk, C. H. (1992), ''Revising and Implementing the 1987 U.S. Standard Classification System,'' paper presented at the Seventh International Round Table on Business Survey Frames, Copenhagen, Denmark.

Popkin, J. (1991), ''Monitoring Economic Performance in the 21st Century: Measurement Needs and Issues,'' paper presented at the International Conference on the Classification of Economic Activity, Williamsburg, VA.

U.S. Department of Labor (1987), *1987 SIC Manual Conversion Tables*, Washington, DC: U.S. Bureau of Labor Statistics.

U.S. Department of Labor (1989), *Employment Data Under the New Standard Industrial Classification, First Quarter 1988*, Report 772, Washington, DC: U.S. Bureau of Labor Statistics.

Sample Design and Selection

CHAPTER EIGHT

Selecting Samples from List Frames of Businesses

Richard S. Sigman and Nash J. Monsour[1]
U.S. Bureau of the Census

Most samples for business surveys are selected from lists. Developing the associated sample designs can be challenging because business populations can have the following characteristics:

- *Skewness.* A small number of businesses account for a large proportion of the population total.
- *Dynamic membership.* Businesses are created, go out of business, change their type or level of activity, or change their identity through mergers, acquisitions, or divestitures.
- *Inter-business relationships.* Businesses may be related to each other in that they are owned by the same legal entity, they employ the same accountant, or their activities are combined in the same set of financial records.

In conjunction with the estimators and population attributes, the sample design determines a survey's operating characteristics such as cost, variance, and respondent burden.

This chapter discusses sample designs commonly used in business surveys. The statistical framework we employ is randomization- or design-based. This means that bias and variance are defined with respect to the distributions of

[1]The authors gratefully acknowledge contributions from Esbjörn Ohlsson, K. P. Srinath, and Ronald M. Carpenter. The authors also thank Larry Ernst and Cary Isaki for helpful comments. The views expressed are the authors' and do not necessarily reflect those of the U.S. Bureau of the Census.

Business Survey Methods, Edited by Cox, Binder, Chinnappa, Christianson, Colledge, Kott.
ISBN 0-471-59852-6 © 1995 John Wiley & Sons, Inc.

potential samples rather than the distribution of potential population data values given the sample. Model-assisted, design-based sampling methods are discussed as well (Rao and Bellhouse 1989), but readers interested in purely model-based approaches are referred to Chapter 30 and to Royall and Herson (1973a, 1973b) and Kirkendall (1992). We begin by discussing single-stage, stratified, simple random sampling without replacement and then move on to Poisson sampling and other probability proportional to size (*pps*) sampling designs. We conclude by introducing rotation designs. Questions relating to sampling a dynamic population are treated in Chapters 9 and 10.

8.1 DESIGN ISSUES IN STRATIFIED SAMPLING

Many business surveys employ stratified, simple random sampling without replacement (*srswor*). Questions that need to be answered with this (or an equivalent) design include:

- What should the sampling unit be?
- How should strata be constructed?
- How should the sample be allocated to strata?
- What type of estimator should be used?

Chapters 2 and 3 address the definition of the sampling unit. We discuss allocation in Section 8.2 and stratum construction in Section 8.3. Chapter 25 discusses weighting and estimation for annual and subannual business surveys. We discuss the type of estimator only as it relates to allocation and stratum construction.

We restrict our attention to estimators that are nearly unbiased and have variances V' of the form

$$V' \approx V_0 + V = V_0 + \sum_{h=1}^{H} \frac{V_h^2}{n_h}, \qquad (8.1)$$

where n_h is the size of the sample selected from stratum h ($h = 1, 2, \ldots, H$) and $V_0, V_1, \ldots, V_H$ are constants. This class of estimators is quite large, however, as it includes expansion estimators, difference estimators, regression estimators, and ratio estimators.

The choice of survey designs also involves cost considerations. We assume that the total survey cost C' has the form

$$C' \approx C_0 + C = C_0 + \sum_{h=1}^{H} C_h n_h, \qquad (8.2)$$

where $C_0, C_1, \ldots, C_H$ are constants. The notation in equations (8.1) and (8.2) is from Kish (1976). The constant C_0 is the fixed cost of the survey which is unaffected by sample size, while C is the survey's variable cost which changes along with the stratum sample sizes. Similarly, V_0 is the fixed component of the estimator's variances, while V is the variable component.

8.2 ALLOCATION OF STRATIFIED SAMPLES

Allocation is the survey designer's specification of stratum sample sizes. In this section we discuss allocation for a survey that collects data for only one item and estimates a single population-level total. Discussions of multivariate and multipurpose allocations follow.

8.2.1 Univariate Allocation

In *proportional allocation*, the size of the sample in each stratum is proportional to the stratum's population size. A survey designer uses proportional allocation when stratum-specific information is lacking on data variability and survey costs. Proportional allocation is rarely used in business surveys, however, because survey values from large businesses are generally more variable than those from small businesses.

Optimal allocation minimizes the product VC with either C or V held fixed. This is equivalent to minimizing variance for a specified cost or minimizing cost for a specified level of accuracy (variance). The global minimum of VC is $[\Sigma(V_h/\sqrt{C_h})]^2$, which occurs when $n_h \propto V_h/\sqrt{C_h}$ (Kokan 1963, Kish 1976). If $C = C' - C_0$ is fixed, we obtain

$$ n_h = \left[\frac{V_h}{\sqrt{C_h}} \right] \left[\frac{C' - C_0}{\sum_k V_k \sqrt{C_k}} \right] ; $$

and if $V = V' - V_0$ is fixed, we obtain

$$ n_h = \left[\frac{V_h}{\sqrt{C_h}} \right] \left[\frac{\sum_k V_k \sqrt{C_k}}{V' - V_0} \right] . $$

Because n_h cannot exceed N_h, however, an iterative sequence of calculations may be needed to obtain all $n_h \leq N_h$. At each step of these calculations, equations (8.1) and (8.2) are redefined so that V' and C reflect variance and variable costs, respectively, for *noncertainty strata* (those in which $n_h < N_h$), whereas C_0 is redefined to capture both the original fixed costs and the additional survey costs for collecting data from the units in *certainty strata* (those in which $n_h = N_h$). Cochran (1977, p. 104) explains this sequence of calculations for the case of equal stratum costs. Mergerson (1988a) provides a corresponding explanation for unequal stratum costs and also (1989) describes a Pascal program for performing univariate optimal allocations.

Neyman allocation attempts to minimize (1) the variance of an estimator given a specified sample size or (2) the sample size for a given level of accuracy. This is equivalent to optimal allocation with $C_0 = 0$ and $C_1 = C_2 = \cdots = C_H = 1$.

In practice, the sample allocation is nearly always only approximately optimal. For one thing, the actual stratum h sample allocation n_h must be an integer, while the optimal allocation n_h^* need not be. Kish (1976) provides the following useful formula for determining the relative *proximation loss*—that is, the relative increase in VC with either V or C fixed:

$$L = \sum_h U_h K_h \sum_h \frac{U_h}{K_h} - 1, \qquad (8.3)$$

where $K_h \propto n_h^*/n_h$ and $U_h = V_h \sqrt{C_h}/\Sigma V_k \sqrt{C_k}$ is a relative measure of the variability and cost associated with stratum h.

For the case where $C_1 = \cdots = C_H = 1$ and $C = \Sigma n_h = n$, Cochran (1977, pp. 115–116) shows that

$$L = \sum_h \left[\frac{n_h g_h^2}{n} \right], \qquad (8.4)$$

where $g_h = (n_h - n_h^*)/n_h$ is the relative difference between the *proximal allocation* n_k and the optimal estimator for stratum h. Consequently, $L \leq g^2$, where g is the largest absolute g_h. For example, if g is 20 percent, then L is 4 percent. Thus, for optimal incremental sample costs across strata and a fixed sample size, moderate departures from optimal sample sizes do not have a major effect on variance. Another reason for departing from an optimal allocation is to reduce the burden on large businesses. Dillard and Ford (1986) investigate such reduced allocations.

8.2.2 Model-Assisted Allocation

The calculation of optimal sample sizes requires that the V_h values be known. This is a third reason why actual (proximal) and optimal sample sizes fail to coincide (Eltinge 1993). Often all the survey designer knows before a sample is drawn is the variability within strata of an auxiliary variable thought to be correlated with the survey variable of interest. Such an auxiliary variable is often referred to as a *measure of size*. For a retail trade survey, the auxiliary variable might be employment or it might be total annual sales from a recent census. For a survey of schools, it might be registered students; for a survey of prisons, inmates; and so forth.

Dayal (1985) discusses the use of an auxiliary variable to calculate proximal allocations. He assumes that there exists a linear model,

$$y_j = \alpha + \beta x_j + x^q e_j, \qquad (8.5)$$

relating the unknown survey value of interest y_j to an auxiliary variable x_j. The parameters α and β are unknown constants, while the e_j are independent random errors satisfying $E_m(e|x) = 0$ and $E_m(e^2|x) = \delta^2$, where E_m denotes expectation with respect to the assumed model.

For the expansion estimator, we obtain

$$\hat{Y}_E = \sum_{h=1}^{H} \frac{N_h}{n_h} \sum_{j \in S_h} y_j, \tag{8.6}$$

where S_h is the set of sampled units from stratum h. Dayal's allocation rule minimizes the model expectation of the (design) variance, what Isaki and Fuller (1982) call the *anticipated variance*. Dayal's rule is

$$n_h \propto \frac{N_h[\beta^2 V_{xh}^2 + \delta^2 M_h(x^{2q})]^{1/2}}{\sqrt{C_h}}, \tag{8.7}$$

where V_{xh} is the population variance of the x_j in stratum h, and $M_h[g(x)]$ is the mean of $g(x_j)$ in stratum h. If $\delta = 0$, then

$$n_h \propto \frac{N_h V_{xh}}{\sqrt{C_h}}, \tag{8.8}$$

which is called *x-optimal allocation*.

In practice, x-optimal allocation is often confused with optimal allocation even though they are not the same when the correlation between x_j and y_j is less than perfect. If the assumed model is correct, and good estimates for β, δ, and q can be obtained (see Harvey 1976), then equation (8.7) will likely be superior to (8.8) in terms of producing a sample allocation that minimizes the variance for fixed survey costs. Because of the flatness of this optimum, however, the superiority of the allocation equation (8.7) over equation (8.8) may be small. By assuming that equation (8.7) yields an optimal allocation, equation (8.3) can be employed to determine the proximation loss resulting from use of equation (8.8).

Godfrey et al. (1984) investigated allocation for the *general difference estimator*:

$$\hat{Y}_{DIFF} = \hat{Y}_E + B(X - \hat{X}_E),$$

where B is a constant. It is not difficult to see this produces a mild generalization of Dayal's result:

$$n_h \propto N_h \frac{[(\beta - B)^2 V_{xh}^2 + \delta^2 M_h(x^{2q})]^{1/2}}{\sqrt{C_h}}. \tag{8.9}$$

When $B = \beta$, this reduces to

$$n_h \propto N_h \frac{[M_h(x^{2q})]^{1/2}}{\sqrt{C_h}} \tag{8.10}$$

In practice, B is often not a constant but a survey-based estimate of β. This is the case for the *combined regression estimator*. When α in the model in equation (8.5) is assumed to equal zero, β can be estimated by

$$B = \frac{\displaystyle\sum_{h=1}^{H} \frac{N_h}{n_h} \sum_{i \in S_h} y_i}{\displaystyle\sum_{h=1}^{H} \frac{N_h}{n_h} \sum_{i \in S_h} x_i}.$$

The result is the *combined ratio estimator*. With either the combined ratio or the combined regression estimator, the model-assisted allocation rule is simply equation (8.10).

8.2.3 Multivariate Surveys

A multivariate survey collects data for two or more items. Each item has a separate variance, so we add a subscript i to the terms in equation (8.1) to index the data item:

$$V_i' \approx V_{i0} + V_i = V_{i0} + \sum_h \frac{V_{ih}^2}{n_h}. \tag{8.11}$$

There are two approaches to allocating a sample for a multivariate survey. One approach is to minimize the product AC, where $A = \Sigma_i A_i V_i$ is a linear combination of the design variances, with either A or C fixed (Kish 1976). Then

$$n_h \propto \left[\frac{\displaystyle\sum_i A_i V_{ih}^2}{C_h} \right]^{1/2}$$

It is often difficult, however, to specify meaningful A_i weights.

The second approach to allocating a sample for a multivariate survey is to specify an upper bound V_i^* for each design variance V_i' and then minimize total variable costs C subject to $V_i' \leq V_i^*$ for all i. Bethel (1989b) invokes the Kuhn–Tucker theorem to show that there exist λ_i such that

$$n_h = \left[\frac{\displaystyle\sum_i \lambda_i V_{ih}^2}{C_h} \right]^{1/2} \tag{8.12}$$

is the desired allocation. He describes a simple computation algorithm for obtaining the λ_i and discusses its convergence properties. Mergerson (1988b) describes an available Turbo Pascal program for implementing Bethel's algorithm. Chromy (1987) develops an alternative algorithm for determining the λ_i in equation (8.12) but does not prove that his algorithm always converges. Chromy's algorithm, which is suitable for automated spreadsheets, consists of the following steps:

- Step 1: Set $\lambda_i = 1$ for all i.
- Step 2: Calculate the n_h using equation (8.12).
- Step 3: Calculate the $V_i = V_i' - V_{i0}$ using equation (8.11).
- Step 4: Calculate revised λ_i, denoted λ_i', using the updating equation: $\lambda_i' = \lambda_i [V_i / (V_i^* - V_{i0})]^2$.

Steps 2 through 4 are repeated over and over again with λ_i' each time replacing λ_i. The minimum cost solution is obtained when $V_i' \leq V_i^*$ and $\lambda_i (V_i' - V_i^*) = 0$ for all i. A result by Causey (1983) can be used to define a stopping rule for Chromy's algorithm. Causey's result is that $\Sigma_i \lambda_i |V_i' - V_i^*|$ is an approximate upper bound on the distance (in terms of survey cost) that a solution is away from the minimum cost solution.

8.2.4 Multipurpose Surveys

In a multipurpose survey, estimates are calculated for the entire population and for particular domains (subpopulations). The allocation that is optimal for the entire population estimate will generally not be optimal for domain estimates. One approach is to recast the problem into a form solvable by one of the allocation methods for multivariate surveys. For example, let y_j denote the survey values of interest and let I_d be an indicator variable for membership in domain d, $d = 1, 2, \ldots, D$. Then, determine the allocation that is optimum for the multivariate data associated with the overall population and the D domains or $[y_j, y_j I_1, y_j I_2, \ldots, y_j I_D]$. This approach is used in allocating the U.S. Monthly Retail Trade Survey. Causey (1983) discusses calculations associated with this approach that take advantage of the creation of artificially zero data in a large number of strata.

One common practice with mutually exclusive domains is to treat the domains as strata and to allocate samples so that the coefficients of variation for the domain estimates are approximately equal. Bankier (1988) discusses a more general allocation that is a compromise between this practice and Neyman allocation for the entire population. A drawback of both these methods for business surveys is that often there are two levels of stratification, but domain estimates are desired for only one of the stratifiers. For example, a business survey may be stratified by industrial classification and size, but domain estimates may be desired only by industrial classification.

8.3 STRATUM CONSTRUCTION

Stratum construction, like new home construction, is driven by customer needs and available resources. Among the reasons for stratum construction are: (1) to obtain domain estimates, (2) to reduce design variances, and (3) to reduce survey costs. The materials used in constructing strata are the available data on the individual business in the sampling frame. These data are referred to as *planning data, control data,* or *stratification variables.* An example of a continuous stratification variable is the measures of size x_j in Section 8.2. Businesses that lack control data for classification may be placed in a separate "unclassified" stratum.

8.3.1 Stratification to Obtain Domain Estimates

The population of a business survey is sometimes composed of mutually exclusive domains of interest. In this situation, it is reasonable for the domains to serve as sampling strata or for the members of each sampling strata to lie entirely within a particular domain. However, for that to happen, the control data must contain (or be capable of producing) a variable that indicates domain membership. If it does not, then alternative methods must be used for stratum construction.

8.3.2 Stratification to Reduce Variance—Univariate Case

In business surveys, survey data from large businesses are usually more variable than those from small businesses. Fortunately, frame data usually contain one or more continuous variables related to business size. Thus, stratifying by size is one way to reduce design variances in business surveys. In this section, we consider the case of one characteristic of interest and one continuous stratification variable. Stratum construction then consists of specifying for each stratum an upper and a lower endpoint for associated values of the stratification variable.

The *cum* $\sqrt{f}$ of Dalenius and Hodges (1959) is one method for obtaining stratum endpoints (see also Heike and Jaspers 1993). The *cum* $\sqrt{f}$ rule creates strata with approximately equal values of $W_h S_h$, where W_h is the proportion of businesses assigned to stratum h, and S_h is the stratification variable's standard deviation in stratum h. For a fixed number of strata, endpoints from the *cum* $\sqrt{f}$ rule result in a lower expansion-estimator variance than that from alternative endpoints if certain conditions are satisfied. These conditions are as follows: (1) The stratification variable is identical to the characteristic of interest, (2) the stratification variable is approximately uniformly distributed between adjacent endpoints, (3) Neyman allocation is used, and (4) the allocation variable is the characteristic of interest.

Cochran (1977, pp. 127–133) provides a detailed description of the *cum* $\sqrt{f}$ rule. He also discusses the effect that the number of strata (constructed with the *cum* $\sqrt{f}$ rule) has on the expansion variance when the stratification variable is used for Neyman allocation. Cochran concludes that there is very little additional variance reduction from having more than six strata unless the correlation between the stratification variable and the characteristic of interest exceeds 0.95. Bethel (1989a) examines the limit of the variance of the expansion estimator as the number of strata constructed with the *cum* $\sqrt{f}$ rule increases. He also examines model-based stratification and allocation rules (see also Wright 1983, Godfrey et al. 1984, Kott 1985).

In deriving the *cum* $\sqrt{f}$ rule, Dalenius and Hodges (1959) assume that the stratification variable is approximately uniformly distributed between adjacent endpoints. The extreme skewness of business populations, however, violates this assumption. When the stratification variable arises from a highly skewed Pareto distribution, Wang and Aggarwal (1984) show that the *cum* $\sqrt{f}$ endpoints are inferior to optimum endpoints expressed in terms of the number of strata and the parameter of the underlying Pareto distribution. Wang and Aggarwal (1984) suggest that their results could be used to stratify actual populations. They discuss how to estimate the needed parameter and provide a graphical aid for obtaining the optimum endpoints.

Nevertheless, a more common method for stratifying a skewed population is to create a certainty stratum of large businesses in which all units are sampled. The *cum* $\sqrt{f}$ rule is then used to construct noncertainty strata. One way to construct the certainty stratum is to define it to be all large business "outliers." These are subjectively identified from listings sorted by the stratification variable. Potential outliers are often added to the certainty stratum based on subject matter knowledge such as businesses with unusually large projected growth.

Hidiroglou (1986) describes an objective procedure for constructing the certainty stratum (the "take all" stratum). A simple random sample is then selected from the businesses outside this certainty stratum. Lavallee and Hidiroglou (1988) extend Hidiroglou's procedure to the selection of a power-allocated stratified sample of noncertainty ("take some") businesses. Chen (1989) discusses the programming of Lavallee and Hidiroglou's algorithm and documents an accompanying FORTRAN program listing.

Detlefsen and Veum (1991) modified Lavallee and Hidiroglou's algorithm to carry out Neyman allocation in the noncertainty strata. However, in applying the modified algorithm to the redesign of the U.S. Monthly Retail Trade Survey they found that the algorithm's convergence was slow (often 50–100 iterations) or nonexistent. Different starting values for the same population resulted in different ending boundaries, and many times the boundaries differed substantially.

8.3.3 Stratification to Decrease Variance—Multivariate Case

For the case of two continuous stratification variables, Kish and Anderson (1978) examined the problem of using stratification variables x_1 and x_2 to stratify a sample to be used to estimate population totals for characteristics y_1 and y_2 (we have suppressed the unit identifier j for convenience.) They assume high correlations between x_1 and y_1 and between x_2 and y_2 and low correlations between x_1 and x_2, between x_1 and y_2, and between x_2, and y_1. Kish and Anderson apply the $cum \sqrt{f}$ rule to the x_1 data to construct L_1 intervals over the range of variable x_1 and apply it to x_2 data to construct L_2 intervals over the range of variable x_2. The two sets of intervals and the bivariate (x_1, x_2) data then define $L = L_1 L_2$ strata. They then use proportional allocation to select a sample from the L strata. (Neyman allocation for both y_1 and y_2 over the L strata is possible, however, using one of the multivariate allocation methods described in Section 8.2.3.) The authors' empirical results suggest that a bivariate stratification provides additional variance reduction over the best univariate stratification and protects against a poor univariate stratification resulting from the choice of an inefficient stratification variable (see also Roshwalb and Regnier 1991).

An alternative approach for using multiple stratification variables is to cluster the stratification data. An automated clustering routine assigns similar data vectors to the same cluster and assigns dissimilar data to different clusters. Businesses are assigned to strata on the basis of the cluster membership of their corresponding stratification data. Jarque (1981) recommends the use of Ward's clustering algorithm or the k-means clustering algorithm for stratum construction because these algorithms minimize a criterion that also measures the effectiveness of multivariate stratification.

Julien and Maranda (1990) discuss the use of clustering to create strata for the list portion of the Canadian National Farm Survey. They had nine stratification variables. Prior to clustering these data, however, they created a certainty stratum using an intuitively based rule similar to the subjective procedure described in Section 8.3.2. The stratification data for farms outside the certainty stratum were then clustered with the SAS procedure FASTCLUS (SAS Institute 1989, pp. 823–850). This produced 250 clusters, which were then combined into a smaller number of clusters with the SAS procedure CLUSTER (SAS Institute 1989, pp. 519–614). FASTCLUS implements the k-means algorithm; whereas CLUSTER implements agglomerative clustering methods, one of which is Ward's algorithm.

8.3.4 Stratification To Reduce Costs

One approach to sample allocation is to minimize variable survey costs subject to item design variances not exceeding specified limits. If this approach is used to determine stratum sample sizes, the stratum construction methods described in Section 8.3.3, which reduce item design variances, will also reduce variable survey costs. In some cases, however, construction of strata can reduce variable survey costs even when stratification has no effect on the design variances. For this to occur, however, it is necessary that different strata have different variable unit costs and that they survey designer uses optimal allocation rather than Neyman allocation.

Although it may be possible to relate variable costs per business to some continuous variable, we are unaware of procedures that partition a continuous variable to create strata that minimize variable costs. Moreover, in most cases, the only available cost information will be the average unit costs within business categories such as geographical area, industrial classification, or type of ownership. Armed with such information, a survey designer could group categories into strata so that the average unit costs are similar within each stratum and dissimilar across strata. This can be done manually when the number of categories is small. When there are a large number of categories, automated clustering procedures can be used to group categories with similar average unit costs.

8.3.5 Systematic Sampling

A popular alternative to *srswor* is systematic sampling. Cochran (1977, pp. 205–232) describes this alternative and its attributes in detail. Stratified systematic sampling often leads to more efficient estimation than stratified *srswor* because it can incorporate an additional level of implicit stratification within explicit strata. A good example of this phenomenon is described in Chapter 12.

8.4 PROBABILITY PROPORTIONAL TO SIZE DESIGNS

Probability proportional to size (*pps*) sampling is sometimes used in place of size stratification especially when estimating price indexes. In what follows, we first discuss Poisson sampling and then alternative without-replacement *pps* designs that, unlike Poisson sampling, have fixed sample sizes. Although we treat *pps* sampling of an unstratified population, the discussion applies equally well to *pps* sampling within strata, which is more common.

8.4.1 Poisson Sampling

The term "Poisson sample" appears to have been coined by Háyek (1960, 1964). To draw a Poisson sample, one begins with a list of N units and conducts N independent random trials. The outcome of the jth trial is that unit j is either included in the sample or it is not. The probability that unit j is selected for the sample is π_j. The sample size n is a random variable and $E(n) = \sum_{j=1}^{N} \pi_j$. The chief advantage of Poisson sampling is that it is easy to coordinate Poisson samples—that is, to minimize or maximize overlap between samples selected from the same population (see Chapter 9). For example, the U.S. Bureau of the Census uses Poisson sampling to select its Annual Survey of Manufactures (ASM). A new ASM sample is selected every 5 years, and for the noncertainty sample, overlap is minimized (Ogus and Clark 1971, Waite and Cole 1980).

An unbiased estimator for a population total Y is the Horvitz–Thompson

(1952) estimator, $\hat{Y}_{HT} = \Sigma_{j \in S} y_j / \pi_j$, where S denotes the sample. It has a design variance of $V' = \Sigma_{j=1}^{N} y_j^2 [(1/\pi_j) - 1]$. An unbiased estimator for V' is $v' = \Sigma_{i \in S} y_j^2 [(1/\pi_j) - 1]/\pi_j$. This simple formula for estimating V' is another advantage of Poisson sampling.

For a given expected sample size $E(n)$, π_j proportional to y_j minimizes V'. In practice, the y_j are not known for all businesses in the population. Nevertheless, suppose each business j has a known measure of size x_j, and let $\pi_j = x_j E(n) / \Sigma^N x_k$ be the selection probability for j. It is clear that the expected sample size is $E(n)$. Moreover, if x_j is proportional to y_j, then V' will be minimized given $E(n)$. An implicit assumption in the above discussion is that $x_j \leq \Sigma_{g=1}^{N} x_g / E(n)$ for all j. Any unit that fails this inequality must be selected with certainty.

When combined with a Horvitz–Thompson estimator, Poisson sampling tends to be less efficient than alternative *pps* designs that have a fixed sample size. This is because there is a variance component associated with the random sample size. However, one can increase the efficiency of a Poisson sample by incorporating auxiliary information into the estimator. For example, the U.S. Annual Survey of Manufactures using a Horvitz–Thompson difference estimator in which the quinquennial Census of Manufacturers provides the auxiliary information.

Another disadvantage of Poisson sampling is that in multivariate surveys, a set of π_j efficient for one survey item of interest may be inefficient for another survey item. By considering a business to be a stratum of size one, this problem is identical to allocating a stratified sample for a multivariate survey. Consequently, the allocation procedures for multivariate surveys described in Section 8.2.3 can be used to determine optimum π_j by changing n_h to π_j and replacing V_{ih}^2 with a prediction of y_{ij}^2 (y_{ij} is data item i for business j) based on the available measures of size.

8.4.2 Other Probability-Proportional-to-Size Sampling Designs

Without-replacement *pps* sampling with a fixed sample size n satisfies the Prob $(i \in S) = \pi_j = nx_j \Sigma_{g=1}^{N} x_g$. An unbiased estimator for a population total Y is the Horvitz–Thompson estimator $\hat{Y}_{HT}$. If x_h is proportional to y_h, then y_h / π_h is a constant, say b, and $V' = \text{Var}(bn) = 0$.

A simple selection procedure is to order the list of units and then systematically sample from the cumulated x_j. (We have assumed that $x_j \leq \Sigma_{g=1}^{N} x_g / n$ for all j). In other words, let $k = X/n$, and chose a real number a at random from the interval $(0, k]$. Unit j is in the sample if $\Sigma_{g=1}^{j-1} x_g < a + (m - 1)k \leq \Sigma_{g=1}^{j} x_g$ for $m = 1, 2, \ldots, n$. A disadvantage of this *systematic pps* procedure is that there is no unbiased estimator for V'. Wolter (1985) discusses applicable variance estimation procedures and how to reduce V' by the choice of ordering variable.

Hanif and Brewer (1980) and Brewer and Hanif (1983) catalog 50 published methods for selecting *pps* samples. Except for systematic *pps* selection, however, the cataloged methods of without-replacement *pps* sample designs with fixed sample sizes are quite complex when $n > 2$. However, in a more recent paper, Sunter (1986) describes a simple list-sequential procedure that is approximately *pps* and can be adjusted to be exactly *pps*. In Chapter 9, Ohlsson describes a simple procedure, which he refers to as "sequential Poisson sampling," that is approximately *pps*. Another sampling method that is approximately *pps* is the Rao–Hartley–Cochran method (Cochran 1977, pp. 266–267), which has the advantage of a simple variance formula.

When the relationship between y_j and x_j satisfy the linear model,

$$y_j = \beta x_j + x_j^q e_j, \tag{8.13}$$

where the e_j are uncorrelated random variables with a common variance, a without-replacement *pps* sampling design coupled with a Horvitz–Thompson estimator is very efficient. When $q = 1$, this strategy is optimal in the sense that, asymptotically, it has the smallest anticipated variance among all nearly unbiased strategies with fixed sample size n. When q in equation (8.13) is arbitrary, then the estimator

$$\hat{Y}_R = X\hat{Y}_{HT}/\hat{X}_{HT} \tag{8.14}$$

and a design where π_j is proportional to x_j^q form the asymptotically optimal strategy (Brewer 1973, 1979; Särndal et al. 1992, pp. 453–454). The disadvantages of these strategies include: (1) complex or nonexistent variance formulas, (2) potential inefficiency when the presumed relationship between y_j and x_j does not exist, and (3) the problem of selecting an efficient measure of size for multivariate surveys.

Due to the complex variance formula, the procedure described in Section 8.4.1 for selecting a Poisson sample for a multivariate survey does not readily extend to other *pps* designs. For the case of a multivariate survey with two characteristics of interest, Pollock (1984) discusses the selection of two *pps* samples from the same population with maximum overlap between the samples. The first and second samples are *pps* with respect to measures of size that are closely related to first and second characteristics of interest, respectively.

8.5 THE USE OF ROTATING PANELS

Sample rotation is used in most periodic surveys. To describe the various types of rotation, standardization of terms is needed. To this end, a *panel* is defined as the collection of all units in the sample reporting for a given period. A period could be a week, a month, a quarter, or whatever. In some periodic surveys, there is a *fixed panel* that does not change except for births and deaths. The exclusive use of a fixed panel produces very efficient estimates of periodic change. Other business surveys employ *rotating panels* to reduce respondent burden.

The use of rotating panels can also lead to more efficient estimates of period totals. Often, however, the largest sampling units need to be included with certainty in a fixed panel, surveyed every period, even when some sample rotation is desired. By definition, only those units from the noncertainty portion of the population that are selected in a given period constitute the rotating panel. Wolter (1979) describes the various forms of rotation sampling; we incorporate much of his terminology. In the remainder of Section 8.5, we discuss rotating panels exclusively unless otherwise indicated.

8.5.1 Types of Rotation

In a *one-level rotation sample*, there may be an overlap in the (noncertainty) sampling units surveyed in time periods $t - 1$ and t, but individual units when reporting for time period t report for that one time period only. The pattern of rotation of the sampling units is an important feature of this type of rotation. The collection of sampling units within a rotating panel that undergo a specific pattern of reporting is referred to as a *rotation group*. The properties of the estimates used with one-level rotation may depend heavily on the pattern of rotation used. Rotating panels usually overlap across periods to obtain better estimates of period-to-period change. They can also be used to improve estimates of population totals. One-level rotation is frequently used in selecting samples from an area frame (Bush and House 1993, Chhikara and Deng 1991) and is employed in the Current Population Survey, a household survey in the United States.

In contrast to one-level rotation, suppose that a unit reporting at time period t reports for multiple periods. If a unit reports for two periods, say t and $t - 1$, this is a *two-level scheme*. If a unit reports for three periods, say t, $t - 1$, and $t - 2$, this is called a *three-level scheme*, and so on. To avoid collecting the same data again for the same unit in a two-level scheme, at least two distinct rotating panels are needed, one for period t and one for period $t - 1$. Likewise, with a three-level scheme, at least three distinct rotating panels are needed.

In multilevel rotation the panels are usually selected so that they do not overlap. This means that estimates derived from the panels are negatively correlated (because the population is finite). Nevertheless, if the population of sampling units is large compared to the size of the panels, the estimates derived from the panels are approximately uncorrelated. By having nonoverlapping panels, the sample units only have to report every m periods if there are m rotating panels.

8.5.2 Estimation

An estimator for a population total in a particular period may be based exclusively on survey values from that period. It is possible, however, to decrease the variance of that estimator by employing information from outside the panel, given a one-level rotation sample with overlapping rotating panels or a multi-level sample with distinct rotating panels. In this section, we follow Wolter (1979) and limit our formal discussion to a two-level rotation scheme. Extensions to other multilevel rotation samples are obvious. Cochran (1977, pp. 351–355) and Cantwell (1990) discuss one-level rotation samples.

Let Y_s be the total one wishes to estimate for period s. The rotating panel reporting in period s reports for period s and retrospectively for the previous period $s - 1$. Let $\hat{Y}'_s$ and $\hat{Y}''_{s-1}$ be the Horvitz–Thompson estimators of the period s total Y_s and the period $s - 1$ total Y_{s-1} obtained from the period s rotating panel. For a particular time period s, two estimators exist for total Y_s: the estimator $\hat{Y}'_s$ from the rotating panel reporting in period s and the estimator $\hat{Y}''_s$ from the rotation panel reporting retrospectively in period $s + 1$. After T survey periods ($t - T + 1$ to t), we have the following vector of observations:

$$\hat{\mathbf{Y}}_t = (\hat{Y}'_t, \hat{Y}''_{t-1}, \hat{Y}'_{t-1}, \hat{Y}''_{t-2}, \hat{Y}'_{t-2}, \ldots, \hat{Y}''_{t-T+1}, \hat{Y}'_{t-T+1}, \hat{Y}''_{t-T}).$$

In this section, we review two types of estimators that are more efficient than a simple combination of $\hat{Y}'_s$ and $\hat{Y}''_s$ for estimating Y_s when $t - T \le s \le t$. The two estimators are the minimum variance linear unbiased (MVLU) estimator and the composite estimator.

Gurney and Daly (1965) and Wolter (1979) solved the problem of finding *minimum variance linear unbiased estimators* for a total. With the above notation, the MVLU estimator for Y_s has the form

$$\hat{Y}^*_s = \sum_{k=0}^{T-1} (\alpha_{s,k} \hat{Y}'_{t-k} + \beta_{s,k} \hat{Y}''_{t-k-1}), \tag{8.15}$$

where $\alpha_{t,k}$ and $\beta_{t,k}$ depend on the covariance matrix V of $\hat{\mathbf{Y}}_t$. Observe that $\hat{Y}^*_s$ depends on the observations before and after period s. In many applications, only the estimate $\hat{Y}^*_t$ for the last observed period is desired. The computation of $\hat{Y}^*_t$ may be prohibitively expensive because it depends on all past observations and on the covariance matrix V.

One way of simplifying the estimation procedure for two-level rotations is to have an iterative estimator $\hat{Y}'''_s$ that depends on the previously obtained estimators for past periods only through $\hat{Y}'''_{s-1}$. This type of estimator first described by Bershad is given by Hansen, et al. (1953, pp. 497–503). This *composite estimator* can take two forms:

$$\hat{Y}'''_s = \hat{Y}'_s + \beta(\hat{Y}'''_{s-1} - \hat{Y}''_{s-1}) \tag{8.16}$$

or

$$\hat{Y}'''_s = \hat{Y}'_s + \frac{\beta(\hat{Y}'''_{s-1} - \hat{Y}''_{s-1})\hat{Y}'_s}{\hat{Y}''_{s-1}} \tag{8.17}$$

depending on alternative simplifying assumptions about V. Equation (8.16) is particularly appropriate when the variance of $\hat{Y}'_s$ is roughly constant across time, while equation (8.17) is appropriate when the relative variance of $\hat{Y}'_s$ is roughly constant. The latter assumption may be more reasonable for estimators that are subject to inflation or growth, like sales in a retail trade survey.

Under covariance structures of $\mathbf{Y}_t$ with high positive correlation, which are common for monthly sales of many retail and wholesale businesses, the variance of $\hat{Y}'''_t$ can be reduced to less than half the variance of $\hat{Y}'_t$ (Wolter 1979).

Another composite estimator can be computed to update the composite estimator $\hat{Y}'''_{s-1}$. It is

$$\hat{Y}^+_{s-1} = (1 - \alpha)\hat{Y}''_{s-1} + \alpha\hat{Y}'''_{s-1}, \qquad (8.18)$$

which can be used in conjunction with equation (8.16) or (8.17).

The choice of coefficients in equations (8.16), (8.17), and (8.18) depends on the covariance structure of $\hat{Y}_t$ and on what variances are to be minimized. For example, we may want to minimize the variances of $\hat{Y}'''_s$, $\hat{Y}^+_{s-1}$, $\hat{Y}'''_s / \hat{Y}^+_{s-1}$, and $\hat{Y}'''_s / \hat{Y}^+_{s-12}$ (assuming the period is a month) simultaneously. This becomes something of a compromise because minimizing one variance does not usually minimize the others.

8.5.3 Large Observation Procedures for Two-Level Rotation

Some variables collected in business surveys are subject to large changes in size, especially at the sampling unit or finer level. For example, individual sampling units can have large increases in the absolute value of a variable such as sales, shipments, or profits. These large increases can seriously increase the relative variances of survey estimates. Woodruff (1963) described two unbiased methods or "large observation procedures" that can modify the increases in relative variances in two-level rotation samples. These methods involve collecting data for sampling units with weighted observations above a cutoff for periods other than those when the units would normally respond, thereby reducing the weighted observations for these units (see also Wolter et al. 1976).

8.5.4 Drawbacks of Rotation

In two-level rotation, each respondent in a rotating panel reports the current period's data and the previous period's data. These data can sometimes be subject to reporting biases. If these biases are identical in both Horvitz–Thompson estimators, $\hat{Y}'_s$ and $\hat{Y}''_{s-1}$, then bias is transferred to the MVLU or composite estimators in (8.15), (8.16), (8.17), and (8.18). In fact, if the relative bias is the same, say b, for $\hat{Y}'_s$ and $\hat{Y}''_{s-1}$ for all periods s, then the relative bias in the MVLU and composite estimators is also b. However, if there is a differential bias in the estimators, $\hat{Y}'_s$ and $\hat{Y}''_{s-1}$, then the resulting differential bias is magnified in the composite estimators.

In some surveys, more downward bias has been noted in the estimator $\hat{Y}'_s$ than in $\hat{Y}''_{s-1}$. This bias is called *early reporting bias* because the bias in $\hat{Y}'_s$ seems to result from the fact that the data for this estimate are being reported for a month s that is just ending or has just ended. The bias tends to be downward because (it is conjectured) respondents are generally conservative in reporting their data before final figures become available. Waite (1974) noted this apparent early reporting bias in the U.S. Monthly Retail Trade Survey. Under a simplifying assumption, he derives the asymptotic effect of this differential downward bias on composite estimators (8.17) and (8.18). For example, if $\hat{Y}''_{s-1}$ is unbiased and $\hat{Y}'_s$ has a downward bias of 0.1 percent for each month s and if the coefficient β in (8.17) is 0.8, then $\hat{Y}'''_s$ will eventually have a downward bias of 0.5 percent. This asymptotic bias is nearly reached in equation (8.13). A similar effect occurs with the estimates in equations (8.15), (8.16), and (8.18).

The problem of differential or early reporting bias is probably the most serious problem with regard to using rotating panels and two-level rotation, and it can occur with any multilevel rotation. There are at least two ways to solve the problem. First, one can delay the reporting for the survey so that respondents have enough time to get their final or book data for the current month, thus eliminating the differential bias. Another method is to use *benchmarking*—that is, forcing the estimates to conform to a more accurate survey conducted on a less frequent basis. For example, sales estimates from a monthly survey can be forced to sum for each year to an annual estimate produced by a more accurate annual survey. See Helfand, et al. (1977) and Monsour and Trager (1979) for discussions of benchmarking.

A problem that affects one-level rotation more than multilevel rotation is time-in-sample bias, which is also referred to as *rotation group bias*. This bias is dependent on the number of times a sampling unit has been surveyed. It has largely been studied for household surveys (e.g., Bailar 1975), but this bias also affects business surveys, especially for units reporting for the first time.

REFERENCES

Bailar, B. A. (1975), "The Effects of Rotation Group Bias on Estimates from Panel Surveys," *Journal of the American Statistical Association*, **70**, pp. 23–30.

Bankier, M. D. (1988), "Power Allocations: Determining Sample Sizes for Subnational Areas," *The American Statistician*, **42**, pp. 174–177.

Bethel, J. (1989a), "Minimum Variance Estimation in Stratified Sampling," *Journal of the American Statistical Association*, **84**, pp. 260–265.

Bethel, J. (1989b), "Sample Allocation in Multivariate Surveys," *Survey Methodology*, **15**, pp. 47–57.

Brewer, K. R. W. (1963), "Ratio Estimation and Finite Populations: Some Results Deducible from the Assumptions of an Underlying Stochastic Process," *Australian Journal of Statistics*, **5**, pp. 93–105.

Brewer, K. R. W. (1979), "A Class of Robust Sampling Designs for Large-Scale Surveys," *Journal of the American Statistical Association*, **74**, pp. 911–915.

Brewer, K. R. W., and M. Hanif (1983), *Sampling with Unequal Probabilities*, New York: Springer-Verlag.

Bush, J., and C. House (1993), "The Area Frame: A Sampling Base for Establishment Surveys," *Proceedings of the International Conference on Establishment Surveys*, Alexandria, VA: American Statistical Association, pp. 335–344.

Cantwell, P. J. (1990), "Variance Formulas for Composite Estimators in Rotation Designs," *Survey Methodology*, **16**, pp. 153–163.

Causey, B. D. (1983), "Computational Aspects of Optimal Allocation in Multivariate Stratified Sampling," *SIAM Journal of Scientific and Statistical Computing*, **4**, pp. 322–329.

Chen, W. (1989), "Stratification of a Population: Programming of Lavallee and Hidiroglou's Algorithm," *Proceedings of the Survey Research Methods Section, American Statistical Association*, pp. 620–624.

Chhikara, R., and L. Deng (1991), "A Multiyear Rotation Design for Sampling in Agricultural Surveys," *Proceedings of the Survey Research Methods Section, American Statistical Association*, pp. 461–465.

Chromy, J. (1987), "Design Optimization with Multiple Objectives," *Proceedings of the Survey Research Methods Section, American Statistical Association*, pp. 194–199.

Cochran, W. G. (1977), *Sampling Techniques*, 3rd ed., New York: Wiley.

Dalenius, T., and J. L. Hodges (1959), "Minimum Variance Stratification," *Journal of the American Statistical Association*, **54**, pp. 88–101.

Dayal, S. (1985), "Allocation of Sample Using Values of Auxiliary Characteristic," *Journal of Statistical Planning and Inference*, **11**, pp. 321–328.

Detlefsen, R. E., and C. S. Veum (1991), "Design Issues for the Retail Trade Sample Surveys of the U.S. Bureau of the Census," *Proceedings of the Survey Research Methods Section, American Statistical Association*, pp. 214–219.

Dillard, D. M., and B. L Ford (1986), "Deviating from the Optimal Allocation to Reduce Burden on Large-Sized Units," *Proceedings of the Survey Research Methods Section, American Statistical Association*, pp. 592–596.

Eltinge, J. (1993), "Using Previous-Survey Data for Optimal Sample Allocation Among Strata in Agricultural Surveys," paper presented in International Conference on Establishment Surveys, Buffalo, NY.

Godrey, J., A. Roshwalb, and R. Wright (1984), "Model-Based Stratification in Inventory Cost Estimation," *Journal of Business and Economic Statistics*, **2**, pp. 1–9.

Gurney, M., and J. F. Daly (1965), "A Multivariate Approach to Estimation in Periodic Sample Surveys," *Proceedings of the Social Statistics Section, American Statistical Association*, pp. 242–257.

Hanif, M., and K. R. W. Brewer. (1980), "Sampling with Unequal Probabilities without Replacement: A Review," *International Statistical Review*, **48**, pp. 317–355.

Hansen, M. H., W. N. Hurwitz, and W. G. Madow (1953), *Sample Survey Methods and Theory*, Vol. 1, New York: Wiley.

Harvey, A. C (1976), "Estimating Regression Models with Multiplicative Heteroscedasticity," *Econometrica*, **44**, pp. 461–465.

Háyek, J. (1960), "Limiting Distributions in Simple Random Sampling from a Finite Population," *Publications of the Mathematical Institute of the Hungarian Academy of Science*, **5**, pp. 361–374.

Háyek, J. (1964), "Asymptotic Theory of Rejective Sampling with Varying Probabilities from a Finite Population," *Annals of Mathematical Statistics*, **35**, pp. 1431–1523.

Heike, H.-D., and W. Jaspers (1993), "Optimum Stratification and Allocation by Arithmetic Resp. Geometric Sequences and Iterative Refinement," *Proceedings of the International Conference on Establishment Surveys*, Alexandria, VA: American Statistical Association, pp. 633–638.

Helfand, S. D., N. J. Monsour, and M. L. Trager (1977), "Historical Revision of Current Business Survey Estimates," *Proceedings of the Business and Economic Statistics Section, American Statistical Association*, pp. 246–250.

Hidiroglou, M. A. (1986), "The Construction of a Self-Representing Stratum of Large Units in Survey Design," *The American Statistician*, **40**, pp. 27–31.

Horvitz, D. G., and D. J. Thompson (1952), "A Generalization of Sampling Without Replacement from a Finite Universe," *Journal of the American Statistical Association*, **47**, pp. 663–685.

Isaki, C. T., and W. A. Fuller (1982), "Survey Design Under the Regression Superpopulation Model," *Journal of the American Statistical Association*, **77**, pp. 89–96.

Jarque, C. M. (1981), "A Solution to the Problem of Optimum Stratification in Multivariate Sampling," *Applied Statistics*, **30**, pp. 163–169.

Julien, C., and F. Maranda (1990), "Sample Design of the 1988 National Farm Survey," *Survey Methodology*, **16**, pp. 117–129.

Kirkendall, N. J. (1992), "When Is Model-Based Sampling Appropriate for EIA Surveys?" *Proceeding of the Survey Research Methods Section, American Statistical Association*, pp. 637–642.

Kish, L. (1976), "Optima and Proxima in Linear Sample Designs," *Journal of the Royal Statistical Society*, Series A, **139**, pp. 80–95.

Kish, L., and D. W. Anderson (1978), "Multivariate and Multipurpose Stratification," *Journal of the American Statistical Association*, **73**, pp. 24–34.

Kokan, A. R. (1963), "Optimum Allocation in Multivariate Surveys," *Journal of the Royal Statistical Society*, Series A, **126**, pp. 557–565.

Kott, P. S. (1985), "A Note on Model-Based Stratification," *Journal of Business and Economic Statistics*, **3**, pp. 284–288.

Lavallee, P., and M. A. Hidiroglou (1988), "On the Stratification of Skewed Populations," *Survey Methodology*, **14**, pp. 33–43.

Mergerson, J. W. (1988a), "Allocations Requiring 100% Sampling in Some Strata," NASS Staff Report Number SSB-88-10, Washington, DC: National Agricultural Statistics Service.

Mergerson, J. W. (1988b), "ALLOC.P: A Multivariate Allocation Program," *The American Statistician*, **42**, p. 85.

Mergerson, J. W. (1989), "A Generalized Univariate Optimal Allocation Program," *The American Statistician*, **43**, p. 128.

Monsour, N. J., and M. L. Trager (1979), "Revision and Benchmarking of Business

Time Series,'' *Proceedings of the Business and Economics Statistics Section, American Statistical Association*, pp. 333–337.

Ogus, J. L., and D. F. Clark (1971), "The Annual Survey of Manufactures: A Report on Methodology," Technical Paper 24, Washington DC: U.S. Bureau of the Census.

Pollock, J. (1984), "PPES Sampling of Two Subdomains with Independent Probabilities," *Proceedings of the Survey Research Methods Section, American Statistical Association*, pp. 223–227.

Rao, J. N. K., and D. R. Bellhouse (1989), "The History and Development of the Theoretical Foundations of Survey Based Estimation and Analysis," *Proceedings of the American Statistical Association: Sesquicentennial Invited Paper Sessions*, pp. 406–428.

Roshwalb, A., and S. Regnier (1991), "Comparison of Two Bivariate Stratification Schemes," *Proceedings of the Survey Research Methods Section, American Statistical Association*, pp. 478–483.

Royall, R. M., and J. Herson (1973a), "Robust Estimation in Finite Populations I," *Journal of the American Statistical Association*, **68**, pp. 880–889.

Royall, R. M., and J. Herson (1973b), "Robust Estimation in Finite Populations II: Stratification on a Size Variable," *Journal of the American Statistical Association*, **68**, pp. 890–893.

Särndal, C.-E. B. Swensson, and J. Wretman (1992), *Model Assisted Survey Sampling*, New York: Springer-Verlag.

SAS Institute Inc. (1989), *SAS/STAT User's Guide*, Vol. 1, Version 6, 4th ed., Cary, NC: SAS Institute.

Sunter, A. (1986),"Solution to the Problem of Unequal Probability Sampling Without Replacement," *International Statistical Review*, **54**, pp. 33–50.

Waite, P. J. (1974), "An Evaluation of Nonsampling Errors in the Monthly Retail Trade Sales Data," *Proceedings of the Business and Economic Statistics Section, American Statistical Association*, pp. 602–607.

Waite, P. J., and S. J. Cole (1980), "Selection of a New Sample Plan for the Annual Survey of Manufactures," *Proceedings of the Social Statistics Section, American Statistical Association*, pp. 307–311.

Wang, M. C., and V. Aggarwal (1984), "Stratification Under a Particular Pareto Distribution," *Communications in Statistics—Theory and Methods*, **13**, pp. 711–735.

Wolter, K. M. (1979), "Composite Estimation in Finite Populations," *Journal of the American Statistical Association*, **74**, pp. 604–613.

Wolter, K. M. (1985), *Introduction to Variance Estimation*, New York: Springer-Verlag.

Wolter, K. M., C. T. Isaki, T. R. Sturdevant, N. J. Monsour, and F. M. Mayes (1976), "Sample Selection and Estimation Aspects of the Census Bureau's Monthly Business Surveys," *Proceedings of the Business and Economic Statistics Section, American Statistical Association*, pp. 99–109.

Woodruff, R. S. (1963), "The Use of Rotating Samples in the Census Bureau's Monthly Surveys," *Journal of the American Statistical Association*, **58**, pp. 454–467.

Wright, R. L. (1983), "Finite Population Sampling with Multivariate Auxiliary Information," *Journal of the American Statistical Association*, **78**, pp. 879–884.

CHAPTER NINE

Coordination of Samples Using Permanent Random Numbers

Esbjörn Ohlsson[1]
Stockholm University and Statistics Sweden

Selecting multiple samples from the same sampling frame presents many problems. Efficient estimation of periodic change often requires the use of fixed panel designs or one-level rotation samples with large overlaps between samples (see Chapter 8 for details). However, most business populations are subject to rapid changes due to births, deaths, splits, mergers, and changes in size or activity. List frames and samples must be periodically updated to reflect such changes.

The desire to draw up-to-date samples may seem incompatible with planned overlaps between successive samples. This chapter discusses the permanent random number (PRN) technique, which provides a simple solution to this problem. Even when frames are updated or surveys redesigned, the PRN technique generally produces sizable overlap between successive samples. For Australia's Commonwealth Bureau of Census and Statistics, Brewer et al. (1972) suggested the use of PRNs in connection with Poisson sampling. Independent of the Australian work, Atmer et al. (1975) introduced a PRN technique for simple random sampling without replacement (*srswor*) at Statistics Sweden.

In most countries, many business surveys sample from the same register or list frame. If sampling is independent, response burden is unlikely to be evenly

[1]The author is indebted to Phillip S. Kott for contributing the ''PRN collocated sampling'' idea and other valuable comments. He thanks Diane Ramsey at Statistics New Zealand, Geoff Lee of the Australian Bureau of Statistics, and Frank Cotton of INSEE (France) for kindly providing information on sampling procedures used at their agencies.

Business Survey Methods, Edited by Cox, Binder, Chinnappa, Christianson, Colledge, Kott.
ISBN 0-471-59852-6 © 1995 John Wiley & Sons, Inc.

spread among the businesses over repeated selections. Because strata sizes are often small, some units will be included in several samples and others in none. A method of reducing the overlap between samples for different surveys (*negative coordination*) is needed to address this problem. A major feature of PRN techniques is their ability to negatively coordinate many surveys in a simple fashion. For data collection, editing, or estimation reasons, sample overlap may need to be maximized for some surveys (*positive coordination*), however. With PRNs both negative and positive coordination can be obtained even when different sampling designs are used.

Other techniques have been proposed for positive and negative coordination of samples. Keyfitz (1951), Kish and Scott (1971), and Causey et al. (1985) discuss non-PRN methods for positive coordination of samples. They focus on unequal probability sampling of one unit per stratum in the first stage of a multistage design and do not address negative coordination. Perry et al. (1993) present a negative coordination method in which all samples are drawn simultaneously. Van Huis et al. (1994) present Statistics Netherlands' new system, called EDS, for negative and positive coordination of samples. The major advantage of the system is that on each occasion the sampling procedure concentrates on the businesses with the least cumulated response burden. EDS is not as flexible as the PRN technique in allowing arbitrary stratification of coordinated samples. In EDS the strata must be formed in a hierarchical fashion from fixed "substrata." EDS also does not include the option of using probability proportional to size sampling.

In this chapter, I describe the PRN techniques for *srswor*, Bernoulli, Poisson, and other probability-proportional-to-size sampling procedures. Two techniques for rotating a PRN sample are also presented. I compare Statistics Sweden's SAMU system (Atmer et al. 1975) for coordinating business samples to synchronized sampling used by the Australian Bureau of Statistics (Hinde and Young 1984). I also describe uses of PRN techniques in New Zealand (Templeton 1990) and France (Cotton 1989). In Chapter 10, Srinath and Carpenter discuss the Canadian system for coordinating a single survey over time.

9.1 THE PERMANENT RANDOM NUMBER TECHNIQUE FOR SIMPLE RANDOM SAMPLING

In this section, I discuss how PRNs are used in selecting a simple random sample without replacement (*srswor*) of size n from a population of size N. This "population" could be the subpopulation associated with a particular sampling stratum.

The following approach leads to an efficient computer algorithm for selecting an *srswor*. Each unit in the list frame is assigned a random number drawn independently from the uniform distribution on the interval [0, 1]. Let X_i denote the random number assigned to unit i. The frame units are sorted in ascending order of the X_i. The sample is composed of the first n units in the ordered list. Ohlsson (1992) presents a formal proof that this technique produces an *srswor*. Fan et al. (1962) describe this technique, which they label "sequential." Following their lead, I refer to this type of selection as *sequential srswor*.

In theory, no duplicates are found among continuously distributed random numbers. In practice, duplicates are possible because the computer's pseudorandom number generator yields random numbers with a finite number of decimals. Duplicates can be avoided by using a congruential generator with a cycle greater than the number of frame units. (See, for example, Morgan 1984, pp. 57–67.) When new units are added to the frame (e.g., births), the assignment of random numbers should start at the place in the cycle where the latest assignment stopped, or else duplicate random numbers may result.

9.1.1 Coordination Over Time

The Swedish SAMU system uses sequential *srswor* with PRNs to coordinate sampling across surveys and over time. (SAMU is an acronym for "coordinated samples" in Swedish.) Atmer et al. (1975) developed the basic technique, referring to it as "JALES," an acronym for its inventors. SAMU permanently associates a random number with each frame record. Units persisting on the list frame, or *persistants*, have the same random number used on every sampling occasion; hence the term *permanent random number* (PRN). When a unit is added to the frame (a birth), a random number is generated and recorded on the frame data record. Deaths, such as closed-down businesses, are withdrawn from the list frame together with their random number.

On each sampling occasion, these permanent random numbers are used to select a new sample by sequential *srswor*. This process yields a fresh *srswor* from an up-to-date list without any complications or compromises. A large amount of overlap with the previous sample is obtained because persistants have the same PRN on both occasions. Not all persistants may be retained in the new sample, because the new sample might get more births than there were deaths in the old sample. This result may be a chance event associated with the random selection, or the population may have experienced more births than deaths. In either case, persistants may leave or enter the new sample. Most frequently, they stay in sample yielding the desired overlap (see Figure 9.1).

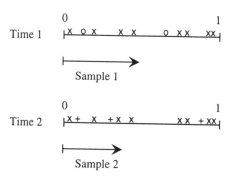

Figure 9.1 Overlap of subsequent samples
($\times$ = persistants, o = deaths, + = births).

The implicit assumption is that births and deaths are not too numerous; in Sweden they comprise less than 15 percent of the list frame.

9.1.2 Coordination of Samples for Different Surveys

Returning to sequential *srswor*, note that the last n units also yield an *srswor*. Indeed, selecting the first n units to the left or to the right of any fixed point a in [0, 1] will yield an *srswor*. The frame is treated as a circular list. If there are not enough points (n) to the right (left) of the starting point a, the selection continues to the right (left) of the point 0 (the point 1), as exemplified in Figure 9.2.

To reduce the overlap between two surveys with desired sample sizes n_1 and n_2, choose two constants a_1 and a_2 in [0, 1]. Then take the units with the n_1 PRNs closest to the right (or left) of a_1 as the first sample and take the ones with the n_2 PRNs to the right (or left) of a_2 as the second sample. If a_1 and a_2 and the sampling directions are chosen properly, the samples will be negatively coordinated. (See Figure 9.3 for an example.) If the population is large enough (i.e., $N \gg n_1 + n_2$), the samples can be made disjoint. On the other hand, when $N < n_1 + n_2$ the samples cannot be disjoint, but their overlap can be reduced. When N is large enough, any number of samples can be disjoined.

Positive coordination of two surveys is maximized by using the same starting point and direction for both samples.

9.1.3 Coordination of Samples with Different Designs

As discussed in Chapter 8, many business surveys use stratified *srswor* designs. It is a simple matter to extend the PRN technique to this environment. For a specific survey, advantages can accrue when the same direction and starting point a are used in all strata.

Positive coordination of two surveys can often be obtained even when different sampling designs are used, which may involve different population definitions and stratification as well as different allocations. For subsequent samples of a recurring survey, this means that the survey may be redesigned and still incorporate an overlap between the old and new sample. When the same starting point is used in all strata, units in the old sample that change strata (due to changes in size or activity, for example) still have a large probability of being included in the new sample.

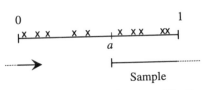

Figure 9.2 Sampling from an arbitrary point a.

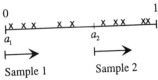

Figure 9.3 Negative coordination of samples.

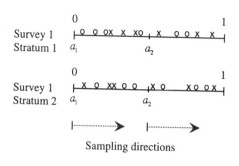

Figure 9.4 Coordination of samples with different stratifications ($\times$ = unit in stratum 1 of survey 2; o = unit in stratum 2 of survey 2).

Similarly, suppose that two surveys use different stratifications of the same population. If the surveys have different starting points or directions, the samples will tend to be negatively coordinated. This is because a "small" random number in one stratum is likely to be "small" in another stratum too, and vice versa (see Figure 9.4). The extent to which negative coordination is attained depends on the starting points and number of sample and population units in each stratum. Similar arguments can be made for coordinated surveys that have different target populations within the same frame.

9.1.4 Sequentially Deleting Out-of-Scope Units

Suppose that the list frame contains many units that are out of scope for a particular survey. Furthermore, suppose that these units can be detected at a cost that is reasonable for the sample, but too large to identify all those on the full list frame. For example, this is the case with the Swedish consumer price index (CPI) sample.

In such a situation, an ordinary *srswor* will have a variable number of in-scope units. With the sequential technique, on the other hand, we can simply continue (sequentially) down the list of random numbers deleting out-of-scope units as they are identified during sampling, until the predetermined number n of in-scope units has been achieved. By the independence of the PRNs, this "net" sample will have the same probability distribution as if the out-of-scope units had not been there. Hence, the result is an *srswor* of size n from the population of in-scope units. The problem with this technique is that the number N of in-scope units on the frame is unknown.

When ratios are being computed, however, knowledge of N is not required. (This is the case for the CPI).

9.2 BERNOULLI SAMPLING

Because Bernoulli sampling is a special case of Poisson sampling, Poisson sampling needs to be introduced. Suppose a sample of size n is desired, in which unit i is included with probability proportional to p_i. The variate p_i may be arbitrary, but is usually some measure of the size of unit i. For simplicity, let the p_i be normed so that

$$\sum_{i=1}^{N} p_i = 1.$$

Let π_i denote the probability of including unit i in the sample. The goal is to have π_i proportional to the size measure p_i for unit i—or more specifically,

$$\pi_i = np_i \qquad (9.1)$$

A procedure that satisfies equation (9.1) will be called a procedure for *strict probability proportional to size (pps) sampling*.

Poisson sampling, a strict *pps* procedure, can be carried out as follows. For each i, let X_i be a uniform random number associated with unit i, and let the X_i's be mutually independent. Choose a starting point a in the interval [0, 1]. Next, pass through the units on the frame, one by one, and include unit i in the sample if and only if

$$a < x_i \leq a + np_i \qquad (9.2)$$

with a "wrap-around" adjustment in case $a + np_i > 1$ (see Figure 9.2). The resulting sample has a random sample size m with expectation n. Obviously, equation (9.1) is satisfied.

9.2.1 Bernoulli Sampling

Bernoulli sampling refers to the special case of Poisson sampling in which $p_i = 1/N$ for all i. The random sample size is a major drawback of Bernoulli sampling. This is especially true for stratified designs where some strata may have small sample sizes allocated. The variability in stratum sample sizes may cause severe deviations from the desired allocation. If N is the stratum size and n the desired stratum sample size, the probability of getting a realized stratum sample of size zero is $(1 - n/N)^N < e^{-n}$. For some strata, this probability may be appreciable.

9.2.2 Coordinated Bernoulli Samples

Brewer et al. (1972) suggested the use of Poisson sampling, and hence Bernoulli sampling, for coordination. As with *srswor*, the idea is to let the X_i be the permanent random numbers. Again, positive or negative coordination is obtained by choosing appropriate starting points a.

When $a = 0$ in Bernoulli sampling and in sequential *srswor*, units having small random numbers X_i are selected. The difference is that the n smallest X_i are selected to get a sequential *srswor*, while in Bernoulli sampling the X_i that

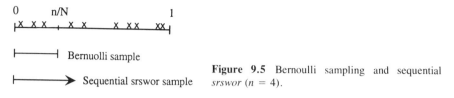

Figure 9.5 Bernoulli sampling and sequential *srswor* ($n = 4$).

fall in the interval $(0, n/N]$ are selected (see Figure 9.5). Bernoulli samples using PRNs have more or less the same coordination properties as sequential *srswor* with PRNs.

In theory, the Bernoulli sampling interval for a survey could be kept fixed from year to year, making the overlap complete among persistants. In practice, N may change between years, resulting in slightly different sampling fractions n/N for some years even when n remains the same. As a result, the overlap of persistants will not be complete but is likely to be larger for Bernoulli sampling than for sequential *srswor*.

When the sum of their selection interval lengths does not exceed one, two or more Bernoulli samples can be made nonoverlapping. In sequential *srswor*, overlaps can only be made improbable, but not absolutely impossible. This advantage of Bernoulli sampling should be balanced against the disadvantage of the random sample size.

It is quite possible to combine the two methods so that an *srswor* sample is coordinated with a Bernoulli sample. This is done by simply using the same PRNs for both surveys.

In Chapter 10, Srinath and Carpenter present modified collocated sampling, a PRN technique related to sequential *srswor* and Bernoulli sampling. This technique solves the problem of providing up-to-date, overlapping samples for a single survey, but is not intended for controlling the sample overlap for different surveys.

9.3 PROBABILITY PROPORTIONAL TO SIZE SAMPLING

When good size data are available for frame units, probability proportional to size sampling may be optimal. In Chapter 8, Sigman and Monsour discuss the advantages and disadvantages of using *pps* sampling for a business survey. In this section, three procedures for *pps* sampling are described as they are used with PRNs. As far as I know, Poisson, sequential Poisson, and collocated sampling are the only *pps* procedures that have been adapted for use with PRNs.

9.3.1 Poisson Sampling

Poisson sampling is a (strict) *pps* design; that is, it satisfies equation (9.1). The coordination of Bernoulli samples is easily generalized to unequal probability Poisson samples by again letting the X_i in equation (9.2) be PRNs.

Coordination of Poisson samples through PRNs was introduced by Brewer et al. (1972) and further discussed by Sunter (1977). The properties of coordinated Bernoulli sampling carry over to the general *pps* case with minor modifications.

The realized sample size m of a Poisson sample is random. One can show that, as in the Bernoulli special case, $\Pr(m = 0) < e^{-n}$. Because the sample size is random, Brewer et al. (1972) recommend that the ratio estimator,

$$\hat{Y}_R = \begin{cases} \dfrac{1}{m} \sum_{i \in S} \dfrac{y_i}{p_i} & \text{if } m > 0 \\ 0 & \text{if } m = 0, \end{cases} \tag{9.3}$$

be used in place of the Horvitz–Thompson estimator $\hat{Y}_{HT} = \Sigma_{i \in S} y_i / \pi_i$. Equation 9.3 is a ratio estimator because m is random. For positive sample sizes ($m > 0$), equation (9.3) is the special case of equation (8.14) in Chapter 8 with $q = 1$.

9.3.2 Sequential Poisson Sampling

Sequential Poisson sampling has been explored as a way to generalize sequential *srswor* to the *pps* case and to obtain a fixed sample size alteration of Poisson sampling (Ohlsson 1990). First, define the normed random numbers ξ_i, where

$$\xi_i = \frac{X_i}{Np_i}. \tag{9.4}$$

Observe that unit i is included in a Poisson sample with $a = 0$ if and only if $X_i \le np_i$, or

$$\xi_i \le \frac{n}{N}.$$

In sequential Poisson sampling, on the other hand, the population is sorted by ξ_i and then the first n units are selected from the ordered list. When the p_i are equal, $\xi_i = X_i$ and sequential Poisson sampling reduces to sequential *srswor*. The coordination of sequential Poisson samples is through the X_i PRNs, while the ξ_i differ from time to time. If a different starting point than $a = 0$ is desired, subtract a from X_i before the calculation of ξ_i in equation (9.4).

Unfortunately, sequential Poisson sampling is not a strict *pps* procedure (see Ohlsson 1990). Exact expressions for the inclusion probabilities of sequential Poisson sampling are not readily obtained, so the (exact) Horvitz–Thompson estimator cannot be used. One approach is to use the estimator $\hat{Y}_R$ in equation (9.3), with m replaced by the fixed sample size n. In a simulation study, I tested this approach using data from the Swedish CPI and a survey of financial accounts (Ohlsson 1990). The inclusion probabilities were very close to the equation (9.1) values. Moreover, $\hat{Y}_R$ under sequential Poisson sampling performed slightly better (in terms of mean square error) than $\hat{Y}_R$ under Poisson sampling.

9.3.3 Collocated Sampling

Another way to reduce, but not quite eliminate, the sample size variation of Poisson sampling is to use *collocated sampling* (Brewer et al. 1972, Brewer et al. 1984). The name of this procedure refers to collocation of the random numbers, not the samples.

Here, the units in the population (or stratum) are ordered at random, giving unit i the rank L_i. Independent of this random ordering, a single random number ε is selected from the uniform [0, 1] distribution. For each unit i, define

$$R_i = \frac{L_i - \varepsilon}{N}. \tag{9.5}$$

A sample of expected size n is now selected by the same inclusion rule as in Poisson sampling in equation (9.2), but with R_i replacing X_i. Because the R_i are equally spaced in the interval [0, 1], collocated sampling "substantially reduces the variability of the sample size and, in particular, reduces the probability of selecting an empty sample very greatly" (Brewer et al. 1984, p. 23). This procedure is strictly *pps*; that is, it satisfies equation (9.1).

Brewer et al. (1972, 1984) suggest that the R_i should play the part of the PRNs for coordination purposes. Although suitable for some applications, this approach cannot coordinate samples that involve different stratifications and/or populations, nor does it allow survey redesign. The reason is that the R_i depend on the actual stratification used (through the stratum size N).

PRN collocated sampling is another approach to consider. The random ordering in the first step of collocated sampling is performed in the same way as in sequential *srswor*. That is, the units are ordered by the size of their permanent random numbers X_i, so that $L_i = \text{Rank}(X_i)$ where X_i is the PRN of unit i. The R_i from equation (9.5) can be regarded as an equal spacing adjustment of the X_i.

The similarities between sequential Poisson and PRN collocated sampling can be seen from their inclusion rules, which can be rewritten as

$$\text{Rank}\left\lceil \frac{X_i}{Np_i} \right\rceil \le n$$

and

$$\frac{\text{Rank}(X_i) - \varepsilon}{Np_i} \le n,$$

respectively. Both procedures are equivalent to sequential *srswor* where the p_i are equal and hence may be viewed as extensions of sequential *srswor* to a *pps* environment.

9.3.4 Coordinated Probability Proportional to Size Samples

The basic features of coordination through PRNs are also valid for *pps* Poisson sampling, sequential Poisson sampling, and PRN collocated sampling. Samples drawn using any of these methods can be coordinated with others, through the use of common PRNs. Therefore, samples can be coordinated that have quite different designs, as regards population delimitation, stratification, allocation, inclusion probabilities, and the choice of one of these three sampling procedures.

A comparison of the coordination properties of sequential Poisson and ordinary *pps* Poisson sampling gives very much the same result as the Section 9.2 comparison in the equal probability case. For example, two or more Poisson samples can be designed to be nonoverlapping when the sum of the maximal π_i (maximal for each sample) is less than 1. For sequential Poisson sampling, an overlap is improbable when this sum is less than 1.

Size measures can vary from one survey period to the other. If just one unit in a stratum changes in sizes, every unit changes its size value p_i because the sum of these values is fixed. Hence, *pps* Poisson sampling will not yield a perfect overlap of persistants. As in the equal probability case, if there are many births and deaths, sequential Poisson sampling will often produce less overlap than Poisson sampling. PRN collocated sampling can be expected to behave more or less as sequential Poisson sampling in these situations.

Poisson and PRN collocated sampling have the advantage of being strictly *pps*. For these two procedures, Brewer and Hanif (1983, p. 83ff) give approximate expressions for the mean square error of $\hat{Y}_R$. Sunter (1986) gives an approximate formula for the bias of $\hat{Y}_R$ in the Poisson case. Sequential Poisson sampling suffers from the lack of analytic results for the bias and variance of $\hat{Y}_R$. On the other hand, sequential Poisson sampling is the only one of these procedures that yields fixed size samples. It also has the advantage of allowing sequential deletion of out-of-scope units.

9.4 SAMPLE ROTATION

Sample rotation can be handled quite easily with the PRN technique. To illustrate the various approaches, I assume that (1) the sample should be rotated every year, and that (2) persistants with selection probabilities of 0.10 or less should only have to stay in sample for 5 years.

9.4.1 The Constant Shift Method

Brewer et al. (1972) suggested the following procedure for rotation of a Poisson sample. Shift the starting points of all surveys a specified distance, say

0.02, to the right between the years. For units with inclusion probability 0.10, the expected rotation will then be 20 percent, and they are out of sample after 5 years. Less rotation will occur for units with larger inclusion probabilities. Small units with inclusion probability less than 0.02 will stay in sample for only 1 year. For the rotation to be meaningful, the distance between the starting points for negatively coordinated surveys must be greater than 0.20, or else units may rotate out of one survey and into another.

For a single survey, the amount of shift may be adjusted so that the overall rotation fraction is 20 percent (or whatever fraction is desired). Variations in the number of years that units are in the sample are hardly acceptable from the respondents' point of view, however. Furthermore, when several surveys need to be coordinated (positively or negatively), the appropriate amount of shift will differ from survey to survey. However, different shifts cannot be used for each survey, because this destroys the coordination after a few years.

9.4.2 The Random Rotation Cohort Method

The random rotation cohort (RRC) method was introduced in the Swedish SAMU system in 1989 (Ohlsson 1990). The RRC method was derived from this observation about the constant shift method. Instead of shifting the starting points of the surveys 0.02 to the right, the same result can be achieved by shifting all the PRNs 0.02 to the left. Monitoring a complex system is easier when starting points are not updated every year.

The *random rotation cohort method* is a variation of the constant shift method in which the starting point of each year's survey is constant over time but the permanent random numbers change. In RRC, each unit in the register is randomly designated to one of five rotation cohorts. This means that each unit has a multinomial trial performed that gives it a 20 percent probability of being included in each rotation cohort. The rotation cohort is permanently associated with the unit. Births are assigned a rotation cohort when they enter the list frame. After the first year, the PRNs of the units in rotation cohort 1 are shifted 0.10 to the left. The next year the PRNs in cohort 2 are shifted 0.10 to the left, and so forth.

For the vast majority of units with selection probabilities less than 0.10, an expected rotation rate of 20 percent will be achieved each year, and the unit can expect to be out of sample after 5 years. This holds true for any survey in the system, irrespective of sampling design. Among units with larger inclusion probabilities, rotation will be slower. It may be a good idea to exclude large units from rotation—that is, those units that have a probability close to 1 of being included in at least one survey. Note that a collection of independent uniform random variables that are shifted this way remain such a collection. Therefore, the RRC rotation does not change the probabilistic features of the PRN sampling procedures.

The RRC method, as well as the constant shift method, can be applied to any PRN sampling technique. In both cases, units for which the inclusion probability is less than 0.10 over the years can be guaranteed to be out of a

Poisson sample after (at most) 5 years. Unfortunately, for sequential *srswor*, sequential Poisson, and PRN collocated sampling, we can only say that when a unit's inclusion probability is considerably less than 0.10, it will very likely be out of sample in the prescribed time. Among units with selection probability less than 0.10, an expected rotation rate of 20 percent each year will occur for all coordinated surveys. In modified collocated sampling for a single survey, introduced in Chapter 10, exact time-in and time-out-of-sample constraints can be specified for the units.

9.5 IMPLEMENTATIONS IN DIFFERENT COUNTRIES

How PRN techniques are applied in a program of surveys can be illustrated by the varied ways Sweden, Australia, New Zealand, and France use PRN sampling techniques for their business surveys. In these countries, PRNs are used predominantly for equal probability sampling.

9.5.1 The Swedish SAMU System

For many business surveys at Statistics Sweden, the sampling frame is the business register. The Swedish register contains all businesses, including enterprises, farms, and institutions (the legal units). Samples are sometimes drawn using combinations of legal units. Another "layer" of the register records the establishments controlled by the legal units.

The SAMU system has been used since 1972 for coordinating samples from the business register. For a description of the early SAMU, see Thulin (1976). About 15 annual and subannual surveys are now selected using SAMU. Most samples are drawn in December, but samples for a few annual surveys are drawn in May. With PRNs, no problem occurs in coordinating the May samples with December's.

In SAMU, stratified samples are drawn using sequential *srswor* with PRNs. The Swedish CPI, which uses sequential Poisson sampling, is an exception. The surveys are grouped into "blocks" for which the same starting point and sampling direction are used. Figure 9.6 shows the starting points and sampling directions of the six blocks in the present SAMU.

When positive coordination is desired, two surveys can be put into the same

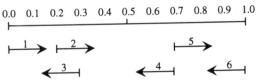

Figure 9.6 Starting points and sampling directions in the interval [0, 1] for the six blocks of surveys in SAMU.

block. When two surveys have different target populations, they can also be put into the same block without risk of sample overlap, thereby facilitating negative coordination with other surveys. Similarly, no risk for overlap exists between Blocks 5 and 6 in Figure 9.6, because Block 6 consists of surveys of the service sector, while the surveys in Block 5 are confined to manufacturers only. Figure 9.6 should hardly be viewed as an ideal placing of samples, but rather the product of 20 years of additions and adjustments.

Built into SAMU are three hierarchical standard stratifications by industry and a default stratification by number of employees. SAMU also contains a default mechanism for sample allocation, yielding Neyman allocation on the control variable "number of employees." In principle, the survey designers are free to use any feasible stratification and allocation. Using the RRC method, the samples are rotated before sampling each December.

On occasion, a unit is included in the sample one year, excluded the next year, and then included again within a year or two. To some extent, this is inescapable with such systems. To a large extent, however, these undesirable shifts in sample inclusion are caused by frequent minor redesigns of the surveys. For instance, mechanical use of the default Neyman allocation system often causes slight reallocations across years. Such minor changes could be eliminated because they have a negligible impact on precision.

SAMU is less efficient in coordinating samples drawn at different levels of the register. For example, take an enterprise with several establishments. The enterprise is assigned the same random number as its largest establishment. The other establishments cannot be linked to their "mother" without destroying the independence of their random numbers. This forces Statistics Sweden to take cluster or two-stage samples of locations when they want the establishment surveys coordinated with samples on the enterprise level.

9.5.2 The Australian Synchronized Sampling System

Before 1982, the Australian Bureau of Statistics (ABS) used equal probability collocated sampling to coordinate two surveys. For rotation, the constant shift method was used. Because of the disadvantages of these methods, a new technique called *synchronized sampling* was introduced. Since 1982, this technique has been used to coordinate several samples. Detailed descriptions of the technique can be found in Hinde and Young (1984) and Australian Bureau of Statistics (1985).

First, let me begin by describing how the technique is used for positive coordination without rotation. As before, the units are assigned permanent random numbers from [0, 1]. At the beginning, a sample of size n is selected to the right of a starting point by using sequential *srswor*. For the next sampling occasion, a selection interval is determined as follows. The end point e is positioned at the PRN of unit number $n + 1$ to the right of the starting point. The starting point is then moved to the right and positioned at the first PRN; let us call this position s. The selection interval is $[s, e)$, including s but not e

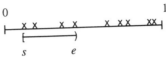

Figure 9.7 A selection interval for synchronized sampling (original starting point is 0, $n = 3$).

(see Figure 9.7). Suppose that a birth or death occurs in the interval or that the sample size is changed between the occasions. The interval is then either extended to the right to include more units or abridged from the left to exclude units until the desired sample size is obtained. The new sample consists of the units in the adjusted selection interval.

Note that the selection interval moves slowly to the right over time. With sequential *srswor* and collocated sampling, a persistant is occasionally removed from the sample one year and then included the next year. The restriction that the endpoints may only move in one direction was introduced to eliminate this annoying effect.

It can be shown by example that this procedure has a "birth bias," in that births have a larger inclusion probability than do persistants. To compensate, the procedure has been modified by introducing a *birth gap*, an open interval between the largest PRN in a stratum and 1.0 that is excluded from the selection interval. Hinde and Young (1984) show that this almost removes the birth bias when there are only births. Furthermore, they show that no selection bias occurs with only deaths and with one death and one birth. The general case with an arbitrary number of births and deaths "was found to be intractable" (p. 15). The second-order inclusion probabilities of the sample were not reported, and it is unknown if the procedure yields an *srswor*.

Negative coordination is achieved by giving each stratum of each survey an individual *overlap range*. The overlap ranges are fixed, disjoint intervals. The selection interval is then, at least in principle, forced to remain within this range. By properly choosing the overlap ranges, negative coordination is obtained, but "the process of allocating the position and length of the overlap ranges for all the strata of all surveys is very complex" (Australian Bureau of Statistics 1985, p. 6.1.5).

To some extent, synchronized sampling allows surveys to use different stratifications. This problem is "solved in practice by improvisation and compromise" (Hinde and Young 1984, p. 26). The overlap ranges differ between strata of a survey. Even two strata that have the same range will eventually have different starting points. Hence, the system does not readily allow positive coordination of surveys with different designs, as opposed to the PRN techniques described earlier. In particular, a large overlap of subsequent samples for the same survey is difficult to achieve when the survey is redesigned. Another consequence of having different starting points in the strata is that a unit that changes stratum is not as likely to remain in sample (or remain out of sample) as when a common starting point is used.

The system also includes the option to rotate samples. Besides the slow movement to the right (described above), voluntary rotation is obtained by letting the selection intervals move to the right within the overlap range. Suppose it is desired that units stay a maximum of R years in sample. The basic idea is to place the starting point as close as possible to the right of the endpoint R years earlier. The actual rotation out of sample will depend on both this rotation technique and the birth rate. If enough units are within the range, each stratum of each survey can rotate at its own rate, which ensures that (small) units only have to stay in sample for (at most) a preset number of years. To the best of my knowledge, the probability distribution of the resulting sample has not been studied.

9.5.3 Permanent Random Number Techniques in New Zealand and France

In 1989, Statistics New Zealand started to use Bernoulli sampling (under the name "Poisson sampling") with permanent random numbers. Surveys using the business register as a frame are being gradually introduced into the system which included five annual and subannual surveys in 1992. The problem of variable sample sizes is handled by choosing large expected sample sizes. So far, the samples have not been rotated. For more information, see Templeton (1990).

In 1989, a PRN technique for business surveys was introduced at the Institute National de la Statistique et des Etudes Economique (INSEE) in France (Cotton 1989). The sampling technique is sequential *srswor*, starting from $a = 0$. Negative coordination is obtained by transforming the random numbers after drawing each sample, rather than by sampling from different starting points. In this transformation, units that have already been drawn are shifted towards the right end of [0, 1] while the other units are shifted to the left. The transformation leaves the joint probability distribution of the PRNs unchanged. The technique can be used for positive or negative coordination.

A special feature is the way the PRN of an enterprise is linked to the PRNs of its establishments. The stated goal is to provide for positive/negative coordination between samples drawn at the enterprise level and the establishment level.

At present, INSEE uses this PRN technique to coordinate samples for different surveys drawn the same year and not for coordination across the years. The reason is that the sampling frame is updated by using information from the samples. The result of such updating can given selection bias in any PRN system.

9.6 CONCLUDING REMARKS

Using PRNs for sample selection is a simple way to obtain negative and positive coordination over both time and space in a dynamic population. The flexibility in allowing different stratification and inclusion probabilities for different surveys makes PRNs suitable for the construction of sampling systems for multiple business surveys.

When using PRNs to coordinate samples over time, information from the surveys must not be used to update the list frame (except for survey information from certainty strata that are completely enumerated). Suppose that the list contains deaths that cannot be detected through the ordinary updating mechanism. Suppose further that all reported deaths in the sample are used to update the list frame one year. If the sample is drawn from the same starting point next year, the new sample will contain fewer deaths than the rest of the frame. As a result, population totals will be overestimated.

REFERENCES

Atmer, J., G. Thulin, and S. Bäcklund (1975), "Coordination of Samples with the JALES Technique," *Statistisk Tidskrift*, **13**, pp. 443–450. (In Swedish with English summary).

Australian Bureau of Statistics (1985), *ABS Computing Network Systems Manual*, **VIII**, unpublished report, Belconnen: Australian Bureau of Statistics.

Brewer, K. R. W., L. J. Early, and M. Hanif (1984), "Poisson, Modified Poisson and Collocated Sampling," *Journal of Statistical Planning and Inference*, **10**, pp. 15–30.

Brewer, K. R. W., L. J. Early, and S. F. Joyce (1972), "Selecting Several Samples from a Single Population," *Australian Journal of Statistics*, **14**, pp. 231–239.

Brewer, K. R. W., and M. Hanif (1983), *Sampling with Unequal Probabilities*, New York: Springer.

Causey, B. D., L. H. Cox, and L. R. Ernst (1985), "Application of Transportation Theory to Statistical Problems," *Journal of the American Statistical Association*, **80**, pp. 903–909.

Cotton, F. (1989), "Use of SIRENE for Enterprise and Establishment Statistical Surveys," paper presented at the 4th International Round Table on Business Survey Frames.

Fan, C. T., M. E. Muller, and I. Rezucha (1962), "Development of Sampling Plans by Using Sequential (Item by Item) Techniques and Digital Computers," *Journal of the American Statistical Association*, **57**, pp. 387–402.

Hinde, R., and D. Young (1984), *Synchronised Sampling and Overlap Control Manual*, unpublished report, Belconnen: Australian Bureau of Statistics.

Keyfitz, N. (1951), "Sampling with Probabilities Proportional to Size: Adjustment for Changes in the Probabilities," *Journal of the American Statistical Association*, **46**, pp. 105–109.

Kish, L., and A. Scott (1971), "Retaining Units after Changing Strata and Probabilities," *Journal of the American Statistical Association*, **66**, pp. 461–470.

Morgan, B. J. T. (1984), *Elements of Simulation*, New York: Chapman and Hall.

Ohlsson, E. (1990), "Sequential Sampling from a Business Register and Its Application to the Swedish Consumer Price Index," R&D Report 1990:6, Stockholm: Statistics Sweden.

Ohlsson, E. (1992) "SAMU—The System for Co-Ordination of Samples from the Business Register at Statistics Sweden—A Methodological Description," R&D Report 1992:18, Stockholm: Statistics Sweden.

Perry, C. R., J. C. Burt, and W. C. Iwig (1993), "Methods of Selecting Samples in Multiple Surveys to Reduce Respondent Burden," *Proceedings of the International Conference on Establishment Surveys*, Alexandria, VA: American Statistical Association, pp. 345–351.

Sunter, A. B. (1977), "Response Burden, Sample Rotation, and Classification Renewal in Economic Surveys," *International Statistical Review*, **45**, pp. 209–222.

Sunter, A. B. (1986), "Implicit Longitudinal Sampling from Administrative Files: A Useful Technique," *Journal of Official Statistics*, **2**, pp. 161–168.

Templeton, R. (1990), "Poisson Meets the New Zealand Business Directory," *The New Zealand Statistician*, **25**, pp. 2–9.

Thulin, G. (1976), "Co-Ordination of Samples in Enterprise Statistics (SAMU). Implementation of the JALES Technique," *Statistisk Tidskrift*, **14**, pp. 85–101. (In Swedish with English summary.)

Van Huis, L. T., C. A. J. Koeijers, and S. J. M. de Ree (1994), "EDS, Sampling System for the Central Business Register at Statistics Netherlands," unpublished manuscript, Voorburg: Statistics Netherlands.

CHAPTER TEN

Sampling Methods for Repeated Business Surveys

K. P. Srinath and Ronald M. Carpenter[1]
Statistics Canada

Business surveys are usually repeated periodically. The period could be a week, a month, a quarter, a year, or some other time period. To increase accuracy and reduce response burden, it may be desirable that the sample change from one survey occasion to the next. In practice, the same sample is often used for several survey occasions. Our focus in this chapter is on sampling from a list frame in a particular stratum for a repeated survey. Every unit in the stratum has an approximately equal probability of selection. Although the composition of the stratum itself may change over time due to births (the introduction of new businesses into the stratum) and deaths (the removal of businesses from the stratum), the definition of the stratum and its desired sampling fraction remain roughly constant.

In Chapter 8, Sigman and Monsour introduce one method of sampling from a list frame for a repeated survey—the rotation group method. They do not, however, address how births should be handled with this method. In Chapter 9, Ohlsson discusses the use of permanent random number sequential sampling. That methodology, which is identical to collocated sampling for a fixed population (Brewer et al. 1972), features a clever technique for handling births in a repeated survey.

In this chapter we discuss the rotation group method in depth. We also review a method Tambay (1988) suggested for handling births with collocated

[1]The authors thank Phillip S. Kott, whose constructive comments led to a number of improvements in the manuscript. The views expressed by the authors do not necessarily reflect the policies of Statistics Canada.

Business Survey Methods, Edited by Cox, Binder, Chinnappa, Christianson, Colledge, Kott.
ISBN 0-471-59852-6 © 1995 John Wiley & Sons, Inc.

sampling. We also introduce a new method of sample selection that combines rotation group and collocated sampling techniques. Finally, we compare the various techniques using a simulation study.

10.1 SELECTION AND ROTATION PROCEDURES

In this section, we consider three methods of sample selection within a particular stratum for a repeated survey with births and deaths. The first method, rotation group sampling, is used in retail and wholesale trade surveys in the United States and Canada. The second, repeated collocated sampling, has been studied internally at Statistics Canada. The third, modified collocated sampling (which we introduce here), combines elements of the first two methods.

10.1.1 Rotation Group Sampling

In rotation group sampling, the population units in each stratum are randomly divided into mutually exclusive rotation groups of equal or nearly equal size. A without-replacement simple random sample (*srswor*) of rotation groups is then selected for the sample.

For the first sampling occasion, the following method is often used to allocate the stratum units to rotation groups. Suppose there are P rotation groups numbered from 1 to P. A random permutation of this ordering is used for allocation and is called the *assign ordering*. The first unit in the population is assigned to the first rotation group in the assign ordering, the second population unit is assigned to the second rotation group in the assign ordering, and so on to the Pth population unit, which is assigned to the Pth rotation group in the assign ordering. The process begins again with the $(P + 1)$th unit assigned to the first rotation group in the assign ordering, the $(P + 2)$th unit assigned to the second rotation group, and so on.

The original numbers of the rotation groups before permutation form what is referred to as the *rotation ordering*. This ordering is used for purposes of selection and rotation of units. For example, rotation groups numbered 1 to p in the rotation ordering are included in the sample on the first occasion; on the second occasion, rotation group 1 rotates out of the sample while the $(p + 1)$th rotation group rotates into the sample.

To illustrate, let the population size N in a stratum be 16, and let the sample size n be equal to 8. Therefore, the sampling fraction f is equal to 0.5. Suppose we want the units to stay in the sample for four occasions; then the total number of rotation groups in the population is $4(1/f)$, which is equal to 8. That is, 16 units in the population are randomly divided into 8 rotation groups numbered 1 to 8, each rotation group containing two units. A random permutation of the numbers 1 to 8 is used for assigning units to the rotation groups. If rotation group 6 in the rotation ordering is the first rotation group in the assign ordering, then the first unit in the population is assigned to the sixth rotation group, and so on. Rotation groups 1 to 4 before permutation are selected in the sample for the first occasion. On the second occasion, rotation group 1 rotates out and group 5 in the rotation ordering rotates in.

Suppose there are N population units in the stratum on the first sampling occasion. If N/P is not an integer, then the P rotation groups will not be of equal size. Thus, the number of units in a sample of p rotation groups will be a random variable with expected value $n = (p/P)N$. The probability of selection for any unit in the population would be p/P. If f is the desired sampling fraction for the stratum, then p and P should be chosen so that f is approximately p/P.

Many surveys that employ rotation group sampling do not use the *unconditional* unit selection probability p/P in estimation. Instead, they rely on *conditional* selection probabilities given realized (actual) sample sizes. Given a realized sample size m in the stratum, each unit has an m/N probability of being in the sample. Conditional probabilities like m/N can easily be calculated for the sampling designs discussed in this chapter.

The following method is used to handle births in some of Statistics Canada's business surveys. Births are assigned to rotation groups using the assign ordering. For example, suppose that on the first sampling occasion the last (Nth) unit in the stratum was assigned to the qth rotation group in the assign ordering. The first new birth for the second sampling occasion would then be assigned to the $(q + 1)$th group in the assign ordering. On each sampling occasion, the rotation group to which the last birth was assigned is noted, and subsequent births are assigned to rotation groups starting with the next rotation group in the assign ordering.

Some business surveys at Statistics Canada use a one-level rotation scheme (see Chapter 8) in which one rotation group is rotated out of the sample and another is rotated into the sample on each new sampling occasion. Considerable effort is extended so that small units that have been in the sample stay out of the sample for the designated time period (Hidiroglou et al. 1991). This can lead to complications such as rotation groups without any members. The interested reader is referred to Hidiroglou and Srinath (1993) for an in-depth discussion of such problems. For our purposes, it suffices to recognize that rotation group sampling readily assures that no rotation group, and thus no sampling unit, is sampled too often due to random chance.

10.1.2 Repeated Collocated Sampling

Chapter 9 discusses collocated sampling in an unequal selection probability context. Below, we describe a procedure first given by Tambay (1988) that uses equal-selection-probability collocated sampling on the first sampling occasion. The handling of births, however, differs from the treatment in Chapter 9. We refer to this procedure as *repeated collocated sampling*.

Arrange the units within the stratum in a random order. A sample selection number, $SSN(i)$, is assigned to the ith ranked unit in the population as follows:

$$SSN(i) = (i - \varepsilon)/N, \qquad (10.1)$$

where ε is a number randomly drawn from the uniform distribution on the unit interval [0, 1]. All sample units whose sample selection numbers lie within the interval [0, f], where f is the desired sampling fraction, are included in the sample.

Rotation of the sample is achieved by simply shifting the sampling interval [0, f]. For example, the sample on the second sampling occasion might be every unit with a sample selection number in the interval [s, $s + f$]. The value of s can be chosen to either eliminate sample overlap (if possible) or to create a degree of overlap that would be near optimal for estimation purposes.

On each sampling occasion, new births are assigned a sample selection number independent of the numbers assigned to older units. This means that on each sampling occasion, the Q new births since the last sampling occasion are assigned sample selection numbers:

$$\text{SSN}(i) = (i - \varepsilon)/Q,$$

where ε may be the same value as in equation (10.1) or may be a new draw from the unit interval.

Equally spacing the new births on the unit interval alters the equal spacing of the stratum units as a whole and may lead to variations in the sample size over time. The permanent random number (PRN) sequential *srswor* (Chapter 9) uses a similar approach that keeps the sample size constant over time. The disadvantage of that approach is that the PRNs of births are not equally spaced on the interval [0, 1]. This may lead to births being over- or underrepresented in the sample due to random chance.

Rotation group sampling methods are conditionally equivalent to simple random sampling without replacement. The same can be said about repeated collocated sampling because births are selected independently each time. Estimating the variance of an estimator based on a repeated collocated sample may not be a trivial matter, however. This is a minor point, because the *srswor* variance estimate can be used to approximate the variance under repeated collocated sampling. If anything, the *srswor* variance estimator will be slightly conservative (because repeated collocated sampling does a better job at proportionally representing the births in the population than does *srswor*).

10.1.3 Modified Collocated Sampling

In this section, we propose a modified collocated sampling approach that is a mixture of rotation group sampling and collocated sampling. Instead of rigidly spacing the sample selection numbers on the unit interval, the selection numbers are assigned as follows. First the interval [0, 1] is divided into N intervals (if N is very large, then it is possible to put a limit on the number of intervals created). A random permutation of the intervals is obtained. The ordering obtained by the random permutation of the numbers 1, 2, . . . , N can be thought of as an assign ordering and can be used to assign random numbers in the corresponding intervals. The first unit in the population is assigned a number randomly selected from a uniform distribution on the first interval in the assign ordering, the second unit is assigned a random number in the second interval in the assign ordering, and so on.

The following example may help clarify this procedure. Let the stratum population size N be 20 and let the desired sample size n be 5. The unit interval [0, 1] is divided into 20 intervals of length 0.05 each. Let the random permutation of the numbers 1 to 20 be 9, 15, 1, 4, 20, 19, 14, 2, 10, 8, 17, 7, 11, 18, 3, 13, 12, 16, 6, and 5.

Now the first unit in the population is assigned a random number between 0.40 and 0.45; that is, the first population unit is assigned to the ninth interval. The second unit in the population gets a random number between 0.70 and 0.75 (the 15th interval), the third unit gets a random number between 0.00 and 0.05, and so on. Each unit gets a unique random number in the interval to which that unit has been assigned as shown in Table 10.1. All units having random numbers in the interval 0–0.25 (the desired sampling fraction, 5/20) are selected in the sample. This means that the units assigned to the first five intervals in the original ordering (namely, units 3, 4, 8, 15, and 20) are in the sample. Sample rotation is achieved in the same manner as described in the previous section.

Table 10.1 Assignment of Random Numbers

Sample Unit	Random Permutation	Corresponding Interval
1	9	0.40–0.45
2	15	0.70–0.75
3	1	0.00–0.05
4	4	0.15–0.20
5	20	0.95–1.00
6	19	0.90–0.95
7	14	0.65–0.70
8	2	0.05–0.10
9	10	0.45–0.50
10	8	0.35–0.40
11	17	0.80–0.85
12	7	0.30–0.35
13	11	0.50–0.55
14	18	0.85–0.90
15	3	0.10–0.15
16	13	0.60–0.65
17	12	0.55–0.60
18	16	0.75–0.80
19	6	0.25–0.30
20	5	0.20–0.25

Births are assigned random numbers in the intervals given by the assign ordering. In the example above, the first birth on the second occasion is assigned to the 9th interval and receives a random number between 0.40 and 0.45, the next birth is assigned to the 15th interval, and so on. Each birth is sequentially assigned to the intervals in the assign ordering. This assignment is independent of the sampling process; any birth in the sampling interval for that occasion is included in the sample.

A disadvantage of this method may be a slight variability of the sampling fraction from one sampling occasion to the next because the sample selection numbers are not equally spaced. The principal advantage of modified collocated sampling over rotation group sampling is realized when more than a single repeated survey is based on the same list frame. Interval selection methods like repeated and modified collocated sampling simplify the coordination of sample selection across surveys (see Chapter 9).

10.1.4 Removal of Deaths

A simple unbiased procedure to remove population deaths (businesses that cease operation) from the sampled and nonsampled parts of the population is to identify them as deaths through an external source that is independent of the survey. Once they are identified as dead units by this external source, the units are dropped from the sampling frame. This method is applicable to the selection and rotation procedures described above. The removal of deaths may cause an imbalance in the number of units in a rotation group in rotation group sampling or in the sampled intervals in the method described in Sections 10.1.2 and 10.1.3.

Deaths identified from the survey itself are treated differently, however. Such dead units remain in the frame. When they are sampled, their survey values are all zero. For a more detailed discussion on the removal of deaths, see Srinath (1987) and Hidiroglou and Srinath (1993).

10.1.5 Changes in Classification

Because of the dynamic nature of the population, changes over time in classification variables such as industry, geography, or size are common in repeated business surveys (Finkner and Nisselson 1978). These changes are observed more often in the sampled units than in the nonsampled portion of the population. Consequently, changes in the frame can be implemented if they are identified by a source that is independent of the sample survey. As we have suggested already, a simple way to implement such changes is to treat the units coming into a stratum as births and to treat the units going out of a stratum as deaths.

This should not cause any problems in rotation group sampling, though the possibility exists of creating an imbalance in the number of units per rotation group. In the procedure described in Sections 10.1.2 and 10.1.3, this method of treating classification changes results in the assignment of new random numbers for the old units. If it is desired to retain the old numbers for sample coordination purposes (see Chapter 9), then the units can be shifted to new strata with their old numbers assuming that all changes in the strata are identified through an external source.

10.1.6 Sequential Sampling

Given the sample selection numbers of Sections 10.1.2 and 10.1.3, one can come closer to the desired sampling fraction by replacing interval sampling with sequential sampling (Chapter 9). Rather than sampling all units with sample selection numbers within an interval, one could sample the first $n \approx fN$ units to the right (or left) of a designated number on the unit interval.

Not only does sequential sampling provide closer to ideal sample sizes, it also removes the need for sample selection numbers to be nearly equally spaced on the unit interval (this is the distinction between the sample selection numbers in this chapter and the permanent random numbers in Chapter 9). The problem with sequential sampling, and the reason it is not used by Statistics Canada, is that it is not conducive to the establishment of strict time-out-of-sample rules for previously sampled units. With interval sampling, we can avoid the selection of certain intervals and thus the units within them. We have no such mechanism with sequential sampling.

10.2 SIMULATION RESULTS

A short scale simulation was carried out to study the long term behavior of (1) the rotation group sampling method, (2) the modified collocated sampling, and (3) Bernoulli sampling (see Chapter 9). Repeated collocated sampling was not included in the study, because it was thought that the results would be similar to modified collocated sampling. The realized sampling fraction, defined as the number of units in the sample divided by the number of units in the population, was compared with the desired sampling fraction to determine the variability of the sampling fractions over time. Also, the realized sampling fraction for births was compared with the expected fraction to determine whether new births were properly represented in each period.

Three population sizes of 50, 500, and 1000 were used. Two sampling rates, 0.10 and 0.20, were considered for a population of 50 while a sampling rate of 0.05 was used for populations of size 500 and 1000. After the initial sample selection, it was assumed that the number of births and deaths on each occasion follow a Poisson distribution with means of 20 and 15, respectively, for populations of size 50 and 500 and means of 200 and 185 for population of size 1000. The deaths are removed from the populations only if they are identified as such by an external source. Though deaths occur each month, the probability of identifying a death through an external source was assumed to be 0.70.

The overall sampling fractions and the sampling fractions just for births were studied assuming a sample rotation for over 50 occasions. The results of the simulation with respect to maintaining a constant sampling fraction are shown in Figures 10.1a to 10.4b. Figures 10.1a, 10.2a, 10.3a, and 10.4a show the overall sampling fraction, whereas Figures 10.1b, 10.2b, 10.3b, and 10.4b show the realized sampling fractions just for births. The solid lines on either side of the reference line at the desired sampling fraction f in each graph represent deviations from f by $\pm 1/N$. The sampling fractions represented by these lines lead to sample sizes within $+$ or -1 unit of the desired sample size. Sampling procedures are judged on the basis of how often the realized sampling fractions fall within these bounds. The means assumed for the occurrence of births and deaths for each population are shown in the figures as E(birth) and E(death).

It is easily seen that Bernoulli sampling has the largest variation in sampling fractions between occasions for all three population sizes. Rotation group sampling and modified collocated sampling provide sampling fractions closer to the desired sampling fraction. For a population of size 50 with a sampling fraction of 0.10, the rotation group method produced a slightly lower sampling fraction than desired for later occasions whereas the modified collocated sampling provided sampling fractions both above and below the desired sampling fraction. This may be due to the fact that all the units in a rotation group rotate out or rotate in, whereas only a fraction of the units in an interval rotate in or rotate out, depending on the shift. With a sampling fraction of 0.20, rotation group sampling provided samples which resulted in sampling fractions closer to the desired sampling fraction. The rotation group method and modified collocated sampling give sampling fractions closer to the expected for populations of size 500 and 1000.

It can be seen from the figures that the sampling fractions for births vary widely for Bernoulli sampling. For larger populations, both the rotation group method and the modified collocated sampling do well in representing births in the sample.

10.3 CONCLUDING REMARKS

If maintaining a roughly constant sampling fraction within a sampling stratum and avoiding undue variation in sample size unrelated to population size are important concerns, then Bernoulli sampling without controls over the allocation of births is undesirable though it is an unbiased procedure for sampling. In addition to causing difficulties in managing data collection, the changes in sample size may result in varying levels of reliability for the estimates. Moreover, the realized sampling fraction for births may not be the same as that for the older units. As a result, there is a possibility of under- or overrepresenting births in the sample on any given occasion.

Rotation group and modified collocated sampling methods provide sampling fractions closer to the desired fractions, especially when the population sizes

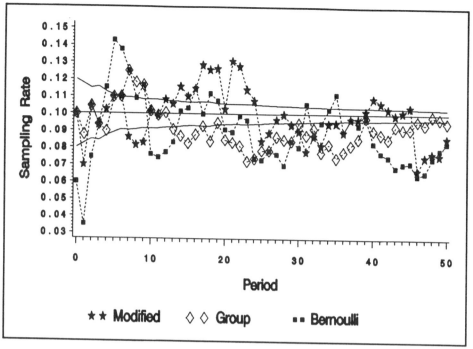

A

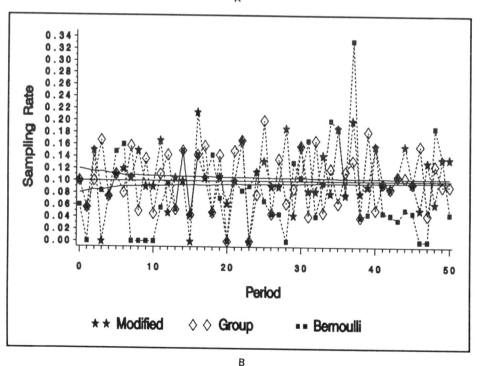

B

Figure 10.1 (A) Overall sampling fraction: Initial population = 50, sampling fraction = 0.10, E(birth) = 20, E(death) = 15, probability {death removed} = 0.70. (B) New births only: Initial population = 50, sampling fraction = 0.10, E(birth) = 20, E(death) = 15, probability {death removed} = 0.70.

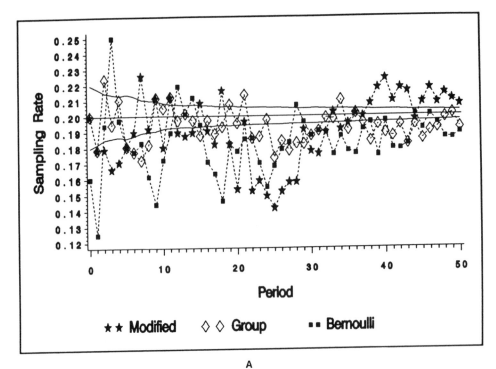

A

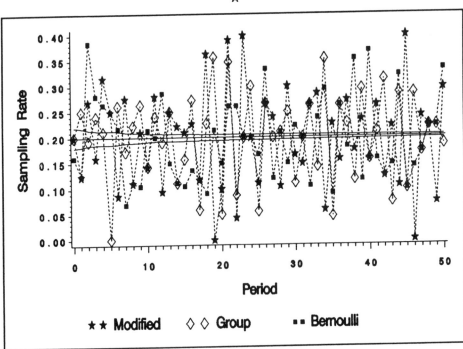

B

Figure 10.2 (**A**) Overall sampling fraction: Initial population = 50, sampling fraction = 0.20, E(birth) = 20, E(death) = 15, probability {death removed} = 0.70. (**B**) New births only: Initial population = 50, sampling fraction = 0.20, E(birth) = 20, E(death) = 15, probability {death removed} = 0.70.

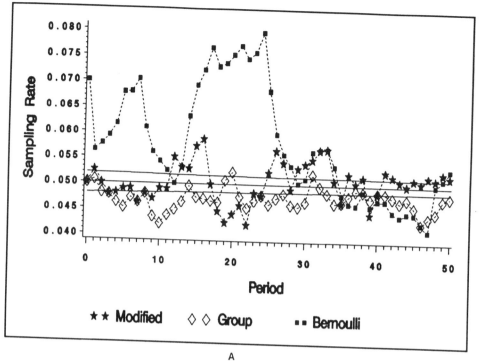

A

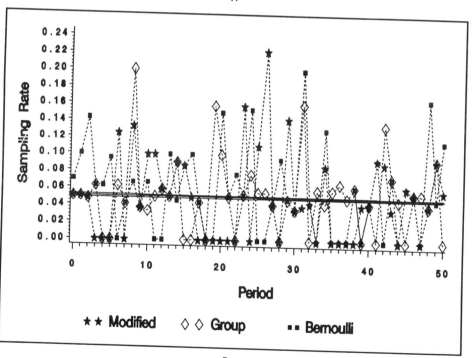

B

Figure 10.3 **(A)** Overall sampling fraction: Initial population = 500, sampling fraction = 0.05, E(birth) = 20, E(death) = 15, probability {death removed} = 0.70. **(B)** New births only: Initial population = 500, sampling fraction = 0.05, E(birth) = 20, E(death) = 15, probability {death removed} = 0.70.

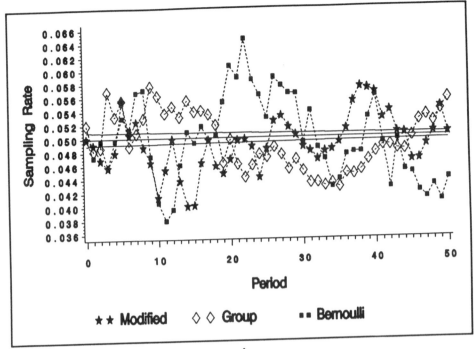

A

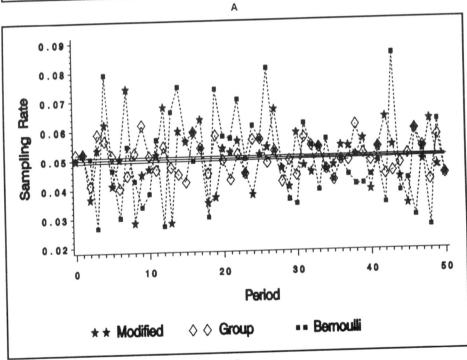

B

Figure 10.4 **(A)** Overall sampling fraction: Initial population = 1000, sampling fraction = 0.05, E(birth) = 200, E(death) = 185, probability {death removed} = 0.70. **(B)** New births only: Initial population = 1000, sampling fraction = 0.05, E(birth) = 200, E(death) = 185, probability {death removed} = 0.70.

are large. Both procedures do not attempt to represent precisely the proportions of the births in the sample because births are spaced by allocation panels or subintervals. One minor disadvantage of the rotation group method is the difficulty of changing the sampling fraction from one sampling occasion to the next, because it requires the creation of additional rotation groups or dropping some rotation groups. The number of rotation groups is directly linked to the sampling fraction and the number of occasions a unit is expected to be in the sample. With repeated collocated sampling, births are sampled independently on each occasion, and therefore better representation for births is provided. But because spacing of the units on the first occasion is directly dependent on the number of units for that occasion, births on subsequent occasions may interfere with obtaining the desired sampling fraction if sampling fractions are later changed.

Modified collocated sampling has many of the desirable attributes of rotation group sampling and repeated collocated sampling, including a mechanism for ensuring that units with small selection probabilities are not sampled too often by sheer random chance. Unlike rotation groups sampling, however, this sampling scheme has no trouble accommodating a change in the desired sampling fraction from one sampling occasion to the next. Moreover, it is conducive to the coordination of several samples sharing the same sampling frame.

REFERENCES

Brewer, K. R. W., L. J. Early, and S. F. Joyce (1972), "Selecting Several Samples from a Single Population," *Australian Journal of Statistics*, **14**, pp. 231–239.

Finkner, A. L., and H. Nisselson (1978), "Some Statistical Problems Associated with Continuing Cross-Sectional Surveys," in *Survey Sampling and Measurements*, N. K. Namboodri (ed.), New York: Academic Press, pp. 45–68.

Hidiroglou, M. A., G. H. Choudhry, and P. Lavallée (1991), "A Sampling and Estimation Methodology for Sub-Annual Business Surveys," *Survey Methodology*, **17**, pp. 195–210.

Hidiroglou, M. A., and K. P. Srinath (1993), "Problems Associated with Designing Sub-Annual Business Surveys," *Journal of Business and Economics Statistics*, **11**, pp. 397–406.

Srinath, K. P. (1987), "Methodological Problems in Designing Continuous Business Surveys: Some Canadian Experiences," *Journal of Official Statistics*, **3**, pp. 283–288.

Tambay, J. L. (1988), Collocated Sampling, Unpublished Manuscript, Ottawa: Statistics Canada.

.

CHAPTER ELEVEN

Multiple-Frame Business Surveys

Phillip S. Kott and Frederic A. Vogel
U.S. *National Agricultural Statistics Service*

Business surveys are usually designed to provide estimates of characteristics such as total sales, expenditures, number of workers, and inventories. Business populations, whether of farms, retail stores, factories, buildings, schools, or governments, often possess common characteristics that affect the choice of sampling frame and overall sample design. Among these characteristics are skewed distributions, diversity of variables of interest, and changing population membership. Basic principles from finite population sampling theory apply regardless of the specific approach used.

A population of units must first be defined (not always a trivial task in business surveys; see Chapters 2 and 3), and a sampling frame must be constructed from which a sample of units can be drawn. The general term *unit* is used to mean a standard statistical unit in the sense of Chapters 2 and 3. In many business surveys, but certainly not all (see Section 11.2.1), the population of units is identical to the population of businesses of interest, where the latter may be composed of enterprises, establishments, or some other business component. Every unit in the population must have a known probability of selection. Ideally, the frame should also be complete; that is, the selection probability of every unit in the target population should be positive.

Because business populations are skewed, efficient sample designs demand the use of list frames that incorporate known or projected measures of size for each population unit (see Chapter 8). Unfortunately, when the variables of interest are diverse and weakly correlated, a single list frame with one measure of size is not sufficient. Moreover, the rapidly changing nature of the population causes any list frame or combination of list frames to become quickly outdated and therefore incomplete in its coverage of the target population.

One way to ensure the completeness of the frame is to use an area frame

Business Survey Methods, Edited by Cox, Binder, Chinnappa, Christianson, Colledge, Kott.
ISBN 0-471-59852-6 © 1995 John Wiley & Sons, Inc.

covering the entire target population so that all population units and all potential future units are contained somewhere within the area frame. Whereas a list frame is a list of the units in the population, an area frame is a collection of geographical areas. In list frame designs for business surveys, the sampling units are usually the businesses themselves, and stratified single-stage sample designs are commonly used to select units directly from the list. In area frame designs, area segments are the sampling units, and these units are often selected using stratified multistage designs. Correspondence rules are needed to link the businesses in the population to the area segments in the frame.

Although area frames ensure complete coverage, they do not generally lead to efficient sampling designs because area segments are essentially clusters of units. Segment sizes (in terms of numbers of constituent units) are usually unequal and unknown at the time of the sample design. In fact, some area segments may contain no businesses at all. Large area sample sizes are needed to overcome problems with populations containing rare variables and with skewed distributions.

Area frames are most useful for general-purpose surveys covering a wide spectrum of items that are fairly evenly distributed geographically or when the sizes (as defined above) of the area segments are available or can be estimated reasonably well. That is why they are used so often in social surveys where population censuses provide adequate measures of the size of an area segment. In addition, an area frame has a long life span. It only needs to be updated when geographical features have changed to the point that it becomes difficult to associate population units with sampled area segments or when updated size information allows the development of improved sampling designs. This occurs often in agricultural surveys with the use of recent aerial photographs and/or satellite data.

Area frame business surveys are generally more expensive than list frame surveys of comparable sample size. This is because area frames are much more costly to develop. Because name and address data are unavailable, sampled businesses from an area frame also have to be personally enumerated, while list frame units can be enumerated less expensively by mail, telephone, or electronic methods. On a per-business basis, then, area frame samples are more expensive. On a per-unit-of-precision basis, area frame samples can be much more expensive when the businesses being surveyed vary in size or are sparsely distributed geographically. In such situations, area frame samples must include more units than the list frame in order to achieve the same precision.

A *multiple frame* survey uses a combination of frames. The primary reason for using multiple frame sampling for business surveys is to offset the weaknesses of each frame with the strengths of another frame. In principle, the theory of multiple frame sampling can be applied to the use of more than one list frame (Bankier 1986). The main focus in this chapter, however, will be on combining list and area sampling frames. Area frame sampling ensures completeness but at greater cost per completed interview, while list frame sampling is less complete but also less costly and more effective for targeting large and/or rare items.

11.1 FUNDAMENTALS OF MULTIPLE FRAME SAMPLING

The list and area frames may be thought of as a set of mutually exclusive primary sampling units. The units in the target population can be mapped to the primary sampling units of a particular frame in a one-to-one or a many-to-one fashion. The mapping need not be complete for each frame; some population units may not be associated with any sampling unit of a particular frame. For simplicity, we say a population unit belongs to a frame when it maps onto a sampling unit in the frame. After one or more stages of random sampling, a sample of population units is drawn independently from each frame. Hereafter in this chapter, the unmodified term "unit" will refer to the population unit.

Two important assumptions are made in this section. They are:

- *Completeness*: Every unit in the target population belongs to at least one frame.
- *Identifiability*: For any sampled unit from any frame, it is possible to determine whether the unit belongs to any other frame.

The completeness assumption is usually satisfied when an area sampling frame is one of the multiple frames. Populations such as farms, for instance, are naturally associated with the land units comprising the area frame. The identifiability assumption, while simple in theory, accounts for many operational difficulties in implementing a multiple frame survey. It is not trivial to ascertain whether a unit sampled from one frame is also contained in another frame (see Section 11.5). Matching and record linkage procedures such as those described by Winkler in Chapter 20 are used to make this determination.

Hartley (1962) developed the basic theory of multiple frame sampling, which was later extended by Cochran (1965). See also Lund (1968) and Hartley (1974). Hartley divided the population into mutually exclusive domains defined by the sampling frames and their intersections. For example, two sampling frames, A and B, would form three possible domains:

- Domain *a* containing units belonging only to Frame A,
- Domain *b* containing units belonging only to Frame B, and
- Domain *ab* containing units belonging to both Frames A and B.

With k frames, there will be $2^k - 1$ domains as the completeness assumption precludes the existence of a domain without any members from at least one frame.

Let us focus attention on the two-frame example to clarify the issues involved in multiple frame estimation. Suppose we are interested in estimating a population total Y. It is possible to decompose Y as

$$Y = Y_a + Y_b + Y_{ab}, \tag{11.1}$$

where Y_d is the population total for domain d ($d = a$, b, or ab). When $d = a$ or b, Y_d can be estimated using the Horvitz–Thompson (1952) estimator

$$\hat{Y}_d = \sum_{i \in S_d} w_i y_i, \tag{11.2}$$

where S_d is the set of sampled units in domain d, w_i is the sampling weight or expansion factor (the inverse of the selection probability), and y_i is the item value for unit i. This estimator is unbiased under the sampling design. The overlap domain ab total can be estimated using either the Frame A or Frame B sample or a combination of the two samples. A continuum of unbiased estimators for the domain ab total Y_{ab} is given by

$$
\begin{aligned}
\hat{Y}_{ab(p)} &= p \sum_{i \in S_{ab}^A} w_i y_i + (1 - p) \sum_{i \in S_{ab}^B} w_i y_i \\
&= p\hat{Y}_{ab}^A + (1 - p)\hat{Y}_{ab}^B,
\end{aligned}
\tag{11.3}
$$

where $0 \le p \le 1$, S_{ab}^D is the set of units in domain ab sampled from Frame D ($D = A$ or B), and $\hat{Y}_{ab}^D$ is the Horvitz–Thompson estimator for Y_{ab} based on the Frame D sample. The limits on p ensure that $\hat{Y}_{ab(p)}$ will be nonnegative whenever the y_i are nonnegative.

Based upon this continuum of unbiased estimators for domain ab, we now have a continuum of unbiased estimators for the overall total Y:

$$\hat{Y}_{(p)} = \hat{Y}_a + \hat{Y}_b + \hat{Y}_{ab(p)}, \tag{11.4}$$

where $0 \le p \le 1$. The sampling design is independent across frames but not necessarily across domains. Consequently, the variance of $\hat{Y}_p$ is

$$
\begin{aligned}
\text{Var}\,(\hat{Y}_{(p)}) = {}& \text{Var}\,(\hat{Y}_a) + \text{Var}\,(\hat{Y}_b) + 2p\,\text{Cov}\,(\hat{Y}_a, \hat{Y}_{ab}^A) \\
& + 2(1 - p)\,\text{Cov}\,(\hat{Y}_b, \hat{Y}_{ab}^B) \\
& + p^2\,\text{Var}\,(\hat{Y}_{ab}^A) + (1 - p)^2\,\text{Var}\,(\hat{Y}_{ab}^B).
\end{aligned}
\tag{11.5}
$$

A value of p can be chosen that minimizes the variance of $\hat{Y}_{(p)}$. The optimal (i.e., variance minimizing) p is

$$p = \frac{\text{Var}\,(\hat{Y}_{ab}^B) - \text{Cov}\,(\hat{Y}_a, \hat{Y}_{ab}^A) + \text{Cov}\,(\hat{Y}_b, \hat{Y}_{ab}^B)}{\text{Var}\,(\hat{Y}_{ab}^B) + \text{Var}\,(\hat{Y}_{ab}^A)} \tag{11.6}$$

If the two covariance terms in equation (11.6) are zero (or equal to each other), p takes on the form:

$$p = \frac{\text{Var}\,(\hat{Y}^B_{ab})}{\text{Var}\,(\hat{Y}^B_{ab})\,+\,\text{Var}\,(\hat{Y}^A_{ab})} \tag{11.7}$$

which always lies between zero and one. As one would expect, the optimal value for p is directly related to the precision of $\hat{Y}^A_{ab}$ relative to that of $\hat{Y}^B_{ab}$. The more relatively precise the Frame A sample is in estimating domain ab, the more weight $\hat{Y}^A_{ab}$ is given in estimating Y_{ab}.

The weights w_i in the above expressions can be either unconditional or conditional. That is to say, they may reflect either the original probabilities of selection or the recomputed selection probabilities within the domains. For example, if the Frame A sample design were simple random sampling without replacement, then the unconditional weight for every sampled frame unit would be N_A/n_A, where N_A and n_A are the respective population and sample sizes for Frame A, while the conditional expansion factors for the sampled units in the intersection of domain d and Frame A would be $N_{A(d)}/n_{A(d)}$, where $N_{A(d)}$ and $n_{A(d)}$ are the respective population and sample sizes in this intersection. For theoretical reasons, estimation with conditional rather than unconditional expansion factors is preferred (Rao 1985). In many applications, however, especially those involving an area frame (where population sizes are often unknown), conditional selection probabilities are either impossible to calculate or impractical. Consequently, unconditional inference has to suffice.

Although one can, in principle, choose p so that the variance (conditional or unconditional) is minimized, this is usually impossible in practice because the component variance and covariance terms in equation (11.5) are unknown. They could be estimated from the sample, but then the choice of p would not really minimize the variance of $\hat{Y}_{(p)}$ but the *estimated* variance of $\hat{Y}_{(p)}$. As a consequence, this estimated variance would be biased downward. Even if the distinction between the variance-minimizing p and the estimated variance-minimizing p could be ignored, the following inconvenience remains: The optimal p could vary from question to question. Armstrong (1979) demonstrates this point using Canadian farm data.

A popular alternative to minimizing variance is to eschew issues of optimality or near-optimality and to fix the value of p in advance, usually at zero or one. One can then estimate the components on the right-hand side of equation (11.5) in an unbiased fashion and create an unbiased estimator for the variance of $\hat{Y}_{(p)}$. When p is preset to zero or one, the overlap domain ab total is only estimated from one frame sample.

A more direct Horvitz–Thompson estimator for Y in equation (11.1) could be developing by treating the samples from the two frames as a single sample and then computing the probability of selection for each sampled unit (the sum of its probability of selection from each frame minus the product of these two terms). The variance of the resultant estimator may be little different from the variance of $\hat{Y}_{(p)}$ when a near-optimal p is used. Moreover, estimating the variance of this alternative is often quite difficult.

One can sometimes improve on $\hat{Y}_{(p)}$ by using auxiliary information. The interested reader is directed to Fuller and Burnmeister (1972), Bosecker and Ford (1976), Bankier (1986), Skinner (1991), and Rao and Skinner (1993).

11.2 THE DOMINANT SPECIAL CASE: ONE LIST FRAME, ONE AREA FRAME

The U.S. Bureau of the Census' 1949 Sample Survey of Retail Stores (Hansen et al. 1953) was one of the first business surveys to use a multiple-frame technique. It employed a single list frame and a single area frame. This two-frame approach is still used in the Bureau's Monthly Retail Trade Survey but the feasibility of dropping the area frame is being investigated (Konschnik et al. 1991). It is fair to say that the multiple-frame approach currently finds its widest use in agricultural surveys. See Vogel (1975) and Julien and Maranda (1990) for discussions of U.S. and Canadian multiple-frame agricultural surveys.

11.2.1 Area Sampling

In Chapter 8, Sigman and Monsour review list sampling procedures for business surveys. The basic concepts for area frame sampling are almost as simple as those of list frame sampling. The total area to be surveyed is first stratified by geography or other known characteristics that relate to the variables of interest (population density for a retail survey, farm density for an agriculture survey). Segments—usually compact blocks of land—are then selected within each stratum using a multistage sampling design.

Gallego and Delincé (1993) discuss sampling designs where points are randomly selected within sampled area segments. In this section, however, we focus on designs for which the last stage of sampling is the selection of area segments. All population units linked to a sampled segment are included in the sample. The area expansion factor or weight for each unit is simply the inverse of the selection probability of the segment with which it is linked. When employing such designs, rules need to be developed to uniquely link (map) the units in the population with the sampled segments.

Although the basic concepts are simple, the successful application of area frame sampling can become complex, especially if used in a multiple-frame environment. The theories and applications of area frame sampling are well documented and included Jessen (1942), King and Jessen (1945), Houseman (1975), Nealon and Cotter (1987), and Bush and House (1993). Fecso et al. (1986) provide references covering area frame sample design issues.

The link between population units and sample segments is complicated in those area frame surveys (e.g., agricultural surveys) where the physical location of businesses (farms) may cross segment boundaries and hence the business (farm) could have multiple opportunities for selection depending upon the sampling and reporting rules adopted. There are three approaches frequently used in this situation.

The *open approach* eliminates multiple selection opportunities by linking each business operation to one segment (usually the location of the headquarters) and then including the operation in the sample only when that segment is selected. Large businesses with complicated corporate or partnership structures can be difficult to link to just one segment in an unambiguous fashion. For example, many medical practices have several doctors with offices in more than one location; identifying the headquarters may not be simple. Complex counting rules must be devised that ensure that such businesses do not have a chance of being linked to more than one area segment.

In the *closed* or *tract approach*, the enumeration unit is defined as the intersection of a business and an area segment (i.e., the portion of the business that lies within the area segment). That is, the business is conceptually divided into sections corresponding to the area segments with which its plants or operations are associated. The business reports for only those portions associated with sampled segments. Response errors occur when the business is not able to measure and report accurately on that portion of its operation that lies within the sample segment's boundaries. The closed approach is statistically robust and relatively efficient for items that are distributed evenly over wide areas, such as crop acreages. For many items, outliers can be controlled by the size of the segment. For example, if the segment size is 600 acres, the maximum acres for any crop in the segment is 600 acres even though a single farm located in the segment may contain many times that number of acres in its entire operation.

Using the open approach, sampled segments will capture fewer units with large item values than with the closed approach. As a result, the open approach is often statistically less efficient than the closed approach. A potential advantage of the open approach, however, is that the respondent may find it easier to report values for the entire business rather than for the portion located within the sample segment. Segment boundaries have little intrinsic meaning for respondents.

The open and closed approaches are very different, but both are designed to eliminate multiple reporting. The third method, the *weighted approach*, allows multiple reporting but weights the data to adjust for this multiple reporting. The definition of the enumeration unit in the weighted approach is the same as in the closed approach: the portion of a business that lies within an area segment. The difference is in the data values attributed to the unit. Under the weighted approach, the operations of the entire business are reported whenever a segment containing one of its operations is selected. The data values for businesses that are the ''parent'' of several units (each in a different area segment) are prorated using *multiplicity adjustment factors* that sum to unity. For agricultural surveys, it is common to use the fraction of the farm's total acreage within a sampled segment as the adjustment factor.

The weighted approach leads to less sampling variability than the open approach because large operations are spread across many segments. A major problem with this approach, especially when applied to agricultural surveys, is that the adjustment factors themselves are prone to measurement errors (Nealon 1984). In principle, the weights for all the enumeration units with the same parent business should sum to unity. In practice, since only one unit is likely to be selected in an area sample, pains must be taken to ensure that the factors used to prorate business values to sampled units are not systematically larger or smaller than they should be.

Nealon (1984) provides a theoretical and empirical examination of these three approaches for area frame and multiple frame surveys conducted by the U.S. National Agricultural Statistics Service (NASS). The results favor the weighted approach. Houseman (1975) provides a more in-depth discussion of the three approaches. Faulkenberry and Garoui (1991) explore additional approaches.

11.2.2 Estimation

Suppose one list frame and one area frame are to be used, and a total is to be estimated. Using the notation of Section 11.1, let Frame A denote the area frame and let Frame B denote the list frame. Observe that domain b (units found only on Frame B) is empty because all units are associated with the area frame. Consequently, $Y_b = \hat{Y}_b = 0$.

In almost all applications of the two-frame design, the list frame sample is used exclusively to represent list frame members (domain ab) whereas area frame selections are only included in estimation when they are not found on the list frame (i.e., they belong to domain a). In this situation, p in equations (11.3) and (11.4) is set equal to zero, rendering $\hat{Y} = \hat{Y}_a + \hat{Y}_{ab}^B$. This value for p, besides simplifying variance estimation, is often close to optimal (depending on the efficiency of the list frame stratification) because the variance of the domain ab list frame estimator $\hat{Y}_{ab}^B$ is almost always considerably smaller than the variance of the area frame estimator $\hat{Y}_{ab}^A$. Consequently, equation (11.7) implies that p should be very small. Depending on the content of the list frame, Cov $(\hat{Y}_a, \hat{Y}_{ab}^A)$ in equation (11.6) may actually be positive because $\hat{Y}_a$ and $\hat{Y}_{ab}^A$ are based on the same sampled segments. This would make the optimal p even smaller. [Note that Cov $(\hat{Y}_b, \hat{Y}_{ab}^B)$ in equation (11.6) is zero because $\hat{Y}_b$ is zero.]

A little renaming to simplify the notation: let $\hat{Y}_L = \hat{Y}_{ab(0)} = \hat{Y}_{ab}^B$, because it is the Horvitz–Thompson expansion from the list frame L of domain ab (the overlap domain); and let $\hat{Y}_N = \hat{Y}_a$, because it is the area expansion for the nonoverlap N domain (using either the closed, open, or weighted approach to unit linkage).

The estimator for the total $\hat{Y} = \hat{Y}_L + \hat{Y}_N$ is called the *screening multiple frame estimator* because units in the area sample are "screened" and only those units in the nonoverlap domain are enumerated. Its variance has the obvious form:

$$\text{Var } (\hat{Y}) = \text{Var } (\hat{Y}_L) + \text{Var } (\hat{Y}_N). \tag{11.8}$$

Estimating Var $(\hat{Y}_L)$ is straightforward when, as is usually the case, $\hat{Y}_L$ is based on a single-stage stratified list sample.

Estimating Var $(\hat{Y}_N)$, however, is more complicated. Recall that segments are selected from the area frame. In many cases, the design is equivalent to stratified simple random sampling (see Kott 1989). In practice, within-strata sampling fractions are so low that the distinction between with and without replacement sampling can be ignored. Suppose there are H strata and n_h sampled segments within stratum h. If X_{hj} is the sum of the expanded values of all nonoverlap units in segment j of stratum h (i.e., $X_{hj} = w_h \Sigma y_i$, where w_h is the weight for all sampled units in the stratum h, y_i is the value of unit i, and the summation is over all nonoverlap units in segment j), then

$$\text{Var}\,(\hat{Y}_N) = \sum_{h=1}^{H} \frac{n_h}{n_h - 1} \left[\sum_{j=1}^{n_h} X_{hj}^2 - \frac{\left(\sum_{j=1}^{n_h} X_{hj}\right)^2}{n_h} \right] \tag{11.9}$$

is an unbiased estimator for Var $(\hat{Y}_N)$ whenever finite population corrections can be ignored. Although equation (11.9) appears straightforward, note that X_{hj} will be zero for many sampled segments either because the segments have no linked units or their linked units are found on the list frame.

11.2.3 Estimating Means, Ratios, and Regression Coefficients

The area sample is composed of segments, whereas the population is composed of businesses. Although this difference poses no problems when the goal is estimating population totals, it can interfere when estimating population ratios (including population means) and regression coefficients.

Estimating population ratios and regression coefficients with a multiple-frame sample is a relatively trivial matter when the open approach to unit linkage in the area frame is used with $p = 0$. When a business selected from the area frame is also on the list frame, the business' area frame data values are ignored in both the numerator and denominator of the ratio estimator. An analogous technique is used for estimating population regression coefficients. As in equation (11.9), however, estimates of variance need to reflect the existence of those sampled segments that do not contribute to the estimation of the parameter itself.

When the weighted or closed approach is used, estimating ratios is again straightforward because either approach can be used to estimate the appropriate total in both the numerator and the denominator. It is unclear, however, how one can estimate population regression coefficients with the closed approach. The weighted approach is discussed below.

The vector of finite population regression coefficients can be written as

$$\mathbf{B} = \left(\sum_{k \in U} \mathbf{x}'_k \mathbf{x}_k \right)^{-1} \sum_{k \in U} \mathbf{x}'_k y_k \qquad (11.10)$$

where k denotes a business, U denotes the target population of businesses, y_k is a value attached to business k, and $\mathbf{x}_k$ is a row vector of values attached to business k. Without loss of generality, we assume that the units of the list frame sample are businesses. In what follows, a business is said to be associated with a sampled unit when it is sampled from the list frame or when it is the parent of a sampled unit from the area frame.

A reasonable estimator for $\mathbf{B}$ that uses the weighted approach to unit linkage is

$$\hat{\mathbf{B}} = \left(\sum_{k \in S} w_k a_k \delta_k \mathbf{x}'_k \mathbf{x}_k \right)^{-1} \sum_{k \in S} w_k a_k \delta_k \mathbf{x}'_k y_k, \qquad (11.11)$$

where S is set of businesses associated with sampled units, w_k is the expansion factor for the unit associated with k, a_k is the multiplicity adjustment factor attributed to the unit associated with k if it is sampled from the area frame (if not, $a_k = 1$), and δ_k is 0 if k is associated with a unit sampled from the area frame that is also on the list frame (if not $\delta_k = 1$).

Two observations about equation (11.11) are in order. First, the set S may include particular businesses more than once due either to list/area overlap or to a particular business being in two sampled area segments. The δ_k and a_k terms assure us that neither the y values nor the $\mathbf{x}$ values for such businesses are double-counted. Second, if we call $g_k = w_k a_k \delta_k$ the multiplicity-adjusted expansion factor, then $\hat{\mathbf{B}}$ can be expressed as

$$\hat{\mathbf{B}} = \left(\sum_{k \in S} g_k \mathbf{x}'_k \mathbf{x}_k \right)^{-1} \sum_{k \in S} g_k \mathbf{x}'_k y_k.$$

In this form, a survey analysis software package such as SUDAAN (Shah et al. 1991) can be used to calculate $\hat{\mathbf{B}}$ and estimate its variance.

11.3 SUBSAMPLING FROM AN AREA FRAME

In this section, we explore two methods of subsampling from an area frame sample. What distinguishes these methods from simple multistage area sampling is that each incorporates the development of a list frame for subsampling. The first method starts with an area sample of primary sampling units (PSUs) and then uses a multiple-frame estimation strategy (list and area) within each PSU. The second method begins with an area sample of units using the weighted method of unit linkage and then draws a subsample of these units. Because subsampling need not be done independently within the original sampling units, this is a two-phase sampling design (see Särndal et al. 1991).

11.3.1 Using Multiple Frame Designs Within Area PSUs

In many cases the cost of developing a list frame is prohibitively high. On the other hand, the variance of an estimator based on a pure area frame sample is prone to be unacceptably large. This was the situation faced by the designers of the U.S. Commercial Buildings Energy Consumption Survey (CBECS) which measures the consumption of energy in commercial buildings (U.S. Energy Information Administration 1989). Their solution was to use a multistage area sample and to create list frames of large commercial buildings within sampled PSUs. This approach reduced the potential size of the expansion factors for large buildings.

Let j denote a PSU, and let Y_j denote the first-stage expanded values of interest from all units in the PSU. A list and area multiple-frame estimation strategy can be constructed to estimate each Y_j. Because the units in CBECS were buildings that never crossed segment boundaries, linking units to the area segments was straightforward. The estimator developed for the CBECS survey did not estimate each Y_j with a member of the family in equation (11.4). Instead, the CBECS estimator has a form equivalent to

$$\hat{Y}_j' = \hat{Y}_j + w_{1j}\left(\sum_{k \in P_j'} y_k - \sum_{k \in S_j'} w_{Lk}y_k\right), \qquad (11.12)$$

where $\hat{Y}_j$ is the screening estimator ($\hat{Y}_{jL} + \hat{Y}_{jN}$) for Y_j, w_{1j} is the first-stage (PSU) expansion factor, P_j' is the set of all units subsampled from the area frame that are also on the list frame in PSU j, S_j' is the set of all units subsampled on both the list and area frames of PSU j, and w_{Lk} is the inverse of the probability that unit k is subsampled from the list frame in j. It is not difficult to show that $\hat{Y}_j'$ is unbiased (Chu 1987).

Let $d_j = \hat{Y}_j' - \hat{Y}_j$. The variance of $\hat{Y}_j'$ conditioned on the first-stage selection of PSU j is

$$\text{Var}_2\,(\hat{Y}_j') = \text{Var}_2\,(\hat{Y}_j) + \text{Var}_2\,(d_j) + 2\,\text{Cov}_2\,(\hat{Y}_j, d_j), \qquad (11.13)$$

where the subscript 2 denotes the conditional nature of the variance (covariance) operator. Because the covariance term in equation (11.13) is likely to be negative, it is possible for $\hat{Y}_j'$ to have less variance than $\hat{Y}_j$.

Now consider unbiased estimators having the general form

$$\hat{Y}_{jc} = \hat{Y}_j + cd_j,$$

where c is a constant. Observe that $\hat{Y}_{jc} = \hat{Y}_j$ when $c = 0$, and $\hat{Y}_{jc} = \hat{Y}_j'$ when $c = 1$. The value of c may be chosen so that the variance of $\hat{Y}_{jc}$ is minimized, although that may not be easy to do in practice.

Suppose there are H strata in the original area frame and n_h sampled PSUs within stratum h. A variance estimator for $\hat{Y}_c = \Sigma_h \Sigma_j \hat{Y}_{jc}$ is

$$\widehat{\text{Var}}\,(\hat{Y}_c) = \sum_{h=1}^{H} \frac{n_h}{n_h - 1} \left[\sum_{j=1}^{n_h} \hat{Y}_{jc}^2 - \frac{\left(\sum_{j=1}^{n_h} \hat{Y}_{jc} \right)^2}{n_h^2} \right]. \qquad (11.14)$$

This estimator is unbiased when the first-stage sample is drawn with replacement and is in many practical applications nearly unbiased when PSUs are selected without replacement. Goldberg and Gargiullo (1988) discuss variance estimation for surveys (like the CBECS) in which some PSUs are selected with certainty.

11.3.2 Using Lists Across Area Segments

Suppose segments have been previously selected from an area frame for a particular multiple-frame (list and area) survey and now a new, smaller survey must be implemented to estimate different population parameters. This happens frequently in agricultural surveys where different commodities are of interest at different times of the year and particular commodities change over time. Suppose further that the same list frame will be employed for the new survey so that the nonoverlap domain is effectively fixed. Although this supposition is unnecessary, it simplifies the exposition. To reduce costs for the new survey, the previous area frame sample can be subsampled for the new survey. Because list sampling is generally more efficient than area sampling, it makes sense to use list sampling techniques to draw the area subsample.

Area frame sampled units can be separated into two groups: those that are found on the list frame and those that are not. As discussed in Section 11.2.2, the values of overlap area frame units are ignored when estimating Y with the screening estimator, $\hat{Y} = \hat{Y}_L + \hat{Y}_N$. This means that these units are attributed values of zero. Because we know their values in advance of subsampling, we can think of any area frame subsample as containing all originally sampled overlap units. The only question then is how to subsample from the sampled nonoverlap area frame units.

Let w_{1i} be the original expansion factor of unit i from the area sample, and let a_i be the adjustment factor attributed to i in the weighted approach to unit linking. In the open and closed approach, a_i is unity. Let y_i be the value of interest for the business associated with i. Define $z_i = w_{1i}a_iy_i$. The goal of subsampling from the area frame sample is to estimate

$$\hat{Y}_N' = \sum_{i \in S_{NA}} z_i, \qquad (11.15)$$

where S_{NA} is the set of nonoverlap units in the original area sample. In the open approach, S_{NA} is restricted to units with headquarters in the original area sample; in the closed approach, S_{NA} is a set of enumeration units.

List sampling designs to estimate parameters like $\hat{Y}_N'$ in equation (11.15) in an unbiased and efficient manner are the object of Chapter 8. There is no requirement that the original area sampling design be reflected in the subsampling design used to estimate $\hat{Y}_N'$. The U.S. National Agricultural Statistics Service, for example, restratifies area-sampled nonoverlap units from its June Agriculture Survey for follow-on surveys in subsequent months. Those units believed to have similar z_i values are placed in the same subsampling stratum irrespective of their original area stratum or their original expansion factor (except so far as w_{1i} impacts on z_i), and a without-replacement simple random sample is drawn within each new stratum.

An attractive alternative to stratified simple random sampling is probability proportional to size (pps) sampling, which can also incorporate stratification. Because there are likely to be a number or values of interest in the new survey and because y_i values are unknown before enumeration anyway, one may choose to treat the $w_{1i}a_i$ as the measures of size for pps sampling. Alternatively, data from the original survey can be used to determine reasonable measures of size.

A Horvitz–Thompson estimator for $\hat{Y}_N'$, and thus Y_N, is

$$\hat{Y}_N = \sum_{i \in S_2} w_{2i} z_i = \sum_{i \in S_2} g_i y_i, \tag{11.16}$$

where w_{2i} is the inverse of unit i's probability of being subsampled given that it is in the original area sample, S_2 is the set of units subsampled from the nonoverlap domain of the area frame, and $g_i = w_{2i} w_{1i} a_i$. Note than when unstratified probability proportional to $w_{1i}a_i$ sampling is used for subsampling, g_i is identical for all subsampled units.

The variance of the two-phase estimator described above is the sum of the variance of $\hat{Y}_N'$ in equation (11.15) as an estimator of Y_N plus the expected value of the variance of $\hat{Y}_N$ in equation (11.16) as an estimator for $\hat{Y}_N'$ (Cochran 1977). An estimator for this variance is more difficult to express. Särndal and Swensson (1987) provide a general formulation.

Kott (1990) explores the important special case where the first phase is a stratified simple random sample of area segments and the second phase is a (re)stratified simple random sample of units. It turns out that if x_{hj} is the sum of the fully expanded values of all nonoverlap units in area segment j of stratum h and there are n_h sampled segments in stratum h ($h = 1, 2, \ldots, H$), then

$$\widehat{\text{Var}}\,(\hat{Y}_N) = \sum_{h=1}^{H} \frac{n_h}{n_h - 1} \left[\frac{\sum_{j=1}^{n_h} x_{hj}^2 - \left(\sum_{j=1}^{n_h} x_{hj} \right)^2}{n_h} \right], \tag{11.17}$$

is biased upward as an estimator for the variance of $\hat{Y}_N$.

It is well known that equation (11.17) is unbiased under two-stage sampling when the first-stage sampling fractions can be ignored. This is because the x_{hj}

are (virtually) independent. In the design discussed in Kott (1990), the x_{hj} tend
to be inversely correlated: When one segment has more units than expected—
and thus a larger x_{hj} value—another segment would tend to have fewer units
than expected. Such inverse correlations tend to depress the variance of $\hat{Y} =
\Sigma_{h=1}^{H} \Sigma_{j=1}^{nh} x_{hj}$ while increasing the value of the right-hand side of equation
(11.17). This informal reasoning suggests that equation (11.17) is likely to
provide a conservative variance estimator for two-phase estimation strategies
more complicated than the one formally treated in Kott (1990).

11.4 PROBLEMS IN SURVEYS WITH OVERLAPPING FRAMES

An area frame provides complete coverage of the population, but it can lead
to inefficient estimation when the target population has a skewed distribution.
List frames, by contrast, provide the means for more efficient estimation but
are often incomplete. When a list of businesses is used along with an area
frame in a multiple-frame environment, the list need not be complete. It should,
however, contain those large and rare businesses that are the bane of area frame
surveys.

11.4.1 List Frame Issues

All businesses on a list frame must be completely identified by name, address,
and so on. Operations that are large, have complex management structures,
and/or are scattered over different locations must be so identified. This is the
identifiability assumption described in Section 11.1. In a two-frame (list and
area) sampling design, one must be able to determine whether a business (or
portion of a business) sampled from the area frame could also have been selected from the list frame; otherwise, intractable nonsampling errors will result. It must be realized, however, that the identifiability requirement can
greatly increase the cost of list development.

  When developing a list frame of businesses, more than one source of business names is often available. The survey designer must decide whether to use
each list as a separate frame in a multiple frame survey or to combine the lists
prior to sampling and to create a single list frame for sampling purposes. Suppose the former approach is chosen with two list frames and a single area
frame. The population can then be divided into four mutually exclusive domains: (1) businesses on neither list frame, (2) businesses on the first list frame
but not the second, (3) businesses on the second list frame but not the first,
and (4) businesses on both list frames.

  One can see that the need to identify all domains when there are two or
more list frames greatly complicates the survey and estimation process. Many
statistical agencies have therefore decided that it is more practical to combine
all lists and remove duplication prior to sampling. This can be a significant

undertaking requiring the use of record linkage methodologies that may be prone to errors. Winkler discusses these methodologies in Chapter 20.

11.4.2 Overlap Detection Issues

Matching names on the list frame to sampled area frame units to determine overlap complicates the survey process. Whether a business is surveyed via the list or the area frame, it may be necessary to determine the primary business name, other business names, and the names of individuals who are associated with the business. Rules of association of individuals' names with businesses must be defined. For example, an area frame unit may contain the residence of Dr. X associated with the medical practice of Drs. X, Y, and Z. Should the association of the business with an area frame unit be based on the residence of each name or on the location of the practice itself? The answer will depend upon the counting rules used to determine the overlap between the frames. This situation becomes more complicated if there is a Dr. T involved in the practice, but the list frame only identifies the practice involving Drs. X, Y, and Z. If counting rules are used that would allow Dr. T to report for the entire practice of Drs. X, Y, and Z, then it must be possible to identify the practice as overlap with the list frame. This situation occurs often in agriculture when an individual associated with a partnership can report for the entire partnership, but the name matching process fails to link it with a name on the list frame, resulting in an upward bias. Vogel (1975) discusses these problems. The survey questionnaire for both frames must be carefully designed to identify and link names with businesses so that the overlap domain can be properly determined. Resources need to be available to reinterview "questionable links" so that the domain determination is correct. Several studies have shown the domain determination to be the single largest source of nonsampling errors in multiple-frame surveys (Nealon 1984).

11.4.3 Estimation Difficulties

A two-frame (list and area) sampling design generally yields more efficient and robust estimators than an area frame design by itself. Nevertheless, outliers can still occur when the list frame is not constructed carefully or is not up to date. A business missing from the list frame that is sampled in the nonoverlap domain of the area frame can have a much larger expansion factor than it would have had as a list selection. (Because the costs associated with an area frame are larger, its sampling fractions are usually smaller). Five such farms in a 1992 U.S. Department of Agriculture survey accounted for 6 percent of the national estimate. In Chapter 26, Lee describes the problems that such outliers pose for business surveys.

Although the desire to eliminate potential outliers provides a powerful argument for including as many business names as possible on the list frame, there are equally compelling reasons for limiting the list frame to only larger businesses. For instance, the smaller the list, the easier it is to check overlap. Moreover, the incremental costs of adding units to the nonoverlap domain are relatively small because (1) the fixed cost of developing an area frame has already been expanded, and (2) each unit linked to a sampled area segment must be contacted to determine its overlap status whether or not it eventually needs to be enumerated.

A multiple-frame design can make it more difficult to measure change over time. Ratio estimators are usually an efficient means of estimating change when a portion of the sample in the reference (denominator) time period remains in the sample in the comparison (numerator) period. Because businesses may move between the overlap and nonoverlap domains as their structures change or as the list frame is updated, the efficiency of ratio estimators of change in multiple-frame surveys is reduced.

Extreme caution must also be exercised in multiple-frame designs to ensure that businesses found through the area frame sample are not added to the list frame during the duration of the design, because such additions can bias estimation (by effectively changing selection probabilities). Such additions to the list frame should only be allowed when an entirely new sample is selected from both frames. The independence of sample selection between the area and list frames must be maintained for the area frame to estimate for the incompleteness of the list frame (Vogel and Rockwell 1977).

11.5 THE FUTURE OF MULTIPLE-FRAME SURVEYS

Multiple-frame surveys are heading in two opposite directions. The first is the elimination of the area frame altogether. Because of the cost of maintaining an area frame, Canada models the incompleteness of its list frame of retail and wholesale business establishments rather than measure list undercoverage with an area sample (Sande 1986). The U.S. Bureau of the Census may soon follow suit and abandon the area sample component of its Monthly Retail Trade Survey (Konschnik et al. 1991). This pattern could be reversed in Italy, however, where lists of businesses are viewed as unreliable. Petrucci and Pratesi (1993) discuss the possibility of introducing multiple-frame business surveys in that country.

The future use of area frames both by themselves and in a multiple frame environment is more secure for surveys of agriculture and construction. Here we find the other direction in which multiple-frame surveys are heading: that of more complicated estimators. NASS is currently experimenting with regression estimation for the second phase of area sampling and with outlier-resistant estimators. One intriguing idea is to employ time-series techniques (see Scott et al. 1977) to dampen the effects of outliers from the area frame. In addition, Chapman (1993) discusses using area cluster sampling within a list frame to reduce the cost of a survey conducted by face-to-face interview. In Chapter 12, Garrett and Harter review an innovative use of area frame methods to update list samples in a repeated survey. Johnson discusses a network sampling design linking workers and employers that starts with an area frame sample in Chapter 13.

REFERENCES

Armstrong, B. (1979), "Test of Multiple Frame Sampling Techniques for Agricultural Surveys: New Brunswick," *Survey Methodology*, **5**, pp. 178–184.

Bankier, M. D. (1986), "Estimators Based on Several Stratified Samples with Applications to Multiple Frame Surveys," *Journal of the American Statistical Association*, **81**, pp. 1074–1079.

Bosecker, R. R., and B. L. Ford (1976), "Multiple Frame Estimation with Stratified Overlap Domain," *Proceedings of the Social Statistics Section, American Statistical Association*, Part I, pp. 219–224.

Bush, J., and C. House (1993), "The Area Frame: A Sampling Base for Establishment Surveys," *Proceedings of the International Conference on Establishment Surveys*, Alexandria, VA: American Statistical Association, pp. 335–344.

Chapman, D. W. (1993), "Cluster Sampling for Personal-Visit Establishment Surveys," *Proceedings of the International Conference on Establishment Surveys*, Alexandria, VA: American Statistical Association, pp. 645–650.

Chu, A. (1987), "Proof that the Assignment of Conditional Weights Will Produce Unbiased Estimates," in L. A. Le Blanc (ed.), *Weighting Procedures for CBECS III*, unpublished report, Rockville, MD: Westat, pp. 12–14.

Cochran J. G. (1977), *Sampling Techniques*, New York: Wiley.

Cochran, R. S. (1965), *Theory and Applications of Multiple Frames Surveys*, unpublished Ph.D. dissertation, Ames: Iowa State University.

Faulkenberry, G. D., and A. Garoui (1991), "Estimation a Population Total Using an Area Frame," *Journal of the American Statistical Association*, **86**, pp. 445–449.

Fecso, R., R. D. Tortora, and F. A. Vogel (1986), "Sampling Frames for Agriculture in the United States," *Journal of Official Statistics*, **2**, pp. 279–292.

Fuller, W. A., and L. F. Burnmeister (1972), "Estimates for Samples Selected from Two Overlapping Frames," *Proceedings of the Social Statistics Section, American Statistical Association*, pp. 245–249.

Gallego, F. J., and J. Delincé (1993), "Sampling Frames Using Area Samples in Europe," *Proceedings of the International Conference on Establishment Surveys*, Alexandria, VA: American Statistical Association, pp. 651–656.

Goldberg, M. L., and P. M. Gargiullo (1988), "Variance Estimation Using Pseudostrata for a List-Supplemented Area Probability Sample," *Proceedings of the Survey Research Methods Section, American Statistical Association*, pp. 479–484.

Hansen, M. H., W. N. Hurwitz, and M. A. Madow (1953), *Sample Survey Methods and Theory*, Vol. 1, New York: Wiley, pp. 516–556.

Hartley, H. O. (1962), "Multiple Frame Surveys," *Proceedings of the Social Statistics Section, American Statistical Association*, pp. 203–206.

Hartley, H. O. (1974), "Multiple Frame Methodology and Selected Applications," *Sankhyá*, Series C, **36**, pp. 99–118.

Horvitz, D. G., and D. J. Thompson (1952), "A Generalization of Sampling Without Replacement from a Finite Universe," *Journal of the American Statistical Association*, **47**, pp. 663–685.

Houseman, E. (1975), *Area Frame Sampling in Agriculture*, SRS Report No. 20, Washington, DC: Statistical Reporting Service, U.S. Department of Agriculture.

Jessen, R. J. (1942), *Statistical Investigation of a Sample Survey for Obtaining Farm Facts*, Research Bulletin 304, Ames: Iowa State College of Agriculture and Mechanical Arts.

Julien, C., and F. Maranda (1990), "Sample Design of the 1988 National Farm Survey," *Survey Methodology*, **16**, pp. 116–129.

King, A. J., and R. J. Jessen (1945), "The Master Sample of Agriculture," *Journal of the American Statistical Association*, **40**, pp. 38–56.

Konschnik, C. A., C. S. King, and S. A. Dahl (1991), "Reassessment of the Use of an Area Sample for the Monthly Retail Trade Survey," *Proceedings of the Survey Research Methods Section, American Statistical Association*, pp. 208–213.

Kott, P. S. (1989), "The Variance of Direct Expansions from a Common Area Sampling Design," *Journal of Official Statistics*, **5**, pp. 183–186.

Kott, P. S. (1990), "Variance Estimation When a First Phase Area Sample Is Restratified," *Survey Methodology*, **16**, pp. 99–104.

Lund, R. E. (1968), "Estimators in Multiple Frame Surveys," *Proceedings of the Social Statistics Section, American Statistical Association*, pp. 282–288.

Nealon, J. P. (1984), *Review of the Multiple and Area Frame Estimators*, SF&SRB Staff Report 80, Washington, DC: Statistical Reporting Service, U.S. Department of Agriculture.

Nealon, J. P., and J. Cotter (1987), *Area Frame Design for Agricultural Surveys*, Washington, DC: U.S. National Agricultural Statistics Service.

Petrucci, A., and M. Pratesi (1993), "Listing Frames and Maps in Area Sampling Surveys on Establishments and Firms," *Proceedings of the International Conference on Establishment Surveys*, Alexandria, VA: American Statistical Association, pp. 554–559.

Rao, J. N. K. (1985), "Conditional Inference in Survey Sampling," *Survey Methodology*, **11**, pp. 15–31.

Rao, J. N. K., and C. J. Skinner (1993), "Estimation in Dual Frame Surveys With Complex Designs," unpublished paper, Ottawa: Carleton University.

Sande, I. (1986), "Business Survey Redesign Project: Frame Options for the Monthly Wholesale Retail Trade Survey Redesign," unpublished memorandum, Ottawa: Statistics Canada.

Särndal, C.-E., and B. Swensson (1987), "A General View of Estimation for Two Phases of Selection with Application to Two-Phase Sampling and Nonresponse," *International Statistical Review*, **55**, pp. 279–294.

Särndal, C.-E., B. Swensson, and J. Wretman (1991), *Model Assisted Survey Sampling*, New York: Springer-Verlag.

Scott, A. J., T. M. F. Smith, and R. G. Jones (1977), "The Application of Time Series Methods to the Analysis of Repeated Surveys," *International Statistical Review*, **45**, pp. 13–28.

Shah, B. V., B. G. Barnwell, P. H. Hunt, and L. M. LaVange (1991), *SUDAAN™ User's Manual*, Research Triangle Park, NC: Research Triangle Institute.

Skinner, C. J. (1991), "On the Efficiency of Raking Ratio Estimation for Multiple Frame Surveys," *Journal of the American Statistical Association*, **86**, pp. 779–784.

U.S. Energy Information Administration (1989), *Commercial Buildings Characteristics 1989*, Washington, DC: U.S. Department of Energy.

Vogel, F. A. (1973), "An Application of a Two-Stage Multiple Frame Sampling Design," *Proceedings of the Social Statistics Section, American Statistical Association*, pp. 617–622.

Vogel, F. A. (1975), "Surveys with Overlapping Frames—Problems in Application," *Proceedings of the Social Statistics Section, American Statistical Association*, pp. 694–699.

Vogel, F. A., and D. Rockwell (1977), *Fiddling with Area Frame Information in List Development and Maintenance*, SF Staff Report 77-01, Washington, DC: Statistical Reporting Service, U.S. Department of Agriculture.

CHAPTER TWELVE

Sample Design Using Peano Key Sequencing in Market Research

Joseph K. Garrett and Rachel M. Harter
Nielsen Marketing Research

Market research surveys have as their objective the collection of information or data that enables efficient marketing of goods and services. Although many market research surveys focus on households, businesses are also surveyed. In this chapter, we present a methodology for sequencing a list of businesses geographically, enabling efficient sample designs and maintenance. While it is illustrated for a market research survey of supermarkets, the methodology has widespread applicability for business surveys in general.

12.1 SAMPLE DESIGN

A. C. Nielsen's establishment-based SCANTRACK® survey of U.S. supermarket trade provides packaged goods manufacturers and retailers with estimates of consumer purchases by product and by week for various subdivisions such as major markets, individual retailers, and custom-defined geographies. In conjunction with a product's sales movement, the survey provides data on weekly promotional activity such as coupons, in-store displays, and newspaper advertisements. Sales data are collected weekly via the in-store checkout optical scanner; promotional data are collected via newspaper subscriptions and store visits by field personnel.

Because SCANTRACK is a periodic survey, the sample must be maintained to minimize bias and adequately reflect the changing population of supermar-

Business Survey Methods, Edited by Cox, Binder, Chinnappa, Christianson, Colledge, Kott.
ISBN 0-471-59852-6 © 1995 John Wiley & Sons, Inc.

kets, while at the same time providing reliable estimates of year-to-year trends. Furthermore, in today's competitive environment, the sample design needs to be conducive to automated maintenance to reduce labor costs. Geographic sequencing of the frame and systematic sampling within strata are integral parts of the SCANTRACK design. Sample maintenance rules are automated using the Peano key concept (Wolter and Harter 1990). In this section, we describe SCANTRACK's sample design. This description includes frame construction, stratification, sample selection, and a comparison of the sample design with other designs in terms of efficiency.

12.1.1 The Frame

The target population for the survey is all U.S. supermarkets located in the 48 contiguous states and the District of Columbia. For the market research industry, a *supermarket* is a grocery store (as defined by the Standard Industrial Classification Manual) with over two million dollars in annual sales. The grocery store population contains roughly 175,000 stores with total annual volume of slightly under 370 billion dollars according to recent Nielsen estimates. Approximately 30,000 of these stores are supermarkets with total annual volume of 305 billion dollars (83 percent). Of the 30,000 supermarkets, about 20,000 are *chain stores* (defined as four or more stores operating under common ownership and operation), while 10,000 are independent operations. Chain store supermarkets comprise roughly 79 percent of supermarket annual volume.

Nielsen maintains a listing of all supermarkets operating in the contiguous United States; this listing forms the basis for SCANTRACK's frame. Lists of stores and their characteristics are derived from Nielsen proprietary data, data supplied by cooperating retailers, commercially available data, and publicly available data from the federal government.

12.1.2 Stratification

As with many sample design stratification strategies (see Chapter 8), the SCANTRACK strata simultaneously allow for estimation at the stratum level, improve the overall efficiency of estimators, and create a building block approach for aggregated levels of estimation. The land area comprising the 48 contiguous states and Washington, DC is stratified into 73 areas defined by counties or county equivalents. These areas form 50 major markets and 23 remaining areas. The 50 markets are fairly large in size and contain major metropolitan areas and their surrounding counties. Markets are defined to correspond to television viewing areas so that advertising and market response can be easily correlated. The 23 remaining areas fill in the geographic voids remaining after market formation. This broad geographic stratification permits market-level estimates to be produced, and it facilitates efficient estimation at

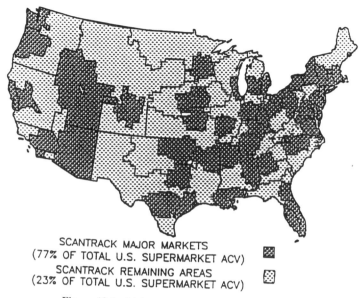

SCANTRACK MAJOR MARKETS
(77% OF TOTAL U.S. SUPERMARKET ACV)
SCANTRACK REMAINING AREAS
(23% OF TOTAL U.S. SUPERMARKET ACV)

Figure 12.1 Major markets and remaining areas.

aggregated levels by stratifying on regional differences related to sales of consumer goods. Figure 12.1 shows the 50 markets and 23 remaining areas.

Within the 50 major market strata, additional stratification by major retailing chains is employed. Retailer stratification within markets allows for efficient market-level estimates by recognizing chain differences, and it permits precise within-market estimates for individual retailer chains. Within major markets, the remaining chains not specifically stratified are placed into one of two size strata. Within each of the 23 remaining areas of the United States, stores are stratified into two size strata. Overall, the frame is stratified into more than 300 retailer-by-geographic strata. On average, a typical major market has separate strata for the four or five major grocery retailers in that market and two remainder strata.

12.1.3 Geographic Sequencing Of Stores Within Strata: Peano Keys

The stratification described in the previous section might be referred to as the *explicit* stratification. Prior to sample selection, the stores within these strata are geographically sequenced. This provides implicit substratification when coupled with the systematic sample selection procedures employed within strata. Careful use of systematic selection within explicit strata often produces additional gains in efficiency when frame records are sequenced based upon correlates of response or important reporting domains (Cochran 1977, Chapter 8). These gains occur because of what might be termed *implicit stratification,* or the stratifying effects implicit in careful use of systematic sampling from an ordered list. For a particular retailer stratum in a given market, this additional level of implicit stratification effectively benefits from demographic and socio-economic variables that affect consumer buying. In this section, we explain the procedure used to geographically sequence stores within a particular re-tailer-by-market stratum.

Stores can be sequenced geographically in many ways. For example, postal ZIP codes could be used to order stores. The geographic sequencing the SCANTRACK survey uses is based on a fractal, space-filling curve called the *Peano key*. In general, the Peano key is a parameter that defines a mapping from R^2 to R^1 such that points or spatial objects in R^2 can be arranged in a unique order in a list. The Peano key for a particular point in R^2 is obtained by interleaving the bits of its coordinates in binary form (Peano 1908, Laurini 1987, Saalfeld et al. 1988).

In this application, the space R^2 is represented by earth's geographic coor-dinate system; the spatial objects are sampling units. For a brief discussion of Peano key formation, let $X = X_k X_{k-1} \ldots X_3 X_2 X_1$ and $Y = Y_k Y_{k-1} \ldots Y_3 Y_2 Y_1$ represent the longitude and latitude of an arbitrary sampling unit in k-digit binary form. Then, the corresponding Peano key is $P = X_k Y_k \ldots X_3 Y_3 X_2 Y_2 X_1 Y_1$. Given latitude–longitude coordinates with k binary digits of accuracy (for any finite k), the spatial point represented by the value of P is actually a square in R^2. As k increases, the sizes of the squares decrease. In fact, as k tends to infinity, the value of P tends to represent a specific point in R^2.

The space-filling curve created by all possible values of the Peano key P is in the shape of a recursive N. Figure 12.2 illustrates the N curve, using a grid of 1024 points. This figure displays the self-similarity feature of fractal im-ages. The N curve passes once and only once through each point in space, with points being defined as squares whose size is determined by the number of digits in the latitude and longitude coordinates. The order of points on the

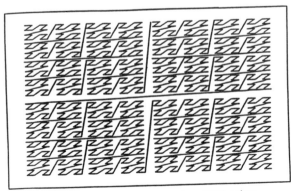

Figure 12.2   Peano order based on 1024 points.

curve (Peano order) largely preserves geographic contiguity. Thus, Peano order facilitates proximity searches. Peano order involves a few geographical discontinuities (e.g., see Figure 12.2) as do most mappings from R^2 to R^1. An alternative ordering based on Hilbert curves may have better contiguity, but the Peano order is easier to compute and to extend to larger geographic areas (Laurini 1987).

In SCANTRACK's actual application of this sequencing, the store's latitude and longitude are transformed, which effectively scales them to the store's county. Table 12.1 illustrates the transformation and the calculation of the Peano key for a specific store. By this procedure, economic establishments are arranged on a list in Peano order within county by means of their latitude and longitude coordinates. Counties within markets are geographically sequenced in a similar manner. Probability samples of establishments can then be drawn systematically from the ordered listings within strata.

12.1.4 Sample Selection Based On Peano Key Ordering

In this section, we describe the use of Peano key sequencing in initial sample selection. For ease of presentation, we develop this for an arbitrary, geographically defined stratum. Let U_i, $i = 1, 2, \ldots, N$, denote the population units, and let P_i represent the Peano key value associated with U_i. Also, assume that the U_i's have been ordered in such a manner that $P_1 < P_2 < \ldots < P_i < \ldots < P_N$.

Table 12.1 Example of a Peano Key Calculation

	Longitude	Latitude
Store	73.2622	40.7824
County Minimum	71.7786	40.5342
County Maximum	73.4971	41.3081

1. Translate and scale store's coordinates using county min/max:

   $(73.2622 - 71.7786)/(73.4971 - 71.7786) \times 2^{14} = 14144$

   $(40.7824 - 40.5342)/(41.3081 - 40.5342) \times 2^{14} = 5255$

2. Express transformed coordinates as binary and interlace:

   Longitude: 1 1 0 1 1 1 0 1 0 0 0 0 0 0

   Latitude: 0 1 0 1 0 0 1 0 0 0 0 1 1 1

   Peano Key: 10110011101001100000000010101

3. Express Peano key in base 10:

   188375061

Note that this implies that the latitude and longitude associated with any given unit are sufficiently detailed so that the P_i's are unique values.

The sampling scheme we employ is single-start systematic selection induced on the geographically ordered sequence $\{U_i\}_{i=1}^N$. Systematic sampling does not require that $N = nk$, with integer values for the sample size n and sampling interval k. Suppose $N = nk + c$, $1 \le c < k$. In this situation, the remaining c units can be assigned to samples 1 through c, yielding c systematic samples of size $n + 1$ and $k - c$ systematic samples of size n. Alternatively, if the sample size n must be fixed, then let $k = N/n$ be a noninteger. Select a number g at random from the uniform distribution on the interval $[0, k]$. The systematic sample consists of the units $\lfloor g + 1 \rfloor$, $\lfloor g + k + 1 \rfloor$, $\lfloor g + 2k + 1 \rfloor$, . . . , $\lfloor g + (n - 1)k + 1 \rfloor$, where $\lfloor n \rfloor$ denotes the largest integer less than or equal to n. This is equivalent to the systematic *pps* methodology of Chapter 8, using the same size measure for all units (e.g., $x_j = 1$). For ease of illustration, we assume that $N = nk$ in this chapter. Then all possible systematic selections of size n (i.e., each selection being an "every kth" systematic selection) can be viewed as in Table 12.2.

The implicit geographic substrata are defined by each sequential set of k units and the geographic areas surrounding them. To formalize this, we define the following concepts. Let P_L and P_U be the lower and upper bounds, respectively, on Peano key values for a particular geographic-by-retailer stratum. Thus, we have $P_L \le P_1 < P_2 < \ldots < P_i < \ldots < P_N \le P_U$. Let $P_{i,i+1}$, $i = 1, 2, \ldots, N - 1$, denote the average of P_i and P_{i+1}. Next, we define n Peano key *zones* and N Peano key *segments* as numerical intervals in the following manner:

$$
\text{Peano key zone } i = \begin{cases} [P_L, \overline{P}_{k,k+1}) & \text{for } i = 1 \\ [\overline{P}_{(i-1)k,(i-1)k+1}, \overline{P}_{ik,ik+1}) & \text{for } i = 2, \ldots, (n-1) \\ [\overline{P}_{(n-1)k,(n-1)k+1}, P_U] & \text{for } i = n \end{cases}
$$

$$
\text{Peano key segment } i = \begin{cases} [P_L, \overline{P}_{1,2}) & \text{for } i = 1 \\ [\overline{P}_{(i-1)i}, \overline{P}_{i,i+1}) & \text{for } i = 2, \ldots, (N-1) \\ [\overline{P}_{N-1,N}, P_U] & \text{for } i = N \end{cases}
$$

By construction, every segment initially contains one and only one population unit. Also by construction, these numeric intervals correspond to a geographic area. The implicit geographic substrata are Peano key zones of k consecutive segments. For illustration, each column in Table 12.2 is a Peano key zone. Systematic selection induced on $\{U_i\}_{i=1}^N$ can be analogously viewed as a systematic selection of Peano key segments, with one Peano key segment being systematically selected per Peano key zone, and with complete store enumeration within each selected segment.

Table 12.2   Systematic Samples for $N = nk$ Case

Systematic Sample	Composition of the Sample				
1	U_1	U_{k+1}	U_{2k+1}	$\cdots$	$U_{(n-1)k+1}$
2	U_2	U_{k+2}	U_{2k+2}	$\cdots$	$U_{(n-1)k+2}$
3	U_3	U_{k+3}	U_{2k+3}	$\cdots$	$U_{(n-1)k+3}$
$\vdots$	$\vdots$	$\vdots$	$\vdots$	$\vdots$	$\vdots$
k	U_k	U_{2k}	U_{3k}	$\cdots$	U_{nk}

We illustrate the stratum universe sequencing and sample selection process with a particular market (New York) and a particular retailer (Pathmark). Table 12.3 shows the current retailer stratification in New York with universe counts and current optimally allocated stratum sample sizes.

Using the store-specific latitude and longitude information, the Peano key value is constructed for each of the 103 Pathmark stores in the stratum. Using single-start systematic sampling, an illustrative sample of $n = 19$ stores selected from this sequenced listing appears as Figure 12.3. The map in Figure 12.3 shows the locations, based on store latitude and longitude, with the various symbols denoting the 19 zones into which the 103 Pathmark stores are divided. Through coding, Figure 12.3 portrays the implicit geographic stratification occurring through systematic selection from this sequenced listing. Because the sampling rate is approximately 1 in 5.42, the coding symbols form zones of five or six stores; the symbol of each selected store is circled for this example. This additional implicit geographic stratification within a retailer

Table 12.3   An Illustration of the Retailer Stratification for New York

Strata	N	ACV (%)	
A&P	226	22	
Grand Union	103	15	
King's	18	3	
Pathmark	103	19	60.5%
First National	25	6	
Shop Rite	110	20	
Waldbaum	106	12	
Remainder	893	49[a]	
Total New York	1584	146	

[a]This includes 38 stores from remaining chain organizations plus 11 stores from independents.
Source: A. C. Nielsen Company.

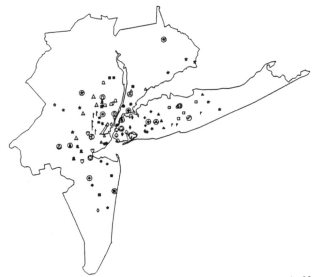

Figure 12.3 Peano key geographic sequencing of Pathmark supermarkets in New York market—an illustrative sample.

stratum brings socioeconomic and demographic variables affecting consumer buying into the design even though these variables are not explicitly accounted for in stratification. The benefits of this implicit stratification are demonstrated in the next section.

12.1.5 Comparison to Other Sampling Schemes

Based on research conducted using retailer-supplied universe data, the sampling scheme illustrated in Section 12.1.4 compares favorably to other common probability-based sampling schemes. Specifically, Nielsen often obtains weekly sales data for a "market basket" of products for *all* stores (not just sample stores) of a particular retailer for an 8- to 10-week period of time. This population information allows comparison of the coefficients of variation of various probability-based sampling strategies of fixed sample size within a retailer stratum. For each retailer studied, coefficients of variation were tabulated for each sampling strategy and product reviewed. Average coefficients of variation were computed across products within retailers, and then averaged by sampling scheme across retailers.

Two estimators were studied: (1) the simple expanded mean and (2) the ratio estimator that uses the auxiliary variable of annual total store sales or all commodity volume (ACV). ACV ratio estimation is employed at the stratum level in the SCANTRACK survey.

Several design strategies are compared in Table 12.4. The code to the strategies appearing in the table is the following:

Table 12.4 Average Coefficients of Variation Under Alternative Sample Designs[a]

	Simple Expanded Mean			Ratio Estimators						
Chain	srswor	Strat. by ACV (Count)	Strat. by ACV (ACV)	srswor	Syst. (ZIP)	Syst. (Peano)	Zone (ZIP)	Zone (Peano)	Strat. by ACV (Count)	Strat. by ACV (ACV)
Jewel—Chicago	0.116	0.085	0.084	0.086	0.077	0.072	0.081	0.078	0.082	0.082
Ralph's—Los Angeles	0.160	0.078	0.112	0.112	0.087	0.107	0.103	0.106	0.106	0.095
Randall's—Houston	0.210	0.164	0.161	0.173	0.163	0.168	0.164	0.163	0.162	0.159
Safeway—Seattle	0.188	0.171	0.171	0.143	0.121	0.139	0.140	0.137	0.143	0.143
Publix—Jacksonville	0.210	0.139	0.143	0.146	0.142	0.141	0.154	0.150	0.137	0.141
Publix—Miami	0.118	0.095	0.095	0.094	0.086	0.092	0.089	0.088	0.092	0.092
Publix—Orlando	0.152	0.121	0.117	0.114	0.108	0.112	0.107	0.109	0.114	0.111
Publix—Tampa	0.118	0.089	0.087	0.089	0.092	0.072	0.101	0.089	0.086	0.084
Pathmark—New York	0.111	0.090	0.087	0.084	0.065	0.067	0.084	0.082	0.084	0.082
Pathmark—Philadelphia	0.150	0.120	0.120	0.113	0.094	0.097	0.104	0.102	0.115	0.114
Average	0.153	0.115	0.118	0.115	0.104	0.107	0.113	0.110	0.112	0.110

[a]Abbreviations in this table are defined in text.

213

- *srswor* = simple random sampling without replacement;
- Strat. by ACV (Count) = stratification of stores by ACV, sample allocation to strata proportional to store counts within strata;
- Strat. by ACV (ACV) = stratified by ACV, allocation to strata proportional to ACV;
- Syst. (ZIP) = systematic selection applied to stores sequenced by ZIP code;
- Syst. (Peano) = systematic selection applied to stores sequenced by Peano key; and
- Zone (ZIP) and Zone (Peano) = random selection of one store per "geographic zone" as induced by the implicit geographic zone stratification of systematic selection when used with either ZIP code or Peano key sequencing.

Table 12.4 presents results of this empirical research. As can be observed, systematic selection applied to geographically sequenced listings coupled with ACV-based ratio estimation produced lower average coefficients of variation. In terms of statistical efficiency (i.e., low sampling variability), Peano key and ZIP code sequencing were roughly equivalent. The results in Table 12.4 indicate that conversion from ZIP code to Peano key sequencing does not degrade statistical quality despite its occasional discontinuities. Peano keys are better-suited for maintenance of initially selected samples through time than ZIP code sequencing. The primary advantage of Peano key ordering over ZIP code ordering is in the fixed nature of latitudes and longitudes. ZIP codes sometimes change, which could rearrange the sequence order and complicate automated maintenance. As a practical matter, then, long-term quality of the sample is more easily retained with Peano key sequencing.

12.2   CONTINUAL QUALITY MAINTENANCE

Maintenance of initially selected samples must ensure that the samples remain representative over time by properly reflecting "births" and "deaths" of establishments in both the sample and the population as they occur. As with any large-scale, probability-based sample, continuous maintenance is necessary for the SCANTRACK sample because the national supermarket universe is not static. In a recent 12-month period, approximately 2200 new supermarkets opened and 2450 existing stores went out of business. Another 170 stores were reclassified during the year. Reclassification can result from a number of changes. Smaller grocery stores enter the survey population when their ACV's surpass the $2-million-per-year threshold which defines a supermarket. The principal objectives of a maintenance system for the SCANTRACK sample are as follows:

- The sample should maintain geographic balance through time.

- The system should maintain sample size through time and be amenable to automation.
- The sample should adhere to principles of probability sampling to avoid bias in estimates of total sales.
- Sample changes should not excessively disturb estimates of year-to-year trends.

See Colledge (1989) and Chapter 5 for general comments on maintenance issues for business surveys.

Geographical balance is a proxy for socioeconomic and demographic balance. Because different neighborhoods have different purchasing patterns, geographical balance is important in achieving an efficient sample design for a wide range of products. Some products are targeted to specific demographic populations, and a representative sample should include stores carrying such products. Furthermore, geographical balance is an important factor in customers' perception of an appropriate sample.

Sample size must also be kept constant. A sample size decrease adversely affects the standard errors of estimators, whereas a sample size increase adversely affects costs. Neither outcome is desirable. Furthermore, contracts with chain organizations specify sample sizes and cooperation payments, and any changes have to be renegotiated. This, too, is undesirable.

12.2.1 Birth and Death Updating

For recurring surveys, the initially selected sample must be properly modified over time to ensure that it accurately represents the population at any given time. Essentially, this implies that population units that come into existence after initial sample selection must be given known, nonzero probabilities of selection. Similarly, proper attention must be given to population units that cease to exist. Often these essentials are incorporated into the basic structure of the sample design, estimation methodology, and frame updating operations.

With the structure defined in Section 12.1.4, we give a rule for incorporating births into the sample. As births arise, they are assigned their own unique Peano key value, and this value is a member of one and only one Peano key segment. Thus, the Peano key permits us to automatically place new units (stores) in their correct and unique positions on the ordered frame.

Rule 1. A birth store is subjected to sampling if and only if its Peano key value is an element of a selected Peano key segment. Birth stores whose Peano key values are elements of nonselected segments are not subjected to sampling, and hence not selected. Subjecting births to sampling implies that all supermarkets within an initially selected segment are given a nonzero probability of being selected, including the initially selected store, subject to the constraint that these nonzero probabilities sum to one.

As the supermarket population changes, we no longer have a one-to-one correspondence of segments to stores. In maintenance procedures, the segment can be viewed as the primary sampling unit. In the case of births, subsampling of stores takes place within the selected segments. In effect, the design is a two-stage cluster design where initially each cluster (segment) has exactly one secondary sampling unit (store), but subsampling is repeated within a cluster whenever it gains a new secondary unit. Initially, the within-cluster variance is zero because the cluster is completely enumerated, but as birth stores are added and subsampling takes place, within-cluster variance is no longer zero.

For establishments that go out of business or otherwise cease to exist, various rules can be given for proper handling. For SCANTRACK, the frame maintenance procedures are such that deaths are known on a population basis, not just those occurring in the sample. In such situations the following rule is unbiased.

Rule 2. Remove all deaths from the sampling frame. If a sample store dies, subject any remaining stores in the segment to probability sampling. If the segment contains no other live stores, then the segment is included with sales of zero.

The birth and death rules are simple and easy to automate. To ensure a constant sample size, Rule 2 might be altered in practice so that if the originally selected segment has no live stores, another segment in the zone is selected, and stores within that segment are subjected to probability sampling. Automated maintenance can be implemented at regularly scheduled intervals such as monthly, quarterly, or semiannually. Whether or not the sample actually changes during maintenance, the weights applied to the sample stores are adjusted to reflect the births and deaths in the universe. (See Chapter 25 for a discussion of calibrating weights to known population totals.) According to the proposed maintenance rules, the systematic cluster sample of segments remains rather stable over time. This stability maintains geographic balance and discourages trend disruptions. New stores are selected with probability consistent with the cluster sample design.

12.3 CONCLUDING REMARKS

In this chapter, we present a sample design and sample maintenance system that uses store location given by latitude and longitude. Building on this, we define a unique set of stores, Peano key segments, which can be viewed as clusters in the standard sample design sense. We illustrate systematic selection from the universe of clusters on a stratum basis and provide empirical evidence of the efficiency of this sampling strategy. Finally, we provide general rules for maintaining sample representativeness over time. We have found, as we transition our SCANTRACK sample to this design, that the Peano system is easy to design, implement, and automate.

The concepts underlying SCANTRACK's sample design and maintenance are not restricted to supermarkets. Peano key sequencing for geographic ordering is suitable for surveys of businesses whose geographic coordinates can be uniquely determined; that is, the businesses stay put, and they are separated enough geographically that their coordinates can be distinguished. Many variations on sample design and maintenance rules can be automated using Peano keys.

REFERENCES

Cochran, W. G. (1977), *Sampling Techniques*, New York: Wiley.

Colledge, M. J. (1989), "Coverage and Classification Maintenance Issues in Economic Surveys," in D. Kasprzyk, G. Duncan, G. Kalton, and M. P. Singh (eds.), *Panel Surveys*, New York: Wiley, pp. 80–107.

Laurini, R. (1987), *Manipulation of Spatial Objects by a Peano Tuple Algebra*, Technical Report CS-TR-1893, College Park: University of Maryland.

Peano, G. (1908), "La Curva di Peano nel Formulario Mathematico," in *Opere Scelte di G. Peano*, Vol. I. Edizioni Cremonesi, Roma (1957), pp. 115–116.

Saalfeld, A., S. Fifield, F. Broome, and D. Meixler (1988), "Area Sampling Strategies and Payoffs Using Modern Geographic Information System Technology," unpublished paper, Washington, DC: U.S. Bureau of the Census.

Wolter, K. M., and R. M. Harter (1990), "Sample Maintenance Based on Peano Keys," *Survey Methodology*, **16**, pp. 181–194.

CHAPTER THIRTEEN

Business Surveys as a Network Sample

Ayah E. Johnson[1]

Henry M. Jackson Foundation for the Advancement of Military Medicine

Surveys generally seek to make estimates about a target population. The individual members of the population whose characteristics are to be measured are called *target statistical units*. In conventional surveys, each of these units is the ultimate sampling unit or is uniquely linked to the sampling unit. Moreover, the target units (or some aggregation thereof) serve as the analysis units. For example, to make estimates of the uninsured in the United States, the target statistical unit may be an individual person, while the sampling unit is a household. Each individual is linked to one and only one household, and health insurance status is determined for all members of sampled households.

Network sampling modifies the above linking procedure. It allows a target statistical unit to be linked to more than one sampling unit. An early example of network sampling is provided by Sirken (1970) who used network sampling as a means of increasing the precision of estimators for rare populations such as individuals with cancer or epilepsy.

This chapter explores the use of network sampling in a different context—namely, the joint examination of two populations composed of different but linked analysis units. Specifically, the characteristics of a population of persons and the businesses to which they are linked are examined. For example, a researcher may be interested in estimating both the number of uninsured

[1]The author wishes to thank Phillip S. Kott of the U.S. National Agricultural Statistics Service and Steven B. Cohen of the U.S. Agency for Health Care Policy and Research (AHCPR) for their helpful comments. This work was done while the author was with AHCPR. The views expressed in this paper are her own and no official endorsement by AHCPR is intended or should be inferred.

Business Survey Methods, Edited by Cox, Binder, Chinnappa, Christianson, Colledge, Kott.
ISBN 0-471-59852-6 © 1995 John Wiley & Sons, Inc.

individuals in the United States and the number and types of employers that do not offer insurance. Both estimates may be needed for a policy decision. Two independent surveys—a household survey of individuals and a business survey—could be carried out to provide the estimates. By contrast, in the network sampling design discussed in this chapter, two separate surveys are conducted, but the business survey is dependent on the household survey. This allows linkage of data from the two surveys for analytical purposes.

This chapter describes a particular network sample of employers and the strategy for obtaining employee and business estimates from that sample. It also demonstrates problems associated with the practical application of network sampling designs. The merits and issues associated with network sampling are discussed in terms of (1) the achievement of the dual objectives for the two analysis levels; (2) the analytical power of the broader database; (3) the sources and magnitude of nonresponse; and (4) the bias, reliability, and accuracy of the resultant estimators.

13.1 SAMPLE DESIGN

Consider a household survey where the target statistical units are persons residing within households in a population of interest, such as the United States. Baseline data on household composition, employment, and other characteristics are collected. Respondents are asked to sign a permission form authorizing the collection agency to approach their employer(s) for additional information. Using the household identification of employers and signed permission forms as a frame, a *follow-back survey* is conducted with the employers of sampled household members. The sample units of the follow-back survey are the business establishments identified by the household respondents.

This network sample can serve as a stand-alone survey for estimating employment or employer characteristics. Moreover, the data collected can include information from the employers that supplements or verifies the sampled individual's data. Just as the original sample of persons, when properly weighted, represents the household population in the United States, the network sample of business establishments, when properly weighted, represents the population of U.S. employers.

The Health Insurance Provider Survey (NMES-HIPS), a component of the 1987 National Medical Expenditure Survey (NMES), was designed using this network sampling methodology. NMES-HIPS respondents were providers of health insurance and included employers, unions, and private insurance carriers identified by household respondents to the primary survey—the household component of the NMES, which I denote as NMES-HHS. This chapter focuses exclusively on the over 10,000 employer establishments in the NMES-HIPS. The term "establishment" is used here in the narrow sense of a business at a single location (see Chapter 3).

13.2 ESTIMATION

One main advantage of the network sample design is its analytical capability of facilitating both person-level and establishment-level analyses. To allow estimation at these two levels, we start with the set of all respondent individuals in the original household sample, called the *primary sample*. Each individual in the primary sample is linked to the establishment(s) the respondent identifies as an employer. This forms a sample of person-establishment pairs which can be enumerated for the follow-back establishment survey. (In practice, of course, an establishment in more than a single sampled pair need only be contacted once to obtain data on the establishment and each linked sampled employee.) Establishments could be subsampled for the follow-back survey, but that was not done for the NMES-HIPS.

The next step involves defining what constitutes a response to the establishment survey. To maximize the usefulness of survey data, the definition of response varies depending on the level of analysis. For establishment-level analyses, a response can be considered complete if all establishment information has been completed. For person-level analyses (of data provided by establishments for linked employees), a response can be considered as complete if all establishments identified by the household respondent provided the required individual-level information. Different sampling weights are derived using these different definitions of completeness that account for nonresponse and for the design's complexity.

13.2.1 Estimation at the Establishment Level

In a conventional establishment survey, the basic sampling unit is the establishment itself, and the unit's sampling weight is simply the inverse of its probability of selection. In this network sampling setting, establishments with multiple employees have an opportunity of selection through each employee. This and the complex design employed in most household surveys make calculation of exact establishment selection probabilities extremely difficult.

A multiplicity adjustment approach is commonly used in this context for computing the sampling weights. Tourangeau (1993) describes its application for NMES-HIPS. The sampling weight for the jth establishment is set equal to

$$w_j = \frac{\sum_{i=1}^{n_j} w_{ij}}{s_j} \tag{13.1}$$

where n_j is the number of individuals in the primary sample linked to the jth establishment, s_j is the number of individuals employed by the jth establishment, and w_{ij} is the sampling weight from the primary survey associated with the ith individual that is employed by the jth establishment.

With these weights, establishment characteristics can be estimated in an unbiased fashion. For example, let y_j denote a value of interest for establishment j, and let $Y = \Sigma^M y_j$, where M is the total number of establishments in the population. The estimator for the total Y based on the network sample is $\hat{Y}_{mult} = \Sigma^m w_j y_j$, where m is the number of establishments in the sample. $\hat{Y}_{mult}$ is unbiased because the expected value of w_j (assuming no nonresponse or measurement error) is one for every establishment j in the population of M establishments [observe that the expected value of the numerator of equation (13.1) equals the denominator]. Estimators such as $\hat{Y}_{mult}$ are called *multiplicity* estimators because they account for the multiplicity of chances that each element j has of being selected for the sample. For a general discussion of multiplicity estimators, see Lessler and Kalsbeek (1992, pp. 94–102). Also of relevance are Sirken (1972), Sirken and Levy (1974), Levy (1977), Levy and Lemeshow (1991, pp. 342-345), Cohen and Farley (1984), and Cooper and Johnson (1993).

13.2.2 Estimation at the Person Level

The employer can be asked to provide individual-level data that are unknown to the individual, such as employer-paid health insurance premiums. To create estimates at the person level, adjustments must be made for nonresponse to the primary survey and the follow-back survey. The first set of nonresponse adjustments adjust for nonresponse to the primary survey and are incorporated into the primary sampling weights [the w_{ij} of equation (13.1)]. For the follow-back survey, the definition of response is slightly more complex because an individual may be linked to more than one establishment. A strict definition of *response* can be adopted: An individual respondent to the primary survey is counted as a respondent for the follow-back data if all establishment(s) linked to that individual respond to the follow-back survey. Thus, if one establishment linked to a primary sampled individual responds to the follow-back survey but another does not, then that individual is considered a follow-back nonrespondent.

To optimize all collected data, imputation may be used in cases with information from some, but not all, of the individual's linked establishments. If the number of cases requiring imputation is small, the gain from this strategy will be minimal. An alternative strategy, followed with the NMES-HIPS, is to use data only for follow-back respondents and to adjust their primary sampling weights to account for nonresponse. See Cox and Cohen (1985) for a general discussion of weighting and imputation for longitudinal surveys such as NMES.

13.3 EVALUATING THE DESIGN OF THE NMES-HIPS

Only a few empirical studies have evaluated the advantages and disadvantages of network sampling. In one such study, Nathan (1976) showed that a slight reduction in mean square error could be achieved and that this reduction was not offset by an increase in response bias. In his study, individuals within sampled households reported about other households, such as those of relatives or neighbors. In this section, the advantages and disadvantages of the network sampling strategy for a joint survey of employers and employees are explored by evaluating the results of NMES-HIPS.

13.3.1 Measurement Errors

Two types of measurement errors encountered by the NMES-HIPS are of particular concern. The first involves inaccurate information provided by household respondents concerning the name, address, and telephone number of their places of employment. In approximately 16 percent of the sampled establishments, nonresponse resulted from omission of identifying information such as location. [Nathan (1976) and Nathan et al. (1977) encountered similar types of measurement error.] The remainder involved cases that were not fielded because they were incorrectly categorized as ineligible. This error resulted from drawing the sample based on preliminary NMES-HHS data before the household database was fully edited.

Table 13.1 indicates nonresponse by establishment size, industry code, and changing employment of the primary survey respondent. As size increased, it became easier to locate and obtain cooperation from establishments. Agriculture, forestry or fishing, and establishments in nonclassifiable industries had the highest nonresponse rate due to errors in the primary survey. Changing jobs also increased nonresponse.

The second type of measurement error occurs when a primary survey respondent and his or her linked establishment(s) provide different responses. For example, a household respondent might report not holding private insurance, while the employer reports otherwise. In such cases, it is difficult to ascertain whose report is accurate. The magnitude of both types of measurement error may be quantified to qualitatively assess the extent of their impact on the estimator of interest. Table 13.2 compares estimates based on primary survey (NMES-HHS) data with estimates based on follow-back data. Estimates based on the primary survey are comparable to those based on the follow-back survey, suggesting that overall the impact of measurement error may be minor.

13.3.2 Bias Resulting from Nonresponse

Surveys need to make a concerted effort—subject to time and cost constraints—to minimize the rate of nonresponse. Nonresponse reduces the efficiency of estimators because it reduces sample size and introduces bias when the variable of interest and survey response/nonresponse behavior are related. Sources of nonresponse to a network sample include: (1) nonresponse to the primary survey, (2) failure to sign permission forms allowing contact with linked estab-

Table 13.1 Establishment's Profile by Response Status

			Nonresponse	
Establishment Characteristic	Eligible	Percent Complete	Percent Not Fielded	Percent Contact-Nonresponse
Total	12,681	73.3	15.9	10.8
Size of Establishment[a]				
< 10	2,960	66.9	19.9	13.2
10–25	2,011	69.7	18.3	12.0
26–100	2,984	72.7	16.1	11.1
101–500	2,545	78.1	13.0	9.0
501+	2,168	79.1	12.5	8.5
Major SIC Code				
Agriculture, Forestry, Fishing	265	69.5	20.9	9.6
Mining	81	78.5	15.2	6.3
Construction	669	69.9	17.9	12.1
Manufacturing	2,324	78.1	12.6	9.3
Transportation,	781	72.2	15.0	12.8
Communications, Utilities,	2,759	73.0	14.4	12.6
Sales				
Finance, Insurance, Real	736	75.3	13.7	10.9
Estate				
Repair Services	736	68.1	15.6	16.3
Personal Services	474	66.4	18.1	15.5
Entertainment, Recreation	150	72.7	15.6	11.7
Professional, Services	2,581	80.5	12.7	6.9
Public Administration	484	77.3	12.3	10.5
Unknown	606	32.0	60.3	7.7
Geographic Location[a]				
Northeast	2,617	70.4	18.2	11.3
Midwest	2,954	75.6	14.4	10.1
South	4,526	73.7	15.6	10.7
West	2,420	70.5	17.6	11.9
More than One Location				
Missing	276	8.1	54.1	37.9
Yes	7,148	73.3	15.9	10.9
No	5,257	77.3	13.7	9.0
Burden Level				
1–5 employees	12,591	72.5	16.7	10.9
5+ employees	90	93.1	0.0	6.7
Job Change During Year				
Yes	2,810	69.4	19.8	10.8
No	9,859	74.5	14.8	10.8
Labor Union Member				
Missing	671	66.8	22.2	11.0
Yes	1,908	76.3	13.9	9.8
No	10,102	73.1	15.9	11.0

[a]Data are missing on establishment characteristics (<1%), and thus the sum of the establishments for this category does not equal the total number of establishments.

Source: Agency for Health Care Policy and Research, 1987 National Medical Expenditure Survey.

Table 13.2 Insurance Status: Comparisons Between NMES-HHS and NMES-HIPS Estimates

Estimated Number of Persons	NMES-HHS				NMES-HIPS				Difference HHS-HIPS
	Sample Size	Total	Standard Error	Design Effect	Sample Size	Total	Standard Error	Design Effect	
Employed	15,217	114,958,334	2,445,353	9.3	9,139	111,339,360	2,371,948	3.25	3,618,974
Privately Insured	24,365	178,669,444	2,759,858	11.34	21,493	175,478,552	3,657,854	7.61	3,190,892
Employed, Private Insurance	12,108	95,368,118	1,772,647	4.99	7,141	89,734,988	2,170,910	4.06	5,633,130
Only Public Insurance	4,784	26,309,715	1,202,842	8.05	7,180	27,274,032	1,157,407	5.19	(964,317)
Uninsured	5,110	34,413,697	1,379,899	7.04	4,186	36,640,271	1,434,002	4.72	(2,226,574)

Source: Agency for Health Care Policy and Research, 1987 National Medical Expenditure Survey.

lishments, (3) failure to provide enough information in the primary survey to contact units in the establishment survey, and (4) establishment nonresponse to the follow-back survey.

For the 1987 NMES-HHS survey, 80 percent of sampled individuals responded. Signed permission forms were obtained from 81 percent of these respondents. Nonresponse to the NMES-HHS stemmed from refusals and from the usual sources of nonresponse to a household survey. Permission-form nonresponse had two major sources: refusal to sign permission forms (9.8 percent) and failure to generate a permission form (4.7 percent). The first source of permission-form nonresponse included individuals who refused to sign any permission forms and individuals who signed some, but not all, forms.

The second source of nonresponse was the result of the primary and follow-back survey timing. Because both surveys collected data for the same time period, delays in fielding the network sample were undesirable. Lists of sample establishments were generated while data from the primary survey were still being collected or edited. One consequence of this timing issue was that not all individuals who should have signed a permission form received one to sign. A second consequence was that sample identification for the follow-back survey was incomplete at the time of data collection because the follow-back relied on NMES-HHS preliminary data. To rectify the problem, several draws from the primary survey (NMES-HHS) were made to update the NMES-HIPS sample listing. However, this procedure did not mitigate the problem of not having all required permission forms. To attenuate nonresponse resulting from lack of signed permission forms, such NMES-HIPS cases were fielded with instructions to collect only data pertaining to establishment characteristics. Without a signed permission form, establishments had to have more than 25 employees to be included in NMES-HIPS. This last restriction was applied to ensure respondent confidentiality.

A third source of nonresponse was incomplete contact information for establishments of interest. Again, the tight time schedule in fielding both the primary and follow-back survey left little leeway to investigate these cases. For primary respondents eligible for NMES-HIPS, the person-level response rate was only 62 percent and the establishment-level response rate was 73 percent.

Despite these multiple levels of nonresponse, it is worth noting that where information was available to field the case, the NMES-HIPS completion rates were about 87 percent. This points out that it is possible to gain a high level of cooperation from establishments once they are part of the network sample.

Analysis of the characteristics of establishments not participating in the NMES-HIPS survey indicated that a significant component of nonresponse was associated with previous steps of the data collection, suggesting the need for a more complete listing of eligible establishments and obtaining signed permission forms (Johnson and Cohen 1992). Another finding was that nonresponse bias is of particular concern when analyses focus on a small, specific subset of the population, such as uninsured persons under the age of 24 who work in establishments with less than 25 employees. Other correlates of non-response to the NMES-HIPS include the size and the location of the establishment and the ability of primary survey individuals to describe their occupation and the industry in which they work.

13.3.3 Enhanced Analytical Capability

Although the potential effects of nonresponse bias must be considered, the network sampling design obtains information not only about establishments but also about individual persons employed by (or retired from) these establishments. Linking the data collected from the primary survey respondent to the individual-level data collected from the establishment provides a more complete profile for the individual. For instance, employees seldom know the cost of health insurance provided by their employer. For the follow-back NMES-HIPS survey, individual-level data included information such as health insurance benefits received from the employer (if any), the name of the insurance plan, the cost of health insurance, the portion paid by the employer and employee respectively, and the level of coverage.

Individual-level NMES-HIPS data were then supplemented by NMES-HHS data to include sociodemographic characteristics, data on utilization of health services, expenditures, sources of payment, medical conditions, job information, and health insurance. For example, Table 13.3 provides estimates of the cost of health insurance premiums by the individual's sociodemographic characteristics. The standard errors and the design effect are computed using SE-SUDAAN (Shah 1981), which adjusts for complex survey design using the Taylor series linearization method.

Using a conventional household survey, data on the total cost of health insurance premiums and the employer share of this cost could not be easily or accurately obtained. By contrast, such data can be obtained from a conventional establishment survey; it would, however, be very difficult to construct individual-level estimates by sociodemographic characteristics. Moreover, establishment-level data could not be used alone to link information on use and expenditures of medical services with the cost of health insurance premiums.

13.3.4 Reliability and Accuracy of Estimates

The reliability of major estimators of interest is quantified by examining the relative standard errors and the size of design effects of estimators both at the person level and at the establishment level. The *design effect* for a survey estimate is defined as the ratio of the variance of the statistic under the actual design divided by the variance that would have been obtained from a simple random sample of the same size. Differing for each design, the design effect represents the cumulative effect of design components such as stratification, unequal weighting, and clustering.

Estimates of design effects and percent relative standard errors (coefficients

Table 13.3 Cost per Individual of Health Insurance Premiums for Employment-Related Policies by Sociodemographic Characteristics[a]

Sociodemographic Characteristic	Policy Holders (Thousands)	Annual Premium		
		Average Cost ($)	Relative Standard Error (%)	Design Effect
Total	68,487	2,048	1.19	2.14
Poverty Status				
Poor	2,138	1,810	6.11	1.33
Near Poor	1,118	1,825	5.90	1.27
Low Income	5,266	1,975	3.05	1.33
Middle Income	24,947	1,973	1.65	1.57
High Income	35,017	2,135	1.72	2.17
Age of Employee				
<25	7,035	1,493	2.76	0.98
25–34	22,650	1,880	1.99	1.78
35–54	30,373	2,271	1.51	1.84
55–64	7,105	2,174	2.50	0.99
65+	1,324	2,095	7.14	1.63
Gender				
Male	40,301	2,272	1.33	1.70
Female	28,186	1,728	1.66	1.63
Race/Ethnicity				
Nonwhite	12,596	1,998	2.29	2.14
White	55,981	2,060	1.32	1.88
Marital Status: Male				
Married	28,539	2,558	1.32	1.69
Divorced	3,157	1,943	5.32	1.29
Separated	906	2,043	8.16	1.44
Never Married	7,388	1,361	3.60	1.45
Married Status: Female				
Married	15,496	1,952	2.12	1.65
Widowed	1,318	1,613	6.84	1.11
Divorced	4,579	1,559	3.64	1.18
Separated	1,032	1,760	6.32	1.26
Never Married	5,712	1,274	3.18	1.29
Health Status				
Excellent/Good	59,078	2,042	1.28	2.07
Fair/Poor	6,681	2,096	3.10	1.50
Region				
Northeast	14,288	2,175	1.82	1.04
North Central	17,732	2,186	2.52	2.33
South	23,162	1,825	1.97	2.26
West	13,305	2,118	2.88	2.37

[a]Estimates were computed for Findings on Employment-Related Insurance; relative standard errors and design effects were added.

Source: Agency for Health Care Policy and Research, 1987 National Medical Expenditure Survey.

of variation) associated with the mean annual cost of health insurance premiums are given in Table 13.3. The design effect for the total population is 2.14. Within each subclass the design effect is usually below 2.0. This relatively low design effect for a household survey is achieved despite the effect of the complex survey design, oversampling in the household survey, and the variability of the analysis weights caused by the multiple adjustments for nonresponse.

To evaluate the accuracy of estimates at the establishment level, Table 13.4 compares NMES-HIPS estimates of the number of establishments with 1987 estimates from the *Statistical Abstract* (U.S. Bureau of the Census 1991, pp. 532). The NMES-HIPS estimate of the number of establishments is 30 percent higher than the estimated number given in the statistical abstract. This difference may be due in part to differences in definitions of the collapsed SIC codes. In addition, agricultural and livestock production and small businesses are not included in the *Statistical Abstract* but are included in NMES-HIPS estimates. In a few instances, NMES-HIPS underestimates the number of establishments. Those are establishments such as (1) mining establishments employing less than 100 individuals, (2) construction establishments employing more than 100 employees; (3) manufacturing establishments employing more than 500 employees; (4) sales establishments employing between 20 and 499 employees; and (5) finance, insurance, or real estate establishments with at least 20 employees. To adjust for this systematic bias, the NMES-HIPS weights were poststratified using the number of establishments published in the statistical abstracts by SIC and by size. The poststratification adjustment factors averaged 0.87 and ranged from 0.40 to 1.57. The relative standard errors are below 30 percent, except for the mining industry with more than 100 employees.

13.4 CONCLUDING REMARKS

Conventional sampling strategy is sensible when one has to produce estimates for a single target population—for example, the estimated number of establishments satisfying a set of conditions. When analytical objectives warrant the joint examination of more than one population, however, network sampling can enhance the analytical power of the collected data and offers flexibility in the type of estimates that can be produced. In this chapter, I explore a particular application of network sampling that links individuals from a household survey with their employers. Other applications using network sampling include linking a sample of students who were in the eighth grade with their math, history, science, and English teachers (Ingels et al. 1992).

For person-level estimates, a network sampling strategy can enhance the information collected during the primary survey with data from the followback survey. It also allows for verification of the information provided by the respondent to the primary survey. Reliability and accuracy of the estimates do not seem significantly affected by this design. However, a mechanism to con-

Table 13.4 Number of Establishments by SIC Code and by Size

Characteristic	Establishment Size	Sample Size	Statistical Abstract	NMES-HIPS Total	Adjustment Factor	After Poststratification[a] Standard Error	After Poststratification[a] Relative Standard Error
Total	All	8,416	5,937,000	8,489,410	0.70	188,207	0.03
Agriculture, Forestry, Fishing		72	76,000	146,139	0.52	13,788	0.18
Mining	1–99	28	32,000	23,701	1.35	8,956	0.28
	100+	34	1,000	1,648	0.61	347	0.35
Construction	1–19	241	509,500	896,895	0.57	61,240	0.12
	20–99	125	43,500	48,098	0.90	4,617	0.11
	100+	74	5,000	4,676	1.07	823	0.16
Manufacturing	1–19	255	243,500	605,128	0.40	23,932	0.10
	20–99	430	89,500	114,353	0.78	5,674	0.06
	100–499	569	32,000	34,570	0.93	2,200	0.07
	500+	457	6,000	5,783	1.04	488	0.08

Transportation, Communications, Utilities	1–19	112	187,000	292,548	0.64	26,929	0.14
	20–99	142	33,000	40,504	0.81	3,809	0.12
	100–499	124	7,000	8,460	0.83	769	0.11
	500+	159	1,000	1,352	0.74	120	0.02
Sales	1–19	846	1,700,000	2,138,359	0.80	92,463	0.05
	20–99	706	255,000	210,259	1.21	12,937	0.05
	100–499	314	26,000	21,501	1.21	1,957	0.08
	500+	122	1,000	1,378	0.73	111	0.11
Finance, Insurance, Real Estate	1–19	190	479,000	596,150	0.80	55,184	0.12
	20–99	124	47,000	39,015	1.20	5,497	0.12
	100–499	99	8,000	6,641	1.20	977	0.12
	500+	125	2,000	1,269	1.58	310	0.16
Services	1–19	941	1,797,000	2,713,596	0.66	101,794	0.06
	20–99	701	148,000	198,997	0.74	7,089	0.05
	100–499	618	30,000	41,755	0.72	1,988	0.05
	500+	615	5,000	6,600	0.76	339	0.07
Nonclassifiable	All	193	173,000	290,036	0.60	35,688	0.21

[a]After poststratification, the control totals are equal to those published in the *Statistical Abstract*.

Source: Agency for Health Care Policy and Research. 1987 National Medical Expenditure Survey.

trol for measurement error during the sample listing should be carefully devised and closely monitored.

Establishment-level estimates indicate that the design strategy should include careful monitoring of data quality to avoid large biases. Further research is needed on the accuracy of establishment-level estimators based on network samples compared to more conventional list samples. In addition, a systematic assessment of the costs associated with a network sampling scheme is needed to compare the approach with more conventional sampling schemes.

REFERENCES

Cox, B. G., and S. B. Cohen (1985), *Methodological Issues for Health Care Surveys*, New York: Marcel Decker.

Cohen, S. B., and P. J. Farley (1984), *Estimation and Sampling Procedures in the NMCES Insurance Surveys*, Instrument and Procedures 3, Washington, DC: National Center for Health Center for Health Services Research.

Cooper, P. F., and A. E. Johnson (1993), *Employment Related Health Insurance in the U.S.: The Health Insurance Provider Survey*, Research Finding 17, Washington, DC: Agency for Health Care Policy and Research.

Ingels, S. J., L. A. Scott, J. T. Lindmark, M. R. Frankel, and S. Wu (1992), *National Education Longitudinal Study of 1988, First Follow-up: Teacher Component Data File User's Manual*, NCES 93-085, Washington, DC: National Center for Education Statistics.

Johnson, A. E., and S. B. Cohen (1992), "Profile of Nonrespondents in a Follow-Back Survey: The Health Insurance Provider Survey," paper presented at the Public Health Association Annual Meeting, Washington, DC.

Lessler, J. T., and W. D. Kalsbeek (1992), *Nonsampling Error in Surveys*, New York: Wiley.

Levy, P. S., and S. Lemeshow (1991), *Sampling of Populations: Methods and Applications*, New York: Wiley.

Levy, P. S. (1977), "Optimum Allocation in Stratified Random Network Sampling for Estimating the Prevalence of Attributes in Rare Populations," *Journal of the American Statistical Association*, **72**, pp. 758–763.

Nathan, G. (1976), "An Empirical Study of Response and Sampling Errors for Multiplicity Estimates with Different Counting Rules," *Journal of the American Statistical Association*, **71**, pp. 808–815.

Nathan, G., U. O. Schmelz, and J. Kenvin (1977), *Multiplicity Study of Marriages and Births in Israel*, DHEW Publication No. (HRA) 77-1344, Series 2, Number 70, Washington, DC: National Center for Health Statistics.

Shah, B. V. (1981), *SESUDAAN: Standard Errors Program for Computing Standardized Rates from Sample Survey Data*, Report No. RTI/5250/00-15, Research Triangle Park, NC: Research Triangle Institute.

Sirken, M. G. (1970), "Household Surveys with Multiplicity," *Journal of the American Statistical Association*, **65**, pp. 257–266.

Sirken, M. G. (1972), "Stratified Sample Surveys with Multiplicity," *Journal of the American Statistical Association*, **67,** pp. 224–227.

Sirken, M. G., and P. S. Levy (1974), "Multiplicity Estimation of Proportions Based on Ratios of Random Variables," *Journal of the American Statistical Association*, **69,** pp. 68–73.

Tourangeau, R. (1993), *Sample Design for the Health Insurance Plan Survey (NMES-HIPS) of the National Medical Expenditure Survey*, NORC/WESTAT Technical Report, Washington, DC: Agency for Health Care Policy and Research.

U.S. Bureau of the Census (1991), *Statistical Abstract of the United States: 1991*, 111 ed., Washington, DC.

Data Collection and Response Quality

CHAPTER FOURTEEN

Issues in Surveying Businesses: An International Survey

Anders Christianson[1]
Statistics Sweden

Robert D. Tortora
U.S. Bureau of the Census

When compared to household surveys, there is little published literature on the measurement of the quality of business surveys. Morgenstern (1950) indicated levels of accuracy for several areas of economic statistics. The general approach is to identify sources of error at a disaggregate level, note how they might cumulate, and obtain differences in aggregate economic statistics when error is included in the estimation process. In a more recent report, the Federal Committee on Statistical Methodology (1988) uses the same initial approach as Morgenstern, identifying six sources of error in business surveys: sampling, specification, coverage, response, nonresponse, and processing error. Sources of each nonsampling error are identified along with specific techniques to control and measure the errors.

Survey models are essential to understand the process generating measurement errors. In Chapter 15, Biemer and Fecso discuss methods to evaluate measurement errors in business surveys using two survey error models. The first is the Hansen et al. (1961) general survey error model, and the second is a cognitive model adapted by Edwards and Cantor (1991) for business surveys.

---

[1]Twenty-one international statistical agencies provided 104 survey questionnaires, which create the basis for this chapter. The questionnaires were accompanied by well-considered comments as well as reprints of articles, memoranda describing issues in further detail, and questionnaires for specific surveys. The agencies took considerable effort to provide answers, and we take this opportunity to express our deep gratitude to all involved.

Business Survey Methods, Edited by Cox, Binder, Chinnappa, Christianson, Colledge, Kott.
ISBN 0-471-59852-6 © 1995 John Wiley & Sons, Inc.

Data collection processes not only need to be evaluated and controlled, but they also need to be improved. Improvements have been achieved for household surveys by applying behavioral science methods. In Chapter 16, Dippo et al. demonstrate how similar methods can be adapted to business surveys.

Mail is the predominant mode of data collection for business surveys, either as a single mode or in combination with other modes like telephone and face-to-face visits. Reasons for this predominance are the perceived lower costs associated with mail surveys and the frequent need for access to information filed in the business' accounting system. Dillman (1978) describes the total design method (TDM) for improving response rates for mail surveys. TDM is adapted to business surveys by mail by Paxson et al. in Chapter 17.

The telephone mode, whether computer-assisted or paper and pencil, is being used much more often for data collection. Some agencies are switching to the telephone as a primary mode, but typically the telephone is used to reduce nonresponse error. Thus, business surveys, while still using mail for a large part of data collection, are quickly becoming multimode surveys. These modes are quite different from those used in household surveys. In Chapter 18, Werking and Clayton describe a 7-year research program on the impact of automated telephone collection technologies for the U.S. Current Employment Statistics Survey. The use of touchtone data collection in tandem with computer-assisted telephone interviews reduced collection costs below that of mail while improving timeliness and control.

Increasingly, businesses are storing information on electronic media. This raises the possibility that businesses might transfer data directly from their electronic files, either to a diskette or through a telecommunications network, thereby reducing respondent burden and costs. Electronic data interchange (EDI), this new and not yet proven mode of data collection, is described by Ambler et al. in Chapter 19, together with its implementation by the U.S. Bureau of the Census.

Chapters 15 to 19 describe these major trends in the development of business surveys. This chapter provides a background for this work by discussing data collection issues and response quality, based upon an international survey of statistical agencies conducting business surveys. Although not probability-based, the international survey suggests that mail may be the predominant mode of data collection. It further suggests that response rates can be increased by personalizing the approach to businesses. Business survey questionnaires do not appear to be redesigned very often; but when they are redesigned, modern methods such as the use of focus groups or cognitive interviews are not used. The participating agencies reported many potential sources of measurement error.

When quality is evaluated, it is done mostly by standard methods such as computation of confidence intervals, nonresponse rates, and comparisons with external sources. Reinterviews and nonresponse follow-up studies to measure bias are rarely performed. The major conclusion drawn from this international survey is that there is a considerable gap between the need for quality evaluation (indicated by the many issues reported) and the evaluation actually performed. This gap limits the ability to accurately evaluate the quality of business surveys.

14.1 ABOUT THE SURVEY

Twenty-one statistical agencies in 16 countries responded to this mail survey. The agency response rate was 75 percent. The survey obtained information on several broad topics: (1) the total number of business surveys and censuses (hereafter called surveys) by mode of data collection, (2) specific data on self-selected agency data collections, and (3) the survey manager's views about quality issues. Caution should be used in making inferences to all government statistical agencies or all business surveys because the survey was not based on a probability sample of agencies or surveys.

The survey had two questionnaires. The first, the agency-level questionnaire, obtained data on the total number of business surveys conducted by each agency. The second questionnaire, where agencies self-selected important data collection efforts, had several parts: scope of the survey, sample design, data collection, nonresponse rates and issues, questionnaire design issues, quality evaluations, and survey research. For this second questionnaire, we asked agencies to choose surveys that were important—that is, played a vital role in quantifying the economic health of a specific sector of the economy. This part of the survey questionnaire was designed to collect data on different types of industries. The agencies provided data on a total of 1387 surveys on the agency-level questionnaire and provided detailed information on 104 important surveys.

14.2 THE AGENCY-LEVEL QUESTIONNAIRES: RESULTS

The agency questionnaire obtained information on the total number of surveys conducted by each agency by data collection mode. When more than one mode was used, the agency was asked to estimate the proportion of data collected by each mode. Of the 1387 surveys reported by the agencies, almost half (663) used mail as the only mode of data collection. Of the other single-mode surveys, the telephone was used for 81 surveys whereas face-to-face visits were used for 50 surveys. Other single-mode data collections were electronic data interchange and secondary uses of registers or administrative records (e.g., data on imports into a country from customs duties or taxes), with 62 and 65 total collections, respectively.

Almost every combination of modes was reported. The most common mixed-mode survey used mail/telephone/face-to-face interviewing (222), with the second most common using mail and telephone (172). Mixed-mode mail/ face-to-face visits and telephone/face-to-face visits accounted for only 24 and

12 surveys, respectively. Thirty-six surveys were identified as using other mode combinations such as mail/registers, mail/EDI, and so forth.

For mixed-mode surveys, the agencies were asked to estimate the percent of data collected by the different modes. Data collected by mail dominates these mixed-mode surveys. Of the 222 surveys using mail/telephone/face-to-face modes, the median percents were 70, 20, and 10 percent, respectively. The corresponding medians for the 172 surveys using mail/telephone were 79 and 21 percent, respectively.

Overall, then, the most striking result for the agency-level questionnaire is the dominance of mail. Nearly 80 percent of business surveys use the mail as a mode of data collection. For mixed-mode mail/telephone or mail/face-to-face surveys, at least 70 percent of the questionnaires are collected by mail. Although we do not know of any attempts to quantify household or person-based government surveys with respect to data collection mode, we hypothesize that only a minority of those surveys use mail. (Of course, some very large population censuses, such as those in the United States, use mail as a primary mode.)

14.3 AGENCY-SELECTED SURVEYS: RESULTS

The 104 agency-selected surveys were distributed by type as follows: 21 industrial; 17 agricultural; 15 retail, wholesale, or service; 12 multipurpose; 11 construction; 8 local government; 6 education; 6 foreign trade; 4 energy; and 4 health care providers. A wide variety of surveys was obtained in an attempt to cover the diverse nature of business surveys.

14.3.1 Modes of Data Collection

Respondents classified the self-selected surveys by mode of data collection. Five modes were identified: mail, telephone, face-to-face, electronic, and secondary use of registers or administrative records. The telephone mode was subdivided into paper and pencil, computer-assisted telephone interviews (CATI), touchtone response, and voice recognition; the face-to-face mode was subdivided into paper and pencil and computer-assisted personal interviews (CAPI). Electronic response included options such as facsimile responses (FAX) and digital files created by responding businesses. Of the 104 surveys, slightly more than half (54) were single-mode data collections. Thirty were mail only, 15 were face-to-face only, 7 used secondary registers or records, and 1 each were solely telephone or electronic.

Fifty of the 104 self-selected surveys used multiple modes of data collection. Table 14.1 provides a summary of these multimode surveys, noting the different mode combinations, the median proportion of data collected by mode, and the range of data collected by mode.

As seen in the agency-level results, mail predominated even in mixed-mode

Table 14.1 Multimode Surveys: Number, Median Percentage, and Ranges of Data Collected by Mode

Multimode Combinations	Median Percentage Collected by Mode	Range of Percentages Collected by Mode
Mail	90	30–90
Telephone ($n = 20$)	10	1–70
Mail	20	10–30
Face-to-Face ($n = 3$)	80	70–90
Mail	50	2–91
Electronic ($n = 7$)	50	4–98
Mail	52	25–80
Secondary ($n = 2$)	50	20–70
Mail	60	2–70
Telephone	30	20–80
Face-to-Face ($n = 5$)	18	5–30
Mail	89	75–95
Telephone	5	3–20
Electronic ($n = 4$)	5	5–9
Mail	61	NA
Telephone	5	NA
Face-to-Face	30	NA
Electronic ($n = 1$)	4	NA
Mail	84	NA
Telephone	13	NA
Electronic	2	NA
Secondary ($n = 1$)	1	NA
Telephone	65	2–96
Face-to-Face ($n = 4$)	35	4–98
Secondary	50	10–90
Electronic ($n = 2$)	50	10–90
Secondary	90	NA
Telephone	5	NA
Face-to-Face ($n = 1$)	5	NA

surveys. Forty-three surveys (86 percent) used mail as a component. When used in combination with other modes, mail usually accounted for the largest proportion of data. Of the 20 mail/telephone mixed-mode surveys, the median percentage collected by mail was 90 percent. In almost all combinations where mail was used, the median percentage obtained by mail was at least 50 percent. Only when used in combination with face-to-face interviewing did mail account for a median percentage of less than 50 percent—then only 20 percent.

Single-mode telephone data collections were rare in this survey (only one reported), but in combination with other modes, the telephone was used for 36 of the 50 mixed-mode data collections. However, only when telephone was mixed with face-to-face interviewing did the telephone account for a majority of completed interviews, with a median of 65 percent. Respondents indicated that the telephone was used for nonresponse follow-up in 17 surveys. (In two cases, respondents indicated that the telephone was used to obtain the name of the appropriate contact person or to schedule an interview.) CATI was more prevalent than CAPI, reported for nine surveys. The median percentage of data collected by CATI was 18 percent, with a range of 3–100 percent. In three cases, survey questionnaires were hand-delivered to respondents and either picked up at a later date or returned by mail.

Ten data collection efforts used three modes of data collection, and two used four modes. In one survey, while the data collection effort involved personal delivery of the questionnaire, respondents supplied data using different modes depending on the data type: accounting data were provided on digital tapes, capital value items were provided in a face-to-face interview, and other data were returned by mail.

Other technologies were identified in only 18 surveys. Bar codes for automated check-in were mentioned in eight surveys. Voice recognition (in an experimental mode), optical mark recognition, on-line editing, digital tapes, and FAX were other new technologies being used.

There were few recent changes in data collection mode. Only eight respondents replied positively to a question about a change in data collection mode occurring after January 1, 1990. Most changes involved moving towards automation: paper-and-pencil/face-to-face interviewing to CAPI, face-to-face interviewing or mail to CATI (mentioned four times), direct automated electronic reporting (mentioned two times), mail replaced by touchtone data entry, and face-to-face interviewing replaced by mail.

With respect to upcoming changes of data collection mode, including new technologies, 86 surveys reported no changes planned by 1994. The types of change for the remaining 18 are summarized in Table 14.2. Using the telephone more was mentioned 12 times (CATI, touchtone, paper and pencil telephone, and voice recognition). Changing to CAPI was not identified as a pos-

Table 14.2 Planned Changes in Data Collection Mode by 1994

Change	Number Mentions
Electronic Response, Including Diskettes	12
Administrative Registers	4
Computer-Assisted Telephone Interviews	4
Touchtone Data Entry	3
Paper-and-Pencil Telephone	3
Voice Recognition	3
Optical Character Recognition	1
Total	30[a]

[a]Adds to more than 18 because some agencies reported more than one planned change.

sible mode change, reflecting the relatively low use of face-to-face interviewing in data collection and the dominant use of less expensive modes.

14.3.2 Approaching the Business

An important part of the data collection process is the approach to the respondent, particularly for mail surveys. The main goal is to get the questionnaire into the hands of the proper person for accurate completion. This often includes going through a "gatekeeper," such as a receptionist, to reach the owner or company official responsible for a particular activity. For mail surveys, Dillman (1978) noted the importance of personalizing the request to get the attention of the respondents (see Chapter 17). To quantify current practice in business surveys, for each agency-selected survey, we asked "Who do you approach?" For all surveys, about two-thirds use a title or contact name in their approach (Table 14.3). On the other hand, for surveys that collect at least 50 percent of data by mail, slightly more than 75 percent personalize the approach by using a title or contact name.

14.3.3 Nonresponse

Nonresponse issues are important for all surveys. Of interest for our survey were the latest (unit) nonresponse rates, trends in nonresponse rates, and issues that affect nonresponse. Respondents were asked to report the unit nonresponse rate obtained the last time the survey or census was conducted. In total, they reported on 19 censuses and 78 surveys. Table 14.4 summarizes these unit nonresponse rates. Readers are cautioned that agencies used their own measures of nonresponse, which must be considered in interpreting these quantitative results.

Respondents were asked to categorize the trend in nonresponse rates over the past 10 years as increasing, unchanged, or decreasing. Table 14.5 summarizes these results. Overall, whether looking at a census or survey, the trend for nonresponse was about the same. The number of censuses (or surveys) with an increasing trend in nonresponse rate were about the same as those with a decreasing trend. However, the majority had an unchanged nonresponse rate over the 10-year period. For those with decreasing trends, reasons such as increased nonresponse follow-up, more automation, improved notification or precanvasing, use of reminders, use of administrative registers, reduction of data content, and later survey data collection were cited by the statistical agencies. For those data collections with unchanged rates, the most common reason cited was a change to mandatory reporting (four times). When respondents mentioned reasons for an increasing trend in nonresponse, they cited nonavailability of accounts during the survey period, longer interviews, sensitive nature of financial data, decreased budgets, strikes, business consolidation or bankruptcy, and inaccurate postal addresses.

Our question about the maximum item nonresponse rates for major variables

Table 14.3 Approach to the Business: Number by Approach for All Surveys and for Surveys That Collect Majority of Data by Mail

Approach	All Surveys	Surveys That Collect Majority of Data by Mail
Company Name Only	20	15
Company Name and Title of Appropriate Official	37	23
Contact Name Identified at First Contact	14	12
Contact Name Identified from Previous Data Collection or Other Source	27	13
Presurvey Notification to Obtain Contact Name	5	—
Not Applicable,[a] Not Answered	6	1
Total	109[b]	64

[a]Data collected from registers or administrative records.
[b]Total adds to more than 104 because some surveys used multiple approaches.

Table 14.4 Most Recent Unit Nonresponse Rates: Medians, Quartiles, and Ranges

Type	Median	Lower Quartile	Upper Quartile	Range
Census ($n = 19$)	3	0	10	0–25
Survey ($n = 85$)	13	6	22	0–60

Table 14.5 Trends in Nonresponse Rates[a] Over the Last 10 Years

Type	Increasing Trend	Unchanged	Decreasing Trend	No Answer
Census ($n = 19$)[a]	4	9	3	3
Survey ($n = 85$)	23	34	25	3

[a]Included in this category are data collection efforts that rely totally on administrative registers.

had a high item nonresponse rate. Only 32 respondents supplied an answer. Of those responding, the median maximum item nonresponse rate was 10 percent with a lower quartile of 3 percent, an upper quartile of 25 percent, and a range of 0–39 percent.

When respondents mentioned timeliness, reasons such as the data collection period being too short (five times) or the survey arriving when requested data were unavailable (five times) were cited (Table 14.6). The most common relevance issue was a general lack of interest on the part of respondents (11 times), but other reasons include the unit not having the requested data (five times), tired respondents dealing with many data requests, and the nonmandatory nature of the request. With request to cost, the most frequent reason cited for nonresponse was that the data were too hard to supply because they either asked for too much or were difficult to obtain (six times). On the other hand, respondents indicated a lack of resources to conduct nonresponse follow-up (four times), a major nonresponse issue.

Table 14.6 Most Important Nonresponse Issues

Issue	Number
Timeliness of Effort	10
Relevance of Data	19
Cost to Survey Organization or Respondent	13
Accuracy of Data or Addresses	20
Size of Business	7
Characteristics of Nonrespondents	5
Coverage of Administrative Registers	2

Table 14.7 Distribution of Nonresponse Rates by Data Collection Mode: Medians, Quartiles, and Ranges

Mode	Median (%)	Lower Quartile (%)	Upper Quartile (%)	Range (%)
Mail ($n = 62^a$)	13	15	25	0–60
Telephone[b] ($n = 8^a$)	6	1	15	0–20
Face-to-Face[b] ($n = 19^a$)	10	3	16	0–40

[a]A data collection reporting 50 percent of data collected by two modes was counted in both modes.
[b]Including CATI and CAPI data collections.

Accuracy issues fell into several categories. Inability on businesses' part to provide accurate data (four times), poor addresses or inability to contact respondents (eight times), and out-of-scope units that were difficult to identify (five times) were major issues. Seven respondents mentioned size as an issue, four citing small businesses and three citing large ones. Finally, five respondents cited the need to identify the characteristics of nonrespondents to account for nonresponse during estimation.

Not surprisingly, surveys using mail had the highest nonresponse rates in all categories of comparison (Table 14.7). The highest median nonresponse rate (16 percent) was associated with the most general approach to the business—namely, use of the company name only (Table 14.8). However, it was closely followed by the most complex approach, using a presurvey notification asking for a company contact name (15 percent median nonresponse rate). The other, more specific approaches were (1) company name and title of an official, (2) contact name from previous data collection of another source, and (3) contact name identified at first contact, which have lower median nonresponse rates of 10 percent, 10 percent, and 9 percent, respectively. However, even

Table 14.8 Respondent Approach by Nonresponse Rate: All Surveys

Approach	Number	Median Nonresponse Rate (%)	Range (%)
Company Name	20	16	2–31
Company Name and Title of Appropriate Official	37	10	0–49
Contact Name Identified at First Contact	14	9	0–25
Contact Name Identified from Previous Data Collection or Other Source	27	10	0–60
Presurvey Notification Asking for Company Contact Name	5	15	7–40

Table 14.9 Respondent Approach by Nonresponse Rate: Surveys Where the Majority of Data Were Collected by Mail

Approach	Number	Median Nonresponse Rate (%)	Range (%)
Company Name	15	20	6–40
Company Name and Title of Appropriate Official	23	10	0–49
Contact Name Identified at First Contact	12	9	0–25
Contact Name Identified from Previous Collection or Other Source	13	10	0–60

these very specific approaches did not guarantee low nonresponse rates because 75 percent of the more specific approaches had higher maximum nonresponse rates than did the most general approach using names only.

Table 14.9 compares respondent approach and nonresponse rate for mail surveys. Mail surveys are defined as those surveys where a majority of the responses are obtained by mail. Also included are those surveys where the response rate is equally divided between mail and some other mode. Again, we see the same overall pattern. The most general approach, company name only, has the highest median nonresponse rate (20 percent). The more specific approaches have a median nonresponse rate of about half of the nonresponse rate of the general approach.

14.3.4 Questionnaire Design

Questions about the process of designing or redesigning the questionnaire was obtained by asking respondents to describe the evolution of the current questionnaire (from January 1990) and plans for future questionnaire development (to December 1994).

Agencies were asked about (1) the use of cognitive psychologists or other specialists, either for the current questionnaire or for coming redesign, and (2) involvement of respondents in questionnaire redesign. Eight respondents did not answer the question because they did not use a questionnaire. The use of cognitive experts in the broad sense used here was rather limited (17 current questionnaires). There were no signs that cognitive experts will be used much more in future redesigns, primarily because there will be little redesign (5 out of 34). Table 14.10 quantifies respondent involvement for questionnaire redesign.

Almost 90 percent of the responses indicated some form of respondent involvement. The most common type of respondent involvement was through informal or ad hoc contact (slightly more than 40 percent of the total). While one-third indicated contacts with organizations when redesigning question-

Table 14.10   Respondent Involvement in Questionnaire Redesign

Classification	Number of Surveys[a]
No Respondents Involved in Redesign	14
Informal or Ad Hoc Contacts with Respondents	57
Contacts with Respondents' Organizations	45
Discussion in Focus Groups	18
Total	134

[a]Adds to more than 104 because multiple respondent contacts were reported.

naires, focus groups were reported only 18 times. For upcoming questionnaire redesigns, focus groups will be used nine times.

Behavioral science methods, like the use of focus groups and cognitive experts, have proven useful for improving questionnaire design (see Chapter 16). These methods could be an important potential means of improving the data collection process. Part of the improvement will be directed at reducing measurement errors associated with the interview and questionnaire.

14.4   MEASUREMENT ISSUES FOR IMPORTANT VARIABLES

The agency-selected survey questionnaires asked for the two most important variables of each survey and for the main measurement problem associated with each variable. The variables fall into seven broad classes: (1) business characteristics; (2) production; (3) expenditures and investments; (4) turnover, revenues, and volume sold; (5) employment and wages; (6) consumption and waste; and (7) transactions.

Business characteristics include variables such as ownership, hospital bed size, floor space, and farm acreage. Examples of measurement problems for farm variables were: (1) determining acres operated whenever multiple parties (landlords, tenants, partners) are involved in one farm arrangement; (2) incorrect reporting of acreage especially by telephone; (3) underreporting of pasture, unimproved, and noncontiguous land; and (4) misclassification of livestock numbers by category.

Production is the major variable for most industry surveys but is also important for other sectors such as construction. Requested by volume or value, production generally comes from the business' accounting system. One major source of error is the discrepancy between the accounting system specifications and questionnaire definitions. As one agency noted, "Industry Gross Product (IGP) is a derived item with several major components. Several of the components are currently the focus of a post-enumeration study. The main problem faced with IGP is that the components are defined for national accounting purposes and these definitions do not always align with definitions used in financial accounting."

Some agencies reported uncertainty about the overall quality of production volume measurements, whereas others identified more specific issues. Some of these issues were: (1) subjective assignment of fair market values for products shipped to plants within the company; (2) reference period differences between businesses and survey; (3) data not available when needed for the survey resulting in rough estimates; (4) small businesses lacking complete or organized accounting records; (5) real estate developers having difficulties in estimating the market prices of development projects and underlying land values; (6) difficulty in measuring quantity change of manufacturing production; (7) difficulty in finding the right person to approach; (8) time for respondents to learn to complete the questionnaire; and (9) for construction production statistics, loan applications for government-granted loans may cause applicants to use loan application rules instead of survey definitions.

Expenditures and investments are frequently the major variables for industry and local government surveys. Specific variables may be current expenditures, fixed capital expenditures, and total government investments for particular uses such as support for education or environmental protection. The measurement problems show many similarities to those for production.

One frequently reported issue for local government surveys was the reconciling of accounting data to economic concepts. Timing issues, such as floating fiscal years, cause problems. Another issue resulted from the conceptual problems for capital investment for farm financial surveys.

Measuring industry investment in environmental protection involves serious classification problems. Investment in new production equipment may be made for cost effectiveness as well as for environmental protection. Classification of such investments into environmental and production categories is based upon judgment. Turnover, revenues, and volume sold are the typical variables for retail, wholesale, and service surveys, although they occur with other surveys.

Specification problems occur, including different time periods for requested data versus the data available in business' accounts. For instance, businesses may report retail instead of total turnover. Other surveys reported irrelevant turnover included in error as well as the exclusion of relevant turnover. Businesses may also revise figures when reporting turnover.

Like wages, salaries, and hours, number of employees is an important variable for business surveys. This variable occurs frequently in industry and multipurpose surveys. For local government and institutional surveys, it is increasingly being collected by EDI.

The number of employees variable may be affected by the fact that increasingly companies are using leased employees. Based on current definitions, these employees are not included because they are contracted through another company. For small companies and farms, owners may erroneously report themselves as employees. Occasional employees may be excluded in error, while employees absent due to illness may not be excluded when they should be. For labor variables, the major issue seems to be the classification of qualifications and the determination of main field of practice.

For wages and hours, the issues revolve around the inclusion of irrelevant salary components, underreporting of special payments, and part-time employees counted as full-time employees. In reporting total income, off-farm income may be excluded from farmers' returns.

Surveys of business consumption of energy, farm consumption of water for irrigation, and industrial waste report the following examples of measurement issues: (1) incorrect energy supplier accounting; (2) deliveries of fuel to gas stations may be reported when survey specifications request that only deliveries to end users be recorded; (3) water used for irrigation is not metered accurately, and therefore the data are estimates only; and (4) waste measurements are based on rough estimates because companies do not record quantities of waste generated.

Typical variables for foreign trade surveys are the characteristics of particular trade transactions. Although measurement of such characteristics seems uncomplicated, underreporting of exports and inadequate coverage of international services can create serious problems for a nation's balance-of-payments accounts (National Research Council 1992). Ryten (1988) analyzes the differences between the origin country's export data and the destination country's import data and their relations to factors such as timing, value, and third country.

Value of transaction, country of origin (for imports), or destination (for exports) are the most common variables measured in foreign trade surveys; all face measurement problems. Examples of such problems are: (1) the insurance and freight components may be improperly omitted; (2) full values may be reported although reporting in thousands is requested; and (3) for the classification of imports, difficulties may occur in coding orders accurately to direct consumption, intermediate use, or capital goods.

14.5 EVALUATION OF ACCURACY

Table 14.11 summarizes the coded responses on how accuracy was evaluated for the 104 surveys. The responses were classified into the table categories, with more than one answer possible per survey. Only special efforts to evaluate

Table 14.11 Evaluation of Accuracy

Type of Evaluation	Number of Times Reported
Comparison with Other Sources	57
Nonresponse Follow-Up Studies	8
Validation/Reinterview	10
Response Analysis Surveys	4
Other	5
No Response	27

quality are reported for this question. Standard measures of quality such as sampling variances and nonresponse rates are not included.

More than one-fourth of the surveys report no special effort to evaluate quality. Half use external source comparisons or checks using benchmark data, other previous estimates, or related statistics not involving additional data collection. Special studies to obtain measures of quality such as nonresponse follow-up studies, reinterviews, and response analysis surveys occurred much less frequently, perhaps because their demand for new data collection makes them more expensive.

As reported in Table 14.11, for 57 surveys comparisons were used to evaluate current estimates. The inventive power of survey designers is easily recognized. For one particular survey, the following comparisons were made to verify accuracy: (1) unit reports and aggregates were edited to ensure consistency and logical relationships between items; (2) aggregates were compared with internal and external data sources, reconciling survey data from the national accounts data; (3) each year's data were evaluated with an emphasis on editing movements between years for unit records and aggregates; and (4) unit record data for the largest businesses were compared with publicly available annual reports. As shown in Table 14.12, the 57 surveys reported 80 different comparisons.

Comparisons with statistics obtained from other sources, within or outside the agency, were most frequently used to check survey results. The external comparisons came from surveys in related fields, administrative data, annual financial reports, association data, or research institute data. Specific examples include: hospital associations; birth registrations; input–output construction tables; reconciliation of data with major trading partners; trade statistics; trade organization data; financial statistics; and for agriculture surveys, comparison with processing statistics such as slaughtering and ginnings data.

Internal consistency checks were used for many surveys. The checks identify outliers and are often used to edit data. Internal checks include standard errors or relative standard errors, unit-by-unit plausibility checks, unit-by-unit checks with previous records for the same unit, unit-by-unit checks with similar administrative data, and agency auditing of financial statements for governments and institutions.

Table 14.12 Evaluation of Quality by Internal or External Comparisons

Type of Comparison	Number of Times Reported
Check with External Sources	43
Check for Internal Consistency	14
Check with Benchmark Data	8
Comparison with Previous Estimates	15

14.5.1　Evaluation of Nonresponse Bias and Coverage Error

Nonresponse is an important source of error for surveys and censuses. Estimation of nonresponse bias can be expensive, and such studies are infrequent. Only eight surveys reported such studies, with two expected in the future. The reported studies fell into three classes: studies for nonresponse occurring for any reason, special studies of refusals or late respondents, and studies using an external data source to compare respondent and nonrespondent characteristics. Three surveys reported regular nonresponse follow-up studies.

Coverage follow-up was more infrequent than nonresponse follow-up. The two examples provided were (1) a supplemental sample of physicians not engaged in office-based practice and (2) an evaluation study for an employment survey.

14.5.2　Recontacts and Reinterview Surveys

Household surveys frequently include reinterview surveys. Their main uses are to measure response variance and to prevent curbstone interviews. Business surveys make little use of reinterview surveys. Generally, reinterview studies are conducted only for a few large surveys, a fact perhaps attributable to cost.

Pafford (1988) describes a large-scale reinterview program. The reinterview program included 1000 respondents interviewed within 10 days of the original data collection. The reinterview's purpose was to estimate biases in grain stock estimates in major agricultural states. Other reinterview studies reported by agencies include reinterviews to check the accuracy of construction data, verification by field visit of a subsample of telephone cases for a trade survey, abstracting a subsample of hospital discharge records, and subsampling physicians to validate interviewer conduct.

One survey reported a more thorough and more detailed reinterview, called a *response analysis survey*, which analyzes respondents' comprehension and interpretation of questions. In one instance, CATI was used to check on the effectiveness of an initial conversion to some other mode of data collection.

14.5.3　Other Validation

For some applications, postcensus comparisons of individual measurements with those obtained by a superior measurement instrument can be less expensive than reinterview surveys. One example is the comparison of expectations with actual realizations for loan amounts, company investments, and foreign trade data.

Agencies also described sources of inaccuracy that were not quantified, such as reporting periods that differed from the survey's specifications and the use of unsatisfactory measurement instruments. Only a few agencies explicitly mentioned using sampling errors or confidence intervals to evaluate quality, perhaps because this was not an explicit answer category on the survey questionnaire.

14.6 IMPROVING SURVEY QUALITY

When comparing measurement issues with the evaluation of accuracy, two questions occur:

- Are the evaluations appropriate in relation to the issues?
- Do the evaluations provide sufficient knowledge about survey quality?

A conceptual model for the interrelations between quality, agency actions, and user perceptions provides a suggested framework for increasing knowledge of survey quality. The model is visualized in Figure 14.1.

Perfect quality is an imaginary concept corresponding to the survey's ideal goal. It is achieved when the survey is relevant to all user needs, obtains true values from the interviews, and publishes the results in time for user decision-making. Although imaginary, it is useful for the survey agency and its users to work to achieve an understanding of perfect quality. Users may stress such aspects of quality as relevance and timeliness, which should then be acknowledged by the agency. On the other hand, users may underestimate the importance of the accuracy aspects of quality.

Perceived quality is an agency's belief about the survey's quality based on data, folklore, or other factors. Perceived quality can change over time. Survey design, survey operations, and quality control measures determine *actual quality*. Sampling, coverage, nonresponse, and measurement errors create discrepancies between perfect and actual quality, thus creating differences between survey estimates and the true values. Those differences are not estimable. Thus, the most important consequence of discrepancy between perfect and actual quality is the uncertainty associated with survey results whenever decisions are based upon them.

The evaluation of quality reflects the agency's quantification of quality. Confidence intervals describe the uncertainty associated with sampling. Other measures of uncertainty for coverage bias, nonresponse bias, and measurement errors are not as easily obtained. Special evaluation studies are expensive and rarely used. Instead, indicators of quality such as nonresponse rates are used.

This suggests a quality hierarchy. In many instances, there is only a vague feeling that "something may be wrong" without real knowledge of what is wrong or how wrong it is. In fact, this perception may be accidentally obtained. For example, an initial use of CATI may result in the discovery of incorrect reporting of retail turnover, or survey units may report two different answers to the same basic question. The next level of the hierarchy is knowledge of sampling errors that are used in confidence intervals as measures of quality and uncertainty. A higher level includes the measurement of indicators of quality as advocated by Groves and Tortora (1991). They argue that weak indicators of nonsampling errors may offer valuable information even if they fail to meet the necessary criteria for inclusion in formal cost–error models. Thus, an agency's perceived quality is the result of the agency's interpretation

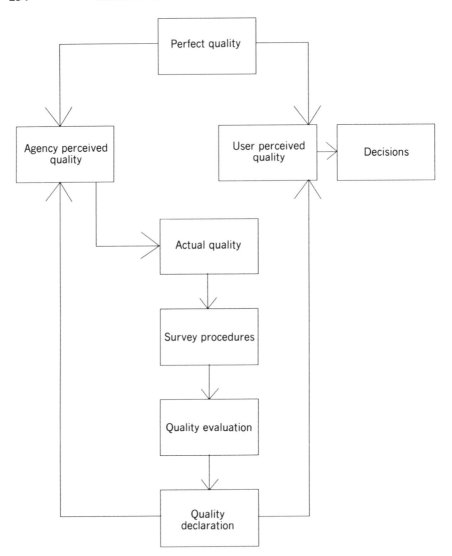

Figure 14.1 Survey quality cycle.

of quality evaluation. The poorer the quality evaluation, the more an agency's perceived quality relies on assumptions.

The quality declaration is an important way for the agency to communicate (internally) to users. Statistics Canada (1992) provides a working model for a quality declaration. The users' perceived quality, informed by quality declaration, provides a basis for decision-making. Iterations of the process can improve overall survey quality.

14.7 CONCLUDING REMARKS

With agencies choosing their important business surveys for this study, surprisingly little is known about data accuracy considering the magnitude of measurement issues raised. In the first edition of *On the Accuracy of Economic Observations*, Morgenstern (1950) called for the improvement of error evaluations. It appears that little progress has been made since then.

There are at least four hypotheses for why quality is poorly evaluated. First, users may not be demanding information about the accuracy of survey estimates. In particular, they may not question whether survey statistics are accurate. Second, some agencies may find it difficult or embarrassing to publish information on survey shortcomings. Third, for repeated surveys, the results of quality evaluation may call for a redesign. Such redesigns cause changes in the time series not attributable to real change, creating user interpretation problems. Fourth, there may be an over-belief in the benefits of editing. In Chapter 21, Granquist argues that agencies spend too much resources on editing, not achieving an acceptable quality increase to support the investment. The international survey supported this view.

The issues we report in this chapter need to be addressed specifically, because general quality control measures or editing are unlikely to resolve these problems. A two-path approach to obtaining better information about accuracy seems appropriate. One path evaluates specific components of accuracy (through special studies). The other path offers regular quantification of accuracy. Together, these paths bring the concepts of quality (actual, agency-perceived, and user-perceived) closer while moving toward perfect quality.

REFERENCES

Dillman, D. (1978), *Mail and Telephone Surveys—The Total Design Method*, New York: Wiley.

Edwards, W. S., and D. Cantor (1991), "Recall Error: Sources and Bias Reduction," in P. P. Biemer, R. M. Groves, L. E. Lyberg, N. A. Mathiowetz, and S. Sudman (eds.), *Measurement Errors in Surveys*, New York: Wiley, pp. 211–236.

Federal Committee on Statistical Methodology (1988), *Quality in Establishment Surveys*, Statistical Policy Working Paper 15, Washington, DC: Office of Management and Budget.

Groves, R. M., and R. D. Tortora (1991), "Developing a System of Indicators for Unmeasured Survey Quality Components," *ISI Booklet*, **II**, pp. 280–295.

Hansen, M. H., W. N. Hurwitz, and M. A. Bershad (1961), "Measurement Errors in Censuses and Surveys," *Bulletin of the International Statistical Institute*, **38**, pp. 359–374.

Morgenstern, O. (1950), *On the Accuracy of Economic Observations*, 1st ed., Princeton, NJ: Princeton University Press.

National Research Council (1992), *Behind the Numbers: U.S. Trade in the World Economy*, ISBN 0-309-04590-8, Washington, DC: National Academy of Sciences.

Pafford, B. (1988), "Quality Assurance Program for the Agricultural Survey Program," *Proceedings of Survey Research Methods Section, American Statistical Association*, pp. 71–74.

Ryten, J. (1988), "Errors in Foreign Trade Statistics," *Survey Methodology*, **14,** pp. 3–18.

Statistics Canada (1992), "Policy on Informing Users of Data Quality and Methodology," Ottawa.

CHAPTER FIFTEEN

Evaluating and Controlling Measurement Error in Business Surveys

Paul P. Biemer
Research Triangle Institute

Ronald S. Fecso
U.S. National Agricultural Statistics Service

Survey estimates are almost never identical to the population value they are trying to measure because of the sampling and nonsampling errors they contain. *Sampling error* is the error due to surveying only a subset of the population rather than conducting a complete census of all businesses in the target population. Even in a complete census of businesses, however, the estimate may still differ considerably from the population value as a result of nonsampling error. *Nonsampling error* is the difference attributable to all sources other than sampling error. Nonsampling errors arise during the planning, conducting, data processing, and final estimation preparation stages of all types of surveys including business surveys.

As shown in Figure 15.1, the sources of nonsampling errors may be classified as specification errors, frame errors, nonresponse errors, processing errors, or measurement errors. *Specification error* occurs when (1) survey concepts are unmeasurable or ill-defined, (2) survey objectives are inadequately specified, or (3) the collected data do not correspond to the specified concepts or target variables. *Frame errors* include erroneous inclusions, omissions, and duplications in the sampling frame or process. Also included are errors in listing subunits in surveys that require subsampling within businesses. *Nonresponse errors* include unit nonresponse, item nonresponse, or incomplete data.

Business Survey Methods, Edited by Cox, Binder, Chinnappa, Christianson, Colledge, Kott.
ISBN 0-471-59852-6 © 1995 John Wiley & Sons, Inc.

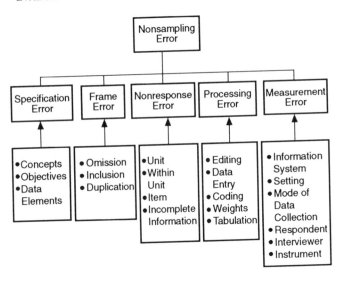

Figure 15.1   Sources of nonsampling error.

Processing errors refer to the errors in post-data collection processes such as coding, keying, editing, weighting, and tabulating the survey data. Finally, *measurement error*, the topic of this chapter, refers to those errors that occur at the time of data collection.

At a recent conference on the subject, Biemer et al. (1991, p. xvii) defined survey measurement error as:

> . . . error in survey responses arising from the method of data collection, the respondent, or the questionnaire (or other instrument). It includes the error in a survey response as a result of respondent confusion, ignorance, carelessness, or dishonesty; the error attributable to the interviewer, perhaps as a consequence of poor or inadequate training, prior expectations regarding respondents' responses, or deliberate errors; and the error attributable to the wording of the questions in the questionnaire, the order or context in which the questions are presented, and the method used to obtain the responses. At the time survey responses are collected, all of these factors may intervene and interact in such a way as to degrade response accuracy.

Five sources of measurement error in business surveys are discussed in this chapter: (1) the survey instrument, (2) the respondent, (3) the information system, (4) the mode of data collection, and (5) the interviewer. The *survey instrument* refers to the survey form or questionnaire and the instructions to the respondent for supplying the requested information. The *respondent* refers to the person(s) whose task it is to supply the requested information, either by accessing the business' information system or by relying on personal or other knowledge. Perhaps the most important feature of business surveys and one that distinguishes these surveys from household surveys is a greater reliance on information or record systems for the required information. These include the business' formal record systems, company or individual files, and informal records. The *mode of data collection* refers to the combination of the communication medium (i.e., telephone, face to face, self-administration, etc.) and the method of data capture (i.e., paper and pencil, computer keyboard, telephone keypad, etc.). Finally, in some business surveys, the *interviewer* may play a key role as the individual who obtains the information from the business and enters the information into the questionnaire. For some surveys, the interviewer may simply be a prerecorded voice over the telephone or may be nonexistent for self-administered surveys.

In this chapter, our primary goals are to review methods for controlling and evaluating measurement error and to study the use of these methods in business surveys. Two general models for the response process are reviewed first. One model is a cognitive model of the response process proposed by Tourangeau (1984) and adapted for business surveys by Edwards and Cantor (1991). The second model is a statistical model for measurement error originally developed by Hansen et al. (1961). Knowledge of both models is needed to understand the evaluation and control techniques that are described subsequently.

15.1 A COGNITIVE MODEL OF SURVEY RESPONSE

Tourangeau (1984) and, more recently, Eisenhower et al. (1992) described five stages of the survey response process for household surveys. As depicted in Figure 15.2, the five stages are: (1) the encoding of information, (2) comprehension of the survey question, (3) retrieval of information, (4) judgment, and (5) communication. These stages reflect an idealized paradigm for response which may be described as follows:

- In formulating an answer for a survey question, the respondent must first have the knowledge, belief, or attitude required to provide a valid response (*encoding*).

- There must be a shared meaning among the researcher, the interviewer, and the respondent with respect to each word in the question as well as the question as a whole (*comprehension*).

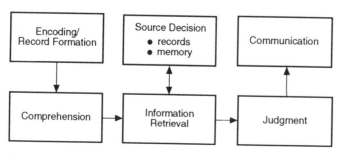

Figure 15.2 Stages of the response process. (Adapted from Edwards and Cantor 1991.)

- To respond to questions concerning past events or behaviors, the respondent attempts to retrieve the required information from memory (*retrieval of information*).
- Once information has been retrieved, the respondent decides how to respond appropriately, taking into account risk, benefit, available answer choices, and so on (*judgment*).
- Finally, the respondent communicates the response to the interviewer (*communication*).

For a single question, the respondent may cycle through some or all of the stages repeatedly as the interviewer continues to probe for a satisfactory response.

As Edwards and Cantor (1991) recognized, this response paradigm does not describe a typical response process for a business survey. Business surveys rely much more extensively on information (record) systems than do household surveys. As an example, when a business survey respondent accesses company records to provide a response for a question such as "the number of full-time employees during the period (date) to (date)," "encoding" and "retrieval of information" take on new meaning. The respondents know where to obtain the required data. The cognitive encoding of this basic procedural information is not enough, however, because to be retrieved by the respondent the required information must be entered in the company's record system. This stage of the response process, which is analogous to cognitive encoding in the household survey model, is called *record formation* by Edwards and Cantor. The retrieval of survey information may require the respondent to invest considerable effort to locate and access the appropriate record sources. The respondent may have several options for responding to the request, and the accuracy of the information may vary considerably with the effort required to access each potential source. The traditional household response model does not explicitly account for this *source decision* stage of the process. To account for these differences between the household and business surveys' paradigms, Edwards and Cantor proposed a revised response model for business surveys that explicitly accounts for the respondent's reliance on information systems rather than on individual recall to respond to business survey questions (see Figure 15.2).

An important benefit of response models is that they allow survey methodologists to decompose the response process into smaller steps that may be treated separately in survey design and evaluation. Let us now consider how the response model can assist the control and evaluation of measurement errors is business surveys.

15.1.1 Encoding/Record Formation

All business survey questions, to some extent, rely on the knowledge, attitudes, and beliefs of the respondent. Thus, careless or uncontrolled choice of the business respondent can be an important source of measurement error. The survey procedures should clearly specify the minimum qualifications the respondent should possess to ensure the accuracy of the respondent's reports. An important aspect of the interviewer's or the addressee's (for mail surveys) task, then, is to identify a respondent meeting these minimum qualifications. In some cases, multiple persons within a business may serve as respondents for various parts of the questionnaire. Separate questionnaires may be needed for each respondent, or, alternatively, a single questionnaire may be used with separate sections corresponding to the change in respondents. For surveys that rely on information contained in the business information system, knowledge of the information system, as well as of the terms and definitions of the data items, are essential qualifications for the respondent. Ultimately, however, the choice of respondent is the business' decision.

The record formation process is also subject to a number of error sources which may undermine data integrity. Some record systems allow input from multiple sources with little or no validation of input. As an example, university student enrollment records may be updated by each college's administrative assistant. Errors are corrected as the system is used, but no formal system for error detection may exist. Another common problem in using administrative record systems to obtain survey data is the inconsistency between the record items and the survey questions. The record system item may include unwanted data components or exclude components that are part of the survey question. The record may reflect a time frame different from the survey's. The burden of adjusting for these differences may be so great that few respondents do so.

Incompatibilities between the survey and the business record system are best remedied by revising the survey item definition because respondents are unlikely to revise their record systems. In a study of the U.S. Current Employment Statistics Survey, Ponikowski and Meily (1989) reported that 59 percent of the businesses did not adhere to the survey's employment definition. The most common problem cited was the inclusion of employees on unpaid vacation. Only 56 percent of the respondents who made this error agreed to adjust their figures to correct the error for future surveys.

15.1.2 Comprehension and Retrieval of Information

The next two stages in the response process are comprehension of the question and the retrieval of required information, either from memory or from records. In evaluating the accuracy of the 1977 U.S. Economic Censuses, Corby (1984) reported that these two stages explained most error found in the censuses' data.

Comprehension
Comprehension of the survey question is particularly problematic in business surveys. Often, the questions contain "legalese" and technical jargon which are not understood by some respondents. Furthermore, the questions may re-

quire that the respondent aggregate data values across time or other units, and it may be unclear what items should be included or excluded. The Economic Censuses study provides many examples of this. For annual payroll in the 1977 U.S. Census of Manufactures, the amount of erroneous inclusion and exclusions totaled $3.7 billion, or about 2 percent of the census total. Erroneous exclusion of vacation pay accounted for almost one-third of this error. In a study conducted by the U.S. National Agricultural Statistics Service (O'Connor 1993), respondents were asked to define "calf," a common term in agricultural surveys. Over 30 different meanings were recorded, including: not yet weaned, up to 800 pounds, under 600 pounds, anything under a yearling, under 300 pounds, everything except stock cows, 8 months or less, and not sure.

Retrieval of Information

Following comprehension, the next stage in the response process is the retrieval of information. Ideally, the respondent accesses records that contain the required information. If no records exist or if retrieving information from records is too difficult, the respondent may attempt to recall the information or to "estimate" the correct response. Krosnick and Alwin (1987) suggest that some respondents *satisfice*; that is, they exert minimal effort in responding to questions, often providing answers that are "good enough" rather than precise responses. This theory may operate both for the retrieval of information from memory and for the retrieval from records. In the 1977 Census of Retail Trade, estimation error accounted for 75 percent of the error in the reported value of employment (Corby 1984). Groves (1989) and Eisenhower et al. (1991) provide excellent reviews of the literature on recall error.

15.1.3 Judgment and Communication

Following comprehension and information retrieval, the last two stages of the response process involves the respondent's judgment of the appropriate response to make and communication of that response to the interviewer or onto the questionnaire for self-administered forms. For closed-ended questions, the respondent decides which response category fits the information retrieved in the previous stage. For open-ended responses, the respondent formulates his/her own response. The respondent may judge a response as posing an economic or social risk or threat and alter his/her response accordingly.

Response judgment and communication may be influenced by the characteristics of the interviewer and the interview mode as well as by the questionnaire. For self-administered surveys, the mechanism by which a respondent records his/her response, such as a laptop computer or a telephone keypad, can exacerbate communication problems if it is faulty or inefficient. Deliberate misreporting arising from a perceived economic or legal risk may be somewhat abated by emphasizing the neutrality of the survey sponsor. An interviewer who probes completely and persistently until an acceptable response is obtained can counteract some of the effects of satisficing.

15.2 A STATISTICAL MODEL FOR SURVEY RESPONSE

The cognitive model of survey response aids in understanding and investigating the causes of response error in surveys. By considering each stage of the response process in the survey design, the problems associated with the successful completion of each stage can be identified and mitigated. However, the response model focuses primarily on the respondent. Other factors that induce measurement error, such as the interviewer, the mode of data collection, and the questionnaire, are represented in the model only through their effects on the respondent. In this section, we consider an alternative model that explicitly represents these other error sources. Such a model is useful for measuring how much of the total error in a survey estimate is attributable to the interviewer, mode, and so on, without regard to which stages of the response process are affected.

15.2.1 Sources of Measurement Error

The statistical model postulates that the measurement error in an estimator is attributable to six sources: (1) the information system, (2) the respondent, (3) the mode of data collection, (4) the interviewer, (5) the survey questionnaire, and (6) the interview setting. As depicted in Figure 15.3, the respondent is the link between the interviewer (or questionnaire in self-administered surveys) and the business information system. Communication through this link may be distorted by the mode of data collection. Each error source may give rise to errors in reported values. Incorrect data may reside in the business' information system; these erroneous data will be duplicated in the survey reports.

The respondent may answer in error at each of the five stages of the response process. Errors may be attributable to mode of data collection in that data collected by one mode are more accurate than those of another mode. Although

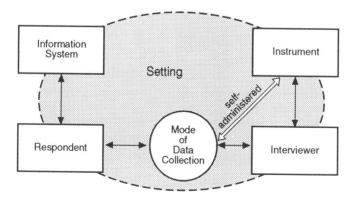

Figure 15.3 Sources of measurement error.

the mode does not commit errors in the way that respondents and interviewers do, the mode causes an error by interfering with or hampering the response process. Likewise, the questionnaire may cause errors through its construction, the position of individual questions, its layout, and so on. Finally, the interview setting may contribute to measurement error through its effects on the response process. The interview setting includes such factors as survey sponsorship, location of the interview, degree of confidentiality and privacy perceived by the respondent, and so on. Focusing on these error sources and the stages of the response process makes it possible to isolate the causes of measurement error and thereby to understand the actions needed to eliminate the causes. In the next two sections, we introduce statistical methods that allow the survey methodologist to partition the total error into error components that correspond to each source and to estimate their magnitudes.

15.2.2 An Introduction to Statistical Modeling

This section provides a background in the concepts and notation of measurement error modeling. These ideas serve as the foundation of the error evaluation and control techniques discussed in Section 15.3. For other introductions to this topic, interested readers should refer to Groves (1989), Biemer and Stokes (1991), and Lessler and Kalsbeek (1992).

Our approach to measurement error modeling is an adaption of earlier work by Hansen et al. (1953) and by Sukhatme and Seth (1952). The simplest form of the model specifies that a single observation y_j from a randomly selected respondent j is the sum of two terms: a true value μ_j and an error ε_j. Mathematically, this may be written as

$$y_j = \mu_j + \varepsilon_j, \tag{15.1}$$

where $\varepsilon_j \sim (0, \sigma_j^2)$, $\mu_j \sim (\mu, \sigma_\mu^2)$, and all covariances between the terms on the right are zero. Under this model, the variance of the mean $\bar{y}$ of a sample of n observations, ignoring the finite population correction factor, is

$$\text{Var}(\bar{y}) = \frac{\sigma_\mu^2}{n} + \frac{\sigma_\varepsilon^2}{n}, \tag{15.2}$$

where σ_μ^2 is the finite population variance of the true values and $\sigma_\varepsilon^2 = \text{E}(\sigma_j^2)$ is the finite population mean of the individual variances σ_j^2. The term σ_ε^2 is often referred to as the *simple response variance* (SRV).

Finally, we define the *reliability ratio R* as

$$R = \frac{\sigma_\mu^2}{\sigma_\mu^2 + \sigma_\varepsilon^2}. \tag{15.3}$$

Because $\text{Var}(\bar{y}) = R^{-1} \text{Var}(\bar{\mu})$, where $\bar{\mu}$ is the sample mean of the true values, R determines the increase in variance of the sample mean or total due to measurement error. The reliability ratio is widely used as a measure of the stability of the response process. A ratio of $R = 1$ indicates total reliability ($\sigma_\varepsilon^2 = 0$) in the measure, whereas a ratio approaching $R = 0$ indicates lack of response stability; that is, the variability among the true values μ_j in the population is small relative to the variability attributable to the response process.

A number of extensions of this simple model serve useful purposes in exploring measurement error. Two extensions that we consider are the extensions: (1) to accommodate remeasurement of the same sample unit and (2) to incorporate multiple sources of measurement error.

Extension for Reinterview Data

Some methods for evaluating measurement error require at least two measurements on some sample units. Let $y_{j\alpha}$ denote the observation on unit j on the αth occasion. Extending equation (15.1), we have

$$y_{j\alpha} = \mu_j + \varepsilon_{j\alpha}, \tag{15.4}$$

where $\varepsilon_{j\alpha} \sim (0, \sigma_{j\alpha}^2)$, and we assume that the covariance involving the terms on the right are zero both between and within trials.

A special case of equation (15.4) assumes that the second trial produces responses that are distributed identically to the responses of the first trial; that is, $\sigma_{j\alpha}^2 = \sigma_j^2$ where $\alpha = 1, 2$. This model is usually assumed for reinterview studies where the second interview is intended as an independent replication of the first interview. Under these assumptions, an estimator of the simple response variance is

$$\hat{\sigma}_\varepsilon^2 = \frac{\sum_j (y_{j1} - y_{j2})^2}{2n} \tag{15.5}$$

and an estimator of R is $(1 - \hat{\sigma}_\varepsilon^2/\hat{\sigma}_y^2)$, where $\hat{\sigma}_y^2$ is the average of

$$\hat{\sigma}_{y\alpha}^2 = \frac{\sum_j (y_{j\alpha} - \bar{y}_\alpha)^2}{n - 1} \quad \text{for } \alpha = 1, 2.$$

The assumptions for model (15.4) may not always be satisfied in practice.

Another use of model (15.4) is in estimating measurement error bias using record check studies or true value reinterviews. For this purpose, the assumptions associated with the error term $\varepsilon_{j\alpha}$ are altered. Instead, we assume $\varepsilon_{j1} \sim (B_j, \sigma_j^2)$ and $\varepsilon_{j2} \equiv 0$. That is, we assume that the error associated with the first observation has a nonzero mean (referred to as *measurement bias*) denoted by B_j and that the second observation has no error. Under this model we obtain

$$E(\bar{y}_1 - \bar{y}_2) = B, \tag{15.6}$$

where $B = E(B_j)$, the measurement bias in the estimator $\bar{y}_1$. Thus, the aim of studies that employ model (15.4) under this set of assumptions is to estimate B. Under the model assumptions, the bias can be estimated by

$$\hat{B} = \bar{y}_1 - \bar{y}_2. \tag{15.7}$$

Again, there may be severe difficulties in designing evaluation studies that satisfy the "true value" assumptions.

Extensions to Incorporate Error Sources

The statistical models discussed previously do not explicitly provide for partitioning the error variance or bias into separate components that correspond to the five error sources (see Figure 15.3). For example, suppose we wish to investigate the contribution that interviewers make to the error variance. To accomplish this, let y_{ij} denote the response obtained from the jth respondent in the ith interviewer's assignment. Then partition the error as $\varepsilon_{ij} = b_i + e_{ij}$, where b_i is a random error variable associated with interviewer i and e_{ij} is the difference $e_{ij} = \varepsilon_{ij} - b_i$. Substituting this error term into equation (15.1), we have

$$y_{ij} = \mu_{ij} + b_i + e_{ij}, \tag{15.8}$$

where $b_i \sim (0, \sigma_b^2)$, $e_{ij} \sim (0, \sigma_e^2)$, and, as before, all terms on the right are uncorrelated. If we further assume that interviewer assignments are *interpenetrated* (i.e., each assignment is a random subsample of the initial sample), then the model can be rewritten as the traditional analysis of variance (ANOVA) model:

$$y_{ij} = \mu + b_i + e'_{ij}, \tag{15.9}$$

where $e'_{ij} = e_{ij} + (\mu_{ij} - \mu)$, $e'_{ij} \sim (0, \sigma_{e'}^2)$, and μ is a fixed constant. Using ANOVA methods, the interviewer variance σ_b^2 can be estimated as a function of the between and within interviewer's sums of squares (see Groves 1989, p. 318).

A useful measure of the contribution of interviewers to the error variance is

$$\rho_y = \frac{\sigma_b^2}{\sigma_b^2 + \sigma_{e'}^2}, \tag{15.10}$$

referred to as the *intra-interviewer correlation*. An estimator of ρ_y is also available from the analysis of variance of interviewers. Model (15.9) can be extended to include other sources of error that may be viewed as random effects such as coders, editors, and supervisors.

When the model effects associated with an error source are fixed effects, a slightly different model is more appropriate. For example, we may have three modes of data collection or two alternative question wordings. Here, we may be interested in the response reliability under the alternative mode or question versions as well as their contributions to response bias.

A model appropriate for investigating mode effects may be created by re-writing model (15.8) as

$$y_{ij} = \mu_{ij} + M_i + e_{ij}, \tag{15.11}$$

where now y_{ij} is the response of the jth respondent assigned to the ith mode and M_i is a constant effect associated with the ith mode. Again, by randomly assigning respondents to modes, we may invoke the ANOVA model for ana-lyzing the mode's "biases." Under this model, the mode differential biases, $M_i - M_{i'}$, $i \neq i'$, are estimable, but the biases themselves are not estimable unless one assumes that $M_i = 0$ for some i. As an example, in a study of the effects of face-to-face interviewing, telephone interviewing, and mail self-administered methods, it might be appropriate to assume that the bias associ-ated with face-to-face interviewing is zero for some characteristics. In this case, the mode biases associated with the other two modes are estimable.

Finally, we may be interested in comparing the reliability ratios R_i associ-ated with alternative modes or questionnaire versions. For this objective, we extend model (15.11) for repeated measurements in analogy to model (15.4) as follows:

$$y_{ij\alpha} = \mu_{ij} + M_i + \varepsilon_{ij\alpha}, \tag{15.12}$$

and we assume $\varepsilon_{ij\alpha} \sim (0, \sigma_i^2)$. The reinterview methods described earlier may be used to obtain estimates of σ_i^2 and R_i.

15.3 TECHNIQUES FOR ERROR EVALUATION AND CONTROL

In this section, we describe techniques for evaluating and controlling mea-surement error components and illustrate their use. The methods discussed are: (1) cognitive laboratory methods, (2) experimental design, (3) observational studies, (4) administrative record check studies, (5) true value reinterview studies, (6) replicated reinterview studies, and (7) external and internal con-sistency studies.

15.3.1 Cognitive Laboratory Methods

Cognitive methods encompass a wide array of exploratory techniques for (1) investigating one or more components of the cognitive response model or (2) identifying the types of errors that may be introduced during data collection. In cognitive laboratory methods, the objective is to obtain information on the causes of error and the ways in which alternative survey methods or design features affect response. Laboratory studies usually involve small sample sizes (less than 100 subjects) and are conducted at a small number of sites where survey conditions can be tightly controlled. Nonrandom or restricted random samples are used to identify test subjects. Hence, inferences cannot be made about the magnitude of measurement error components. For inferring error magnitudes, it has been found that field studies, observational studies, or embedded survey experiments are much more effective.

Forsyth and Lessler (1991) summarize the cognitive laboratory methods that have been used in demographic surveys. Documented applications of cognitive laboratory methods to business survey methodology are relatively uncommon. In Chapter 16, Dippo et al. describe recent research by the U.S. Bureau of Labor Statistics in applying these methods to business surveys.

15.3.2 Experimental Design and Observational Studies

Whether it be an investigation of alternative interview modes, a test of alternative question wordings, or the estimation of interview effects, randomized experimental design has played a pivotal role in the development of measurement error theory and methods. In an early use of interpenetrating designs, Mahalonobis (1946) randomized interviewer assignments to obtain evidence of interviewer biases for an agricultural survey. Hansen et al. (1961), Fellegi (1964), and Bailar and Dalenius (1969) formalized the method of interpenetration. The Section 15.2.2 discussion of the use of ANOVA for estimating interviewer variance is a special case of this more general methodology.

The use of randomized experimental designs in full-scale surveys is referred to as an *embedded experiment*. Experimental designs have also been used in small-scale field tests and in cognitive laboratory experiments. In laboratories, the experimental manipulation of survey conditions is easier, and designs with complex treatment structures are possible. For example, O'Reilly et al. (1994) describe a laboratory study employing an incomplete block, repeated measures design in which three alternative modes of data collection were tested.

In the field, treatments are typically simpler, manipulating only one or two experimental conditions simultaneously. However, the randomization of experimental units may be considerably more complex. As an example, in a test of two types of interviewer training and two data collection modes, interviewer training types may be randomly assigned to primary sampling units whereas the interview mode may be randomly assigned to businesses within the interviewer assignments.

In many situations, the required randomizations and experimental manipulations would be too expensive, too impractical, or simply impossible to perform; "nonexperimental" or observational studies are conducted instead. For instance, to answer questions about a possible causal relationship between the respondent's position within the business and the accuracy of responses to a mail questionnaire, an experimental study needs to control who in the business completes the questionnaire for a large number of businesses over a considerable time period. Because of the difficulties of controlling the assignment of respondents as well as the management of costs, an experimental study is infeasible. Furthermore, the effort exerted by the researchers to control the type of respondent might affect the results of such a study. Although the researcher may instruct the business as to who should complete the instrument, the business makes the ultimate decision.

In an observational study, the survey proceeds normally without experimentally manipulating the type of respondent. Rather, when the returns are received, the researcher observes the position of the respondent in the business and relates this variable to response accuracy in much the same way as in an experimental analysis. For example, we may have observed that response accuracy tends to be better for members of the business' clerical staff than for members of the management team. However, we should then ask whether this result has as much to do with the business' size as it does with the respondent's position. In larger businesses, special clerical positions may be established to provide accounting information for regulatory purposes, whereas in smaller businesses where such managerial luxuries are not affordable, the task falls to the busy proprietor or his/her spouse. To examine this possibility and similar ones, explanatory variables such as business size are incorporated in the analysis to eliminate these extraneous factors as causes of the observed relationship. Of course, there is always the risk that an important explanatory variable will escape the researcher's attention or will be unavailable for use in analysis; in such a case, conclusions regarding causality will be misleading. The literature on causal modeling provides methods and principles for making causal inferences using observational data. (See Stolzenberg and Land 1983, for a review.)

15.3.3 Administrative Record Check and True Value Reinterview Studies

While cognitive methods provide valuable insights into the response process, an estimate of the magnitude of measurement bias is often desirable. Such estimates require that the true value of the business be known. With such information, faulty data items can be identified, measurement variance and bias can be estimated, and, in some cases, the sources and causes of error can be determined. Record check and true value reinterview studies are the two most widely used techniques for estimating measurement bias.

Administrative Record Check Studies
Because of the lack of administrative data for the characteristics typically measured in business surveys and the expense of obtaining records from official sources, record check studies are relatively infrequent. In record check studies, survey responses are compared to administrative records such as income or sales tax reports, licensing information, or other government data. Privacy limitations and differences in record keeping and reporting requirements often limit the ability to find suitable administrative sources. In one type of record

check study, a sample of businesses is drawn from the frame and then the corresponding administrative records are located. Alternatively, we may start with a sample of records and then interview the corresponding businesses. The latter is usually more efficient because the precision for rare items is better controlled. While selecting records from accountant's files and other financial services may be inexpensive, the coverage of the population can be poor. For example, Willimack (1989) found that only about 11 percent of farmers reported using financial services.

An implicit assumption of the administrative record check method is that the records contain accurate information (i.e., true values) for the survey characteristics of interest. Under this assumption, estimates of B can be obtained as in equation (15.7). However, three problems typically plague the method and limit the usefulness of the records data:

- The time periods for the record data and the survey data may not coincide.
- The characteristic(s) being reported in the record system may not exactly agree with the characteristic(s) being measured in the surveys.
- To save evaluation study costs, the records study may be confined to a restricted geographic area and inferences beyond this area may be invalid.

The Federal Committee on Statistical Methodology (1983) provides guidance on methodological requirements to conduct record checks. Furthermore, Ponikowski and Meily (1989) describe an examination of data on employment, earnings, and hours worked reported by businesses in the U.S. Current Employment Statistics Survey.

True Value Reinterview
Accessing administrative records is often not feasible because of availability, cost, privacy concerns, or other reasons. In these cases, a *reinterview study* may be appropriate, in which a second interview is conducted in an attempt to obtain the true values of survey variables. In the reinterview, more time may be taken and greater attention given to the reporting task, in an effort to obtain highly accurate data and to reconcile originally reported data against a reinterview report. Reinterviews may require that respondents access company records for "book values" whenever they are available. Yet many small businesses keep records informally in a "shoebox" file. Willimack (1989) found that 25 percent of U.S. farm operators used such informal records, whereas only 3 percent used a computer and 16 percent used a workbook.

In household surveys, it is well known that obtaining a true value during reinterview is very difficult. Indeed, for variables such as attitudinal views, true values may not exist. Business surveys, while more factually based, still present challenges for reinterviews. For example, Van Nest (1985) found that when computer records did not contain breakdowns such as expenditures by new versus used equipment, true values for the detailed data could not be obtained.

Reinterview designs vary depending on the assumptions that are most reasonable for the particular data and survey. Because the reinterview is often combined with interviewer performance evaluations, the reinterviewer is often the supervisor of the original interviewer. Forsman and Schreiner (1991) discuss difficulties associated with this practice. Alternatively, more experienced interviewers can be used provided that they do not reinterview their own cases. Although reinterviews conducted by telephone are less costly, some data are best obtained in a face-to-face interview to ensure that records are consulted. Cantwell et al. (1992) list questions that need to be resolved when designing a true value reinterview, including:

- How should the reinterview be introduced to the respondent?
- Should there be prior notification of reinterview?
- How soon after the original response should the reinterview occur?
- Should question wording be identical?
- How much burden should be placed on respondents?
- Should proxy responses be allowed in reinterview?
- Should the respondent or the reinterviewer decide if the original response or the reinterview response (or neither) is correct?

The U.S. Bureau of the Census used true value reinterviews to study the accuracy of data from the economic censuses (Corby 1984, 1985). For this study, the characteristic of interest was operating expenses, including components such as cost for purchased advertising, unemployment compensation paid, and so on. Measurement biases were estimated for each component and then combined for later use in correcting the reported data. An interesting aspect of the study was the classification of the reinterview responses into three categories: (1) values the reinterview respondent obtained directly from company records, (2) estimated values deemed reliable by the reinterviewer, and (3) estimated values deemed unreliable by the reinterviewer. For unreliable items, the respondent was asked to provide a range of values within which the true value could fall.

Another example is provided by Fecso and Pafford (1988) from a reinterview study used to measure bias for a U.S. National Agricultural Statistics Service (NASS) survey. Until recently, the detailed nature of acreage, stocks, and livestock inventory items led NASS to rely on face-to-face interviews to collect data from farmers. Like many survey organizations faced with cost containment and the need to publish results quickly, NASS has switched to more extensive use of computer-assisted telephone interviews (CATI) to collect data. Researchers were initially concerned that telephone responses would be inaccurate, not only because of the detailed nature of the data required but also because the centralized telephone staff lack the field staff's familiarity with farm terms. A face-to-face interviewer, recruited from the local farm community, might have more knowledge of farming.

The focus in the study was to estimate the measurement bias by treating the final reconciled response between the CATI and independent face-to-face reinterview response as the "truth." To obtain "truth" measures, experienced supervisory field enumerators reinterviewed approximately 1000 farm operations for the 1986 December Agricultural Survey. The Fecso and Pafford (1988) study of corn and soybean stocks in three states indicated that the difference in the original CATI and final reconciled responses (the bias) was significant for all but one item (soybean stocks in Indiana). The direction of the bias indicated that use of CATI tended to lead to underestimates of stocks of corn and soybeans.

In the process of reconciliation, the reasons for differences were collected. Table 15.1 indicates that an overwhelming percent of differences, 41.1 percent, was related to definitional problems (bias-related discrepancies), and not to those of simple response variance (random fluctuation). Examples of these definitional problems are erroneously included rented-out bins or grain belonging to someone else, confusion about reporting government reserve grains, and the mistaken inclusion of bins on a relative's farm. Definitional discrepancies were the largest contributors to bias. In contrast, differences due to rounding and estimation tended to reflect random variation and contributed little to overall bias.

Fecso and Pafford (1988) suggest that the bias in the survey estimate generated from use of the CATI might be reduced through a revised questionnaire design, improved training, or a shift back to more face-to-face interviews. Considering time and budget constraints, additional face-to-face interviews are unlikely. Thus, the suggested alternative was to use reinterview techniques to monitor bias over time and to determine whether the bias has been reduced through improvement in questionnaires and/or training. If large discrepancies continued, the estimates for grain stocks could be adjusted for bias through a continuing reinterview program.

15.3.4 Replicated Reinterview Studies for Measuring Reliability

In equation (15.3), the reliability ratio R was defined as the ratio of the variance of the true values to the total variance with measurement variance included. Estimation of R typically requires replicated measurements of a subsample of units via a reinterview study. For the estimator in equation (15.5) to be unbiased for σ_ε^2, two assumptions must hold for the reinterview survey:

- **Assumption 1:** The mean and variance of the response error associated with the reinterview and the original interview are identical, that is, $\varepsilon_{\alpha i}$ ~ $(B_i, \sigma_\varepsilon^2)$ for $\alpha = 1, 2$.
- **Assumption 2:** The covariance between the errors ε_{1i} and ε_{2i} is zero.

Assumption 1 requires that the reinterview survey create the same conditions that existed in the original interview—that is, the same questions, the

Table 15.1 Reason for Differences in CATI and Reinterview Responses for Corn Stocks in Minnesota—December 1987

Reason	Number	Percent of Total
Definitional	37	41.1
Estimated/Rounding	28	31.1
Other	25	27.8
Total	90	100.0

Source: Fecso and Pafford (1988).

same type of interviewers, the same mode of interview, the same respondent rules, and so on. However, because reinterview surveys are conducted at a later point in time, the survey questions may need modification to account for the time difference. Furthermore, to save cost and respondent burden, only a portion of the original questionnaire may be used. If the original survey was conducted in person, the reinterview may not be affordable unless it is conducted by a less expensive mode. Other changes in the reinterview procedures may be needed to save costs. The consequences of these changes may affect the response error distribution, thus invalidating Assumption 1. In household surveys, this assumption has been tested and it appears achievable in most situations (Forsman and Schreiner 1991, Groves 1989). For business surveys, reported research on reinterview assumptions is scant, yet Assumption 1 seems plausible for a well-designed business reinterview survey.

Assumption 2, however, is much more problematic. Errors between trials may be correlated due to a number of factors. Respondents may recall their original responses and simply repeat them in the reinterview without regard to accuracy. If the error is recorded in the business' record system, the error will be repeated if the same record system is accessed in reinterview. Respondents may also repeat the same response process to arrive at an answer. Thus, the errors in the two responses are likely of similar magnitude and direction, inducing a positive correlation between the errors. For household surveys, O'Muircheartaigh (1991) reports correlations as high as 57 percent in an unemployment classification. However, little is known regarding the between-trial correlation for business surveys. Bushery et al. (1992) discuss the difficulties of implementing a reinterview by mail when the original survey was conducted by mail. Some respondents retain a copy of their original responses and transcribe these responses to the reinterview form.

For a between-trial correlation of τ, the relative bias in the estimate of the simple response variance in equation (15.5) is $-\tau$. Thus, a positive correlation results in a downward bias in the estimates of reliability. More research is needed to study the magnitude and direction of τ for characteristics measured in business reinterview surveys.

15.3.5   Internal and External Consistency Studies

Two techniques for assessing measurement error (at times in combination with other sources of nonsampling error) are external and internal consistency checks. *External consistency studies* compare the estimate with estimates from another source, done most effectively when compared as a data series. *Internal consistency studies* include analysis of edit output and analysis allowed by the use of rotating panel designs.

External Consistency Studies

The comparison of several series of estimates can provide an indication that measurement error may exist. If sampling error is small enough, a difference in level between the series is an indication of nonsampling error. If one series is considered very accurate and another series is parallel but at a different level, the difference is a crude estimate of overall bias. In crop yield surveys conducted in the United States, time series produced by different measurement methods (counts from fields versus farmer reporting techniques) often track with one another, but level differences of 10 percent occur (Fecso 1991). Unfortunately, determining the contribution to total error due to measurement error is not possible without other sources of information. In an example of an external series comparison, Silverberg (1983) examined several series of estimates of retail gasoline prices in the United States. He found that all the national estimates of retail motor gasoline prices were within 5 percent of the Energy Information Administration series, but sizable differences existed for wholesale prices in some series. Furthermore, Silverberg found that internal assessments were more informative than external comparisons.

Internal Consistency Studies

Internal consistency studies are often exploratory in nature. Editing activities may lead to the most familiar forms of internal consistency study. A detailed description of editing processes can be found in Chapter 21. In a study of U.S. federal business surveys (Federal Committee on Statistical Methodology 1988), almost all the surveys studied used practices such as analyst review of the data, edits for reasonableness, and follow-up procedures. About 75 percent of these surveys compute edit failure rates, and about half compute interviewer error rates as an indirect technique to measure response error. The information about the relative magnitude of potential measurement error gleaned from analysis of error rates and from respondent and interviewer feedback can target areas in need of direct measurement.

For most surveys, these edit failure rates relate to procedures based on experience and logical relationships between data items, with range checks to detect outliers. Most statistically based procedures to identify unusual data are available. Such techniques can be useful in associating potential causes of measurement error with the errors. Barnett (1983) discusses statistically based methods for detecting outliers in data sets. Silverberg (1983) presents a graphical technique for examining changes over time in the distribution of company-reported prices for gasoline. Temporal comparison of response distributions is useful in detecting unexpected changes to conditions impacting response or, as Silverberg found, verifying expected changes such as greater price variability during the 1979 oil crisis.

Repeated use of the same business and data items in periodic surveys provides an opportunity to measure validity and reliability of the data. Saris and Andrews (1991) and Munck (1991) discuss the use of structural modeling in evaluating measurement errors. Reiser et al. (1992) apply such models in studying measurement error in a corn yield survey in the United States. Variables that were easy for the enumerator to count, such as number of stalks, were found to be highly reliable, but some measurements, such as length of the ear, were not.

In research on the use of historic data, methodology similar to a reinterview approach, Pafford (1986) studied the effect of using farmer-reported planted acreage from a spring survey during the fall interview. Differences in planted acreage between the spring and fall survey were largest in the control treatment where prior responses were not used. When spring-reported data were worded directly into the fall question, the differences were smallest. The control group had more instances of unreported acreage in both survey periods, whereas few reports of zero acreage occurred when the spring report was nonzero and incorporated into the fall question. While interviewers had no problems using prior data and response variance between surveys decreased, a recommendation to use historic data could not be made because there was no way to determine which treatment had lower response bias. Similar results were found in a later study of grain stocks (Pafford 1988).

15.3.6 Quality Standards

Many researchers view the use of standards as fundamental for quality improvement. *Standards* reflect principles for guiding and assessing the quality of an activity. In a survey environment, standards are the guide to the process that keeps an organization focused on the essential conditions and procedures that could produce unwanted changes in measurements. Morganstein and Hansen (1990) indicate that without clearly stated standards that are followed without compromise, increased and unnecessary variability will result. Areas where standards are useful in controlling measurement error include (1) questionnaire design; (2) editing; and (3) interviewer recruitment, training, and supervision.

Freedman (1990) notes that standards alone cannot improve data quality. Enforcement, education, and evaluation are also necessary, along with the commitment of everyone in the survey organization. Furthermore, the organization must decide where standards are needed. Accuracy, consistency, efficiency, and clarity can be used as guiding principles. Some standards apply regardless of the subject matter covered in the survey, such as pretesting all questionnaires to ensure clear wording of questions and the respondent's ability to answer questions. Others depend upon the circumstances of the survey, such as the need for interviewers not to know the survey's sponsor when the data are used as evidence in judicial actions (Morgan 1990). U.S. organizations with very detailed standards for instrument design and collection include the Energy Information Administration (1989) and the National Center for Education Statistics (Cooperative Education Data Collection and Reporting Standards Project Task Force 1991).

The Energy Information Administration (EIA) uses a quality audit to check for compliance with standards. The audits focus on determination of the soundness of the data system and its ability to produce data that meet standards (acceptable quality). Survey processes are reviewed and tested, and documentation is reviewed. Checklists ensure coverage of all standards and record deviations. Offices are responsible for implementing the recommendations made by the standards review team when deficiencies are found.

While enforcement may be necessary to ensure that the standards are followed as intended, auditors may be resented within the organization. For quality standards to work, a cooperative environment built upon trust, voluntary self-improvement, and an emphasis on organizational goals and missions are necessary. Such efforts are the emphasis of the growing movement from inspection sampling of final products toward quality improvement strategies.

15.4 CONCLUDING REMARKS

In this chapter, we overview methods for evaluating measurement errors and describe how they are applied to business surveys. We present measurement error models that guide efforts to improve data quality. Error evaluation methods, such as cognitive laboratory techniques, are applied in the survey's design stage. Reinterviews, experimental designs, and observational studies are applied during survey execution; the results are not known until after the survey is completed. Finally, post-survey methods such as internal and external validity studies and administrative record check studies are another possibility.

One area not addressed in this chapter is evaluation methods that may be applied *during* data collection to provide timely information on data quality as it is collected. The need for such methods is obvious. To be controlled, survey errors must be measured continuously during the course of the survey. When information on data quality is available on a real-time basis, survey managers can determine the degree to which errors are being controlled and can intervene if action is needed to reduce errors. Such methods are called *process quality control methods.*

Usually, a component of measurement error cannot be directly estimated and monitored continuously during data collection. Rather some *indicator* of data quality—response rate, edit failure rate, item nonresponse rate, and so on—is measured and monitored. A *quality indicator* may be any routinely observable quantity that is correlated with a component of error. As an example, interviewer edit failure rates may be correlated with interviewer variance or bias. The rationale is that keeping the quality indicators within acceptable limits will also control the corresponding mean squared error component, resulting in higher-quality data. Unfortunately, there is no assurance of this; studies that directly measure bias and variance components are still needed to determine the true level of data quality. Furthermore, many serious measurement errors can only be detected using direct approaches. In continuing surveys, evaluating the magnitude of measurement errors periodically (say once every 2–3 years) may be adequate to ensure that routine quality control practices are controlling errors at the desired levels. For one-time surveys, direct evaluation studies may be expensive but still necessary for two purposes: (1) to provide data users with an accurate assessment of the data's true information content and (2) to obtain information on survey error that can be used in future surveys. There is a vast literature on methods for quality improvement in industrial or service settings, but only a few articles specifically for survey operations (e.g., Fecso 1989, Colledge and March 1992).

While a variety of methods are available for the control of data quality, the uses made of the methodology vary. The Federal Committee on Statistical Methodology (1988) profiled the use of various measurements and control procedures for measurement error in business surveys conducted by the U.S. Federal Government. A majority of the surveys reported the use of editing, analyst review, and the production of edit failure rates. Useful techniques such as reinterviews, record check studies, and cognitive studies were reported as rarely used. In general, indirect measures rather than direct measures of error components are used. The lack of standard approaches to incorporate error measurement strategies into surveys may have stimulated the recent interest in total quality management approaches in survey organizations. Federal agencies are further motivated to use such approaches because recent administrations have encouraged such thinking government-wide.

A total quality strategy of management emphasizes a focus on continuous improvement in all aspects of an organization. The framework for such a strategy includes the following elements (Fecso 1993):

- identify customers and their needs,
- move from only measuring quality to improving it,
- base decisions on analyses of data,
- anticipate and accept change,
- emphasize an interorganizational team approach to problem solving,
- ensure that staff members are aware and involved in the organization's quality goals, and
- accomplish the above activities with top management leadership.

While the survey research literature has many reports on the use of total quality techniques in various situations or in units of a survey organization, studies about the effectiveness of the efforts as a strategy in survey organizations are scarce. Most survey organizations appear to be in the development stages of implementation during which they focus on training and team formation. In subsequent stages, we would expect open discussions leading to agreement on the most important quality problems in the organization, further use of real-time measurements during the survey process, increasing communication of the results with those involved in the operations, and delegation of responsibility for taking actions to correct problems.

Quality profiles can be very helpful in quality improvement efforts, yet profiles that consist only of indirect measures of data quality can be misleading (Bailar 1983). Use of the models and techniques described in this chapter are also necessary for error evaluations and data quality improvement. The Federal Committee on Statistical Methodology's 1988 report convincingly demonstrates the need for frequent use of the evaluation methods with federally sponsored business surveys. Studies in the private sector that directly evaluate the components of measurement error are seldom reported and should be in even greater demand.

REFERENCES

Bailar, B. A. (1983), "Error Profiles: Uses and Abuses," in T. Wright (ed.), *Statistical Methods and the Improvements of Data Quality*, Orlando, FL: Academic Press, pp. 117–130.

Bailar, B. A., and T. Dalenius (1969), "Estimating the Response Variance Components of the U.S. Bureau of the Census' Survey Model," *Sankhā B*, pp. 341–360.

Barnett, V. (1983), "Principles and Methods for Handling Outliers in Data Sets," in T. Wright (ed.), *Statistical Methods and the Improvement of Data Quality*, Orlando, FL: Academic Press, pp. 131–158.

Biemer, P., and S. L. Stokes (1991), "Approaches to the Modeling of Measurement Errors," in P. P. Biemer, R. M. Groves, L. E. Lyberg, N. A. Mathiowetz, and S. Sudman (eds.), *Measurement Errors in Surveys*, New York: Wiley, pp. 487–516.

Biemer, P. P., R. M. Groves, L. E. Lyberg, N. A. Mathiowetz, and S. Sudman (eds.) (1991), *Measurement Errors in Surveys*, New York: Wiley.

Bushery, J. M., D. Royce, and D. Kasprzyk (1992), "The Schools and Staffing Survey: How Reinterview Measures Data Quality," *Proceedings of the Survey Research Methods Section, American Statistical Association*, pp. 458–463.

Cantwell, P. J., J. M. Bushery, and P. P. Biemer (1992), "Toward a Quality Improvement System for Field Interviewing: Putting Content Reinterview into Perspective," *Proceedings of the Survey Research Methods Section, American Statistical Association*, pp. 74–83.

Colledge, M., and M. March (1992), "Quality Management: Development of a Framework for a Statistical Agency," *Proceedings of the Annual Research Conference*, Washington, DC: U.S. Bureau of the Census, pp. 121–136.

Cooperative Education Data Collection and Reporting (CEDCAR) Standards Project Task Force (1991), *Standards for Education Data Collection and Reporting*, Washington, DC: National Center for Education Statistics.

Corby, C. (1984), *Content Evaluation of the 1977 Economic Census*, Statistical Research Division Report Series, Washington, DC: U.S. Bureau of the Census.

Corby, C. (1985), "Content Evaluation of the 1982 Economic Censuses Petroleum Distributors," *Statistical Research Division Report Series*, Washington, DC: U.S. Bureau of the Census.

Edwards, W. S., and D. Cantor (1991), "Toward a Response Model in Establishment Surveys," in P. P. Biemer, R. M. Groves, L. E. Lyberg, N. A. Mathiowetz, and S. Sudman (eds.) *Measurement Errors in Surveys*, New York: Wiley, pp. 211–236.

Eisenhower, D., N. A. Mathiowetz, and D. Morganstein (1991), "Recall Error: Sources and Bias Reduction," in P. P. Biemer, R. M. Groves, L. E. Lyberg, N. A. Mathiowetz, and S. Sudman (eds.), *Measurement Errors in Surveys*, New York: Wiley, pp. 127–144.

Energy Information Administration (1989), "Energy Information Administration Standards Manual," Washington, DC: U.S. Department of Energy.

Fecso, R. (1989), "What Is Survey Quality: Back to the Future," *Proceedings of the Survey Research Methods Section, American Statistical Association*, pp. 88–96.

Fecso, R. (1991), "A Review of Errors of Direct Observation in Crop Yield Surveys," in P. P. Biemer, R. M. Groves, L. E. Lyberg, N. A. Mathiowetz, and S. Sudman (eds.), *Measurement Errors in Surveys*, New York: Wiley, pp. 327–346.

Fecso, R. S. (ed.) (1993), *Quality in Student Financial Aid Programs: A New Approach*, Washington, DC: National Academy Press.

Fecso, R., and B. Pafford (1988), "Response Errors in Establishment Surveys with an Example from an Agribusiness Survey," *Proceedings of the Survey Research Methods Section, American Statistical Association*, pp. 315–320.

Federal Committee on Statistical Methodology (1983), *Approaches to Developing Questionnaires*, Statistical Policy Working Paper 10, Washington, DC: U.S. Office of Management and Budget.

Federal Committee on Statistical Methodology (1988), *Quality in Establishment Surveys*, Statistical Policy Working Paper 15, Washington, DC: U.S. Office of Management and Budget.

Federal Committee on Statistical Methodology (1990), *Data Editing in Federal Statistical Agencies*, Statistical Policy Working Paper 18, Washington, DC: U.S. Office of Management and Budget.

Fellegi, I. P. (1964), "Response Variance and Its Estimation," *Journal of the American Statistical Association*, **59**, pp. 1016–1041.

Forsman, G., and I. Schreiner (1991), "The Design and Analysis of Reinterview: An Overview," in P. P. Biemer, R. M. Groves, L. E. Lyberg, N. A. Mathiowetz, and S. Sudman (eds.), *Measurement Errors in Surveys*, New York: Wiley, pp. 279–302.

Forsyth, B. H., and J. T. Lessler (1991). "Cognitive Laboratory Methods: A Taxonomy," in P. P. Biemer, R. M. Groves, L. E. Lyberg, N. A. Mathiowetz, and

S. Sudman (eds.), *Measurement Errors in Surveys*, New York: Wiley, pp. 211–233.

Fowler, F. J. (1991), "Reducing Interviewer-Related Error Through Interviewer Training, Supervision, and Other Means," in P. P. Biemer, R. M. Groves, L. E. Lyberg, N. A. Mathiowetz, and S. Sudman (eds.), *Measurement Errors in Surveys*, New York: Wiley, pp. 259–278.

Freedman, S. R. (1990), "Quality in Federal Surveys: Do Standards Really Matter?" *Proceedings of the Survey Research Methods Section, American Statistical Association*, pp. 11–17.

Groves, R. M. (1989), *Survey Errors and Survey Costs*, New York: Wiley.

Groves, R. M. (1990), "On the Path to Quality Improvement in Social Measurement: Developing Indicators of Survey Errors and Survey Costs," *Proceedings of the Survey Research Methods Section, American Statistical Association*, pp. 1–10.

Hansen, M. H., W. N. Hurwitz, and M. A. Bershad (1961), "Measurement Errors in Censuses and Surveys," *Bulletin of the International Statistical Institute*, **38**, pp. 359–374.

Hansen, M. H., W. N. Hurwitz, E. S. Marks, and W. G. Madow (1953), *Survey Methods and Theory, Vol. I: Methods and Applications, Vol. II: Theory*, New York: Wiley.

Krosnick, J. A., and D. F. Alwin (1987), "An Evaluation of a Cognitive Theory of Response-Order Effects in Survey Measurement," *Public Opinion Quarterly*, **51**, pp. 201–219.

Lessler, J. T., and W. D. Kalsbeek (1992), *Nonsampling Errors in Surveys*, New York: Wiley.

Mahalanobis, P. C. (1946), "Recent Experiments in Statistical Sampling in the Indian Statistical Institute," *Journal of the Royal Statistical Society*, **109**, pp. 325–378.

Morgan, F. W. (1990), "Judicial Standards for Survey Research: An Update and Guidelines," *Journal of Marketing*, **54**, pp. 59–70.

Morganstein, D. R., and M. H. Hansen (1990), "Survey Operations Processes: The Key to Quality Improvement," in G. E. Liepins and V. R. R. Uppuluri (eds.), *Data Quality Control: Theory and Pragmatics*, New York: Marcel Dekker, pp. 91–104.

Munck, I. M. E. (1991), "Path Analysis of Cross-National Data Taking Measurement Errors into Account," in P. P. Biemer, R. M. Groves, L. E. Lyberg, N. A. Mathiowetz, and S. Sudman (eds.), *Measurement Errors in Surveys*, New York: Wiley, pp. 599–616.

O'Connor, T. P. (1993), "An Analysis of the Cognitive Aspects of the January, 1993 Cattle on Feed Quality Assessment Survey," Washington, DC: National Agricultural Statistics Service.

O'Muircheartaigh, C. O. (1991), "Simple Response Variance: Estimation and Determinants," in P. P. Biemer, R. M. Groves, L. E. Lyberg, N. A. Mathiowetz, and S. Sudman (eds.) *Measurement Errors in Surveys*, New York: Wiley, pp. 551–574.

O'Reilly, J., M. Hubbard, J. Lessler, P. Biemer, and C. Turner (1994), "Audio and Video Computer-Assisted Self-Interviewing: Preliminary Tests of New Technologies for Data Collection," *Journal of Official Statistics*, **10**, pp. 199–214.

Pafford, B. V. (1986), "The Influence of Using Previous Survey Data in the 1986 April ISP Grain Stock Survey," Washington, DC: National Agricultural Statistical Service.

Pafford, B. (1988), "Use of Reinterview Techniques for Quality Assurance: The Measurement of Response Error in the Collection of December 1987 Quarterly Grain Stocks Data Using CATI," Washington, DC: National Agricultural Statistics Service.

Ponikowski, C. H., and S. A. Meily (1989), "Controlling Response Error in an Establishment Survey," *Proceedings of the Survey Research Methods Section, American Statistical Association*, pp. 258–263.

Reiser, M., R. Fecso, and M. Chua (1992) "Some Aspects of Measurement Error in the United States Objective Yield Survey," *Journal of Official Statistics*, **8**, pp. 351–375.

Saris, W. E., and F. M. Andrews (1991) "Evaluation of Measurement Instruments Using a Structural Modeling Approach," in P. P. Biemer, R. M. Groves, L. E. Lyberg, N. A. Mathiowetz, and S. Sudman (eds.), *Measurement Errors in Surveys*, New York: Wiley, pp. 575–598.

Silverberg, A. R. (1983), "An Approach to an Evaluation of the Quality of Motor Gasoline Prices," in T. Wright (eds.), *Statistical Methods and the Improvements of Data Quality*, Orlando, FL: Academic Press, pp. 297–319.

Stolzenberg, R. M., and K. C. Land (1983). "Causal Modeling and Survey Research," in P. H. Rossi, J. D. Wright, and A. B. Anderson (eds.), *Handbook of Survey Research*, San Diego, CA: Academic Press, pp. 613–672.

Sukhatme, P. V., and G. R. Seth (1952), "Nonsampling Errors in Surveys," *Journal of Indian Society of Agricultural Statistics*, **5**, pp. 5–41.

Tourangeau, R. (1984), "Cognitive Sciences and Survey Methods," in T. Jabine, E. Loftus, M. Straf, J. Tanur, and R. Tourangeau (eds.), *Cognitive Aspects of Survey Methodology: Building a Bridge Between Disciplines*, Washington, DC: National Academy of Science, pp. 73–100.

Van Nest, J. G. (1985), *Content Evaluation Pilot Study*, Statistical Research Division Report Series, Washington, DC: U.S. Bureau of the Census.

Willimack, D. K. (1989), "The Financial Record-Keeping Practices of U.S. Farm Operators and Their Relationship to Selected Operator Characteristics," paper presented at the American Agricultural Economics Association annual meetings, Baton Rouge, LA.

CHAPTER SIXTEEN

Designing the Data Collection Process

Cathryn S. Dippo, Young I. Chun, and Joan Sander[1]
U.S. Bureau of Labor Statistics

The way survey researchers think about data collection has changed over the last decade. Not only is there a greater recognition of the effect that data collection can have on the quality of the survey estimates, but now resources are being expended to understand and improve collection. The last two international conferences on survey research methods were dedicated to sharing information related to the quality of the data collection process (Groves et al. 1988; Biemer et al. 1991).

Measurement error research now goes beyond answering the question "How do we measure it?" to "How do we understand the causes of measurement error and use that information to improve data quality?" This is illustrated by our changed perception of the role of computer-assisted survey information collection (CASIC). In the past, survey researchers focused on CASIC's ability to control interviewer errors related to skip patterns and control response errors recognized through data inconsistencies. Now, we focus on CASIC technology as a tool for aiding the respondent and interviewer in completing their tasks with efficiency and accuracy (Dippo et al. 1994). This change parallels the change in thinking about quality, from a paradigm based on control to one based on process improvement.

In this chapter, we discuss strategies to identify the causes of measurement error in business surveys and how this information can be used to improve the data collection process. Some of these methods have been in use for many years; others have been adapted from the fields of psychology, sociology, and

[1]The authors are grateful to the Behavioral Science Research Staff of the U.S. Bureau of Labor Statistics for their help in preparing this chapter. Views expressed are those of the authors and do not necessarily reflect those of the U.S. Bureau of Labor Statistics.

Business Survey Methods, Edited by Cox, Binder, Chinnappa, Christianson, Colledge, Kott.
ISBN 0-471-59852-6 © 1995 John Wiley & Sons, Inc.

anthropology. Some provide field-based *quantitative* information; others are laboratory-based and produce *qualitative* information. We describe these methods and present examples of their application for business surveys at the U.S. Bureau of Labor Statistics (BLS). Because data collection is one part of the larger, interrelated production process for survey estimates, we follow the survey model, beginning with defining survey concepts and ending with the processing of data into estimates. We conclude by addressing the limitations of these methods and recommend an integrated approach for reducing measurement errors.

16.1 STRATEGIES AND METHODS

Historically, data collection design strategies for business surveys concentrated on specifying data products and the underlying definitions and concepts. Business questionnaires were more like forms than scripts. Their design focused on the technical specification of the required data; respondents were viewed as professionals easily providing data from available records. The complexity of the measured concepts should dictate decisions on the data collection mode. Until rather recently, however, these decisions were often based on costs or timeliness. The most viable options were mail or face-to-face visits.

Thus, when the Federal Committee on Statistical Methodology (FCSM 1988) examined data collection designs for U.S. business surveys, they focused on specification error and the use of techniques such as requirements reviews, respondent consultations, questionnaire review by expert panels, questionnaire pretests, and cognitive studies. The subcommittee found that of 55 business surveys in nine profiled agencies, only 50–60 percent used these techniques on a regular basis. Approximately 60 percent used questionnaire pretesting on an irregular basis, but only 30 percent pretested regularly. Few surveys used cognitive studies. More recently, Christianson and Tortora (see Chapter 14) found little evidence of the use of behavioral science methods in their study of 1387 business surveys in 16 countries.

In examining practices to control response error, FCSM found that few surveys used methods involving direct contact with respondents, such as cognitive investigations or studies of record-keeping practices. Commonly used methods were expert review of micro-level data and other data editing methods that are post-collection and that involve no respondent contact and whose results have little integration into process design. In summary, until recently, little quantitative information was available on overall data quality for business surveys. Informal, ad hoc procedures were more prevalent than research using scientifically designed studies to design and evaluate the data collection process.

Another FCSM subcommittee listed methods and tools for developing, testing, and evaluating questionnaires (FCSM 1983), including unstructured individual interviewing, qualitative group interviews, participant observation, pilot and split-sample testing, respondent and interviewer debriefings, and record checks. Most methods are applicable to the measurement process as a whole and are not limited to the questionnaire only. Except for the first few, the methods are field-based.

Recently, greater emphasis has been placed on laboratory-based methods due to the "Advanced Research Seminar on Cognitive Aspects of Survey Methodology," sponsored by the U.S. Committee on National Statistics (Jabine et al. 1984). The seminar led the way to the establishment of cognitive research laboratories at the National Center for Health Statistics, the Bureau of Labor Statistics, and the Bureau of the Census. In addition, Statistics Canada has created a Questionnaire Design Resource Centre (Gower 1991). See Forsyth and Lessler (1991) for a taxonomy of cognitive laboratory methods, such as focus groups, cognitive forms appraisals, think-aloud interviews, paraphrasing, interactional behavior coding, and memory cueing and rating tasks.

The methods described in the 1983 FCSM report are applicable to business surveys, as are the cognitive, laboratory-based methods discussed by Forsyth and Lessler. Until recently, however, published applications for business surveys have been rare. Most applications describe the use of focus groups and think-aloud interviews to determine problems that respondents are having with questionnaires received in the mail or mailouts followed by the telephone interviews (Bureau 1991; Cox et al. 1989; DeMaio and Jenkins 1991; Jenkins 1992; Statistics Canada 1992).

In contrast, much has been written about data collection mode research for BLS's Current Employment Statistics (CES) program, a large monthly business survey. Initial research focused on the use of computer-assisted telephone interviewing (CATI). Although CATI was found to be useful in improving response rates and data timeliness (Werking et al. 1986), it was more expensive than mail (Clayton and Harrell 1989). Thus, subsequent research focused on computer-assisted self-response using touch tone data entry and voice recognition by telephone (Clayton and Winter 1992). (For a more in-depth discussion of how to improve data quality through the use of automated telephone collection methods, see Chapter 18.)

An important result has been the discovery that CASIC provides a cost-effective, field-based method for identifying sources of error in the measurement process. Each step of the mode-directed research has included a response analysis survey conducted using CATI. The surveys have focused on determining the causes of measurement error (Ponikowski and Meily 1989; Rosen et al. 1991).

One important difference in applying cognitive laboratory methods to business surveys as opposed to household surveys relates to the respondent. Respondents in business surveys are employees who provide information as part of their job duties. In general, they are unable to come to a laboratory. Thus, laboratory methods must be administered in a field environment. Although not a difficult task, researcher visits to companies are more expensive than reimbursing volunteers with $15 or $20 each to cover their travel expenses. Gower and Nargundkar (1991) discuss other differences in applying laboratory-based methods to household and business surveys.

Primarily through work in progress on BLS surveys, we next illustrate how many of these laboratory-based methods can be applied to business surveys. An early BLS application of laboratory-based methodology was for a one-time-only business survey conducted to determine the extent of employer drug-testing and employee assistance programs (Palmisano 1988). Having shown the benefits of applying behavioral science methods to business surveys, additional research and more long-term projects were initiated. Many projects involved a long-term goal of converting BLS face-to-face data collection to CASIC.

16.2 EXAMPLES OF BEHAVIORAL SCIENCE METHODS

BLS is the only U.S. statistical agency that has field staff collecting data in person from businesses from all sectors of the economy. In most cases, the business data collection process differs significantly from that for households. The data being collected are often conceptually complex, resulting in a need for in-person interviews. BLS interviewers usually have at least a bachelor's degree in economics and are given extensive training in the concepts being measured and forms to complete. For example, the Occupational Compensation Survey Program (OCSP) trains interviewers in the definitions of over 50 occupations with occupational levels ranging from I (novice in the job) to VIII (most expert in the job). While on site, OCSP interviewers must identify which of the approximately 150 occupation-level combinations exist at the business.

Data collection should be viewed not as a process unto itself, but within the full context of the entire survey. Thus, the applications we present include examples from conceptualization through data processing.

16.2.1 Survey Concepts

The concepts used in business surveys are rich in meaning; many terms have multiple meanings. For example, "layoff" may mean anything from "being fired" to "receiving unemployment insurance." Operationalizing concepts rich in meaning into indicators (i.e., questions) is not an easy task (Babbie 1973). Conceptualizing, testing concepts, and operationalizing them in business surveys require rigorous research using techniques such as focus groups, multi-dimensional scaling, and think-aloud techniques.

Focus Groups
Focus groups are conducted to develop, assess, and clarify survey concepts and their indicators, to evaluate questions and instructions, and to identify errors or burdens associated with understanding and answering questions and retrieving data. Usually, 6–12 people participate in the structured group dis-

cussion facilitated by a moderator who prepares a protocol in advance. Members of a focus group communicate informally with one another and generate and assess face-validity measures of survey questions (Desvousges and Frey 1989, Krueger 1988, Merton 1987, Merton and Kendall 1946, Morgan 1988). In a relatively short time, the group dynamics elicit more information about survey concepts and their measurements than could be obtained by separate interviews with each individual. Often the focus groups are audio- or video-taped. The qualitative results provide useful insights by exposing how respondents might respond to questions and the reasoning behind their responses. Recently, research has begun on how to make quantitative estimates from focus groups (Dietz 1992).

To illustrate the use of focus groups in defining concepts, consider the Employment Turnover and Job Openings (ETJO) pilot survey. Two focus groups assessed proposed concepts relevant to job separations, new hires, average wage of new hires, job openings, and the duration of job openings (Phipps et al. 1992). Following a preset protocol, personnel specialists were asked to give examples of hard-to-fill jobs and occupations where labor shortages might exist and then about variables that might be used to measure labor shortage. The ETJO survey team's proposed variables were then presented for comparison with the participants' proposed variables. The focus groups generated many of the same variables as the ETJO survey team did, including separations, new hires, and job openings. They also suggested other measures of labor shortage such as costs of filling a job and funding training programs, but were uncertain about whether businesses kept these records.

Focus groups have also assisted in evaluating concepts for BLS user surveys. These surveys have been used to assess users' satisfaction with the quality of BLS's data service and guide program managers in making decisions. After discussing their most recent experience in contacting BLS, focus group members discussed what information could be used to identify satisfaction with a statistical agency. Then, six constructs of service and data quality were presented (tangibles, data quality, dependability, responsiveness, assurance, and empathy), along with 18 questions designed to obtain data on the six indicators. Most focus group members reported that all 18 questions adequately measured the six indicators. The validity of each question was suspect, however. For example, a question intended to measure dependability was understood as one measuring reliability. As a result, tests were later conducted to assess the validity of each question item, to identify questions with low validity, and to find approaches to improve them. Redrafting based on findings from these tests significantly increased the validity of questions designed to measure the quality of data and service. Focus groups also revealed the extent to which respondents understand terms such as often, one time, bulletin, and news release.

Multidimensional Scaling

Multidimensional scaling (MDS) techniques allow cognitive psychologists to understand how people mentally structure knowledge and then make decisions based on those structures (Ramsay 1978). BLS is using MDS techniques in research designed to understand "job match" decisions made by interviewers in the OCSP. Interviewers match job definitions with jobs found in sample establishments, and then they collect wage data for the matched jobs. As noted earlier, interviewers are trained and well-educated (all hold at least a bachelor's degree in economics or a related field), but the matching task is difficult and often performed inconsistently. It is difficult to discern what parts of the job match decision cause confusion for the interviewers. The job match decision is based, in part, on the interviewer's conceptual understanding of occupations and the organizations that provide these jobs. Therefore, an investigation into how interviewers structure their understanding of occupations and industries is a first step towards improving the quality of matching.

The MDS techniques allow researchers to graphically portray the interviewer's mental representation of jobs, and they demonstrate how these jobs are similar or dissimilar based on salient properties the interviewer uses to classify them (Cantor et al. 1985). For example, these properties might include tasks, education, and salary. The interviewers sort the jobs in pairs according to their similarity, and a similarity matrix is created from each subject's sorting. Using software for MDS analysis, researchers may discover that the job properties are arrayed along several dimensions (e.g., supervisory responsibilities) according to their similarity to one another. The closer the properties are positioned to each other on the dimension, the more similar they are on that dimension. There is no guarantee that MDS produces meaningful dimensions, but if it does, it can provide insight as to why one interviewer decides that a position may be matched to the official accountant definition while another matches the same job to the budget analyst definition.

Think-Aloud Techniques
Think-aloud techniques encourage respondents to verbalize the thoughts they have or the thought processes they use when answering questions (Ericsson and Simon 1984, FCSM 1991, Palmisano 1988). In the *concurrent think-aloud interview*, the respondent verbalizes this information while answering the question. In contrast, the *retrospective think-aloud interview* first obtains the respondent's answer and then later obtains the explanation of processes used to generate responses. The concurrent think-aloud interview obtains the sequence of information attended to by the respondent without altering the cognitive process, whereas the retrospective interview gets the respondent to retrieve the trace of the process (Ericsson and Simon 1984). Typically, field tests are used to obtain think-aloud responses to business questionnaires, so researchers can observe in a natural setting the interactions of business respondents with the interviewer and the questionnaire. Respondents at the work sites are first briefed about the purpose of the think-aloud and are occasionally probed when answering survey questions. The proceedings of the think-aloud interviews are taped for later analysis. Findings are qualitatively analyzed and used in designing or redesigning the questionnaire.

Think-aloud techniques often detect different conceptual problems than those found in focus groups. Thus, both techniques are useful in combination. For example, retrospective interviews were conducted with nine respondents to test findings from the focus groups for one of BLS' user surveys. Two major sources of measurement errors were identified in the think-aloud interviews, and approaches to resolve these were introduced. First, most respondents expressed difficulties in identifying the specific Employment and Unemployment Statistics (EUS) programs they used during the past year. Respondents cited their lack of recognition of BLS acronyms and the indistinguishable descriptions of some programs. To resolve this difficulty, examples of each program's major publications were added to the questionnaire. Second, most think-aloud respondents reported difficulty in ranking the importance of six factors evaluating quality of service. Several respondents suggested that the first two or three factors perceived as important could be used to anchor their judgment of service quality. Thus, the question was redesigned to ask respondents to rank only the three most important factors.

16.2.2 Questionnaire Development

Historically, writing questions for surveys has been considered an art, in that each questionnaire was a hand-crafted original (Payne 1951; Converse and Presser 1986). With increased concerns about the cognitive aspects of survey methodology and the application of laboratory-based research techniques, questionnaire development and testing is moving towards becoming a science. Questionnaire wording, layout, and formatting can be tested using both qualitative and quantitative tools until the data desired by the analyst corresponds to the data provided by users (i.e., respondents and interviewers).

Respondents have different degrees of cognitive ability for information processing. Perception, experience, emotional state, and motivation can result in respondents' assigning different meanings to the same stimuli (FCSM 1991, Groves 1989). Some respondents engage in *satisficing* behavior; that is, they provide an adequate but incomplete answer when a more substantial cognitive effort is needed to answer a question completely and accurately (Blair and Chun 1992, Krosnick 1991, Simon 1957, Simon and Stedry 1968). In a business mail survey, where the questionnaire is the primary instrument for data capture, questionnaire layout and format play a significant role in motivating respondents to complete the form. The presence of visual stimuli in the self-administered business survey elicits psychological and emotional appeals and influences the cognitive process of retrieving data (FCSM 1991).

Focus groups, think-aloud techniques, and rating tasks are useful tools for identifying the causes of measurement errors related to the questionnaire. However, identifying causes does not necessarily provide the prescription for improving data quality. Field testing of alternatives is often necessary, especially when the goal is to increase response.

Think-Aloud Techniques

The retrospective think-aloud technique was used to understand the effect of different layouts for the Nonwage Cash Payments Pilot Survey (Phipps 1990). Instructions and definitions are typically included in mail surveys to aid the respondent in comprehending the questions. This study hypothesized that when instructions and definitions are in a location that is completely different from that of the questions, respondents make more mistakes. For the test, these instructions, definitions, and examples were on the back of a one-page questionnaire for which two different layouts were used. In one layout, respondents were first asked to provide annual nonwage cash payment and annual payroll totals and then (if the business made nonwage cash payments) to complete a set of yes/no questions. In contrast, the set of yes/no questions were placed first on the alternative form, followed by questions about the nonwage cash payment and payroll totals at the bottom of the page. Respondents receiving the second layout were less likely to provide annual payroll data. Apparently, they overlooked the question about annual payroll data, did not understand that the question was to be answered, or overlooked the instructions on the back of the questionnaire. Insights from this study were applied to designing the questionnaire for the ETJO pilot survey (Phipps et al. 1992).

A think-aloud protocol also identified two possible sources of measurement error in the mail questionnaire for the Survey of Occupational Illness and Injuries. First, respondents for businesses with multiple work sites or seasonal variations in payroll and hours worked experienced serious difficulties in answering "How many employees worked for your business in an average month?" and "Did any employees work more or fewer hours than expected in 1990?" due to a lack of clarity in the meaning of terms such as "average month" and "expected." Unclear instructions and visual perceptions were another source of measurement error. Respondents failed to notice instructions indicating the option of attaching a substitute form with the requested information rather than manually completing each individual lost-workday case report form. This failure was attributed to the shading of the instructions, which some respondents assumed to mean "For Office Use Only." Other respondents reported that the wording of the instruction was unclear.

Rating Tasks

Rating tasks can be designed to evaluate respondents' overall ability to understand and use questionnaires. For example, after think-aloud interviews or focus groups, respondents can be asked to evaluate on a rating scale [say, from 1 (meaning "very easy") to 7 (meaning "very difficult")] how easy it was to understand and answer a questionnaire. Such a procedure was used in redesigning the mail questionnaire for the Survey of Occupational Illness and Injuries. Analysis of responses to the rating scales suggested that respondents believed that the documents were difficult to understand and would be difficult to use. Needed revisions were made by defining concepts, providing or rephrasing instructions, changing fonts to clarify words and phrases, providing an example, and changing skip patterns.

Field Testing of Alternatives
Often, laboratory-based research suggests several questionnaire approaches that might reduce measurement error. Further laboratory testing of the alternatives may indicate that some or all of the alternatives appear to reduce error. Often, the viable alternatives must be field-tested to collect enough data to make a choice or to verify that improvements observed in the laboratory occur within the natural survey setting.

If nonresponse is being investigated, alternative procedures to improve response must be field-tested to determine if they work. From the theories of perceptual and motivational psychology, one might hypothesize that a colored questionnaire is more easily identifiable in an office setting, and color could have visual appeal for some business respondents. Using a meta-analysis, Fox et al. (1988) found that response rates increased using a green instead of a white questionnaire. For the Occupational Employment Statistics Wage Pilot Survey (OESWPS), Phipps et al. (1991) conducted a split-panel experiment to examine the effect of colored questionnaires on response. Four industries were chosen for OESWPS in which 15 states participated. The sample was split so that half the establishments received a green questionnaire and the remaining ones received a white one. The researchers concluded that green questionnaires produced a higher response rate than white ones in both the follow-up mailings and telephone follow-up. Overall, the green questionnaires had a statistically significant 3.5 percent higher response than the white ones.

Another study is now underway to compare the use of telephone and certified mail as nonresponse follow-up procedures. The mode of interviews in the follow-up procedure is an important factor in converting refusals and in increasing the response rate while maintaining data quality. In the EUS user survey, half the hard-core nonrespondents remaining after a second follow-up are randomly selected to receive the third package by certified mail, and the remaining half are prompted by the telephone. The hypothesis is that the continued efforts of the survey organization to obtain a response motivate the respondent to participate in the survey. A competing hypothesis is that telephones have become more frequent channels of communication in mundane business life, so that hard-core holdouts who reject the mail request may react more favorably to telephone follow-up. The split-panel comparison should aid in evaluating the effects of certified mail versus telephone prompting in a business setting.

Field-Based Respondent Debriefing
Respondent debriefings can enhance the usefulness of field tests of alternative questionnaires. The debriefing may consist of either (1) follow-up questions at the end of the interview or (2) questions administered at a later point in time. The questions may probe about the respondent's use of records, understanding of instructions, comprehension of terms, time required to respond, procedures followed to provide responses, and so forth. A respondent debriefing in the Redesigned Occupational Safety and Health Survey (ROSH) field test was administered via the combined use of a follow-up mail questionnaire and a fol-

low-up telephone call. Both techniques proved invaluable in providing information for survey redesign that could not have been obtained otherwise (American Institutes for Research 1992). Based upon a probability sample, the debriefing yielded quantitative results that could be generalized to the target population. Usually, subjects involved in laboratory research, such as focus groups and think-aloud interviews, are not derived from a representative sample, making inferences to a larger population statistically invalid. Thus, appropriately designed field-based respondent debriefings can provide data that permit a quantitative analysis of hypotheses developed from the qualitative data obtained through focus groups and think-aloud interviews.

16.2.3 Interviewer Training

The collection of business survey data by face-to-face visits requires that interviewers have *complex learning skills*. Complex learning skills are characterized by the coordination of perceptual input with motor responses (Ellis 1972). Training design has advanced by teaching complex learning skills with a combination of cognitive and behavioral approaches to learning, supplemented by intermittent evaluation and feedback on skill performance (Tannenbaum and Yukl 1992). In this regard, BLS instituted a new interviewer training process for the Occupational Compensation Survey Program (Committee on Training Manual 1991). The strategy is called a *data collection certification program*.

The goal of the certification program is to promote the consistent collection of high-quality data. To achieve that goal, performance standards were established for interviewers via a task analysis of the interviewer's job from start to conclusion of the interview. Next, training components (field observation, classroom training, on-the-job training, feedback from reinterviews) were designed so that interviewers would achieve a minimum competency rating on each standard. Each component assists the interviewer's progression through the various stages of complex learning skill acquisition, from knowledge about facts or what to do (*declarative knowledge*) through knowledge about how to do things (*knowledge compilation*) to automatic performance (*procedural knowledge*) (see Ackerman and Kyllonen 1991).

The first training component in certification facilitates the interviewer's achievement of the initial stage of skill acquisition, namely, declarative knowledge. During this stage, trainees come to understand what is required of them. Trainees learning to conduct face-to-face business interviews must encode orientation information, including the survey's purpose, the correct way to fill out survey forms, appropriate techniques to make appointments with respondents, and other key orientation facts required when confronting a new task. Learning a complex task can be facilitated by helping the trainee develop a global, mental conceptualization of the whole task (Burke and Day 1986). Hence, the observation training component was designed to help trainees build a broad cognitive framework for the task and to develop a set of field examples that clarify concepts presented in the next phase of training, namely, the class-room.

Classroom training allows the trainers to organize and explain the parts of the task that fit into the trainee's cognitive framework. Novice interviewers who have completed classroom training are still at the declarative knowledge stage, and their performance may be slow, effortful, and error-prone. Trainees must move beyond trial-and-error task engagement to be minimally competent to perform their job. Therefore, OCSP created the on-the-job training (OJT) component to assist interviewers with knowledge compilation, which is the next stage in the acquisition of complex learning skills.

The knowledge compilation stage has been termed the *associative stage* of skill acquisition (Fitts and Posner 1967). During this stage, trainees strengthen the associations between stimuli and appropriate response patterns. To be certified to collect OCSP survey data, novice interviewers must demonstrate minimal competence on all performance standards for six interviews. The OJT trainers rate the interviewers on their ability to perform certain competencies, providing immediate verbal and written feedback to them after the interviews. During the course of OJT, trainers observe that the interviewers' full attention to task completion is no longer necessary, because they begin to internalize "how things are done" so that they could explain the task to someone else (*conscious competence*). Certified interviewers are then allowed to collect data on their own.

The final component of certification is a reinterview program that assesses experienced interviewers' performance by the same standards they met when initially certified. Using an intermittent schedule, OCSP data reviewers contact a random sample of respondents for which an interviewer completed data collection. The reviewer reinterviews the respondent and notes discrepancies and problems. The reinterview process focuses on *all* of the information collected for the OCSP surveys, including wages collected for the jobs and even the firm's demographic information. Regional personnel then notify the interviewer of problems and, if necessary, provide remedial training. At this stage, interviewers have reached the final level of complex learning skill acquisition: proceduralized knowledge or *unconscious competence*. These interviewers no longer need declarative knowledge to perform their task adequately (Shiffrin and Schneider 1977). Though they can now perform their task effortlessly, they may no longer be able to explain the process to someone else. In the longer run, this loss of memory about how to accomplish a task has been linked to deterioration of performance over time (Tannenbaum and Yukl 1992). Continuing monitoring and feedback procedures like those involved in the reinterview program are necessary to prevent this decline in performance.

16.2.4 Data Collection

Over the last decade, the expanding use of computer technology has resulted in major structural changes to data collection methods. Eventually, almost all interviewers will use a computer to aid them in collecting data, either in face-to-face visits or over the telephone. Moreover, respondents to mail surveys will report their data using touch tone data entry, voice recognition, or electronic data interchange (see Chapters 18 and 19).

Initial research for the Consumer Price Index (CPI) program concentrated on the acceptability of computers by interviewers and on machine design (Couper et al. 1990). In the laboratory, interviewers tested lifting various weights and the use of differently styled machines while standing and sitting. Interviewers found it easier to use the the pen-based machines when standing, which is the most prevalent position used by field interviewers when collecting consumer prices. Moreover, to avoid fatigue, the machine should weigh less than 4.5 pounds.

More recently, an interagency research group has been established in the United States to identify research issues related to computer-assisted survey interviewing. The CASIC group, which includes researchers from BLS, the Bureau of the Census, the National Agricultural Statistics Service, and other U.S. statistical agencies, plans collaborative CASIC research to prevent duplication of efforts across agencies and to facilitate resource sharing. A subcommittee is investigating *human-computer interface*—that is, the effect of instrument design, screen design, and input mechanisms on overall data quality and on the interviewer's accuracy and speed of data entry. Research topics particularly relevant to business surveys are the examination of the effects of user-initiated telephone data entry, touch-tone entry systems, and different data input modes such as pen or voice recognition systems.

Earlier we described how MDS techniques might address the problem of job match inconsistencies among interviewers. The MDS study is one of three studies now underway to improve the job match process in wage programs. While the MDS study investigates how field interviewers mentally perceive the concepts of ''job'' and ''industry,'' the second study depicts interviewers' mental processing during the job match decision. In controlled interview settings, one expert and one less experienced interviewer are videotaped while conducting interviews with two different businesses. Immediately after the interview, the interviewer views the tape and is asked to ''think aloud,'' reporting her or his thought processes during the interview. This procedure allows researchers to determine how interviewers arrive at their decisions and to decide if categorization bias is present.

The third study uses data from the Job Match Validation Program to explore sources of error associated with particular job descriptions. For a sample of job descriptions, BLS researchers compute how often interviewers match definitions to jobs incorrectly, how often they do not match definitions to jobs when they should, and the overall errors per use of each definition. These errors will be subjected to a factor analysis to identify job descriptions and levels that share underlying factors. The descriptions will then be examined for factors that lead to confusing associations among similar jobs and job levels. The outcome should be guidelines that emphasize the distinct factors of similar job definitions and the deletion of features that cause inappropriate mental associations among field interviewers.

16.2.5 Post-Interview Procedures

Efforts to seek the causes of measurement error and to improve the data collection process should continue after data collection. Many business surveys are repeated or longitudinal, so continuous improvement is needed. Even special, one-time surveys should include a post-interview research component to prepare for similar survey problems in future studies. Post-interview research investigations can concentrate on macro-level or micro-level data.

The response analysis survey (RAS) is a post-interview research tool for investigating the quality of macro-level data. Although reinterview or recontact surveys are more prevalent for household surveys, they have been used in business surveys (Scott 1980). The modification of the recontact survey into the RAS is a change in data collection strategy which began with business surveys and is now being adopted for housing unit surveys.

Two post-interview research tools for investigating the quality of micro-level data—job match validation and expert systems—are described below. Another historical difference between business and household surveys has been in the amount of hands-on data validation and editing. Typically, considerably more resources are expended in the examination of individual responses for business surveys. Researchers working on business surveys are leading the development of new strategies for data editing (see Chapters 21 to 23).

Response Analysis Surveys

Often BLS evaluates the quality of business data by recontacting a random sample of approximately 150–500 respondents and asking them to complete a retrospective set of questions as part of a *response analysis survey*. Typically, a brief telephone reinterview is conducted shortly afterwards with a sample of respondents to the initial survey questionnaire. The reinterview identifies the sources of information (memory vs. records); evaluates respondents' comprehension and interpretation of questions, instructions, terminology, and definitions; and estimates the time and effort involved in answering the questionnaire. The Employment Turnover and Job Openings RAS (Phipps et al. 1992) indicated that small establishments (those with 49 or less employees) used memory while the medium (50–249) and large (250+) establishments depended upon more than one source to answer the ETJO questions. The RAS also unraveled respondents' difficulties in following multiple reference periods, in understanding the ambiguous meaning of inter-establishment transfers, and in estimating wages under nonstandard pay schemes. Respondents had the most difficulty with the reference period of the last business day of the month for reporting job openings, while more respondents correctly reported separations and new hires for the entire reference month. Some respondents undercounted internal transfers when calculating employee separations and new hires, whereas the number of laid-off employees appeared to be correct. (This result suggested that the instructions for internal transfers need to be included in the column heading, as do those for layoffs.) The motor freight transportation and warehousing industry had the most difficulty with reporting the average hourly wage of new hires; employees are paid by the mile, percent of load, or percent of revenue or commission. In addition, the RAS indicated that large establishments spent five times as much time to complete the survey questionnaire as did the small establishments and about twice as much time as the medium-sized establishments.

Job Match Validation

To ensure the quality of job match data for the Occupational Compensation Survey Program, BLS instituted a pilot job match validation program in 1983. Interviewers with various levels of experience are selected from the eight regional offices each year. One of their survey forms is randomly selected for review by regional staff every 2 weeks during the course of the survey. The reviewer telephones the respondent to discuss job match decisions made by the interviewer. During the call, the reviewer asks about jobs that were: (1) matched on the schedule but seem inappropriate, (2) probably overlooked by the interviewer and could be potential matches, and (3) identified by the interviewer but not matched. In addition, the reviewer validates a subsample of job matches that appear to be correct and queries the respondent about possible reasons for job match errors (Morton 1986). Finally, the reviewer shares the review with the interviewer who conducted the initial interview. The regional reviewer then completes a report that includes sources, reasons, and frequencies of interviewer errors in that region. Data from the eight regional reports are then compiled, analyzed, and published shortly after the survey results are released.

OCSP has identified six broad categories of error sources: training, job descriptions, interviewing techniques, survey instructions, administrative, and other (Morton 1986). Some error sources are continuously addressed through the immediate feedback from regional staff to interviewers; others require changes in materials and procedures by national office staff. For example, national training staff can modify their training courses according to the learning and skill problems that are causing the most difficulty in the field.

Findings from the job match validation indicate that the overall error rate for job matching has been relatively low. Since 1983, the error rate has ranged from 5 to 8 percent; some new jobs such as computer systems analyst have caused unusually high error frequencies in the last few years (Cohen 1990). Special efforts are underway to determine the sources of new error. Once sources are found, steps will be initiated to improve data quality. In summary, a major advantage of this strategy to reduce measurement error is that it provides immediate feedback to field interviewers about their performance while the interview is still fresh in their minds. In addition, the job match validation program is a cyclical improvement process that identifies new errors and tracks changes in the frequency of measurement error over time.

Expert Systems

Expert systems are computer programs that simulate the decision processes of an expert in a particular task. Expertise is encoded as a series of if–then rules. These rules are then applied to solve a problem. For example, an expert system has been developed to assist physicians with diagnosing infectious diseases. The physician inputs data about the characteristics of the bacterium, patient health history, and so on. The expert system integrates these facts to decide what the likely alternatives are and what antibiotic should destroy it (Charniak and McDermott 1985).

An expert system is in the early stages of development to assist CPI commodity analysts with their review of interviewers' decisions about substitutions for goods and services included in the index. Sometimes interviewers are unable to find a previously priced good or service that they must price for the current month. This can happen if the item has been discontinued by the supplier or if the business no longer carries it. In such a case, the interviewer must find an adequate substitute for the item, with the same major characteristics as the original item. For example, the major characteristics of a package of Dove Bars might be that they (1) are a package of three bars, (2) are milk chocolate with vanilla ice cream, and (3) weigh 8 ounces. Characteristics of this sort appear on a checklist which interviewers use to select comparable substitutions. Potential substitutions are later evaluated by commodity analysts by comparing the checklist characteristics of the original and substitute products. (There are 719 checklists used in the monthly or bimonthly collection of prices for approximately 100,000 items in the CPI.)

As one might imagine, the substitution decisions made by interviewers and analysts can disagree. Visual search errors may be made in reading product descriptions. Knowledge possessed may not be applied or may be applied in error. Therefore, an expert system can allow a commodity analyst to evaluate the substitution decisions made by interviewers without errors of memory, logic, or perception. Rules for this expert system are formatted so that if certain item characteristics are present, an appropriate substitution may be made. The program does not take the place of the analyst, but serves as a tool for making the final decision about the adequacy of a substitution (Conrad et al. 1992). Eventually, the rules could become an integral part of the software used for actual data collection by interviewers using pen-based machines to aid them in finding the best substitute.

16.3 CONCLUDING REMARKS

Until recently, little research has been conducted on the causes of measurement error in business surveys; the strategies used in designing data collection systems were relatively ad hoc. With the establishment of behavioral science laboratories at statistical agencies, laboratory-based research techniques are being introduced to improve data collection procedures for business (and housing unit) surveys. Because these techniques are often qualitative, they must be used carefully and preferably in conjunction with methods that yield quantitative information on the change in measurement error resulting from improvements to the data collection process. Focus groups and think-aloud interviews with small convenience samples are appropriate for identifying possible causes of measurement error, but split-panel experiments and response analysis surveys with larger probability samples can verify that the resulting changes to data collection procedures do, in fact, reduce measurement error. Of course, these probability-sample-based methods are not without their limitations (e.g., the inability to control potentially confounding variables, costs, and respondent burden). Thus, we encourage designers of data collection methods and procedures to develop an integrated approach to research (Copeland and Rothgeb 1990, Esposito et al. 1991).

One way to facilitate the development of an integrated approach to improve the quality of business survey data is to establish interdisciplinary teams that include behavioral scientists. Behavioral scientists bring new strategies and methods to the survey researcher's tool box. At the same time, old strategies, such as the use of experts, experimental field-testing, and reinterview surveys, can be improved to fit the new paradigm. Together, these new and improved strategies for investigating the causes of measurement errors and strategies to reduce them are quite effective. At the end of this decade, our knowledge of measurement errors in business surveys will be more extensive than now.

REFERENCES

Ackerman, P., and P. Kyllonen (1991), "Trainee Characteristics," in J. Morrison (ed.), *Training for Performance: Principles of Applied Human Learning*, New York: Wiley, pp. 193–229.

American Institutes for Research (1992), "Redesigning and Testing the 1992 Survey of Occupational Injuries and Illnesses," unpublished final report, Washington, DC.

Babbie, E. R. (1973), *Survey Research Methods*, Belmont, CA: Wadsworth Publishing Company.

Biemer, P. P., R. M. Groves, L. E. Lyberg, N. A. Mathiowetz, and S. Sudman (1991), *Measurement Errors in Surveys*, New York: Wiley.

Blair, J., and Y. Chun (1992), "Quality of Data from Converted Refusals in Telephone Surveys," paper presented at the Annual Meeting of the American Association of Public Opinion Research, St. Petersburg, FL.

Bureau, M. (1991), "Experience with the Use of Cognitive Methods in Designing Business Survey Questionnaires," *Proceedings of the Survey Research Methods Section, American Statistical Association*, pp. 713–717.

Burke, M., and R. Day (1986), "A Cumulative Study of the Effectiveness of Managerial Training," *Journal of Applied Psychology*, **71,** pp. 232–245.

Cantor, D., J. Brown, and L. Groat (1985), "A Multiple Sorting Procedure for Studying Conceptual Systems," in M. Brenner, J. Brown, and D. Danter (eds.), *The*

Research Interview: Uses and Approaches, Orlando, FL: Academic Press, pp. 79–114.

Charniak, E., and D. McDermott (1985), *Introduction to Artificial Intelligence*, Reading, MA: Addison-Wesley.

Clayton, R. L., and L. Harrell, Jr. (1989), "Developing a Cost Model for Alternative Data Collection Methods: Mail, CATI and TDE," *Proceedings of the Survey Research Methods Section, American Statistical Association*, pp. 264–269.

Clayton, R. L., and D. L. S. Winter (1992), "Speech Data Entry: Results of a Test of Voice Recognition for Survey Data Collection," *Journal of Official Statistics*, **8**, pp. 377–388.

Cohen, S. (1990), "Capturing Causes of Interviewer Error in Wage Surveys," paper presented at the International Conference on Measurement Errors in Surveys, Tucson, AZ.

Committee on Training Methodology (1991), *CPI/OCSP Introductory Training Manuals*, Washington, DC: U.S. Bureau of Labor Statistics, Office of Field Operations.

Conrad, F., R. Kamalich, J. Longacre, and D. Barry (1992), "An Expert System for Reviewing Commodity Substitutions in the Consumer Price Index," paper presented at the Annual Conference on Computing for the Social Sciences, Ann Arbor, MI.

Converse, J. M., and S. Presser (1986), *Survey Questions: Handcrafting the Standardized Questionnaire*, Beverly Hills, CA: Sage Publications.

Copeland, K. R., and J. M. Rothgeb (1990), "Testing Alternative Questionnaires for the Current Population Survey," *Proceedings of the Survey Research Methods Section, American Statistical Association*, pp. 63–71.

Couper, M., R. Groves, and C. Jacobs (1990), "Building Predictive Models of CAPI Acceptance in a Field Interviewing Staff," *Proceedings of the Annual Research Conference*, Washington, DC: U.S. Bureau of the Census, pp. 685–702.

Cox, B. G., G. E. Elliehausen, and J. D. Wolken (1989), "Surveying Small Businesses About Their Finances," *Proceedings of the Survey Research Methods Section, American Statistical Association*, pp. 553–557.

DeMaio, T. J., and C. R. Jenkins (1991), "Questionnaire Research in the Census of Construction Industries," *Proceedings of the Survey Research Methods Section, American Statistical Association*, pp. 496–501.

Desvousges, W. H., and J. H. Frey (1989), "Integrating Focus Groups and Surveys: Examples from Environmental Risk Studies," *Journal of Official Statistics*, **5**, pp. 349–363.

Dietz, S. K. (1992), "Quantitative Estimation Using Focus Groups," paper presented at the Annual Meeting of the American Association of Public Opinion Research, St. Petersburg, FL.

Dippo, C., A. Polivka, K. Creighton, D. Kostanich, and J. Rothgeb (1994), "Redesigning a Questionnaire for Computer-Assisted Data Collection: the Current Population Survey Experience," in house report, Washington, DC: U.S. Bureau of Labor Statistics.

Ellis, H. (1972), *Fundamentals of Human Learning and Cognition*, Dubuque, IA: William C. Brown Co.

Ericsson, K. A., and H. A. Simon (1984), *Protocol Analysis: Verbal Reports as Data*, Cambridge, MA: The Massachusetts Institute of Technology Press.

Esposito, J. L., P. C. Campanelli, J. M. Rothgeb, and A. E. Polivka (1991), "Determining Which Questions Are Best: Methodologies for Evaluating Survey Questions," *Proceedings of the Survey Research Methods Section, American Statistical Association*, pp. 46–55.

Federal Committee on Statistical Methodology (1983), *Approaches to Developing Questionnaires*, Statistical Policy Working Paper 10, Washington, DC: U.S. Office of Management and Budget.

Federal Committee on Statistical Methodology (1988), *Quality in Establishment Surveys*, Statistical Policy Working Paper 15, Washington, DC: U.S. Office of Management and Budget.

Federal Committee on Statistical Methodology (1991), *Seminar on Quality of Federal Data*, Statistical Policy Working Paper 20, Washington, DC: U.S. Office of Management and Budget.

Fitts, P., and M. Posner (1967), *Human Performance*, Belmont, CA: Brooks/Cole.

Forsyth, B. H., and J. T. Lessler (1991), "Cognitive Laboratory Methods: A Taxonomy," in P. P. Biemer, R. M. Groves, L. E. Lyberg, N. A. Mathiouetz, and S. Sudman (eds.), *Measurement Errors in Surveys*, New York: Wiley, pp. 393–418.

Fox, R. J., M. R. Crask, and J. Kim (1988), "Mail Survey Response Rate: A Meta-Analysis of Selected Techniques for Inducing Response," *Public Opinion Quarterly*, **52**, pp. 467–491.

Gower, A. R. (1991). "The Questionnaire Design Resource Centre's Role in Questionnaire Research and Development at Statistics Canada," presented at the 48th Session of the International Statistical Institute, Cairo, Egypt.

Gower, A. R., and M. S. Nargundkar (1991), "Cognitive Aspects of Questionnaire Design: Business Surveys versus Household Surveys," *Proceedings of the Annual Research Conference*, Washington, D.C.: U. S. Bureau of the Census, pp. 299–312.

Groves, R., P. P. Biemer, L. E. Lyberg, J. T. Massey, W. L. Nicholls II, and J. Waksberg (1988), *Telephone Survey Methodology*, New York: Wiley.

Groves, R. M. (1989), *Survey Errors and Survey Costs*, New York: Wiley.

Jabine, T. B., M. L. Straf, J. M. Tanur, and R. Tourangeau (1984), *Cognitive Aspects of Survey Methodology: Building a Bridge Between Disciplines*, Washington, DC: National Academy Press.

Jenkins, C. R. (1992), "Questionnaire Research in the Schools and Staffing Survey: A Cognitive Approach," *Proceedings of the Survey Research Methods Section, American Statistical Association*, pp. 434–439.

Krosnick, J. A. (1991), "Response Strategies for Coping with the Cognitive Demands of Attitude Measures in Surveys," *Applied Cognitive Psychology*, **5**, pp. 213–236.

Krueger, R. A. (1988), *Focus Groups: A Practical Guide for Applied Research*, Newbury Park, CA: Sage Publications.

Merton, R. K. (1987), "The Focused Interview and Focus Groups: Continuities and Discontinuities," *Public Opinion Quarterly*, **51**, pp. 550–566.

Merton, R. K., and P. L. Kendall (1946), "The Focused Interview," *American Journal of Sociology*, **51**, pp. 541–557.

Morgan, D. L. (1988), *Focus Groups as Qualitative Research*, Newbury Park, CA: Sage Publications.

Morton, J. (1986), "Quality Control in the PATC Survey: a Report on the Job Match Validation Experience," internal report prepared for the Office of Compensation and Working Conditions, Washington, DC: U.S. Bureau of Labor Statistics.

Palmisano, M. (1988), "The Application of Cognitive Survey Methodology to an Establishment Survey Field Test," *Proceedings of the Survey Research Methods Section, American Statistical Association*, pp. 179–184.

Payne, S. L. (1951), *The Art of Asking Questions*, Princeton, NJ: Princeton University Press.

Phipps, P. (1990), "Applying Cognitive Techniques to an Establishment Mail Survey," *Proceedings of the Survey Research Methods Section, American Statistical Association*, pp. 608–612.

Phipps, P., S. Butani, and Y. Chun (1992), "Designing Establishment Survey Questionnaires," paper presented at the Annual Meeting of the American Association of Public Opinion Research, St. Petersburg, FL.

Phipps, P., K. W. Robertson, and K. G. Keel (1991), "Does Questionnaire Color Affect Survey Response Rates?" *Proceedings of the Survey Research Methods Section, American Statistical Association*, pp. 484–489.

Ponikowski, C. H., and S. A. Meily (1989), "Controlling Response Error in an Establishment Survey," *Proceedings of the Survey Research Methods Section, American Statistical Association*, pp. 258–263.

Ramsay, J. O. (1978), *Multiscale*, Chicago: National Educational Resources.

Rosen, R. J., R. L. Clayton, and T. B. Rubino, Jr. (1991), "Controlling Nonresponse in an Establishment Survey," *Proceedings of the Survey Research Methods Section, American Statistical Association*, pp. 587–592.

Scott, S. (1980), "Reinterview Methods in an Establishment Survey of Job Openings," *Proceedings of the Survey Research Methods Section, American Statistical Association*, pp. 445–449.

Shiffrin, R. M., and W. Schneider (1977), "Controlled and Automatic Human Information Processing: Perceptual Learning, Automatic Attending, and a General Theory," *Psychological Review*, **84**, pp. 127–190.

Simon, H. A. (1957), *Models of Man: Social and Rational, Mathematical Essays on Rational Human Behavior in a Social Setting*, New York: Wiley.

Simon, H. A., and A. C. Stedry (1968), "Psychology and Economics," in G. Lindzey and E. Aronson (eds.), *Handbook of Social Psychology*, 2nd ed., Vol. 5, Reading, MA: Addison–Wesley, pp. 269–314.

Statistics Canada (1992), "Final Report on Focus Groups and Personal Interviews with Farm Operators," *Farm Financial Survey*, Questionnaire Design Resource Centre Methodology Branch.

Tannenbaum, S., and G. Yukl (1992), "Training and Development in Work Organizations," *Annual Review in Psychology*, **43**, pp. 399–441.

Werking, G. A., A. R. Tupek, C. H. Ponikowski, and R. J. Rosen (1986), "A CATI Feasibility Study for a Monthly Establishment Survey," *Proceedings of the Survey Research Methods Section, American Statistical Association*, pp. 639–644.

CHAPTER SEVENTEEN

Improving Response to Business Mail Surveys

M. Chris Paxson
Washington State University

Don A. Dillman
Washington State University and U.S. Bureau of the Census

John Tarnai
Washington State University

Large numbers of business mail surveys are conducted each year with varying results. Christianson and Tortora report in Chapter 14 that of more than 1300 surveys implemented in 16 countries, half were single-mode mail surveys and another third were mixed-mode surveys with mail as the predominant mode. Some obtain acceptably high response rates, whereas response is intolerably low for others. This chapter examines the difficulties of achieving high response to business mail surveys with a view toward assessing the state of knowledge that exists.

During the last two decades, much has been learned about how to improve response for mail surveys of individual persons. Clearly, certain core procedures produce reasonably high response rates for surveys of individuals and households. Yet there is no core set of agreed-upon procedures for business mail surveys, not only because of the ways business and individual person surveys differ, but also because of different survey challenges.

This chapter summarizes what is known about achieving response to individual-person mail surveys. It then discusses how business surveys differ from surveys of individuals and why some procedures commonly used with individual-person surveys may not be applicable for businesses. Finally, the chapter

Business Survey Methods, Edited by Cox, Binder, Chinnappa, Christianson, Colledge, Kott.
ISBN 0-471-59852-6 © 1995 John Wiley & Sons, Inc.

compares the procedures and experiences of two survey organizations. The first organization is the Social and Economic Sciences Research Center at Washington State University, which regularly conducts business surveys using procedures modeled closely after those found effective in individual-person surveys. The second organization is the U.S. Bureau of the Census, which regularly conducts large-scale mail surveys and uses somewhat different procedures.

Based on this comparison, we identify important issues that must be addressed by survey organizations, if response rates comparable to those expected in individual-person surveys are to be attained in business surveys. Our overall goals are to assess the current state of knowledge about attaining high response rates to business mail surveys and to suggest how response might be improved.

17.1 RESPONSE RATE RESEARCH FOR INDIVIDUAL-PERSON SURVEYS

Research literature on improving survey response rates differs, but certain techniques have emerged as important in achieving high mail-survey response rates. These techniques include follow-up contacts and replacement questionnaires, prior notice, certified or special delivery postage, financial incentives, sponsorship, personalization, content of correspondence, and salience of the topic to the respondent (Dillman 1991). The hundreds of experiments and analyses of response-inducing techniques reported in the literature defy easy summary, but few would doubt that the single most powerful inducement to response is number of contacts. The use of financial incentives is second in importance.

Achieving the highest possible response depends upon more than simply applying these techniques. How they are applied also makes a difference. The Total Design Method (TDM) attempts to combine numerous elements in a way that favorably influences respondents and maximizes the likelihood of response (Dillman 1978). This technique is based upon the assumption that individuals are most likely to respond when they perceive that the eventual rewards of responding outweigh the costs of doing so.

Dozens of prescribed details assist in achieving this favorable respondent perception and subsequent cooperation. The details include (1) a carefully timed sequence of four first-class mailings, the last of which is certified, (2) personalized correspondence, and (3) an attractive booklet questionnaire designed to increase the perceived salience of the questions and ease of responding (Dillman 1978). Numerous other details of preparing, sending, and retrieving questionnaires are also specified as a part of the TDM. Updated details include use of modern word processing equipment for questionnaire design, stamped return envelopes, and elimination of the certified mail contact in favor of a telephone or other special mail contact (Dillman 1991).

James and Bolstein (1990, 1992) and Johnson and McLaughlin (1990) demonstrated that even when TDM details are followed, response from individuals can still be improved with the use of financial incentives. Use of the complete TDM plus financial incentives provides a means for consistently achieving response rates of 70 percent and higher in surveys of the general public. Higher rates can often be achieved for specialized groups with higher educational levels.

Past research also suggests that there is no single "magic bullet" to secure high response. Research findings are contradictory and somewhat confusing. Contrary to the results of other research studies, Childers and Ferrell (1979) and Jobber and Sanderson (1981) found no significant improvements in response rates from techniques such as reduced questionnaire length and prior notification. Despite these caveats, high response rates appear to result from combining multiple techniques. For example, a series of experiments conducted by the U.S. Bureau of the Census on household responses to Census questionnaires showed that the use of five techniques together could improve response from about 40 percent to about 70 percent. The amount of response improvement achieved by each of the five techniques in controlled experiments was 8.0 percent for respondent-friendly construction and a shorter form, 10.4 percent for use of a replacement form mailing, and 12.7 percent for a prenotice and reminder (Dillman et al. 1993a; 1993b). This research also demonstrated that each of the paired techniques (e.g., prenotice and reminder) independently contributed to the overall response. In summary, reasonably high response rates can be obtained to individual-person surveys through the use and integration of multiple techniques.

17.2 HOW BUSINESS SURVEYS DIFFER

Applying the TDM or similar comprehensive procedures to the conduct of business surveys may be difficult for a number of reasons:

- Businesses are often hard to define. Sending a questionnaire to a named business does not mean that it reaches the appropriate survey respondent. What telephone directories or tax records identify as a business may be only a subunit of what the survey researcher views as the business entity of interest.

- Addresses can be problematic. It is often unclear who should respond to the survey, and therefore to whom to address correspondence. Lists used to draw samples of businesses often do not provide names of owners and management personnel. Even if a name is provided, it may not be that of the appropriate individual to receive the questionnaire.

- Businesses often have gatekeepers who decide whether survey requests should be given to the appropriate person for response. In some businesses, gatekeepers are receptionists or secretaries. In other businesses, correspondence not addressed to a specific individual is sent to an executive or simply to an office (e.g., of a tax accountant).

- Some businesses have policies about responding to surveys. There may be a policy against responding to any survey, or there may be a process for deciding which surveys should receive a response.
- The questions asked in business surveys are often difficult to answer, requiring record checks or even compilation of information. By their nature, business surveys often ask for information that is not immediately known to the respondent, whereas individual-person surveys ask more easily answered questions about personal characteristics or opinions.

Based on these differences, the techniques found effective for individual-person surveys may not work as well for business surveys. For example, in business surveys it is difficult to personalize correspondence and other individual contacts. Repeated contacts may be delivered to different individuals. In other cases, correspondence may need to be aimed at several handlers of the desired information, each of whom have different concerns, making it difficult to be concise and targeted.

Under the best of circumstances, respondents need time to compile the needed information for surveys, so repeated contacts over a short time period may be inappropriate. The task of responding may appear even more laborious than it actually is for most businesses because of detailed directions and definitions needed for compiling complex information under all possible situations. Financial incentives also present problems because it may be unclear who gets them, and business policies may prohibit their acceptance. A business may find that accepting a financial incentive creates more costs than its monetary value or the costs of responding to the survey.

To elaborate on these dilemmas and potential methods for overcoming them, we report the experiences of two markedly different survey organizations: the Social and Economic Sciences Research Center (SESRC), which conducts surveys for university, government, and business clients and the U.S. Bureau of the Census, one of the largest survey organizations in the world. SESRC adheres to a comprehensive survey response model to the extent that resources allow, whereas the Census Bureau uses a somewhat different approach.

17.2.1 Social and Economic Sciences Research Center Surveys (SESRC)

Table 17.1 provides a summary of response rates and procedures for SESRC business mail surveys over the 3-year period from 1989 to 1992. SESRC attempts to use a comprehensive TDM approach to mail surveys. Implementation strategies are decided in conjunction with each client; frequently, the full TDM cannot be used because of client funding constraints or preferences. Therefore, the implementation treatments for studies presented in Table 17.1 do not represent a random assignment of procedures to projects. Nonetheless, several interesting differences emerge from an examination of Table 17.1.

The average response rate across the 26 surveys is 51 percent. However,

Table 17.1   Response Rates for Business Mail Surveys Conducted by the Social and Economic Sciences Research Center, November 1989 to November 1992

Survey Number	Date, Population, Topic	Sample Size	Questionnaire Pages	Response Rate (%)	Procedures
1	11/89, County Businesses, Hazardous Waste Management	1040	4	38	Owner/Manager, Postcard Follow-Up
2	1/89, County Businesses, Hazardous Waste Management	967	4	42	Owner/Manager, Postcard Follow-Up
3	1/89, County Businesses, Hazardous Waste Management	952	4	37	Owner/Manager, Postcard Follow-Up
4	2/91, County Businesses, Hazardous Waste Management	1020	4	28	Owner/Manager, Postcard Follow-Up
5	8/90, Nationwide, Apple Storage Operators	1156	8	28	Owner/Manager, Postcard Follow-Up
6	10/88, County Businesses	902	8	59	Owner/Manager, Three Contacts
7	11/89, Washington State Businesses, Waste Management Practices	1033	8	49	Owner/Manager, Three Contacts
8	3/90, Nationwide Apple Growers, Production Practices	2798	8	49	Owner/Manager, Three Contacts
9	10/89, Oregon Businesses, Parental Leave Policies	938	12	38	Owner/Manager, Three Contacts
10	1/90, Wisconsin Businesses, Parental Leave Policies	1056	12	36	Owner/Manager, Three Contacts
11	2/90, Rhode Island Businesses, Parental Leave Policies	1099	12	29	Owner/Manager, Three Contacts
12	2/90, Minnesota Businesses, Parental Leave Policies	1463	12	40	Owner/Manager, Three Contacts
13	5/90, State Businesses, Changes in the Minimum Wage Law	3029	12	43	Owner/Manager, Three Contacts
14	5/90, State Businesses, Training and Retraining Issues	9829	16	43	Owner/Manager, Three Contacts
15	3/91, Nationwide, Apple Growers' Pesticide Use	2739	20	33	Owner/Manager, Three Contacts

(Continued)

Table 17.1   *(Continued)*

Survey Number	Date, Population, Topic	Sample Size	Questionnaire Pages	Response Rate (%)	Procedures
16	1/92, Washington State Businesses, Satisfaction with Revenue Policies	3740	12	56	Owner/Manager, TDM with Telephone Follow-Up
17	3/90, County Farms, Management of Hazardous Waste Law	466	12	34	Named Individual, Postcard Follow-Up
18	11/89, County Farms, Management of Hazardous Waste	385	12	54	Named Individual, Postcard Follow-Up
19	9/89, State Businesses, Large Generators of Hazardous Waste	518	8	73	Named Individual, Three Contacts
20	9/89, State Businesses, Generators of Small Amounts of Hazardous Waste	763	8	71	Named Individual, Three Contacts
21	4/91, Nationwide Cooperative Extension Directors	74	12	86	Named Individual, Three Contacts
22	4/90, K–12 Public School Superintendents, International Issues	294	16	67	Named Individual, Three Contacts
23	10/89, Nationwide, Internationalization of Universities	238	20	76	Named Individual, Three Contacts
24	11/92, Nationwide, Dental Offices, Malpractice Issues	4300	16	75	Named Individual, TDM with $5 Incentive
25	4/90, Nationwide, Hotel Operators	790	8	95	Named Individual, TDM with Telephone Follow-Up
26	5/92, Washington State Agency Executives	120	8	90	Named Individual, TDM with Telephone Follow-Up

the range in response rates is from 28 percent to 95 percent. Listed first in the table are all of the surveys that could not be addressed to a named individual and instead were sent simply to the company, in some cases with the notation "owner/manager" preceding the company name. The average response rate for these surveys was 40 percent, compared to 72 percent for those addressed to a named individual.

Five surveys (numbers 1–5) addressed to owner/manager included only an original mailing, plus a follow-up postcard. Most surveys used a questionnaire produced by the client; the typical booklet construction of the TDM was not used. These surveys achieved an average response rate of only 37 percent.

Ten surveys (numbers 6–15) addressed to owner/manager used normal TDM booklet construction for the questionnaires and three contacts by mail and achieved an average response rate of 42 percent. These surveys were generally longer and more complex than the ones for which only follow-up postcards were used, a fact probably reflected in the limited increase in the response rate. The TDM survey that was addressed to owner/manager and used a telephone follow-up contact (number 16) attained a response rate of 56 percent.

In contrast to these surveys, four surveys (number 19–22) with a named individual, TDM questionnaire construction used, and three contacts had an average response rate of 74 percent. In length and complexity, their question-aires were comparable to those of the 10 surveys (numbers 6–12) that used the same number of contacts (3) but did not mail to a named person; these 10 surveys achieved only a 42 percent response. Note that lower response rates also occurred for two surveys (numbers 17–18) that had a named individual and that used TDM questionnaire construction but only a postcard follow-up.

Finally, 84 percent response was achieved for four surveys sent to named individuals and using the three mail-contact procedures as well as including either a $5.00 incentive or telephone follow-up (numbers 23–26). A more detailed discussion of these surveys is warranted.

The Internationalization of Universities Survey was sent to the presidents of the 238 largest universities in the United States, with the request that the person "most responsible for decisions about international activities at your university" be asked to complete and return the questionnaire. Recipients were also asked to give us the person's name so that he/she could be followed up. If no name was received from the president's office, follow-up telephone calls were made to obtain the name of the designated respondent so that all follow-ups could be sent to the appropriate person.

The nationwide Hotel Operators Survey was also sent to the president or chief executive officer (CEO) with a request that the questionnaire be completed by him/her or given to an appropriate person for completion. When no response was received, the president or CEO was telephoned and given the option to complete and mail back the questionnaire or to provide answers over the telephone. Nearly three-fourths of those contacted chose to complete the questionnaire by telephone.

In the Survey of Washington State Agencies, agency directors were mailed a questionnaire with a request that they complete the questionnaire or forward it to an appropriate person. Nonrespondents were contacted by telephone. The Survey of Dental Offices did not use a telephone follow-up, but did use a $5.00 incentive to increase response rates.

Generally, the conclusion to be drawn from these results is that high response to voluntary business surveys is possible. Furthermore, reasonably high response rates can be obtained when one has the name of an individual to whom correspondence should be addressed and when general procedures found successful with individual-person surveys are augmented with procedures to get past gatekeepers.

These latter results contradict much of the published literature about business surveys that suggest that high response rates cannot be achieved. For example, Paxson (1992) found an average response rate of 21 percent for 183 studies identified in a search of academic and trade literatures published since 1990. Nearly all of these studies used no follow-up procedures.

17.2.2 U.S. Bureau of the Census Surveys

Table 17.2 lists mail surveys regularly conducted by the U.S. Bureau of the Census. Some surveys are conducted monthly, while others are decennial. Some involve millions of businesses, while others involve only a few respondents. Many surveys contact the same respondents repeatedly. In addition, the questionnaires vary in complexity from one survey to another.

Response rates for these business mail surveys are fairly high overall, averaging 79 percent and ranging from 57 percent to 96 percent. Although the surveys differ considerably in their response-inducing procedures, many use multiple contacts by mail, followed by attempts to interview nonrespondents by telephone. In some instances, telephone attempts are followed by certified letter.

The switch to telephone is especially important for surveys that must be completed. For example, the construction surveys use one mailing, followed by telephone contacts within 7 days. For one of the surveys, response was obtained entirely by telephone. For several of the business surveys, a substantial proportion of the response (shown in parentheses under the overall response rates) was completed by telephone.

The mandatory reporting requirement is a second factor that appears to encourage response. Response to the mandatory surveys is 84 percent versus 69 percent for voluntary surveys. A number of experiments conducted by the Census Bureau leave little doubt that mandatory reporting is an important contributor to overall response. Informing respondents that response is mandatory adds about 20 percentage points to overall response (Worden and Hoy 1992, Tulp et al. 1991). Significantly, this increment of response occurs in the high response range (i.e., improving response rates from 60 percent to 80 percent or higher) where increases in response rate are generally difficult to achieve.

A number of procedures used for the Census Bureau surveys contrast sharply with those found effective for individual-person surveys. Some surveys, par-

Table 17.2   Selected Business Mail Surveys Conducted by the U.S. Bureau of the Census

Survey Number	Survey	Approximate Sample Size	Mandatory	Estimated Response Rate (%)	Procedures
1	1987 Census of Agriculture	4,100,000	Yes	86	Multiple Mail Contacts, Some Telephone, Some Certified Mail
2	1988 Census of Horticultural Specialties	24,590	No	75	Multiple Mail Contacts, Some Telephone, Some Certified Mail
3	Agricultural Sample Surveys	Varies	Yes	78–85	Multiple Mail Contacts, Some Telephone, Some Certified Mail
4	1987 Agricultural Nonresponse Survey	27,100	Yes	85	Multiple Mail Contacts, Some Telephone, Some Certified Mail
5	1987 Agricultural Classification Error Survey	15,300	Yes	85	Multiple Contacts, Then Telephone
6	Multifamily Residential Housing	1,700	No	57	One Mailing, Telephone Contact in 7 Days
7	Government-Owned Projects	6,000	No	96	One Mailing, Telephone Contact in 7 Days
8	Private Non-Residential Housing	4,700	No	66	One Mailing, Telephone Contact in 7 Days
9	Advance Monthly Retail Trade	3,250	No	60 (45)[a]	One Mailing, Then Telephone Contact
10	Monthly Retail Trade and Retail Inventory	3,000	No	63 (22)[a]	One Mailing, Then Telephone, Then Field Representative
11	Monthly Wholesale Trade	2,800	No	67 (18)[a]	One Mailing, Then Telephone Contact
12	Annual Retail Trade	20,000	Yes	90 (20)[a]	Two Mailings, Then Certified Letter, Then Telephone Contact

(Continued)

Table 17.2 *(Continued)*

Survey Number	Survey	Approximate Sample Size	Mandatory	Estimated Response Rate (%)	Procedures
13	Annual Trade	4,500	Yes	93 (17)[a]	Two Mailings, Then Certified Letter, Then Telephone, Then Certified Again
14	Service Annual	33,000	Yes	70 (10)	Two Mail Contacts, Then Telephone Contact
15	Motor Freight Transportation	1,500	Yes	88 (8)[a]	Two Mail Contacts, Then Telephone Contact
16	Warehousing Communication Services	1,200	Yes	94 (9)[a]	Two Mail Contacts, Then Telephone Contact
17	Truck Inventory and Use	135,000	Yes	82 (16)[a]	Mail Contact, Then Telephone Contact
18	Nationwide Truck Activity	44,000	No	70	Multiple Mail Contacts
19	Commodity Flow	197,000	Yes	70	Multiple Mail Contacts
20	Assets and Expenditures	52,000	Yes	87	Four Mail Contacts, Then Telephone Contact

[a]Portion of response obtained over the telephone.

ticularly the censuses, must send questionnaires by bulk rate rather than by first-class mail. While this lowers survey costs, it also lengthens the response period, sometimes substantially. Correspondence usually consists of form letters, rather than letters addressed specifically to the business or an owner/manager, and often look like a mass mailing. Later mailings often have rather intense, insistent wording, informing recipients that their response is required by federal law.

The questionnaires are often quite complex, with information sheets enclosed to provide definitions and explain how to handle certain situations. The questionnaires also have a "form" appearance, with boxes separating questions, special notices, and return address information. For many years, color has been used for areas of the questionnaire that ask questions, with white spaces reserved for respondents' answers. Nonetheless, many of the forms appear fairly difficult to complete and may require the specialized knowledge of a person (or office) in some businesses, as well as consultation with records.

Considering the high response achieved in the Census Bureau surveys, it appears that these procedures are hardly an inhibitor to response. Major changes in the factors seem unlikely to yield significant improvement in response rates.

The mandatory requirement and the telephone follow-up appear to compensate for any significant response-rate increment that could be achieved by efforts to redesign the forms in a user-friendly manner, to personalize letters, or to switch to first-class mailings. The principal gain likely to be achieved through such procedures would be to decrease the data collection period by increasing the speed of response. The effects on quality of response and costs from implementing such changes are unknown.

Current efforts to improve response focus on ways of making it easier for respondents to answer, for example, accepting facsimile responses, using automated touchtone entry systems for reporting data, or using toll-free 800 numbers to give answers to an interviewer (Wallace 1992). The overall effects of such procedures on response rates are not yet known.

To summarize, Census Bureau business mail surveys follow a different model than that found effective for individual-person mail surveys. This approach results in high response rates. The Census Bureau takes full advantage of the ability to use mandatory authority and the fact that most businesses are listed in telephone directories, where telephone numbers can be located. In contrast to households, business employees are unlikely to disconnect a call or refuse to talk to an interviewer. Not only do businesses wish to avoid such unprofessional behavior, they also wish to avoid the considerable risk of legal and other costs associated with not responding to the Federal Government. In effect, the anticipated costs of not responding to the survey could be far greater than the costs of responding, the opposite of the situation that applies to non-government-sponsored surveys.

17.3 CONCLUDING REMARKS

In this chapter, we juxtapose procedures that are generally effective when used to conduct individual-person surveys, contrasting the effects of these procedures when used by the SESRC and the U.S. Bureau of the Census for business surveys. This comparison leads to several conclusions.

First, the procedures used by the Census Bureau follow a model that, in part, is available only to certain government surveys. The high level of response is achieved principally by using mandatory authority and telephone follow-up. Only the latter technique is available for use by nongovernmental organizations.

Second, the results of nonmandatory Census Bureau surveys and the SESRC surveys suggest that relatively high response to business mail surveys can be obtained, even without the use of mandatory authority. For individual-person surveys other than the Census Bureau's, identifying a person to whom the survey can be addressed appears to be quite successful. Christianson and Tortora's findings in Chapter 14 corroborate this notion. In Chapter 6, Pietsch suggests using the profiling technique to improve response rates for business surveys. Profiling involves contacting large businesses to gain better insight into their structure. In a sense, obtaining a name allows one to turn the contact process for a business survey into an individual-person survey. More intensive techniques, such as telephone calls and certified mail, seem essential for pushing response rates closer to those obtained in Census Bureau surveys.

Third, the results presented here suggest that low response rates (below 50 percent) result more from not using available knowledge about obtaining high response than from inherent features of business surveys. The main suggestion for organizations without governmental mandatory authority are therefore to send business surveys to named individuals and to use telephone follow-up methods to encourage response or obtain the needed data.

This analysis leaves some important questions unanswered. One concerns the response-rate implications of exercising mandatory authority. Some statisticians oppose the use of mandatory authority as a means of encouraging response to business surveys as coercive. Use of such authority may reduce the quality of the reported data (Worden and Hamilton 1989).

Another question is whether the use of telephone follow ups introduces the possibility of mode effects. An expanding literature suggests that questions may be answered differently over the telephone than on a self-administered form (Bishop et al. 1988, Ayidiya and McClendon 1990, Tarnai and Dillman 1992). Whether such effects occur for business surveys is unknown. Because the telephone is an important means of improving response for mail surveys, this issue is important.

The combination of telephone interviews in conjunction with mandatory response may invite rough estimations or guessing which could skew survey results. The effects of mandatory reporting and mode effects should be examined simultaneously so that their combined influence on response quality can be determined. These studies should determine whether biases in survey results are greater or less than those resulting from imputation of missing data.

Another unanswered question is how the use of impersonal mailings, bulk rate mail, form-like construction, and other procedures affect respondent efforts to answer accurately. If requests for information appear important and are easy to understand, the quality of answers might improve.

Another question is whether first class mail, personalized mailings, and respondent-friendly questionnaire construction are cost-effective measures. Some Census Bureau business surveys have an extremely long response period and vary in follow-up methodology. After a few months, surveyors switch from mail procedures to telephone or, in some cases, face-to-face follow-up. The SESRC findings that response can be improved by a contact name could have potential implications for Census Bureau surveys. It may be possible to call businesses prior to conducting some mail surveys, obtaining the name of the individual to whom the questionnaire should be sent. Such a telephone call could serve the function of a mailed ''prenotice'' and also facilitate a quicker follow-up. The cost of finding a person to whom business surveys can be sent may be more than offset by gains in mailback response rates that lower the substantial costs of collecting information by telephone from nonrespondents. Such research should be a high priority.

REFERENCES

Ayidiya, S. A., and M. J. McClendon (1990), "Response Effects in Mail Surveys," *Public Opinion Quarterly*, **54,** pp. 229–247.

Bishop, G. G., H. J. Hippler, N. Schwarz, and Strack (1988), "A Comparison of Response Effects in Self-Administered and Telephone Surveys," in R. M. Groves, P. B. Biemer, L. E. Lyberg, J. T. Massey, W. L. Nicholls, and J. Waksberg (eds.), *Telephone Survey Methodology*, New York: Wiley, pp. 321–340.

Childers, T. L., and O. C. Ferrell (1979), "Response Rates and Perceived Question-naire Length in Mail Surveys," *Journal of Marketing Research*, **16,** pp. 429–431.

Dillman, D. A. (1978), *Mail and Telephone Surveys: The Total Design Method*, New York: Wiley.

Dillman, D. A. (1991), "The Design and Administration of Mail Surveys," *Annual Review of Sociology*, **17,** pp. 225–249.

Dillman, D. A., J. Clark, and M. Sinclair (1993a), "How Prenotice Letters, Stamped Return Envelopes and Reminder Postcards Affect Mailback Response Rates for Census Questionnaires," *Proceedings of the Annual Research Conference*, Washington, DC: U.S. Bureau of the Census, pp. 37–48.

Dillman, D. A., M. Sinclair, and J. Clark (1993b), "Effects of Questionnaire Length, Respondent-Friendly Design, and a Difficult Question on Response Rates for Oc-cupant-Addressed Census Mail Surveys," *Public Opinion Quarterly*, **57,** pp. 289–304.

James, J. M., and R. Bolstein (1990), "The Effects of Monetary Incentives and Fol-low-up Mailings on Response Rate and Response Quality in Mail Surveys," *Public Opinion Quarterly*, **54,** pp. 356–361.

James, J. M., and R. Bolstein (1992), "Large Monetary Incentives and Their Effect on Mail Survey Response Rates," *Public Opinion Quarterly*, **56,** pp. 442–453.

Jobber, D., and S. Sanderson (1981), "The Effects of a Prior Letter and Colored Questionnaire Paper on Mail Survey Response Rates," *Journal of the Market Research Society*, **25,** pp. 339–349.

Johnson, T., and S. McLaughlin (1990), *GMAT Registrant Survey Design Report*, Los Angeles Graduate Admission Council.

Paxson, M. C. (1992), *Unpublished Data: Response Rates for 183 Studies*, Pullman, WA: Department of Hotel and Restaurant Administration, Washington State Uni-versity.

Tarnai, J., and D. A. Dillman (1992), "Questionnaire Context as a Source of Response Differences in Mail and Telephone Surveys," in N. Schwarz and S. Sudman (eds.), *Context Effects in Social and Psychological Research*, New York: Springer-Ver-lag, pp. 115–129.

Tulp, D. R., Jr., C. E. Hoy, G. L. Kusch, and S. J. Cole (1991), "Nonresponse

Under Mandatory vs. Voluntary Reporting in the 1989 Survey of Pollution Abatement Costs and Expenditures (PACE)," *Proceedings of the Survey Research Methods Section, American Statistics Association*, pp. 272–277.

Wallace, M. E. (1992), "Response Improvement Intitiatives for Voluntary Surveys by Business Division within the Census Bureau," prepared for October 22–23 Joint Census Advisory Committee Meeting, Washington, DC.

Worden, G., and H. Hamilton (1989), "The Use of Mandatory Reporting Authority to Improve the Quality of Statistics," prepared for April 13–14 Joint Census Advisory Committee Meeting, Washington, DC.

Worden, G., and E Hoy (1992), "Summary of Nonresponse Studies Conducted by Industry Division, 1989–91," unpublished paper, Washington, DC: U.S. Bureau of the Census.

CHAPTER EIGHTEEN

Automated Telephone Methods for Business Surveys

George S. Werking and Richard L. Clayton[1]

U.S. Bureau of Labor Statistics

The 1980s brought many dramatic technological changes to the workplace, most notably widespread use of microcomputers, telecommunications, and electronic information exchange. These new technologies offer statistical agencies new opportunities for improving the timeliness and quality of the data collected for business surveys. Though inexpensive, data collection by mail severely limits the timeliness of published estimates and thus lessens their usefulness for economic analysis. In contrast, automated telephone collection approaches afford almost instantaneous access to business data. Some automation options even offer this improved timeliness and control at lower cost than traditional mail collection. This chapter describes the U.S. Bureau of Labor Statistics' (BLS) 7-year research program into automated telephone collection techniques for the Current Employment Statistics (CES) survey. The chapter also summarizes significant results from a 5-year, large-scale implementation program for the chosen automated collection methods.

18.1 AUTOMATED DATA COLLECTION: AN ISSUE OF TIMELINESS

Large-scale national business surveys traditionally have relied on mail as the primary mode of data collection. Because the requested data often require em-

---

[1]The authors gratefully acknowledge the contributions of the State Employment Security Agencies which conducted many of the early research studies, and they also acknowledge the U.S. Bureau of Labor Statistics staff at the regional offices, the Data Collection Centers, and the Office of Employment and Unemployment Statistics. The views expressed are those of the authors and do not necessarily reflect those of the U.S. Bureau of Labor Statistics.

Business Survey Methods, Edited by Cox, Binder, Chinnappa, Christianson, Colledge, Kott.
ISBN 0-471-59852-6 © 1995 John Wiley & Sons, Inc.

ployers to refer to their firm's records, mail collection has been viewed as a logical and cost-effective approach. For example, a typical collection procedure in the 1970s consisted of an initial mailing followed by additional mailings to nonrespondents at 3- to 4-week intervals and usually concluded with a telephone contact to delinquent respondents. (See Chapter 14 for a survey of international approaches and Federal Committee on Statistical Methodology 1988 for U.S. practices.) As noted in Chapter 17, with such special efforts, acceptable response rates can be obtained for mail surveys, particularly for mandatory surveys. However, in terms of data timeliness, these mail collection procedures are badly flawed, often requiring many months before data collection can be finalized and estimates produced.

To address this timeliness limitation, survey agencies often publish *preliminary estimates* that use only those responses available by an initial closing date and then follow the preliminary estimates with *final estimates* several months later when all data have been received. Creating both preliminary and final estimates introduces an additional problem; sometimes quite substantial revisions to the published preliminary data are needed. Thus, mail business surveys face a somewhat intractable tradeoff between timeliness and accuracy. Fortunately, new technologies offer a wide range of options to address this issue.

18.1.1 New Computer-Based Collection Methods

Survey agencies' approaches to interviewer-collected data and mail-collected data have changed as the result of the microcomputer-based collection methods that emerged in the 1980s. In this section, we describe each method; more extensive descriptions of these methods are given by the Federal Committee on Statistical Methodology (1990) and Groves et al. (1988).

Computer-assisted telephone interviewing (CATI) offers a structured approach to traditional telephone data collection. Under CATI the survey questionnaire and full editing capabilities reside on the computer. The CATI interviewer accesses the next scheduled sample case and autodials the respondent. A computer scheduling algorithm usually controls scheduling of cases along with callbacks. The interviewer takes the respondent through a computer-controlled sequence of questions, resolving edit failures on a flow basis, and then schedules any subsequent callbacks that may be required.

Computer-assisted personal interviewing (CAPI) provides a structured approach to face-to-face data collection. In this case, the questionnaire and editing capabilities reside on a portable laptop computer that interviewers carry with them. All other interview functions are similar to CATI. Normally, the data are electronically transmitted on a daily basis from the field site to the survey agency's computer.

Touchtone data entry (TDE) offers an alternative to mail collection. Under TDE, the respondent uses a touchtone telephone to call a toll-free number connected to the survey agency's touchtone computer system and to activate an interview session. Again, the questionnaire resides on the computer in the form of prerecorded questions that are read to the respondent. The respondent enters numeric responses by pressing the touchtone phone buttons. Each answer is then read back for respondent verification. Limited editing is also possible under this method; however, survey agencies are currently relying on interviewer follow-up calls for edit reconciliation. TDE benefits include the elimination of mail handling activities, mail delays, and key entry activities.

Voice recognition (VR) is a variation of the TDE collection process where the respondent no longer needs a touchtone phone. The VR system is speaker-independent and accepts continuous speech; it recognizes digits zero through nine and "yes" and "no". The respondent's verbal answer is translated by the VR system; the answer is then repeated by the VR system for respondent verification.

Facsimile transmission (FAX) also offers an alternative to mail collection. Questionnaires and respondent data are electronically transmitted, bypassing mail delays. In addition, intelligent character recognition (ICR) systems are becoming available to translate the incoming FAX responses and thus eliminate the need for paper output and subsequent key entry.

Electronic data interchange (EDI) offers the ability to collect large volumes of data from businesses. As described in Chapter 19, respondents extract the needed data in a prespecified format from their computer databases and initiate electronic transmissions to the survey agency. Basic EDI systems simply accept data files; respondent recontacts for edit reconciliation are done later by interviewers. More sophisticated EDI systems offer direct on-line editing by the respondent.

In the future *E-mail transmission* will offer a full electronic replacement for mail, TDE, VR, and FAX. Under systems such as Internet, questionnaires for periodic surveys and subsequent nonresponse prompting and reconciliation can be sent directly to the respondent's E-mail address. The respondent enters the required information and forwards the response to the statistical agency's E-mail address.

These new collection methods have several advantages in common. They create a paperless collection environment where all questionnaires and reported data reside in electronic form only, thus eliminating labor-intensive mail-handling activities. All respondent-reported data, whether collected by interviewers or entered directly by respondents, are transmitted electronically across telephone lines, thus eliminating the time delays of mail transmission and key entry. Scheduling of visits/calls and questionnaire disposition is also computer-based, allowing instantaneous status reports on data collection progress and upcoming workload. Computer-controlled branching in the questionnaire and extensive on-line editing with the respondent provide higher-quality microdata and reduce the volume and associated cost of subsequent edit reconciliation follow-ups.

18.1.2 Automated Data Collection for Business Versus Household Surveys

It is useful to contrast the different direction that automated collection is taking for the business survey environment versus the household survey environment.

Traditionally, there have been significant distinctions between business and household survey environments (see Chapter 1). These distinctive features include population skewedness, sampling plans, respondent profile, data collection approaches, and nonsampling error sources. The distinctions for data collection will grow markedly in the future.

In the past, large-scale national household surveys relied upon face-to-face visits to ensure high response rates and to control response errors due to comprehension and recall problems and the effects of proxy respondents. Automated data collection for household surveys has led to CAPI collection with a far stronger reliance on cognitive approaches for questionnaire development to control respondent and interviewer error. However, automation leaves the underlying cost structure for household surveys little changed, assuming that hardware can be amortized effectively over time.

In contrast, business surveys are moving in quite different directions. Business survey population distributions are highly skewed so that a small portion of the population units can account for a vast amount of the item of interest. For instance, 2 percent of U.S. businesses account for over 50 percent of the total employment in the United States. This high concentration allows for cost-effective targeting, allowing a mixed-mode collection approach incorporating a range of new automated collection methods. With the advent of personal computers (PCs), FAX, and electronic data transmission, virtually all employers can soon be expected to have strong technology capabilities for reporting economic data. The new electronic transmission capability of employers, including TDE, FAX, and (more recently) E-mail and EDI, provide a wide range of cost-effective automated collection approaches for business surveys.

New opportunities for controlling data quality in business surveys are also emerging. In growing numbers, employers are using sophisticated off-the-shelf software for tasks such as producing payroll data or are contracting for centralized services. Because response error at the business level rests more on the software systems that produce the data than on specific individuals, a one-time validation of major software packages can have a significant, ongoing impact on the quality of economic data. (For an in-depth discussion of measurement error, see Chapter 15.) Under the new automated data collection methods of TDE, VR, FAX, and EDI, the cost of obtaining economic data may well be lowered, while at the same time significantly increasing the timeliness and the quality of the reported data.

18.2 CES RESEARCH PROGRAM

With 380,000 establishments interviewed monthly, the Current Employment Statistics survey is the largest sample survey in the United States. CES data are used as input for many of the nation's economic indicators. On the first Friday of each month, BLS releases data on the employment situation for the previous month. On release day, the Commissioner of Labor Statistics appears before the Joint Economic Committee of Congress and provides a detailed analysis of the current month's data and trends; data are made available to the news media and the financial and business communities at the same time. This closely watched set of statistics covers employment, hours, and earnings by industries. These statistics are the earliest indicator available on the previous month's economic activity and are used to gauge the health of the U.S. economy.

Although the timeliness and accuracy of its employment statistics are essential in analyzing current economic conditions, the CES survey has relied on mail data collection since its inception in the early 1900s. This collection process results in the publication of preliminary estimates based upon the 50 percent of sample returns available from the first 2 weeks of data collection, followed by final estimates 2 months later when all sample returns are in house. Producing both preliminary and final estimates for a given month periodically results in substantial revisions to the initial estimates. These revisions not only affect the basic CES statistics but also affect other statistics that use CES estimates as input.

18.2.1 Research Goals

In the early 1980s, BLS initiated a 7-year research effort into the causes of late response and into alternative collection methods that could increase response for the preliminary estimates and thereby reduce the size and frequency of large revisions to preliminary estimates. The focus of the survey research centered around three basic questions:

- Does the establishment have data available in time to respond to the publication deadline for the preliminary estimates?
- Are there data collection methods that ensure an 80 to 90 percent response rate under tight time constraints?
- Can the cost of these data collection methods be controlled at about the same level as the current mail collection costs?

At the conclusion of the research program, a mixed-mode, CATI/TDE collection approach emerged that satisfied the response rate and cost constraints for CES. The following sections describe these PC-based data collection methods, the research tests, response rate results, and the cost analysis.

18.2.2 Data Collection Methods

The CES survey is a federal and state cooperative program where the states are the collection agents. Additionally, the data collection time period is limited by the publication deadline for the preliminary estimates. The CES survey's reference date is the payroll period containing the 12th day of the month; only 2.5 weeks are available to collect, keypunch, edit, tabulate, validate, and publish the data. To meet this tight time constraint, a collection method must obtain the required data as soon as they become available to the business. Four data collection methods were studied.

CES Mail Procedures

The basic CES questionnaire has changed little over the past 50 years. It is a single-page, mail-shuttle form that provides space for the employer to record 12 months of data. Five basic data items are collected each month: all employees, women workers, production (or nonsupervisory) workers, hours, and earnings. The earlier months of employer responses serve as a quality check for the employer, provide information in cases where the firm's respondent changes, and assist with BLS' edit reconciliation callbacks. The employer receives the questionnaire in the mail on or about the 12th of each month (i.e., the survey's reference date) and fills in the row of data items corresponding to the current month. Once completed, the employer mails the form back to the state agency where it is keypunched and edited. The form is then filed so that it can be returned to the employer for entering the next month's report. Currently, this process yields a 50 percent response rate in the 2.5 weeks available for the preliminary estimates.

CATI Procedures

Under CATI collection, the employer is mailed the CES questionnaire at the beginning of the year and retains it for recording monthly data throughout the year. Each month, as their payroll data become available, employers receive an advance notice postcard prompting them to fill in the data items for that month and reminding them of the time of their prearranged CATI call from the state agency. The interviewer then calls the respondent, collects and edits the data under CATI, and arranges a time for the next month's call. The direct contact between respondent and interviewer provides an opportunity to address respondent issues, including questions on data definitions, uses of the reported data, and the monthly reporting time frame.

TDE Procedures

Under TDE reporting, the employer does the same activities as under CATI, with one exception. Instead of waiting for the state agency to call, the employer calls a toll-free number connected to a touchtone PC located at the state agency. The employer then enters the data items following the prompts in the automated CES interview. Each data item entered by the employer is read back to the respondent for verification. Similar procedures are used with the voice recognition collection system.

18.2.3 Research Test: 1984–1990

BLS began developing a PC-based CATI system in 1983 for use in a two-state test that began in 1984. The University of California at Berkeley's CATI system was selected for the test and was subsequently used throughout the research effort. A small random sample of 200 units was initially selected in each state, and collection procedures and systems were refined over the next 7 years. The initial research tests were highly successful in the response rates they achieved, and the tests were expanded to nine states in 1986 and then to 14 states in 1988.

The composition of the test sample changed in 1986. The initial random samples of existing CES reporters contained a mixture of timely respondents along with late respondents. Because timely respondents needed no improvement in response rates, subsequent research tests focused only on random samples of habitually late CES respondents (e.g., those units that had a past history where the average response rate was under 20 percent for the preliminary estimates publication deadline). Thus, the success of CATI was measured in terms of its ability to move samples of habitually late CES reporting units to a stable, ongoing, 80 to 90 percent response rate. By the end of the CATI research phase in 1990, BLS was collecting over 5000 interviews monthly with CATI and had conducted well over a quarter of a million CATI interviews.

While CATI was proving to be highly successful in improving response rates, by 1985 it became clear that ongoing large-scale CATI collection would be far more expensive than mail collection. At this time, a separate path of telephone collection research was started which focused on reducing the cost of CATI while still maintaining the high monthly response rates it achieved.

By 1985, many U.S. banks were operating a version of touchtone entry verification for check cashing at drive-in windows. BLS identified a PC-based touchtone reporting system suitable for survey research testing. By 1986, BLS was conducting a small two-state test of touchtone data entry for collecting data. TDE was not viewed as a direct replacement for mail or as a competitive method to be tested against CATI. CATI's role was to take habitually late mail respondents and turn them into timely respondents through personal contact and an educational process, while TDE's role was to take these timely CATI respondents and maintain their response rates at the same high level, but at a greatly reduced unit cost (Ponikowski and Meily 1988). In the subsequent 5 years of data collection research, TDE proved to be a very successful and reliable method of telephone data collection. The research phase for TDE was concluded in 1990, with response rates averaging 82 percent under TDE across 14 states; in total, BLS collected over 100,000 schedules during the TDE research phase using this new automated reporting method.

In 1988, as a natural follow-up to TDE, BLS began conducting small research tests of a new voice recognition reporting system. As described in Clayton and Winter (1992), preliminary results for VR reporting replicated the same high monthly response rates achieved under TDE but, with the important advantage that respondents found VR reporting more natural and generally preferred it over TDE. At this time (mid-1994), the cost of the VR hardware is approximately four times that of TDE; however, as the initial costs of VR hardware drops, this collection method should become a viable replacement for TDE. Over-the-telephone VR technology is also rapidly improving, allowing verbal entry of multiple digits ("four thousand, three hundred, and six") and small vocabularies which will open up many other survey applications.

18.2.4 Research Results

Over a 7-year period of research tests, BLS has established that payroll data are available in most firms prior to the publication deadline for the preliminary estimates. BLS has also determined that CATI collection can take traditionally late mail respondents (e.g., 0 to 20 percent response rate for preliminary estimates) and, within 6 months, turn them into timely respondents with response rates of 82 to 84 percent (Table 18.1). Over the research period, the response rates were remarkably stable as the CATI sample was expanded from 400 units to 5000 units and the number of participating states was increased from 2 to 14 states. The research results also indicate that the CATI response rate can be maintained in the targeted range of 80 to 90 percent over long periods of time (Werking et al. 1986).

The principal factor limiting a respondent's ability to make the publication deadline was found to be the length of the firm's pay period (Figure 18.1). Employer pay periods are generally weekly, biweekly, semimonthly, or monthly. Weekly and semimonthly payrolls can almost always be collected in time for publication with biweekly pay periods available most of the time; however, many monthly payroll systems close out well after the preliminary estimate's publication deadline. Because survey publication timing is critical and cannot be altered, monthly payrolls will continue to be a large factor in limiting the final CATI response rate.

Several other important results have come out of the CATI research. Under CATI, approximately 60 percent of the respondents have their data available on the prearranged date for the first call, with the remaining 40 percent using the first call for prompting (Table 18.1). This rate has varied little across states or over the years of testing. A small test was conducted to see whether an advance postcard mailed shortly before the prearranged CATI contact would reduce the number of callbacks required. This test showed the unexpected benefit of a 4 percentage point increase in the CATI response rate along with a 3 percentage point reduction in the number of required callbacks and a corresponding reduction in costs.

The average time for a CATI interview depends on the number of items to be collected, the time efficiency of the interview instrument, and the experience of the data collector. The average time for a CATI call (Table 18.1) was reduced by one-third as the CATI instrument was streamlined and as interviewers gained experience. Another important concern in the testing was the effect of CATI on sample attrition. There was some concern that employers would not want to be bothered repeatedly by telephone contacts and would drop out of the program. However, the sample attrition rate for CATI was

Table 18.1 Research Summary

Collection Method	Item	1984	1985	1986	1987	1988	1989	1990
CATI	Units	400	400	2000	3000	5000	5000	5000
	Response Rate (%)	83	84	82	84	83	84	82
	Percent Call Back (%)	44	42	40	41	42	41	41
	Average Minutes	5.6	5.6	5.0	4.8	4.4	3.5	3.8
TDE	Units				400	600	2000	5000
	Response Rates (%)				78	80	84	82
	Percent Call Back (%)				45	45	43	40
	Average Minutes				1.8	1.8	1.7	1.7

Source: Werking and Clayton 1991, p. 7.

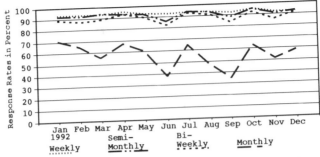

Figure 18.1 CES CATI first closing performance by length of pay period, January 1992–December 1992.

about one-third of that for mail, with almost no loss of large businesses. In summary, CATI provided the high 80 to 90 percent response rate required for the preliminary estimates while enjoying broad support from respondents. As described in Clayton and Winter (1990), these results represented a major breakthrough in performance level over traditional mail methods. (See Juran et al. 1979, for a discussion of quality performance measures.)

As noted earlier, the increased cost associated with CATI collection led BLS to initiate research into touchtone collection. As shown in Table 18.1, 4 years of research testing demonstrated TDE's ability to take timely CATI reporters having 82 to 84 percent response rates and to maintain these high rates under completely automated TDE reporting (Werking et al. 1988a). The cost savings under TDE collection versus CATI collection are significant. A major concern for TDE collection was that TDE respondents might tend to call during the same time period. This would generate busy signals and require an excessive number of touchtone PCs to handle peak-load reporting. This was not the case, however. While the touchtone PC's are on-line 24 hours a day, most calls are more or less uniformly distributed between 8 A.M. and 5 P.M. (Figure 18.2). TDE respondents tend to require the same proportion of prompt calls as CATI respondents—approximately 40 percent. Methods are being tested to reduce the TDE prompting workload. An advantage for the respondent is that TDE collection requires only one-half the time of a CATI interview. The average TDE interview lasts only 1 minute and 45 seconds. Additionally, touchtone phones are widely available at most establishments; current estimates indicate that over 93 percent of employers could report under touchtone data collection and that the remainder could report by VR. While TDE reporting offers many advantages to the survey agency, its strongest feature is respondent acceptance; respondents' reaction to touchtone reporting has been very positive because of its speed and convenience.

18.2.5 Cost Analysis

With performance testing and respondent acceptance for CATI and TDE proving to be highly successful, the final phase of research shifted to analyzing the

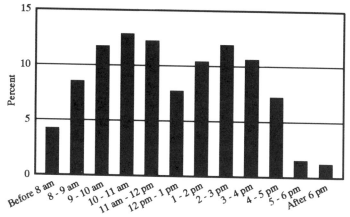

Figure 18.2 Touchtone data entry for distribution of TDE calls by time of day, March–May 1993.

transitional costs of CATI and the ongoing costs of TDE collection. As described in Clayton and Harrell (1989), the major labor and nonlabor cost categories were studied for mail, CATI, and TDE collection (Table 18.2). The study looked not only at estimates of current cost but also at projected costs over the next 10 years, using the current rate of increase for the major cost items. The basic cost factors for CATI are identical to TDE except CATI requires a large interviewer labor cost that is eliminated under TDE. Because CATI plays only a 6-month transitional role (e.g., moving late responding mail

Table 18.2 Data Collection Costs[a]

Cost Category	Mail	CATI	TDE and VR
Labor			
Mailout	↑		
Mail Return	↑		
Data Entry	↑	↑	
Edit Reconciliation	↑	↑	↑
Nonresponse Follow-Up	↑		↑
Nonlabor			
Postage	↑	↑	↑
Telephones		↓	↓
Microcomputers		↓	↓

	Recent Annual Price Change Factors
Labor	+3.6% Employment Cost Index, state and local government
Postage	+5.0% U.S. Postal Service
Telephone	−1.8% Consumer Price Index-U, intrastate toll calls
Microcomputers	−20.9% Producer Price Index: PCs and workstations

[a]Arrows show direction of recent price changes

units to on-time responding CATI units) prior to conversion to ongoing TDE, the cost analysis focused on the ongoing cost tradeoffs between mail versus TDE data collection.

For labor categories, the mail operations of monthly mailout, mailback, check-in, and forms control are replaced by a single annual mailout operation under TDE, thus eliminating a large monthly clerical operation. The batch keypunching, keypunch validation, and forms control operations under mail are completely eliminated under TDE, where the respondent touchtone enters the individual firm's data and validates each entry. Telephone nonresponse follow-up is more feasible under TDE than mail. Under TDE, an accurate up-to-the-minute list can be generated of respondents who have not called in data; this list can then be used for telephone prompting. Under mail, telephone follow-up of apparent nonrespondents is awkward because survey staff do not know the exact status of the respondent's form (i.e., has not been completed, currently in the mail, in the state check-in process, or at keypunch). In addition, respondents who recently returned their form tend to resent the additional reminder for an activity they perceive as complete. Because of the voluntary nature of the program and the uncertainty of a respondent's response status, telephone prompting under mail is only used for critical, large employers.

No significant cost savings occur for edit reconciliation because the number of edit failures under TDE remain at about the same level as under mail. This also is true for postcard reminders where the number of postcards used under mail collection for late respondents is approximately the same as the number used under TDE, where respondents receive an advance postcard notice to touchtone their data by the due date.

In the nonlabor categories, postage costs under mail (currently 58 cents per unit) are replaced by telephone charges and the amortized cost of the TDE machine (together currently 46 cents per unit). Postage is a continually increasing cost with an annual price increase of approximately 5 percent. Postage costs are driven by annual labor cost increases ($+3.6$ percent) and by fuel costs (also generally increasing), with labor accounting for over 80 percent of total postage costs. In contrast, under TDE, the cost of telephone calls has been decreasing in recent years (-1.8 percent), along with the cost of microcomputers (-20.5 percent).

Demonstrable cost savings result in shifting from mail to TDE. Perhaps more importantly, under a 10-year projection of future costs of these two collection methods (Figure 18.3), these savings grow substantially. The CES survey will use future cost savings from TDE to help offset full nonresponse prompting activities.

Several major conclusions concerning TDE reporting have emerged from the performance and cost analysis review. First, the traditional view that mail is the least expensive collection data method is no longer true. The major technological breakthroughs of the 1980s in automated telephone collection not only reduce collection costs below that of mail but also improve timeliness and control over the collection process. Additionally, over the next 5 to 10

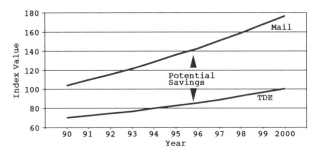

Figure 18.3 Estimated unit costs by mode: 1990–2000.

years, the cost of mail will become even less cost competitive with these high-technology/low-labor collection approaches due to increasing labor and postage costs. Second, the transition of respondents from mail to TDE appears to cause little disruption to monthly reporting. In BLS' follow-up interviews with respondents who converted to TDE, respondents had very little trouble adapting to this new method of reporting. Virtually all respondents completed their first-month TDE report accurately and without assistance, with many respondents commenting on TDE's ease of reporting. Third, TDE can be viewed as a reliable replacement method for conventional mail data collection. Over the past 4 years of collection, no major equipment failure problems or disruptions of the collection process have occurred. Minor equipment problems have been easily resolved using a back-up PC. In addition, current back-up protection for the state TDE collection process uses a call-forwarding option to reroute calls to a central site should major problems occur.

18.3 CATI/TDE IMPLEMENTATION

By 1990, BLS had completed a successful research program and had sustained high performance levels over a 7-year period. Research results suggested that substantial cost-effective quality improvements could be made in the area of automated data collection. However, no matter how extensive the research phase is, unexpected differences always occur between research results and the actual long-term results achieved under full-scale implementation in a production environment. This turned out to be particularly true, in a very positive way, for CATI collection.

In 1991, funding was obtained to begin a full-scale implementation of CATI/TDE for all employers with 50 or more employees. This conversion covered over 100,000 of the 380,000 monthly units and represented over 85 percent of the sample's unweighted employment. The conversion was to take place over a 5-year period concluding in 1995. One very important difference was introduced into the implementation plan; the CATI conversion process was shifted from a decentralized state environment to a centralized environment, using two newly established CATI collection centers. These centers were designed to provide a 6-month CATI transition in moving employers from mail to ongoing TDE collection.

18.3.1 Centralized CATI Collection Centers

There has always been controversy about how best to implement CATI: Should CATI be run out of large centralized facilities or run out of smaller localized (decentralized) field operations? In a sharp departure from BLS's decentralized state collection environment where the research was conducted, BLS created two central CATI centers for the implementation phase. This was done because of the tight time constraints for implementing the project and because it was not cost effective to implement CATI collection in all 50 states for a temporary 5-year period. The establishment of the centers created additional systems development workload but resulted in major performance improvements for CATI collection.

BLS brought two regional office CATI collection centers into full production in 1992 (Donohue et al. 1992). These centers are staffed with 20 interviewers each and, at full capacity, can handle approximately 30,000 CATI conversion cases per year. The shift to the central collection centers required a number of systems changes. The CATI system, originally designed to be a stand-alone PC-based system for the states, was rewritten to support a LAN environment for the centers. Data linkages were established between the centers and each state to handle a monthly transfer of up to 2400 mail units from the states to the centers for CATI conversion and to handle a corresponding return of 2400 units that had completed the 6-month CATI conversion process and were now being moved to ongoing TDE collection in the states. In addition to the transfer activities for the incoming and outgoing units, the centers are responsible for collecting data for approximately 15,000 CATI units each month. To provide BLS and the states with immediate access to the collected data, a central-site data access system was developed. Under this system, the collection centers transfer CATI-collected data to the central site where the data are then partitioned into separate individual state mailboxes for immediate access by the states and by BLS. While the creation of the regional collection centers has required a great deal of effort, the centers have proven to be a very successful and cost-effective alternative for the CATI conversion process.

CATI had always been viewed as the most critical component because its purpose was to establish the high response rates that TDE would attempt to maintain. While the average CATI response rate over the 7-year research period was approximately 83 percent, there was a wide 30 percent variation in state response rates, which ranged from a low of 65 percent up to a high of 95 percent. There had always been a certain amount of anecdotal evidence to support the contention by some localized collection sites that their respondents were more difficult to obtain responses from than respondents in other geographic areas and, thus, to support a belief that there was a natural variation in geographic response rates, which is inherent in any survey environment.

However, this geographic variation in response rates essentially vanished, once the collection was centralized. Over a 3-year period of operating experience in the CATI centers, the average response rate was raised to 92 percent (Table 18.3), approximately 10 percent higher than under the decentralized state environment. Of equal importance, the geographic variation among states dropped from 30 percentage points (e.g., 65 percent to 95 percent) under the decentralized state environment to approximately 3 percentage points (e.g., 89 percent to 92 percent) under a centralized CATI center environment. This result provides support for the view that existing differential geographic response rates are predominantly an artifact of current organizational structures and not an inherent constraint of the survey environment itself.

18.3.2 TDE Operations in the States

TDE provides the cornerstone for production data collection for over 100,000 of the most critical units in the CES sample. Over the past 7 years, over one million schedules have been collected under TDE, and, in many ways, TDE has proven very successful under the implementation phase, offering these cost-effective advantages over mail:

- It is widely and readily accepted by the respondents.
- It provides instantaneous data to the survey agency.
- The TDE hardware costs have dropped to one-fourth of the original cost during the past 5 years of research testing.
- TDE has maintained a strong systems reliability record throughout its research and production use.
- The overall collection cost per unit under TDE continues to drop compared to the escalating cost of mail.

At this time, TDE has not been as responsive to improvement efforts as CATI has under the implementation phase.

Ongoing TDE response rates have remained at or a little below the response rates achieved under the research phase (Table 18.3). This has remained true even though the CATI response rates that feed into TDE have increased. Thus, while TDE has not lost in terms of response rate, it has failed to gain from the extra improvements in the centralized CATI process. Several factors contribute to the lower TDE response rate: the loss of the personal contact under CATI, the periodic turnover in contact persons at the business, and the lack of systematic nonresponse prompting.

The most important aspect of TDE collection is the development of an effective, ongoing, nonresponse-prompting operation to ensure high response rates. Nonresponse prompting is a critical component and one of the most difficult for the states to control because it generates a large peak end-of-the-month workload and is one of the few staff-intensive activities under TDE. Implementation tests are currently being conducted in five states to centralize

Table 18.3 Implementation Summary

Collection Method	Item	1991	1992	1993	1994	1995
CATI	Units	4,000	9,000	20,000	30,000	30,000
	Response rate (%)	90	92	92	92	92
	Percent call back (%)	43	43	42	40	40
	Average minutes	4.4	4.0	4.1	4.1	4.1
TDE	Units	11,000	20,000	40,000	70,000	100,000
	Response rates (%)	79	80	80	80	80
	Percent call back (%)	39	38	35	35	35
	Average minutes	1.8	1.9	2.1	2.0	2.0

Source: Werking and Clayton 1991, p. 7.

the prompting activities to ensure that all prompt calls can be completed by the closing due date. As discussed by Clayton et al. (1993), test states experienced a 10 percent increase in response rates, bringing their TDE response rates to 80 percent or better.

To control the nonresponse-prompting workload and to minimize the volume of nonresponse prompting required, a scheduling algorithm was developed to balance the data flow workload by assigning a due date to each TDE reporter. The due date is calculated based on the respondent's payroll type (i.e., weekly, biweekly, semimonthly, monthly) and their reporting experience under CATI collection. To ensure that respondents are aware of the survey's reporting dates, the due date is printed beside each month on the respondent's questionnaire and also printed on the TDE advance postcard notice sent each month. If the respondent's TDE report has not been received, nonresponse prompting calls take place 2 days after the respondent's due date. Based on simulation studies using historical data (see Rosen et al. 1991), the TDE prompting workload was expected to be reduced from the current level of 40 percent to approximately 25 percent and to be spread uniformly across the last week of collection (Figure 18.4). However, even after the implementation of these improvements, the prompting rate was reduced only to about the 35 percent level.

A large-scale research study is currently being conducted to test the effectiveness of using a FAX nonresponse prompt notice as a replacement for the telephone contact. These notices are sent on the evening of the normally scheduled prompt day. The FAX notices are autodialed and transmitted using a specially programmed PC containing a FAX board; if the FAX prompt is unsuccessful, the respondent receives a follow-up prompt call from the interviewer. If successful, this approach could significantly reduce the staffing cost for telephone nonresponse prompting and more easily accommodate an ongoing, non-

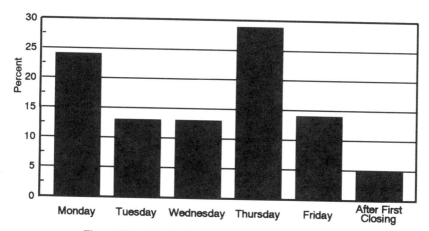

Figure 18.4 Distribution of nonresponse prompts by day.

response prompt rate of 35 percent. As described by Rosen and Clayton (1992), FAX prompts yield response rates in the same range as the prompt calls placed by interviewers.

18.4 EDI AND MULTIUNIT EMPLOYERS

Business frames generally contain both single-unit firms [i.e., the specific employer identification number (EIN) only appears on one record in the frame] and multiunit firms (e.g., more than one record on the frame with the same EIN as with chains of grocery stores). While the automated data collection methods of CATI and TDE are appropriate for single-unit employers or multiunit employers with less than 7 units, these methods are not appropriate for larger multiunit employers. Large multiunit employers represent approximately 10 percent of the population units and 25 percent of total employment and can present a special collection problem for economic surveys.

The CES survey is currently set up for worksite sample selection. Often several worksites are selected within multiunit companies. The employer requirement to file multiple CES reports each month is complicated by the ongoing sample rotation process that may increase, decrease, or change the specific worksites to be filed. This filing process became further complicated during the CATI/TDE research phase when some employers found themselves filing some reports by mail, some by CATI, and some by TDE.

During this same time period, an increasing number of large multiunit employers began requesting that BLS accept automated monthly files of all worksites versus the subset of sample sites. With the high degree of automation and data transfer capabilities between worksites and the central office, multiunit employers became increasingly interested in centralizing and simplifying their response burden. The need to automate the collection for multiunit employers along with the request from these employers for simplified reporting approaches led BLS towards an EDI approach for this important segment of the population.

BLS is developing a pilot study using an EDI collection center for conversion and ongoing collection and edit reconciliation of multiunit employer data. Procedures are being developed for the direct conversion of some large multiunit employers from decentralized unit-by-unit mail collection to centralized electronic reporting of all worksite data to the EDI collection center. An initial personal visit with the multiunit employers will establish contacts, reporting procedures, timing, and, most importantly, data accuracy. To minimize the center's future edit reconciliation workload, each firm's payroll processing system will go through a systems certification process for each CES data element to ensure that the automated files sent to BLS each month conform to CES definitions. BLS has developed an EDI format that will be requested from the employer; however, we also will accept data in any format and use preprocessing software for conversion to the standard format. If successful, the collection center will be expanded in 1995.

18.5 USING AUTOMATED COLLECTION METHODS FOR IMPROVING DATA QUALITY

Automated telephone methods have proved useful not only in the area of data collection but also for data quality. CATI and CAPI provide the ability to have a complex, structured, probing, reinterview questionnaire for identifying major sources of response error. During the 1980s, BLS conducted a number of reinterview surveys termed *response analysis surveys* (RAS) using the automated collection methods of CATI and CAPI. While the CES survey has extensive monthly edits, including range checks, internal consistency checks, and longitudinal edit checks, these standard edits generally will not identify systematic and consistent reporting errors. Detailed reinterview studies are required to provide a broader profile of survey response error. (For a more detailed discussion of response error control, see Chapter 16).

In 1984, over 400 CAPI, face-to-face RAS reinterviews were conducted; these were followed by over 3500 CATI RAS reinterviews during the next 3 years (see Werking et al. 1988b). Additionally, during the CATI/TDE conversion process, each of the 100,000 units received a CATI RAS reinterview for employment data. The information provided by the RAS reinterviews has been used to correct the individual employer's reporting and, more importantly, to improve questionnaire wording, definitions, and instructions.

RAS activities are being extended into an important additional area. The 1980s introduced the widespread use of PCs and off-the-shelf standardized software. In a separate study, BLS identified a number of firms that are now offering PC-based payroll software systems or payroll services or both (Werking et al. 1993). These software systems and services currently cover over 25 percent of the nation's employment. This level of coverage offers a unique opportunity to control a significant portion of potential response error at the source. BLS is conducting RAS visits with up to 200 major firms in this area and will work with many of these firms to complete a systems validation for the CES data items (Werking et al. 1993). At the conclusion of this process, these payroll software systems will produce accurate, well-labeled data for employer use in reporting BLS payroll data and also for future EDI collection.

18.6 CONCLUDING REMARKS

The new collection methods described in this chapter—CATI, TDE, VR, FAX, and EDI—represent high-performance and generally low-cost collection methods. High response rates and low attrition are the result of focusing not only on the technology but also on the survey respondent as the customer in the collection process. These new methods would not have been successful if the Bureau had not constantly strived to involve the respondent more fully in the development of procedures, systems, and survey materials. Throughout the research and implementation, respondent reaction was repeatedly sought for each method and then used to improve and refocus the procedures and systems.

The growing number of collection methods offers the respondent a choice in how to provide data appropriate to their office environment and level of technology.

The 1990s will provide even greater opportunities for using electronic technology such as E-mail and EDI to improve data collection timeliness and quality at lower costs. As employers adopt electronic communication during the 1990s, the new electronic approaches of E-mail/Internet will quickly replace the 1980s state-of-the-art telephone-based collection methods of CATI, TDE, and FAX.

Technology has rapidly changed the way business is conducted in the workplace; however, the data collection approach of our surveys often date back to their inception. Our cost assumptions and cost studies are usually outdated and often simplistic in approach. Because data collection generally represents the largest part of a survey's cost, it is usually well-entrenched in the agency's organizational structure and can be quite difficult to restructure to accommodate large-scale change. Within this survey environment, we will face the major challenges and opportunities of the 1990s. These challenges will determine for the future the cost and quality competitiveness of our economic programs and our survey agencies.

REFERENCES

Clayton, R. L., and L. J. Harrell, Jr. (1989), "Developing a Cost Model of Alternative Data Collection Methods: MAIL, CATI, and TDE," *Proceedings of the Survey Research Methods Section, American Statistical Association*, pp. 264–269.

Clayton, R. L., and D. L. S. Winter (1990), "Applying Juran Concepts to the Improvement of Data Collection in the Current Employment Statistics Survey," *Conference Proceedings Impro'90, Juran Institute Conference*, pp. 981–1011.

Clayton, R. L., and D. L. S. Winter (1992), "Speech Data Entry: Results of the First Test of Voice Recognition for Survey Data Collection," *Journal of Official Statistics*, **8**, pp. 377–388.

Clayton, R. L., R. J. Rosen, and T. B. Rubino, Jr. (1993), "Nonresponse Prompting Behavior in an Establishment Survey," *Proceedings of the Survey Research Methods Section, American Statistical Association*, pp. 850–854.

Donohue, K. R., R. L. Clayton, and G. S. Werking (1992), "Integrating CATI Centers in a Decentralized Data Collection Environment," *Proceedings of the Survey Research Methods Section, American Statistical Association*, pp. 143–147.

Federal Committee on Statistical Methodology (1988), *Quality in Establishment Surveys*, Statistical Policy Working Paper 15, Washington, DC: Office of Management and Budget.

Federal Committee on Statistical Methodology (1990), *Computer Assisted Survey Information Collection*, Statistical Policy Working Paper 19, Washington, DC: Office of Management and Budget.

Groves, R., P. P. Biemer, L. E. Lyberg, J. T. Massey, W. L. Nicholls II, and J. Waksberg (1988), *Telephone Survey Methodology*, New York: Wiley.

Juran, J. M., F. N. Gryna, Jr., and R. S. Bingham, Jr. (1979), *Quality Control Handbook*, 3rd edition, New York: McGraw Hill, pp. 2–14.

Lyberg, L., and D. Kasprzyk (1991), "Data Collection Methods and Measurement Error: An Overview," in P. P. Biemer, R. M. Groves, L. E. Lyberg, N. A. Mathiowetz, and S. Sudman (eds.), *Measurement Errors in Surveys*, New York: Wiley, pp. 237–257.

Phipps, P. A., and A. R. Tupek (1991), "Assessing Measurement Errors in a Touchtone Recognition Survey," *Survey Methodology*, **17**, pp. 15–26.

Ponikowski, C. H., and S. A. Meily (1988), "Use of Touchtone Recognition Technology in Establishment Survey Data Collection," paper presented at the First Annual Field Technologies Conference, St. Petersburg, Florida.

Rosen, R. J., R. L. Clayton, and T. B. Rubino (1991), "Controlling Nonresponse in an Establishment Survey," *Proceedings of the Survey Research Methods Section, American Statistical Association*, pp. 587–591.

Rosen, R. J., and R. L. Clayton (1992), "An Operational Test of FAX for Data Collection," *Proceedings of the Survey Research Methods Section*, American Statistical Association, pp. 602–607.

Werking, G. S., A. R. Tupek, C. Ponikowski, and R. J. Rosen (1986), "A CATI Feasibility Study for a Monthly Establishment Survey," *Proceedings of the Survey Research Methods Section, American Statistical Association*, pp. 639–643.

Werking, G. S., R. L. Clayton, R. J. Rosen and D. L. S. Winter (1988a), "Conversion from Mail to CATI in the Current Employment Statistics Program," *Proceedings of the Survey Research Methods Section, American Statistical Association*, pp. 431–436.

Werking, G. S., A. R. Tupek, and R. L. Clayton (1988b), "CATI and Touchtone Self-Response Applications for Establishment Surveys," *Journal of Official Statistics*, **4**, pp. 349–362.

Werking, G. S., and R. L. Clayton (1991), "Enhancing Data Quality Through the Use of Mixed Mode Collection," *Survey Methodology*, **17**, pp. 3–14.

Werking, G. S., P. M. Getz, and R. L. Clayton, (1993) "New Approaches to Controlling Response Error in Establishment Universe and Sample Data," *Proceedings of the Survey Research Methods Section, American Statistical Association*, pp. 753–757.

CHAPTER NINETEEN

Electronic Data Interchange

Carole A. Ambler, Stanley M. Hyman, and Thomas L. Mesenbourg
U. S. Bureau of the Census

Electronic exchange of information is becoming the norm for the business world. Electronic mail and information retrieval services are rapidly replacing reports, letters, and other documents. Many businesses are moving towards a paperless environment. *Electronic Data Interchange* (EDI) offers businesses an electronic means to exchange standard information, and it is rapidly replacing order forms, shipping notes, and other documents. In this chapter, we describe initiatives implemented by the U. S. Bureau of the Census to develop an EDI reporting capability for its economic censuses and surveys.

19.1 WHAT IS EDI?

EDI is the electronic transfer of business transaction information in a standard format between business partners. EDI grew out of businesses' need to compete in an increasingly complex and changing international economy. When computers talk to one another without human intervention, errors are reduced; physical handling of paper is virtually eliminated; information is instantly available; decisions can be made easily and quickly; and for businesses, delivery of goods and services can be made within days. Everyday business transactions such as buying, shipping, selling, billing, and payments are completed in minutes rather than weeks or months. Businesses realize considerable savings using EDI for transactions.

EDI has become the norm for buying, selling, shipping, and inventory control for numerous large businesses. Smaller companies that wish to continue

Business Survey Methods, Edited by Cox, Binder, Chinnappa, Christianson, Colledge, Kott.
ISBN 0-471-59852-6 © 1995 John Wiley & Sons, Inc.

doing business with these large companies also are being required to establish EDI programs.

Industry experts project rapid growth in EDI usage during the 1990s. As international competition in the domestic marketplace becomes more intense, businesses continue to seek methods to cut costs and streamline operations. EDI offers a means to accomplish both tasks. The Data Interchange Standards Association (1990) and General Electric Information Services (1985) both provide detailed explanations of EDI in booklets available to interested EDI users.

19.2 WHY USE EDI IN ECONOMIC CENSUSES?

The U. S. economy has been changing rapidly. Technological advancements, globalization, and the increased importance of the service sector are just a few of the dynamic forces affecting the economy of today and the 21st century. Demands for data to measure the activities of this changing economy are also growing.

While data demands grow, sources of supply are in some ways shrinking. A large portion of the service sector consists of large numbers of small businesses. Recordkeeping for small businesses can be particularly burdensome, and small businesses may not maintain the kinds of records needed for government statistical reporting. Many industries engaged in finance, communications, insurance, real estate, transportation, and public utilities are unaccustomed to reporting data to government statistical agencies, especially economic census data. Some form of electronic reporting could benefit these small businesses and relieve them of the paperwork burden of reporting census data.

The problems of reporting detailed statistical information to the government are not however, limited to the service sector or to small businesses. Large businesses also are finding it increasingly difficult to report detailed information, especially at the establishment (physical location) level. Resources for reporting statistical information to the government have been cut, because businesses have reduced staff formerly devoted to external reporting.

Large multinational businesses, however, must maintain sophisticated internal accounting systems for their own reporting requirements. EDI offers businesses the opportunity to retrieve information electronically from these internal systems and to forward that information directly to the government through a communications network.

The U.S. Bureau of the Census (1990) conducted the 1989 Recordkeeping Practices Survey to determine the kinds of information businesses could report and how census reporting could be made easier. The survey showed the increasing interest of businesses in electronic reporting. Businesses mentioned EDI frequently as a means to reduce reporting burden on companies, both large and small. For small businesses, a diskette-based reporting system needs to be implemented that includes translation software on the diskette format the data reported by the business into the EDI standard.

For the 1992 Economic Census, the Census Bureau developed and implemented an electronic data interchange reporting program for large businesses. Over the past 200 years, collecting economic data has ranged from collection via personal interview, to mail collection, to a limited program of magnetic tape reporting. For the 1992 Economic Census, another step forward is the collection of information via EDI. As businesses automate, the technology will be extended to other surveys.

19.3 WHAT ARE THE ECONOMIC CENSUSES?

The Economic Censuses are the cornerstones for all economic data published by the Census Bureau and provide the benchmark and sampling frame for many current surveys. They provide a detailed snapshot of the economy at a particular point in time; the censuses are conducted in years ending in 2 and 7. The 1992 Economic Censuses cover economic activity of U.S. businesses for calendar year 1992. Forms were mailed in December 1992, and data collection activities extended until Fall 1993. The first reports from these censuses were released in late 1993.

Response to the censuses is required by U.S. law. All data provided by companies are confidential and can be published as statistical totals only. Detailed economic data, including information on over 7000 manufactured products and 2000 merchandise, commodity, and receipts lines, must be reported for each physical location (establishment) owned and operated by the business. The 1992 Economic Censuses forms were mailed to approximately 4.2 million establishments. About 1.2 million of these establishments are owned by some 165,000 enterprises. The remaining 3 million businesses are single establishment firms.

Even though the censuses cover over 15 million establishments, forms were mailed to only 4.2 million establishments. The information for some 2 million small employers and approximately 9 million firms with no paid employees was obtained from the administrative records of other government agencies in lieu of direct reporting.

To reduce the reporting burden of companies, the Census Bureau tailors the forms to individual industries or groups of industries. For 1992, almost 500 different forms collected data on some 950 industries. The forms included common data items such as sales/shipments/receipts, payroll, and employment. Other data items are tailored to individual industries or groups of industries.

19.4 OTHER ECONOMIC PROGRAMS OF THE CENSUS BUREAU

Over 125 separate monthly, quarterly, and annual surveys are conducted by the U.S. Bureau of the Census. These current surveys are designed to provide broad economic measures such as total sales, inventories, or orders for an industry or group of industries. These surveys generally collect much less detail than the censuses, are more frequent and timely, and (if conducted on a quarterly or monthly basis) often are adjusted for seasonal variation. Many of the current programs of the Census Bureau provide data that are designated as principal economic indicators.

The Census Bureau also compiles the nation's foreign trade statistics. The merchandise trade balance is among the most important data series available to government and industry analysts. To publish these data, the Bureau processes approximately 1 million import records and 1 million export records each month.

19.5 ELECTRONIC DATA COLLECTION INITIATIVES

Many Census Bureau programs are focusing on electronic reporting. Increasingly, businesses want to move away from a paper reporting system. Many businesses are willing to make the investment necessary to move towards electronic reporting of information.

19.5.1 Foreign Trade Statistics

Until recently, import data were compiled from trade documents submitted to the U.S. Customs Service for each import and export transaction. In 1986, some import brokers began reporting data electronically to the Customs Service using Customs' internally developed Automated Broker Interface (ABI). By the end of 1992, over 91 percent of all import transactions were collected electronically using ABI.

Exporters have, until recently, continued to report information on exports on Census Form 7525, the Shipper's Export Declaration (SED). These SEDs are collected at the ports by the Customs Service and forwarded to the Bureau's processing office in Jeffersonville, Indiana for clerical processing and data entry.

Currently, three methods of electronic reporting are offered to exporters: magnetic cartridge, floppy disk, and direct computer-to-computer transmission. To date, over 100 major exporters, freight forwarders, and carriers report in one of these three ways. They account for over 20 percent of all transactions processed each month.

19.5.2 Governments Division Statistics

The Government Division collects and publishes data relating to state and local governments. Over the past several years, more and more of the information collected from the state governments has been received electronically. For example, data for the Survey of Elementary and Secondary School Finances are provided via magnetic tape annually through the 50 state education agencies and cover all the financial transactions of approximately 13,000–14,000 local school districts. These districts account for about 50 percent of all local government financial activities.

In the Survey of State Government Finances, 15 states report via magnetic tapes. Information is reported for the financial activities of the states' general funds. For the Annual Survey of General Purpose Governments Finances, the Census Bureau works with state agencies that collect data on the financial transactions of local governments to ensure that data are collected which satisfy both the states' requirements and ours. In 18 states, data are transmitted electronically in the form of data tapes or diskettes. The data from these sources account for approximately 70 percent of the Bureau's general local government data collection.

19.5.3 The Economic Censuses

In the 1972 Economic Censuses, as the demand from data users for more detailed information grew, many large companies were finding it increasingly difficult and costly to provide the kinds of detailed data requested in the censuses for each of their establishments. That year, for the first time, the Census Bureau accepted computer printouts from a few large businesses in lieu of their completing paper report forms. Because of the cost savings to businesses to report in this manner, the program was expanded in 1977 and 1982 to include a few additional businesses, although the mail-out/mail-back method of collecting data remained the basic collection method.

By the 1982 censuses, businesses began expressing interest in filing census forms in an electronic format. While the idea of electronic filing for businesses was explored within the Census Bureau during the 1982 planning phase, it was not until 1987 that the first magnetic tape reporting program was developed.

By 1987, the Bureau was ready to accept limited tape reporting from some of the very large businesses. A number of large retail businesses (generally those with 1000 establishments or more—both those that had previously reported via computer printout and those that had not) were offered the opportunity to report their 1987 economic census data on magnetic tape. Only 10 businesses participated in this pilot program, but those businesses accounted for over 25,000 establishments for which it was not necessary to key or process data in any manner once the magnetic tape was received and read.

Approximately 80 businesses are participating in the magnetic tape reporting program for 1992. While magnetic tape reporting continues to be an option for electronic reporting, an increased interest in using EDI is anticipated because of the growth and benefits of EDI and the increased usage of EDI in everyday transactions.

19.6 DEVELOPING AN EDI PROGRAM

In the fall of 1990, the Census Bureau began to develop an EDI program for the economic area and in particular the 1992 Economic Censuses. At the time there were no statistical applications of EDI, but a number of large businesses expressed particular interest in this technology. EDI offers businesses the opportunity to reduce reporting burden and offers the Census Bureau the opportunity to reduce paper handling, data entry, and other resource-intensive operations, particularly in the Economic Censuses where large enterprises with thousands of establishments are required to file detailed information for each of their physical locations. Following is a discussion of the development of the Census EDI Program.

19.6.1 Organizing

In the fall of 1990, staff attended a number of seminars and EDI programs and met with representatives of businesses using EDI to learn more about developing and implementing an EDI program. In addition, an EDI Working Group was established that included staff from several Census Bureau divisions. Each member of this group played a key role in the successful implementation of EDI. Team members include representatives from (1) the Business Division, responsible for the 1992 Census of Retail Trade; (2) the Economic Programming Division, responsible for all EDI programming-related activities; (3) the Economic Census and Surveys Division, responsible for planning and coordinating the agriculture and economic censuses, the economic census centralized processing functions, and the annual Company Organization Survey; and (4) the Telecommunications Staff, responsible for Census Bureau data and voice communications.

The EDI team's first task was to determine the feasibility of developing and implementing an EDI program for the economic censuses and to identify the key activities associated with EDI implementation. Time was short because it was necessary to have an EDI program in place by Fall 1992 to ensure complete testing of the system before the beginning of data collection for the censuses. There was no time to acquire the expertise necessary to successfully complete each of the required activities in the time frame available for implementation of EDI for the 1992 censuses, so the decision was made to contract out some of the work.

Many activities required immediate attention. There was no existing standard that could be used for reporting economic statistical information to the Census Bureau. In this context, a *standard* is an agreed-upon format for exchanging data. Development of that standard was a top priority. At the same time, it was necessary to determine what industries or groups of industries should be included in an EDI program. An early caution from EDI experts was not to try to do too much, too soon; that is, start small and build the EDI program as expertise develops during the implementation process. The Bureau decided in November 1990 to move ahead with a limited EDI program for the 1992 Economic Censuses. The implementation plan was conservative; that is, implement EDI reporting on a limited, incremental basis in the 1992 censuses, then build on that experience to expand electronic reporting in post-census current surveys and in the 1997 Economic Censuses.

It was necessary to complete 10 activities for the successful implementation of the EDI program. These activities were (1) determine the scope of the program, (2) develop a standard, (3) create an implementation plan, (4) determine hardware requirements, (5) select a translation software package, (6) determine communications requirements, (7) develop an application interface, (8) elicit industry cooperation, (9) develop an implementation guide, and (10) procure the necessary hardware and software.

19.6.1 Determining the Scope of the EDI Program

EDI experts recommended a limited EDI program be established for the 1992 Economic Censuses. With that decision made, it was necessary to select the industries for the 1992 program. Criteria used for that selection were the following:

- *Industry usage of EDI:* Industries using EDI would be more likely to have the resources in place to implement EDI for census reporting.
- *Homogeneity of the business's operations:* Businesses operating a significant number of establishments classified in a limited number of industries make EDI implementation simpler and more attractive, as compared to a highly diversified business structure with establishments operating in hundreds of different industries. Also, the more homogeneous the business, the easier it is to explain and describe data content and filing requirements.
- *Number of establishments operated by the business:* The more establishments the business operates, the greater the potential benefits of electronic reporting to the business and to the Census Bureau.

The U.S. retail trade sector met all three of these criteria. Many leading retailers have extensive EDI experience.

In fact, this industry has been a pioneer in developing and using EDI. Retailers have established a separate EDI group, the Voluntary Interindustry Communications Standards (VICS) committee, to facilitate the development of a standard. There are a number of large retailers operating thousands of individual establishments classified in a limited number of industries. The relatively successful experience with retailers reporting on computer tape in the 1987 censuses also was a factor in deciding to offer EDI reporting to large retailers in the 1992 censuses.

19.6.2 Developing the Standard

Within the world community, there are two major standards used for transmitting information via EDI: (1) the Accredited Standards Committee (ACS) X12 standard, developed and maintained by the American National Standards Institute's (ANSI) X12 Committee; and (2) the United Nations Electronic Data

Interchange for Administration, Commerce, and Transport (UN/EDIFACT), which operates under the auspices of the United Nations Economic Commission for Europe (ECE) (see Economic Commission for Europe 1991 for a complete explanation of this EDI standard). Businesses in the United States are using the X12 standard for conducting domestic business and are using the UN/EDIFACT standard for international EDI transactions.

The X12 Committee was chartered in 1979 to develop uniform standards for interindustry electronic interchange of business transactions. Since that time, a number of subcommittees have been established within the ASC X12 Committee to work with specific groups and on specific standards. One of these subcommittees is the X12G (government) subcommittee that is responsible for developing standards for government use. Most domestic businesses use the domestic X12 standard for many of their day-to-day transactions, the determining factor in developing and X12 standard for census data collection.

There are two ways to initiate the development of a new standard for EDI. Any organization can submit a work request explaining the need for an EDI standard to the Data Interchange Standards Association (DISA), the secretariat for the X12 Committee. DISA will log the request and forward it to the Technical Assessment Subcommittee (TAS). TAS then assigns and forwards the work request to the appropriate development subcommittee, usually with comments.

The development committee determines whether an existing standard can be used, a new standard is required, or no action is needed. If it is determined that a new standard needs to be developed, the development subcommittee creates a project proposal.

This process can take a year or longer. Given the Census Bureau's extremely tight time frame, an alternative approach was needed. To shorten the process, a project proposal was submitted directly to the X12G Subcommittee at the February 1991 X12 Committee meetings. Approval was received at that meeting to develop a government statistical reporting *transaction set*, the X12 terminology for an EDI standard. To meet the time schedule for collection of 1992 census data, the X12 Subcommittee granted preliminary approval of the transaction set at the June 1991 Committee meetings.

To ensure that this time schedule was met, the Census Bureau contracted the development of the transaction set to the Price Waterhouse EDI Consulting Group and worked closely with a group of large retailers through the VICS group. Their comments were invaluable in ensuring that the transaction set was technically correct and usable by the retailers.

Approval of Transaction Set 152, Statistical Government Information, at the June 1991 meeting was the first of many steps. Other activities included TAS approval in July 1991, approval by the X12G Subcommittee and recommendation for full X12 Committee vote in October 1991, and final approval in February 1992.

Transaction Set 152 is generic and can be used for any program collecting statistical data (U.S. Bureau of the Census 1992). In developing the standard, many existing X12 data elements and segments were used. This greatly simplified the approval process. Basically, the transaction set requires a name, address, an identifier for the unit reporting, an identifier for the data collected, and p᾿ ᾿vision for that data as a dollar amount, quantity, or percent.

19.6.3 Implementation Plan

Price Waterhouse's contract included the creation of an implementation plan for Census Bureau use. That was completed in July 1991. That document detailed the process for standards development and approval, provided a management overview of EDI, described how to map data received in the EDI standard into a file format, and outlined the steps necessary for successful implementation of EDI. As a result of that plan, the EDI team identified all of the key activities, established schedules, and assigned responsibilities. The following discussion details those activities.

19.6.4 Hardware

Two stand-alone 486 personal computers (PCs) send, receive, and translate EDI messages. These machines are physically separated from the cluster of DEC computers used for census processing. The PCs are configured with a tape drive so EDI data can be read to tape as a backup. Data are transferred from the PC to the DEC processing system using commercial software. This is done only when all communications links to the outside have been disconnected.

19.6.5 Translation Software

Translation software "translates" the EDI standard received from the business into a flat file that can then be used to map to the census processing format. Similarly, any information generated by the Census Bureau must be converted into the EDI standard for transmission to the appropriate business. Besides formatting data into the EDI standard, the translation software also tracks outbound documents until an electronic acknowledgment is received.

Users may develop their own translation software or purchase it commercially. There are a number of businesses writing translation software, and most EDI users purchase commercially available software. When Transaction Set 152 was approved as a draft standard, translation software companies provided their customers the module needed for using the census standard.

19.6.6 Communications

Most businesses using EDI contract with a *Value Added Network* (VAN), a third-party network service that provides communications, mailboxes, and gateways to other networks used by businesses. EDI World, Inc. (1993) in its

February issue provides detailed information on VANs. Security considerations have thus far precluded the Census Bureau from using a VAN. To send and receive EDI messages, Census uses a toll-free telephone number that is directly connected to our stand-alone computer. The translation software acknowledges receipt of transmissions and sends other messages to *trading partners*, EDI terminology for businesses exchanging information via EDI. As the EDI program expands, however, use of a toll-free number for communications will become more cumbersome and difficult.

Some businesses have indicated that they prefer to send a magnetic tape containing the EDI messages. This tape would be loaded onto a 486 PC, where the data would be translated and formatted for loading onto the DEC processing system.

19.6.7 Application Interface

Although the translation software converts the EDI data into a flat file, some programming is required to map the data into the format used for Census processing. Some translation software provides mapping facilities that ease this process. The application interface for Transaction Set 152 has been written by the Census Bureau.

19.6.8 Industry Cooperation—Trading Partnerships

The success of the EDI initiative depended to a large extent upon the cooperation of a number of large retailers. It is necessary for businesses to invest the necessary time and effort to convert to the EDI standard. As a first step in gaining that support, the development phase of the transaction set included a number of businesses. In May 1991, the VICS committee scheduled a special session at their spring user meeting to brief key EDI contacts from many of the largest retailers about the 1992 Census of Retail Trade and Bureau objectives regarding EDI. Representatives of six major retailers accounting for thousands of establishments attended and generally were receptive though noncommittal.

In most businesses, EDI responsibilities are separate and distinct from the accounting groups that usually complete census forms. Because many of these retailers have a number of EDI initiatives underway or on the planning board, it was necessary to pull both groups together in meetings to provide to each a clearer understanding of both census reporting and the opportunities afforded by reporting via EDI.

During the summer of 1991, other ways to promote interest in EDI reporting were investigated. An article in the August 1991 issue of *EDI World* described the current status of Bureau activities and invited retailers to contact the Census Bureau for additional details (Sykes 1991). Earlier, letters were mailed to approximately 250 retailers inviting them to report electronically for 1992 using either magnetic tape or EDI.

By the summer of 1992, several very large retailers agreed to report using the EDI standard. Most encouraging are (1) the support that business has shown towards the Census effort to develop an EDI program and (2) the desire to move toward using EDI in census reporting even if they were unable, because of previous commitments, to do so in 1992.

19.6.9 Implementation Guide

An *implementation guide* (1) provides trading partners a comprehensive document that describes in detail how the transaction set will be used to handle the data requested and (2) identifies special requirements and code lists. The Census Bureau has developed an implementation guide that is available for current EDI trading partners, future trading partners, and those interested in learning more about the Census Bureau and EDI.

19.6.10 Procurement of Hardware and Software

Early on, it was decided to use a PC to send and receive EDI messages. The use of a PC significantly cuts the cost for hardware and software. Translation software for a PC that meets all of the Census Bureau's requirements is relatively inexpensive (about $3000 in 1992).

19.7 THE FUTURE—UN/EDIFACT OR X12?

The ASC X12 Committee was chartered to develop uniform standards for electronic interchange of business information. To date, over 150 standards (transaction sets) such as order placement, shipping and receiving, and invoicing have been developed. There are over 1200 data elements and 500 data segments used in these transaction sets. As many as 20,000 companies are using X12 standards as a way of conducting business within the United States, and the number of companies is increasing.

By 1985, the international business community recognized EDI as the wave of the future and moved to begin the development of international EDI standards for the interchange of international business transactions. The United Nations supported the development of international standards; and in 1986 the acronym *EDIFACT*, which stands for Electronic Data Interchange for Administration, Commerce, and Transport, was approved by the UN/ECE/WP.4. By 1987, Rapporteurs for North America, Western Europe, and Eastern Europe had been appointed, the EDIFACT syntax had been adopted, and standards (message) development had begun. Currently 34 messages are approved for use, with many more in the development stage.

UN/EDIFACT has grown rapidly in the past few years. There are now five regional EDIFACT boards, with the addition of Australia/New Zealand and Japan/Singapore. The North American Board, now known as the Pan American EDIFACT Board (PAEB), has been expanded to include all of the Americas. The structure within each of these boards differs; but each has a Rapporteur appointed by the United Nations, who is responsible for initiating and coordinating UN/EDIFACT development work in his/her geographical area of jurisdiction. The Rapporteurs meet twice yearly, where work groups also convene to work on message development.

The PAEB serves as the coordinating body of the Pan American national EDI standards organizations and provides a forum for Pan American representation and consensus to the U.N. Working Group. Until June 1992, the PAEB was a standing committee of the X12 Committee. At the June X12 meetings, however, it was voted to separate the PAEB from X12.

There has been increasing concern within the United States about the continued development and maintenance of two EDI standards. The X12 Committee established the EDIFACT/X12 Alignment Committee to study the problem. Because the syntax for the two standards is so different, the Alignment Committee concluded there could not be a convergence between the two. Rather, at the February 1992 X12 meetings the EDIFACT/X12 Alignment Committee proposed that X12 adopt EDIFACT, after release of X12 Version 4 in 1997.

The effect of this resolution is to establish EDIFACT as the syntax for the development of X12 transaction sets after 1994/1995. No X12 syntax would be approved after that time. Businesses could continue to use the existing transaction sets; but as more international standards are developed, it is believed there will be a steady migration to the UN/EDIFACT standard.

The Census Bureau developed an X12 transaction set for use in EDI reporting, because Census trading partners are U.S. businesses that are currently using X12 for EDI. However, the Census Bureau recognizes the changing environment for EDI and the move within X12 toward the EDIFACT syntax.

In May 1992, at the request of the Office of Management and Budget, the agency which coordinates the work of the various statistical agencies within the U.S. Government, the Census Bureau hosted an International Conference on EDI. Attending were representatives from groups around the world working on UN/EDIFACT message development. The conference provided a forum for the exchange of information on X12 and EDIFACT statistical messages.

EDIFACT statistical messages are currently being developed within the Western European EDIFACT Board's statistical message development group. This group (WE/EB-MD6, Message Development Group 6, Statistics) is headed by a representative from EUROSTAT. The objective of the group is to develop standardized messages for statistical information. Five subgroups have been established within the MD6 Working Group to develop statistical messages:

- WG1: exchanges of aggregated statistical data,
- WG3: statistical aspects of code lists (statistical nomenclature),

- WG4: new methods of collection of basic statistical information (raw statistics),
- WG5: internatinoal trade statistics, and
- WG6: balance of payment statistics.

Transaction Set 152 is the only X12 EDI transaction set available for collecting raw economic statistical information. There is no equivalent EDIFACT message. Therefore, there is, much interest in Transaction Set 152.

The Census Bureau now proposes to convert the X12 Transaction Set 152 into a UN/EDIFACT message so that U.S. businesses can use either standard to transmit information to the Bureau. As a start, at the September 1992 X12 Committee meetings the Census Bureau submitted a new project proposal for converting Transaction Set 152 to a UN/EDIFACT message. That was approved. The Bureau is working with the Western European MD6 Working Group on Statistics to convert the transaction set to an international message.

EDIFACT is the wave of the future. Plans are to accept data in either X12 or EDIFACT format at the Census Bureau continues to explore methods of reducing the reporting burden for U.S. businesses.

19.9 CONCLUDING REMARKS

For 1993, EDI was implemented for the Company Organization Survey (COS). This survey, which is conducted annually, asks multi-establishment businesses to provide updated name and address information, number of employees, and payroll for each of their locations. Letters have been mailed to approximately 500 large businesses in this survey, inquiring about their interest in EDI. Respondents in some current business surveys are also being queried to determine their interest in EDI. Plans are to move rapidly towards implementing EDI in all programs where there is business interest.

Implementing EDI on a limited scale in the 1992 Census of Retail Trade has provided the experience needed to expand electronic reporting significantly into many current surveys. Plans are to participate in the development of EDIFACT messages in anticipation of the transition to this standard in the mid-1990s.

It is important to look back over the development and implementation process and reflect on the lessons learned. For the Census Bureau, one of the most important lessons is to start modestly; that is, do not try to do too much, too soon. Move slowly, gain experience, and then expand. Next, it is important to contract whenever possible. Rather than trying to develop EDI expertise immediately, contract with the experts to get started.

Work with the industry for which you are developing the program. This is a critical point. Business involvement in developing the transaction set was instrumental in convincing businesses to adopt EDI reporting in the censuses.

One also needs to sell the benefits of an EDI program. The decision to use an EDI reporting standard requires significant corporate resources. A key to convincing large retailers to report using EDI was bringing together the EDI experts and the accountants to discuss EDI benefits. Finally, follow through by working closely with trading partners. By limiting the number of EDI reporters, it is possible to provide the "hands on" assistance needed to make implementation successful.

REFERENCES

Data Interchange Standards Association, Inc. (1990), "An Introduction to Electronic Data Interchange," Alexandria, VA.

EDI World, Inc. (1993), *EDI World*, **3**, 4, Hollywood, FL.

Sykes II, J. D. (1991), *EDI World*, **1**, 8, Hollywood, FL, pp. 16, 37.

General Electric Information Services (1985), "Introduction to Electronic Data Interchange," Rockville, MD: General Electric Company.

Economic Commission for Europe (1991), "Introduction to UN/EDIFACT," Joint Rapporteurs' Teams, New York: United Nations.

U.S. Bureau of the Census (1990), "1989 Recordkeeping Practices Survey," Washington, DC.

U.S. Bureau of the Census (1992), "A Guide to the Implementation of Statistical Government Information Transaction Set 152," Washington, DC.

PART D

Data Processing

CHAPTER TWENTY

Matching and Record Linkage

William E. Winkler[1]

U.S. Bureau of the Census

Matching has a long history of uses for statistical surveys and administrative data files. Business registers of names, addresses, and other information such as total sales are constructed by combining tax, employment, or other administrative databases (see Chapter 2). Surveys of retail establishments or farms often combine results from an area frame and a list frame. To produce a combined estimator, units must be identified from the area frame sample that are also found on the list frame (see Chapter 11). To estimate the size of a population via capture–recapture techniques, units common to two or more independent listings must be accurately determined (Sekar and Deming 1949; Scheuren 1983; Winkler 1989b). Samples must be drawn appropriately to estimate overlap (Deming and Gleser 1959).

Rather than develop a special survey to collect data for policy decisions, it is sometimes more appropriate to match data from administrative data sources. An economist, for instance, might wish to link a list of companies and the energy resources they consume with a comparable list of companies and the types, quantities, and dollar amounts of the goods they produce. There are potential advantages to using administrative data in analyses. Administrative data sources may contain greater amounts of data and that data may be more accurate due to improvements over time. In addition, virtually all cost of data collection is borne by the administrative programs, and respondent burden associated with a special survey is eliminated. Brackstone (1987) discusses these and other advantages of administrative sources as a substitute for surveys. Methods of adjusting analyses for matching error in merged databases are also available (Neter et al. 1965, Scheuren and Winkler 1993).

[1]The author appreciates many useful comments by Brenda G. Cox, the section editor, and an anonymous reviewer. The opinions expressed are those of the author and not necessarily those of the U.S. Bureau of the Census.

Business Survey Methods, Edited by Cox, Binder, Chinnappa, Christianson, Colledge, Kott. ISBN 0-471-59852-6 © 1995 John Wiley & Sons, Inc.

This chapter addresses exact matching in contrast to statistical matching (Federal Committee on Statistical Methodology 1980). An *exact match* is a linkage of data for the same unit (e.g., business) from different files; linkages for units that are not the same occur only because of error. Exact matching uses identifiers such as name, address, or tax unit number. *Statistical matching*, on the other hand, attempts to link files that have few units in common. Linkages are based on similar characteristics rather than unique identifying information, and strong assumptions about joint relationships are made. Linked records need not correspond to the same unit.

Increasingly, computers are used for exact matching to reduce or eliminate manual review and to make results more easily reproducible. Computer matching has the advantages of allowing central supervision of processing, better quality control, speed, consistency, and reproducibility of results. When two records have sufficient information for making decisions about whether the records represent the same unit, humans can exhibit considerable ingenuity by accounting for unusual typographical errors, abbreviations, and missing data. For all but the most difficult situations, however, modern computerized record linkage can achieve results at least as good as a highly trained clerk. When two records have missing or contradictory name or address information, then the records can only be correctly matched if additional information is obtained. For those cases when additional information cannot be adjoined to files automatically, humans are often superior to computer matching algorithms because they can deal with a variety of inconsistent situations.

In the past, most record linkage has been done manually or via elementary but ad hoc computerized rules. This chapter focuses on computer matching techniques that are based on formal mathematical models subject to testing via statistical and other accepted methods. A description is provided of how aspects of name, address, and other file information affect development of automated procedures. The algorithms I describe are based on optimal decision rules that Fellegi and Sunter (1969) developed for methods first introduced by Newcombe et al. (1959). Multidisciplinary in scope, these automated record linkage approaches involve (1) string comparator metrics, search strategies, and name and address parsing/standardization from computer science; (2) discriminatory decision rules, error rate estimation, and iterative fitting procedures from statistics; and (3) linear programming methods from operations research. This chapter contains many examples because its purpose is to provide background for practitioners. While proper theory plays an important role in modern record linkage, my intent is to summarize theoretical ideas rather than rigorously develop them. The seminal paper by Fellegi and Sunter (1969) is still the best reference on theory and related computational methods.

20.1 TERMINOLOGY AND DEFINITION OF ERRORS

Much theoretical work and associated software development for matching and record linkage have been done by different groups working in relative isolation, resulting in varied terminology across groups. In this chapter I use terminology consistent with Newcombe (Newcombe et al. 1959; Newcombe 1988) and Fellegi and Sunter (1969).

In the product $\mathbf{A} \times \mathbf{B}$ of files A and B, a *match* is an $a_i b_j$ pair that represents the same business entity and a *nonmatch* is a pair that represents two different entities. Within a single list, a *duplicate* is a record that represents the same business entity as another record in the same list. Rather than consider all pairs in $\mathbf{A} \times \mathbf{B}$, attention is sometimes restricted to those pairs that agree on certain identifiers or *blocking criteria*. Blocking criteria are also called *pockets* or *sort keys*. For instance, instead of making detailed comparisons of all 90 billion pairs from two lists of 300,000 records representing all businesses in a particular state, it may be reasonable to limit comparisons to the set of 30 million pairs that agree on U.S. Postal ZIP code. Errors of omission can result from use of such blocking criteria; *missed matches* are those false nonmatches that do not agree on a set of blocking criteria.

A *record linkage decision rule* is a rule that designates a pair either as a link, a possible link, or a nonlink. *Possible links* are those pairs for which the identifying data are insufficient to decide if the pair is a match. Typically, clerks review possible links and determine their match status. In a list of farms, name information alone is not sufficient for deciding whether "John K Smith, Jr, Rural Route 1" and "John Smith, Rural Route 1" represent the same operation. The second "John Smith" may be the same person as "John K Smith, Jr" or may be his father or grandfather. Mistakes can and do occur in matching. *False matches* are those nonmatches that are erroneously designated as links by a decision rule. *False nonmatches* are either (1) matches designated as nonlinks by the decision rule as it is applied to a set of pairs or (2) missed matches that are not in the set of pairs to which the decision rule is applied. Generally, *link/nonlink* refers to designations under decision rules and *match/nonmatch* refers to true status.

Matching variables are common identifiers (such as name, address, annual receipts, or tax code number) that are used to identify matches. Where possible, a business name such as "John K Smith Company" is parsed or separated into components such as first name "John," initial "K," surname "Smith," and business key word "Company." The parse allows better comparison of names and hence improves matching accuracy. Similarly, an address such as "1423 East Main Road" might be parsed into location number "1423," direction "East," street name "Main," and street type "Road." Matching variables do not necessarily uniquely identify matches. For instance, in constructing a frame of a city's retail establishments, name information such as "Hamburger Heaven" may not allow proper linkage if "Hamburger Heaven" has several locations. The addition of address information can sometimes help, but not if many businesses have different addresses on different lists. In such a situation there is insufficient information to separate new units from existing units that have different mailing addresses associated with them. The *matching weight* or *score* is a number assigned to a pair that simplifies assignment of link and nonlink status via decision rules. A procedure, or matching variable, has more *distinguishing power* if it is better able to delineate matches and nonmatches than another.

20.2 IMPROVED COMPUTER-ASSISTED MATCHING METHODS

Historically, record linkage has been assigned to clerks who reviewed the lists, obtained additional information when matching information was missing or contradictory, and made linkage decisions following established rules. Typically these lists were sorted alphabetically by name or address characteristics to simplify the review process. If a name contained an unusual typographical variation, the clerks might not find its matches. For large files, matches could be separated by several pages of printouts, so that some matches might be missed. Even after extensive training, the clerks' matching decisions were not always consistent. All work required extensive review. Each major update required training the clerical staff again.

On the other hand, development of computer matching software can require person-years of time from proficient computer scientists. Existing software may not work optimally on files having characteristics significantly different from those for which they were developed. The advantages of automated methods far outweigh these disadvantages. In situations for which good identifiers are available, computer algorithms are fast, accurate, and yield reproducible results. Search strategies can be far faster and more effective than those applied by clerks. As an example, the best computer algorithms allow searches using spelling variations of key identifiers. Computer algorithms can also account for the relative distinguishing power of combinations of matching fields as input files vary. In particular, the algorithms can deal with the relative frequency that combinations of identifiers occur.

As an adjunct to computer operations, clerical review is still needed to deal with pairs having significant amounts of missing information, typographical errors, or contradictory information. Even then, using the computer to bring pairs together and having computer-assisted methods of review at terminals is more efficient than manual review of printouts.

By contrasting the creation of mailing lists for the U.S. Census of Agriculture in 1987 and 1992, the following example dramatically illustrates how enhanced computer matching techniques can reduce costs and improve quality. Absolute numbers are comparable because 1987 proportions were multiplied by the 1992 base of six million. To produce the address list, duplicates were identified in six million records taken from 12 different sources. Before 1982, listings were reviewed manually and an unknown proportion of duplicates remained in files.

In 1987, the development of effective name parsing and adequate address parsing software allowed creation of an ad hoc computer algorithm for automatically designating links and creating subsets for efficient clerical review. Within pairs of records agreeing on ZIP code, the ad hoc computer algorithm used surname-based information, the first character of the first name, and numeric address information to designate 6.6 percent (396,000) of the records as duplicates and 28.9 percent as possible duplicates to be clerically reviewed. About 14,000 person-hours (as many as 75 clerks for 3 months) were used in this clerical review, and an additional 450,000 duplicates (7.5 percent) were identified. Many duplicates were not located, compromising subsequent estimates based on the list.

In 1992, Fellegi–Sunter algorithms were developed that used effective computer algorithms for dealing with typographical errors. The computer software designated 12.8 percent of the file as duplicates and another 19.7 percent as needing clerical review. About 6500 person-hours were used and an additional 486,000 duplicates (8.1%) were identified. Even without further clerical review, the 1992 computer procedures identified almost as many duplicates as the 1987 combination of computer and clerical procedures. The cost of software development was $110,000 in 1992. The rates of duplicates identified by computer plus clerical procedures were 14.1 percent in 1987 and 20.9 percent in 1992. The 1992 computer procedures lasted 22 days; in contrast, the 1987 computer plus clerical procedure needed 3 months.

20.3 STANDARDIZATION AND PARSING

Appropriate parsing of name and address components is crucial for computerized record linkage. Without it, many true matches would erroneously be designated as nonlinks because identifying information could not be adequately compared. For specific types of business lists, the drastic effect of parsing failure has been quantified (Winkler 1985b, 1986). DeGuire (1988) presents concepts needed for parsing and standardizing addresses; name parsing requires similar concepts.

20.3.1 Standardization of Names and Addresses

The basic ideas of *standardization* are to (1) replace the many spelling variations of commonly occurring words with standard spellings such as fixed abbreviations or spellings and (2) use key words found during standardization as hints for parsing subroutines. In standardizing names, words of little distinguishing power such as "Corporation" or "Limited" are replaced with consistent abbreviations such as "CORP" and "LTD," respectively. First name spelling variations such as "Rob" and "Bobbie" might be replaced with a consistent, assumed, original spelling such as "Robert" or an identifying root word such as "Robt" because "Bobbie" could refer to a woman with "Roberta" as her legal first name. The purpose of name standardization is to allow name-parsing software to work better, by presenting names consistently and by separating out name components that have little value in matching. When business-associated words such as "Company" or "Incorporated" are encountered, flags are set that force entrance into different name-parsing routines than would be used otherwise.

Standardization of addresses operates like standardization of names. Words such as "Road" or "Rural Route" are typically replaced by appropriate abbreviations. For instance, when a variant of "Rural Route" is encountered, a flag is set that forces parsing into routines different from routines associated with house-number/street-name addresses. When reference lists containing city, state or province, and postal codes are available from the national postal service or another source, then city names in address lists can be placed in a standard form that is consistent with the reference list.

20.3.2 Parsing of Names and Addresses

Parsing divides a free-form name field into a common set of components that can be compared. Parsing algorithms often use hints based on words that have been standardized. For instance, words such as "CORP" or "CO" might cause parsing algorithms to enter different subroutines than words such as "MRS" or "DR." In the examples of Table 20.1, "Smith" is the name component with the most identifying information. PRE refers to a prefix, POST1 and POST2 refer to postfixes, and BUS1 and BUS2 refer to commonly occurring words associated with businesses. While exact, character-by-character comparison of the standardized but unparsed names would yield no matches, use of the subcomponent last name "Smith" might help designate some pairs as links. Parsing algorithms are available that deal with either last-name-first types of names such as "John Smith" or last-name-last types such as "Smith, John." None are available that can accurately parse both types of names in a single file.

Humans can easily compare many types of addresses because they can associate corresponding subcomponents in free-form addresses. To be most effective, matching software requires address subcomponents to be in identified locations. As the examples in Table 20.2 show, parsing software divides a free-form address field into a set of corresponding components in specific locations on the data record.

20.3.3 Examples of Names

The main difficulty with business names is that even when they are properly parsed, the identifying information may be indeterminate. In each example of Table 20.3, the pairs refer to the same business entity in a survey frame. Alternatively, in Table 20.4, each pair refers to different business entities that have similar names. Because the name information in Tables 20.3 and 20.4 may be insufficient for accurately determining match status, address information or other identifying characteristics may have to be obtained via clerical review. If the additional address information is indeterminate, then at least one establishment in each pair may have to be contacted.

Table 20.1 Examples of Name Parsing

	Parsed							
Standardized	PRE	FIRST	MIDDLE	LAST	POST1	POST2	BUS1	BUS2
DR John J Smith MD	DR	John	J	Smith	MD			
Smith DRY FRM				Smith			DRY	FRM
Smith & Son ENTP				Smith		Son	ENTP	

Table 20.2 Examples of Address Parsing

	Parsed									
Standardized	Pre2	Hsnm	Stnm	RR	Box	Post1	Post2	Unit1	Unit2	Bldg
16 W Main ST APT 16	W	16	Main			ST		16		
RR 2 BX 215				2	215					
Fuller BLDG SUITE 405									405	Fuller
14588 HWY 16 W		14588	HWY				W			

Table 20.3 Names Referring to the Same Business Entities

Name	Reason
John J Smith ABC Fuel Oil	One list has owner name while the other list has business entity name.
John J Smith, Inc. J J Smith Enterprises	Either name may be used by the business.
Four Star Fuel, Exxon Distrib. Four Star Fuel	Independent fuel oil dealer is associated with major oil company.
Peter Knox Dairy Farm Peter J Knox	One list has establishment name while the other has owner name.

Table 20.4 Names Referring to Different Businesses

Name	Reason
John J Smith Smith Fuel	Similar initials or names but different companies
ABC Fuel ABC Plumbing	Same as previous
North Star Fuel, Exxon Distrib. Exxon	Independent affiliate and company with which affiliated

20.4 MATCHING DECISION RULES

For many projects, automated matching decision rules are developed using ad hoc, intuitive approaches. For instance, the decision rule might be as follows:

- If the pair agrees on a specific three characteristics or agrees on four or more within a set of five characteristics, designate the pair as a link.
- If the pair agrees on a specific two characteristics, designate the pair as a possible link.
- Otherwise, designate the pair as a nonlink.

Ad hoc rules are easily developed and may yield good results. The disadvantage is that ad hoc rules may not be applicable to pairs that are different from those used in defining the rule. Users seldom evaluate ad hoc rules with respect to false match and false nonmatch rates.

In the 1950s, Newcombe et al. (1959) introduced concepts of record linkage that were formalized in the mathematical model of Fellegi and Sunter (1969). Computer scientists independently rediscovered the model (Cooper and Maron 1978, Van Rijsbergen et al. 1981, Yu et al. 1982) and showed that the model's decision rules work best among a variety of rules based on competing mathematical models. Fellegi and Sunter's ideas are a landmark in record linkage theory because they introduce many ways of computing key parameters needed for the matching process. Their paper provides (1) methods of estimating outcome probabilities that do not rely on intuition or past experience, (2) estimates of error rates that do not require manual intervention, and (3) automatic threshold choice based on estimated error rates. In my view the best way to build record linkage strategies is to start with formal mathematical techniques based on the Fellegi–Sunter model and then make ad hoc adjustments only as necessary. The adjustments may be likened to the manner in which early regression procedures were informally modified to deal with outliers and collinearity.

20.4.1 Crucial Likelihood Ratio

The record linkage process attempts to classify pairs in a product space $A \times B$ from two files A and B into M, the set of true matches, and U, the set of true nonmatches. Fellegi and Sunter (1969) considered ratios of probabilities of the form

$$R = \frac{P(\gamma \in \Gamma \mid M)}{P(\gamma \in \Gamma \mid U)}, \tag{20.1}$$

where γ is an arbitrary agreement pattern in a comparison space Γ. For instance, Γ might consist of eight patterns representing simple agreement on the largest name component, street name, and street number. Alternatively, each $\gamma \in \Gamma$ might additionally account for the relative frequency with which specific values of name components such as "Smith," "Zabrinsky," "AAA," and "Capitol" occur.

20.4.2 Theoretical Decision Rule

The decision rule is equivalent to the one originally given by Fellegi and Sunter [1969, equation (19)]. In the following, r represents an arbitrary pair, $\gamma \in \Gamma$ is the agreement pattern associated with r, and R is the ratio corresponding to r that is given by equation (20.1). The decision rule d provides three designated statuses for pairs and is given by:

$$d(r) = \begin{cases} \text{link} & \text{if } R > UPPER \\ \text{possible link} & \text{if } LOWER \leq R \leq UPPER \\ \text{nonlink} & \text{if } R < LOWER. \end{cases} \tag{20.2}$$

The cutoff thresholds *UPPER* and *LOWER* are determined by *a priori* error bounds on false matches and false nonmatches. Rule 20.2 agrees with intuition. If $\gamma \in \Gamma$ consists primarily of agreements, then it is intuitive that $\gamma \in \Gamma$ would be more likely to occur among matches than nonmatches and ratio (20.1) would be large. On the other hand, if $\gamma \in \Gamma$ consists primarily of disagreements, then ratio (20.1) would be small.

Fellegi and Sunter (1969) showed that rule (20.2) is optimal in that for any pair of fixed upper bounds on the rates of false matches and false nonmatches, the clerical review region is minimized over all decision rules on the same comparison space Γ. The theory holds on any subset such as pairs agreeing on a postal code, street name, or part of a name field. The ratio R or any monotonically increasing transformation of it (such as given by a logarithm) is referred to as a *matching weight* or *total agreement weight*. In actual applications, the optimality of rule (20.2) is heavily dependent on the accuracy of the estimated probabilities in equation (20.1). The probabilities in equation (20.1) are called *matching parameters*.

20.4.3 Basic Parameter Estimation Under the Independence Assumption

Fellegi and Sunter (1969) were the first to observe that certain parameters needed for rule (20.2) could be obtained directly from observed data if certain simplifying assumptions were made. For each $\gamma \in \Gamma$, they considered

$$P(\gamma) = P(\gamma \mid M)P(M) + P(\gamma \mid U)P(U) \qquad (20.3)$$

and noted that the proportion of pairs with $\gamma \in \Gamma$ could be computed directly from available data. If $\gamma \in \Gamma$ consists of a simple agree/disagree pattern associated with three variables satisfying the conditional independence assumption that there exist vector constants (marginal probabilities) $m \equiv (m_1, m_2, \cdots, m_K)$ and $u \equiv (u_1, u_2, \cdots, u_K)$ such that, for all $\gamma \in \Gamma$,

$$P(\gamma \mid M) = \prod_{i=1}^{K} m_i^{\gamma^i}(1 - m_i)^{(1 - \gamma^i)} \text{ and } P(\gamma \mid U) = \prod_{i=1}^{K} u_i^{\gamma^i}(1 - u_i)^{(1 - \gamma^i)},$$

$$(20.4)$$

then Fellegi and Sunter provide the seven solutions for the seven distinct equations associated with equation (20.3).

If $\gamma \in \Gamma$ represents more than three variables, then it is possible to apply general equation-solving techniques such as the method of moments (e.g., Hogg and Craig 1978, pp. 205–206). Because the method of moments has shown numerical instability in some record linkage applications (Jaro 1989) and with general mixture distributions (Titterington et al. 1988, p. 71), maximum-likelihood-based methods such as the Expectation-Maximization (EM) algorithm (Dempster et al. 1977, Wu 1983, Meng and Rubin 1993) may be preferred.

The EM algorithm has been used in a variety of record linkage situations. In each, it converged rapidly to unique limiting solutions over different starting points (Thibaudeau 1989; Winkler 1989a, 1992). The major difficulty with the parameter-estimation techniques (EM or an alternative such as method of moments) is that they may yield solutions that partition the set of pairs into two sets that differ substantially from the desired sets of true matches and true nonmatches. In contrast to other methods, the EM algorithm converges slowly and is stable numerically (Meng and Rubin 1993).

20.4.4 Adjustment for Relative Frequency

Newcombe et al. (1959) introduced methods for using the specific values or relative frequencies of occurrence of fields such as surname. The intuitive idea is that if surnames such as "Vijayan" occur less often than surnames such as "Smith," then "Vijayan" has more distinguishing power. A variant of Newcombe's ideas was later mathematically formalized by Fellegi and Sunter (1969; see also Winkler 1988, 1989c for extensions). Copas and Hilton (1990) introduced a new theoretical approach that, in special cases, has aspects of the Newcombe's approach; it has not yet applied in a record linkage system. While the value-specific approach can be used for any matching field, strong assumptions must be made about independence between agreement on specific value states of one field versus agreement on other fields.

The concepts of Fellegi and Sunter (1969, pp. 1192–1194) describe the problem well. To simplify the ideas, files A and B are assumed to contain no duplicates. The true frequencies of specific values of a string such as first name in files A and B, respectively, are given by

$$f_1, f_2, \cdots, f_m; \sum_{j=1}^{m} f_j = N_A$$

and

$$g_1, g_2, \cdots, g_m; \sum_{j=1}^{m} g_j = N_B.$$

If the mth string, say "Smith," occurs f_m times in File A and g_m times in File B, then pairs agree on "Smith" $f_m g_m$ times in $\mathbf{A} \times \mathbf{B}.$ The corresponding true frequencies in M are given by

$$h_1, h_2, \cdots, h_m; \sum_{j=1}^{m} h_j = N_M.$$

Note that $h_j \leq \min (f_j, g_j)$, where $j = 1, 2, \cdots, m$. For some implementations, h_j is assumed to equal the minimum, and $P(\text{agree } j\text{th value of string} \mid M)$ = h_j/N_M and $P(\text{agree } j\text{th value of string} \mid U) = (f_j g_j - h_j)/(N_A \cdot N_B - N_M).$ In practice, observed values rather than true values must be used. The variants of how the h_j frequencies are computed involve differences in how typographical errors are modeled, what simplifying assumptions are made, and how frequency weights are scaled to simple agree/disagree probabilities (Newcombe 1988; Fellegi and Sunter 1969; Winkler 1988, 1989c). As originally shown by Fellegi and Sunter (1969), the scaling can be thought of as a means of adjusting for typographical error. The scaling is

$$P(\text{agree on string} \mid M = \sum_{j=1}^{m} P(\text{agree on } j\text{th value of string} \mid M),$$

where the probability on the left is estimated via the EM algorithm or another method. With minor restrictions, the ideas of Winkler (1989c) include those of Fellegi and Sunter (1969), Newcombe (1988, pp. 88–89), and Rogot et al. (1986) as special cases.

In some situations, the frequency tables are created "on-the-fly" using the files actually being matched (Winkler 1989c); in others, the frequency tables are created *a priori* using large reference files. The advantage of on-the-fly tables is that they can use different relative frequencies in different geographic regions; for instance, Hispanic surnames in Los Angeles, Houston, or Miami and French surnames in Montreal. The disadvantage of on-the-fly tables is that they must be based on files that cover a large percentage of the target population. If the data files contain samples from a population, then the frequency weights should reflect the appropriate population frequencies. For instance, if two small lists of companies in a city are used and "George Jones, Inc" occurs once on each list, then a pair should not be designated as a link using name information only. Corroborating information such as address should also be used because the name "George Jones, Inc" may not uniquely identify the establishment.

20.4.5 Jaro String Comparator Metrics for Typographical Error

Jaro (1989) introduced methods for dealing with typographical error such as "Smith" versus "Smoth." Jaro's procedure consists of two steps. First, a string comparator returns a value based on counting insertions, deletions, transpositions, and string length. Second, the value is used to adjust a total agreement weight downward toward the total disagreement weight. Jaro's string comparator was extended by making agreement in the first few characters of the string more important than agreement on the last few (Winkler 1990b). As Table 20.5 illustrates, the original Jaro comparator and the Winkler-enhanced comparator yield a more refined scale for describing the effects of typographical error than do standard computer science methods such as the Damerau–Levenstein metric (Winkler 1985a, 1990b).

Jaro's original weight-adjustment strategy was based on a single adjustment function developed via ad hoc methods. Using calibration files having true matching status, Jaro's strategy has been extended by applying crude statistical curve fitting techniques to define several adjustment functions. Different curves were developed for first names, last names, street names, and house numbers.

Table 20.5 Comparison of String Comparators Rescaled Between 0 and 1

┌────── Strings ──────┐		Winkler	Jaro	Damerau–Levenstein
billy	billy	1.000	1.000	1.000
billy	bill	0.967	0.933	0.800
billy	blily	0.947	0.933	0.600
massie	massey	0.944	0.889	0.600
yvette	yevett	0.911	0.889	0.600
billy	bolly	0.893	0.867	0.600
dwayne	duane	0.858	0.822	0.400
dixon	dickson	0.853	0.791	0.200
billy	susan	0.000	0.000	0.000

When used in actual matching contexts, the new set of curves and enhanced string comparator improve matching efficacy when compared to the original Jaro methods (Winkler 1990b). With general business lists, the same set of curves could be used or new curves could be developed. In a large experiment using files for which true matching status was known, Belin (1993) examined effects of different parameter-estimation methods, uses of value-specific weights, applications of different blocking criteria, and adjustments using different string comparators. Belin demonstrated that the original Jaro string comparator and the Winkler extensions were the two best ways of improving matching efficacy in files for which identifying fields had significant percentages of minor typographical errors.

20.4.6 General Parameter Estimation

Two difficulties arise in applying the EM procedures of Section 20.4.3. The first is that the independence assumption is often false (Smith and Newcombe 1975, Winkler 1989b). The second is that, due to model misspecification, EM or other fitting procedures may not naturally partition the set of pairs into the desired sets of matches M and nonmatches U.

To account for dependencies between the agreements of different matching fields, an extension of an EM-type algorithm due to Haberman (1975, see also Winkler 1989a) can be applied. Because many more parameters are associated with general interaction models than with independence models, only a fraction of all interactions may be fit. For instance, if there are 10 matching variables, the degrees of freedom are only sufficient to fit all three-way interactions (e.g., Bishop et al. 1975, Haberman 1979); with fewer matching variables, it may be necessary to fit various subsets of the three-way interactions.

To address the natural partitioning problem, $\mathbf{A} \times \mathbf{B}$ is partitioned into three sets of pairs C_1, C_2, and C_3 using an equation analogous to (20.3). The EM procedures are then divided into three-class or two-class procedures. When appropriate, two of the three classes are combined into a set that represents either M or U. The remaining class represents the complement. When both name and address information is used for matching, the two-class EM tends to divide a set of pairs into those agreeing on address information and those disagreeing. If address information associated with many pairs is indeterminate (e.g., Rural Route 1 or Highway 65 West), the three-class EM can yield a proper partition because it tends to divide the set of pairs into (1) matches at the same address, (2) nonmatches at the same address, and (3) nonmatches at different addresses.

The general EM algorithm is far slower than the independent EM algorithm because the M step is no longer in closed form. Convergence is speeded up by using variants of the Expectation-Conditional Maximization (ECM) and Multicycle ECM (MCECM) Algorithm (Meng and Rubin 1993, Winkler 1989a). The difficulty with general EM procedures is that different starting points often yield different limiting solutions. However, if the starting point is relatively close to the solution given by the independent EM algorithm, then the limiting solution is generally unique (Winkler 1992). The independent EM algorithm often provides starting points that are suitable for the general EM algorithm.

Figures 20.1–20.8 illustrate that the automatic EM-based parameter-estimation procedures can yield dramatic improvements. Because there were no available business files for which true matching status was known, files of individuals having name, address, and demographic characteristics such as age, race, and sex were used. Each figure contains a plot of the estimated cumulative distribution curve via equation (20.2) versus the truth that is given by the 45-degree line. Figures 20.1–20.4 for matches and Figures 20.5–20.8 for nonmatches successively display fits according to (1) iterative refinement (e.g., Newcombe 1988, pp. 65–66), (2) three-class, independent EM, (3) three-class, selected interaction EM, and (4) three-class, three-way interaction EM with

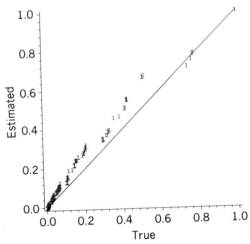

Figure 20.1 Estimates vs. truth, cumulative distribution of matches—two-class, iterative.

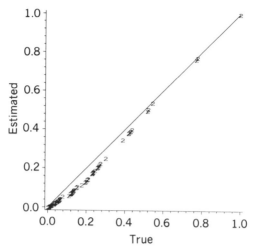

Figure 20.2 Estimates vs. truth, cumulative distribution of matches—three-class, independent EM.

convex constraints. *Iterative refinement* involves the successive manual review of sets of pairs and the reestimation of probabilities given a match under the independence assumption. Iterative refinement is chosen as a reference point (Figures 20.1 and 20.4) because it yields reasonably good matching decision rules (e.g., Newcombe 1988; Winkler 1990b). The algorithm for fitting selected interactions is due to Armstrong (1992). The EM algorithm with convex constraints that predispose a solution to the proper region of the parameter

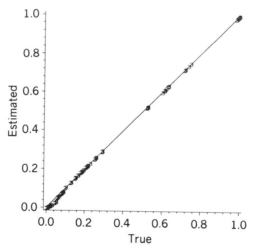

Figure 20.3 Estimates vs. truth, cumulative distribution of matches—three-class, selected interaction EM.

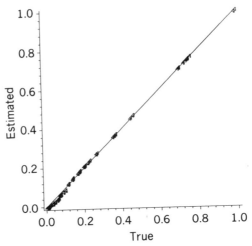

Figure 20.4 Estimates vs. truth, cumulative distribution of matches—three-class, three-way interaction EM, convex.

space is due to Winkler (1989a; also 1992, 1993b). All three-way interactions are used in the last model.

The basic reason that iterative refinement and three-class independent EM perform poorly is that independence does not hold. Three-class independent EM yields results that are closer to the truth because it divides the set of pairs that agree on address into those agreeing on name and demographic information and those that disagree. Thus, nonmatches such as husband–wife and

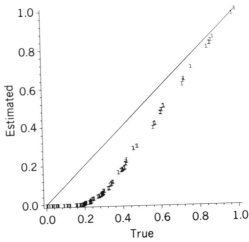

Figure 20.5 Estimates vs. truth, cumulative distribution of nonmatches—two-class, iterative.

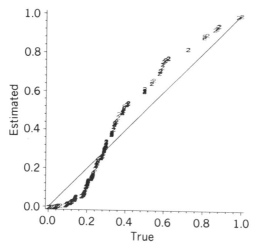

Figure 20.6 Estimates vs. truth, cumulative distribution of nonmatches—three-class, independent EM.

brother–sister pairs are separated from matches such as husband–husband and wife–wife. As shown by Thibaudeau (1993) with these data, departures from independence are moderate among matches whereas departures from independence among nonmatches (such as the husband–wife and brother–sister pairs at the same address) are quite dramatic.

The selected interaction EM does well (Figures 20.3 and 20.7) because true

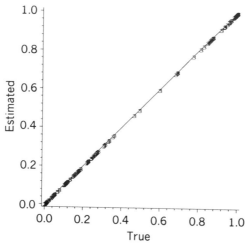

Figure 20.7 Estimates vs. truth, cumulative distribution of nonmatches—three-class, selected interaction EM.

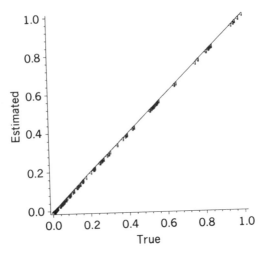

Figure 20.8 Estimates vs. truth, cumulative distribution of nonmatches—three-class, three-way interaction EM, convex.

matching status is used to determine the interactions that must be included. It is unreasonable to expect that true matching status will be available for many matching situations or that the exact set of interactions that were developed for one application will be suitable for use in another. Furthermore, loglinear modeling in latent-class situations is more difficult than for basic loglinear situations where such modeling is known to be difficult (e.g., Bishop et al. 1975). To alleviate the situation, it may be suitable to take a model having all three-way interactions and use convex constraints that bound some probabilities. The bounds would be based on similar matching situations. The all three-way interaction model without convex constraints does not provide accurate fits (Winkler 1992). If the convex constraints are chosen properly, then the three-way interaction EM with convex constraints provides fits (Figures 20.4 and 20.8) that are nearly as good as those obtained with the selected interaction EM (Winkler 1993b).

20.5 EVALUATING THE QUALITY OF LISTS

The quality of lists is primarily determined by how useful the available variables are for matching. For large files, the first concern is how effective common identifiers (blocking criteria) are at reducing the set of pairs to a manageable size. The effectiveness of blocking criteria is also determined by the estimated number of missed matches. Applying a greater number of matching variables generally improves matching efficacy. Name information generally provides more distinguishing power than receipts, sales, or address information. Parameter estimates must be as good as possible. Improving parameter estimates can reduce clerical review regions by as much as 90 percent.

20.5.1 Quality of Blocking Criteria

While use of blocking criteria facilitates the matching process by reducing the number of pairs to be considered, it can increase the number of false nonmatches because some pairs disagree on the blocking criteria. The following describes an investigation of how well different sets of blocking criteria yield sets of pairs containing all matches (Winkler 1984, 1985b). The sets of pairs were constructed from 11 U.S. Energy Information Administration (EIA) lists and 47 state and industry lists containing 176,000 records. Within the set of pairs from the original set of files, name and address information allowed 110,000 matches to be identified. From the remaining 66,000 records, there were 3050 matches having similar names and addresses and 8510 matches having either a different name or a different address. The remaining 11,560 matches (18 percent of the 66,000 records) were identified via intensive manual review and were used in analyzing various blocking criteria.

In the subsequent analysis, only the 3050 matches having similar names and addresses were considered. In the blocking criteria displayed in Table 20.6, NAME represents an unparsed name field. Only the first few characters from different fields were used. These criteria were the best subset of several hundred criteria that were considered for blocking a list of sellers of petroleum products (Winkler 1984). Table 20.7 illustrates that for certain sets of lists it is quite

Table 20.6 Blocking Criteria

1. 3 digits ZIP code, 4 characters NAME
2. 5 digits ZIP code, 6 characters STREET
3. 10 digits TELEPHONE
4. 3 digits ZIP code, 4 characters of largest substring in NAME
5. 10 characters NAME

Table 20.7 Incremental Decrease in False Nonmatches—Each Set Consists of Pairs in the Union of Sets Agreeing on Blocking Criteria

Group of Criteria	Rate of False Nonmatches	Matches/ Incremental Increase	Nonmatches/ Incremental Increase
1	45.5	1460/NA	727/NA
1–2	15.1	2495/1035	1109/289
1–3	3.7	2908/413	1233/124
1–4	1.3	2991/83	1494/261
1–5	0.7	3007/16	5857/4363

difficult to produce groups of blocking criteria that give a set of pairs that include all matches. With the union of pairs based on the best two sets of criteria, 15.1 percent of the matches were dropped from further consideration; with three, 3.7 percent. The last (fifth) criterion was not useful because it enlarged the set of pairs with only 16 additional matches while adding 4363 nonmatches.

20.5.2 Estimation of False Nonmatches Not Agreeing on Multiple Blocking Criteria

If estimates of the numbers of missed matches are needed, then lists can be sampled directly. Even with very large sample sizes, the estimated standard deviation of the error rate estimate often exceeds the estimate (Deming and Gleser 1959). If samples are not used, then following the suggestion of Scheuren (1983), capture-recapture techniques as in Sekar and Deming (1949; see also Bishop et al. 1975, Chapter 6) can be applied to the set of pairs captured by the first four sets of blocking criteria of Section 20.5.1 (Winkler 1987). The best-fitting loglinear model yields the 95 percent confidence interval (27,160). The interval, which represents between 1 and 5 percent of true matches, contains the 50 matches that were known to be missed by the blocking criteria and found via intense clerical review.

20.5.3 Number of Matching Variables

As the number of matching variables increases, the ability to distinguish matches usually increases. For instance, with name information alone, it may only be feasible to create subsets of pairs that are held for clerical review. With name and address information, a substantial number of the matches can be correctly distinguished. With name, address, and financial information (such as receipts or income), it may be possible to distinguish most matches automatically.

Exceptions occur if some matching variables have extreme typographical variations and/or are correlated with other matching variables. For instance, consider the following. Two name fields are available for each record of the pairs. The first is a general business name that typically agrees among matches. The second name field in one record corresponds to the owner of a particular business license (e.g., in some states, all fuel storage facilities must be licensed) and in the other record the name field corresponds to the accounting entity that keeps financial records. While the owner of a particular business license will sometimes correspond to the financial person (owner of a gasoline service station), the two names will often disagree among true matches. When both name fields are used in software that assumes that agreements are uncorrelated, contradictory information can cause loss of distinguishing power. Expedient solutions are to drop the contradictory information in the second name field or to alleviate the problem via custom software modifications.

Table 20.8 Examples of Agricultural Names

John A Smith
John A and Mary B Smith
John A Smith and Robert Jones
Smith Dairy Farm

20.5.4 Relative Distinguishing Power of Matching Variables

Without a unique identifier such as a verified employer identification number (EIN), the name field typically has more distinguishing power than other fields such as address. The ability of name information to distinguish pairs can vary dramatically from one set of pairs to another. For instance, in one situation properly parsed name information, when combined with other information, may produce good automatic decision rules; in other situations it may not.

As an example of the first situation, consider the 1992 U.S. Census of Agriculture in which name parsing software was optimized to try to find surnames (or suitable surrogates) and first names. Because the overwhelming majority of farming operations have names of the form given in Table 20.8, the resultant parsed names will likely all have "Smith" as a surname that will yield good distinguishing power when combined with address information. The exception can occur when two names containing "Smith" have the same address. A similar situation occurs with the 1992 match of the Standard Statistical Establishment List (SSEL) of U.S. businesses with a list of small nonemployers from an Internal Revenue Service (IRS) 1040C file of records for which EIN was unavailable.

General business lists can signify the second situation of the poor decision rule because of the ways in which the name field can be represented. For instance, the same business entity may appear in the following forms given in Table 20.9. Even if name parsing software can properly represent name components, it may be difficult to use the components to distinguish matches. If the name information and clerical-review status were retained, then clerical review could be reduced during future updates. Each business could be represented by a unique record that has pointers to significant name variations of matches and nonmatches along with match status. If a potential update record

Table 20.9 Examples of Business Names That Are Difficult to Compare

John A Smith and Son Manufacturing Company, Incorporated
John Smith Co
John Smith Manufacturing
J A S Inc.
John Smith and Son

is initially designated as a possible link because of a name variation, then the associated name variations could be searched to decide whether a record with a name similar to the potential update record had previously been clerically reviewed. If it had, then the prior follow-up results could be used to determine whether the new record is a match.

20.5.5 Good Matching Variables But Unsuitable Parameter Estimates

Even when name and other matching variables can be properly parsed and have agreeing components, automatic parameter estimation software may not yield good parameter estimates because the lists have little overlap or because model assumptions in the parameter-estimation software are incorrect. In either situation, matching parameters are usually estimated via an iterative procedure involving manual review. Generally, matching personnel start with an initial set of parameters. The personnel review a moderately large sample of matching results and estimate new parameters via ad hoc means. The review-reestimation process is repeated until matching personnel are satisfied that parameters and matching results will not improve much.

The most straightforward means of parameter reestimation is the iterative refinement procedure of Statistics Canada (e.g., Newcombe 1988, pp. 65–66; Statistics Canada 1983; Jaro 1992). After each review and clerical resolution of match results, marginal probabilities given a match are reestimated and matching (under the independence assumption) is repeated. Marginal probabilities given a nonmatch are held as constant because they are approximated by probabilities of random agreement over the entire set of pairs. If the proportion of nonmatches within the set of pairs is very high, then the random-agreement approximation is valid because decision rules using the random agreement probabilities are virtually the same as decision rules using true marginal probabilities given a nonmatch.

For the 1992 U.S. Census of Agriculture, initial estimates obtained via the independent EM algorithm were replaced by refined estimates that accounted for lack of independence. The refined estimates were determined by reviewing a large sample of pairs, creating adjusted probability estimates, and repeating the process. For instance, if two records simultaneously agreed on surname and first name, their matching weight was adjusted upward from the independent weight.

20.6 ESTIMATION OF ERROR RATES AND ADJUSTMENT FOR MATCHING ERROR

Fellegi and Sunter (1969) introduced methods for automatically estimating error rates when the conditional independence assumption (20.4) is valid. Their methods do not involve sampling and can be extended to more general situations. This section provides different methods for estimating error rates within a set of pairs than those given in Section 20.4.6. Estimation of false non-matches due to pairs missed because of disagreement on blocking criteria is covered in Section 20.5. This section also describes new work that investigates how statistical analyses can be adjusted for matching error.

20.6.1 Sampling and Clerical Review

Estimates of the number of false matches and nonmatches can be obtained by reviewing a sample of pairs designated as links and nonlinks. Sample size can be minimized by concentrating the sample in weight ranges in which error is likely to take place. Using a weighting strategy that yields good distinguishing power with rule (20.2), most error among computer-designated links and non-links occurs among weights that are close to the thresholds *UPPER* and *LOWER*. Within the set of possible links that are clerically designated as links and nonlinks, simple random samples can be used. While the amount of manual review needed for confirming or correcting the link–nonlink designations can require substantial resources, reasonable estimates within the fixed set of pairs can be obtained. An alternative to sampling is to develop effective statistical models that allow automatic estimation of error rates. At present, such methods are the subject of much research and should show improvements in the future.

20.6.2 Rubin–Belin Estimation

Rubin and Belin (1991) developed a method of estimating matching error rates when the curves (ratio R versus frequency) for matches and nonmatches are somewhat separated and the failure of the independence assumption is not too severe. Their method is applicable to weighting curves R obtained via a one-to-one matching rule (Jaro 1989) and to which a number of ad hoc adjustments are made (Winkler 1990b). The one-to-one matching rule can dramatically improve matching performance because it can eliminate nonmatches such as husband–wife or brother–sister pairs that agree on address information. Without one-to-one matching, such pairs receive sufficiently high weights to be designated as possible links.

To model the shape of the curves of matches and nonmatches, Rubin and Belin require true matching status for a representative set of pairs. For a variety of basic settings, the procedure yields reasonably accurate estimates of error rates and is not highly dependent on *a priori* curve shape parameters (Rubin and Belin 1991; Scheuren and Winkler 1993; Winkler 1992). The SEM algorithm of Meng and Rubin (1991) is used to get 95 percent confidence intervals for the estimates.

While the Rubin–Belin procedures were developed using files of individuals (for which true match status was known), I expect that the procedures are also applicable for files of businesses. When one-to-one matching is used, the Rubin and Belin method can give better error rate estimates than a modified version of the Winkler method given in Section 20.4.6 (e.g., Winkler 1992). If one-to-one matching is not used, then the Winkler method can yield accurate parameter estimates whereas the Rubin–Belin method cannot be applied because the curves associated with matches and nonmatches are not sufficiently separated.

20.6.3 Scheuren–Winkler Adjustment of Statistical Analyses

Linking information that resides in separate files can be useful for analysis and policy decisions. For instance, an economist might wish to evaluate energy policy by matching a file with fuel and commodity information for businesses against a file with the values and types of goods produced by the businesses. If the wrong businesses are matched, then analyses based on the linked files can yield erroneous conclusions. Scheuren and Winkler (1993) introduced a method of adjusting statistical analyses for matching error. If the probability distributions for matches and nonmatches are accurately estimated, then the adjustment method is valid in simple cases where one variable is taken from each file. Accurate estimates can sometimes be obtained via the method of Rubin and Belin (1991). Empirical applications have been performed for ordinary linear regression models (Winkler and Scheuren 1991) and for simple loglinear models (Winkler 1991). Extensions to situations of more than one variable from each file are under investigation.

20.7 COMPUTING RESOURCES AND AUTOMATION

Many large record linkage projects require new software or substantial modification of existing software. The chief difficulty with these projects is developing the highly skilled programmers required for the task. Few programmers have the aptitude or are allowed the years needed to acquire proficiency in advanced algorithm development and the multi-language, multi-machine approaches needed to modify and enhance existing software. For example, a government agency may use software that another agency spent several years developing in PL/I because PL/I is the only language their programmers know. Possibly more appropriate software written in C may not be used because the same programmers do not know how to compile and run C programs. The same PL/I programmers may not have the skills that allow them to make major modifications in PL/I software that they did not write or to port new algorithms in other languages to PL/I.

A secondary concern is lack of appropriate, general-purpose software. In many situations for which name, address, and other comparable information are available, existing matching software will work well if names and addresses can be parsed correctly. Directly comparable information might consist of receipts for comparable time periods. Nondirectly comparable information might consist of receipts in one source and sales in another. To use such data, custom software modifications have to be added to software. The advantage of some existing software is that, without modification, they often parse a substantial percentage of the records in files.

20.7.1 Need for General Name-Parsing Software and What Is Available

At present, the only general-purpose business-name-parsing software that has been used by an assortment of agencies is the NSKGEN software from Statistics Canada. The software is written in a combination of PL/I and IBM Assembly language. NSKGEN software is primarily intended to create search keys that bring appropriate pairs of records together. Because it does a good job of parsing and standardizing names, it has been used for record linkage (Winkler 1986, 1987). I recently wrote general business-name-parsing software that was used in a match of the U.S. SSEL list of business establishments with the U.S. IRS 1040C list that contains many small establishments (Winkler 1993a). The software achieves better than a 99 percent parsing rate with an error rate of less than 0.2 percent with these lists. It has not yet been tested on a variety of general lists. The code is ANSI-standard C and, upon recompilation, runs on a number of computers. While name parsing software is written and used by commercial firms, the associated source code is generally considered proprietary.

20.7.2 Need for General Address-Parsing Software and What Is Available

Statistics Canada has the ASKGEN package (again written in PL/I and IBM Assembly language) which does a good job of parsing addresses (Winkler 1986, 1987). ASKGEN has recently been superseded by Postal Address Analysis System (PAAS) software. PAAS has not yet been used at a variety of agencies but, with limitations, has been used in creating an address register for the 1991 Canadian Census. The limitations were that most of the source address lists required special preprocessing to put individual addresses in a form more suitable for input to PAAS software (Swain et al. 1992). In addition to working on English-type addresses, the ASKGEN and PAAS software works on French-type addresses such as "16 Rue de la Place."

At the U.S. Bureau of the Census, address-parsing software has been written in ANSI-standard C and, upon recompilation, currently runs on an assortment of computers. The software has been incorporated in all major Census Bureau geocoding systems, has been used for the 1992 U.S. Census of Agriculture, and was used in several projects involving the 1992 U.S. SSEL. As with name-parsing software, source code for commercial address-parsing software is generally considered proprietary.

20.7.3 Matching Software

At present, I am unaware of any general software packages that have been specifically developed for matching lists of businesses. While the ASKGEN and NSKGEN standardization packages were used with the Canadian Business Register in 1984, associated matching was based on search keys generated through compression and standardization of corporate names. One-to-many matches were reviewed by clerks who selected the best match with the help of interactive computer software. At the U.S. Bureau of the Census, I have been involved with the development of software for large projects in which the Fellegi–Sunter model was initially used and a number of ad hoc modifications were made to deal with name-parsing failure, address-parsing failure, sparse and missing data, and data situations unique to the files being matched. In every case, the ad hoc modifications improved matching performance substantially over performance that would have been available from the software alone. The recent projects were the 1992 U.S. Census of Agriculture, the 1993 match of the SSEL file of U.S. businesses with the IRS 1040C list of nonemployers, and the 1993 matching of successive years' SSEL files and the unduplication of individual years' files. The latter two projects used files from 1992. A set of software for agricultural lists and several packages for files of individuals are described below.

The U.S. Department of Agriculture (1980) has a system for matching lists of agricultural businesses, which was written in FORTRAN for IBM mainframes in 1979 and has never been updated. Name-parsing software is available as part of the system. The software applies Fellegi–Sunter matching to the subsets of pairs corresponding to individuals. The remaining records that are identified as corresponding to partnerships and corporations are matched clerically when an exact character-by-character match fails. If the pairs of businesses generally have names that allow them to be represented in forms similar to the ways that files of individuals have their names represented, then matching software (or modifications of it) designed for files of individuals can be used.

While the ASKGEN and NSKGEN packages from Statistics Canada have been given out to individuals for use on IBM mainframes, associated documentation does not cover installation or details of the algorithms. To a lesser extent, the lack of detailed documentation is also true for the USDA system. The software packages require systems analysts and matching experts for installation and use.

General matching software has only been used on files of individuals due to the difficulties of name and address standardization and consistency in business files. Available systems are Statistics Canada's GRLS system (Hill 1991, Nuyens 1993), the system for the U.S. Census (Winkler 1990a), Jaro's commercial system (Jaro 1992), and University of California's CAMLIS system. None of the systems provides name- or address-parsing software. Only the Winkler system is free and, upon recompilation, runs on a large collection of computers. Source code is available with the GRLS system and the Winkler system. The GRLS system has the best documentation.

20.8 CONCLUDING REMARKS

This chapter provides background on how the Fellegi–Sunter model of record linkage is used in developing automated matching software for business lists. The presentation shows how a variety of existing techniques have been created to alleviate specific problems due to name- and/or address-parsing failure or inappropriateness of assumptions used in simplifying computation associated with the Fellegi–Sunter model. Much research is needed to improve record linkage of business lists. The challenges facing agencies and individuals are great because substantial time and resources are needed for (1) creating and enhancing general name and address parsing/software; (2) performing, circulating, and publishing methodological studies; and (3) generalizing and adding features to existing matching software that improve its effectiveness when applied to business lists.

REFERENCES

Armstrong, J. A. (1992), "Error Rate Estimation for Record Linkage: Some Recent Developments," in *Proceedings of the Workshop on Statistical Issues in Public Policy Analysis*, Carleton University.

Belin, T. R. (1993), "Evaluation of Sources of Variation in Record Linkage Through a Factorial Experiment," *Survey Methodology*, **19**, pp. 13–29.

Bishop, Y. M. M., S. E. Fienberg, and P. W. Holland (1975), *Discrete Multivariate Analysis*, Cambridge, MA: MIT Press.

Brackstone, G. J. (1987), "Issues in the Use of Administrative Records for Administrative Purposes," *Survey Methodology*, **13**, pp. 29–43.

Cooper, W. S., and M. E. Maron (1978), "Foundations of Probabilistic and Utility-Theoretic Indexing," *Journal of the Association for Computing Machinery*, **25**, pp. 67–80.

Copas, J. R., and F. J. Hilton (1990), "Record Linkage: Statistical Models for Matching Computer Records," *Journal of the Royal Statistical Society*, Series A, **153**, pp. 287–320.

DeGuire, Y. (1988), "Postal Address Analysis," *Survey Methodology*, **14**, pp. 317–325.

Deming, W. E., and G. J. Gleser (1959), "On the Problem of Matching Lists by Samples," *Journal of the American Statistical Association*, **54**, pp. 403–415.

Dempster, A. P., N. M. Laird, and D. B. Rubin (1977), "Maximum Likelihood from Incomplete Data via the EM Algorithm," *Journal of the Royal Statistical Society*, Series B, **39**, pp. 1–38.

Federal Committee on Statistical Methodology (1980), *Report on Exact and Statistical*

Matching Techniques, Statistical Policy Working Paper 5, Washington, DC: U.S. Office of Management and Budget.

Fellegi, I. P., and A. B. Sunter (1969), "A Theory for Record Linkage," *Journal of the American Statistical Association*, **64**, pp. 1183–1210.

Haberman, S. J. (1975), "Iterative Scaling for Log-Linear Model for Frequency Tables Derived by Indirect Observation," *Proceedings of the Statistical Computing Section, American Statistical Association*, pp. 45–50.

Haberman, S. (1979), *Analysis of Qualitative Data*, New York: Academic Press.

Hill, T. (1991), "GRLS-V2, Release of 22 May 1991," unpublished report, Ottawa: Statistics Canada.

Hogg, R. V., and A. T. Craig (1978), *Introduction to Mathematical Statistics*, 4th ed., New York: Wiley.

Jaro, M. A. (1989), "Advances in Record-Linkage Methodology as Applied to Matching the 1985 Census of Tampa, Florida," *Journal of the American Statistical Association*, **89**, pp. 414–420.

Jaro, M. A. (1992), "AUTOMATCH Record Linkage System," unpublished, Silver Spring, MD

Meng, X., and D. B. Rubin (1991), "Using EM to Obtain Asymptotic Variance-Covariance Matrices: The SEM Algorithm," *Journal of the American Statistical Association*, **86**, pp. 899–909.

Meng, X., and D. B. Rubin (1993), "Maximum Likelihood via the ECM Algorithm: A General Framework," *Biometrika*, **80**, pp. 267–278.

Neter, J., E. S. Maynes, and R. Ramanathan (1965), "The Effect of Mismatching on the Measurement of Response Errors," *Journal of the American Statistical Association*, **60**, pp. 1005–1027.

Newcombe, H. B. (1988), *Handbook of Record Linkage: Methods for Health and Statistical Studies, Administration, and Business*, Oxford: Oxford University Press.

Newcombe, H. B., J. M. Kennedy, S. J. Axford, and A. P. James (1959), "Automatic Linkage of Vital Records," *Science*, **130**, pp. 954–959.

Nuyens, C. (1993), "Generalized Record Linkage at Statistics Canada," *Proceedings of the International Conference on Establishment Surveys*, Alexandria, VA: American Statistical Association, pp. 926–930.

Rogot, E., P. Sorlie, and N. Johnson (1986), "Probabilistic Methods of Matching Census Samples to the National Death Index," *Journal of Chronic Disease*, **39**, pp. 719–734.

Rubin, D. B., and T. R. Belin (1991), "Recent Developments in Calibrating Error Rates for Computer Matching," *Proceedings of the Annual Research Conference*, Washington, DC: U.S. Bureau of the Census, pp. 657–668.

Scheuren, F. (1983), "Design and Estimation for Large Federal Surveys Using Administrative Records," *Proceedings of the Survey Research Methods Section, American Statistical Association*, pp. 377–381.

Scheuren, F., and W. E. Winkler (1993), "Regression Analysis of Data Files That Are Computer Matched," *Survey Methodology*, **19**, pp. 39–58.

Sekar, C. C., and W. E. Deming (1949), "On a Method of Estimating Birth and Death

Rates and the Extent of Registration,'' *Journal of the American Statistical Association*, **44**, pp. 101–115.

Smith, M. E., and H. B. Newcombe (1975), ''Methods of Computer Linkage of Hospital Admission-Separation Records into Cumulative Health Histories,'' *Methods of Information in Medicine*, **14**, pp. 118–125.

Statistics Canada (1983), ''Generalized Iterative Record Linkage System,'' unpublished report, Ottawa: Systems Development Division.

Swain, L., J. D. Drew, B. LaFrance, and K. Lance (1992), ''The Creation of a Residential Address Register for Coverage Improvement in the 1991 Canadian Census,'' *Survey Methodology*, **18**, pp. 127–141.

Thibaudeau, Y. (1989), ''Fitting Log-Linear Models When Some Dichotomous Variables Are Unobservable,'' *Proceedings of the Section on Statistical Computing, American Statistical Association*, pp. 283–288.

Thibaudeau, Y. (1993), ''The Discrimination Power of Dependency Structures in Record Linkage,'' *Survey Methodology*, **19**, pp. 31–38.

Titterington, D. M., A. F. M. Smith, and U. E. Makov (1988), *Statistical Analysis of Finite Mixture Distributions*, New York: Wiley.

U.S. Department of Agriculture (1980), ''Record Linkage System Documentation,'' unpublished report, Washington, DC: National Agricultural Statistics Service.

Van Rijsbergen, C. J., D. J. Harper, and M. F. Porter (1981), ''The Selection of Good Search Terms,'' *Information Processing and Management*, **17**, pp. 77–91.

Winkler, W. E. (1984), ''Exact Matching Using Elementary Techniques,'' technical report, Washington DC: U.S. Energy Information Administration.

Winkler, W. E. (1985a), ''Preprocessing of Lists and String Comparison,'' in W. Alvey and B. Kilss (eds.), *Record Linkage Techniques—1985*, U.S. Internal Revenue Service, Publication 1299 (2-86), pp. 181–187.

Winkler, W. E. (1985b), ''Exact Matching Lists of Businesses: Blocking, Subfield Identification, Information Theory,'' in W. Alvey and B. Kilss (eds.), *Record Linkage Techniques—1985*, U.S. Internal Revenue Service, Publication 1299 (2-86), pp. 227–241.

Winkler, W. E. (1986), ''Record Linkage of Business Lists,'' technical report, Washington, DC: U.S. Energy Information Administration.

Winkler, W. E. (1987), ''An Application of the Fellegi-Sunter Model of Record Linkage to Business Lists,'' technical report, Washington, DC: U.S. Energy Information Administration.

Winkler, W. E. (1988), ''Using the EM Algorithm for Weight Computation in the Fellegi-Sunter Model of Record Linkage,'' *Proceedings of the Survey Research Methods Section, American Statistical Association*, pp. 667–671.

Winkler, W. E. (1989a), ''Near Automatic Weight Computation in the Fellegi–Sunter Model of Record Linkage,'' *Proceedings of the Annual Research Conference*, Washington, DC: U.S. Bureau of the Census, pp. 145–155.

Winkler, W. E. (1989b), ''Methods for Adjusting for Lack of Independence in an Application of the Fellegi-Sunter Model of Record Linkage,'' *Survey Methodology*, **15**, pp. 101–117.

Winkler, W. E. (1989c), ''Frequency-Based Matching in the Fellegi–Sunter Model of

Record Linkage," *Proceedings of the Survey Research Methods Section, American Statistical Association*, pp. 778–783.

Winkler, W. E. (1990a), "Documentation of Record-Linkage Software," unpublished report, Washington, DC: U.S. Bureau of the Census.

Winkler, W. E. (1990b), "String Comparator Metrics and Enhanced Decision Rules in the Fellegi–Sunter Model of Record Linkage," *Proceedings of the Survey Research Methods Section, American Statistical Association*, pp. 354–359.

Winkler, W. E. (1991), "Error Model for Analysis of Computer Linked Files," *Proceedings of the Survey Research Methods Section, American Statistical Association*, pp. 472–477.

Winkler, W. E. (1992), "Comparative Analysis of Record Linkage Decision Rules," *Proceedings of the Survey Research Methods Section, American Statistical Association*, pp. 829–834.

Winkler, W. E. (1993a), "Business Name Parsing and Standardization Software," unpublished report, Washington, DC: U.S. Bureau of the Census.

Winkler, W. E. (1993b), "Improved Decision Rules in the Fellegi-Sunter Model of Record Linkage," *Proceedings of the Survey Research Methods Section, American Statistical Association*, pp. 274–279.

Winkler, W. E., and F. Scheuren (1991), "How Matching Error Affects Regression Analysis: Exploratory and Confirmatory Results," technical report, Washington, DC: U.S. Bureau of the Census.

Wu, C. F. J. (1983), "On the Convergence Properties of the EM Algorithm," *Annals of Statistics*, **11**, pp. 95–103.

Yu, C. T., K. Lam, and G. Salton (1982), "Term Weighting in Information Retrieval Using the Term Precision Model," *Journal of the Association for Computing Machinery*, **29**, pp. 152–170.

CHAPTER TWENTY-ONE

Improving the Traditional Editing Process

Leopold Granquist[1]
Statistics Sweden

Errors in survey data distort estimates, complicate further processing, and de-crease user confidence. *Editing* is the procedure for detecting and adjusting individual errors in data records resulting from data collection and capture. The checks for identifying missing, erroneous, or suspicious values in com-puter-assisted editing are called *edit rules* or simply *edits*. *Imputation* refers to the replacement of missing data using logical edits or statistical procedures (see Chapter 22). *Editing change* is used throughout the chapter to refer to the situation where an item (question) value is adjusted as a consequence of action taken when an error is identified.

Nowadays, editing (and imputation) is a computer-assisted procedure. The editing potential of computers was quickly recognized by statistical agencies. The early development of methods for automatic editing and imputation is described by Nordbotten (1963), who gives an excellent overview of many automatic editing and imputation methods. The stated aim of computerizing the editing process was to reach higher quality at lower costs, with a focus on detecting "possible errors" that could not be identified by manual editing. Subject-matter experts checked what could be verified by various edits with rather narrow bounds. More errors were detected at the price of an increasing number of unnecessary flags. The edits brought no reduction in costs and no definitive improvement in quality.

[1]The author gratefully acknowledges the helpful suggestions of the following individuals in im-proving the presentation of this chapter: Brenda G. Cox, the section editor; Anders Christianson, Chris Denell, and Lars Lyberg of Statistics Sweden; John Kovar of Statistics Canada; and Mark Pierzchala of the U.S. National Agricultural Statistics Service.

Business Survey Methods, Edited by Cox, Binder, Chinnappa, Christianson, Colledge, Kott.
ISBN 0-471-59852-6 © 1995 John Wiley & Sons, Inc.

When improving quality, crucial items to evaluate are the design of the edits and the control of the manual review. When these processes are not properly designed, an edit bias and an editor bias may be introduced. This chapter focuses on improving the efficiency of the editing process in identifying suspicious data. It also discusses basic issues in designing editing processes and offers suggestions for improving data quality in business surveys. For a discussion of editing systems and imputation methods, see Chapters 22 and 23. Editing rules per se are not treated in this chapter because they are targeted on the particular error types affecting the items of the individual survey questionnaire. Every survey is unique in this respect.

21.1 WHAT IS EDITING?

In editing systems, error detection is automatic; data identified as erroneous may be referred to subject-matter staff or may be automatically imputed. According to the Federal Committee on Statistical Methodology (FCSM 1990, p. 2), at least 60 percent of all editing systems in the United States can be characterized as "automated edit checking, all error correction done by analysts/clerks." For business surveys in particular, almost all editing systems forward data records that fail edit to subject-matter staff. That is why I refer to this type of system as a "traditional editing" system. In this section, traditional editing systems are described to introduce editing concepts and to serve as a basis for discussing the editing of business survey data.

21.1.1 The Traditional Editing Process

In traditional editing processes, erroneous or suspected data are identified by computer software based upon edits provided by subject-matter experts. The software produces error messages, which are reviewed by clerks who enter new values when they detect errors. Schematically, the process may be divided into three subprocesses: input editing, machine editing, and output editing.

Input editing covers all editing activities until the data are keyed or captured. When *manual editing* is used before data capture, clerks review the hard-copy questionnaires, code the responses manually, ensure that the unit is in scope for the survey, and verify that the questionnaire is filled out properly. The data are then captured without further editing; this is referred to as "data entry heads down." When *data entry editing* or *integrated editing* is used, the responses are coded and edited when captured; this is referred to as "data entry heads up," which means that the data entry computer edits the answers online. The editor, interviewer, or respondent immediately reviews data that fail edits, and he/she changes responses to get the record accepted or filed for further expert review.

In *machine editing*, the captured data are machine-reviewed by an error-detecting program which is again provided with checks by subject-matter experts; this usually produces error messages in a batch mode. This process is typically followed by a manual review of the messages. The reviewer has to check the questionnaire and often must recontact the respondent to correct the data. As a result, filing and retrieval of questionnaires can become an important and cumbersome subprocess. Changes are entered in batch or on-line, and the changed records are reviewed, usually by the same error-detecting program. In *batch updating*, the changed records are checked in batch mode; when a questionnaire fails an edit, it usually has to be manually reviewed once again, a process called *cycling*. It is not unusual for the average number of cycles per record to exceed two. A record may pass up to 20 times through the machine (Boucher 1991). Different editors may be involved, and several contacts with the respondent may occur. In *on-line updating*, changes are entered on-line; when the record fails an edit, it is reviewed and adjusted by the same editor until it is accepted.

The process known as *macroediting, output editing,* or *output review* is carried out on aggregates of all collected and processed questionnaires to ensure that no serious errors are left in the data. Output review edits are designed to detect potential outliers or substantial errors remaining in individual records. Granquist (1984a) and Chinnappa et al. (1990) suggest that macroediting should be used for finding error sources, systematic measurement errors, or potential problems in the data collection process. Because it is part of data analysis, it is carried out by subject-matter analysts.

21.1.2 Editing Business Surveys

In most countries, business surveys are periodic mail surveys collecting mostly quantitative data (see Chapter 14). Survey designers expect respondents to be knowledgeable and to understand the concepts underlying survey questions. Historic data about the business are often available to the statistical agency from a previous survey or from administrative data sources. Responses to items of interest often have highly *skewed distributions*, which means that a small number of units contribute substantial amounts to the total estimate. Strong correlations usually exist between current and previously reported values of the same item and between many pairs of items. Practically all respondents in business surveys the world over obtain their data directly from accounting records (Linacre 1991, Werking et al. 1988, Corby 1984). The telephone is used for follow-up actions to such an extent that mail surveys include considerable telephone data collection (Linacre 1991, Boucher 1991, Hedlin 1993).

21.1.3 Errors and Edits

Edits are often classified into validity, consistency, and statistical edits. Another classification is by fatal edits versus query or suspicious edits. *Fatal edits* identify data that are clearly erroneous, whereas *query edits* point to data that have a high probability of being erroneous. In business surveys, failed edits very often require recontacts with respondents.

Validity checks (range checks) and consistency checks are generally used to detect fatal errors. Even when fatal errors have only a minor impact on the accuracy of aggregates, the editing process should remove them. Otherwise, users may lose confidence in the data. Fatal errors are easy to detect but not always easy to correct. For example, missing values due to lack of reporting capacity or to difficulties in understanding the questions may cause serious problems for the editor.

Query edits are based upon relationships between two or more questionnaire items or from one collection period to another. Values of an item or function of items outside an acceptance region are flagged as *suspicious data* for further manual review. The acceptance region, usually an interval, is very often set in advance on rather subjective grounds and applied to the whole data set. A very common check is the *ratio check*, $a < X/Y < b$, where X is an item of the current period and Y is another item or item's value in a previous period. When the limits a and b are derived from a statistical analysis, the edits are called *statistical edits*.

21.1.4 Definitions of Editing

At present, there is no generally adopted or accepted formal definition of editing. Operational or formal definitions or indicated aims of editing processes, systems, procedures, and so forth, indicate different goals of the editing operation ranging from "· · · conforming data to the desired format or units" (Pritzker et al. 1965, p. 443) to "· · · verifying the quality of the figures and signaling whether coverage is drifting, definitions are no longer relevant, etc." (Chinnappa et al. 1990, §6.1). In generalized editing and imputation systems based on the Fellegi–Holt methodology, the goal of editing is limited to removing fatal errors (Fellegi and Holt 1976). For this chapter, the most appropriate definition of editing is given by the FCSM (1990, p. 7): "Procedure(s) designed and used for detecting erroneous and/or questionable survey data (survey response data or identification type data) with the goal of correcting (manually and/or via electronic means) as much of the erroneous data (not necessarily all of the questioned data) as possible, usually prior to data imputation and summary procedures."

21.2 PROBLEMS WITH EDITING

Editing has always been expensive. In spite of efforts to rationalize the processes, computers have not decreased the cost of editing. Computers are frequently used to apply more edits than data editors could apply. These additional edits have resulted in an increasing number of questionnaires needing manual review. This has caused *overediting*; that is, the share of resources and time dedicated to editing is not justified by the resulting improvements in data quality.

Three types of costs are connected with editing: producer costs, respondent costs, and losses in timeliness. In the United States, Pritzker et al. (1965, p. 447) notes that "Rough estimates of editing and correction or imputation costs as percent of total costs for the 1963 Manufactures Census, the 1963 Business Census, and the Export Statistics are 18, 18, and 16 percent respectively." At Statistics Sweden, the mean editing cost of all surveys as a percentage of the total survey cost was 40 percent in 1975 (Granquist 1982). The FCSM (1990) found that in the United States the median editing cost as a percentage of total survey costs was also 40 percent for business surveys. The FCSM also noted that there is an opportunity cost of editing: The time spent on editing leaves less time for other tasks, such as converting nonrespondents. Linacre and Trewin (1989) also raised this issue in discussing methods to more rationally allocate resources to data collection and processing operations.

The high query rates of editing processes cause considerable costs to respondents as well, who have to discuss questioned data with the editor and reconcile problematic responses. Respondent costs are of the same order as the producer query costs. A bad will cost is also involved, which can be high when many queries leave data unchanged or only insignificantly changed.

The cited references clearly indicate that business surveys have heavy editing costs (30 to 40 percent of total costs). The aim of the following review of published evaluations of editing systems and methods is to show that (1) editing processes commonly used in business surveys do not produce a corresponding increase in data quality, (2) business survey data are often overedited, and (3) certain kinds of errors may be unaffected by the editing.

Often, much of data editing's impact is attributable to only a small percentage of the total edits. In evaluating the editing of the 1982 U.S. Economic Censuses, Greenberg and Petkunas (1986) found that approximately 5 percent of the editing changes accounted for over 90 percent of the total change in survey estimates; the graphs of the changes of the economic items were almost identical for the first 5 percent versus all editing changes. In her study of Sweden's Annual Survey of Financial Accounts, Wahlström (1990) found that if only the 2 percent largest changes had been made in editing value added, the resulting estimate would be 98 percent of the published total. Hedlin (1993) found that the largest 8 percent of changes accounted for 95 percent of the total change in turnover for Sweden's Annual Survey of Manufacturing. In his study of the impact of microediting for Statistics Canada's Annual Survey of Manufactures, Boucher (1991) found that when 50 percent of the largest changes have been made for shipments, nothing more was to be gained on the total level.

Furthermore, most clerical actions caused by query edits leave the original values unchanged. For example, the *hit rate*, defined as the ratio of the number of editing changes to the number of flags, was found to be 23 percent for query edits in the Australian Retail Census (Australian Bureau of Statistics 1987). Hit rates between 28 and 47 percent were reported by Lindström (1991). The low hit rate of range checks with predetermined bounds prompted the U.S. Department of Agriculture to develop more efficient edits for its Livestock Slaughter Survey. Editing costs were reduced by approximately 75 percent (Mazur 1990).

Editing may also hide serious data collection problems and give a false impression of respondents' reporting capacity. The following two examples are given by the Australian Bureau of Statistics (1987). In reviewing the purchase section of the Australian Retail Census, manual editors systematically deducted a small amount from "purchase of goods" and inserted it in the previously vacant "purchases of wrapping and packaging materials" to avoid reviewing another error message from subsequent machine editing. This process is referred to as *creative editing* because the manual editors invent the editing procedures. Creative editing also occurred for manufactured commodities. The classification was too detailed for respondents, who collapsed items and reported them on the most relevant commodity items. Manual editors compared the current responses with data reported in the previous period, and then they deducted amounts from the reported values and allocated them to commodities for which no current values were reported.

Survey designers or managers are inclined to believe that the tighter the acceptance bounds of query edits, the better the resulting data quality. Indications that this assumption does not always hold true were detected as early as 1965. Using data from the 1962 Norwegian Annual Survey of Manufacturing and Mining with artificially generated errors, Nordbotten (1965) implemented a series of simulation studies of an experimental automatic editing and imputation system. Values flagged by ratio edits were automatically imputed by a hot deck imputation method. Nordbotten found that wider limits yielded higher quality because too narrow limits incorrectly identified too much valid data as erroneous and then replaced these data with erroneous values. Even those broader limits were considered too narrow by Fellegi (1965) in his discussion of Nordbotten's simulation studies.

Corby (1984) provides another example of an editing process yielding lower quality in survey estimates. Her paper is an admirable, detailed report of a reinterview evaluation study of the 1977 U.S. Economic Censuses. In this study, interviewers visited a sample of businesses that responded to the census. By reviewing their record-keeping system, the interviewers collected accurate data, identified errors in reported data, and found out whether correct data could have been provided. For each census item, estimated totals were compared for reinterview data, tabulation data, and reported data. Corby found that census editing had changed the reported data (1) too far but in the right direction for four questions, (2) in the right direction but not sufficiently for six questions, and (3) in the wrong direction for six questions. Furthermore, many errors canceled each other out. The conclusion I draw from that observation is that when reported data contain both negative and positive errors, the edits have to address both types of errors equally. Otherwise, the edits will introduce bias, irrespective of the skill of editors in finding accurate data to replace the flagged data.

Range checks cannot detect small errors and systematic errors consistently reported in repetitive surveys, irrespective of how narrow the bounds are. Such errors occur whenever discrepancies occur between the survey definitions and the definitions used in the firm's accounting system. Werking et al. (1988, p. 353) present the following example: "A range check for average hourly earnings for production workers in a particular industry would most likely not detect the exclusion of both overtime hours and pay." Werking et al. report on a continuous response analysis survey (RAS) designed to focus on differences in definitions and how to get firms to apply the survey's definitions. In this study, production worker earnings were found to be significantly affected by nonapplication of the survey's definition. When changed to the survey's definition, the estimate for RAS units became 10.7 (standard error 3.2) as compared to 1.6 for the control group.

21.3 POSSIBLE SOLUTIONS TO COMMON PROBLEMS

The cost–benefit efficiency of traditional editing processes should be of major concern to survey managers. It is imperative that the sensitivity of commonly used query edits be improved (1) to detect significant errors, (2) to develop new edits targeted on serious data errors, (3) to find new ways of detecting errors that cannot be identified by edits, and (4) to improve the questionnaire to prevent errors from arising. In reviewing evaluation reports from different statistical agencies all over the world, I have found that these problems affect most editing processes of business surveys. The aim of this section is to describe methods for evaluating present processes and edits, to suggest how to limit manual editing to records that have a significant impact on estimates, and to emphasize the need for improving questionnaire design.

21.3.1 Evaluation Methods

The first and easiest way of obtaining data on problem areas in collecting and processing survey data is to analyze the changes made to the original keyed-in data. All that is necessary is to (1) key the original responses, (2) save the captured data in a file, and (3) construct a change file by matching the captured data with the edited, cleaned file of finally tabulated data for each question. By ordering the changes by descending absolute magnitude, the cumulative impact of the editing changes to the total editing change or to the estimated item total can be displayed in graphs and tables (see Greenberg and Petkunas 1986). Boucher (1991) and Hedlin (1993) demonstrate the usefulness of this technique for identifying problem areas such as questionnaire problems, missing values, and respondent errors.

When forms are reviewed before data capture or when general editing changes are noted on questionnaires, evaluation studies can be carried out by selecting a sample of forms and analyzing the effect of the editing procedures on individual data items. When the analysis is carried out by an expert, it is possible to obtain detailed information on respondent errors, hit rates of query edits, and, above all, the likely accuracy of the editing changes (Linacre and Trewin 1989).

A widely used technique of evaluating new editing methods is to simulate the new process using a raw data file from the survey. By replacing values flagged according to the new method with the values from the tabulation file containing the data edited by the alternative editing process, it becomes possible to compare estimates from this newly edited file with estimates from the tabulation file (Anderson 1989a, Granquist 1991, Latouche and Berthelot 1992). This easy and inexpensive way of testing new methods or of evaluating a set of edits can be interpreted as studying what effect the changes make in the current process and what impact the remaining changes have.

Indirect measures of the quality of the editing process can be obtained by conducting surveys that compare survey item definitions with the record definitions of accounting systems, as done by Werking et al. (1988). The aim of such studies is to obtain indicators of the incoming quality of responses and to determine the extent to which the editing process can identify response errors caused by different definitions. Of course, the best but most expensive way of evaluating an editing process is to conduct a reinterview study as reported by Corby (1984).

21.3.2 Developing Improved Edits

Care must be taken in designing the query edits for a business survey because otherwise the query edits may have a low hit rate and the changes may be insignificant (and hence not cost effective). This section presents methods to increase the efficiency of query edits for individual items. It focuses on providing individual range edits with efficient bounds.

Tuning, Coordinating, and Targeting Edits

To avoid unnecessary flags from query edits in a current editing system, the first step is to evaluate the query edits using the simulation method outlined in Section 21.3.1 and coordinate the query edits for related items. Relaxing acceptance limits is an obvious measure to decrease the risk of incorrectly identifying valid data as outliers or as erroneous. Anderson (1989a) found that extending the acceptance limits of the range ratio and flagging only extreme outliers for manual review brought 75 percent savings in clerical resources without loss of quality. Checks may also be unnecessary in the sense that flagged erroneous data of one check form a subset of the flagged erroneous data of another check for the item. In the revision of edits in the Swedish Monthly Survey on Employment and Wages (SEW), some edits were deleted for that reason, which contributed to a notable reduction in unnecessary flags (Granquist 1990). A factor contributing to the low hit rate may be the interaction between edits when two or more range checks are applied to many items.

For example, in an earnings survey, "wages" may be flagged by the ratio checks "wages/hours worked" and "wages at time t/wages at time $(t - 1)$" when at least one ratio falls outside the acceptance limits. Even with limits extended to be approximately equal to the extreme outlier limits for the checks, SEW simulations indicate that for both "wages" and "hours" it might be sufficient to use only the check "wages/hours."

Another method which will be prominent in the near future is targeting the edits to specific, serious error types for the survey. I have found only one example so far, which is to be expected. Such edits depend heavily on the specific survey environment and on the types of errors affecting the incoming data. In the Livestock Slaughter Survey at the U.S. Department of Agriculture, range edits with predetermined bounds could not identify *inliers*—that is, the same values reported every period. The main goal of the project, reported by Mazur (1990), was to create a statistical edit, unique for each plant, using the plant's historic data to define edit limits. Tukey's biweight (Hoaglin et al. 1983) was chosen and resulted in a 75 percent cost reduction for editing.

Macroediting

The use of output editing methods in machine editing was launched by Statistics Sweden under the name of macroediting (see Granquist 1991); using this approach clerical review can be reduced as much as 35 to 80 percent without losses in timeliness and quality. *Macroediting* is a procedure for identifying suspicious data in individual records by editing according to the potential impact on survey estimates. It is considered to be a statistical way of providing microediting checks with efficient acceptance limits. The acceptance regions are determined solely by the distributions of the edit values, either manually or automatically. The methods bring priority thinking to verification. The principle is to start with the worst observation and stop when further review has no impact. Because the methods are based on distributions of the incoming data, they cannot be applied in data entry editing but only in machine editing as generally used for microedits. This fact has no impact on timeliness because the merit of the method is that only a few data values are identified as outliers. Granquist (1991) illustrates the principles of these macroediting methods.

The *aggregate method* carries out checks, first on aggregates and then on the individual records of suspicious (flagged) aggregates. The acceptance limits are manually set by reviewing lists of sorted edit values (calculated from reported weighted data). The lists of the sorted values can also be used directly as a basis for reviewing observations manually (if identifiers are printed out together with observations). Some obvious improvements are to provide statistics such as the median, quartiles, the range or the interquartile range, and graphs (e.g., box plots) of the distribution of the sorted values. Using box plots for determining acceptance bounds is also suggested by Anderson (1989b).

The *top-down method* sorts the values of the edits (which are functions of the weighted, keyed-in values) and starts the manual review from the top or the bottom of the list and continues until there is no noticeable effect on estimates. Anderson (1989a) claims that the top down method is the most efficient output editing method in use at the Australian Bureau of Statistics.

Hidiroglou and Berthelot (1986) suggested the *statistical edits method* to remedy shortcomings connected with the traditional ratio method. To prevent the bounds from being influenced by single outliers, the robust median, quartiles, and interquartile ranges parameters are used instead of the mean and standard deviation. The acceptance bounds are automatically calculated from the reported data in the statistical edits method and from the weighted reported values in the Swedish version of the method. It is called the *Hidiroglou–Berthelot method* by Granquist (1991) and has been successfully used at Statistics Sweden.

Graphical Methods

Graphical editing has been considered by Statistics Sweden (Granquist 1991) and the Australian Bureau of Statistics (Hughes et al. 1990) and has been put into practice at the U.S. Bureau of Labor Statistics (Esposito et al. 1994) and at Statistics New Zealand (Houston and Bruce 1993). *Graphical editing* is an efficient, powerful, on-line output editing method that can significantly reduce clerical effort without a measurable loss in data quality (Hughes et al. 1990). ARIES has increased productivity by 50 percent in the output editing of the U.S. Current Employment Statistics Survey (Esposito et al. 1994, Esposito et al. 1993).

Graphical methods are user-friendly, parameter-free, easy to understand, and flexible. Using graphics is perhaps the best way of implementing the ideas underlying macroediting methods. The idea is to start with the worst values and stop when further editing has no notable impact on estimates. Here, the reviewer selects the worst cases by looking at the edit values visualized in scatter plots and then interactively retrieves the corresponding record. To guide the reviewer in choosing records for manual review, acceptance regions of any shape can be provided and overlaid on the scatter plot. A change is entered interactively, and statistics on the impact of the change are displayed. The method is also useful as tool to provide query edits with efficient acceptance limits (Hughes et al. 1990) and, consequently, for developing and evaluating edits. See Chapter 23 for further discussion of automated systems for graphical editing.

21.3.3 Improving Questionnaire Design

Linacre and Trewin (1989) present data on fatal errors from a study of the editing process for the Australian Retail Census (see also Australian Bureau of Statistics 1987). Rates for item nonresponse and form/system design errors were both about 30 percent. These error rates are considered representative of business surveys in Australia. Boucher (1991) and Hedlin (1993) found similar error rates for item nonresponse and questionnaire design faults. All three papers state that form design problems are responsible for a significant number of respondent errors and conclude that improving questionnaire design would improve the quality of incoming data. Note that high item nonresponse rates may be an indicator of questionnaire/system design problems as well; respondents may not be able to answer the question because they do not understand it or do not have the required data in their accounting systems.

Nontrivial examples of questionable form design are difficult to find. However, Hedlin (1993) presents the following example for a fairly common question sequence. Business questions often have sections where a total item is subsequently subdivided into a number of detailed, subtotal items. These subtotal items are as important as the total item. The respondent is supposed to provide the detailed data, calculate the sum of the subtotal values, and compare the sum with the value of the total item, which should be available in his or her accounting system. Thus, the respondent can check that he or she has reported correctly. Of course, this sum check is also included in the edits, as in the Swedish Annual Survey of Manufacturing (SASM). Hedlin (1993) found that of all the 26,000 changes made in the 1990 SASM, almost 9000 concerned the cost section of the questionnaire. In 20 cases only, the questionnaire design seemed to have detected partial nonresponse error. Evidently, such a design has to be seriously questioned because it seems to cause many unnecessary edits, and it may give rise to creative editing. Hedlin (1993) and Linacre and Trewin (1989) concluded that most clerical changes could have been avoided by deleting the total item from the questionnaire and letting the data entry computer do the totaling.

21.4 METHODS FOR RATIONALIZING EDITING

In this section, I indicate ways of rationalizing not only the editing process but the whole data collection process by focusing on improving incoming data quality and hence the overall quality of survey data. The following means of improving traditional editing are discussed and advocated: (1) moving most of the editing as close as possible to the collection of data, (2) limiting manual follow-up to those flagged records with the heaviest potential impact on estimates, and (3) applying a total quality management (TQM) approach to data editing.

21.4.1 Moving Editing Closer to Data Collection

Edits targeted on easily detected error types should be moved from machine editing to input editing, as advocated by Granquist (1984a), Linacre (1991), and Latouche and Berthelot (1992). Editing can then be integrated into data capture and coding in an interactive mode. Contact is established with the data source, which in mail surveys is the completed questionnaire. Integrated editing eliminates manual input editing and questionnaire filing and retrieval, giving the following advantages: (1) the coding of answers can be supported and the codes can be immediately checked, (2) keying errors may be detected and corrected at the moment they are committed, (3) survey staff are responsible for the whole processing task and can improve their part of the survey process, and (4) unnecessary edits can be avoided by letting the computer total, convert units, and so on. Note that Blaise, the globally distributed input editing system, was developed by Statistics Netherlands just to accomplish data entry editing (Bethlehem et al. 1989). Statistics Netherlands' business surveys processed with Blaise apply only data entry editing. Compared to traditional machine editing, input editing reduces costs by 20 to 50 percent (Statistics Sweden 1990). For further discussion of integrated editing and Blaise, see Chapter 23.

In computer-assisted telephone interviewing (CATI) and computer-assisted personal interviewing (CAPI), data are captured and checked with the respondent. The edits can then reduce the number of fatal errors committed by the respondent and the interviewer. For example, Tortora (1985) found that CATI reduced fatal errors by 77 percent in a U.S. agricultural survey. In computer-assisted self-interview collection modes, data are checked by the respondent. The design of the edits should then bring the respondent to a deeper understanding of the definitions of the survey questions, thus improving incoming data quality. The respondent should also be able to comment upon his/her answers to avoid future follow-ups and to provide information on problems with his/her reporting capacity or other problems with the survey questionnaire. For further discussion of implications of computer-assisted survey collection methods on incoming data quality, see Weeks (1992).

21.4.2 Identifying Records Requiring Manual Review

Independently of one another, Latouche and Berthelot (1992) and McDavitt et al. (1992) elaborated upon the idea of using a score function to identify records requiring manual review in a traditional editing process. Both reports state that the methods can be extended to other surveys. Kovar (1991) uses the term *selective microediting* for the Latouche–Berthelot method of limiting the manual review of flagged records to the most important ones. McDavitt et al. use the term *significance editing*.

In this process, each failed item is given a score based on (1) the size of the unit, (2) the analysis weight, (3) the size of the suspicious value for the item (in selective editing), (4) the size of the expected value of the failed item (in significance editing), and (5) the relative importance of the flagged item (in selective editing). A global score is then calculated for the record by summing the item scores in the selective editing method and by taking the maximum of the item scores in the significance editing method. Only those records for which the score exceeds a predetermined value are manually reviewed. The advantage of determining the threshold value in advance (based upon past experience) is that editing can be carried out immediately, unit by unit, by the data entry computer. The main difference between the methods is that the score in selective editing depends on the value of the failed item, whereas in significance editing it depends on the expected change that results from reviewing a unit's values. This change is given by the difference between the raw data and an expected value for the data.

For the Australian Weekly Survey, reviewing only 40 percent of the failed records had no impact on estimates at the industry level. Eliminating this review should decrease the necessary worker years by 4 years (McDavitt et al. 1992). The selective editing method was evaluated using data from the Canadian Annual Retail Survey. The experiment indicates that only a third of the records flagged by a typical editing process needs manual review (Latouche and Berthelot 1992).

21.4.3 A Total Quality Management Approach

Survey quality is heavily related to the quality of incoming data. For business surveys, this means that respondents must have an in-depth understanding of the questions, including their underlying concepts and definitions, and also have the relevant data available in their accounting systems. A major issue is to ensure that respondents really report the requested data rather than substituting a similar data item from their various accounting systems. When survey definitions differ from the accounting system's, response errors arise that can seldom be identified by editing (Linacre 1991, Werking et al. 1988). Thus, it is important to get the data right the first time. For correct reporting, respondents have to know the conceptual differences between the survey items and the items in their accounting system and must be able to estimate those items that are lacking or that have different definitions within their system. Accordingly, the survey instrument and the editing process should account for possible conceptual differences between survey items and accounting records. Reviewing definitions and collecting data on these differences should be a principal task of the quality assurance process in business surveys. When contacting respondents regarding suspicious or erroneous data, the aim should be more on finding out why data are suspicious or erroneous rather than on correcting them. This means that training staff to find differences and the reasons for these differences is fundamental for improving incoming data quality.

Granquist (1984b), Ferguson (1989), and the FCSM (1990) emphasize the need for the editing process to collect accounting and auditing information. By capturing raw (unedited) data and keeping track of the changes and the reasons for them (audit trail), it is possible to not only estimate the impact of editing and imputation but also to evaluate the performance of the survey questionnaire. One of the two basic elements in the TQM approach to survey processing is to analyze such data and feed back the findings for improvement in staff training, system design, and form design. The other element is the principle of getting the data right the first time.

Thus, in reviewing conceptual differences in editing as a means of doing it right the first time, it is natural to adopt a TQM approach to editing. The

Australian Bureau of Statistics (ABS) has done so. Their approach is outlined by Linacre (1991), who gives a subordinate quality assurance role to editing, though it plays an important role in furnishing data on the survey process. ABS now devotes considerable efforts to staff training, particularly in telephone interviewing. ABS has also established a questionnaire design unit for business surveys.

21.5 CONCLUDING REMARKS

The traditional editing process (automatic error detection, all error messages reviewed by analysts/clerks) is reviewed in this chapter on the basis of a number of evaluation studies. Most were carried out recently by statistical agencies in Australia, Canada, Europe, and the United States. The efficiency of commonly used error detection methods is emphasized. The main editing problems are identified, and possible solutions are suggested. Considerable resources can be saved without loss of quality or timeliness by improved edits, procedures, and form design. There is scope for further resource savings by refining and combining the suggested methods as experience is gained and by developing new means, methods, and procedures.

Reducing overediting may be beneficial to quality. Resources thus released can then be used for controlling the editing process and for addressing those quality issues that are beyond the reach of editing procedures. The opportunity cost of traditional editing may be high. Pullum et al. (1986, p. 313) noted that "Editing is not properly described as the correction of errors; conversely a good many errors of many kinds will not even be touched by the editing process." For examples, see Christianson's and Tortora's presentation of measurement issues for business surveys (Chapter 14). Research is needed on how to allocate resources within the quality assurance process, where editing is but one component. Furthermore, the commonly used edits can only detect randomly occurring errors, not consistently reported errors, which may have a greater impact on estimates. To detect and to analyze possible error types and to find appropriate edits to handle them is an important task for survey designers.

The proper action to take after the detection of errors or suspicious data is a more difficult problem than detecting outliers. Survey managers are inclined to believe that the editors will always find the true value for flagged data, an assumption that does not hold true. For example, Corby (1984) and Linacre and Trewin (1989) found that many editing changes are incorrect. Editing systems commonly produce so many error messages from query edits that the surveys may be characterized as mail surveys followed by telephone surveys. Thus, the editors also become telephone interviewers, needing adequate training and monitoring as in regular telephone interview surveys. Furthermore, they need a deep knowledge of the subject-matter area and survey concepts to obtain accurate values from the respondents. The machine edits may suffer serious errors from the intervention of clerks who feel it is their duty to ensure that the data pass the machine edits (Boucher 1991). This means that data may be fabricated to bypass the editing system, a problem here termed creative editing. An absolute requirement for editing systems is the collection of indications on limits to reporting capacity and the prevention of creative editing. This may be accomplished by requiring that the exact reason for every data change be registered. Note that the latter rule also applies to manual imputation of missing values, a procedure that should be as monitored and audited as automatic imputation (see Chapter 22). Thus, controlling and monitoring manual review is an important issue in designing efficient editing systems.

Editing done after data collection cannot improve data quality other than by removing certain types of errors. Because data quality for a survey is heavily related to quality of incoming data, editing can contribute to quality if it focuses on improving the quality of incoming data. Consequently, the overall conclusion of this chapter is to advocate the TQM approach to editing and to assert that the main role of editing in business surveys should be to review possible conceptual differences between survey and respondent record items.

REFERENCES

Anderson, K. (1989a), "Output Edit Study, Average Weekly Earnings," unpublished draft, Belconnen: Australian Bureau of Statistics.

Anderson, K. (1989b), "Enhancing Clerical Cost-Effectiveness in the Average Weekly Earnings," Unpublished report, Belconnen: Australian Bureau of Statistics.

Australian Bureau of Statistics (1987), "1985–86 Retail Census Edit Evaluation Study," unpublished report, Belconnen.

Bethlehem, J. G., A. J. Hundepool, M. H. Schuerhoff, and L. F. M. Vermeulen (1989), *BLAISE 2.0 An Introduction*, Voorburg: Statistics Netherlands.

Boucher, L. (1991), "Micro-Editing for the Annual Survey of Manufactures. What Is the Value-Added?," *Proceedings of the Annual Research Conference*, Washington, DC: U.S. Bureau of the Census, pp. 765–781.

Chinnappa, N., R. Collins, J. F. Gosselin, T. S. Murray, and S. Simard (1990), "Macro-Editing at Statistics Canada. A Status Report," unpublished report, Ottawa: Statistics Canada.

Corby, C. (1984), *Content Evaluation of the 1977 Economic Censuses*, SRD Research Report No. CENSUS/SRD/RR-84/29, Washington, DC: U.S. Bureau of the Census.

Esposito, R., J. K. Fox, D. Lin, and K. Tidemann (1994), "ARIES—A Visual Patch in the Investigation of Statistical Data," *Journal of Computational and Graphical Statistics*, **3**, pp. 113–125.

Esposito, R., D.-Y. Lin, and K. Tidemann (1993), "The ARIES Review System in the BLS Current Employment Statistics Program," *Proceedings of the International Conference on Establishment Surveys*, Alexandria, VA: American Statistical Association, pp. 843–847.

Federal Committee on Statistical Methodology (1990), *Data Editing in Federal Statis-*

tical Agencies, Statistical Policy Working Paper 18, Washington, DC: U.S. Office of Management and Budget.

Fellegi, I. P. (1965), "Invited Discussion Paper of the I. S. I. Session on: Automatic Detection and Correction of Errors in Data Processing on Electronic Computers," *Proceedings of the International Statistical Institute Meetings*, pp. 468–470.

Fellegi, I. P., and D. Holt (1976), "A Systematic Approach to Automatic Edit and Imputation," *Journal of the American Statistical Association*, **71**, pp. 17–35.

Ferguson, D. P. (1989), "Review of Methods and Software Used in Data Editing," UNDP/ECE, Statistical Computing Project 2, Data Editing Working Paper No. 33, New York: United Nations.

Granquist, L. (1982), "On Generalized Editing Programs and the Solution of the Data Quality Problems," UNDP/ECE, Statistical Computing Project, Data Editing Joint Group, Working Paper No. 17, New York: United Nations.

Granquist, L. (1984a), "On the Role of Editing," *Statistisk Tidskrift*, **2**, pp. 105–118.

Granquist, L. (1984b), "Data Editing and Its Impact on the Further Processing of Statistical Data," *Proceedings of the Workshop on the SCP, Invited Papers*, UNDP/ECE, Statistical Computing Project: ECE/UNDP/SCP/H.3, pp. 25–45.

Granquist, L. (1990), "Data Editing and Quality," *Revista Brasileira de Estatistica*, **51**, pp. 21–51.

Granquist, L. (1991), "Macro-Editing—A Review of Some Methods for Rationalizing the Editing of Survey Data," *Statistical Journal*, **8**, pp. 137–154.

Greenberg, B., and T. Petkunas (1986), *An Evaluation of Edit and Imputation Procedures Used in the 1982 Economic Censuses in Business Division*, 1982 Economic Censuses and Census of Government Evaluation Studies, Washington, DC: U.S. Bureau of the Census, pp. 85–98.

Hedlin, D. (1993), "A Comparison of Raw and Edited Data of the Manufacturing Survey" unpublished report, Stockholm: Statistics Sweden.

Hidiroglou, M. A., and J. M. Berthelot (1986), "Statistical Editing and Imputation for Periodic Business Surveys," *Survey Methodology*, **12**, pp. 73–83.

Hoaglin, D. C., F. Mosteller, and J. W. Tukey (1983), *Understanding Robust and Exploratory Data Analysis*, New York: Wiley.

Houston, G., and A. G. Bruce (1993), "*gred*: Interactive Graphical Editing for Business Surveys," *Journal of Official Statistics*, **9**, pp. 81–90.

Hughes, P. J., I. McDermid, and S. J. Linacre (1990), "The Use of Graphical Methods in Editing," *Proceedings of the Annual Research Conference*, Washington, DC: U.S. Bureau of the Census, pp. 538–550.

Kovar, J. G. (1991), "The Impact of Selective Editing on Data Quality," Working Paper No. 5 presented at the Conference of European Statisticians, Work Session on Statistical Data Editing, Geneva, Switzerland.

Latouche, M., and J. M. Berthelot (1992), "Use of a Score Function to Prioritize and Limit Recontacts in Editing Business Surveys," *Journal of Official Statistics*, **8**, pp. 389–400.

Linacre, S. J., and D. J. Trewin (1989), "Evaluation of Errors and Appropriate Resource Allocation in Economic Collections," *Proceedings of the Annual Research Conference*, Washington, DC: U.S. Bureau of the Census, pp. 197–209.

Linacre, S. J. (1991), "Approaches to Quality Assurance in ABS Business Surveys,"

Invited Papers Booklet, 48th International Statistical Institute Session, **2,** pp. 297–321.

Lindström, K. (1991), "A Macro-Editing Application Developed for PC-SAS," *Statistical Journal,* **8,** pp. 155–165.

McDavitt, C., D. Lawrence, and K. Farwell (1992), "The AWE Significance Editing Study," unpublished report, Belconnen: Australian Bureau of Statistics.

Mazur, C. (1990), "A Statistical Edit for Livestock Slaughter Data," SRB Research Report No. SRB-90-01, Washington, DC: U.S. National Agricultural Statistics Service.

Nordbotten, S. (1963), "Automatic Editing of Individual Statistical Observations," Conference of European Statisticians Statistical Standards and Studies No. 2, New York: U.N. Statistical Commission for Europe.

Nordbotten, S. (1965), "The Efficiency of Automatic Detection and Correction of Errors in Individual Observations as Compared with Other Means for Improving the Quality of Statistics," *Proceedings of the International Statistical Institute Meetings,* pp. 417–441.

Pritzker, L., J. Ogus, and M. H. Hansen (1965), "Computer Editing Methods—Some Applications and Results," *Proceedings of the International Statistical Institute Meetings,* pp. 442–465.

Pullum, T. W., T. Harpham, and N. Ozsever (1986), "The Machine Editing of Large Sample Surveys: The Experience of the World Fertility Survey," *International Statistical Review,* **54,** pp. 311–326.

Statistics Sweden (1990), *Computer-Assisted Data Collection: Production Test in the Labour Force Surveys August 1989–January 1990,* R&D Report 1990:11, Stockholm: Statistics Sweden.

Tortora, R. D. (1985), "CATI in an Agricultural Statistical Agency," *Journal of Official Statistics,* **1,** pp. 301–314.

Wahlström, C. (1990), "The Effects of Editing—A Study on the Annual Survey of Financial Accounts in Sweden," unpublished report, F-Method No. 27, 1990-02-26, Stockholm: Statistics Sweden (in Swedish).

Weeks, M. F. (1992), "Computer-Assisted Survey Information Collection: A Review of CASIC Methods and Their Implications for Survey Operations," *Journal of Official Statistics,* **8,** pp. 445–465.

Werking, G., A. Tupek, and R. Clayton (1988), "CATI and Touchtone Self-Response Applications for Establishment Surveys," *Journal of Official Statistics,* **4,** pp. 349–362.

CHAPTER TWENTY-TWO

Imputation of Business Survey Data

John G. Kovar and Patricia J. Whitridge[1]
Statistics Canada

Nonresponse to business surveys is an ongoing problem for statistical agencies responsible for the collection and dissemination of economic data. Economic survey data are often highly skewed and quantitative in nature; this poses unique challenges in dealing with missing data. Because nonrespondents may not behave in the same manner as respondents in terms of characteristics of interest, special care must be taken to avoid potential biases that can result from inappropriate treatment of missing or inconsistent data. While imputation is but one method of dealing with incomplete data, its use is widespread. In this chapter we discuss reasons for and against using imputation techniques to replace missing survey data.

The sources of nonresponse are numerous and varied. Complete questionnaires may be missing because of the interviewer's inability to contact the business or the business' unwillingness to participate in the survey. We refer to such nonresponse as *total* or *unit nonresponse*. Unfortunately, filled-out and returned questionnaires are often incomplete. This may be because the respondent (i.e., the person or persons the business assigned to complete the survey) is not sufficiently knowledgeable or is unable or unwilling to respond to some questions. Missing, invalid, or inconsistent responses also may be obtained because of misinterpretation of concepts, inaccuracy, or inadvertent omissions. This type of nonresponse is referred to as *item nonresponse*.

Both total and item nonresponse may or may not occur at random; this assessment is difficult to make. In most cases, however, nonresponse is not random: large businesses often have better reporting arrangements, small farms

[1]The authors are grateful for the many useful comments provided by Brenda G. Cox in her role as section editor.

Business Survey Methods, Edited by Cox, Binder, Chinnappa, Christianson, Colledge, Kott.
ISBN 0-471-59852-6 © 1995 John Wiley & Sons, Inc.

tend not to appreciate fully the utility of surveys, and nonprofit institutions may feel they do not have the resources or time to comply. We distinguish here between two types of nonrandom nonresponse: ignorable and nonignorable. Nonrandom nonresponse is deemed *ignorable* if it depends only on the observed variables on the data file. Nonresponse that depends systematically on the actual items subject to nonresponse is deemed *nonignorable*. (See Rubin 1987 for more rigorous definitions.) The choice of imputation method depends on the practitioner's suspicions about the ignorability of the nonresponse mechanism, despite the fact that ignorability is undetectable given the data set in question.

In most instances, ignoring missing data and basing analyses on responding units only lead to serious biases, since respondents rarely form a random subsample of the entire sample. Two broad options exist for minimizing the bias caused by nonresponse. For item nonresponse, the missing values are often replaced by feasible data values to create a completed data file for analyses. This technique is referred to as *imputation*. For total nonresponse, weight adjustment is likely the best approach for nonresponse (see Chapter 25). Weight adjustment could be used for item nonresponse on a variable-by-variable basis. This is cumbersome, because different weights are required for each variable, making imputation the preferable approach. In fact, some business surveys even use imputation in cases of total nonresponse. This would generally be done only if there existed sufficient auxiliary information from other sources or past occasions.

Imputation techniques for business surveys have largely evolved from methods used in social and demographic settings (Madow et al. 1983). Of interest in this chapter is the adaptation of such methods to the skewed, quantitative data of business surveys. Other methods have been developed predominantly within the economic survey framework. These methods are also discussed in the chapter. Methods that find their domain of application almost exclusively within the social survey arena are not discussed. This chapter addresses imputation issues from the business survey perspective, realizing that many issues are common to both economic and social science areas.

Historically, imputation has been a manual process based on subject matter experience and knowledge. Due to advances in technology, imputation processes have become increasingly automated, yielding more objective and reproducible results. Because the computer can maintain useful data concerning the process, automated imputation methods are also easier to monitor and evaluate. Such computer-facilitated imputation methods and associated software are widely used and accepted.

However, the availability of automatic imputation systems may tempt the practitioner to impute in nontraditional situations, as in the case of total nonresponse, the case of nonresponse by design (e.g., two-phase sampling), or cases where respondents are not required to report certain characteristics. The latter scenario can lead to high rates of incomplete data, as is sometimes observed with administrative databases. The use of imputation on a large-scale basis is known as *mass imputation*. In these situations, weighting should be considered very carefully as an alternative to imputation. Its statistical justifiability is usually preferable to the operational convenience of a completed data set. On the other hand, where the propensity for nonresponse can cause serious biases and where nonresponse is not random but is ignorable, imputation can be useful in correcting the bias (Colledge et al. 1978).

Once a data set has been completed by imputation, care must be taken that it is not treated as "clean." Imputed values must be properly flagged and the methods and sources clearly identified. Some analysts may choose to perform their own adjustments for nonresponse, which can be done only if the imputed data are identified. While imputed data sets serve well to produce ad hoc estimates of means and totals at various levels of aggregation, their use for data analysis is not without difficulty, in particular with complex survey designs. Analyses can be misleading if the imputed values are treated as observed data, because variances and covariances may be seriously underestimated. Recent techniques to remedy the problem of variance underestimation include multiple imputation, model-assisted methods, and jackknife procedures.

Because the true data values are unknown for nonrespondents, the resulting imputation-revised data sets are inherently unverifiable. Nonetheless, it is important to provide measures of quality and indications of success or failure. The reason behind the imputation should be taken into account in addition to considering what specific fields to revise. Measures of quality should evaluate the imputation process itself and the impact of the imputation on the data set.

It has been said that the best treatment for nonresponse is not to have any. We take the more pragmatic approach and attempt to show that imputation, despite all of its shortcomings, can be a useful and practical tool. For further insight, readers should refer to the papers by Kalton and Kasprzyk (1986) and Sande (1979, 1982). Extensive bibliographies on the topic are provided by Kalton (1983), Madow et al. (1983), Bogestrom et al. (1983), Little (1988), Pierzchala (1990), and Kovar et al. (1992).

22.1 INCOMPLETE DATA

Business survey data are rarely complete. Because economic data are often more skewed than social data, significant levels of nonresponse are unacceptable. In some business surveys, the skewness combined with large weights leads to situations where even a few nonrespondents can have a large impact on estimates (see Chapter 26). Whether nonresponse is due to refusals, to noncontacts, or to simple inability to answer; whole questionnaires, parts of questionnaires, or individual questions may have missing responses. Furthermore, for some data items the responses may be deleted if they are deemed incorrect as the result of edit failure. From this point on, we refer to all such cases as *missing data*.

Clearly, missing data items cannot be assumed to have zero as their response, though some survey processing systems fill the missing fields with zeros. This practice complicates the process, in that true zeros must be distinguished from missing values later on. We recommend that the processing system provide a data file where missing values are clearly identified.

One technique for dealing with missing data is the *do-nothing option* that simply leaves missing data alone. That is, the values are identified as "missing" on the data file, allowing the analyst to deal with the problem in a locally optimal fashion. The analyst can make a number of assumptions about the missing fields and then build the uncertainty into the econometric model. However, for simple tabulations and analyses, this approach has two flaws. First, different analysts, making different assumptions, would generate different tabulations, yielding inconsistent (if not contradictory) results. This includes those analysts that adjust for nonresponse, as well as those who (perhaps dangerously) base their analyses on responding units only. For example, one could estimate the total by multiplying the mean of the responding unit values by the population size (equivalent to mean imputation) or by simply adding up the responding unit values (equivalent to imputation of zeros). Besides problems in interpretation, such inconsistencies undermine the credibility of the data and, by extension, the data collecting agency. Secondly, the treatment of missing data might often be performed based on incomplete knowledge. The further the data are removed from the collecting agency and the more aggregated they become, the more serious the problem. Confidentiality concerns often dictate that crucial information be suppressed. In the extreme, data are tabulated and disseminated with an "unknown" category. In this instance, users are left to their own devices in interpreting this category. Leaving missing data as missing is more appropriate when the data are being disseminated in their entirety as a microdata file. Indications of how to deal with the missing data must then be provided by the data producers. It is the data collecting agency that best knows the data and all of their limitations, particularly because the agency often has access to related data from the survey frame and other sources.

Another option is to deal with missing data at the estimation stage, through the use of analysis weights. Usually this method is reserved for total nonresponse, though in theory it can be used for item nonresponse as well. It is generally agreed that weighting is preferable to imputation in cases of total nonresponse. Some business surveys are exceptions to this rule, since, if available, quantitative frame data can be better used through imputation to adjust for ignorable nonresponse. Weight adjustment can be done at various levels, including stratum, domain, or unit levels (Little and Rubin 1987); in all cases, the idea is to increase the weights of the respondents to account for the nonrespondents. Raking ratio estimation and poststratification can also be used to advantage when external information is available. The literature on weight adjustment for nonresponse is plentiful; the reader may refer to Chapter 25 and to Chapman (1976), Oh and Scheuren (1983), and Little (1988). For more recent discussion aimed at nonresponse adjustments at the variable level, refer to Yansaneh and Eltinge (1993). Weight adjustment for nonresponse has the desirable property that it is usually theoretically tractable, and thus can be evaluated rigorously. Furthermore, if carried out properly, it can go far in eliminating biases due to uneven distribution of nonrespondents in the population. On the other hand, ad hoc tabulation requests may be more difficult to supply using weight adjustments. If the weights have to be modified, the resulting tabulations may be inconsistent, leading to credibility problems. A further complication arises when weighting is used to adjust for total nonresponse in longitudinal surveys, in that the same unit may have different weights on different occasions, making longitudinal analysis more complicated. In fact, the very concept of total versus item nonresponse comes into question as respondents reply to some, but not all, survey occasions.

The third option is for the data collecting agency to impute for the missing item data. That is, the agency may insert plausible values in place of the missing values, so that internally consistent records are created, and a complete rectangular data file is obtained. Imputation is most feasible for item nonresponse, though it is often used for total nonresponse in business surveys. For example, as mentioned above, imputing data for large nonresponding units may be preferable to weight adjustment, since the collecting agency very often has good auxiliary information about large businesses. In the case of item nonresponse, imputation can make use of additional information about the unit available through other data items. In either case, we assume that auxiliary data are available and thus restrict the remaining discussion to methods of imputation for item nonresponse only. With imputed data files, estimation is greatly simplified and ad hoc tabulations can be produced quickly and consistently. Good imputation techniques can be used to preserve known relationships between variables, address systematic biases, and reduce nonresponse bias. In most cases, imputation can be objective and reproducible. On the other hand, imputed data files can give the users a false impression of quality (Sande 1979; Granquist 1984, 1990). Furthermore, the danger exists that imputation will destroy reported data to create consistent records that fit preconceived models that the analyst will rediscover (Kovar 1991).

22.2 IMPUTATION METHODS

Imputation methods can be classified into two broad categories: deterministic and stochastic (Kalton and Kasprzyk 1986). Some methods of imputation can be labeled *deterministic* because, given the sample of respondents, the imputed values are determined uniquely. Other methods are nondeterministic, or *stochastic*, because the imputed values are subject to some degree of randomness. Deterministic methods include logical imputation, historical imputation, mean imputation, sequential (ordered) hot deck methods, ratio and regression imputation, and nearest-neighbor imputation. These methods can be further divided into (1) those methods that rely solely on deducing the imputed value from data available for the nonrespondent and other auxiliary data (logical and historical) and (2) those methods that make use of the observed data of other responding units for the given survey. Use of current observed data can be made directly by transferring data from a chosen donor record (hot deck and nearest neighbor) or by means of models (ratio and regression). Examples of stochastic methods are the random hot deck, regression with random residuals, and, in fact, any deterministic method with random residuals added.

Many methods can be applied to the entire data set at once or independently within *imputation classes*. For example, one might choose to impute independently in each province or within different standard industrial classifications (SICs). Imputation classes are used to reduce the impact of nonresponse bias and to improve the accuracy of the imputation. The classes are constructed using control variables available for all units, so that the variables to be imputed are homogeneous within classes. That is, if the data were available, respondents and nonrespondents within each class would have similar responses. Often, a proxy for the propensity to respond can be used as a control variable. For example, small and large businesses often respond at different rates. Constructing imputation classes based on a measure of business size can ensure that all sizes are properly represented. In practice, the imputation methods are rarely used in exclusion, because no method is perfect for all situations. The imputation system often comprises several methods to be used in a predefined sequence. Thus, if one method fails, a backup strategy is in place. For example, a monthly survey might use historical imputation as its preferred method but for newly rotated units might instead use mean value imputation.

22.2.1 Deterministic Methods

Often performed during data collection or at the early stages of imputation, *logical* (or *deductive*) *imputation* refers to any method that establishes the value of a missing item with certainty by means of logical constraints and the reported values for that data record. Examples include cases when only one subcomponent of a total is missing or when some reported values correspond to the extremes of the edit bounds. For example, barring negative values, if the edits specify that expenses are to be less than income, then when an income of zero is reported, zero expenses can be deduced. The number of logical relationships between questionnaire items is generally large for economic surveys. However, logical imputation should be based only on exact relationships, not on approximate ones. For example, while it may be acceptable to impute zero wages and salaries to records with zero employees, it would be unwise to do the opposite and impute zero employees to records with zero wages and salaries, since the employees may be taking some form of leave without pay. If used properly, logical imputation can be a useful tool in the early stages of data grooming and is often performed as part of the editing process (see Chapter 21).

Historical imputation is most useful in repeated economic surveys, particularly for variables that tend to be stable over time (e.g., number of employees). Essentially, it consists of using values reported by the same unit on previous occasions to impute for missing data for the current occasion. Clearly, this method attenuates the size of trends and the incidence of change. Variants of the method would adjust the previous values by some measure of trend, often based on other (reported) variables on the record or within the imputation class. The method is most effective when the relationship between occasions is stronger than the relationship between units. When this is the case and when previous values are available, historical imputation can be very appropriate, because it is relatively unaffected by the nonresponse mechanism.

Mean value imputation replaces missing values with the mean of the reported values for that imputation class, and therefore it should only be used for quantitative variables. This method is often used as a last resort. It preserves the respondent means, but it destroys distributions and multivariate relationships, by creating an artificial spike at the imputation class mean value. Mean value imputation performs poorly when nonresponse is not random, unless the imputation classes are chosen exceptionally well. If the same classes are used, it yields the same results for means and totals as weighting class adjustment. However, for statistics other than means and totals, the results can be disastrous. For this reason, mean value imputation is not a method of choice, but it can be a backup strategy when other imputation techniques fail, provided that sufficiently fine imputation classes are used.

Sequential hot deck methods replace the missing data item by a corresponding value from the last responding unit preceding it in the data file. There exist several variants on the method (Kalton 1983, Sande 1983). In general, the data file is processed sequentially, alternately either storing values of a clean record for later use or using previously stored values to impute a record with missing data items. Sequential hot deck methods have their origins in social and demographic survey applications. Their adaptation to economic surveys has usually taken advantage of the continuous nature of the data. While in social applications the data files are usually divided into imputation classes, economic surveys often order the files according to some measure of size (possibly within broader imputation classes) or according to geography. For example, under the hypothesis that neighboring farms tend to behave alike, ordering an agricultural data file geographically increases the likelihood of donors and recipients being spatially close. Care must be taken when imputing from sorted files that a systematic bias is not introduced by forcing the donors to be always smaller or larger than the recipients, depending on the direction of processing. Colledge et al. (1978) addressed this problem by modifying the sequential nature of the procedure so that donors that lie on either side of the recipient would be considered. The choice of classing versus sorting variables needs to be made carefully. In general, continuous variables that measure the size of the business are usually better for sorting. Discrete variables related more to the type of the business are better reserved for classing, especially when it is important that class boundaries not be crossed during imputation.

An often cited disadvantage of sequential hot deck imputation is that the same donors can be used many times (Kalton and Kasprzyk 1986) whenever a "run" of nonrespondents is encountered. This can create artificial spikes in continuous distributions. To remedy this problem, some hot deck systems store three or four good records at a time and cycle through them whenever two or more recipients have to be imputed in a row.

An appealing property of the sequential hot deck method is that actual, observed data are used for the imputation. Thus, no invalid values can be imputed, as can be the case with mean, ratio, or regression methods (e.g., imputing nonintegral values); additionally, the distributions are reasonably well preserved, as are means and totals. However, imputed values may be inconsistent when combined with existing responses. The robustness of the sequential hot deck against nonrandom nonresponse is directly related to the choice of the imputation classes and sorting variables. The sequential hot deck has become a popular imputation technique, since it is easily incorporated into existing data validation programs and allows for data adjustment during and after imputation. However, herein also lies its weakness: Many imputation systems incorporating sequential hot deck methods can become very large and therefore difficult and costly to maintain.

Nearest-neighbor imputation also uses data from clean records to impute missing values of recipient records. The difference is that the donors are chosen in such a way that some measure of distance between the donor and the recipient is minimized. The distance measure is not a spatial one, but rather some multivariate measure based on the reported data; thus, the method is more appropriate for use with continuous data. The nearest-neighbor method shares some of the sequential hot deck and regression imputation properties. Like the hot deck method, it uses actual, observed data and tends to preserve distributions. (For this reason, it is sometimes classified as a hot deck method.) Because missing variables for a given record are usually imputed from the same donor record, multivariate relationships are better preserved. The nearest-neighbor method can control the effect of nonresponse bias due to nonrandom but ignorable nonresponse. However, like the hot deck, it can use donors repeatedly when the nonresponse rate within an imputation class is high. Systems that incorporate the nearest-neighbor method often allow for parametric specification of the minimum number of donors within an imputation class that must be available before an imputation is performed (Kovar and Whitridge 1990). Alternately, the number of times a record has been used as a donor can be incorporated as a penalty measure into the distance function (Colledge et al. 1978). In this latter case, however, unlike the pure nearest-neighbor method, the implementation is sensitive to file ordering.

Ratio and regression imputation methods make use of auxiliary variable(s), replacing the missing values with the corresponding ratio or regression predicted value. These are excellent imputation methods for business surveys, especially in cases where the auxiliary variable is both (1) highly correlated with the variable to be imputed and (2) available for all (or at least a high proportion) of the sampled units. For these methods to be effective, the response variable needs to be continuous, and for the ratio imputation method even the auxiliary variable should be continuous. The independent regression variables may be continuous or discrete, making use of dummy variables in the discrete case. In most business surveys, ratio-type relationships between questionnaire items are abundant and should be put to good use. This is particularly true if it is suspected that the propensity of item response is correlated with the independent variable, as in cases of nonrandom but ignorable nonresponse. Estimates of means and totals based on the ratio or regression imputed data sets are the same as those that would be obtained using weight-adjusted ratio or regression estimators, provided that the same classes were used. The methods perform well in cases of random nonresponse, as well as in nonrandom but ignorable nonresponse situations (i.e., cases where the nonresponse propensity is related to the auxiliary variables used by the ratio or regression). Despite the advantages, some distributional problems persist in cases where the independent (auxiliary) variables can be identical for several units, since the imputed values will be the same in those cases. In addition, time and effort are required to develop the model, which then must be verified or adjusted regularly.

22.2.2 Stochastic Methods

Deterministic methods often reduce the variation of the variable of interest and sometimes distort distributions. Most stochastic imputation techniques have been introduced in an attempt to preserve the distribution and variability of the data set (Little 1988); many are variations of deterministic methods. As will be seen in Section 22.4, all imputation procedures introduce an extra component of variation that must be considered in estimation. The use of stochastic methods in itself is not sufficient to establish correct variance estimates (Särndal 1990).

Following the framework introduced by Kalton and Kasprzyk (1986), we let $\hat{y}_{mi}$ be the imputed value for the ith missing observation. A large number of imputation methods can be approximately described by the general model

$$\hat{y}_{mi} = b_{r0} + \sum_j b_{rj} x_{mij} + \hat{e}_{mi},$$

where x_{mij} are the values of the auxiliary variables (indexed by j) for the ith observation, b_{r0} and b_{rj} are the coefficients of a regression between y and x based on the responding units, and the $\hat{e}_{mi}$ are residuals chosen in a prespecified manner.

Setting the residuals to zero, the ratio and regression methods fit into this framework in an obvious way. By letting the x variables be dummy variables representing the imputation classes, class mean imputation is described. The *random hot deck method*, in which donors are chosen completely at random with replacement from the entire sample of respondents within imputation classes, is represented in this framework by adding to the class mean a residual equal to the difference of the donor value and the class mean. Similarly, many other imputation methods can be explained within this framework.

Kalton and Kasprzyk classify imputation methods as deterministic or stochastic according to whether or not the residuals are set to zero. For example, the above-described mean imputation, ratio imputation, and regression imputation are considered deterministic. The random hot deck, on the other hand, is classified as stochastic. Clearly, any deterministic method can be made stochastic by adding judiciously chosen residuals that satisfy the usual condition of zero expectation. In particular, regression imputation with the addition of a randomly chosen residual from the set of observed residuals has been studied by Lee et al. (1991). To the best of our knowledge, no such method is widely used, perhaps because it decreases the precision of estimated means, increases the complexity of software, and would not by itself account for the variance due to imputation.

The random hot deck method, even with the use of imputation classes, is used less frequently in economic surveys than in social surveys. This infrequent use is the result of the perception that some information will be lost by not exploiting the continuous nature of the data as with the ordered, sequential hot deck. Unlike the sequential hot deck, however, the random hot deck makes all clean records eligible for selection as donors at any one time. Because of this, the distributions are better preserved, and multiple use of donors is limited. Cox (1980) proposed an alternative known as the *weighted random hot deck method* that controls the number of times a donor is used and that ensures that all donors have a nonzero probability of selection. Survey weights are included in this hot deck method so that for each imputation class, weighted estimates of means using imputed data are the same in expectation as weighted estimates of means using respondents only. The weighted hot deck procedure is the imputation analog to weighting class adjustment (Folsom 1981).

22.2.3 Implementation

Many imputation systems have been developed in the last three decades. These range from highly customized and survey-specific systems to fairly generalized and reusable software (see Chapter 23). Initially, the imputation systems were just automating the manual, sequential, "detect and correct" rules, often taking advantage of the computer only to add more rules. The resultant systems were often so large and complex that no record could pass all edits. In the mid-1970s, Fellegi and Holt (1976) proposed that edit and imputation systems should adhere to four principles. A number of automatic imputation systems have since been developed adhering to the Fellegi–Holt principles. The early efforts concentrated on applications with qualitative data and gave rise to systems such as the Canadian CANEDIT, the Hungarian AERO, and the Spanish DIA. [Refer to Economic Commission for Europe (1992) for more information on these systems and a bibliography.] In the mid- to late 1980s, two systems that deal with quantitative economic data emerged. These are the U.S. Bureau of the Census' SPEER (Greenberg 1987, Greenberg and Petkunas 1990, Draper et al. 1990) and Statistics Canada's GEIS (Kovar et al. 1988a, 1988b). The availability of such generalized software has broadened the potential scope for imputation and may tempt practitioners to use imputation for less conventional applications, such as mass imputation.

22.3 MASS IMPUTATION

To save resources, many statistical agencies are resorting to two-phase sampling of administrative records. Administrative record data for the first-phase sample are used to select an efficient subsample for which additional information is collected in a sample survey. Classical estimates based upon the subsample require the derivation of secondary weights. An alternative is to impute the missing parts of the nonsampled primary units to create a complete rectangular file (Whitridge et al. 1990b); this technique is known at Statistics Canada as mass imputation. The sampling rate for the second phase is often between 10 and 30 percent, although it can vary. The amount of missing data is quite high: 70–90 percent. Classical imputation methods, designed for low nonresponse rates, must therefore be used with caution.

Since 1978, Statistics Canada has applied mass imputation techniques to Census of Construction data (Colledge et al. 1978). This survey uses a sample of income tax records for the first-phase information, then selects a subsample to collect detailed second-phase information. In this example, approximately 30 percent of the original tax sample is subsampled; thus, 70 percent is imputed. Similar methods were applied by Hinkins and Scheuren (1986) for a sample of U.S. corporate tax returns, and more recently by Clogg et al. (1991) for industry and occupation classification. These examples had imputation rates between 75 and 90 percent.

Mass imputation can also be useful when large amounts of data are missing for operational reasons. For example, Statistics Canada uses agricultural income tax data to produce balance sheet estimates. However, farmers are not legally required to file a balance sheet. Thus, the resulting data set is missing large amounts of data for many farmers, but the missing farmers do not form a random subsample of all farms.

For many imputation methods, there exists a corresponding weight adjustment method (Folsom 1981; Whitridge et al. 1990a). For example, in the case of simple random sampling, if the sample mean is used for imputation, then mass imputation is equivalent to the direct expansion estimator. Using a ratio estimator with auxiliary data to mass impute for variables in the subsample would be equivalent to using the ratio estimator at the estimation stage. A more complex imputation method, such as nearest-neighbor imputation, is implicitly equivalent to an expansion estimator with variable weights, wherein each

IMPUTATION OF BUSINESS SURVEY DATA

record is assigned a weight corresponding to the number of times it was used as a donor. Weighting has the distinct advantage that it is based on classical statistical theory and hence it is easily defensible, and its properties and behavior are well known. Weighting is a simple and efficient way of estimating for information of a subsample that is relatively independent of the sample. Weighting is to be recommended when estimation of variances, covariances, and correlations is routinely considered by users. However, mass imputation has a place in survey processing for cases where quick, ad hoc estimates are needed or where second-phase sample weights are difficult to calculate, as is the case when information is missing for operational reasons.

Originally, mass imputation techniques were defended using missing-at-random arguments, since it is known that most imputation methods perform well under such conditions. That is, since the subsample was chosen at random, it was argued that not-chosen units form a random sample of missing records. In addition, the equivalence between weighting and mass imputation is comforting. However, further studies (Colledge et al. 1978) suggest that mass imputation can be useful in the presence of ignorable nonresponse for nonrandom subsamples in that it may help attenuate the bias. These results were further borne out by Michaud (1987). For nonrandom, ignorable nonresponse, mass imputation may be preferable to weighting, since it makes more extensive use of auxiliary information when imputing the missing second-phase segments. The choice of imputation method is hence very important. Mass imputation might also be recommended for situations where there are high correlations between variables in the sample and those in the subsample. Depending upon the imputation method chosen, mass imputation can better preserve the relationships between variables, because it can make more extensive use of auxiliary variables. For example, nearest-neighbor imputation lends itself to mass imputation, since it can take advantage of the multivariate relationships between variables in the sample and those in the subsample. By contrast, in the presence of nonignorable nonresponse, most imputation methods perform poorly in traditional settings (Rancourt et al. 1992), let alone in cases of mass imputation.

Typically, for mass imputation applications, large amounts of data are imputed from a small subsample. Since this is the case, it is especially important that all imputed data be flagged and that data users be aware of the imputation. In the extreme case, two separate files could be kept: the complete rectangular file for tabulation and the original file for analysis. With such a large amount of imputed data, the evaluation of the impact of imputation becomes critical. Methodology to evaluate the imputation must be developed carefully. When the "nonresponse" is incurred by design and the underlying model is known, simulation studies can be helpful in evaluating whether mass imputation is feasible. The 1977 National Medical Care Expenditure Survey attempted to apply mass imputation to two-phase sample data from individual persons and their medical care providers in the United States. Mass imputation was successful in imputing unreported medical visits, but not expenditure or diagnosis data (Cox and Folsom 1981; Williams and Folsom 1981; Cox and Cohen 1985, pp. 150–189). Further examination revealed that significant biases could be introduced for variables that were uncontrolled by the imputation process. It was concluded that mass imputation was not feasible for this application.

22.4 DATA ANALYSIS

Rightly or wrongly, imputation methods are often devised with the aim of predicting the correct response. As a result, the imputed data sets provide good estimates of means and totals. With care, the distributions can also be preserved reasonably well. The situation is not as favorable when it comes to estimates of variances and correlations. Numerous studies (e.g., Santos 1981; Kalton and Kasprzyk 1982, 1986; Little 1986) show that imputation tends to attenuate correlations between the imputed variables to various degrees, though the situation is improved if good auxiliary variables are used at the imputation stage. If standard formulas are used, treating imputed values as observed values underestimates variances of estimated means and totals (Rubin 1978). This leads to confidence intervals that are too short and to the tendency to declare significance when none exists. The problem becomes more serious as the proportion of missing items increases (Särndal 1990). For stochastic imputation methods, there exists an associated model for nonresponse, which can be used to calculate a model-based estimate of the variance.

It can be shown that the true variance of the estimator of the mean, V_{Tot}, can be written as $V_{\text{Tot}} = V_{\text{Sam}} + V_{\text{Imp}} + V_{\text{Mix}}$ (Särndal 1990), where V_{Sam} is the sampling variance component, V_{Imp} is the variance introduced by the imputation method in question, and V_{Mix} is a covariance between V_{Sam} and V_{Imp}, which in most cases is negligible or zero. An estimator of V_{Sam} could be obtained by adding a term to the usual variance formula to correct for the fact that the sampling variance component is understated when the data set contains imputed values. This adjustment would depend on the method used, being relatively high in the case of mean imputation and negligible in the case of the random hot deck method. Unfortunately, interest rarely lies in estimating what the variance would have been, had there been no nonresponse; rather, we are interested in the variance of the estimates based on the present, imputed file or V_{Tot}. To estimate V_{Tot}, the additional component of variance due to the imputation mechanism V_{Imp} must be estimated. Empirical evidence suggests that, depending on the imputation method, the underestimation when using the usual variance formulas can be of the order of 2 to 10 percent in the case of 5 percent nonresponse rate, but as high as 10 to 50 percent in the case of 30 percent nonresponse (Kovar and Chen 1994). Three general methods that estimate the variance due to imputation are presented next.

Rubin (1977, 1978, 1986, 1987) proposed the technique of *multiple imputation* to estimate the variance due to imputation by replicating the process a number of times and estimating the between-replicate variation. A number of variants of this approach have been put forth in the literature. The method is theoretically defensible, though care must be taken that the imputation method be proper in Rubin's (1987) sense. Rubin defines *improper* methods as those that do not show enough variability between replicates to provide appropriate variance estimates. (As an example, even the random hot deck method is improper.) Operationally, the multiple imputation method is unattractive because it incurs high computer costs, necessitates the maintenance of multiple files, and complicates data dissemination. The relatively high cost of imputation, even for moderately large files, tempts practitioners of multiple imputation to use a low number of replicates. This results in poor precision for the estimated variance due to imputation. Furthermore, recent results suggest that multiple imputation can produce inconsistent variance estimates when the inference and imputation classes cross (Fay 1992, 1993). As noted by Kalton and Kasprzyk (1986, p. 13), "... it is questionable whether the multiple imputation approach is feasible for routine analyses. It may be best reserved for special studies."

More recently, Särndal (1990) outlined model-assisted estimators of variance, while Rao and Shao (1992) proposed a method that corrects the usual jackknife variance estimator. These methods are appealing in that only the imputed file (with the imputed fields flagged) is required for variance estimation. Särndal's model-assisted approach requires different variance estimators for each imputation method, but yields consistent variance estimates provided that the model holds. Several estimators have been proposed and empirically evaluated with very positive results (Lee et al. 1991). On the other hand, the Rao and Shao adjusted jackknife method requires implementation of only one estimator, though the temporary adjustment of the imputed values depends on the imputation method. The actual adjustments for several imputation methods can be found in Rao (1992) and in Rao and Shao (1992). The method is design-consistent (p-consistent) under uniform nonresponse irrespective of the model, as well as design-model-unbiased (pm-unbiased) under the usual linear model and any ignorable nonresponse mechanism (Rao 1992). Empirical results show that with uniform nonresponse, the adjusted jackknife method along with any of the studied imputation methods essentially eliminates any underestimation of the variance, for simple as well as complex survey designs (Kovar and Chen 1994). For nonrandom but ignorable nonresponse, the adjusted jackknife with the ratio imputation method performs equally well. Rancourt et al. (1993) have recently applied the adjusted jackknife estimator when more than one imputation method is used for the same data set.

In cases where the nonresponse mechanism is not ignorable, all imputation methods tend to produce severely biased point estimates. As such, variance estimation is of minimal interest, because the real problem lies in estimating the mean squared error. More attention needs to be concentrated on improving the point estimates and reducing their biases. Some preliminary results on this front have been put forth by Rancourt et al. (1992). The performance of these three techniques is less than satisfactory for nearest-neighbor imputation, except under ideal conditions—that is, when the auxiliary variables are highly correlated with the response variables and nonresponse is uniform (Kovar and Chen 1994).

22.5 EVALUATION OF IMPUTATION

The success or failure of the imputation process is difficult to evaluate, but audit trails and status reports can help. Theoretical results are scarce, though Little and Rubin (1987) provide some, along with good references. Once an imputation strategy has been implemented, it is very important to evaluate the impact the imputation has on the final data. Despite the fact that naively calculated estimates of sampling variances seem to decrease after imputation, this does not indicate an improvement in data quality. The basic objective of imputation is to reduce the bias due to nonresponse. If bias is indeed reduced and the increase in variance due to imputation does not offset this reduction, then the quality of survey estimates is improved in the mean square error sense. However, there are other aspects to consider when evaluating the impact of imputation (Sande 1982).

Performance measures or *evaluation statistics* provide a standard set of measures that can be used by a manager to plan, monitor, and control all aspects of the survey process. Such statistics can be used not only to provide information about the quality of the data, but also to provide a basis for improvements for the next survey occasion, as described in Chapter 21. Performance measures at the edit and imputation stages reflect the quality of the data being edited, the quality of the edits themselves, and the magnitude of the changes brought about by imputation. They assist in evaluating the difference between the data that are actually collected and the imputed data. Any changes made to the data, such as those due to imputation, should be flagged. Complete documentation of all survey processes is essential for later evaluation.

Sande (1981) presents descriptive statistics that monitor the edit and imputation process. These statistics include counts of the number of times certain fields were identified for imputation, the number of times they were imputed, and comprehensive statistics specific to the nearest-neighbor imputation method (e.g., distance between donor and recipient, transformed values of matching fields, etc.). Other performance measures for imputation might include counts of the number of records that required at least one imputation, the percentages with which certain methods were used on given fields, the number of records with a given number of fields imputed, and specific counts that depend on the imputation methods chosen. For example, if donor imputation is used, then the identifier of the donor should be retained so that statistics about how many times certain donors were used can be examined. Once these performance measures have been gathered during the imputation process, it is important that they be analyzed as fully as possible. These imputation performance measures reflect not only the data, but the survey design as well. For example, very high imputation rates for a certain field may mean that the question was poorly understood by respondents and may suggest the need for improvements in the questionnaire design or the data collection and training procedures. Quality assurance principles need to be applied at all steps when establishing an imputation strategy. Statistics such as these assist in monitoring the quality of the imputation process.

Beyond producing tables of counts of changes in data values due to imputation, it is important to evaluate and validate the final data themselves. To do this, statisticians must consider evaluation tables that consist of the number of times a reported value was increased, decreased, or remained unchanged, and they must evaluate the corresponding estimates. Typically, such tables would be produced at several different levels of aggregation, perhaps corresponding to estimates to be tabulated. When the estimates appear suspicious, effort can be concentrated on large or important units which have a significant impact on the estimates, rather than manually verifying all imputed records, which can be labor-intensive and expensive. Such a technique uses a macro approach rather than the more commonly adopted micro approach, looking at the impact of imputation on the estimates, rather than on specific records. This method involves the same principles that form the basis for the selective editing method discussed in Chapter 21. It ensures that the global estimates will be reasonable even though all underlying data relationships might not hold for all individual records. One way to implement this approach is to produce and examine tables of the records with the highest values for specific fields or to examine the records with the largest weighted contribution to estimates.

Another strategy to evaluate imputed data would be to compare them against expected results. This evaluation could involve external data, perhaps from administrative sources or other surveys in a comparative exercise. The most effective method of evaluating the imputation is probably through follow-up reinterview studies. Respondents could be contacted to resolve inconsistent responses and to complete missing responses. The analyst could then estimate the bias and hence the mean square error under alternate imputation approaches. Attempts to evaluate imputation through simulation studies are usually of questionable value, since too many assumptions must be made with respect to the nonresponse mechanism. This is in direct contrast to the mass imputation situation where data are missing by design.

Evaluating the impact of imputation is often difficult, since it requires a prespecified notion of the true values as well as knowledge about acceptable imputation rates. These are very subjective measures. Acceptable imputation rates depend upon the response rates, the reasons for imputation, and which specific fields are being imputed. For certain key variables that are always reported, the imputation rate should be low, since any imputation represents an actual change to reported data. However, for variables that tend to be missing on many questionnaires, higher imputation rates may be acceptable. The reason for imputation, be it due to nonresponse or to inconsistent data, should be considered when the impact of imputation is being evaluated, since it will have different effects upon the data. For example, imputing for nonresponse will result in a positive change to the unweighted estimates of totals, since all fields are necessarily increased (assuming positive data). Imputing for inconsistent data should not have a large effect upon the estimates themselves, since changes can either increase or decrease the values. In the end, once the imputation-revised data have been evaluated, it is important to step back and consider the costs and benefits of the imputation exercise. Was there a decrease in the nonresponse bias, and at what cost did it occur in terms of resources and increase in total variance?

22.6 CONCLUDING REMARKS

All business surveys suffer from the effects of nonresponse. Whether the non-response is due to errors, misconceptions, noncontacts, or refusals, it is rarely random. Systematic nonresponse patterns can be responsible for serious biases in survey estimates. Methods of dealing with such biases include weight adjustment and imputation. In this chapter, we describe common imputation methods and provide suggestions about the appropriateness of their use in various situations. Most methods produce imputed data files from which good, first-order estimates can be produced. However, some methods, such as mean value imputation, should be used with caution because of their negative impact on distributional properties and multivariate relationships. On the other hand, methods such as the ratio, regression, and nearest-neighbor imputation that make use of auxiliary information are well-suited to business survey data and can go far in reducing nonresponse bias. The type of auxiliary information and its quality need to be considered when choosing the appropriate imputation method. The assumptions regarding the randomness of the underlying response mechanism, though usually unverifiable, must also be evaluated and the robustness of the imputation method against departures from these assumptions must be considered.

The availability of an imputation-revised data file facilitates the production of consistent, ad hoc tabulations. However, caution must be exercised when large-scale imputation is attempted and when data analyses using imputed files are performed. We provide some guidance on the suitability of the mass imputation technique and caution would-be practitioners against its use without a proper evaluation. While mass-imputed data sets may be operationally more convenient, traditional weighting methods are generally preferable. However, when frequent ad hoc estimates have to be produced in a consistent fashion, a mass-imputed data file may serve the purpose well. When the units to be imputed are ignorably nonrandom, mass imputation may in fact help eliminate biases that weighting would not.

Using imputed data files for data analyses, significance testing, and variance estimation, in particular, poses new challenges and requires the analyst's attention. We present a brief overview of several recently developed techniques for this purpose, including multiple imputation, model-assisted methods, and adjusted jackknife techniques.

Finally, the necessity of flagging all imputed data cannot be overstressed. All imputation methods fabricate data to some extent, and the user must be aware of this. Only the data collecting agency can annotate the data properly! The maintenance of accurate audit trails and the complete documentation of the entire imputation process is essential for ultimate success.

REFERENCES

Bogestrom, B., M. Larsson, and L. Lyberg (1983), "Bibliography on Nonresponse and Related Topics," in W. G. Madow, I. Olkin, and D. B. Rubin (eds.), *Incomplete Data in Sample Surveys, Vol. 2: Theory and Bibliographies*, New York: Academic Press, pp. 479–567.

Chapman, D. W. (1976), "A Survey of Nonresponse Imputation Procedures," *Proceedings of the Social Statistics Section, American Statistical Association*, pp. 245–251.

Clogg, C., D. Rubin, N. Schenker, B. Schultz, and L. Weidman (1991), "Multiple Imputation of Industry and Occupation Codes in Census Public-Use Samples Using Bayesian Logistic Regression," *Journal of the American Statistical Association*, **86**, pp. 68–78.

Colledge, M. J., J. H. Johnson, R. Paré, and I. G. Sande (1978), "Large Scale Imputation of Survey Data," *Survey Methodology*, **4**, pp. 203–224.

Cox, B. G. (1980), "The Weighted Sequential Hot Deck Imputation Procedure," *Proceedings of the Survey Research Methods Section, American Statistical Association*, pp. 721–726.

Cox, B. G., and S. B. Cohen (1985), *Methodological Issues for Health Care Surveys*, New York: Marcel Dekker.

Cox, B. G., and R. E. Folsom (1981), "An Evaluation of Weighted Hot Deck Imputation for Unreported Health Care Visits," *Proceedings of the Survey Research Methods Section, American Statistical Association*, pp. 412–417.

Draper, L., B. Greenberg, and T. Petkunas (1990), "On-Line Capabilities in SPEER," *Proceedings of Symposium 90: Measurement and Improvement of Data Quality*, Ottawa: Statistics Canada, pp. 235–243.

Economic Commission for Europe (1992), *Statistical Data Editing Methods and Techniques*, Vol. 1, New York: United Nations, Conference of European Statisticians.

Fay, R. E., III (1992), "When Are Inferences from Multiple Imputation Valid?" *Proceedings of the Survey Research Methods Section, American Statistical Association*, pp. 227–232.

Fay, R. E., III (1993), "Valid Inferences From Imputed Survey Data," *Proceedings of the Survey Research Methods Section, American Statistical Association*, pp. 41–48.

Fellegi, I. P., and D. Holt (1976), "A Systematic Approach to Automatic Edit and Imputation," *Journal of the American Statistical Association*, **71**, pp. 17–35.

Folsom, R. E. (1981), "The Equivalence of Generalized Double Sampling Regression Estimators, Weight Adjustments and Randomized Hot-Deck Imputations," *Proceedings of the Survey Research Methods Section, American Statistical Association*, pp. 400–405.

Granquist, L. (1984), "On the Role of Editing," *Statistisk Tidskrift*, **2**, pp. 105–118.

Granquist, L. (1990), "A Review of Some Macro-Editing Methods for Rationalizing the Editing Process," *Proceedings of Symposium 90: Measurement and Improvement of Data Quality*, Ottawa: Statistics Canada, pp. 225–234.

Greenberg, B. (1987), "Edit and Imputation: A Discussion," *Proceedings of the Annual Research Conference*, Washington, DC: U.S. Bureau of the Census, pp. 204–210.

Greenberg, B., and T. Petkunas (1990), "SPEER (Structured Program for Economic Editing and Referrals)," *Proceedings of the Survey Research Methods Section, American Statistical Association*, pp. 95–104.

Hinkins, S., and F. Scheuren (1986), "Hot Deck Imputation Procedure Applied to a Double Sampling Design," *Survey Methodology*, **12**, pp. 181–196.

Kalton, G. (1983), *Compensating for Missing Survey Data*, Ann Arbor: Survey Research Center, University of Michigan.

Kalton, G., and D. Kasprzyk (1982), "Imputing for Missing Survey Responses," *Proceedings of the Survey Research Methods Section, American Statistical Association*, pp. 22–31.

Kalton, G., and D. Kasprzyk (1986), "The Treatment of Missing Survey Data," *Survey Methodology*, **12**, pp. 1–16.

Kovar, J. G. (1991), "The Impact of Selective Editing on Data Quality," Working Paper No. 5, Work Session on Statistical Data Editing, Conference of European Statisticians, Geneva, Switzerland.

Kovar, J. G., and E. J. Chen (1994), "Jackknife Variance Estimation of Imputed Survey Data," *Survey Methodology*, **20**, pp. 45–52.

Kovar, J. G., J. MacMillan, and P. Whitridge (1988a), "Overview and Strategy for the Generalized Edit and Imputation System," Methodology Branch Working Paper No. BSMD 88-007E/F, Ottawa: Statistics Canada.

Kovar, J. G., J. Mayda, K. Dumičić, and S. Dumičić (1992), "Selected Bibliography on Data Editing and Imputation and Related Topics," Working Paper No. 12, Work Session on Statistical Data Editing, Conference of European Statisticians, Washington, DC.

Kovar, J. G., and P. Whitridge (1990), "Generalized Edit and Imputation System: Overview and Applications," *Revista Brasileira de Estatística*, **51**, pp. 85–100.

Kovar, J. G., P. Whitridge, and J. MacMillan (1988b), "Generalized Edit and Imputation System for Economic Surveys at Statistics Canada," *Proceedings of the Survey Research Methods Section, American Statistical Association*, pp. 627–630.

Lee, H., E. Rancourt, and C.-E. Särndal (1991), "Experiments with Variance Estimation from Survey Data with Imputed Values," *Proceedings of the Survey Research Methods Section, American Statistical Association*, pp. 690–695.

Little, R. J. A. (1986), "Survey Nonresponse Adjustments for Estimates of Means," *International Statistical Review*, **54**, pp. 139–157.

Little, R. J. A. (1988), "Missing-Data Adjustments in Large Surveys," (With Discussion), *Journal of Business and Economic Statistics*, **6**, pp. 287–301.

Little, R. J. A., and D. B. Rubin (1987), *Statistical Analysis with Missing Data*, New York: Wiley.

Madow, W. G., I. Olkin, and D. B. Rubin (1983), *Incomplete Data in Sample Surveys, Vol. 2: Theory and Bibliographies*, New York: Academic Press.

Michaud, S. (1987), "Weighting vs Imputation: a Simulation Study," *Proceedings of the Survey Research Methods Section, American Statistical Association*, pp. 157–161.

Oh, H. L., and F. J. Scheuren (1983), "Weighting Adjustment for Unit Nonresponse," in W. G. Madow, I. Olkin, and D. B. Rubin (eds.), *Incomplete Data in Sample Surveys, Vol. 2: Theory and Bibliographies*, New York: Academic Press, pp. 143–184.

Pierzchala, M. (1990), "A Review of the State of the Art in Automated Data Editing and Imputation," *Journal of Official Statistics*, **6**, pp. 355–377.

Rancourt, E., H. Lee, and C.-E. Särndal (1992), "Bias Corrections for Survey Estimates from Data with Imputed Values for Nonignorable Nonresponse," *Proceedings of the Annual Research Conference*, Washington, DC: U.S. Bureau of the Census, pp. 523–539.

Rancourt, E., H. Lee, and C.-E. Särndal (1993), "Variance Estimation Under More than One Imputation Method," *Proceedings of the International Conference on Establishment Surveys*, Alexandria, VA: American Statistical Association, pp. 374–379.

Rao, J. N. K. (1992), "Jackknife Variance Estimation Under Imputation for Missing Survey Data," unpublished report, Ottawa: Statistics Canada.

Rao, J. N. K., and J. Shao (1992), "Jackknife Variance Estimation with Survey Data Under Hot Deck Imputation," *Biometrika*, **79**, pp. 811–822.

Rubin, D. B. (1977), "Formalizing Subjective Notions About the Effect of Nonrespondents in Sample Surveys," *Journal of the American Statistical Association*, **72**, pp. 538–543.

Rubin, D. B. (1978), "Multiple Imputations in Sample Surveys—A Phenomenological Bayesian Approach to Nonresponse," *Proceedings of the Survey Research Methods Section, American Statistical Association*, pp. 20–34.

Rubin, D. B. (1986), "Basic Ideas of Multiple Imputation for Nonresponse," *Survey Methodology*, **12**, pp. 37–47.

Rubin, D. B. (1987), *Multiple Imputation for Nonresponse in Surveys*, New York: Wiley.

Sande, G. (1981), "Descriptive Statistics Used in Monitoring Edit and Imputation Process," paper presented at the Computer Science and Statistics 13th Symposium on the Interface, Pittsburgh, PA.

Sande, I. G. (1979), "A Personal View of Hot-Deck Imputation Procedures," *Survey Methodology*, **5**, pp. 238–258.

Sande, I. G. (1982), "Imputation in Surveys: Coping with Reality," *American Statistician*, **36**, pp. 145–152.

Sande, I. G. (1983), "Hot-Deck Imputation Procedures," in W. G. Madow and I. Olkin (eds.), *Incomplete Data in Sample Surveys, Vol. 3: Proceedings of the Symposium*, New York: Academic Press.

Santos, R. L. (1981), "Effects of Imputation on Regression Coefficients," *Proceed-

ings of the Survey Methods Section, American Statistical Association, pp. 140–145.

Särndal, C.-E. (1990), "Methods for Estimating the Precision of Survey Estimates when Imputation Has Been Used," *Proceedings of Symposium 90: Measurement and Improvement of Data Quality*, Ottawa: Statistics Canada, pp. 337–350.

Whitridge, P., M. Bureau, and J. G. Kovar (1990a), "Mass Imputation at Statistics Canada," *Proceedings of the Annual Research Conference*, Washington, DC: U.S. Bureau of the Census, pp. 666–675.

Whitridge, P., M. Bureau, and J. G. Kovar (1990b), "Use of Mass Imputation to Estimate for Subsample Variables," *Proceedings of the Business and Economic Statistics Section, American Statistical Association*, pp. 132–137.

Williams, R. L., and R. E. Folsom (1981), "Weighted Hot Deck Imputation of Medical Expenditures Based Upon a Record Check Subsample," *Proceedings of the Survey Research Methods Section, American Statistical Association*, pp. 406–411.

Yansaneh, I. S., and J. L. Eltinge (1993), "Construction of Adjustment Cells Based on Surrogate Items or Estimated Response Propensities," *Proceedings of the Survey Research Methods Section, American Statistical Association*, pp. 538–543.

CHAPTER TWENTY-THREE

Editing Systems and Software

Mark Pierzchala

U.S. National Agricultural Statistics Service

Because software change so quickly, it is difficult to track in a chapter such as this one. In addition, the marketplace for data editing software systems is too small to merit discussion in the monthly or weekly trade press. Systems do not perform all the same functions, and selection may be complicated. The best way to learn about a particular system is to obtain a copy of the software and test it, attend presentations, visit the development site, and question current users. This chapter describes a few general approaches, why they are taken, how to evaluate available software, and the results of documented implementations. The information presented here can also be used as a starting point for development of software if suitable systems are unavailable. Note that the mention or exclusion of a system in this chapter does not infer recommendation or censure, nor does it signify general availability. Some systems in use are not mentioned.

23.1 HISTORIC AND FUTURE DEVELOPMENT OF EDITING SYSTEMS

Almost all editing and imputation techniques are manifested through computer systems. For many business surveys, data problems must still be detected and manually corrected in a post-collection editing state. The treatment of each questionnaire can be time-consuming and often must be done by highly skilled and highly paid subject-matter specialists. Based on its nonprobability sample of United States surveys of all types, the Federal Committee on Statistical Methodology (1990) found that all error correction was done by analysts or

Business Survey Methods, Edited by Cox, Binder, Chinnappa, Christianson, Colledge, Kott.
ISBN 0-471-59852-6 © 1995 John Wiley & Sons, Inc.

clerks in 60 percent of the surveys. For some surveys, that respondent may be contacted many times over long periods of time, increasing respondent and agency burden. In the collection agency, either many people are assigned to the job, the job is done inadequately, or well-designed computer tools and editing strategies allow the agency to achieve or maintain quality estimates while making the editing process more productive.

23.1.1 Historic Justification

Stuart (1966) provides an excellent summary of an early experience with a mainframe batch editing system at the U.S. Bureau of Labor Statistics. The reasons he gives for moving from hand-based editing to computer-based editing are much the same as those given today: (1) to speed up data review, (2) to relieve the human data editor from tedium, (3) to do things that people cannot do such as apply the same edit logic consistently to all forms, (4) to automate imputation techniques, (5) to allow more edits to be applied, and (6) to focus staff time on records of high impact or with high probability of error. Stuart elaborates on edit reviews that the computer can do, such as checks for internal (micro) consistency, external (macro) consistency (e.g., behavior of a firm against other firms in a class), and matching reports. Stuart also mentions the possible integration of the edit review with analysis and summary and the need for a general system that eases the production of edit review programs. This short descriptive paper is worth reading because the topics discussed are still of major concern.

The early to mid-1970s saw the advent of computer-assisted survey interview collection methods (see Chapter 18). This process began with computer-assisted telephone interviewing (CATI). Later, advances in portable computing facilitated the introduction of computer-assisted personal interviewing (CAPI). Recently, the availability of computers in business offices has allowed the development of electronic self-reporting systems (Heath 1993, Hundepool 1993). New technologies have led to new issues, including: which edits and how many edits to apply during the interview as opposed to post-collection review (Eklund 1993); how to design computer screen interfaces for interviewers, respondents, and data editors; and how to handle mixed-mode data collection (Woelfle 1993).

23.1.2 Major Trends in Software Development

Current design opportunities for editing systems stem from advances in personal computer hardware, database technology, modular and generalized system development methods, networking, graphical computer screens displays, and telecommunications. Some trends in editing system development can be discerned by inspecting available systems and by reviewing the literature. Most trends listed below already appear in two or more systems:

- More edits are invoked in the data collection stage by computer assisted interviewing (CAI) or electronic self-reporting. This reduces callbacks and the need to review data after collection.

- Direct computer-to-computer data transfer from business to statistical agency (Electronic Data Interchange) is being tested. Telephone touch-tone collection is operational in a few surveys, as is reception of data by facsimile. Data collection by automated voice recognition is being explored. These technologies speed up data collection and are more convenient for the respondent, but for most of them, edits are not invoked interactively during data collection.

- If human intervention is required for post-collection data review, interactive approaches are being adopted. This speeds up review and allows forms to be cleaned in one treatment, reducing cyclical processing inherent in inappropriate batch methods.

- Batch processing is reserved for file-based activities that do not require human intervention—for example, finding edit failures or making computations.

- Some batch-based generalized edit and imputation (Fellegi and Holt) systems perform tasks previously reserved for human review including automatically finding errors, choosing items for deletion, and effecting changes.

- Top-down (macro- or statistical) editing is very useful where relatively few offending reports must be found in a massive file and treated quickly.

- Integration of survey tasks within one system speeds applications development and maintenance, reduces or eliminates some tasks, and facilitates data flow between tasks. For example, computer-assisted survey interview collection reduces or eliminates the need for data entry.

- *Meta-data* (descriptive information about survey data) are formalized and generated to document a survey and also to ease data transfer from one system to another. In the future, some editing systems will write directly (not export) to different file formats so different packages can process the same data set.

- Where data are collected on paper, editing is done during or after data entry, but not before data entry. This allows the agency to determine the impact of editing on estimates.

- Systems are designed to plan and manage multimode data collection (e.g., a combination of paper and computer-assisted data collection) and to optimize resource allocation between modes.

- *Selective editing* strategies (manual treatment of some reports, automated treatment of others, with automated determination of which report gets which treatment) are being successfully tested and used, as well as strategies for selective sampling and follow-up to reduce respondent burden.

- In the future, cross-survey coordination will be implemented to avoid asking the same questions of a business in different surveys.
- Generalized systems are built to avoid duplication of programming effort as large development costs are spread among many surveys. They are constructed in modules to make maintenance and updating easier.
- Systems are becoming more portable with the advent of the microcomputer, and widely used operating systems are making it easier to investigate the usefulness of systems and approaches to problems.
- Systems are more polished because the end users must interact directly with them. As a result, end users are becoming more demanding.
- Systems can be implemented by nonprogrammers. This gives more control to subject-matter specialists and methodologists.

Many positive results are already beginning to appear through the early implementation of these trends, some of which are discussed below. The coming years should see even more improvements in productivity, timeliness, and data quality.

23.2 FUNCTIONS AND FEATURES OF EDITING SYSTEMS

When an organization buys or develops an editing system, it needs to have a methodical way of evaluating it. There should be two parts of the evaluation: (1) a profile of the functionality of the system and (2) an *honest* statement of agency needs regarding specific features.

23.2.1 Functional Profile of Editing Systems

There are many functions that an editing system might be called upon to perform and numerous ways to perform the same function. No one system performs all functions, so a profile characterizes what a system does and how it does it. Systems often perform complementary tasks. For example, in Statistics Canada, the DC2 system (an interactive integration system) can be used for data entry, data collection, and interactive editing of reports of major impact while without human intervention GEIS (an automated batch system) can then clear up reports where editing has minimal impact on statistical aggregates.

Table 23.1 displays functional profiles of three types of systems representing much development work in the past years. These systems are interactive integration, Fellegi and Holt (1976), and top-down (macro- or statistical) editing systems. An *interactive integration system* performs many survey tasks, combining data editing with functions such as data entry and interviewing. Data flow smoothly from one process to another, and programming code is applied across modules, for activities such as interviewing, interactive editing, and data entry instruments. *Fellegi and Holt systems* automate tasks formerly

Table 23.1 Survey Function Versus Type of System

Function	System Type		
	Interactive-Integration	Fellegi and Holt	Top-Down, (Macro- or Statistical)
Pre-Survey Edit Rule Analysis		Y	
High-Speed Data Entry	Y		
Computer-Assisted Survey Interview Collection	Y		
Coding	Y^a		
Interactive Edit	Y	Y^b	Y
Automated Item Deletion and Replacement		Y	
Model-Based Imputation	Y	Y	
Donor Imputation		Y^b	
Graphical Outlier Inspection			Y
Tabular Outlier Inspection			Y^b
View of Editing Effects on Estimates			Y
Weighting	Y^a		
Tabulation	Y^a		
Generalized Meta-Data Export to Other Software	Y^b		

[a]Included within the system itself or in a generalized family of software.
[b]Not all systems of this type have this capability.

reserved for people. Where records have small impact on aggregated estimates, the system is trusted to not only detect errors but also to decide the appropriate action and then take it. All actions comply with stated methodological principles. A *top-down editing system* (also known as *macro-* or *statistical editing*) allows the editor to look at the data beginning with preliminary calculations of aggregates and to interactively trace suspicious values at the aggregate level to individual reports. See Granquist (1987) for a description of the main features of such systems.

For business surveys, examples of interactive integration systems include Blaise from Statistics Netherlands and DC2 from Statistics Canada. Examples of Fellegi and Holt systems include GEIS from Statistics Canada and SPEER from the U.S. Bureau of the Census. Examples of macroediting systems include ARIES from the U.S. Bureau of Labor Statistics and the *gred* system from Statistics New Zealand. Other systems do not fit into this functional categorization. For example, the Personal Computer Electronic Data Reporting Option (PEDRO) system from the U.S. Department of Energy is primarily used to collect data through respondent electronic self-reporting (Heath 1993).

23.2.1 Lists of Editing System Features

Once it is decided what kind of system is required, then it is necessary to build a list of required and desired features. The Federal Committee on Statistical Methodology (FCSM 1990) presents one such list of editing system features. Their list was originally based on work by Cotton (1988) in evaluating four editing systems. The FCSM list covers features for all kinds of systems for all kinds of surveys. An advantage of FCSM's list is that it also reports on editing in U.S. statistical agencies along with a glossary of terms, a discussion of the role of editing systems, case studies, and descriptions of editing systems. The FCSM list and glossary include brief explanations of the more esoteric words and phrases that infest this field of study. The FCSM list is meant to be modified for specific agency needs. For example, it was a starting point for the U.S. National Agricultural Statistics Service (NASS 1992) list of features for interactive survey processing including data editing, computer-assisted interviewing, data entry, and survey management.

23.2.2 Use of Lists of Features of Editing Systems

The first step in modifying an existing list is to evaluate the agency's survey program, both current and expected. The evaluation should focus on the agency's difficult surveys, noting specifically the unique aspects of such surveys. For example, the surveys may require a complicated data set structure, production of many similar but different questionnaire versions, or extensive referral to external data files for line-by-line data checking. A good place to start is to gather a representative sample of questionnaires, making sure that all difficult data collection and processing situations are covered.

There should be a vision of where the agency is heading with its survey data processing system. The vision should encompass methodological and productivity goals and make some statement about how people are to work and how data are to be processed and passed to analysis and summary. This vision establishes the relative importance of each feature. One strength of NASS's (1992) list of features is that each feature's relative priority is stated.

The more detailed the list becomes, the more chance there is for disagreement within the evaluation committee. On the other hand, it is desirable to be as complete as possible; no one wants a system that cannot handle an important survey due to lack of a few critical features. NASS's list has 400 items and is a compromise between thoroughness and workability. It should be seen as very positive if a candidate system has much more capability than required by the list of features. It is almost impossible to anticipate every situation the agency and its surveys may encounter. The more powerful the system's language, data handling, and utilities, the better suited it will be for unanticipated needs.

Using the Profile and Lists

Editing systems that are under consideration for procurement should be pre-screened against a functional profile and the major headings in the list of features. The objective is to avoid spending too much time on systems that clearly do not meet specifications. Further evaluation of an editing system against the list should be done by agency staff with experience in the system. This may mean that the agency has to invest in training in the system. However, major capabilities or limitations can be missed if the evaluation is done by hearsay or by relying too heavily on the developer for information. To the extent possible, the system should be programmed for the most difficult agency surveys or at least for the most difficult parts of them. The system under consideration should be evaluated against all features on the list. Most features will require a simple "yes" or "no"; others may require comment as to how well the system satisfies the criterion.

Other nontechnical evaluation considerations are (historical) rate of system development, responsiveness of the developer to clients' needs including support, stability of the developer, and the system's adaptability for future needs. Finally, the system must be evaluated as to whether it can handle the agency's survey program as a whole. For example, even if a system can handle each of the agency's surveys one at a time, it may not be able to handle all of them together. This can occur if development and maintenance time for each survey is excessive due to a weak programming language that precludes the possibility of bringing more surveys into the system.

23.2.3 Major Features of Business Survey's Editing Systems

About 250 features are listed in the FCSM (1990) list of editing system features under eight major headings: general features; survey management; systems items; edit writing; types of edits; edit rule analysis; data review and correction; and support, updates, and training.

General Features

General features are characteristics that may immediately eliminate systems because they cannot meet basic needs. These features include types of data handled (categorical, numeric, text), operating systems, kinds of computer platform (mainframe, PC, local area network), other software needed (database, compiler), functionality (data collection, editing, analysis, summary, imputation), how it edits (batch, interactive), and availability (public domain, sold, licensed). Because no system handles all tasks of all surveys and operates on all platforms and operating systems, choices have to be made.

Platform considerations become less important as computer capabilities increase and costs decline. The computing paradigm is shifting from mainframe file-based activity to desktop record-based activity (Bethlehem and Keller 1991). Even the underappreciated DOS operating system on a local area network has enough power to handle a business survey as long as the system is targeted for that environment. For example, an interactive editing system handles a survey one record at a time instead of the whole file at a time, as a batch editing system would. A personal computer can handle one record at a time for any survey. Interactive systems tend to be based on microcomputers or workstation platforms, though there may be a connection to a database server.

Because they work on large files and automate tasks previously reserved for humans, Fellegi and Holt systems are used mostly with mainframes, though they can also be used with workstations. Fellegi and Holt systems are restricted as to type of data that can be handled. For example, both the GEIS and SPEER systems operate on continuous data but not on categorical data because of the mathematical algorithms used for edit rule and error analysis.

Survey Management
An editing system must be able to handle survey management needs or it must be embedded within other systems that provide this capability. If the system handles survey management, there is less work in survey preparation and in production. Survey management features include production of various reports such as lists of questionnaires not yet received, validation of identification variables against a sample master, lists of records that need further treatment, and a statement about the impact of editing and imputation on survey estimates. In addition, the survey manager should be able to respond to reports with action—for example, to call up a record and fix it, to add or delete records, to send a form back for follow-up, to mail out reminders, and to broaden or tighten edit limits.

Systems Items
System features include types of data storage file formats, whether data can be read in or out and with which formats, how the system is put together (modular or not), whether it is interpreted or compiled, whether audit trails and log files are kept, whether it can read external data for edit check bounds, and whether it allows calls to subroutines or has a macro capability. A well-designed and well-executed system can be adapted to new surveys much more quickly than can a sloppily built one.

Edit Writing
Edit writing is important to the subject-matter specialist in preparing the survey's analysis system. Important features are whether a specialist can enter edits directly into the system or whether a programmer must do it instead. The former capability allows the subject-matter specialist greater control. Formatting features for data display are of concern, such as whether historical calculated values can appear in an edit and edit message, whether it is possible to choose variables to be flagged, whether edit messages are in a spoken language or in a code (e.g., "Harvested Acres Exceeds Planted Acres" versus "Error 333"), and whether the edit writer has control over the user interface on a computer screen.

Types of Edits
Types of edits allowed in a system are as important as the data types allowed. Some features are easily understood, such as whether the system allows edits

to contain complex conditional statements, edit checks against historical data, and levels of edit failure (e.g., fatal versus suspicious). Definitions of other features are more obscure—for example, ratio edits, linear edits, and consistency checks (FCSM 1990). Systems that cannot handle all edits must be embedded in, or used with, other systems, for example, to perform an additivity check. However, there may be enough value added by the use of such a system (e.g., excellent imputation options) to make its use worthwhile.

Fellegi and Holt systems are constrained as to the type of edits they can handle. GEIS in Statistics Canada is based on linear edits (mathematical representation is a straight line or plane), whereas the SPEER system is based on ratio edits. However, these systems are not as constrained as it might seem, because it is usually possible to recast other edits in an appropriate way. These limitations are a consequence of the mathematical bases of these systems.

Edit Rule Analysis
Edit rule analysis is unique to Fellegi and Holt systems. It gives the subject-matter specialist the ability to evaluate the edits by checking for redundancy or contradictions, by evaluating unstated edits implied by the explicit edits, and by checking the most extreme data that can pass edits without invoking an edit failure. In Fellegi and Holt methodology, a set of edits are jointly used to define a valid range for imputations. The edit rule analysis is necessary to ensure that a good record can be defined. If not, every record will be in error.

Data Review and Correction
Covered here is how a person interacts with the system to review data and make corrections, if applicable. For example, he or she might need to decide whether changes are made interactively on a computer screen or on paper first. Also covered is which kinds of imputation are allowed, such as automated or manual updates, and whether formula-based and/or donor-based methods are allowed for automated imputations (see Chapter 22).

Support, Updates, and Training
If the system is available to external users, the purchaser must determine if it is supported and if the vendor can provided needed features, training, and documentation. The purchaser should also determine whether the system is being continually updated and with what speed. Vendor commitment to using the system in its own organization is important, as is the kind of laboratory the vendor organization represents. This includes the number and variety of surveys the vendor handles and how successfully they are implemented.

23.3 GENERALIZED SYSTEMS

Many statistical agencies have spent considerable resources in developing generalized systems. In generalized systems, conventions are standardized, programmed, tested, and then applied to many different surveys. The system has to be adapted for each survey, but common system tasks are already handled, eliminating duplication of effort. In this section, I describe in greater detail the three types of generalized systems introduced in Section 23.2. Fellegi and Holt editing systems are designed to perform much or all subject matter review and to use the subject matter specialist in preparing the system to automatically handle erroneous and suspicious forms. All this is done according to a set of guiding methodological principles. Interactive integration systems integrate editing with data collection and high-speed data entry, among other things. The integration is twofold, first in producing instruments to handle related tasks, and second in using those instruments to execute the tasks. The idea is to state specifications once and then to use them in different ways. Advanced top-down (macro- or statistical) editing systems employ interactive graphical or tabular query and tracing to identify and treat offending reports.

23.3.1 Fellegi and Holt Systems

Fellegi and Holt (1976) proposed an editing methodology whereby a computer program in batch mode inspects forms, detects problems, determines which items to replace, and then makes imputations. Certain methodological principles are enforced by such programs: (1) each record satisfies all edits, (2) the fewest possible changes are made, (3) editing and imputation must be part of the same process, and (4) imputations must retain the structure of the data. Two statistical bureaus have applied the principles to the treatment of continuous data. At Statistics Canada, the Generalized Edit and Imputation System (GEIS) is now in operation and was used to process the 1991 Census of Agriculture, among other surveys. The SPEER system from the U.S. Bureau of the Census has been used on several economic surveys since the early 1980s. These systems are described in Pierzchala (1990a, 1990b) and in FCSM (1990).

Role of the Subject-Matter Specialist
Fellegi and Holt systems require subject-matter specialists to shift much of their work from post-collection hand processing to pre-collection specification of edit and imputation rules. Both GEIS and SPEER have an edit analysis capability that gives the specialist another perspective on edit rule specification. For example, edit analysis in GEIS consists of finding conflicting or redundant edit rules, generating implied edits from two or more explicitly stated edits, and creating extremal records. These *extremal records* are artificially generated records that illustrate the worst possible records that can pass through the system untouched. The implied edits allow the specialist to determine if the edits constrain the data in an unintended way. After iterative specification, the resulting edit rules are automatically applied following data collection. Though one goal of the Fellegi and Holt approach is to eliminate post-collection review by subject-matter specialists, this is not fully accomplished in business surveys. In the GEIS system, data coding and callbacks are carried out

before the system is applied. When GEIS is applied, almost all remaining problems are cleared up in batch mode. However, there may be a few records that GEIS cannot handle, which must be reviewed by specialists (Legault and Roumelis 1992). The SPEER system has combined the Fellegi and Holt methodology with interactive processing whereby the specialist can inspect and treat any referred forms. The value of these systems is that they treat each form consistently, providing agency-sanctioned imputation options and standardized procedures both within and between surveys, while adhering to stated methodological principles.

GEIS in Statistics Canada

GEIS may be regarded as an imputation system based on user-specified edits (Kovar 1993). It can be adapted relatively quickly to most new applications, though there will usually be pre-GEIS and post-GEIS modules to build and run. Imputations are objective, reproducible, and include options such as hot deck and model-based imputation methods that can incorporate trend adjustments. GEIS produces many reports and audit trails, which measure the impact of imputation. Statistics Canada uses the system in several different ways. For example, it is used in mass imputation for administrative data subsamples, in selective editing, and in evaluating different imputation strategies.

Mass imputation is used at Statistics Canada when incomplete administrative data are available for the entire population and a subsample of administrative records is surveyed to gather information for missing fields (Kovar 1993). Unsampled records are then mass-imputed to generate complete rectangular data sets, which is especially useful for ad hoc data requests (see Chapter 22). Mass imputation applications include the Census of Construction, the Annual Motor Carrier Freight Survey, Agriculture Tax Data program, and the Agriculture Whole Farm Data Base Project.

Selective editing is a strategy useful in highly skewed populations where relatively few reports have a large impact on estimates. The high impact reports are dealt with manually, whereas GEIS deals with smaller firms of lesser impact. A score function is defined to channel reports to the appropriate processing method. Statistics Canada uses GEIS in a selective editing strategy for the Annual Survey of Manufactures (Boucher et al. 1993), and is considering a similar use for the Survey of Employment, Payrolls and Hours. The system is also being used in the Motor Carrier Freight Survey and in the income tax data acquisition program.

The largest application Statistics Canada has made of GEIS to date was for the 1991 Agricultural Census (Legault and Roumelis 1992). GEIS found quality donor records for imputation that had a reasonable impact on estimates while ensuring that imputations satisfied all edits simultaneously. it was felt that edit and imputation methodology were considerably improved over previous censuses. On the other hand, to treat 280,000 records, there were about 3500 job submissions for GEIS and 1500 submissions for surrounding jobs. Much of this disparity was due to breaking up the form into 16 edit groups (due to limitations of numbers of variables) while the population was divided into 52 imputation regions (similar farms within regions). Streamlining job submission procedures was not straightforward because users wanted to look at output between each submission. Steps taken included: (1) data preparation, (2) fine-tuning edits (to avoid unwanted effects of imputation and to speed up performance), (3) simultaneous edit and imputation in GEIS, (4) post-GEIS processing treatment of the records GEIS cannot handle, and (5) imputation of total refusals. Computer use increased somewhat over the 1986 census, while there was a slight decrease in the use of human resources. The effort made for the 1991 census should be paid back in future censuses, because the same modules can be used again.

SPEER in the U.S. Bureau of the Census

SPEER is used in several large U.S. business surveys and censuses, including the Annual Survey of Manufactures, the Census of Manufactures, the Enterprise Summary Report and the Auxiliary Establishment Report of the Economic Censuses, and the Census of Construction (Draper et al. 1990). At the Census Bureau, the SPEER system does not necessarily reduce the amount of human resources spent on editing a survey, but it allows the specialists to see more forms within the allotted time. SPEER refers selected records to workstations where they can be treated interactively. For selected records, data editors review actions taken in batch processing by SPEER and either confirm the validity of those actions or override them. Because economic data are highly skewed and imputations can have a large impact on some statistics, review capability is needed. The on-line capabilities of SPEER are based on the same programs as the batch modules and ensure that editor actions are edited according to Fellegi and Holt principles as data are changed. The implementation of on-line capability addresses a particularly difficult processing situation at the U.S. Bureau of the Census where paper questionnaires are received in Jeffersonville, Indiana and edited in Suitland, Maryland. In the past, this situation led to delays. By implementing an interactive capability, the batch cyclic process is reduced or eliminated. The SPEER system has also been used for data entry of late-arriving questionnaires in the Annual Survey of Manufactures. It is possible to edit such data as they are entered.

23.3.2 Interactive Integration Systems

Computer-assisted survey interview collection and post-collection editing apply the same (or at least very similar) routing and edit strictures to a questionnaire. Some high-speed data entry approaches also dynamically route the data entry clerk through a complicated questionnaire. Systems that can perform these functions offer the possibility of saving resources both in producing instruments and in completing production tasks. The Blaise System at Statistics Netherlands and the DC2 system at Statistics Canada were developed with the goal of integrating several different survey tasks including computer-assisted survey interview collection, high-speed data entry, interactive editing, and survey management, as well as other tasks. They maximize the use of networked workstations. Blaise is described by Bethlehem (1987), Keller et al. (1990), Bethlehem and Keller (1991), Pierzchala (1990a, 1990b, 1991), and FCSM (1990). Hundepool (1993) describes the use of Blaise in business surveys in the Netherlands. Woelfle (1993) describes the use of DC2 in multimode data collection efforts at Statistics Canada.

Interactive Editing

Blaise, SPEER, DC2, PEDRO, ARIES, and *gred* are examples of interactive editing systems. A major advantage of interactive editing is that the editor gets immediate feedback about the results of edit actions. Questionnaires are treated just once, cleaned, and then stored. This eliminates the cyclical nature of inappropriate batch-oriented editing where some edit actions may engender other edit failures causing the specialist to revisit the same questionnaire days or weeks later. The dynamic computer screen can be used to display comments, historical information, and other auxiliary information. Another advantage of interactive editing is that it is a powerful way to edit data in that it allows data to be key punched without a previous hand edit. In paper-based batch methods, the failure to hand edit before capture results in too many error messages appearing on the printout.

Two NASS studies have confirmed that interactive editing is more productive than batch paper methods. In the December 1989 Agricultural Survey in Wisconsin, there was a 13 percent time savings for interactive methods without preliminary hand edit over batch methods with preliminary hand edit (Pierzchala 1991). In the 1993 June Area Frame Survey in Indiana, there was a 9 percent time decrease when the preliminary hand edit was partially removed and a 16 percent time decrease when the hand edit was totally removed (Eklund 1993). However, the Indiana staff were uncomfortable with totally removing the preliminary hand edit due to the form's complexity. Further work must be done to define a minimal preliminary hand edit for this difficult survey.

Where it works, the capture of raw data increases productivity by cutting down on separate reviews and providing a more automated way of evaluating survey performance. If data are hand-edited before data capture, then much information about poorer performing parts of the survey are lost and must be gathered again through special inspections of paper forms.

In the U.S. Energy Information Administration, the use of the interactive electronic self-reporting system PEDRO has reduced error rates by as much as 60 to 70 percent. For all petroleum supply forms, the number of verification calls has fallen by more than 80 percent (Heath 1993).

Results at Statistics Netherlands

For Statistics Netherlands, the change from traditional batch editing procedures to Blaise interactive editing and from mainframe processing to local area network (LAN) processing was used as an opportunity to change procedures

8

(Bethlehem and Keller 1991). The goal was to eliminate steps in data processing. As a result, for most business surveys, specialists now enter data while editing it. Data entry is slower than before; however, overall processing time is reduced. Statistics Netherlands managed to absorb a 20 percent budget cut over a 4-year period recently without a cut in survey program (Bethlehem 1992, 1993) and a further 13 percent cut with a minimal loss of program. Keller et al. (1990) note that the change to decentralized processing has led to a 25 percent reduction in Automation Department staff because standard data processing tools decrease the need for tailor-made programs. Surveys experienced a large reduction in processing time, therefore improving timeliness while at the same improving data quality because of broad use of computer-assisted survey interview collection methods, including electronic self-reporting in municipalities, fire brigades, and trading firms in the Netherlands (Hundepool 1993). Automation infrastructure costs go down each year. Statistics Netherlands spent 25 million guilders on automation infrastructure in 1987, 20 million in 1990, and only 16 million guilders (excluding personnel costs) in 1993. It is difficult to project the savings other agencies might experience, because there is no way of knowing relative efficiencies of each organization and their unique operating conditions. However, these data are very suggestive that aggressive, focused, and well-planned technological innovations based on well-developed standard software can result in substantial savings while increasing data quality, reducing respondent burden, and improving timeliness.

23.3.3 Interactive Top-Down Graphical and Tabular Query

Top-down (macro- or statistical) editing has long been part of statistical agencies' procedures, but is often found under different names such as "data listings," "statistical analysis," and so on. These edits have usually been part of batch paper processes and have included much tedious inspection of charts and tables. Significant value is added when the macro-statistical inspection is performed interactively in a top-down method which uses a graphical or tabular computer screen display. For example, if an aggregate value is outside historically determined bounds, the problem can be traced quickly to the offending report. The analyst can see the effect of changing the weight or of correcting the data item. Care must be taken so that the process cannot be abused. The data editor is given so much power that it would be very easy to "cook" the data and make the data appear consistent and plausible all the way down to the level of the individual report! (See Chapter 21 for a discussion of creative editing.) The U.S. Bureau of Labor Statistics (BLS) has taken steps to ensure that editor actions do not inappropriately affect important labor statistics. For example, audit trails are kept of editor actions and each editor is allowed to see only a specific part of the overall data file. The Automated Review of Industry Employment Statistics (ARIES) system from the U.S. Bureau of Labor Statistics and the *gred* system from Statistics New Zealand put these concepts into practice.

ARIES from the U.S. Bureau of Labor Statistics

The ARIES system was developed by the U.S. Bureau of Labor Statistics for the macro-treatment of data from the Current Employment Statistics (CES) survey. It greatly eases the review of data from about 300,000 reports in about 1600 basic estimating cells and about 1000 aggregate cells in an extremely tight time frame (Esposito et al. 1993). The system uses graphical and tabular representations of the data at various levels of aggregation. An anomaly map gives an analyst a visual representation of an industry. The *anomaly map* is a tree of cells portrayed in a circular form in which the outer cells represent basic estimating cells and the inner circles represent data cells that are progressively more aggregated as they approach the center. The cells are connected by lines radiating outward from the center. Colors are used to flag large changes from prior periods' data from the aggregate cells to the basic estimation cells. In this way, the analyst can quickly trace movements at the aggregate level to one or several cells at a lower level. A mouse is used to click on nodes, the effect of which is to pop up historical data for that node. By such maneuvers, the analyst can trace offending data back to the micro level. There are also tabular ways to review data in the ARIES system. ARIES has resulted in savings of 14 million lines of computer printout per year (80 percent savings) and has increased productivity by 50 percent, while being well-received by the analysts using it. ARIES is a dedicated package meant to operate only on the CES survey; however, it furnishes a good model for other systems.

gred System from Statistics New Zealand

Graphical editor or *gred* is a general-purpose graphical editing system for business surveys currently under development in New Zealand (Houston and Bruce 1992). Through the use of box plots, *gred* displays unit records for a single variable for several periods at one time (Houston 1993). It is thereby possible to click on a data point with a mouse and pop up a graph showing estimates for an industry with and without the contribution of a particular unit. Thus, *gred* is not used to change data, but may be used to indicate when a manual adjustment of a firm's weight is needed. It can also assist in monitoring surveys.

23.4 CONCLUDING REMARKS

No automated editing system totally replaces the need for human treatment of forms. People are still better than the computer at certain tasks; for example, recognizing problems that the computer has not been programmed to find and applying subject-matter knowledge. On the other hand, computers are fast, consistent, and can accurately compare data across all forms. People and systems interact: People enter data into the system while editing them, as in computer-assisted survey interview collection; the system is used to enhance human review, as in interactive editing; or the system is applied only after people have treated an important subset of the forms. For example, it is easy to find offending reports in a massive file using interactive graphical and tabular query methods. Upon locating such reports, an individual can apply a range of imputation procedures according to specified priorities. Thus, there is not a necessary conflict between human and machine editing. The latter can and should make the former more satisfactory on methodological, personal, and productivity grounds.

REFERENCES

Bethlehem, J. G. (1987), "The Data Editing Research Project of the Netherlands Central Bureau of Statistics," *Proceedings of the Annual Research Conference*, Washington, DC: U.S. Bureau of the Census, pp. 194–203.

Bethlehem, J. G. (1992), letter to the author.

Bethlehem, J. G. (1993), letter to the author.

Bethlehem, J. G., and W. J. Keller (1991), "The Blaise System for Integrated Survey Processing," *Survey Methodology*, **17**, pp. 43–56.

Boucher, L., J.-P. Simard, and J.-F. Gosselin (1993), "Macro-Editing, A Case Study: Selective Editing for the Annual Survey of Manufactures Conducted by Statistics Canada," *Proceedings of the International Conference on Establishment Surveys*, Alexandria, VA: American Statistical Association, pp. 362–367.

Cotton, P. (1988), "A Comparison of Software for Editing Survey and Census Data," *Proceedings of Symposium 88, The Impact of High Technology on Survey Taking*, Ottawa: Statistics Canada, pp. 211–241.

Draper, L., T. Petkunas, and B. Greenberg (1990), "On-Line Capabilities in SPEER," *Proceedings of Symposium 90, Measurement and Improvement of Data Quality*, Ottawa: Statistics Canada, pp. 235–243.

Eklund, B. (1993), "Computer Assisted Personal Interviewing (CAPI) Costs Versus Paper and Pencil Costs," *Proceedings of the International Conference on Establishment Surveys*, Alexandria, VA: American Statistical Association, pp. 425–429.

Esposito, R., D. Lin, and K. Tidemann (1993), "The ARIES System in the BLS Current Employment Statistics Program," *Proceedings of the International Conference on Establishment Surveys*, Alexandria, VA: American Statistical Association, pp. 843–847.

Federal Committee on Statistical Methodology (1990), "Data Editing in Federal Statistical Agencies," Statistical Policy Working Paper 18, Washington, DC: U.S. Office of Management and Budget.

Fellegi, I. P., and D. Holt (1976), "A Systematic Approach to Automatic Edit and Imputation," *Journal of the American Statistical Association*, **71**, pp. 17–35.

Granquist, L. (1987), "A Report of the Main Features of a Macro-Editing Procedure Which Is Used in Statistics Sweden for Detecting Errors in Individual Observations," paper presented at the Data Editing Joint Group, Stockholm: Statistics Sweden.

Heath, C. C. (1993), "Personal Computer Electronic Data Reporting Option

(PEDRO)," *Proceedings of the International Conference on Establishment Surveys,* Alexandria, VA: American Statistical Association, pp. 173–176.

Houston, G. (1993), Letter to the author.

Houston, G., and A. Bruce (1992), "Graphical Editing for Business and Economic Surveys," unpublished report, Wellington: Statistics New Zealand.

Hundepool, A. (1993), "Automation in Survey Processing," *Proceedings of the International Conference on Establishment Surveys,* Alexandria, VA: American Statistical Association, pp. 167–172.

Keller, W. J. J. G. Bethlehem, and K.-J. Metz (1990), "The Impact of Microcomputers on Survey Processing at the Netherlands Central Bureau of Statistics," *Proceedings of the Annual Research Conference,* Washington, DC: U.S. Bureau of the Census, pp. 637–645.

Kovar, J. G. (1993), "Use of Generalized Systems in Editing of Economic Survey Data," *Bulletin of the International Statistical Institute: Proceedings of the 49th Session,* Contributed Papers, Book 2, pp. 77–78.

Legault, S., and D. Roumelis (1992), "The Use of the Generalized Edit and Imputation System (GEIS) for the 1991 Census of Agriculture," Working Paper No. BSMD-92-010E, Ottawa: Statistics Canada, Methodology Branch.

National Agricultural Statistics Service (1992), "Criteria for the Evaluation of Interactive Survey Software," Report of the Interactive Survey Software Committee, Washington, DC: U.S. Department of Agriculture.

Pierzchala, M. (1990a), "A Review of the State of the Art in Automated Data Editing and Imputation," *Journal of Official Statistics,* **6,** pp. 355–377.

Pierzchala, M. (1990b), "A Review of Three Editing and Imputation Systems," *Proceedings of the Survey Research Methods Section, American Statistical Association,* pp. 111–120.

Pierzchala, M. (1991), "One Agency's Experience with the Blaise System for Integrated Survey Processing," *Proceedings of the Survey Research Methods Section, American Statistical Association,* pp. 767–772.

Stuart, W. J. (1966), "Computer Editing Survey Data—Five Years of Experience in BLS Manpower Surveys," *Journal of the American Statistical Association,* **61,** pp. 375–383.

Woelfle, L. (1993), "Mixed Mode Data Collection and Data Processing for Economic Surveys," paper presented at the International Conference on Establishment Surveys, Buffalo, NY.

CHAPTER TWENTY-FOUR

Protecting Confidentiality in Business Surveys

Lawrence H. Cox[1]
U.S. Environmental Protection Agency

The raw material of business surveys is information on individual establishments or other business entities. Data for some statistical purposes can be obtained from publicly available sources such as land tax records or stockholders' reports or from administrative records such as income tax returns. However, many surveys require direct data collection from individual respondents. The respondent in business surveys is either (1) a farm; (2) an establishment such as a store, a plant, or a regional branch or company; or (3) an organizational unit such as a hospital, a prison, or a school.

In obtaining privileged information, the statistical agency often gives the respondent an assurance of confidentiality. The *confidentiality assurance* is a promise that the data collector will not release any data records, aggregated data, or other information from which individual business information could be revealed. Confidentiality assurance is regarded as an ethical and sound practice among statistical agencies. It is ethical in view of the confidentiality pledge made (American Statistical Association 1983), and it is sound because confidentiality assurances contribute to timely, complete, and accurate response.

The statistical agency upholds the confidentiality pledge by means of mechanisms for confidentiality protection, including staff integrity and procedures to ensure the physical security of raw data. Confidentiality protection has been a persistent concern of statistical agencies for decades (Federal Committee on

---

[1]The author developed the expressed views as the result of research begun at the U.S. Bureau of the Census. These views are not intended to represent the policies or practices of the U.S. Environmental Protection Agency or the Census Bureau.

Business Survey Methods, Edited by Cox, Binder, Chinnappa, Christianson, Colledge, Kott.
ISBN 0-471-59852-6 © 1995 John Wiley & Sons, Inc.

Statistical Methodology 1978, 1994; National Research Council 1993). This chapter presents techniques for confidentiality protection that reduce the ability of a third party to determine confidential data from data products released by the statistical agency. This process, once called *disclosure control*, is currently referred to as *disclosure limitation*.

Disclosure limitation reduces the amount of data released to the public or a third party. It is a technically and operationally complex aspect of a business survey that can conflict with the agency's responsibility to release complete, timely, and useful data products (Cox 1980). The agency must balance confidentiality protection with legitimate societal needs for data. This balance has been referred to as *the right to privacy versus the need to know* (Barabba and Kaplan 1975). Consequently, the confidentiality assurance is not absolute: The statistical agency promises to protect the respondent's confidentiality up to prevailing methodological and operational constraints. The trend within statistical agencies has been towards increased confidentiality protection as more powerful methods become available.

Participation in some business surveys is voluntary; participation in others is required by law (Federal Committee on Statistical Methodology 1978, 1994). In both cases, the interaction between the respondent and the statistical agency can be described as a *social contract* between two parties: In return for divulging privileged information that is intended to improve social or economic understanding and planning, the respondent is assured that the statistical agency will take steps to prevent discovery of this information by a third party.

Confidentiality assurance policies and practices differ between statistical agencies and sometimes between surveys within the same agency. For surveys conducted by the U.S. Bureau of the Census under its enabling legislation, Title 13 of the U.S. Code, confidentiality protection is provided to all respondents in both voluntary and mandatory surveys of nongovernmental units (Cox et al. 1986b). Confidentiality protection by the U.S. Bureau of Labor Statistics is established not by law but by longstanding agency policy (Plewes 1985). The confidentiality protection policies and practices of the major U.S. statistical agencies and programs are summarized by the Federal Committee on Statistical Methodology (1978, 1994).

This chapter focuses on confidentiality protection in statistical tables (*tabulations*) of business data. Business surveys are distinct from social surveys of persons or households. This is convenient for confidentiality protection purposes, because the issues and technical problems in confidentiality protection for business data are different from those in person or household data, for several reasons:

- Populations of businesses are typically sparse and exhibit more skewed distributions. This makes certain respondents readily identifiable in tabulations, even when the released data are highly aggregated. For example, a large automobile manufacturer, oil company, hospital, or university may be easily recognized in tabulations at state or even national levels.

- Although highly visible persons such as politicians or entertainment figures can exhibit identifiability problems, sampling designs for most social surveys make it unlikely that detailed data will be collected for any one individual. If collected, it may be possible to purge or modify such data without serious harm to overall data accuracy and completeness. By contrast, in business surveys, the most important businesses are usually the largest, most salient respondents, who typically are sampled with certainty and for whom data accuracy and completeness is essential.

- With some exceptions, confidentiality concerns in social surveys tend to focus on persons and households and not on other groupings of individuals. However, groupings of businesses such as aggregations of establishments to companies or enterprises are both analytically meaningful and confidentiality-sensitive. This creates difficult problems in confidentiality protection for business data. For example, a company may be represented heavily enough in each of two disjoint cells to force the cell union to be a disclosure cell (*multi-cell disclosure*).

Because of the importance and identifiability of large businesses, disclosure limitation methods such as data rounding and data perturbation prove ineffective for tables of aggregate business data (Cox et al. 1986a, 1986b). Consequently, the standard method for protecting respondent confidentiality in tables of aggregate business data is cell suppression.

Cell suppression examines the linear relationships between cell values imposed by the aggregation structure. Based on linear equations, Fellegi (1972) provided the first published framework for cell suppression. Cox (1980) devised mathematical methods for cell suppression and presented the architecture for a computer system to perform cell suppression for the U.S. Economic Censuses. Sande (1984) describes methods used by Statistics Canada based on general linear programming. This chapter focuses on cell suppression and mathematical methods to perform cell suppression based on a specialized linear program known as a mathematical network. Comparisons are made with the methods of Cox and Sande.

Network models for cell suppression were introduced in Cox et al. (1986a). Cox (1987b) provided a network model for cell suppression that formed the basis for the automated system used for confidentiality protection in the 1987 U.S. Economic Censuses. Further work on network suppression models and related linear programming formulations at the U.S. Bureau of the Census were implemented in the 1992 U.S. Economic Censuses. Network methods generalize to other confidentiality protection problems (Cox et al. 1986a) and to more general statistical problems (Causey et al. 1985) beyond the scope of this chapter.

The confidentiality problems associated with microdata drawn from business surveys are significant, with the result that few public-use microdata files exist for business surveys. This problem is also discussed.

24.1 ORGANIZING AND PRESENTING BUSINESS DATA: STATISTICAL TABLES

Statistical tables provide a convenient, familiar format to represent additive relations between quantities. For example, retail sales for a geographic area such as a county equal the sum of retail sales for "establishments with payroll" plus those for "establishments without payroll," total value of shipments for manufacturers in a major industry group (two-digit standard industry code, or SIC, group) equals the sum of shipments for industries comprising the two-digit group, and value of construction work done for private (i.e., nongovernmental) sponsors in the United States equals the sum of private work done in individual states.

If U.S. Economic Censuses tables were simple one- and two-dimensional tables that did not intersect, then published data could be viewed as printed pages of two-dimensional statistical tables and—at least conceptually—confidentiality protection could be done by hand on each table or page independently. Indeed, this was how confidentiality was protected prior to computers and sophisticated mathematical algorithms. However, Economic Censuses' outputs are not simple, because they involve intersecting tables and tables higher than two dimensions. Even the "simple" paradigm of single pages of data in two-dimensional tables requires moderately complex mathematics to resolve. More sophisticated methods and organization are required. Over the past 20 years, mathematically based software tools for confidentiality protection have been developed and used by the U.S. Bureau of the Census, Statistics Canada, and other agencies.

The principal outputs of the U.S. Economic Censuses are statistical tables presenting totals or averages of an economic quantity over sets of establishments called *tabulation cells*. For example, a tabulation cell in the U.S. Census of Construction Industries could comprise all establishments with payroll, engaged in construction of single-family houses (SIC 1521), and located in New York State. Economic statistics for this cell for a base period such as 1992 include value of construction work done, number of persons employed, and total payroll for construction workers. This cell is denoted {SIC 1521, New York, payroll.}

Tabulation cells are usually related hierarchically. For example, the cells {SIC 1521, New York, payroll} and {SIC 1522 = other residential contractors, New York, payroll} comprise the cell {SIC 152 = general contractors—residential buildings, New York, payroll}. This cell and other cells comprise higher-level cells such as {SIC 15 = building construction—general contractors and operative builders, New York, payroll}, {SIC 152, U.S., payroll}, or {SIC 152, New York, all establishments}, which in turn combine with other cells to form even higher-level cells.

In the U.S. Economic Censuses, hierarchical relations are usually based on standard industry code (SIC), geography, or hierarchical aggregation relations between data items (e.g., total payroll = payroll for construction workers + payroll for other workers). Each hierarchical relationship between cells corresponds naturally to an additive, linear equation between the corresponding cell values. The system of equations corresponding to the full tabulation structure is large and contains many variables of the form $x_{\{SIC, Geographic\ Area, Data\ Item\}}$. Each variable occurs in several equations (three in this example). In general, if a variable occurs in d such equations, the variable is contained in a d-dimensional statistical table.

Confidentiality protection aims at preventing the exact or approximate disclosure of data for individual establishments and meaningful groups of establishments such as multiestablishment companies. To do so completely, it is necessary to deal with all additive relations from establishment to company to enterprise, all additive relations from establishment to tabulation cell levels, and all aggregation equations between tabulation cells from lower to higher levels of aggregation. In the past, this was too complicated and computationally demanding. Instead, the basic unit of analysis in confidentiality protection was the tabulation cell rather than the individual establishment. Cells were examined individually to identify the *disclosure cells*—those cells for which publication would result in divulging confidential information pertaining to a respondent. This examination is referred to as *primary disclosure analysis.*

Because tabulation cells are interrelated by aggregation equations, disclosure limitation must be performed within the framework of the linear system of aggregation equations. For example, if publishing the value of the cell {SIC 1521, New York, payroll} would cause disclosure of confidential data for one of the cell's respondents, then each of the equations

$$V(\{SIC\ 152,\ New\ York,\ payroll\}) = V(\{SIC\ 1521,\ New\ York,\ payroll\})$$
$$+ V(\{SIC\ 1522,\ New\ York,\ payroll\}),$$
$$V(\{SIC\ 1521,\ U.S.\ payroll\}) = \sum_{state} V(\{SIC\ 1521,\ state,\ payroll\}),$$

and

$$V(\{SIC\ 1521,\ New\ York,\ all\ establishments\})$$
$$= V(\{SIC\ 1521,\ New\ York,\ payroll\})$$
$$+ V(\{SIC\ 1521,\ New\ York,\ no\ payroll\}),$$

must be analyzed, where $V(*)$ denotes the value of the corresponding cell. Moreover, because other cells in these relations are involved in other equations, for example,

$$V(\{SIC\ 1522,\ New\ York,\ all\ establishments\})$$
$$= V(\{SIC\ 1522,\ New\ York,\ payroll\})$$
$$+ V(\{SIC\ 1522,\ New\ York,\ no\ payroll\}),$$

it is necessary to analyze systems of aggregation equations.

The U.S. Economic Censuses involve large and frequently complex systems of aggregation equations. Beginning with the 1967 Census of Manufactures, disclosure limitation for U.S. Economic Censuses has been automated and organized around two- and three-dimensional tables related hierarchically through the aggregation structure. This has been accomplished by using data structures and data management techniques based on mathematical lattices. The lattice methodology was originally developed by James P. Corbett. Mathematical algorithms for disclosure limitation incorporating the lattice organization were introduced in the 1977 Economic Censuses (Cox 1980). Development and refinement of the disclosure limitation methodology and data management system for U.S. Economic Censuses have continued during each successive 5-year economic censuses cycle.

Although improvements have been made in each economic censuses cycle, the organization of the disclosure limitation system for the U.S. Economic Censuses can be achieved as follows. The disclosure limitation is performed using cell suppression applied to individual two- and three-dimensional tables in a top-down manner, from higher to lower levels of aggregation (Cox 1980). The data management system carried information on what cells have been suppressed and what cells remain unprotected between aggregation levels and identifies potential lapses in disclosure protection between levels (Cox and Reinwald 1979).

A crucial process in the U.S. system is the identification, analysis, and organization of the aggregation structure into a well-defined sequence of one-, two-, and three-dimensional tables (Cox 1980, Cox and Reinwald 1979). The dimensionality of U.S. Economic Censuses tables are as follows. The Census of Manufactures has two-dimensional tables that can be treated using theoretically based, computationally efficient, disclosure-limitation algorithms. The Census of Retail Trade has a two-tier, three-dimensional structure that can be disclosure-limited using modified two-dimensional methods. The Census of Wholesale Trade has three-dimensional tables of various sizes. The Census of Mineral Industries has tables of various dimensions that are geographically sparse. The Census of Construction Industries involves three-dimensional tables and interrelated sets of three-dimensional tables. The Census Bureau has also applied its census-based methods for disclosure limitation beyond the censuses to other economic surveys, such as the Survey of Minority Owned Businesses.

Data management in the U.S. system relies on theories and operational methods developed for lattices of two-dimensional tables (Cox 1980, Cox 1987b, Cox and Reinwald 1979). Applications covered by these methods include the U.S. Census of Manufactures, for which a typical data value would be total value of shipments over a data cell defined by a unique {SIC, geography} pair (e.g., total value of shipments for automobile manufacturers in Michigan). Although the Census of Manufactures covers many industries and many geographic relations, the predominantly two-dimensional table structure ensures a close relationship between theory and application.

Tabular data are typically presented in two-dimensional tables, and it might seem that the disclosure limitation problem is inherently two-dimensional. It is not. Although tables are often treated two-dimensionally in economic analyses, for confidentiality protection each cell must be examined within its ambient tabular structure, which may be higher-dimensional or may involve interrelated sets of tables. Any change made to the value or publication status of an individual tabulation cell must be balanced by a similar change to at least one other cell in each aggregation to which the original cell contributes. For example, an {SIC, Geographic Area, Employment Size Class} cell from the Census of Wholesale Trade or an {SIC, Geographic Area, Payroll Type} cell from the Census of Construction Industries is contained in a three-dimensional table because each contributes to three independent aggregation equations.

In some cases, certain data may not be published for reasons based on subject matter or operational constraints. For example, data for all establishments and data for establishments with payroll may be published, while data for establishments without payroll may be omitted because of unreliability resulting from low response rate or large standard errors. In performing disclosure limitation, it is vital to include all unpublished data in the analysis because these data are derivable by subtraction from published data.

Statistics Canada adopted a different approach to organizing the disclosure-limitation process for its business censuses and surveys. Instead of using a table-based structure organized within a lattice, the Canadian system incorporates all tabulation equations within a single system of linear aggregation equations, and it uses general linear programming to perform the suppression. This methodology has been organized as CONFID, a suite of computer software that has been successfully implemented at Statistics Canada (Sande 1984, Robertson 1993) and subsequently adopted by the U.S. Energy Information Administration (Federal Committee on Statistical Methodology 1994).

24.2 WHAT IS DISCLOSURE?: ACHIEVING AN OPERATIONAL DEFINITION

Statistical disclosure occurs when confidential data are divulged through the release of data products. In some cases, the disclosure may be *exact disclosure*—divulging the precise value of respondent data. Exact disclosure results, for example, when data for a published cell are known to represent data for only one respondent, in which case the respondent's confidential data are divulged to all data users, or when the cell comprises only two respondents, in which case data for either respondent can be determined by the other. Exact disclosure is a clear breach of the confidentiality pledge; statistical agencies are vigorous in limiting exact disclosure. In most cases, however, close approximation of confidential data is considered equally serious. This is called *approximate* or *interval disclosure*—the ability of a third party to deduce a narrow interval estimate of the respondent's confidential data from the published data. For example, a published cell value may represent data for many respondents, but the total contribution of all others may be only a small percentage of the value of a single largest respondent, referred to as the *dominant respondent*.

Typically, the concept of disclosure evolves in a statistical agency over time into a shared sense of what does and does not constitute a breach of the confidentiality pledge. The confidentiality policies and practices of the organization are based upon these experiences. The task for the statistical agency then is to operationalize these policies into unambiguous, quantitative *disclosure rules* that distinguish disclosure from nondisclosure on a case-by-case basis (e.g., cell by cell or data item by data item). This step is quantitative and it must be approached quantitatively.

Disclosure rules differ depending upon the type of data being released—microdata, frequency count tables, or tables of aggregate data such as business data. Microdata are not released from most business surveys, including the U.S. Economic Censuses. Frequency count tables of businesses are released from business surveys, but often without disclosure limitation due to easy access to this information from publicly available sources (such as the telephone Yellow Pages or other directories). The principal outputs of business surveys are statistical tables of aggregate business data such as number of employees, payroll, receipts, and investments in equipment and research and development. It is the individual business data underlying the released tabulations that must be disclosure-limited by the statistical agency.

Disclosure limitation for tables of aggregate data is accomplished by a technique called *cell suppression*. Cell suppression is an algebraic technique for removing selected cell values from publication. It is focused at the tabulation cell level and performed within the ambient system of aggregation equations. The remainder of this chapter focuses on cell suppression techniques for business data. For discussion of confidentiality concerns for microdata and frequency count data, see Cox et al. (1986b), Cox (1987a), Duncan and Lambert (1986), and Duncan and Pearson (1991).

In the U.S. Economic Censuses, disclosure is considered to be interval disclosure of establishment or enterprise data and is defined as follows. A cell is a *primary disclosure cell* if the combined value of the n respondents contributing the largest amounts to the cell value accounts for more than k percent of the total cell value $V(X)$; any other cell is a *nondisclosure cell*. For example, a cell X comprising respondent contributions $a = 60$, $b = 16$, and $c = d = 12$, under a dominance rule with $n = 3$ and $k = 80$, would be a disclosure cell. This rule, called the (n,k) rule, is an example of a *dominance disclosure rule*. As an additional safeguard to confidentiality, the Census Bureau keeps the values of the parameters n and k confidential.

Disclosure analysis begins by applying the disclosure rule to all cells in the aggregation lattice, whether published or unpublished, to identify the primary disclosure cells. The next step is to quantify the concept of "narrow" interval estimate for each disclosure cell—that is, to associate to each primary disclosure cell X a *disclosure interval*, which is the smallest interval $I(X)$ containing the cell value $V(X)$ whose release in place of $V(X)$ would not constitute disclosure. To determine $I(X)$ for the disclosure cell X above, one reasons as follows. If there were a cell Y whose three largest respondents contributed a, b, and c as in X above, but also included additional, smaller respondents contributing a total of 22 units, then

$$V(Y) = 110 = (a + b + c)/0.80$$

$$= \text{(sum of largest } n \text{ respondent values)}/(k/100).$$

Under the (n,k) rule, Y would just barely be a nondisclosure cell. Thus, the upper limit of the disclosure interval for X should be no more than 110, and for similar reasons should be no less than 110; therefore, the upper limit of $I(X)$ should equal 110. Using a dominance rule, there is no natural value for the lower limit of the disclosure interval. A symmetric lower limit is usually chosen, although nonsymmetric intervals offer more flexibility during the cell suppression phase (Kelly et al. 1992). The symmetric disclosure interval for the cell X above is [90,110]. In general, the symmetric disclosure interval of a disclosure cell under the (n,k) rule is:

$$[2V(X) - (100/k)(x_1 + \cdots + x_n), (100/k)(x_1 + \cdots + x_n)],$$

where x_i denotes the contribution to the cell value of the i-th largest respondent (Cox 1981).

Another disclosure rule is the *p-percent percentage* rule by which X is defined to be a disclosure cell if publishing $V(X)$ would enable a respondent or third party to estimate the contribution of another respondent to within p-percent of its value. The greatest threat is posed when the second largest respondent in the cell attempts to estimate the contribution of the largest respondent. Assuming that the cell comprises m respondents, disclosure is equivalent to the condition: $x_3 + \cdots + x_m < (p/100)x_1$.

For the cell X in the preceding example to be a disclosure cell, p would have to exceed 40 percent. Values for p are usually not this large in practice. The symmetric disclosure interval for a disclosure cell under the p-percent rule is $[x_1 + x_2 - (p/100)x_1, x_1 + x_2 + (p/100)x_1]$.

A disclosure rule operationalizes the statistical agency's notion of what constitutes disclosure. A good disclosure rule measures disclosure, quantifies the minimum amount needed to protect the cell via complementary suppression, and facilitates straightforward, unambiguous identification of a disclosure interval for each disclosure cell X failing the rule. The (n,k) and p-percent disclosure rules enjoy these properties. The class of rules that exhibit these properties—and therefore are suitable candidates for disclosure rules—is the class of *subadditive linear sensitivity measures*. The principal advantage of these measures is that they ensure that the disjoint union of two nondisclosure cells remains a nondisclosure cell, a rule failing this property would be ambiguous. Another useful property is that they can be used both to identify disclosure cells X and to compute the disclosure interval $I(X)$ (Cox 1981), as illustrated in the preceding text.

The first step in disclosure limitation for business tabular data is primary disclosure analysis. Using the sensitivity measure, each disclosure cell X is identified and its disclosure interval $I(X)$ computed. The second step is to protect the disclosure cells by suppressing them from publication, together with additional, appropriately selected nondisclosure cells called *complementary suppressions*. The complementary cells are chosen to provide complete confidentiality protection without unnecessary suppression of data. *Complete confidentiality protection* at the cell level can be characterized as follows: For each disclosure cell or cell combination X within the system of equations between the suppressed cells defined by the tabulation structure, the disclosure interval $I(X)$ is a subinterval of the best interval estimate of the cell value $V(X)$.

24.3 CONFIDENTIALITY PROTECTION FOR TABULAR BUSINESS DATA: CELL SUPPRESSION

Tabulation cells are interrelated by aggregation equations. Three ways to obscure (or *mask*) $V(X)$ are as follows: (1) Alter the published values to obscure $V(X)$ sufficiently while maintaining consistency with the aggregation equations, (2) publish only interval estimates of all or some cell values, and (3) do not publish (*suppress*) the cell X together with additional cells to ensure that $V(X)$ is sufficiently obscured (i.e., that no interval estimate of $V(X)$ finer than $I(X)$ can be derived from the published tables).

The preferred disclosure limitation method for business tabulations is cell suppression. The techniques described in this chapter could be taken one step further to replace suppressed cells by nonsensitive interval values or ranges. If acceptable to users, this method could overcome some of the difficulties of using suppressed data encountered by users, particularly users with less sophisticated software tools (Cox 1993b).

Cell suppression is performed as follows. Primary disclosure cells X are suppressed from publication; that is, the corresponding cell value is not shown in the tables but is replaced by a symbol such as "*D*," indicating that disclosure limitation has been performed. Due to linear relations between cell values, additional nondisclosure cells usually most be suppressed to provide the required interval protection to X. These cells, the *complementary suppressions*, are also represented by "*D*" in the released data. They must be carefully chosen so as to provide needed protection at minimal disruption to the completeness and usefulness of the released data. In some circumstances, such as when data items are correlated, primary disclosure analysis and cell suppression may be performed on a single item and the resulting suppression pattern is carried forward to other data items. This technique, known as *key item suppression*,

Table 24.1

x_{11}	x_{12}	x_{13}	10	20	80
10	x_{22}	x_{23}	5	15	60
x_{31}	10	10	x_{34}	10	90
x_{41}	5	15	x_{44}	5	40
75	35	65	45	50	270

is performed by some, but not all, U.S. Economic Censuses (Cox and Reinwald 1979).

Verifying that complementary suppression has been adequately performed is naturally a linear programming problem, as follows. Replace each suppressed cell variable x_{ij}. For each primary disclosure cell X, compute $\min(x_{ij})$ and $\max(x_{ij})$ over the linear system defined by the aggregation equations and the variables corresponding to the suppressed cells. Confidentiality protection is complete if and only if for each primary disclosure X, the interval $I(X)$ is contained in the interval $[\min (x_{ij}), \max(x_{ij})]$.

The fictitious Table 24.1 could represent any two-dimensional business-based table, except that the symbols D have been replaced by unique variables x_{ij} representing the suppressed cells.

By itself, Table 24.1 provides no information on which suppressed cells are disclosure cells and which are complementary suppressions. However, because the corresponding table of business counts is often published without suppression, any cell comprising only one or two businesses would be a disclosure cell. Assume that the $(1,1)$ cell is a disclosure cell and that its disclosure interval is $[15,25]$. It appears that this cell is well-protected—nestled among three suppressions along both its row and column. Using the row and column equations and assuming that zero cells are not suppressed, it appears that $0 < x_{11} < 50$ is the best interval estimate of this cell, well beyond the disclosure interval. However, using linear programming, $\min(x_{11}) = \max(x_{11}) = 20$ can be derived, so that $x_{11} = 20$ is revealed—exact disclosure. Another suppression pattern must be chosen. Two possibilities would be to enhance the current pattern by suppressing either the $(1,4)$ or the $(4,2)$ cells.

This example illustrates that the complementary suppression process is not complete until its adequacy is validated by means of a *disclosure audit*—a linear programming analysis to compute upper and lower bounds for each disclosure cell. Cell suppression methods based on mathematical programming models are usually *self-auditing*. This motivates the study of mathematical and especially linear programming models for complementary suppression.

24.4 PROPERTIES OF MATHEMATICAL NETWORKS

It is beyond the scope of this chapter to discuss the theory of networks or how networks operate computationally. The reader is referred to Dantzig (1963) and other standard texts on linear programming and network optimization for this information. The following facts about networks are important and will be used:

- Networks represent the aggregation relations along rows and columns in two-dimensional tables; each internal or totals cell is represented by an arc corresponding to a unique algebraic variable; the network ensures additivity to totals among table cells.
- Networks are part of linear programming; they are used to minimize a linear cost function of variables corresponding to the internal and totals cells.
- Networks can be used to solve optimization problems involving binary (0–1) variables.
- Networks are extremely efficient computationally; large problems can be solved extremely fast by standard algorithms and software.

The last two properties, in combination, enable the use of networks to solve specialized, nonlinear problems.

Aggregation equations represent additive relations between cell values and are very specialized equations; aggregation equations can be written using only the coefficients 0 and 1. Mathematical networks take advantage of this specialized equation structure for the case of two-dimensional tables. Consider Table 24.2, ignoring, for the time being, that certain cell values are underlined.

Networks correspond naturally to two-dimensional tables, as follows. Draw a point (called a *node*) to represent each row and each column—including the totals' row and totals' column—of Table 24.2. Begin at the node representing the last column—the totals' column of row totals adding to the grand total. The 270 units of the grand total is represented as a *network flow* from the totals' column node to each of the row nodes in amounts 80, 60, 90, and 40, respectively. Represent each of these flows by a directed line segment called a *directed arc* from the totals' column node to the corresponding row node. The value of this flow equals the corresponding row total (e.g., the flow between the totals' column node and the first row node equals 80 units). Next, draw a directed arc from each row node to each column node. Set the flow along the arc between the nodes representing row *i* and column *j* equal to the value of the table cell in position (*i*, *j*) (e.g., the arc from the row 2 to the column 3 nodes will carry a flow of 20 units). Now, draw a directed arc from each

Table 24.2

20	10	20	10	20	80
10	10	20	5	15	60
40	10	10	20	10	90
5	5	15	10	5	40
75	35	65	45	50	270

column node to the totals' row node—the node representing the row of column totals adding to the grand total. Set the flow along each arc equal to the corresponding column total. The 270 units of the table have thereby been distributed along the network: first to the row nodes, from the row nodes to the column nodes, and from the column nodes to the totals' row node. Finally, the total flow is collected at the grand total: Draw an arc from the totals' row node to the totals' column node carrying a flow equal to the grand total.

Table 24.2 is represented as a network of 11 nodes and 30 arcs. In general, a table containing m internal rows and q internal columns ($m = 4$ and $q = 5$ in Table 24.2) is represented by a network containing m + 1 + q + 1 nodes and $(m + 1)(q + 1)$ arcs. Each arc represents a unique internal or totals cell, and each node represents a unique row, column, or totals equation. This is the mathematical network depicted graphically in Figure 24.1.

The network represented in Figure 24.1 represents the current status of Table 24.2. Networks are suited naturally to *equation balancing* in two-dimensional tables—that is, to reallocating values within the table while maintaining the additive relations and nonnegativity constraints of the table. To perform equation balancing in Table 24.2, the Figure 24.1 network is generalized; for each arc in the Figure 24.1 network, create an additional, oppositely directed arc. This defines a new network represented in Figure 24.2. Only selected arcs are drawn to avoid clutter.

In the Figure 24.2 network, the original or *forward arcs* are used to add flow to the corresponding table cell; the new or *backward arcs* are used to subtract flow from this cell. Flows along backward arcs are nonnegative but are "negatively" directed. Thus, for example, flows x_{ij} in the positive direction and y_{ij} in the negative direction between row node i ($i = 1, 2, \cdots, m + 1$) and

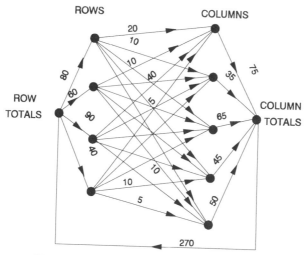

Figure 24.1 Network representation of Table 24.2

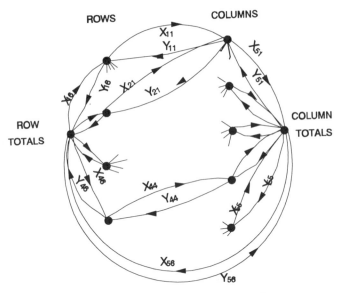

Figure 24.2 Equation balancing network

column node j ($j = 1, 2, \cdots, q + 1$) are equivalent to a flow of $|x_{ij} - y_{ij}|$ in the dominant direction and a zero flow in the opposite direction. The mathematical design of the network ensures that all flows are nonnegative. The network structure represented in Figure 24.2 ensures that all additive table relations between cells and totals and totals and the grand total remain in balance in response to adjustments to individual flows. Adjustments that do not maintain this balance are not permitted. The network structure allows totals entries to be treated in the same manner as internal entries and to be involved in the balancing or not, depending upon the presence or absence of backward arcs for totals.

The Figure 24.2 network is used to balance equations in Table 24.2 as follows. Network flows corresponding to desired adjustments to the values of selected cells in Table 24.2 are defined and provided to the Figure 24.2 network. The Figure 24.2 network is then *solved*—that is, is permitted to balance the table in response to the adjustments. The flows defining this *network solution* are precisely the net changes to Table 24.2 necessary to rebalance it to accommodate the original adjustments. The network either balances Table 24.2 or reports that no solution exists.

For example, assume that Table 24.2 represents population estimates derived from sample data and that a better estimate of the grand total cell, say 243, becomes available (for example, from newly acquired census data). To allocate the loss of $27 = 270 - 243$ units of flow in the original network, set $x_{m+1, q+1} = x_{56} = 0$ and $y_{m+1, q+1} = y_{56} = 27$ and allow the Figure 24.2 network to rebalance the table.

Table 24.3

<u>30</u>	10	<u>15</u>	<u>5</u>	20	80
<u>0</u>	10	30	5	15	60
40	10	<u>0</u>	30	10	90
5	5	20	5	5	40
75	35	65	45	50	270

Capacity constraints are limits on flows along network arcs, and they can be imposed to limit the amount by which any table value can change. Continuing the example, if in addition it is desirable to confine deviations to within 20 percent of the original values a_{ij} of Table 24.2, simply impose the capacity constraints on the Figure 24.2 network: $x_{ij}, y_{ij} \leq 0.20a_{ij}$ ($i = 1, 2, \cdots, m + 1$ and $j = 1, 2, \cdots, q + 1$). There are many balanced tables that solve this problem. To achieve no more than 10 percent deviation, there is only one solution—reduce all values of Table 24.2 by 10 percent. There is no table that will achieve less than 10 percent deviation throughout. In each case, the network either produces a solution or advises that no solution exists. In general, the capacity constraints $y_{ij} \leq a_{ij}$ are imposed to preserve nonnegativity of cell values.

As a second example, assume that the totals entries of Table 24.2 are fixed and, for example, are benchmark data from a census or previous survey, but that the internal cells values are estimates. Suppose that new information makes it desirable to replace the values $a_{11} = a_{23} = a_{34} = 20$ and $a_{44} = 10$ with $a_{11} = a_{23} = a_{34} = 30$ and $a_{44} = 5$, while keeping the values of the totals entries fixed. This could be done by setting $x_{11} = x_{23} = x_{34} = 10$ (= $30 - 20$), $y_{44} = 5$ (= $10 - 5$), $y_{11} = y_{23} = y_{34} = x_{44} = 0$ and all $x_{5j} = y_{5j} = x_{i6} = y_{i6} = 0$, and then solving the network. Table 24.3 represents one solution to this problem. Note that all but the *underlined* cell values remain unchanged.

To reduce the underlined cells of value 20 by 10 units and increase the underlined cell of value 10 by 5 units in Table 24.2, a flow involving the same five additional cells as in Table 24.3 would suffice (see Table 24.4). Therefore, the network has identified a set of five cells that, in combination with the originally selected cells, can be used to increase or decrease the value of each of the four originally selected cells by as much as 50 percent.

Table 24.4

<u>10</u>	10	<u>25</u>	15	20	80
<u>20</u>	10	10	5	15	60
40	10	<u>20</u>	10	10	90
5	5	10	<u>15</u>	5	40
75	35	65	45	50	270

24.5 USING MATHEMATICAL NETWORKS TO SELECT COMPLEMENTARY SUPPRESSIONS IN TWO-DIMENSIONAL TABLES

The preceding example can be used to illustrate the application of mathematical networks to the complementary cell suppression problem for two-dimensional tables. Let Table 24.2 represent a hypothetical table for which the underlined cells are disclosure cells. Assume that 50 percent disclosure protection is required above and below the value of each disclosure cell, so that the disclosure intervals are [10,30] for the three disclosure cells of value 20 and [5,15] for the disclosure cell of value 10. The disclosure cells are suppressed. This is equivalent to replacing them by variables in the system of aggregation equations defining Table 24.2. Complementary suppressions must be selected, which means that additional constant values must be replaced by variables in the system of aggregation equations to prevent a third party from obtaining a linear programming estimate of any of the four underlined disclosure cells of Table 24.2 that is finer than its disclosure interval. This can be accomplished using networks, as follows.

Consider Tables 24.3 and 24.4. Replace each underlined cell by the symbol D to create the table with suppressions, Table 24.5. Each of the four underlined primary suppressions of Table 24.2 has been suppressed, along with five additional underlined complementary cells. Any of Tables 24.2, 24.3 and 24.4 is a solution to Table 24.5 and is a table of values consistent with the system of aggregation equations defined by the suppressed and unsuppressed entries of Table 24.5. Table 24.3 is a solution that meets the upper bounds of the disclosure intervals for the three larger disclosure cells and the lower bound for the disclosure interval for the smaller disclosure cell. Table 24.4 is a solution for which the lower bounds for the three larger disclosure cells and the upper bound for the smaller cell are met. Therefore, for each disclosure cell X and its corresponding variable x, within the system of aggregation equations defined by Table 24.5 the conditions $\max(x) \geq \max(I(X))$ and $\min(x) \leq \min(I(X))$ are satisfied. Thus Table 24.5, which was constructed using network methods, provides adequate disclosure protection to the businesses contributing to Table 24.2.

Table 24.5 is not the only protection-adequate pattern. Using the Figure 24.5 network, other patterns could be found (Cox 1987b). Is Table 24.5 the best solution to the complementary cell suppression problem of Table 24.2? Is

Table 24.5

D	10	D	D	20	80
D	10	D	5	15	60
40	10	D	D	10	90
5	5	D	D	5	40
75	35	65	45	50	270

Table 24.6

D	10	20	D	20	80
D	10	D	5	15	60
40	10	D	D	10	90
D	5	15	D	5	40
75	35	65	45	50	270

it a good solution? That depends on how "best" and "good" are defined and measured. It is instructive to analyze Table 24.5 carefully.

Typically, the number and total value of suppressed cells are used to evaluate cell suppression patterns. Table 24.5 involves five complementary suppressions and 65 total units of complementary suppression. These suppressions succeed in protecting the primary disclosure cells to within $0 \le x_{11}, x_{23}$, $x_{34} \le 30, 0 \le x_{44} \le 25$. Within a single table, lower estimates of 0 are typical and often cannot be improved. The upper estimates on the three larger disclosure cells are tight [i.e., $\max(x) = \max(I(X))$], which is desirable, on the whole. The upper estimate on the smaller disclosure cell is not tight. However, rarely are all estimates tight. The question is, is the suppression pattern of Table 24.5 in some sense optimal, or can it be improved?

To illustrate what a good or improved suppression pattern might be, consider Table 24.6. Table 24.6 is virtually ideal. It involves the fewest possible complementary suppressions (four). This is because there are unprotected suppressions in four distinct rows of Table 24.2. Table 24.6 involves the least total complementary suppressions (35). This is because at least 10 units of complementary suppression are required in the first three rows and at least 5 units are required in the last row to meet the disclosure interval requirements. The table provides tight upper limits on all four disclosure cells. This can be verified by linear analysis.

The confluence of these three conditions is rare, and it is uncommon that a pattern will satisfy any two simultaneously. However, this example illustrates how networks could be used as a framework for examining and achieving these criteria. Some of what is reported here is research in progress (see, for instance, Cox 1993c), and these and other open research problems are examined later.

24.6 USING GENERAL LINEAR PROGRAMMING TO SELECT COMPLEMENTARY SUPPRESSIONS

The cell suppression system for the 1987–1992 U.S. Economic Censuses (henceforth referred to as CENSUS) is based on mathematical networks. CENSUS uses networks to select complementary suppressions to protect disclosure cells one at a time on a table-by-table basis, where the term table can refer to a single, two-dimensional table, a set of hierarchically related two-dimensional tables, or a three-dimensional table treated as a stack of two-dimensional tables (Cox 1993a). CENSUS can be said to perform complementary suppression *locally*, meaning that only equations involved in the immediate table or set of tables are included in the complementary suppression selection process.

Developed at Statistics Canada by Sande (1984) and later refined by Robertson (1993), the CONFID system uses general linear programming to select complementary suppressions *globally*. This means that all equations in the tabulation lattice are considered simultaneously when selecting complementary suppressions to protect a disclosure cell. Cox (1993b) compares the two systems.

The similarities between CENSUS and CONFID are substantial. Both employ subadditive linear sensitivity measures to define disclosure. Both systems process in a top-down manner through the tabulation lattice to select and protect disclosure cells one at a time. Both deal with multicell disclosure by combining direct intervention with the underlying theory. Both optimize mathematical cost functions that attempt to represent the utility of candidate cells for complementary suppression, but are able to do so only indirectly. Through use of the cost function or by other means, both systems can encourage or discourage the use of selected cells for complementary suppression. This is important, for example, when a preliminary publication has already been released, and care is needed to ensure that cells suppressed in the preliminary publication are not effectively disclosed through nonsuppression in the final publication. This capability is also useful to ensure that zero cells are not used as complementary suppressions or to encourage the use of nonpublished cells as complementary suppressions.

On the other hand, the differences between CENSUS and CONFID are revealing. Once a disclosure cell is protected by CONFID, it is fully protected because all aggregation equations have been included in the suppression analysis. CENSUS incorporates only a subset of the aggregation equations—those corresponding to a single table or interrelated sets of tables—into the cell suppression process, and therefore it may create gaps in disclosure protection that may not be detected by the lattice-based data management system. CONFID requires far more computer memory and processing than does CENSUS, and therefore it is likely to be infeasible or impractical to apply to applications as large as the U.S. Economic Censuses. CENSUS is also more portable and applicable to independent sets of smaller problems, such as might be encountered in small surveys or by organizations with specialized statistical reporting programs.

24.7 BALANCING CONFIDENTIALITY PROTECTION AND DATA USEFULNESS

Balancing confidentiality protection and data usefulness can be understood by an examination of three operational alternatives: the 1977–1982 and 1987–

1992 U.S. systems and the Statistics Canada CONFID system. The most common measures of data loss due to cell suppression are *total data value suppressed* and *total number of cells suppressed*. The 1987–1992 CENSUS attempts to minimize total value suppressed by using a unit cost that is a linear function of the cell value. The previous 1977–1982 U.S. Economic Censuses systems sought to minimize the total number of cells suppressed through a combinatorial algorithm. Because there are typically many minimum-number-of-suppression patterns, the 1977–1982 system used a secondary objective consisting of a cost function linear in the cell value to select among the minimum-number-of-suppressions patterns one of least total-value-suppressed. A mathematical search algorithm called "branch-and-bound" was used to perform this optimization. CONFID employs a unit cost that is a logarithmic function of the cell value and finds solutions that lie somewhere between the minimum value suppressed and minimum number of suppressions criteria.

Factors affecting data usefulness are sufficiently varied in that no single measure of data loss will be superior in all circumstances. The importance of individual cells can vary within or between tables and obviously can vary considerably between data users. However, as intuitive notions tend to focus on either total value suppressed or total number of suppressions or both, it is instructive to examine the 1987–1992 Economic Censuses network model in more detail and compare it to the methods underlying the 1977–1982 Economic Censuses system.

24.7.1 Minimizing the Total Value Suppressed

The confidentiality protection systems for the 1987 and 1992 U.S. Economic Censuses are based on network formulations of the complementary cell suppression problem. The use of a network is motivated in part by a desire to minimize the total value suppressed. The network design for the 1987 system is essentially that for a single two-dimensional table and is displayed in Figure 24.2.

The use of a network for complementary cell suppression ensures that suppression audit problems such as that of Table 24.1 do not occur; a disclosure cell will be fully protected within the set of aggregation equations comprising the network. Under 1987 U.S. methodology, for each disclosure cell X, two network solutions must be computed to construct a complementary suppression pattern that protects X: One network ensures that the required upper bound is achieved while a second network achieves the lower bound. The cell requiring the most protection is processed (protected) first, and the composite suppression pattern resulting from each pair of network solutions becomes the input to the problem of protecting the next cell to be protected, until all disclosure cells are protected. At each stage, disclosure cells yet to be protected (but in effect already suppressed) are given priority through the cost function selected in protecting the current cell. In the 1992 system, restrictive upper and lower capacities are used to protect each successive disclosure cell in one, rather than two, network solutions whenever possible. For present purposes, it suffices to consider both systems as based on the network represented in Figure 24.2.

The largest-to-smallest processing sequence is probably the most efficient sequence for a cell-by-cell based confidentiality protection system, because the typically larger cells taken as complementary suppressions early in the sequence are available to contribute to protecting subsequently processed smaller disclosure cells. The CONFID system uses an analogous processing sequence based upon the amount of protection required (Robertson 1993). After one or more disclosure cells are protected, it is desirable to reexamine the entire pattern to eliminate superfluous suppressions, called *oversuppression*. Sullivan and Rowe (1992) and Sullivan (1993) present methods aimed at reducing oversuppression after each successive cell is protected. A method that chose all complementary suppressions for a table or set of tables in one network solution would be much more efficient computationally and would reduce oversuppression.

Although the network design of the 1987–1992 U.S. cell suppression systems was motivated by the desire to minimize the total value of suppressed cells, cost functions that are used such as $\Sigma_{i,j} a_{ij}(x_{ij} + y_{ij})$ fail to do so; these cost functions are only a surrogate for the true cost function (total value suppressed), and often approximate it poorly. This is because the x_{ij} and y_{ij} represent flows through an arc in the Figure 24.2 network corresponding to the table cell (i, j). Other than being limited by the capacity constraints $y_{ij} \leq a_{ij}$, flows through a cell are not directly related to the cell value a_{ij}. Therefore, the product terms in this cost function do not measure data loss (i.e., total value suppressed).

Under the network model and cost function of CENSUS, if there is any flow through a cell, regardless of the amount of flow or the amount relative to the cell value, then the cell is suppressed. Replacing the network by an equivalent integer programming formulation for which $x_{ij} = 1$ or $y_{ij} = 1$ means "suppress cell (i, j)" would avoid this problem in theory, but would render the problem computationally infeasible in all but the smallest examples. Cox (1993c) replaces this integer program by a mathematical network in the two-dimensional case.

Under CENSUS, a large, important cell could be suppressed even though only a fraction of its value is needed to protect a disclosure cell. This type of oversuppression can usually be avoided by adjusting the cost function to favor cells of more appropriate size. Another way to mitigate the effects of suppressing large cells is to publish ranges for suppressed cells. The published range is precisely that which a sophisticated user could compute via linear programming analysis. Publishing ranges thus would put the less sophisticated user on an equal footing in having maximal nonconfidential access to the data.

The advantages of the CENSUS network model for cell suppression are computational speed, ability to deal with sets of hierarchically related tables, assurance of disclosure protection for single cells, and self-audit. The disadvantage is that the desired cost function—total value suppressed—is not optimized by the linear cost function, thereby resulting in oversuppression.

24.7.2 Minimizing the Number of Cells Suppressed

The automated confidentiality protection system used in the 1977–1982 U.S. Economic Censuses was based on a combinatorial procedure for minimizing the number of cells suppressed along rows and columns in a single two-dimensional table. Usually there is more than one minimum-number-of-suppressions pattern, in which case one involving the minimum total value suppressed is chosen (Cox 1980). This method is referred to as *INTRA*. INTRA is based upon a theorem that identifies the minimum number of additional suppressed cells needed to ensure that at least one additional suppression is taken in each row or column requiring suppression. This theoretical minimum provides protection *geometrically* in the sense that it protects disclosure cells completely within its individual row and column equations. However, depending upon the arrangement of cell values in the table, INTRA could fail to provide adequate protection in linear combinations of row and column equations. In practice, however, INTRA did successfully protect cells in the vast majority of actual cases (Cox and Reinwald 1979). The example associated with Table 24.1 illustrates a case in which the INTRA algorithms could fail; the (1,1) cell is protected geometrically but not completely.

The INTRA algorithm first examines patterns involving the minimum number of cells that provide sufficient protection geometrically in all rows and columns. For r = number of rows requiring additional suppression and c = number of columns requiring additional suppression, then in the general case in which $M = \max\{r, c\} \geq 2$, the minimum number of suppressions needed to provide row and column protection geometrically equals M. Among all minimum-number-of-suppressions patterns, one of minimum total-cell-value-suppressed is computed using the branch-and-bound optimization procedure (Cox 1980). If no minimum-number-of-suppressions pattern provides complete numerical protection, then one of maximal incremental protection is selected using branch-and-bound, and protection requirements for the table are reduced accordingly and the suppression process is repeated on the resulting table. A heuristic method for applying INTRA to three-dimensional tables—treating a three-dimensional table as a stack of two-dimensional tables—was successfully implemented for the 1977–1982 Economic Censuses.

The advantages of INTRA are that it controls the number of cells suppressed and, as a secondary objective, controls total value suppressed. Controlling the number of cells suppressed is important because suppressions cascade from higher to lower levels of the aggregation lattice. INTRA also constructs a suppression pattern for the entire table simultaneously, rather than one cell at a time. This reduces oversuppression and is more efficient computationally. A shortcoming of INTRA is that it is based on row and column disclosure, and therefore can miss cell combinations resulting from linear combinations of row and column equations such as the x_{11} cell of Table 24.1. Also, INTRA is not self-auditing. However, in the 1977–1982 Economic Censuses, INTRA was augmented with audit and control modules that dealt effectively with problems not covered by the underlying theory (Cox and Reinwald 1979).

24.7.3 Combining the INTRA and CENSUS Methodologies

Both the network-based CENSUS and the combinatorial INTRA methodologies are designed according to mathematical theories. That is an asset for implementation, analysis, and extendibility. However, in some cases the solutions provided by either method depend heavily upon the geometrical and numerical distribution of cell values. For example, the suppression pattern of Table 24.6 is desirable because geometrical and numerical requirements for protection are well-aligned. Numerically adequate choices of complementary suppressions are available in the right locations. The (1,4) cell provides precisely the 10 units needed to protect the (1,1) cell and does not create the need for additional suppression in column 4. The (2,1), (3,3), and (4,1) cells provide precisely the protection needed in rows 2, 3, and 4 and offer sufficient protection in their respective columns. The pattern does not create any geometrically isolated cells such as the (1,1) cell of Table 24.1. Conversely, if any or all of these cells had been small cells, the suppression pattern of Table 24.6 would not provide sufficient protection. In particular, if any of the complementary cells were zero cells, the pattern would be inadmissible. Thus, the geometrical and numerical distribution of cell values in Table 24.2 is desirable. INTRA is designed to select the optimal pattern: Table 24.6. Assuming that CENSUS actually selects minimum value suppressed at each iteration, and that the (1,1) cell is protected first, CENSUS could also produce Table 24.6. However, it might have selected the complementary suppression pattern (1,2), (2,1), (3,3), and (4,3) under the same conditions or if the (3,4) cell were protected first.

The INTRA methodology computes a complementary suppression pattern for the entire table in one iteration, whereas the CENSUS method proceeds iteratively from one disclosure cell to the next. For this reason, INTRA tends to reduce oversuppression and its results are independent of processing sequence. Although the 1977–1982 system did have procedures for resolving isolated cell and other problems, these procedures are outside the underlying theory and design of the cell suppression software, necessitating iteration of the basic method. Isolated cell problems do not occur within the CENSUS network-based methodology.

INTRA achieves its desired optimum—minimum number of cells suppressed—on a row-and-column basis. Computational experience demonstrated that INTRA tends to provide complete protection in the vast majority of cases. The CENSUS methodology uses a surrogate cost function to minimize total value suppressed, and often this approximation can be poor. Both methods are cell-based rather than establishment- or enterprise-based, and consequently they exhibit problems when the same enterprise is represented in two cells contributing to an aggregate. Both methods are limited to two-dimensional tables, although each has performed adequately when applied to three-dimensional tables treated as stacks of two-dimensional tables. Overall, both methodologies have very strong points.

Recent results of Cox (1993a, 1993c) offer reason to believe that it may be possible to combine the strong points of both methods. The INTRA minimum-number-of-suppressions theory for a single two-dimensional table can be represented as a network in the general case $M = \max\{r, c\} \geq 2$, as follows. Consider Table 24.2. Additional suppression is required in $r = 4$ rows and $c = 2$ columns. Because $r \geq c$ and $r > 1$, and complementary suppressions are restricted to the internal cells of the table only, Table 24.2 can be taken to represent the general case of the theory. The complementary suppression problem for Table 24.2 can be represented by the network illustrated in Figure 24.3. (Node requirements are represented as fixed flows along arcs for the nonmathematical reader.)

The construction of the network displayed in Figure 24.3 can be illustrated using Table 24.2. For Table 24.2, $r \geq c$; otherwise, the role of rows and columns below must be reversed. Similar to the network displayed in Figure 24.1, the Figure 24.3 network contains no backward arcs. However, in place of a feedback arc from the terminal to the initial node, the Figure 24.3 network contains a new terminal node and an arc from the old terminal node to it. Arcs corresponding to suppressed cells or zero cells are removed from the network (or, equivalently, capacitated $x_{ij} \leq 0$). Arcs corresponding to all other internal table cells are capacitated $x_{ij} \leq 1$. For each of the r ($= 4$ in Table 24.2) rows

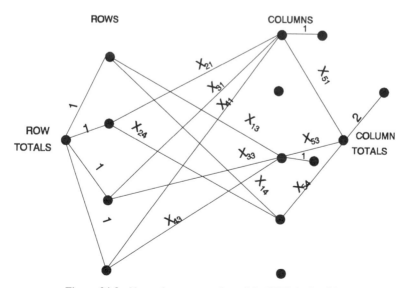

Figure 24.3 Network representation of the INTRA algorithm

requiring complementary suppression, flow along the corresponding row-total arc $x_{i,q+1}$ $(= x_{i6})$ is forced to equal one, and flow along all other row-total arcs is forced to equal zero. For each of the c $(= 2$ in Table 24.2) columns requiring complementary suppression, a new node and an arc to it from the corresponding column-total node are created, and flow along the new arc is forced to equal one. For each column-total arc, flow is forced equal to zero if the column contains no suppressions (so that no further suppressions along this column will be made); otherwise the arc flow is represented by an (uncapacitated) variable $x_{m+1,j}$ $(= x_{5j})$. Finally, the flow along the new arc from the old to the new terminal node is forced to equal $r - c$ $(= 2$ in Table 24.2).

In the network displayed in Figure 24.3, all totals flows are integer. A powerful theoretical property of networks is that, if all totals flows are integer, then any minimum-cost network flow will have integer values for all flows x_{ij} (Dantzig 1963). In view of the capacities $0 \leq x_{ij} \leq 1$, each x-value in a minimum-cost solution of the Figure 24.3 network must equal either zero or one. Cox (1993a, 1993c) used this property to formulate the nonlinear binary problem of selecting complementary suppressions in a two-dimensional table as a network flow problem that is linear and solvable by standard, computationally efficient means. The formulation is as follows. To compute a complementary suppression pattern, solve the network represented in Figure 24.3 and interpret $x_{ij} = 0$ to mean "do not suppress internal cell (i, j)" and $x_{ij} = 1$ to mean "select internal cell (i, j) as a complementary suppression." In Table 24.2, $r = 4$ complementary suppressions are selected, one in each row, with at least one selected in each of the $c = 2$ columns requiring additional suppression (columns 1 and 3).

Any solution of the Figure 24.3 network provides a complementary suppression pattern consistent with the geometrical minimum-number-of-suppressions theory of Cox (1980). The minimum-cost feature of network optimization can be used to increase the likelihood that the selected pattern is also numerically sufficient, thereby avoiding unnecessary additional iteration. This can be done by stratifying arc costs, as follows. Let $a =$ table total $(= 270$ in Table 24.2), and set arc costs c_{ij} as follows:

$$c_{ij} = \begin{cases} a_{ij} & \text{if suppressing } (i, j) \text{ protects row } i \text{ and column } j, \\ aa_{ij} & \text{if suppressing } (i, j) \text{ protects one but not both, and} \\ a^2 a_{ij} & \text{if suppressing } (i, j) \text{ protects neither.} \end{cases}$$

Note that if all cells (i, j) provide sufficient disclosure protection in both rows and columns, then the cost function $f = \Sigma_{i,j} c_{ij} x_{ij}$ is precisely the minimum total value suppressed. For Table 24.2, any solution of the Figure 24.3 network yields a complementary suppression pattern of minimal total number of suppressions. Using the cost function f, a pattern of minimum total value suppressed can be selected (Table 24.6).

24.8 GENERALIZING NETWORKS TO COMPLEX AND HIGHER-DIMENSIONAL TABLES

The network cell suppression system for the 1992 U.S. Economic Censuses incorporates sets of two-dimensional tables related hierarchically along one of the dimensions into a single network. The principal advantage is the enlargement of the set of aggregation equations considered as each disclosure cell is protected. For example, the 1987 system constructed a separate network for each table defined by county within state and by a three-digit SIC within a parent two-digit SIC; the 1992 network relates all of the corresponding SIC tables—six-digit SICs adding to four-digit SICs, four-digit adding to three-digit, three-digit adding to two-digit, and two-digit adding to county and state all-industry totals—along the SIC dimension within a single network. This is more efficient computationally and reduces oversuppression as the selection of complementary suppressions is made over a larger set of equations.

It would be desirable to extend network cell suppression methods to more general aggregation systems; there are obstacles, however. Cox and George (1989) present an algorithm for organizing sets of two-dimensional tables related additively along either the row or column dimension into a single network, but demonstrate by counterexample the impossibility of doing so along both rows and columns in general. Sande (1987) characterizes limitations of network formulations in terms of the geometry of lattices. Ernst (1989) demonstrates by counterexample the impossibility of representing a general three- or higher-dimensional table as a network.

The 1977–1992 U.S. Economic Censuses systems relied on the underlying two-dimensional methodologies for tables of all dimensions, handling the third-dimension heuristically by stacking. Research on genuine three-dimensional cell suppression methods would be useful.

The advantage of general linear programming is that it can be applied to aggregation structures of varying dimensions. Its disadvantage lies in limitations on the size of aggregation structure that it can accommodate. General linear programming formulations such as those of Sande (1984) and Robertson (1993) are worth investigation in the U.S. context. Another approach is to continue to use local cell suppression methods and the lattice, but to employ general linear programming formulations in all but the strictly two-dimensional case (Zayatz 1993).

24.9 RELEASING PUBLIC-USE MICRODATA FROM BUSINESS SURVEYS

Since the 1960s, U.S. statistical agencies have released computer files containing individual respondent data—*microdata*—from demographic censuses and surveys for public use (Federal Committee on Statistical Methodology 1978). However, very few microdata have been released from business surveys. This is because respondents to business surveys, especially large businesses or businesses sampled with certainty, usually exhibit high identifiability. Disclosure limitation methods that work well for demographic data are ineffective when applied to business data. Investigations into creating disclosure-limited public-use microdata files from business surveys have been unsuccessful, either because identifiability could not be reduced sufficiently or because the disclosure-limited microdata file did not accurately reflect econometric relationships between the original data (McGuckin and Nguyen 1990).

The underlying problem is that it is necessary to make sizeable changes to the microdata to mask the identity of respondents (particularly salient respondents). Methods that are routinely and successfully used to disclosure-limit data in social surveys are ineffective for business data. These methods include removing salient respondents from the microdata file, altering their data by rounding or error inoculation, and replacing large values for sensitive variables by an upper limit or *topcode*. One problem is that the largest and most salient respondents in business surveys are often the most important. Statistical agencies typically sample these respondents with certainty and expend considerable effort in obtaining and verifying information collected from them. Therefore, it would be counterproductive to eliminate them from a microdata file or to alter their data severely.

Ideas that promised acceptable compromise between data confidentiality and data usefulness for business microdata such as releasing *microaggregated data—*pseudo-microrecords formed by averaging values between original microrecords with similar characteristics—failed to produce usable data products in the manufacturing sector (Govoni and Waite 1985). Public-use microdata files have been released from the 1987 U.S. Census of Agriculture and further release from the 1992 Census of Agriculture is anticipated. These microdata are displayed only for highly aggregated levels of geography and microrecords are limited to a relatively small number of well-aggregated data items.

Prospects for public-use business microdata files are few, except in sectors involving relatively large or homogeneous respondent populations. The U.S. Census of Agriculture experience is encouraging because it is based on a limited, cautious format for microdata release that can be studied and perhaps expanded over time. An emerging potential methodology for public-use business microdata involves using multiple imputation to create pseudo-microdata (Rubin 1993).

A promising administrative avenue for releasing business microdata is expanded user access mechanisms (McGuckin 1993). Such mechanisms include privileged user status for researchers (as done in Germany) and controlled release of application-specific slices of larger business microdata files. Other enabling factors for public-use business microdata are expanded use of waivers of confidentiality protection from salient respondents and delaying release of microdata files until data confidentiality has substantially decayed. Another administrative (in some cases, legislative) solution is for the statistical agency to share the responsibility for nondisclosure with data users and to impose legal penalties for uses of microdata aimed at identifying microdata respondents. In most statistical agencies, these latter solutions would necessitate substantial changes to long-standing confidentiality policies. A different approach has been used successfully in Britain; Millward (1994) describes a labor relations survey where data are collected with the assurance of confidentiality.

24.10 SELECTED RESEARCH PROBLEMS

Among many important and interesting problems, three problems should contribute to the quality of results of disclosure protection methods for complementary cell suppression, and one developing area of general research should strengthen the underlying theory and methods.

24.10.1 Integrate Minimum Suppression and Minimum Value Suppressed Methods

Data analysts and users are not in agreement as to whether it is more desirable to minimize total value suppressed or to minimize number of cells suppressed to enhance data quality. It is likely that a combination of both criteria would be superior to either criterion, because a suppression pattern that balances these two objectives would be likely to avoid worst-case scenarios. Because minimum-flow networks as currently implemented do not necessarily produce minimum-value suppression patterns, incorporating controls on the number of cells suppressed could improve the quality of results and the computational efficiency of such network solutions. A place to start would be to complete the theoretical model of the minimum-number-of-suppressions network of Section 24.7.3, extending it to hierarchically related sets of tables and investigating its incorporation within current network models.

It would be worthwhile to investigate the extent to which CENSUS network-based methods could be combined with CONFID general linear programming-based methods. CENSUS and CONFID provide the only fully developed methodology and automated systems for cell suppression in current use. Both systems represent extensive long-term research, development, and institutional investments. A detailed methodological and functional comparison of the two systems would likely produce insights into improved combined methods.

24.10.2 Extend Network Cell Suppression Models and Incorporate Side Constraints

Cell suppression problems in three dimensions can be modeled as networks with a quadratic side constraint, as follows. Viewing the three-dimensional table as a stack of two-dimensional tables, include the side constraint given by "sum of squared differences between sum of entries and total entry along each three-dimensional stack equation equals zero.'' Cox et al. (1986a) provide preliminary work on incorporating some equations [but by virtue of Ernst (1989) not *all* three-dimensional equations] within a two-dimensional network. Opportunities to incorporate additional equations into two-dimensional networks and to use side constraints to move beyond two dimensions in network modeling are worth exploring. Three-dimensional networks could also be modeled as interrelated sets of two-dimensional networks. Research leading to methods for solving generalized and interrelated networks would be beneficial.

24.10.3 Protect Establishment and Enterprise Data Directly

Although disclosure is defined at the establishment or enterprise level, in practice, disclosure limitation has been focused at the tabulation cell level. Cells are aggregates of business data, so this approach is in general more conservative and is acceptable from a confidentiality standpoint. It is also necessary operationally, at least under the limitations of current theory and computational methods. However, were methods available focusing directly on business-level data, they would likely incur less removal and distortion of nonconfidential information. These methods also would likely deal directly with the problem of establishments from the same enterprise, combining across tabulation cells to produce multicell disclosure.

Disclosure limitation focused directly at the establishment or enterprise level could be accomplished by building the equation system representing the tabulation structure up directly from the establishment level to the enterprise level and from there to the cell level. The contribution of each establishment to each cell could be represented by a unique variable, and establishment contributions to cells could be aggregated to obtain enterprise-level contributions to published cells. The operational definition of disclosure then would be a *p*-percent rule that prohibits the release of any aggregate that approximates tabulated data within a fixed percentage for an establishment or for an enterprise. Complementary suppression is performed on the combined enterprise and cell systems of aggregation equations. This approach could reduce the amount of suppression needed overall. It would add an additional dimension to the problem—the enterprise dimension, which could be theoretically or computationally infeasible. But this is worth investigating, both with network-based models and with the more open equation structure of the CONFID system.

24.10.4 Use Parallel Computing

Parallel computing offers tremendous promise for many large, computationally intensive problems. For the complementary cell suppression problem, parallel computing might lead to more efficient network algorithms or to methods for integrating solutions of multiple networks to solve three- and higher-dimensional cell suppression problems. It could facilitate solving more complex problems, such as if the establishment-to-enterprise level of aggregations were incorporated into the analysis.

One area in which parallel computation might be particularly useful is that of using branch-and-bound methods to solve overlapping network structure resulting from three-dimensional tables or two-dimensional tables subject to both row and column restraints. Another area is that of achieving more globally optimal complementary suppression patterns by processing different sections of a tabulation hierarchy as separate, linked networks but in parallel.

REFERENCES

American Statistical Association (1983), "Ethical Guidelines for Statistical Practice: Report of the Ad Hoc Committee on Professional Ethics" (with discussion), *The American Statistician*, **37**, pp. 5–20.

Barabba, V., and D. Kaplan (1975), "U.S. Census Bureau Techniques to Prevent Disclosure—The Right to Privacy Versus the Need to Know," paper presented at the 40th Session International Statistical Institute, Warsaw.

Causey, B., L. Cox, and L. Ernst (1985), "Applications of Transportation Theory to Statistical Problems," *Journal of the American Statistical Association*, **80**, pp. 903–909.

Cox, L. (1980), "Suppression Methodology and Statistical Disclosure Control," *Journal of the American Statistical Association*, **75**, pp. 377–385.

Cox, L. (1981), "Linear Sensitivity Measures and Statistical Disclosure Control," *Journal of Statistical Planning and Inference*, **5**, pp. 153–164.

Cox, L. (1987a), "A Constructive Procedure for Unbiased Controlled Rounding," *Journal of the American Statistical Association*, **82**, pp. 520–524.

Cox, L. (1987b), "New Results in Disclosure Avoidance for Tabulations," *Proceedings of the International Statistical Institute*, pp. 83–84.

Cox, L. (1993a), "Solving Confidentiality Protection Problems in Tabulations Using Network Optimization: A Network Model for Cell Suppression in U.S. Economic Censuses," *Proceedings of the International Seminar on Statistical Confidentiality*, Luxembourg: EUROSTAT, pp. 229–245.

Cox, L. (1993b), "Discussion," *Proceedings of the Annual Research Conference*, Washington, DC: Bureau of the Census, pp. 132–135.

Cox, L. (1993c), "Network Models for Complementary Cell Suppression," unpublished manuscript, Research Triangle Park, NC: U.S. Environmental Protection Agency.

Cox, L., J. Fagan, B. Greenberg, and R. Hemmig (1986a), "Research at the Census Bureau into Disclosure Avoidance Techniques for Tabular Data," *Proceedings of the Survey Research Methods Section, American Statistical Association*, pp. 388–393.

Cox, L., S. K. McDonald, and D. Nelson (1986b), "Confidentiality Issues at the U.S. Bureau of the Census," *Journal of Official Statistics*, **2**, pp. 135–160.

Cox, L., and J. George (1989), "Controlled Rounding for Tables with Subtotals," *Annals of Operations Research*, **20**, pp. 141–157.

Cox, L., and L. Reinwald (1979), "U.S. Bureau of the Census Disclosure Analysis System for Economic Censuses and Surveys," paper presented at the 16th Session of the Working Party on Electronic Data Processing, Economic Commission for Europe and the Conference of European Statisticians, Geneva.

Dantzig, G. (1963), *Linear Programming and Extensions*, Princeton, NJ: Princeton University Press.

Duncan, G., and D. Lambert (1986), "Disclosure-Limited Data Dissemination" (with discussion), *Journal of the American Statistical Association*, **81**, pp. 10–28.

Duncan, G., and R. Pearson (1991), "Improving Access to Data While Protecting Confidentiality: Prospects for the Future" (with Discussion), *Statistical Science*, **6**, pp. 219–232.

Ernst, L. (1989), "Further Applications of Linear Programming to Sampling Problems," unpublished report, SRD Report: Census/SRD/RR-89/05, Washington, DC: U.S. Census Bureau.

Federal Committee on Statistical Methodology (1978), *Report on Statistical Disclosure and Disclosure-Avoidance Techniques*, Statistical Policy Working Paper 2, Washington DC: U.S. Office of Management and Budget.

Federal Committee on Statistical Methodology (1994), *Report on Statistical Disclosure Limitation Methodology*, Statistical Policy Working Paper 22, Washington, DC: U.S. Office of Management and Budget.

Fellegi, I. (1972), "On the Question of Statistical Confidentiality," *Journal of the American Statistical Association*, **67**, pp. 7–18.

Govoni, J., and J. Waite (1985), "Development of a Public Use File for Manufacturing," *Proceedings of the Survey Research Methods Section, American Statistical Association*, pp. 300–302.

Kelly, J., B. Golden, and A. Assad (1992), "Cell Suppression: Disclosure Protection for Sensitive Tabular Data," *NETWORKS*, **22**, pp. 397–417.

McGuckin, R. (1993), "Analytic Use of Economic Microdata: A Model for Researcher Access with Confidentiality Protection," *Proceedings of the International Seminar on Statistical Confidentiality*, Luxembourg: EUROSTAT, pp. 83–97.

McGuckin, R., and S. Nguyen (1990), "Public Use Microdata: Disclosure and Usefulness," *Journal of Economic Measurement*, **16**, pp. 19–40.

Millward, N. (1994), "Established Surveys in Britain: A Boon to Labour Economists," *Proceedings of the International Conference on Establishment Surveys*, Alexandria, VA: American Statistical Association, pp. 266–274.

National Research Council (1993), *Private Lives and Public Policies*, Washington, DC: National Academy Press.

Plewes, T. (1985), "Confidentiality: Principles and Practice," *Proceedings of the Annual Research Conference*, Washington, DC: U.S. Bureau of the Census, pp. 219–226.

Robertson, D. (1993), "Cell Suppression at Statistics Canada," *Proceedings of the Annual Research Conference*, Washington, DC: U.S. Bureau of the Census, pp. 107–131.

Rubin, D. (1993), "Discussion: Statistical Disclosure Limitation," *Journal of Official Statistics*, **9**, pp. 461–468.

Sande, G. (1984), "Automated Cell Suppression to Preserve Confidentiality of Business Statistics," *Statistical Journal of the United Nations: ECE*, **2**, pp. 33–41.

Sande, G. (1987), "Discussion," *Proceedings of the Annual Research Conference*, Washington, DC: U.S. Bureau of the Census, pp. 275–276.

Sullivan, C. M. (1993), "A Comparison of Adjustment Techniques to Supplement a Network Flow Disclosure Avoidance System," *Proceedings of the International Conference on Establishment Surveys*, Alexandria, VA: American Statistical Association, pp. 582–587.

Sullivan, C. M., and E. G. Rowe (1992), "A Data Structure and Integer Programming Technique to Facilitate Cell Suppression Strategies," *Proceedings of the Survey Research Methods Section, American Statistical Association*, pp. 685–690.

Zayatz, L. (1993), "Using Linear Programming Methodology for Disclosure Avoidance Purposes," *Proceedings of the International Seminar on Statistical Confidentiality*, Luxembourg: EUROSTAT, pp. 341–351.

Weighting and Estimation

CHAPTER TWENTY-FIVE

Weighting and Estimation in Business Surveys

Michael A. Hidiroglou
Statistics Canada

Carl-Erik Särndal
University of Montreal

David A. Binder
Statistics Canada

Government agencies such as Statistics Canada conduct business surveys to estimate means, totals, and ratios and change over time in these parameters. Annual surveys obtain structural information about business finances, production, employment, and ownership. Subannual surveys focus on timely measurement of economic trends. Subannual estimates are published for the nation and major geographic and industrial categories. Annual estimates are published at much finer levels of detail; demand coupled with data capacity and agency resources are the limiting factors.

For characteristics of interest, estimation focuses on the production of unbiased (or nearly unbiased) point estimates such as means, totals, and ratios and their associated measures of precision. The precision of these estimates is measured using the *coefficient of variation*, calculated as the estimated standard deviation of the point estimate divided by the point estimate and expressed as a percentage. The associated weighting and estimation procedures reflect the sampling design. Point estimates are generated for specific domains of interest. A *domain* may be the entire population or any specified subpopulation for which separate estimates are designed. The domains may completely overlap with the sampling strata or may cut across them (see Chapter 8). This

Business Survey Methods, Edited by Cox, Binder, Chinnappa, Christianson, Colledge, Kott.
ISBN 0-471-59852-6 © 1995 John Wiley & Sons, Inc.

chapter describes point estimation and variance estimation procedures for sample designs commonly used for annual and subannual business surveys.

Auxiliary data are sometimes used to improve the efficiency of the estimates. *Auxiliary data* are derived from sources outside the survey such as regularly updated administrative sources or annual totals from a larger independent survey. When the auxiliary data are correlated with the variable(s) of interest, the additional information can be incorporated into the estimation process using ratio, poststratification, regression, and raking ratio methods. These estimators can be derived by creating "new" weights as close as possible to the original design weights according to a specified metric or distance function. The application of these new weights to the auxiliary variables results in sample totals that equal the known auxiliary totals. We describe this approach following the methodology developed by Särndal et al. (1992) and by Deville and Särndal (1992). This approach is illustrated for several weight adjustment procedures commonly used for business surveys.

When estimating level or change for subannual surveys, methodological issues arise such as (1) how to handle unit and item nonresponse and (2) the proper computation of variances of ratios of two subannual estimates (trends) given that the frame and the sample change over time. These issues are also discussed in this chapter.

Computation of the variance and the coefficient of variation for the ratio of two monthly levels (trends) requires an estimate of the covariance between the two estimated levels. The chapter presents a method for computing this covariance assuming that (1) the population may change between the two occasions due to births and deaths; (2) the sample composition may also change due to births, deaths, and rotation; and (3) no auxiliary information is incorporated into the estimation process. This covariance is also useful for computing coefficients of variation of aggregates such as annual totals and for benchmark adjustments of estimated subannual totals to known annual totals.

25.1 POPULATION, SAMPLE, AND MODEL GROUPS

We begin the discussion by establishing basic concepts that are used throughout the chapter. Let $U = \{1, 2, \ldots, k, \ldots, N\}$ denote the universe of N units of a finite population of business entities. Let y_k denote the value for the kth population unit of a variable of interest y. Estimating population totals is typically an important objective for business surveys; denote the population total of y as $Y = \Sigma_U y_k$.[1]

Suppose a sample survey is conducted to estimate a population total Y. Let s denote a probability sample of units drawn from the universe U by a given sample design. The probability that unit k is selected for the sample s is $\pi_k = P(k \in s)$. The probability of unit k and unit l both appearing in the sample is $\pi_{kl} = P(k \,\&\, l \in s)$. We assume that the π_k and π_{kl} are known and positive.

---

[1]If A is a set of units, we write Σ_A to denote $\Sigma_{k \in A}$, for example, $Y = \Sigma_{k \in U} y_k = \Sigma_U y_k$.

The *Horvitz–Thompson* (1952) *estimator* of the population total Y is

$$\hat{Y}_\pi = \sum_s a_k y_k$$

where a_k is the *sampling weight* of the kth unit or $a_k = 1/\pi_k$. Typically, business surveys select their samples from a list frame using stratified simple random sampling with n_h units selected from the N_h total units in stratum h, $h = 1, 2, \ldots, H$. In this case, $a_k = 1/f_h$ for all k in stratum h, where $f_h = n_h/N_h$ is the sampling fraction in the stratum.

The Horvitz–Thompson estimator can often be improved though the use of auxiliary data. These *auxiliary data* are known totals for one or more auxiliary variables. These totals may be known for the entire population or for specified subpopulations. In this chapter, we describe how business surveys can use auxiliary data to improve the precision of survey estimates. The objective is to incorporate the information as efficiently as possible in estimation. The overall benefit derived from the use of auxiliary variables depends upon the extent to which they are correlated with the target variables for which estimates are desired.

Model groups are subpopulations for which one or more auxiliary variable totals are known. We assume that the universe U consists of a set of mutually exclusive and exhaustive subpopulations. That is, there are P model groups U_p that partition U, where $U_p \subseteq U$ and $p = 1, 2, \ldots, P$. These groups could correspond to the original sampling strata or a regrouping of them, or, more generally, to an arbitrary partition of the population for which auxiliary totals are available. Note that auxiliary totals may be available at finer levels than the chosen partition. However, the chosen partition represents a compromise such that no model group has an extremely small number of sampled units. When $P = 1$, the entire population is the only model group. The auxiliary totals are then known only for the entire population. This case leads to combined-type estimators. When $P = H$ and the groups are the original sampling strata, we obtain separate-type estimators. Cases that pool the strata into a number of groups $1 < P < H$ lead to new estimators that are compromises between the combined and separate type. These groupings are set up so as to minimize accumulation bias. An arbitrary partitioning of the population leads to poststratified-type estimators.

The sample s, drawn from population U with the given sampling design, also can be partitioned into P disjoint subsamples based upon model group membership. That is, $s = \bigcup_{p=1}^{P} s_p$, where $s_p = s \cap U_p$ is the subset of the sample that falls in the p-th model group.

25.2 WEIGHTING OF OBSERVATIONS USING KNOWN AUXILIARY TOTALS

We next describe a general weighting procedure for surveys that incorporates auxiliary information. For the kth sample unit associated with U_p, let $\mathbf{x}_{pk}$ be a vector of auxiliary measurements. More specifically, we call U_p a *model group* if:

- the auxiliary vector $\mathbf{x}_{pk}$ can be observed for every sample unit k contained in subpopulation U_p (i.e., for every $k \in s_p = s \cap U_p$), and
- the total over the subpopulation U_p is known for each auxiliary variable (i.e., the vector $\mathbf{X}_p = \Sigma_{U_p} \mathbf{x}_{pk}$ can be quantified).

For every sample unit $k \in s$, we assume that the model group to which k belongs can be identified and that the measurement (y_k, $\mathbf{x}_{pk}$) can be obtained, so that a linear regression of y on $\mathbf{x}_p$ can be fit for each model group. Note that $\mathbf{x}_{pk}$ need be known only for units selected for the sample and hence can be derived from survey responses.

The index p is added to the vector $\mathbf{x}_{pk}$ of auxiliary measurements for sample unit k because (1) the number of auxiliary variables may differ across model groups and (2) the auxiliary variables being measured may be different across model groups. As an example of the latter, we could have only one auxiliary measurement for each model group, but in model group U_1 we might have x_{1k} = gross business income of company k, in model group U_2 we might have x_{2k} = number of employees of company k, and so on. For this example, total gross business income must be known for the subpopulation of business that define U_1, total employees must be known for the subpopulation of businesses that define U_2, and so on.

Ideally, $\mathbf{x}_p$ is a good predictor of the variable of interest y within the model group U_p. For example, knowledge of the group counts N_p can improve the precision of survey estimates. In this case, the known auxiliary information for model group U_p is the total number of units belonging to model group U_p or N_p. We have $x_{pk} = 1$ for all $k \in U_p$ and $X_p = \Sigma_{U_p} x_{pk} = N_p$. Here, the model groups correspond to poststrata (see Section 25.2.2). Two procedures for taking into account auxiliary data for groups are next provided. These are the regression approach and the calibration approach. We first proceed with the regression approach.

25.2.1 Regression Approach

This approach is based on fitting a regression model within each model group. For the pth model group, consider the regression model stating that

$$y_k = \mathbf{x}'_{pk} \boldsymbol{\beta}_p + \varepsilon_k \qquad \text{for } k \in U_p \qquad (25.1)$$

where $E_\xi(\varepsilon_k) = 0$, $\text{Var}_\xi(\varepsilon_k) = c_k \sigma^2$, and $\text{Cov}_\xi(\varepsilon_k, \varepsilon_l) = 0$ for all $k \neq l$, where the subscript ξ indicates moments with respect to the regression model, and $\mathbf{x}_{pk}$ is the auxiliary variable vector for which the total $\mathbf{X}_p = \Sigma_{U_p} \mathbf{x}_{pk}$ is known. The known constants c_k are determined by the variance structure of the assumed underlying regression model given by equation (25.1). Here the super population regression parameters β_p is estimated from the finite population parameter $\mathbf{B}_p$, using the estimate from sample s or $\hat{\mathbf{B}}_p$. The population regression vector $\mathbf{B}_p$ is associated with the best fit (in the sense of generalized least squares) when all units in U_p are observed. The sample estimate $\hat{\mathbf{B}}_p$ is defined as the solution of

$$\sum_{s_p} \frac{a_k \mathbf{x}_{pk} \mathbf{x}'_{pk}}{c_k} \hat{\mathbf{B}}_p = \sum_{s_p} \frac{a_k \mathbf{x}_{pk} y_k}{c_k}$$

This represents the system of normal equations when the data $\{(y_k, \mathbf{x}_{pk}): k \in s_p\}$ are used to fit model (25.1). The weights a_k in this system of equations serve the purpose of making $\hat{\mathbf{B}}_p$ a design-consistent estimator of the population regression coefficient vector $\mathbf{B}_p$.

The total for model group p or $Y_p = \Sigma_{U_p} y_k$ is estimated by $\hat{Y}_{p\pi} + (\mathbf{X}_p - \hat{\mathbf{X}}_{p\pi})' \hat{\mathbf{B}}_p$, which is the sum of the Horvitz-Thompson estimator $\hat{Y}_{p\pi} = \Sigma_{s_p} a_k y_k$ and a regression adjustment $(\mathbf{X}_p - \hat{\mathbf{X}}_{p\pi})' \hat{\mathbf{B}}_p$. Here, $\mathbf{X}_p = \Sigma_{U_p} \mathbf{x}_{pk}$ is the known auxiliary total for the model group U_p, and $\hat{\mathbf{X}}_{p\pi} = \Sigma_{s_p} a_k \mathbf{x}_{pk}$ is the Horvitz-Thompson estimator[2] of the known auxiliary group total $\mathbf{X}_p$. To obtain the *generalized regression* (GREG) *estimator* of the entire population total, we sum over model groups, that is

$$\hat{Y}_{\text{GREG}} = \sum_{p=1}^{P} [\hat{Y}_{p\pi} + (\mathbf{X}_p - \hat{\mathbf{X}}_{p\pi})' \hat{\mathbf{B}}_p]. \tag{25.2}$$

If this estimator is written as a weighted linear sum over the sample or $\Sigma_s w_k y_k$, it is easy to verify that the weight w_k is precisely $w_k = a_k g_k$, where g_k is given by

$$g_k = 1 + (\mathbf{X}_p - \hat{\mathbf{X}}_{p\pi})' \left(\sum_{s_p} \frac{a_k \mathbf{x}_{pk} \mathbf{x}'_{pk}}{c_k} \right)^{-1} \frac{\mathbf{x}_{pk}}{c_k}. \tag{25.3}$$

The system of g-weights, calculated from equation (25.3) for $p = 1, 2, \ldots , P$, incorporates the auxiliary information associated with the particular model groups used in the estimation. We rewrite (25.2) as

$$\hat{Y}_{\text{GREG}} = \sum_{p=1}^{P} \sum_{s_p} a_k g_k y_k. \tag{25.4}$$

The fit also produces the regression residuals $e_k = y_k - \mathbf{x}'_{pk} \hat{\mathbf{B}}_p$ for $k \in s_p$. These are needed in computing $\hat{V}(\hat{Y}_{\text{GREG}})$, the estimated variance of $\hat{Y}_{\text{GREG}}$ or

---

[2] In this chapter, estimators identified by a "^" and the subscript "π" are Horvitz-Thompson estimators.

$\hat{V}$ for short. This variance estimator is given by

$$\hat{V} = \sum\sum_s \frac{\Delta_{kl}}{\pi_{kl}} \frac{g_k e_k}{\pi_k} \frac{g_l e_l}{\pi_l}, \tag{25.5}$$

where $\Delta_{kl} = \pi_{kl} - \pi_k \pi_l$, $\pi_{kk} = \pi_k$, and $\sum\sum_s$ is a compact notation for the double sum $\sum_{k \in s} \sum_{l \in s}$. The theoretical justification for g-weighting the residuals in the variance estimator (25.5) is given by Särndal et al. (1989).

When the π_{kl} are all different, equation (25.5) is computed as a double sum for many probability-proportional-to-size, without-replacement designs. For many other sampling schemes, it is possible to reduce the right-hand side of equation (25.5) to a computationally more suitable form. For example, consider stratified simple random sampling without replacement (*srswor*). For this case, $s = \cup_{h=1}^{H} s_h$, where s_h is a *srswor* drawn from the hth stratum. Here, equation (25.5) becomes

$$\hat{V} = \sum_{h=1}^{H} N_h^2 \frac{1 - f_h}{n_h} \sum_{s_h} \frac{(g_k e_k - \overline{ge}_h)^2}{n_h - 1},$$

where $\overline{ge}_h = \sum_{s_h} g_k e_k / n_h$ and $f_h = n_h / N_h$ is the sampling fraction for the hth stratum. For example, for *srswor* this variance estimator becomes

$$\hat{V} = N^2 \frac{1 - f}{n} \sum_s \frac{(g_k e_k)^2}{n - 1},$$

where $f = n/N$ and $\overline{ge} = \sum_s g_k e_k / n$ is assumed to be equal to zero. Note that $\overline{ge} = 0$ when the c_k in the model's variance structure satisfy $c_k = \lambda' \mathbf{x}_k$ for all k and for some constant vector λ. For instance, under the homoscedastic variance structure that has $\text{Var}_\xi(\varepsilon_k) = \sigma^2$ for all k, a regression model that contains an intercept term implies that $\overline{ge} = 0$.

As a standard measure of precision, survey organizations often use the estimated design-based coefficient of variation (*CV*). With the aid of $\hat{V}$ given by equation (25.5), the coefficient of variation is calculated for the generalized regression estimator (25.2), as

$$\widehat{CV} = \frac{(\hat{V})^{1/2}}{\hat{Y}_{GREG}}. \tag{25.6}$$

The following example illustrates the use of the regression estimator.

Example 1: Regression Estimation for the Canadian Survey of Employment, Payrolls and Hours
The Survey of Employment, Payrolls and Hours (SEPH) is conducted monthly by Statistics Canada. As the name suggests, a number of important variables

are collected by this survey including payrolls, employment, and number of hours worked. SEPH uses a rotating, stratified simple random sampling design with stratification by province, standard industrial classification (SIC), and employment size groups (Schiopu-Kratina and Srinath 1991).

For several years, the Horvitz-Thompson estimator has been used to produce estimates of totals for characteristics of interest. With the availability of well-correlated auxiliary variables from administrative files, alternate procedures have been explored for possible implementation later. The auxiliary data are monthly remittance data that Revenue Canada makes available to Statistics Canada. Correlations are high between monthly remittances (x) and important SEPH survey variables such as payroll and employment (y). These correlations are somewhat reduced by irregular reporting of remittances over time.

Lee and Croal (1989) studied the use of these auxiliary data in the regression estimator. They concluded that the regression estimator with monthly remittances of payroll deductions as the auxiliary variable would lead to considerable improvement over the Horvitz-Thompson estimator for the smaller SEPH strata. However, regression estimation was found suitable in a given province by SIC2 (an aggregation of the four-digit SIC) group only when two conditions were satisfied: (1) the sample size was at least 10 and (2) the correlation between number of employees and monthly remittances exceeded a specified minimum level.

For the larger strata, the Horvitz-Thompson estimator was retained. Regression estimation was used with model groups defined at the province by the SIC2 level for the smaller strata. For each model group, a simple regression model $y_k = \alpha + \beta x_k + \varepsilon_k$ was fit. (The model group subscript p is dropped for convenience.) The estimator of the y-total at the province by SIC2 level can be described as a *combined regression estimator* because the strata are combined in fitting the model. For a given province, let h be the size group index used for stratification, let i be the SIC2 category index, and let s_{hi} be the sample in the hith size by the SIC2 group. The sampling weights are $a_k = N_{hi}/n_{hi}$ for all k in stratum hi, where N_{hi} and n_{hi} are the population and sample sizes respectively, in stratum hi.

The population regression coefficient B is estimated by

$$\hat{B} = \frac{\sum_h \sum_i \sum_{k \in s_{hi}} a_k (y_k - \tilde{y}_s)(x_k - \tilde{x}_s)}{\sum_h \sum_i \sum_{k \in s_{hi}} a_k (x_k - \tilde{x}_s)^2},$$

where $\tilde{y}_s = \hat{Y}_\pi / \hat{N}_\pi$ with $\hat{N}_\pi = \sum_h \sum_i \sum_{k \in s_{hi}} a_k$; the expression for $\tilde{x}_s$ is analogous. The regression estimator of the Y in a given province by SIC2 group is

$$\hat{Y}_{\text{REG}} = \hat{Y}_\pi + \hat{B}(X - \hat{X}_\pi)$$

where $\hat{Y}_\pi = \sum_h \sum_i \sum_{k \in s_{hi}} a_k y_k$ and $\hat{X}_\pi = \sum_h \sum_i \sum_{k \in s_{hi}} a_k x_k$ are Horvitz-Thompson estimators of the totals Y and X, respectively. The g-weights are given by

$$g_k = 1 + \frac{(X - \hat{X}_\pi)(x_k - \tilde{x}_s)}{\sum_h \sum_i \sum_{k \in s_{hi}} a_k (x_k - \tilde{x}_s)^2}.$$

Testing of this estimator has indicated that large efficiency gains can be obtained compared to the Horvitz-Thompson estimator.

25.2.2 Poststratification

Poststratification is a special case of the regression approach. It is commonly used in large-scale surveys to increase the efficiency of the estimators (Holt and Smith 1979, Rao 1985, Särndal and Hidiroglou 1989, Valliant 1993). Poststratification can produce considerable variance reduction compared to ordinary Horvitz-Thompson estimation. The classic textbook example refers to the case where the sample is drawn with *srswor*. The poststratified estimator then realizes large efficiency gains when the poststrata means are widely dispersed. Statisticians have also found the traditional poststratified estimator appealing because of its favorable properties from a conditional inference perspective (Holt and Smith, 1979).

A commonly encountered form of the *poststratified estimator* is

$$\hat{Y}_{\text{POST1}} = \sum_{p=1}^{P} N_p \tilde{y}_{s_p}, \tag{25.7}$$

where $\tilde{y}_{s_p} = \sum_{s_p} a_k y_k / \hat{N}_p$ with $\hat{N}_p = \sum_{s_p} a_k$ and $s_p = s \cap U_p$.

The poststratified estimator is a special case of equation (25.2). It is derived from a model that is the special case of equation (25.1) such that $x_{pk} = 1$ for all $k \in U_p$. That is, the model is

$$y_k = \beta_p + \varepsilon_k \qquad \text{for } k \in U_p, \tag{25.8}$$

where $E_\xi(\varepsilon_k) = 0$, $\text{Var}_\xi(\varepsilon_k) = \sigma_p^2$, and $\text{Cov}_\xi(\varepsilon_k, \varepsilon_l) = 0$ for $k \neq l$. The model groups are called *poststrata* in this case. The required auxiliary information are the poststratum counts $N_p = \sum_{U_p} x_{pk}$ for $p = 1, 2, \ldots, P$. The general formula (25.2) then yields the poststratified estimator (25.7). The variance estimator for $\hat{Y}_{\text{POST1}}$ is obtained from equation (25.5) by setting $e_k = y_k - \tilde{y}_{s_p}$ for $k \in s_p$. The g-weights are $g_k = N_p/\hat{N}_p$ for all $k \in s_p$.

In business surveys, poststratification will often be used when more accurate industrial or size classifications become available. Consider a survey in which there are sampling strata based on an "old" classification as well as poststrata based on an up-to-date classification. Suppose the sampling fraction n_h/N_h is used in stratum h. Then $a_k = N_h/n_h$ for all k in stratum h. Usually the poststrata will cut across the sampling strata. Let N_{hp} be the population frequency in stratum h and poststratum p. Define the stratum total

$$N_{h.} = \sum_{p=1}^{P} N_{hp}$$

and the poststratum total

$$N_{.p} = \sum_{h=1}^{H} N_{hp}.$$

Assume that the auxiliary information is available across the strata within each poststratum so that the counts $N_{.p}$ are accurately known. The estimator (25.7) becomes

$$\hat{Y}_{POST2} = \sum_{p=1}^{P} N_{.p} \frac{\sum_{h=1}^{H} N_{h.} n_{hp} \bar{y}_{shp}/n_{h.}}{\sum_{h=1}^{H} N_{h.} n_{hp}/n_{h.}}. \tag{25.9}$$

The advantage with estimator (25.9) compared to the Horvitz-Thompson estimator $\sum_{h=1}^{H} N_{h.} \bar{y}_{sh.}$ is that if the strata are considerably out of date due to classification changes for many businesses, then estimator (25.9) will have considerably lower variance. Note that the cell counts N_{hp} may be known.

However, a different case arises if the auxiliary information is available at the cell level, so that the N_{hp} are known and sufficiently large. (If some of them are small, we recommend that $\hat{Y}_{POST2}$ be used.) The model associated with this case is

$$y_k = \beta_{hp} + \varepsilon_k$$

for units k in cell hp. Still assuming *srswor* within strata, we obtain from expression (25.2) an alternative poststratified estimator,

$$\hat{Y}_{POST3} = \sum_{h=1}^{H} \sum_{p=1}^{P} N_{hp} \bar{y}_{shp}. \tag{25.10}$$

An example of the use of $\hat{Y}_{POST3}$ would be a case where the sampling strata correspond to old industry by size categories and where the poststrata correspond to new industry by size categories. Estimates of variance for $\hat{Y}_{POST2}$ and $\hat{Y}_{POST3}$ are obtained from expression (25.5). Residuals and g-weights which enter into this expression are computed using the appropriate regression model. Next, we illustrate this procedure using an example from the SEPH.

Example 25.2: Poststratification in the SEPH
Before October 1990, the SEPH sample was drawn from a frame based on
Statistics Canada's old Business Register. In October 1990, the new Central
Frame Data Base began to be used as a frame. Several differences exist between the old and the new frame, one of them being a change in classification
system. Units on the old frame were coded using the 1970 Standard Industrial
Classifications. On the new frame, coding was based on the 1980 SICs. The
size codes and geographical codes on the new frame were also more up-to-date
than those on the old frame.

A sample was drawn from this new frame in October 1990. An analysis
showed that the sample was not representative with respect to changes in size
codes between the old and the new frame. The sample showed that 1.8 percent
of the units had an increase in their size code, while in fact 2.3 percent of the
new frame's units had an increase in size code assigned. The sample units with
an increase in their size code were smaller (in terms of employment) than the
average for units in their new size group. Because the smaller units were underrepresented and the Horvitz-Thompson estimator was used, the estimates
had a conditional upward bias. The number of units on the new frame whose
size code decreased represented 1.9 percent of the units on the frame. Based
on the sample, it was estimated that 6.1 percent of the units on the frame had
a decrease in their size code. These units had larger-than-average employment
in their new size group. Because these units were overpresented, this resulted
in a conditional upward bias in the estimates.

Consequently, it was decided to produce estimates based on poststratification, using the new population characteristics. The units in each of the three
smaller strata (based on the new classification) were poststratified. The
poststratification was based on a comparison of the old register September 1990
size code and the new October 1990 size code. Units were assigned to the
poststrata according to whether their size code had increased, decreased, or
remained the same for each SIC and size group at the Canada level. More
details of the poststratification methodology are given in Gossen and Latouche
(1992).

25.2.3 Calibration Approach

The regression approach in Section 25.2.1 is one way to incorporate auxiliary
data into the estimates. An alternative is the *calibration approach*, which finds
new weights w_k, as close as possible to the original weights a_k, that satisfy the
constraints

$$\sum_{S_p} w_k \mathbf{x}_{pk} = \mathbf{X}_p \qquad (25.11)$$

where $\mathbf{X}_p$ is the known auxiliary total for the model group U_p. In other words,
we require the weights w_k to reproduce $\mathbf{X}_p$, group by group, in such a way that
the weighted x-total over the sample gives the known group total $\mathbf{X}_p$. The

weights $w_k = a_k g_k$ with g_k given by (25.3) do in fact have this property. From this point on, we treat the whole population U as the only model group, bearing in mind that calibration could be done within each model group U_p independently.

The calibration approach, as given by Deville and Särndal (1992), leads to a class of estimators which contains the generalized regression estimator (25.2) as a special case. It also permits finding new weights that are bounded by a lower and an upper limit. For example, negative weights can be ruled out.

A metric must be specified to quantify the distance between w_k and a_k. Several possible distance functions can be considered. We provide two such functions. The first one, the *generalized least squares distance function*,

$$F\left(\frac{w_k}{a_k}\right) = \left(\frac{w_k}{a_k} - 1\right)^2 \bigg/ 2$$

leads to the regression estimator given in Section 25.2.1. The second one, the *raking ratio distance function*,

$$F\left(\frac{w_k}{a_k}\right) = \frac{w_k}{a_k} \log\left(\frac{w_k}{a_k}\right) - \frac{w_k}{a_k} + 1$$

leads to raking-ratio like estimators as given by Deming and Stephan (1940). Other distance functions considered by Deville and Särndal (1992) guarantee weights bounded by a lower limit and an upper limit. Thus, negative weights and very large positive weights can be eliminated. The computation of variances for the resulting estimators make use of residuals obtained by fitting y on $\mathbf{x}$ as in equation (25.1).

A requirement on a distance function $F(w_k/a_k)$ is that $F(1) = 0$, so that when $w_k = a_k$, the distance is zero. Let $f(z) = F'(z)$, the first derivative of F. We require $f(1) = 0$. We minimize the total weighted distance for the sample $\Sigma_s \, a_k c_k F(w_k/a_k)$ subject to constraint (25.11). The additional weight c_k accounts for heteroscedastic residuals resulting from fitting y on $\mathbf{x}$. The uniform weighting $c_k = 1$ is likely to dominate in applications. We minimize

$$\sum_s a_k c_k F\left(\frac{w_k}{a_k}\right) - \lambda'\left(\sum_s w_k \mathbf{x}_k - \mathbf{X}\right) \tag{25.12}$$

with respect to the w_k, where the vector λ is a Lagrange multiplier. Differentiating with respect to w_k, equating the result to zero, and solving for w_k, we get $w_k = a_k g(\lambda' \mathbf{x}_k/c_k)$ where $g = f^{-1}$ is the inverse function of f. To compute the weights, we first obtain the value of λ by solving the system of calibration equations obtained from expression (25.12):

$$\sum_s a_k g\left(\frac{\lambda' \mathbf{x}_k}{c_k}\right) \mathbf{x}_k = \mathbf{X}.$$

For the GLS distance function, this leads to $w_k = a_k g_k$, where the g_k are the g-weights for the GREG estimator given by equation (25.1).

For the raking ratio distance function $g(u) = e^u$, the calibration equations (25.12) can be solved iteratively. For example, the program CALMAR (Deville et al. 1993) solves the calibration equations by Newton's method and calculates the w_k weights for several distance functions, including the generalized least squares and the raking ratio functions. Other programs serving a similar purpose are M-WEIGHT (Huang and Fuller 1978) and BASCULA (Göttgens et al. 1991).

Calibration theory can be applied when the auxiliary information consists of known marginal counts in a frequency table of any number of dimensions. The family of distance functions then leads to "generalized raking ratio" estimators. When the raking ratio distance function is used, we obtain the raking ratio estimator given by Deming and Stephan (1940). We illustrate the procedure for a two-way table with r rows and c columns where the $r + c$ marginal population counts are $N_{i+} = \Sigma_{j=1}^{c} N_{ij}$; $i = 1, 2, \ldots, r$ and $N_{+j} = \Sigma_{i=1}^{r} N_{ij}$; $j = 1, 2, \ldots, c$. Here the N_{ij}'s represent the unknown $r \times c$ cell counts. The corresponding $\mathbf{x}_k$ vector can be written as $\mathbf{x}_k = (\delta_{1.k}, \delta_{2.k}, \ldots, \delta_{r.k}, \delta_{.1k}, \delta_{.2k}, \ldots, \delta_{.ck})$ where $\delta_{i.k} = 1$ if unit k belongs to row i and 0 otherwise. Similarly, $\delta_{.jk} = 1$ if unit k belongs to column j and 0 otherwise. In this case, $\mathbf{X} = \Sigma_U \mathbf{x}_k = (N_{1+}, N_{2+}, \ldots, N_{r+}, N_{+1}, N_{+2}, \ldots, N_{+c})'$, corresponding to the vector of known population marginal counts used in the calibration.

If we let $\lambda = (\rho_1, \rho_2, \ldots, \rho_r, \gamma_1, \gamma_2, \ldots, \gamma_c)'$, then $\lambda' \mathbf{x}_k = \rho_i + \gamma_j$ if unit k belongs to cell (i, j). Further, if we let $c_k = 1$ for all k, then $g(\lambda' \mathbf{x}_k / c_k) = g(\rho_i + \gamma_j)$, and equation (25.12) leads to the following system of equations to be solved for the ρ_i and γ_j:

$$\sum_{j=1}^{c} \hat{N}_{ij} g(\rho_i + \gamma_j) = N_{i+}, \quad i = 1, 2, \ldots, r,$$

$$\sum_{i=1}^{r} \hat{N}_{ij} g(\rho_i + \gamma_j) = N_{+j}, \quad j = 1, 2, \ldots, c, \qquad (25.13)$$

where $\hat{N}_{ij} = \Sigma_{s_{ij}} a_k$ is the ordinary sample weighted count estimate for cell (i, j) and s_{ij} represents the part of the sample falling in cell (i, j). Solving for λ, as in Deville and Särndal (1992), we obtain the *generalized raking ratio (RR) estimator*:

$$\hat{Y}_{RR1} = \sum_{i=1}^{r} \sum_{j=1}^{c} \hat{N}_{ij}^{W} \bar{y}_{s_{ij}} \qquad (25.14)$$

where $\hat{N}_{ij}^{W} = \hat{N}_{ij} g(\rho_i + \gamma_j)$ is the revised cell count estimate for cell (i, j) and $\bar{y}_{s_{ij}} = \Sigma_{s_{ij}} a_k y_k / \Sigma_{s_{ij}} a_k$. Note that the g-weights are given by $g_k = g(\rho_i + \gamma_j) = \hat{N}_{ij}^{W} / \hat{N}_{ij}$ for all units k in cell (i, j). In the special case where the raking ratio distance function is used, so that $g(\rho_i + \gamma_j) = \exp(\rho_i + \gamma_j)$, then the $\hat{N}_{ij}^w$ in equation (25.14) can be obtained by the iterative proportional fitting procedure of Deming and Stephan (1940). However, Newton's method converges more rapidly.

Example 25.3: Raking Ratio Estimation in the Canadian Retail Trade Survey

Raking ratio estimation was investigated for the Canadian Monthly Retail Trade Survey. The survey is based upon a stratified sample of companies, where the strata are defined by province by industry by size. Here we look at estimation for a given province U, which also defines a model group. (The province index is suppressed in the following presentation.) The objective is to estimate the province total $\Sigma_U y_k$. The industry categories are indexed $i = 1, 2, \ldots, r$, and the size classes are indexed $j = 1, 2, \ldots, c$. Also, let x be an auxiliary variable for which there exist known industry totals X_{i+}, $i = 1, 2, \ldots, r$, and known size class totals X_{+j}; $j = 1, 2, \ldots, c$. (The industry-by-size totals X_{ij}, although known, were not used in the calibration because individual cells could be very small.) The **x**-vector in (25.12) is of dimension $r + c$ and given by

$$\mathbf{x}_k' = (x_{1.k}, x_{2.k}, \ldots, x_{r.k}, x_{.1k}, x_{.2k}, \ldots, x_{.ck}),$$

where $x_{i.k} = x_k$ if k is in the i-th industry and $x_{i.k} = 0$ otherwise, and $x_{.jk} = x_k$ if k is in the jth size class and $x_{.jk} = 0$ otherwise. Let $\lambda = (\rho_1, \rho_2, \ldots, \rho_r, \gamma_1, \gamma_2, \ldots, \gamma_c)'$. We have $\lambda' \mathbf{x}_k = (\rho_i + \gamma_j) \mathbf{x}_k$ for $k \in U_{ij}$. Note that $\Sigma_U \mathbf{x}_k = (X_{1+}, X_{2+}, \ldots, X_{r+}, X_{+1}, X_{+2}, \ldots, X_{+c})$, corresponding exactly to the information used for calibration in this example. If we assume that $c_k = 1$ for all k, then the resulting system of equations obtained from (25.12) is as (25.13) with the N's being replaced with X's. The model formulation (25.1) for this case takes the form

$$y_k = (\rho_i + \gamma_j)x_k + \varepsilon_k$$

for $k \in U_{ij}$, where we assume that $E_\xi(\varepsilon_k) = 0$, $\mathrm{Var}_\xi(\varepsilon_k) = \sigma^2$, and $\mathrm{Cov}_\xi(\varepsilon_k, \varepsilon_l) = 0$ for all $k \neq l \in U$. The g-weights are then readily obtained by solving the above system of equations. Using these g-weights, $\hat{X}_{RR,ij} = \Sigma_{s_{ij}} a_k g_k x_k$ is the estimate of the cell total X_{ij} obtained from raking. The corresponding estimator for the characteristic of interest y for cell (i, j) is $\hat{Y}_{RR,ij} = \Sigma_{s_{ij}} a_k g_k y_k$.

Another complicating factor to be incorporated in the estimation, however, is that weight adjustment is needed for frame deaths. We multiply each g-weight in cell (i, j) by $\hat{N}_{ij}/N_{ij}^*$, where $\hat{N}_{ij} = \Sigma_{s_{ij}} a_k$ is an estimate of the actual number, N_{ij}, of units in cell (i, j), and N_{ij}^* is the number of units according to the frame; if deaths have occurred in the cell, then $N_{ij} \leq N_{ij}^*$. The final weight for every $k \in U_{ij}$ is $\tilde{g}_k = \{\hat{N}_{ij}/N_{ij}^*\}\{\hat{X}_{RR,ij}/(\Sigma_{s_{ij}} a_k x_k)\}$, and the resulting estimator of the province total is

$$\hat{Y}_{RR,ij} = \sum_{i=1}^{r} \sum_{j=1}^{c} \sum_{s_{ij}} a_k \tilde{g}_k y_k.$$

This estimator was found to be a considerable improvement on the standard Horvitz-Thompson estimator. The coefficient of variation was reduced from 0.08 percent of 0.05 percent at the Canada level.

25.2.4 Nonresponse

Even with reasonable efforts to obtain full response to a survey, a certain amount of nonresponse is inevitable. In some cases there is complete nonresponse (*unit nonresponse*), whereas in other cases some, but not all, variables of interest are missing (*partial nonresponse*). Typically, partial nonresponse is treated using imputation (see Chapter 22). We concentrate here on the case of unit nonresponse. Even for total nonresponse, we often have useful auxiliary information to improve the estimation. This includes classification, geographical, size, and other information on the sampling frame. This information can be used to reduce nonresponse bias.

Typically, all nonresponse treatments are equivalent to some form of weighting adjustment. Often for stratified simple random samples, for instance, nonresponse can be assumed to be random within strata. This assumption leads to substitution of n_{hr}, the count of stratum h sample respondents, for n_h in calculating the stratum h sampling fraction. This is equivalent to using a weight adjustment of n_h/n_{hr} for strata h sample respondents. It is also equivalent to poststratification where the poststrata correspond to the original design strata. As another example, repeated surveys commonly use ratio imputation, where the ratio is applied to some historical value for the nonresponding unit. The ratio is estimated from those units that have responded in the current period and for which previous period information is available. Again this can be reformulated as an adjustment to the weights.

In discussing nonresponse weighting adjustments, Oh and Scheuren (1983) advocate treating the response mechanism as another stage in the overall probability sample design. This approach is also suggested by Särndal and Swensson (1987), who consider the response mechanism as if it were a second phase of sampling. A simple and effective sampling model is to assume Bernoulli sampling (independent and identically distributed outcomes) for the response outcomes within weighting classes. This leads to reweighing each weighting class according to the estimated response probability. Särndal and Swensson (1987) discuss variance estimation under this scenario using theory from two-phase sampling. Oh and Scheuren (1983) also point out that this reweighing is analogous to poststratification adjustments.

Bethlehem (1988) argues that nonresponse bias will be reduced when $\hat{Y}_{GREG}$ in equation (25.4) is used with the inclusion probabilities adjusted according to the response probabilities. In fact, if the regression coefficients are the same for both respondents and nonrespondents, the bias vanishes. This phenomenon was also pointed out by Thomsen (1973) for the poststratification estimator $\hat{Y}_{\text{POST 1}}$.

This result is important in practice because many surveys ignore the problem of modeling the nonresponse mechanism. Instead, they assume that homogeneous response probabilities exist for the complete sample. Hence, such surveys have to rely on the bias reduction achievable using regression estimators such as $\hat{Y}_{\text{GREG}}$. As Little (1986) points out, explicit modeling of the response probabilities can lead to an inflated variance for the estimate. Little suggests looking at the relationship of the variables of interest to the estimated response probability and using an empirical Bayes approach to estimation. This is not a common procedure in practice. This would lead to different weights for different variables, a practice that is avoided for large-scale surveys.

Modeling for response probability is useful when there is doubt about the validity of the regression model for the respondent portion of the sample and for the nonrespondents. It is also a useful tool to verify that the simpler approaches can lead to satisfactory results. In general, logistic regression methods to estimate the response probability provide a rich class of models, including Bernoulli sampling within weighting classes as special cases. These models are particularly useful when the auxiliary data provides good discriminating power for the response probabilities. Folsom (1991) and Iannacchione et al. (1991) describe use of design weights when fitting these models for the response probabilities.

25.3 ESTIMATING DOMAIN TOTALS

Domains are subpopulations for which point estimates of totals, means, or other parameters are required, together with the corresponding precision measures. Denote by $s_{(d)} = s \cap U_{(d)}$ the part of the sample s that falls in the domain $U_{(d)}$. Except in rare, controlled situations such as when $U_{(d)}$ is identical to a stratum, the size of $s_{(d)}$ will be random. The theory given in Section 25.2 can easily accommodate domain estimation.

The y-data observed within the domain are $\{ y_k : k \in s_{(d)} \}$. In many cases, this information can be supplemented with auxiliary information to produce estimates with better precision. Here, we consider estimation of the following kind. Suppose that $\mathbf{x}_{pk}$ is an auxiliary vector whose total $\mathbf{X}_p = \Sigma_{U_p} \mathbf{x}_{pk}$ is known for specified model groups U_p that partition the population U. We use the data $\{ (y_k, x_k) : k \in s_{(d)} \}$ and the $\mathbf{X}_p$ totals to estimate the domain total $Y_{(d)} = \Sigma_{U_{(d)}} y_k$.

A domain $U_{(d)}$ may be related to the model groups in a variety of ways. For example, suppose that a national survey requires estimates for census divisions. From the entire population, suppose a national sample s is drawn. We may have:

1. domain = census division = model group, and
2. domain = census division; model group = entire country.

In case 1, an auxiliary total is known for the domain itself, whereas in case 2, the known auxiliary total is for the entire population. Between these extremes, there can be intermediate cases such as:

3. domain = census division; a model group exists for the region that contains the census division, and
4. domain = census division; two nonoverlapping model groups exist with the census division contained in the union of the two groups.

Although the auxiliary information in cases 3 and 4 is not about the domains themselves, the information is too valuable to waste. The question is how best to use it in estimating for the domain.

In many applications there are D domains $U_{(d)}$, $d = 1, 2, \ldots, D$, that form a partition of U. Then the whole sample s can be correspondingly partitioned as

$$s = \bigcup_{d=1}^{D} s_{(d)}.$$

Sample cells also can be identified. The dp-th sample cell is defined as $s_{(d)p} = s \cap U_{(d)} \cap U_p = s_p \cap U_{(d)}$, the set of sample units that falls in the domain $U_{(d)}$ and in the model group U_p.

Now we turn to estimation of the domain total $Y_{(d)} = \Sigma_{U_{(d)}} y_k$. A standard device in domain estimation is to introduce a *domain variable*, denoted $y_{(d)}$, whose value for the kth unit is defined as

$$y_{(d)k} = \begin{cases} y_k & \text{if } k \in U_{(d)}, \\ 0 & \text{if } k \notin U_{(d)}. \end{cases} \tag{25.15}$$

The domain total $Y_{(d)}$ can then be written as the total over the entire population U of the domain variable $y_{(d)}$, that is,

$$Y_{(d)} = \sum_{U} y_{(d)k}.$$

Estevao et al. (1995) suggest the following design-based generalized regression estimation procedure for $Y_{(d)}$. A supply of g-weights g_k is first computed according to equation (25.3) for each model group p. The weights $a_k g_k$ are then applied to the observed y_k-values in the domain to calculate the generalized regression domain estimator

$$\hat{Y}_{(d)\text{GREG}} = \sum_{p=1}^{P} \sum_{s_p} a_k g_k y_{(d)k} \tag{25.16}$$

Note that the g-weights are functions of auxiliary totals known at some level of aggregation (the model group level); this may be a coarser level than the domain level. One simple way to describe (equation) (25.16) is to say that it is given by equation (25.4) except that y_k is replaced by the domain variable value $y_{(d)k}$. Another description is as follows, assuming that all g-weights have been calculated in advance according to equation (25.3):

- Identify the *intersecting model groups* for the domain $U_{(d)}$—that is, the model groups U_p such that $U_{(d)} \cap U_p$ is nonempty.
- If U_p is an intersecting model group for $U_{(d)}$, apply the weight $a_k g_k$ to the value $y_{(d)k}$ and sum over the elements $k \in s_p = s \cap U_p$.
- Sum over all intersecting model groups; the result is the generalized regression domain estimator $\hat{Y}_{(d)\text{GREG}}$ given by equation (25.16).

The concept of intersecting model groups is important for estimating the variance of $\hat{Y}_{(d)\text{GREG}}$. The variance estimator $\hat{V}(\hat{Y}_{(d)\text{GREG}})$ is denoted $\hat{V}_{(d)}$ for short. It is calculated as

$$\hat{V}_{(d)} = \sum\sum_s \frac{\Delta_{kl}}{\pi_{kl}} \frac{g_k e_{(d)k}}{\pi_k} \frac{g_l e_{(d)l}}{\pi_l}, \tag{25.17}$$

where $e_{(d)k} = y_{(d)k} - \mathbf{x}'_{pk}\hat{\mathbf{B}}_{(d)p}$ for $k \in s$. It is simply the variance estimator (25.5), where y_k has been replaced by the domain variable value $y_{(d)k}$. Note that this replacement entails replacing $\hat{\mathbf{B}}_p$ by $\hat{\mathbf{B}}_{(d)p}$, defined as the solution of

$$\sum_{s_p} \frac{a_k \mathbf{x}'_{pk} \mathbf{x}_{pk}}{c_k} \hat{\mathbf{B}}_{(d)p} = \sum_{s_p} \frac{a_k \mathbf{x}'_{pk} y_{(d)k}}{c_k}. \tag{25.18}$$

Three different types of residuals $e_{(d)k}$ enter into the computation of equation (25.17). The first two types occur for sample units k belonging to intersecting model groups; the third type occurs for sample units k belonging to nonintersecting model groups. More specifically, with $s_p = s \cap U_p$ we have

$$e_{(d)k} = \begin{cases} y_k - \mathbf{x}'_{pk}\hat{\mathbf{B}}_{(d)p} & \text{if } k \in s_p, \ U_{(d)} \cap U_p \text{ is nonempty, and } k \in U_{(d)}; \\ -\mathbf{x}'_{pk}\hat{\mathbf{B}}_{(d)p} & \text{if } k \in s_p, \ U_{(d)} \cap U_p \text{ is nonempty, and } k \notin U_{(d)}; \\ 0 & \text{if } k \in s_p, \ U_{(d)} \cap U_p \text{ is empty.} \end{cases}$$

$$\tag{25.19}$$

The fact that $e_{(d)k}$ is zero for all k in nonintersecting model groups simplifies the calculation of $\hat{V}_{(d)}$. For example, if the sample s is drawn by *srswor*, then equation (25.17) becomes

$$\hat{V}_{(d)} = N^2 \frac{1-f}{n} \sum_s \frac{(g_k e_{(d)k} - \overline{ge}_{(d)})^2}{n-1},$$

where $\overline{ge}_{(d)} = \Sigma_s g_k e_{(d)k}/n$.

The design-based coefficient of variation is computed in a manner completely analogous to equation (25.6), namely,

$$\widehat{CV}_{(d)} = \frac{[\hat{V}_{(d)}]^{1/2}}{\hat{Y}_{(d),\text{GREG}}}$$

In practice, it is important to calculate $\widehat{CV}_{(d)}$ for all domains of interest. In some domains, $\widehat{CV}_{(d)}$ may exceed what is considered the maximum acceptable level for release, for example, $\widehat{CV}_{(d)} > 25$ percent. This may occur if the domain contains few observations and/or the auxiliary information is not strong enough. If the decision is taken to suppress the publication of the estimates $\hat{Y}_{(d)\text{GREG}}$ for some or all domains, one may consider non-design-based estimation techniques, such as synthetic estimation (see Chapter 27). However, the release of any non-design-based point and variance estimates should be accompanied by an explanation that nonstandard methods have been used.

25.3.1 Computational Issues

The computations for a domain mimic the computations carried out for the entire population. For point estimation, replacing y_k by $y_{(d)k}$ for $k \in s$ implies that equation (25.4) turns into equation (25.16). For variance estimation, replacing y_k by $y_{(d)k}$ for $k \in s$ automatically implies replacing e_k by $e_{(d)k}$ for $k \in s$, and equation (25.5) turns into equation (25.17). In other words, the computation of the domain estimator (25.16) and the corresponding variance estimator (25.17) is handled formally by replacing the y-variable of interest by the domain variable $y_{(d)}$ defined by equation (25.15). Computational simplicity is thereby gained.

The normal equations (25.18) correspond formally to the fit of the regression of the domain-specific dependent variable $y_{(d)}$ on the predictor $\mathbf{x}_p$, using the sample observations from the p-th group. This fit may be mediocre because $y_{(d)}$ is not a natural dependent variable: It equals the y-variable inside the domain but is always equal to zero outside. However, we are not primarily interested in the goodness of the fit at the domain level. Instead, the primary objective is to be able to work with g-weights that (1) yield additive domain estimates (see Case 4), and (2) remain unchanged from one domain to another. This approach has computational advantages and permits the calculation of domain estimates other than those officially reported by the organization. For alternative domain estimators, see Särndal et al. (1992, p. 408).

25.3.2 Statistical Properties

The reason that "close estimates" are obtained for domains hinges on the property of design consistency. It is known that $\hat{Y}_{GREG}$ given by equation (25.4) is a design-consistent estimator of the entire population total Y. Regardless of the configuration of finite population values ($y_1, y_2, \ldots, y_k, \ldots, y_N$), $\hat{Y}_{GREG}$ will be near Y with a high probability when the sample size is large. This is because g_k tends to 1 for large samples. Thus, the property holds in particular for the domain-specific vector ($y_{(d)1}, y_{(d)2}, \ldots, y_{(d)k}, \ldots, y_{(d)N}$). So $\hat{Y}_{(d)GREG}$ given by equation (25.16) is a design-consistent estimator of the domain total $Y_{(d)}$. Similarly, for variance estimation, $\hat{V}$ given by equation (25.7) is a design-consistent variance estimator. That is, whatever the configuration of y-values, $\hat{V}$ will be near the variance of $\hat{Y}_{GREG}$ with a high probability when the sample size is large. It follows that if the formula $\hat{V}$ is calculated for the domain-specific vector ($y_{(d)1}, y_{(d)2}, \ldots, y_{(d)k}, \ldots, y_{(d)N}$), which gives the result $\hat{V}_{(d)}$ in equation (25.17), then we have a design consistent variance estimator for $\hat{Y}_{(d)GREG}$.

Suppose we seek to estimate the total for each of D domains $U_{(d)}$, $d = 1$, $2, \ldots, D$, forming a partition of U. Then $\hat{Y}_{GREG} = \Sigma_{d=1}^{D} \hat{Y}_{(d)GREG}$ where $\hat{Y}_{GREG}$ and $\hat{Y}_{(d)GREG}$ are given by equations (25.4) and (25.16), respectively. This says that the sum of the domain estimates is equal to the estimate made with the same auxiliary information for the entire population. This additivity property is built into the estimates because it is often required by users of official statistics. It follows easily from the fact that $\Sigma_{d=1}^{D} y_{(d)k} = y_k$, for all $k \in U$. Other useful estimators satisfying this property are given by Särndal et al. (1992, pp. 397–413). Next we illustrate the use of these ideas in estimating economic production.

Example 25.3: Two-Phase Sampling of Tax Records for Economic Surveys
Annual estimates of Canadian economic production are obtained by combining estimates from two sources, namely, a large business portion and a small business portion. All large businesses are directly surveyed. For small businesses, a two-phase sample of tax records is selected, as described by Choudhry et al. (1989) and Armstrong et al. (1993). Important features of the design are:

- Bernoulli sampling is used in each phase.
- The sample drawn in each phase is poststratified.
- Estimates are constructed for the business population and for various domains of interest defined by SIC, province, revenue, and assets.

For each of the two phases, there is a sampling weight and a g-weight. These weights are essential for point estimation as well as for the corresponding variance estimation.

The first-phase sample, denoted s_1, is a stratified random sample of tax filers selected from a frame created using Revenue Canada administrative files. First-phase strata are defined by province by industry group (SIC2 or SIC3) by size.

To carry out the Bernoulli sampling, each tax filer is assigned a permanent random number in the (0, 1) interval (see Chapter 9). This number stays the same from year to year. The first-phase selection probabilities π_{1k} may be updated from one year to the next to reflect changes in strata composition and to accommodate births. The first-phase sample is longitudinal—that is, maintained over time. Bernoulli sampling facilities the selection of such a sample. Tax filers may be added to the first-phase sample each year to improve precision and the replace previously selected tax filers no longer in-scope.

Let U_p, $p = 1, 2, \ldots, P$, denote a set of first-phase poststrata. These poststrata are typically defined by collapsing first-phase sampling strata. Denote by N_p the known number of tax filers in poststratum U_p. The first-phase weight for tax filer k is

$$w_{1k} = a_{1k} g_{1k} = \frac{1}{\pi_{1k}} \frac{N_p}{\hat{N}_p}$$

for $k \in s_{1p} = s_1 \cap U_p$, where $a_{1k} = 1/\pi_{1k}$ is the first-phase sampling weight of unit k, and $g_{1k} = N_p/\hat{N}_p$ with $\hat{N}_p = \Sigma_{s_{1p}} 1/\pi_{1k}$ is the first-phase g-weight for every $k \in s_{1p}$. The realized sample size in a poststratum is random because of the Bernoulli sampling feature, and the g-weights provide variance stabilization.

Estimates are required for the business population, not tax filers. Some businesses are partnerships, and the estimators require an adjustment by which partnership data are linked to tax filer data. This technicality need not occupy us in this example because our primary aim is to illustrate domain estimation with auxiliary information. In the formulas shown below, each tax filer is treated as a business entity.

The second-phase sample, denotes s_2, is a stratified random subsample; $s_2 \subset s_1$. Second-phase strata are defined by province by SIC4 by size. The SIC4 codes are assigned by Statistics Canada to the first-phase sample. Let π_{2k} be the second-phase inclusion probability of unit k.

Let U_q, $q = 1, 2, \ldots, Q$, denote the set of second-phase poststrata. Second-phase poststrata are defined by province by SIC4 by size. Denote by N_q the unknown number of tax filers in the second-phase poststratum U_q. This number can be estimated in two ways. Using the first-phase sample, we get the estimate $\hat{N}_q = \Sigma_{s_{1q}} w_{1k}$, where $s_{1q} = s_1 \cap U_q$, and w_{1k} is the first-phase weight. Using only units in the second-phase sample, we get the alternative estimate $\tilde{N}_q = \Sigma_{s_{2q}} w_{1k} a_{2k}$. Here $s_{2q} = s_2 \cap U_q$, and $a_{2k} = 1/\pi_{2k}$ is the second-phase sampling weight of unit k. The second-phase weight for tax filer k is then $w_{2k} = a_{2k} g_{2k} = (1/\pi_{2k})(\hat{N}_q/\tilde{N}_q)$ for $k \in s_{2q} = s_2 \cap U_q$, where $g_{2k} = \hat{N}_q/\tilde{N}_q$ is the second-phase g-weight for every $k \in s_{2q}$. Note that the weights are calibrated at the first phase so that $\Sigma_{s_{1p}} w_{1k} = N_p$ for $p = 1, 2, \ldots, P$. Moreover, they are conditionally calibrated at the second-phase level, given s_{1q}, so that $\Sigma_{s_{2q}} w_{1k} w_{2k} = \Sigma_{s_{1q}} w_{1k} = \hat{N}_q$ for $q = 1, 2, \ldots, Q$. The total weight of unit k, denoted w_k, is $w_k = w_{1k} w_{2k}$. $Y_{(d)}$, the y-total for domain $U_{(d)}$, is estimated as

$$\hat{Y}_{(d)} = \sum_{k \in s_2} w_k y_{(d)k} = \sum_p \sum_q \frac{N_p}{\hat{N}_p} \frac{\hat{N}_q}{\tilde{N}_q} \sum_{k \in s_{2pq}} a_{1k} a_{2k} y_{(d)k},$$

where $s_{2pq} = s_2 \cap U_p \cap U_q$, $y_{(d)k} = y_k$ if $k \in U_{(d)}$ and $y_{(d)k} = 0$ otherwise. The corresponding variance estimator is given by

$$\hat{V}_{(d)} = \sum_p \sum_q \left(\frac{N_p}{\hat{N}_p}\right)^2 \left(\frac{\hat{N}_q}{\tilde{N}_q}\right)^2 \sum_{k \in s_{2pq}} a_{1k}(a_{1k} - 1)a_{2k}(e_{(d)p \cdot k})^2$$

$$+ \sum_p \sum_q \left(\frac{N_p}{\hat{N}_p}\right)^2 \left(\frac{\hat{N}_q}{\tilde{N}_q}\right)^2 \sum_{k \in s_{2pq}} (a_{1k})^2 a_{2k}(a_{2k} - 1)(e_{(d)q \cdot k})^2.$$

The residuals in this expression are generalizations to two-phase sampling of the residuals (25.19). We have

$$e_{(d)p \cdot k} = y_{(d)k} - \tilde{y}_{(d)p} \qquad \text{for } k \in s_{2p} = s_2 \cap U_p$$

where

$$\tilde{y}_{(d)p} = \frac{\displaystyle\sum_{s_{2p}} w_k y_{(d)k}}{\displaystyle\sum_{s_{2p}} w_k},$$

and

$$e_{(d) \cdot qk} = y_{(d)k} - \tilde{y}_{(d)q} \qquad \text{for } k \in s_{2q} = s_2 \cap U_q$$

where

$$\tilde{y}_{(d)q} = \frac{\displaystyle\sum_{s_{2q}} w_k y_{(d)k}}{\displaystyle\sum_{s_{2q}} w_k}.$$

The variance estimator $\hat{V}_{(d)}$ can be viewed as an extension to two phases of formula (25.17), applied here to the case of Bernoulli sampling in each of the two phases. Armstrong and St. Jean (1993) have extended the above methodology to incorporate several control totals for each group.

25.4 ESTIMATING OVER TIME

The data for economic time series are usually in the form of monthly, quarterly, or annual values. Two widely used measures that summarize these data are level and change. *Level* is a synonym for total. *Change* may be defined as the difference between, or the ratio of, two totals at two different time periods. The estimation of level variances is given in the previous sections. For estimation of the variance for change, covariances must be computed between estimates for the two time periods of interest. These covariances reflect the changing nature of the universe (births and deaths) as well as that of the sample (births, deaths and rotation). Tam (1984) provided expressions for covariances under repetitive sampling designs, for simple random sampling of the same finite population over time. Hidiroglou and Laniel (1986) extended Tam's results, for rotating samples (based on a stratified simple random sample clustered sampling design) in a changing population. For a given time period t and domain d, let the estimator of domain total be

$$\hat{Y}_d(t) = \sum_{h=1}^{H} \sum_{k=1}^{n_h(t)} \frac{N_h(t)}{n_h(t)} y_{(d)hk}(t),$$

where $N_h(t)$ and $n_h(t)$, respectively, represent the number of units in the population and sample at time t; $y_{(d)hk}(t)$ is as defined in equation (25.15). The variance for $\hat{Y}_d(t)$ is estimated as shown in Section 25.3. The estimated covariance between the estimates $\hat{Y}_d(t)$ and $\hat{Y}_d(s)$, $s < t$, is given by

$$
\begin{aligned}
\widehat{\mathrm{Cov}}(\hat{Y}_d(t), \hat{Y}_{(d)}(s)) = & \sum_{h=1}^{L} \left(1 - \frac{n_h(t)n_h(s)}{N_h(t)N_h(s)} \frac{N_h(t,s)}{n_h(t,s)} \right) \\
& \cdot \frac{n_h(t,s)}{n_h(t,s)-1} \sum_{k=1}^{n_h(t,s)} (z_{(d)hk}(t) - \bar{z}_{(d)} \cdot (t))(z_{(d)hk}(s) \\
& - \bar{z}_{(d)\cdot}(s)),
\end{aligned}
$$

where $N_h(t,s)$ and $n_h(t,s)$ represent the number of population units and sampled units, respectively, present in the sample at both times t and s. Here

$$z_{(d)hk}(t) = \frac{N_h(t)}{n_h(t)} y_{(d)hk}(t)$$

and $\bar{z}_{(d)\cdot}(t)$ is the sample mean of $z_{(d)hk}(t)$ based on $n_h(t)$ observations. The former results extended one of Tam's (1984) sampling scenarios. Laniel (1988) provided an additional extension to another one of Tam's (1984) sampling scenarios.

Now, these covariances enter into the estimation of the variance for change as follows. The difference between estimated totals for two time periods t and s is

$$\hat{D}_d(t, s) = \hat{Y}_d(t) - \hat{Y}_d(s),$$

and its estimated variance is

$$\hat{V}(\hat{D}(t, s)) = \hat{V}(\hat{Y}_d(t)) - 2\,\widehat{\text{Cov}}\,(\hat{Y}_d(t), \hat{Y}_d(s)) + \hat{V}(\hat{Y}_d(s)).$$

For the ratio $\hat{R}_d(t, s) = \hat{Y}_d(t)/\hat{Y}_d(s)$, the estimated variance is

$$\hat{V}[\hat{R}_d(t, s)] = \frac{\hat{V}[\hat{D}_d(t, s)]}{\hat{Y}_d(s)^2}.$$

The computation of these covariances is also required for procedures such as composite estimation which exploits the time-series aspect of successive survey estimates. The covariances are required for the computation of optimal weighting factors that combine estimates of totals over time, as well as their estimated variances. These methods hinge on the idea that if there is good correlation for data for common units between successive survey occasions, then the reliability of estimators that pool those data over time will be an improvement over those that do not (see Chapter 8).

A given economic series may be measured both subannually and annually via different data collection mechanisms. These two sources of data will most likely differ if the subannual data are aggregated to produce annual data. Traditionally, the problem of adjusting subannual data obtained from a given source to agree with annual data from a different source is referred to as *benchmarking*. The annual source is considered as exact or binding. As a concrete example, Statistics Canada publishes monthly estimates of the retail sales for a number of industries in that sector. In addition, Statistics Canada conducts a separate annual survey, from which the totals of yearly retail sales are obtained.

If the annual source is considered as binding, then benchmarking revises the subannual series so that the resulting series adds up (over a given time period) to a given annual series figure (benchmark) for the same time period. This problem has been studied by Denton (1971), Helfand et al. (1977), Monsour and Trager (1979), Fernandez (1981), and Cholette (1984). Assuming monthly figures, these authors revise the monthly figures using constrained minimization of a quadratic form in the differences between revised and unrevised series. Their objective is to minimize month-to-month changes and the distortion of the data's seasonal pattern.

These authors' approach does not recognize that: (1) both series (subannual and annual) are subject to sampling and nonsampling error and (2) the subannual series may be biased. Recognizing that the series are subject to error, Hillmer and Trabelsi (1987) used time-series methods to produce a solution. They showed that the resulting benchmark estimates have smaller mean squared error than do the pre-benchmark estimates (see also Binder et al. 1993). Extending the Hillmer–Trabelsi procedure to include the possibility of biased subannual series Laniel and Fyfe (1989) obtained the benchmarked series by applying least squares theory to a system of equations involving the models to represent the stochastic nature of the subannual series. The Gauss–Newton procedure was used to solve this problem, taking into account the nonbinding constraints.

25.5 CONCLUDING REMARKS

In this chapter, we describe a number of weighting and estimation methods applicable to business surveys. We show how auxiliary data can be integrated within one framework of estimation. Several examples of commonly used estimators are displayed using this general framework. The important problem of domain estimation is linked to the general framework. The estimation of the covariance of an estimated population total at two different points in time when both the population and sample composition may have changed is discussed. The computation of this covariance is useful for estimates that are based on benchmarking or composite estimation procedures.

REFERENCES

Armstrong, J., and H. St-Jean (1993), "Generalized Regression Estimation for a Two-Phase Sample of Tax Records," *Proceedings of the International Conference on Establishment Surveys*, Alexandria, VA: American Statistical Association, pp. 402–407.

Armstrong, J., C. Block, and K. P. Srinath (1993), "Two-Phase Sampling of Tax Records for Business Surveys," *Journal of Business and Economic Statistics*, **11**, pp. 407–416.

Bethlehem, J. G. (1988), "Reduction of Nonresponse Bias Through Regression Estimation," *Journal of Official Statistics*, **4**, pp. 251–260.

Binder, D. A., S. R. Bleuer, and J. P. Dick (1993), "Time Series Methods Applied to Survey Data," *Bulletin of the International Statistical Institute*, **49**, pp. 327–344.

Cholette, P. A. (1984), "Adjusting Sub-Annual Series to Yearly Benchmarks," *Survey Methodology Journal*, **10**, pp. 35–49.

Choudhry, G. H., P. Lavallée, and M. A. Hidiroglou (1989), "Two-Phase Sample Design for Tax Data," *Proceedings of the Survey Research Methods Section, American Statistical Association*, pp. 646–651.

Deming, W. E., and F. F. Stephan (1940), "On a Least Squares Adjustment of a Sampled Frequency Table when the Expected Marginal Totals are Known," *Annals of Mathematical Statistics*, **11**, pp. 427–444.

Denton, F. T (1971), "Adjustment on Monthly or Quarterly Series to Annual Totals: An Approach Based on Quadratic Minimization," *Journal of the American Statistical Association*, **46**, pp. 99–102.

Deville, J.-C., and C.-E. Särndal (1992), "Calibration Estimators in Survey Sampling," *Journal of the American Statistical Association*, **87**, pp. 376–382.

Deville, J.-C., C.-E. Särndal, and O. Sautory (1993), "Generalized Raking Procedures in Survey Sampling," *Journal of the American Statistical Association*, **88**, pp. 1013–1020.

Estevao, V., M. A. Hidiroglou, and C.-E. Särndal (1995), "Requirements on a Generalized Estimation System at Statistics Canada," *Journal of Official Statistics*, **10**, in press.

Fernandez, R. B. (1981), "A Methodological Note on the Estimation of Time Series," *Review of Economics and Statistics*, **63**, pp. 471–476.

Folsom, R. E. (1991), "Experimental and Logistic Weight Adjustments for Sampling and Nonresponse Error Adjustment," *Proceedings of the Social Research Section, American Statistical Association*, pp. 197–202.

Gossen, M., and M. Latouche (1992), "Post-Stratification to Reduce Sample Bias in an Establishment Survey," *Proceedings of the Business and Economic Statistics Section, American Statistical Association*, pp. 209–214.

Göttgens, R., B. Vellen, M. Odekerken, and L. Hofman (1991), "Bascula, Version 1.0. A Weighting Package Under MS-DOS, User Manual," CBS-Report, Voorburg: Statistics Netherlands.

Helfand, S. D., N. J. Monsour, and M. L. Trager (1977), "Historical Revision of Current Business Survey Estimates," *Proceedings of the Business and Economic Statistics Section, American Statistical Association*, pp. 246–250.

Hidiroglou, M. A., and N. Laniel (1986), "Specifications for the Estimation System of the Monthly Wholesale and Retail Trade Survey," internal document, Ottawa: Statistics Canada.

Hidiroglou, M. A., G. H. Choudhry, and P. Lavallée (1991), "A Sampling and Estimation Methodology for Sub-Annual Business Surveys," *Survey Methodology*, **17**, pp. 195–210.

Hillmer, S. C., and A. Trabelsi (1987), "Benchmarking of Economic Time Series," *Journal of the American Statistical Association*, **82**, pp. 1064–1071.

Holt, D., and T. M. F. Smith (1979), "Post-Stratification," *Journal of the Royal Statistical Society, Sec A.*, **142**, pp. 33–46.

Horvitz, D. G., and D. J. Thompson (1952), "A Generalization of Sampling Without Replacement from a Finite Universe," *Journal of the American Statistical Association*, **47**, pp. 663–685.

Huang, E., and W. A. Fuller (1978), "Nonnegative Regression Estimation for Sample Survey Data," *Proceedings of the Social Statistics Section, American Statistical Association*, pp. 300–305.

Iannacchione, V. G., J. G. Milne, and R. E. Folsom (1991), "Response Probability Weight Adjustments Using Logistic Regression," *Proceedings of the Survey Research Methods Section, American Statistical Association*, pp. 637–642.

Laniel, N. (1988), "Variances for a Rotating Sample from a Changing Population," *Proceedings of the Business and Economic Statistics Section, American Statistical Association*, pp. 246–250.

Lee, H., and J. Croal (1989), "A Simulation Study of Various Estimators Which Use Auxiliary Data in an Establishment Survey," *Proceedings of the Survey Research Methods Section, American Statistical Association*, pp. 336–341.

Little, R. J. A. (1986), "Survey Nonresponse Adjustments for Estimates of Means," *International Statistical Review*, **54**, 139–157.

Monsour, N. J., and M. L. Trager (1979), "Revision and Benchmarking of Business Time Series," *Proceedings of the Business and Economic Statistics Section, American Statistical Association*, pp. 333–337.

Oh, H. L., and F. Scheuren (1983), "Weighting Adjustments for Unit Nonresponse," In W. G. Madow, I. Olkin, and D. Rubin (eds.) *Incomplete Data in Sample Surveys*, Volume 2: *Theory and Bibliographies*. New York: Academic Press. pp. 143–184.

Rao, J. N. K. (1985), "Conditional Inference in Survey Sampling," *Survey Methodology*, **11**, pp. 15–31.

Särndal, C.-E., and B. Swensson (1987), "A General View of Estimation for Two Phases of Selection with Applications to Two-Phase Sampling and Non-Response," *International Statistical Review*, **55**, pp. 279–294.

Särndal, C.-E., and M. A. Hidiroglou (1989), "Small Domain Estimation: A Conditional Analysis," *Journal of the American Statistical Association*, **84**, pp. 266–275.

Särndal, C.-E., B. Swensson, and J. H. Wretman (1989), "The Weighted Residual Technique for Estimating the Variance of the General Regression Estimator of the Finite Population Total," *Biometrika*, **76**, pp. 527–537.

Särndal, C.-E., B. Swensson, and J. H. Wretman (1992), *Model Assisted Survey Sampling*: New York: Springer-Verlag.

Schiopu-Kratina, I., and K. P. Srinath (1991), "Sample Rotation and Estimation in the Survey of Employment, Payrolls and Hours," *Survey Methodology*, **17**, pp. 79–90.

Tam, S. M. (1984), "On Covariances from Overlapping Samples," *The American Statistician*, **38**, pp. 288–292.

Thomsen, I. (1973), "A Note on the Efficiency of Weighting Subclass Means to Reduce the Effects of Non-Response when Analyzing Survey Data," *Statistisk Tidskrift*, **11**, pp. 278–283.

Valliant, R. (1993), "Poststratification and Conditional Variance Estimation," *Journal of the American Statistical Association*, **88**, pp. 89–96.

CHAPTER TWENTY-SIX

Outliers in Business Surveys

Hyunshik Lee[1]
Statistics Canada

Outliers are a well-known problem for almost all types of statistics. Chambers (1986, p. 1063) called outliers a "perennial problem for applied survey statisticians." In statistical disciplines other than survey sampling, the sample is assumed to have been generated from a model or a population that follows a certain parametric distribution. Outliers are regarded as having been derived from a source other than the presumed model or population. This notion is reflected in the names used for outliers, such as discordant observations, rogue values, contaminants, and so on.

Outliers are quite a different concept in design-based survey sampling. There, samples are selected from finite populations and outliers are legitimate values for the units selected from the population under study. However, they may be *extreme values* isolated from the bulk of the data. Some reported values in sample surveys are not extreme but can greatly influence the estimate because of their large sampling weight. In this case, they are not outliers in the sense of extreme values situated far away from the bulk of the data, but they are *influential observations* in the sense that their inclusion or exclusion affects the estimate greatly. Extreme values are not necessarily influential when they have small weights. The distinction between extreme values and influential values is very useful in sample surveys. Other authors share this view (see, for instance, Gambino 1987, Srinath 1987, and Bruce 1991). Both extreme and influential values are outliers in one sense or another.

The influence of an observation varies depending on the estimator used. For

[1]The author thanks Michael A. Hidiroglou and David A. Binder of Statistics Canada, Beat Hulliger of the Swiss Federal Statistical Office, and Louis-Paul Rivest of Laval University for their helpful comments.

Business Survey Methods, Edited by Cox, Binder, Chinnappa, Christianson, Colledge, Kott. ISBN 0-471-59852-6 © 1995 John Wiley & Sons, Inc.

the expansion estimator, an influential observation is an extreme value based on its weighted value. Here the expansion estimator includes all unbiased "blow-up"-type estimators, including the Horvitz–Thompson (1952) estimator. An influential observation for the ratio estimator is an observation that greatly influences the estimation of the ratio. An influential observation for the expansion estimator may not be influential for the ratio estimator and vice versa. In this chapter the term *outlier* is used to mean either an outlying (extreme) observation or an influential observation with respect to the particular estimator used.

In business surveys, outliers are both common and difficult to treat, especially in a design-based framework. For instance, business populations often have highly skewed distributions with respect to variates of interest. In such situations, inclusion or exclusion of large units influences the expansion estimator so much that it can become unreliable. The problem is aggravated when large or medium-sized observations have large sampling weights. For this reason, samples for skewed populations are usually designed so that large units are selected with certainty or with large probabilities so that relatively small sampling weights are assigned to them. Stratification or probability proportional to size (*pps*) sampling is often employed for this purpose (see Chapter 8).

Even though an efficient sample design can minimize outlier problems, it cannot eliminate them. Outliers often occur no matter how carefully the design stage attempts to prevent them. For instance, most surveys are multivariate and only key variables are used for stratification or for obtaining the size measures for *pps* sampling. If the stratification or *pps* sampling is not efficient for non-key variables, outliers inevitably occur in the estimates for these variables. Moreover, frame data such as size measures may decay over time, producing a potential for size misclassification and associated outliers. Because outliers can occur in both tails, small values can also be considered outliers. In a repeated survey, for instance, outliers could occur at either tail when change from one period to another is estimated using differences or ratios in the values of common units.

Some authors classify outliers in sample surveys as *representative outliers* if they are correctly recorded and represent other population units similar in value to the observed outliers. *Nonrepresentative outliers* are those outliers that are incorrectly recorded or unique in the sense that there is no other unit like them (Chambers 1986; Hulliger 1992). Errors that lead to nonrepresentative outliers should be detected and corrected at the editing stage as discussed in Chapter 21. In this chapter, I assume that the data are clean and hence that only genuine outliers must be detected and treated. These outliers may be representative or nonrepresentative.

There are two main concerns in dealing with outliers: the efficiency of point estimates and the validity of inferences. The former is concerned with an estimator's mean squared error (MSE), and the latter involves its confidence interval. Though both topics are equally important, in this chapter I focus on the efficiency or MSE aspect of the outlier problem.

There is a vast amount of literature on outliers for parametric cases or infinite populations. For instance, Hawkins (1980) and Barnett and Lewis (1984) describe outlier detection and treatment for samples from populations with parametric distributions. For regression analysis, Belsley et al. (1980) and Cook and Weisberg (1982) review methods of outlier detection and treatment. Beckman and Cook (1983) provide a historical review of outliers in statistics including Bayesian and robust regression methods. Huber (1981), Hampel et al. (1986), and Rousseeuw and Leroy (1987) present robust estimation theory.

By contrast, less has been written about outliers in survey sampling. In some situations, sample surveys can use methods from other branches of statistics. This situation occurs when (1) the population can be approximated by a parametric model or distribution and (2) either simple random sampling is employed to select the sample or a model-based approach to surveys is used which ignores the sample design.

Outliers are usually dealt with in two steps, namely, detection and treatment. In this chapter, I discuss the detection and treatment of outliers as well as robust estimation in sample surveys, reviewing the literature for these areas. I also describe methods for infinite skewed populations. The outlier problem in estimation is essentially a problem of *robust estimation* (i.e., creating an estimation procedure insensitive to the presence of outliers). Robust estimation can be implemented by treating identified outliers or by direct application of robust estimation techniques such as the M-estimator. Much research remains to be done as I indicate in this chapter.

Although outliers are more common in business surveys, the methods described in this chapter may also be applied to household surveys. Populations that are highly skewed are more susceptible to outlier problems.

26.1 OUTLIER DETECTION

Several factors affect outlier methodology for sample surveys. These factors are as follows: (1) no distributional assumption is usually made (an exception would be a superpopulation framework with a parametric model; (2) sampling units may be correlated and are often selected with unequal probabilities resulting in unequal sampling weights; and (3) in many cases, survey populations are skewed. In large-scale surveys, outliers are detected at a subpopulation level. I use the term *outlier domain* to mean the subpopulation where outlier detection (and/or treatment, depending on the application) is performed.

Traditionally, outliers are detected using their relative distances from the center of the data. Let $y_1, y_2, \ldots, y_n$ be ordered observations in an outlier domain and let m and s be estimates of the location and scale, respectively. Then the *relative distance* of y_i is defined as the absolute value of the distance between y_i and the location estimate m divided by the scale estimate s or

$$d_i = \frac{|y_i - m|}{s}. \tag{26.1}$$

If this measure exceeds a predetermined cutoff value c, then the observation is considered to be an outlier. Alternatively, an interval given by $(m - c_l s, m + c_u s)$ is used where c_l and c_u are predetermined values. If the population is skewed, unequal values of c_l and c_u are used. Observations falling outside of this interval are declared as outliers. This interval is referred to as the *tolerance interval*.

The sample mean and variance are the statistics most frequently used to estimate location and scale. Hence, it is natural to use them in calculating the relative distance d_i. When the sample mean and standard deviation are used to estimate location and scale, the relative distance is quite efficient for detecting a single outlier. In this case, the single outlier is the observation with the largest d-value. However, use of the mean and standard deviation to define the relative distance d_i is not efficient when multiple outliers are present because the sample mean and variance are themselves very sensitive to outliers. The sample mean is shifted towards the outliers if they are clustered on one side and the sample variance is greatly inflated. Therefore, the d-values of outliers appear rather small and the procedure fails to detect them. This problem is called the *masking effect*.

To avoid this problem, robust estimates should be used for the location and scale. In the robust estimation literature, the ability of an estimator to be resistant to outliers is measured by the *breakdown point*, the smallest fraction of outliers in the sample that can cause the estimator to "break down." To define the breakdown point of an estimator G, consider a sample $\mathbf{Z} = \{y_1, y_2, \ldots, y_n\}$ and a contaminated sample $\mathbf{Z}'$ that is obtained by replacing l values of $\mathbf{Z}$ with l arbitrary values. Define the maximum bias caused by such contamination by $b(l; G, \mathbf{Z}) = \sup_{\mathbf{Z}'} \| G(\mathbf{Z}') - G(\mathbf{Z}) \|$. The (finite-sample) breakdown point of G defined in Donoho and Huber (1983; see also Rousseeuw and Leroy 1987, pp. 9–10) is then given by

$$\varepsilon_n^*(G, \mathbf{Z}) = \min \left\{ \frac{l}{n} \mid b(l; G, \mathbf{Z}) = \infty \right\}.$$

The larger the measure is, the more capable the estimator is for handling outliers before it "breaks" down.

For instance, the sample mean and variance have a breakdown point of $1/n$. This means that a single contamination can cause arbitrarily large bias. The same is true for outlier detection procedures based on the sample mean and variance. On the other hand, the sample median has the maximum breakdown point of 0.5. For a high breakdown point scale estimate, the *median absolute deviation* (MAD) is frequently used in the robust statistics literature (Andrews et al. 1972, Huber 1981, Hampel et al. 1986). The MAD is the median of absolute deviations from the sample median. It is formally defined as

$$\text{MAD} = \underset{i}{\text{median}} \{ | y_i - \underset{j}{\text{median}} (y_j) | \}.$$

The sample median and MAD could be good candidates for m and s in definition (26.1). However, the MAD is not commonly used for sample surveys. Instead, the lower and upper interquartile ranges (the definitions are given shortly) are used for their robustness, simplicity, and nonparametric character. This is changing, however, due to emerging techniques based on the M-estimation technique which often employ MAD as the scale estimate. For simple random sampling without replacement (*srswor*), the breakdown point of the interquartile ranges is 0.25, which is much smaller than that of the MAD (0.5). In sample surveys, however, we normally expect only a small fraction of outliers so the interquartile ranges should have a sufficiently large breakdown point. Otherwise, the sample should be redesigned to minimize the occurrence of outliers.

Now, I turn to the definition of finite population quantiles. Let the indicator function $I_{k,y} = 1$ if $y_k \leq y$ and 0 otherwise. The population distribution function is denoted by $F(y) = (1/N)\Sigma_U I_{k,y}$ where Σ_U denotes summation over the finite population U. Then the *θ-th population quantile* is defined by $Q_\theta = F^{-1}(\theta)$. $F(y)$ can be estimated by the sample distribution function (Särndal et al. 1992) given by $\hat{F}(y) = (\Sigma_s \pi_i^{-1}I_{k,y})/\Sigma_s \pi_i^{-1}$, where Σ_s denotes summation over the sample s. Although this estimator may be biased, it is consistent. Using the sample distribution function, the θ-th population quantile can then be estimated as

$$q_\theta = \hat{F}^{-1}(\theta). \tag{26.2}$$

When the inverse function at a given θ is an interval, the midpoint of the interval is used. The sample median and the first and third sample quartiles are obtained by taking $\theta = 0.5, 0.25$, and 0.75 in equation (26.2).

The lower and upper interquartile ranges, defined by $d_l = q_{0.5} - q_{0.25}$ and $d_u = q_{0.75} - q_{0.5}$, respectively, are used for the scale measures. Using these scale measures, the tolerance interval is constructed as

$$(q_{0.5} - c_l d_l, q_{0.5} + c_u d_u)$$

with some predetermined values for c_l and c_u. These bounds can be chosen by examining past data or using past experience. Any observation falling outside of the tolerance interval is identified to be an outlier. This method will be referred to as the *quartile method*. Note that, in general, the interval is not symmetric, that is, $c_l d_l \neq c_u d_u$, reflecting the skewness of the data. Also, the interval can be made virtually one-sided by choosing an arbitrarily large value for c_l or c_u or replacing one of the limits by the smallest or the largest possible value. In fact, a one-sided interval is often used because survey data can be bounded on one side.

When this method is applied to real data, either d_l or d_u may be too small. When the data points are clustered together, observations with a rather small deviation from the median may be identified as outliers. To avoid this problem,

d_l or d_u are modified as follows:

$$d_l = \max (q_{0.5} - q_{0.25}, |Aq_{0.5}|)$$
$$d_u = \max (q_{0.75} - q_{0.5}, |Aq_{0.5}|)$$

for $0 \leq A \leq 1$. The value of 0.05 for A is adequate in most applications. However, the impact of a particular value should be carefully considered before choosing the value.

For trend data, Hidiroglou and Berthelot (1986) proposed a method similar to the quartile method. However, their method is applied to transformed data of period-to-period change ratios. These change ratios are usually skewed and more variable for units with small y-values. Therefore, if the quartile method is applied directly to the ratios, the ratios with small y-values are more likely to be identified as outliers. However, the units with large y-values contribute more to the trend and thus should get more attention. This is called the *size masking effect*.

To neutralize the size masking effect, Hidiroglou and Berthelot (1986) used a two-step transformation. To reduce the skewness of the r_i's, the ratios are first transformed as follows:

$$s_i = \begin{cases} 1 - \dfrac{q_{0.5}}{r_i} & \text{if } 0 < r_i < q_{0.5}, \\[2mm] \dfrac{r_i}{q_{0.5}} - 1 & \text{if } r_i \geq q_{0.5}, \end{cases}$$

where $r_i = y_i(t + 1)/y_i(t)$ and $y_i(t)$ is the y-value of unit i at time t. Here, $q_{0.5}$ is the median of the r_i's. The second step of the transformation incorporates the magnitude of the data and is defined by

$$E_i = s_i \{\max [y_i(t), y_i(t + 1)]\}^V$$

for $0 \leq V \leq 1$. The E_i is referred to as the *effect* of unit i. The exponent V provides control over the importance of the magnitude of the data. A large value of V increases the importance of large y-values and thus further reduces the size masking effect. If $V = 0$, nothing is done to reduce the size masking effect. The quartile method is applied to E_i's to detect trend outliers.

To detect influential observations for the expansion estimator, the quartile method is often applied to weighted data. The method effectively detects extreme values in weighted data. The quartile method has been widely used at Statistics Canada for either level or trend data (Lee 1991b, Lee et al. 1992). Some authors (Smith 1987, Bruce 1991) advocate the use of the *leave-one-out* (or, more generally, the leave-k-out) diagnostic for both level and change. In Bruce (1991), the leave-one-out method is defined by $[\hat{G} - \hat{G}^*(i)]/\hat{G}$, where $\hat{G}$ is the full sample estimate and $\hat{G}^*(i)$ is calculated excluding the ith unit. If $\hat{G}$ is not robust, this method suffers from the masking effect problem.

At the editing stage, it is common practice to ignore the sampling weight when the quartile method is used. In this case, only unweighted extreme values are detected. The quartile method using unweighted data has been implemented in the Generalized Edit and Imputation System (Cotton 1991).

Besides these formal methods, informal procedures are sometimes used. For example, percentage contributions of weighted values to the estimate of the population total are compared with a predetermined allowable percentage to screen out influential observations. Occasionally, even more informal procedures are used such as having a certain number of the largest weighted values inspected by subject-matter experts (Earwaker 1987, Maranda 1989).

Lacking a distributional assumption in sample surveys makes it difficult to test for outliers. If the survey population follows a specific parametric distribution, it may be possible to develop a statistical test for detecting outliers as in Fuller (1991) and Porter (1980). The task of developing a test for outliers will not be trivial when the sample design is complex. Further research is required.

26.1.1 Outliers in Ratio and Regression Estimators

For infinite populations the literature on regression analysis proposes many methods for detection of regression outliers (Belsley et al. 1980, Cook and Weisberg 1982, Beckman and Cook 1983). Ghangurde (1989a) studied Cook's distance in the context of ratio estimation assuming simple random sampling. *Cook's distance* measures the influence of a unit in the least squares estimate of the regression coefficient (Cook 1979). It is large if (1) the point has a large residual or (2) it is a leverage point. A *leverage point* is an outlier in the value of the independent variable. Since the method is based on a non-robust least squares estimator, it suffers from the masking effect.

To avoid masking effects, a robust estimator such as the M-estimator (Huber 1981, Hampel et al. 1986) can be used to detect regression outliers. For robust regression and outlier detection, Koenker and Bassett (1978) proposed the use of L_1-regression, which minimizes the sum of absolute deviations. These methods are not good in detecting leverage points; some authors advocate using a generalized M-estimator (Hampel et al. 1986).

For the multiple regression estimator, the breakdown point aspect of the method must be considered. The regression M-estimators described in Section 26.2.2 have a breakdown point of at most $1/p$, where p is the dimension of β. Rousseeuw (1984) proposed the *least median of squares* (LMS) *estimator* which minimizes the median of the squared deviations and has a very high breakdown point of $[(n/2) - p + 2]/n$. Rousseeuw and van Zomeren (1990) suggested using the LMS estimator along with a robust Mahalanobis distance to detect both ordinary regression outliers and leverage points. The normalized residuals, calculated from the robust regression equation and robust scale estimate, are then examined to detect regression outliers. This can be done when the robust estimates are calculated. The method is computer-intensive and thus is not practical for large data sets.

These techniques could be applicable for equal probability samples, though their properties as well as their adaptation to complex samples have not been studied. Detection of outliers for ratio and regression estimators is a new area which requires further research.

26.2 OUTLIER TREATMENT AND ROBUST ESTIMATION

The specific treatment of outliers depends upon the stage of survey processing where they are identified and corrective action taken. Outliers detected at the editing stage are treated in various ways. In a manual editing system, outliers are examined and corrected if they resulted from errors. In an automated editing system such as the Generalized Edit and Imputation System, detected outliers are often imputed (see Chapter 23). In some cases, subjective judgment plays a vital role because outliers are left untreated if deemed unimportant.

At the estimation stage, the treatment procedures are more formal; some procedures are based upon objective criteria such as MSE-efficiency. Outlier treatment in estimation normally introduces bias (i.e., deviation of the estimator's expected value over repeated samples from the true parameter value). When outliers are present, the objective is to reduce the variance of an unreliable unbiased or nearly unbiased estimator. Such reduction of variance may incur a small bias. For this reason, in this chapter the efficiency of an estimator refers to its MSE-efficiency.

There are basically three approaches to the formal treatment of outliers in estimation of finite population quantities: (1) changing the values of outliers by Winsorization or trimming, (2) reducing the weights of outliers (in many cases, the sampling weights), and (3) using robust estimation techniques such as M-estimation. While trimming is rarely used in business surveys for treating outliers, Winsorization is frequently used. Winsorization has been used mostly for simple random sampling because of the difficulty of extending the concept of order statistics in the case of unequal probability sampling. Weight reduction methods are a preferred treatment in sample surveys, and important research has been done in this area. The third approach is a fairly recent development. In the following discussion, the population is the domain where outlier treatment is to be applied.

26.2.1 Methods Based on Winsorization and Weight Reduction

Because most business surveys mainly deal with nonnegative variables that are skewed to the right, this discussion concentrates on nonnegative, right-skewed y-variables. First, I consider the application of these methods to the simple expansion estimator with *srswor* for the population total Y, which is denoted by $\hat{Y}_E = (N/n) \Sigma_s y_i$, where N is the population size. If these outlier domains are sampling strata, the cases I describe include the stratified *srswor*.

Assume that the sample observations y_i, $i = 1, 2, \ldots, n$, are ordered in ascending order. If the k largest y-values are considered as outliers, the *one-sided, k-times Winsorized estimator* is defined by replacing these outlier values by y_{n-k}. That is,

$$\hat{Y}_{W1} = \frac{N}{n} \left(\sum_{i=1}^{n-k} y_i + ky_{n-k} \right). \tag{26.3a}$$

A variant of the above Winsorized estimator is obtained by replacing the k outliers by a cutoff point t, that is, $y_{n-k} < t \leq y_{n-k+1}$. This type of estimator is called the *censored estimator*. The cutoff point t is the threshold for identifying the outliers. The estimator obtained this way is given by

$$\hat{Y}_{W2} = \frac{N}{n} \left(\sum_{i=1}^{n-k} y_i + kt \right). \tag{26.3b}$$

Instead of altering y-values, an estimator can reduce the weights of the k outliers to $r < N/n$. Several such estimators have been proposed. An early version was proposed by Bershad (1960) and is defined by

$$\hat{Y}_{R1} = \frac{N}{n} \sum_{i=1}^{n-k} y_i + r \sum_{i=n-k+1}^{n} y_i.$$

In this form, the weights for nonoutliers are kept intact, while the weights for outliers are reduced to r. A more general form of the weight reduction method is given:

$$\hat{Y}_{R2} = \frac{N - rk}{n - k} \sum_{i=1}^{n-k} y_i + r \sum_{i=n-k+1}^{n} y_i. \tag{26.4}$$

The principle in this formulation is that the sum of weights must equal N. This estimator is called the *reweighted estimator*. Dividing these estimators by the population size N, we obtain estimators for the population mean. (I use the equation numbers for the population total to refer to the estimators for the population mean.)

Searls (1966) proposed estimator (26.4) for the mean of an infinite skewed population. Searls proved that estimator (26.3b) with an optimal cutoff is more efficient than the ordinary sample mean. In the context of finite skewed populations, Rivest and Hurtubise (1993) investigated Searls' optimal Winsorization and provided a method by which the optimal cutoff can be determined if the population distribution function $F(y)$ is known or a large amount of past data is available.

Ernst (1980) investigated the efficiencies of seven estimators for the mean of an exponential population. These include estimators (26.3a), (26.3b), and

(26.4). Ernst showed that estimator (26.3b) with optimal choice of t performed the best, followed closely by estimator (26.4) with the optimal weight reduction factor r.

Fuller (1991) proposed the Winsorized mean for the Weibull distribution. Let the Weibull density be given by

$$f(y; \alpha, \lambda) = \begin{cases} \alpha\lambda^{-1}y^{\alpha-1}\exp(-\lambda^{-1}y^{\alpha}) & \text{if } y > 0, \\ 0 & \text{otherwise,} \end{cases}$$

where $\lambda > 0$ and $\alpha > 0$. The once-Winsorized mean is (1) more efficient than the sample mean if the shape parameter $\gamma = 1/\alpha > 1$, (2) just as efficient if $\gamma = 1$ (the distribution becomes the exponential), and (3) less efficient if $\gamma < 1$. Fuller also proposed an adaptive method referred to as the test-and-estimate procedure (also called the preliminary test estimator). Whether the sample mean or the Winsorized mean should be used is determined based on the sample data. A Monte Carlo simulation using two real populations and a Weibull distribution showed large gains achieved by the once- and twice-Winsorized means and by the test-and-estimate procedure over the sample mean, especially for small samples. Porter (1980) adapted the test procedure to examine whether the largest two observations are outliers and to modify them if the test result so indicates.

Rivest (1993a) established the superiority of the Winsorized mean for a large family of skewed distributions characterized as having a decreasing failure rate; that is, $f(y)/[1 - F(y)]$ decreases for large y where $f(y)$ and $F(y)$ are the density and distribution functions. Rivest's formulation is different from either the (26.3a) or (26.3b) estimator in that y_n is replaced with a convex combination of adjacent extreme values; that is, $\tau y_{n-1} + (1 - \tau)y_n$ for $\tau \in (0, 1]$ or y_n and y_{n-1} are replaced with $(\tau - 1)y_{n-2} + (2 - \tau)y_{n-1}$ for $\tau \in (1, 2]$. The optimal value τ_0 depends on the distribution. For example, $\tau_0 = 0.5$ for the exponential distribution and $\tau_0 = 1 - 0.5\alpha$ for the Weibull distribution. Rivest concluded that the once-Winsorized mean ($\tau = 1$) provides the largest efficiency for all practical purposes. However, for moderate skewness, the same efficiency with less bias can be achieved with $\tau = 0.75$. He provided a nearly unbiased MSE estimator for the τ-Winsorized mean, given by

$$\frac{S^2}{n} - \frac{\tau}{n^2}\left(y_n + y_{n-1} - 2\frac{\hat{Y}_{W1}}{N}\right)(y_n - 3y_{n-1} + 2y_{n-2})$$
$$- \frac{\tau - \tau^2}{n^2}(y_n - y_{n-1})^2,$$

where S^2 is the sample variance of the original sample and $\hat{Y}_{W1}/N$ is the once-Winsorized mean. In a Monte Carlo experiment, estimator (26.3b) with optimal or near-optimal cutoff performed best and had less bias than other Winsorized estimators (Rivest 1993b). The test-and-estimate procedure performed very well when no information other than the sample was used, which indicates that adapting the sample information in estimation can be beneficial if no prior knowledge about the population distribution is available.

For a finite population but assuming a superpopulation model, Fuller (1991) proposed the following model-based estimator for the total:

$$\hat{Y}_F = \sum_{i=1}^{n} y_i + (N - n)\hat{\mu}, \tag{26.5}$$

where $\hat{\mu}$ is the minimum MSE estimator for the superpopulation mean. Fuller proved that this estimator has the minimum MSE for estimation of the total. When the superpopulation is the Weibull distribution, $\hat{\mu}$ may be replaced by the Winsorized mean or the estimate obtained by the test-and-estimate procedure. Thorburn (1993) proposed a similar approach for a superpopulation that follows the lognormal distribution. He used logarithmic transformation of y-values to get a more efficient predictor $\hat{\mu}$ for unobserved y-values.

In dealing with outliers, the model-based approach allows us to get around the difficulty of unequal sampling weights for complex designs. However, this approach has been criticized as vulnerable to model failures (Rao 1978, Hansen et al. 1983). This brings up another important concept of robustness, namely, robustness against model failure. This chapter does not explore this topic, however.

I turn to the weight reduction method. To use the estimator given by equation (26.4), we have to determine the reduced weight r for the k outliers. Assuming that observed outliers are self-representative or nonrepresentative but genuine (that is, there are no unsampled outliers in the population), Rao (1971) and Chinnappa (1976) assigned a weight of 1.0 to the k outliers (i.e., $r = 1$). This estimator is referred to as $\hat{Y}_{R2}$. If outliers are representative and if K, the number of outliers in the population, is known, then we can use the poststratified estimator:

$$\hat{Y}_{R3} = \frac{N - K}{n - k} \sum_{i=1}^{n-k} y_i + \frac{K}{k} \sum_{i=n-k+1}^{n} y_i.$$

Note that $\hat{Y}_{R2}$ is a special case of the poststratified estimator with $k = K$. Normally, K is unknown and $\hat{Y}_{R2}$ is often used assuming $k = K$.

Hidiroglou and Srinath (1981) determined the optimal weight of r by minimizing the MSE of estimator (26.4), unconditional and conditional on the number of outliers k in the sample. The optimal weight is a function of several population quantities: the number of outliers in the population, the population means, and the variances of outliers and nonoutliers. These quantities are usually unknown, but approximate values might be obtained from previous surveys or a census. I refer to this reweighted estimator with the optimal weight as $\hat{Y}_{R4}$. Hidiroglou and Srinath also studied another variant of the reweighted estimator (26.4) defined by

$$\hat{Y}_{R5} = \frac{N}{n}\left(1 + \frac{k}{2n}\right)\sum_{i=1}^{n-k} y_i + \frac{N}{n}\left(1 - \frac{n+k}{2n}\right)\sum_{i=n-k+1}^{n} y_i.$$

The weight reduction for this estimator depends on k. When k is small compared to n, the reduced weight for the k outliers is about half of the original weight.

From an extensive numerical study of these estimators under various conditions that affect their performances, Hidiroglou and Srinath noted the following conclusions based on unconditional MSE-efficiencies of the estimators:

- The use of the estimator $\hat{Y}_{R2}$ results in substantial gains in efficiency over $\hat{Y}_{E}$ (the expansion estimator) when the sampling fraction and the number of outliers in the population is small.
- If the sampling fraction and K are moderately large, use of $\hat{Y}_{R5}$ is recommended.
- $\hat{Y}_{R4}$ is best used if the optimal weight reduction factor can be determined.
- If K is known and large, $\hat{Y}_{R3}$ should be used.

It is quite interesting to observe that $\hat{Y}_{R2}$ is almost as good as $\hat{Y}_{R4}$ when K is small.

Ghangurde (1989a, 1989b) proposed an outlier-robust ratio estimator obtained by optimally reducing the weight for outliers. He used the variance-inflation model to derive the form of the estimator under a model-based approach, and then the weight reduction factor was computed by minimizing the design-based MSE. The variance-inflation model is given by

$$y_i = \beta x_i + e_i, \tag{26.6}$$

where $e_i \sim (0, \sigma_1^2 x_i)$, for $i = 1, 2, \ldots, n - k$, and $e_i \sim (0, \sigma_2^2 x_i)$, for $i = n - k + 1, \ldots, n$, with unknown β, σ_1^2, and σ_2^2. Without loss of generality, I assume that the first $n - k$ units are nonoutliers and the remaining k units are outliers with respect to the ratio estimator $\hat{Y}_R = (\bar{y}/\bar{x}) X$. Let $W = \sigma_1^2/\sigma_2^2$, where $0 < W \le 1$. The best linear unbiased estimator of β under the model (26.6) is then given by

$$\hat{\beta}_L = \frac{\sum_{i=1}^{n-k} y_i + W \sum_{i=n-k+1}^{n} y_i}{\sum_{i=1}^{n-k} x_i + W \sum_{i=n-k+1}^{n} x_i}. \tag{26.7}$$

This expression implies that if the variance-inflation factor for the outliers is $1/W$, then the outliers should be down-weighted by W, the reciprocal of the variance-inflation factor. Considering W as a weight reduction factor to be determined, Ghangurde derived the design-based MSE formula for $\hat{\beta}_L$ conditional on k and then obtained the optimal weight reduction factor W_0 by minimizing the conditional MSE. His modified ratio estimator is then defined by

$$\hat{Y}_{RR1} = \hat{\beta}_{L0}X, \qquad (26.8)$$

where $\hat{\beta}_{L0}$ is obtained from equation (26.7) with W replaced by W_0. If we take $x_i = 1$ for all i, then equation (26.8) becomes

$$\hat{Y}_{RR2} = \frac{N\left(\sum\limits_{i=1}^{n-k} y_i + W_0 \sum\limits_{i=n-k+1}^{n} y_i \right)}{n - k + W_0 k}. \qquad (26.9)$$

With the optimal W_0, this estimator becomes exactly the same as $\hat{Y}_{R4}$ obtained by Hidiroglou and Srinath (1981). For this reason, Ghangurde contended that the variance-inflation model provides a theoretical basis to the weight reduction method. For the general case of estimator (26.8), an analytical solution is unavailable. Instead, the optimal W is obtained by minimizing the model variance of $\hat{\beta}_L$ (Ghangurde 1989b).

Departing from the rather restrictive assumption of *srswor*, Dalén (1987) and Tambay (1988) studied an interesting method to treat influential observations in unequal probability sampling. This method is a hybrid of Winsorization by a cutoff and the weight reduction method. Let w_i be the weight for unit i. Now define

$$z_i = \begin{cases} w_i y_i & \text{if } w_i y_i < T, \\ T + (y_i - T/w_i) & \text{otherwise,} \end{cases} \qquad (26.10)$$

where T is the threshold for detecting influential observations. When the weighted value $w_i y_i$ exceeds T, the modified value is the sum of the cutoff value T and the value obtained by dividing the excess over T by the weight. Therefore, the excess over the cutoff $(w_i y_i - T)$ gets a reduced weight of 1.0. The total is estimated by the sum of the z_i's. The resulting estimator reduces to Fuller's in equation (26.5) in the case of *srs* if $\hat{\mu}$ is estimated by the Winsorized mean with Winsorization at the cutoff (nT/N) (Hidiroglou 1991). Therefore, this method has a theoretical justification in the case of *srswor* from a finite population that approximately follows the Weibull at the right tail. The estimator performed well in an evaluation study, using samples from the lognormal distribution and real data (Dalén 1987). Tambay (1988) also applied the method to a survey conducted at Statistics Canada and obtained encouraging results.

So far, I have considered one outlier domain. When sampling strata are outlier domains, using optimal procedures within each stratum does not necessarily yield an optimal strategy at the aggregate level. Rivest (1993b) considered the optimization of the stratified Winsorized mean over all strata by minimizing the overall MSE.

Conceptually, the weight reduction method is preferred to Winsorization or other changing-value methods for treating outliers, assuming that the detected outliers represent the conceivable outlier stratum in the population. However, this method poses a serious operational problem for surveys that collect multivariate data. In this case, each variable will have its own outlier-treated weights, which is undesirable in practice. Therefore, outlier detection and treatment are often carried out using only one key variable. As an alternative to this approach, outliers' values can be changed even for the weight reduction method. The result on the estimates is the same, and the original weight can be used for all variables. However, the estimation of the MSE remains an open question. A more desirable approach would be to detect and treat outliers in a multivariate sense, which is a little-explored area in sample surveys.

26.2.2 Methods Based on the M-Estimation Technique

Huber (1964) introduced the M-estimator (M stands for the maximum-likelihood type) as a robust alternative to the sample mean for the distribution that follows a normal in the middle and a double exponential at the tails. In general, an M-estimator for the location parameter θ is defined as a solution that minimizes $\sum_{i=1}^{n} \rho(y_i - \theta)$ with respect to θ for some function ρ, or by an implicit equation

$$\sum_{i=1}^{n} \psi(y_i - \theta) = 0, \qquad (26.11)$$

where $\psi(y - \theta) = (\partial/\partial\theta)\rho(y - \theta)$. This definition includes a class of estimators, one of which is the ordinary least squares estimator obtained by taking $\psi(r) = r$. The Huber M-estimator is defined by the following ψ-function:

$$\psi(t) = \begin{cases} c & \text{if } t > c, \\ t & \text{if } |t| \leq c, \\ -c & \text{if } t < -c. \end{cases} \qquad (26.12)$$

The constant c in the above formula is called the *tuning factor*, and it determines how many outliers are treated. If c is very large, the estimator derived from equation (26.11) is the same as the least squares estimator (i.e., the sample mean), which means that there is no treatment of outliers. Note that the sample median is the limiting case of the Huber M-estimator as $c \to 0$. Because

M-estimators are not usually invariant with respect to scale, a robust scale estimator such as MAD should be used in solving equation (26.11).

Since the publication of Huber (1964), many M-estimators have been proposed and studied. In the famous Princeton study (Andrews et al. 1972), the M-estimators for the location parameter of long-tailed symmetric populations performed very well among other alternatives to the sample mean.

For estimation of a regression model given by

$$y_i = \beta' x_i + e_i, \qquad e_i \sim (0, \sigma^2 v_i), \tag{26.13}$$

where β and x_i are p-column vectors, the M-estimator is defined as a solution with respect to β to the following vector equation:

$$\sum_{i=1}^{n} \psi\{r_i(\beta)\} \, \ddot{x}_i = 0, \tag{26.14}$$

where $r_i(\beta) = (y_i - \beta' x_i)/(\sigma\sqrt{v_i})$ and $\ddot{x}_i = x_i/(\sigma\sqrt{v_i})$. Again the scale parameter σ should be estimated to solve equation (26.14). Because an explicit solution for equations (26.11) and (26.14) is usually not available, an iterative procedure is used to solve the equations.

Motivated by equation (26.12) but constrained by the requirement that the weight of the outliers should not be reduced to a value less than one, Bruce (1991) proposed an estimator for period-to-period change of finite populations obtained by

$$\psi(t) = \begin{cases} (1 - n/N)c + (n/N)t & \text{if } t > c, \\ t & \text{if } |t| \leq c, \\ -(1 - n/N)c - (n/N)t & \text{if } t < -c, \end{cases}$$

which leads to the two-sided version of the estimator given by equation (26.10). Note that the one-sided Winsorized estimator at the cutoff discussed in Section 26.2.1 can be derived by using a modified ψ-function in equation (26.12), that is, $\psi(t) = c$ if $t > c$ and $\psi(t) = t$ if $t < c$.

Chambers (1986) adapted the M-estimation technique to ratio estimation ($p = 1$). The usual model-based ratio estimator for the population total Y is given by

$$\hat{Y}_{LS} = \sum_s y_i + \sum_{U-s} \hat{\beta}_{LS} x_i,$$

where s denotes the sample, $U - s$ is the nonsampled part of the finite population, and $\hat{\beta}_{LS} = (\sum_s x_i y_i/v_i)/(\sum_i x_i^2/v_i)$ is the least squares (LS) estimator for β. To make this unrobust estimator robust, Chambers used the M-estimator for β, say $\hat{\beta}_R$. However, the use of $\hat{\beta}_R$ generally introduces a bias, and a bias correction is incorporated in the estimator. The Chambers estimator is defined as

$$\hat{Y}_C = \sum_s y_i + \hat{\beta}_R \sum_{U-s} x_i + \sum_s z_i \phi[r_i(\hat{\beta}_R)],$$

where $z_i = \ddot{x}_i \Sigma_{U-s} x_j / \Sigma_s(\ddot{x}_j)^2$ and $\phi(r)$ is a bounded odd function. If $\phi(r) = r$, then $\hat{Y}_C$ becomes $\hat{Y}_{LS}$, which is not robust. On the other hand, if $\phi(r) = 0$, no bias is corrected but the estimator is variance-efficient. Thus, using a non-zero-bounded ϕ-function is a compromise that results in a more MSE-efficient estimator than $\hat{Y}_{LS}$. Being a model-based estimator, the Chambers estimator can be applied to any sampling design as it ignores the sampling weights. In a numerical study with real data, the estimator performed very well compared to other model-based and design-based estimators under various sampling designs (Chambers 1986). Gwet and Rivest (1992) observed that the Chambers estimator can be written as $\hat{Y}_C = \Sigma_s y_i + \Sigma_{U-s} \hat{\beta}_M x_i$, where $\hat{\beta}_M$ is another M-estimator for β that can be obtained using the modified residual procedure of Huber (1981) starting with the initial value $\hat{\beta}_R$.

Gwet and Rivest (1992) also studied a robust ratio estimator using a generalized M-estimator of the Schweppe-type (see Hampel et al. 1986, p. 315). Their approach is design-based under simple random sampling. Assuming $v_i = x_i$, their robust ratio estimator is defined as

$$\hat{Y}_{GR} = \hat{\beta}_G \sum_U x_i,$$

where $\hat{\beta}_G$ is a solution to the following equation:

$$\sum_s \frac{\sqrt{x_i}}{h(x_i)} \psi\left(\frac{(y_i - \beta x_i)h(x_i)}{K\sqrt{x_i}}\right) = 0$$

for a bounded odd function ψ and a scaling constant K. This type of estimator can deal with regression outliers as well as leverage points simultaneously for estimation of β. If $h(t)$ is constant, then $\hat{\beta}_G$ becomes an ordinary M-estimator. The robust estimator is biased, but its conditional bias (conditional on the proportion of outliers in the sample) behaves better than that of the ordinary ratio estimator. Gwet and Rivest's numerical results showed superior performance of the robust estimator over the ordinary ratio estimator when outliers are present. Using an estimator for the bias of $\hat{Y}_{GR}$ given by $(N/n)\Sigma_s(y_i - \hat{\beta}_G x_i)$, a fully bias-corrected (BC) estimator can be obtained as

$$\hat{Y}_{BC} = \sum_U \hat{\beta}_G x_i + \frac{N}{n}\sum_s (y_i - \hat{\beta}_G x_i). \tag{26.15}$$

This estimator is inefficient. However, it is useful for comparison with the proposed robust ratio estimators. For instance, Gwet and Rivest (1992) noticed that the Chambers estimator can be expressed as

$$\hat{Y}_C = (1 - f)\hat{Y}_{GR} + f\hat{Y}_{BC},$$

where $f = n/N$ if $\hat{\beta}_G$ is also used in the Chambers' estimator. This implies that the Chambers model-based estimator is a compromise between the Gwet–Rivest estimator that is biased but variance efficient and the estimator given in equation (26.15) that is unbiased but inefficient.

Lee (1990, 1991a) also used an M-estimator to define a robust regression estimator. The estimator has the same form as the Gwet–Rivest estimator and is given by $\hat{Y}_{L1} = \Sigma_U \hat{\beta}'_M \mathbf{x}_i$ where $\hat{\beta}_M$ is an M-estimator for the regression coefficient vector $\beta' = (\beta_1, \beta_2)$ and $\mathbf{x}'_i = (1, x_{i2})$. The generalization to multiple regression is straightforward. The superiority of this estimator over the usual regression estimator was clearly demonstrated in a numerical study using real data (Lee 1990). Taking a model-based approach, Lee (1991a) proposed a robust (multiple) regression estimator using the GM-estimation technique defined as follows:

$$\hat{Y}_{L2} = \sum_s y_i + \sum_{U-s} \hat{\beta}'_G \mathbf{x}_i,$$

where $\hat{\beta}_G$ is implicitly defined as a solution of the vector equation

$$\sum_s w(\mathbf{x}_i)\psi[r_i(\beta)]\mathbf{x}_i = 0$$

for appropriately chosen real-valued functions w and ψ. The preceding equation gives a Mallows-type GM-estimator for β. However, the Schweppe-type GM estimator could have an advantage under a heteroscedastic regression model. The estimator $\hat{Y}_{L2}$ was compared with the usual regression estimators and various versions of robust regression estimators in a simulation study. Real and artificially generated data were used with two sampling designs, namely, *srswor* and stratified *srswor*. It performed the best and its MSE was sometimes only half that of the usual regression estimator. (Note that all numerical results mentioned in this chapter are based on repeated sampling experiments.)

These robust estimators are usually biased and not consistent (in the sense of Isaki and Fuller 1982). If we require consistency as suggested in Lee (1991a), we may study the following:

$$\hat{Y}_{RREG} = \sum_U \hat{\beta}'_R \mathbf{x}_i + \theta \sum_s \frac{1}{\pi_i}(y_i - \hat{\beta}'_R \mathbf{x}_i)$$

with $0 \le \theta \le 1$ and π_i is the inclusion probability of unit i. The value for θ should be chosen by some optimality criterion. To obtain consistency, we need $\theta \to 1$ as $n \to \infty$. Note that this estimator includes as special cases the previously discussed robust estimators defined using the M- or GM-estimation techniques.

Another recent development in this area is the work of Hulliger (1992). He proposed a robust Horvitz–Thompson (HT) estimator using the M-estimation technique. Basu's (1971) elephant example dramatically illustrates the nonrobustness of the HT estimator. To make the HT estimator robust, Hulliger first expressed the estimator as a least squares functional of the sample distribution function of the two variables x and y. This formulation accounts for the sampling weights. The sample distribution function is given by

$$\hat{F}(x, y) = \frac{\sum_s \frac{1}{\pi_i} I(x_i \le x, y_i \le y)}{\sum_s \frac{1}{\pi_i}},$$

where $I(x_i \le x, y_i \le y) = 1$ if $x_i \le x$ and $y_i \le y$, and 0 otherwise. The x-variable here is used as the size measure for pps sampling. Under the model given in equation (26.13) with scalar β and x_i, the least squares estimator for β is defined as a value that minimizes $\int [r(\beta)]^2 \, d\hat{F}(x, y)$, where $r(\beta) = (y - \beta x)/(\sigma\sqrt{v_i})$, or equivalently as a solution of

$$\sum_s \frac{1}{\pi_i} \psi[r_i(\beta)]\ddot{x}_i = 0, \tag{26.16}$$

where $\psi(r) = r$ and $\ddot{x}_i = x_i/(\sigma\sqrt{v_i})$. The above equation is in the same form as equation (26.14) except for the sampling weights $1/\pi_i$. If the model is the usual ratio model (i.e., $v_i = x_i$), then the least squares estimator for β from equation (26.16) is $\hat{\beta}_{LS} = \Sigma_s(\pi_i^{-1}y_i)/\Sigma_s(\pi_i^{-1}x_i)$. If we set $\pi_i = nx_i/\Sigma_U x_i$, then the HT estimator can be written as

$$\hat{Y}_{HT} = \sum_s \frac{y_i}{\pi_i} = \hat{\beta}_{LS} \sum_U x_i.$$

Robustification of the HT estimator is achieved by applying the GM-estimation technique to equation (26.16); $\psi(r) = r$ is replaced by $\eta\{\ddot{x}_i, r_i(\beta)\}$ which is used to obtain a GM-estimator. The robustified HT estimator proposed by Hulliger is given by

$$\hat{Y}_{RHT} = \hat{\beta}_G \sum_U x_i,$$

where $\hat{\beta}_G$ is a solution of the equation

$$\sum_s \frac{1}{\pi_i} \eta[\ddot{x}_i, r_i(\beta)]\ddot{x}_i = 0. \tag{26.17}$$

This definition generates a wide range of estimators depending on the choice of the η-function, including the HT estimator with $\eta(\ddot{x}, r) = r$. If outliers are suspected in the x-values, the Mallows-type $\eta(\ddot{x}, r) = w(\ddot{x})\psi(r)$ is useful as is the Schweppe-type $\eta(\ddot{x}, r) = w(\ddot{x})\psi[r/w(\ddot{x})]$. Otherwise, Huber's ψ-function (26.12) would be a good choice. The formulation given in equation (26.17) can be used to derive a robust estimator for any nonrobust estimator if it can be expressed as a least squares functional of $\hat{F}$. The stratified estimator and the π-weighted ratio and regression estimators are such examples. Variance estimators are also available. The "robustified" estimators obtained in this way, however, are biased and not design-consistent.

Hulliger also provided an analytical tool called the *sensitivity curve* to investigate the robustness of an estimator. It is a finite population analogue to the influence curve for infinite populations (Hampel et al. 1986). Using the sensitivity curve, Hulliger derived the influence of an observed unit i to the HT estimator that is given by $(y_i - \beta_{\mathrm{LS}}x_i)/(\pi_i\Sigma_s\pi_j^{-1})$. The influence of a unit is large if its absolute residual is large. This problem is aggravated if such a unit has a small π_i. The robustified HT estimator is robust to units with outlying residuals. However, it does not provide protection against outlying sampling weights. The Hájek (1971) estimator, given by $\hat{Y}_{\mathrm{H}} = N\hat{Y}_{\mathrm{HT}}/\Sigma_s\pi_i^{-1}$ is less sensitive to the problem of outlying weights (see Särndal et al. 1992, p. 258). It would be interesting to study the estimator given by $\hat{Y}_{\mathrm{MRHT}} = N\hat{Y}_{\mathrm{RHT}}/\Sigma_s\pi_i^{-1}$ which would provide robustness to outliers with large residuals as well as those with large sampling weights.

In general, the robust estimators obtained by the M- or GM-estimation techniques are biased and their efficiencies depend on the choice of ψ-function in equation (26.14) or η-function in equation (26.17). Unless there is good prior knowledge about outliers in the population, the function should be chosen adaptively based on the sample information. In this spirit, Hulliger (1993) also proposed the Minimum Estimated Risk (MER) estimator. He used Huber's ψ-function given in equation (26.12) with the tuning factor determined to minimize an estimated mean square error. By inclusion of the HT estimator as a possible choice, the MER estimator is design consistent. In a simulation study its overall performance was shown to be superior to the HT estimator and comparable or even better than that of the RHT estimator.

Nonlinearity of the robust estimators discussed in this section poses problems for survey practitioners who are accustomed to attaching estimation weights or raising factors to sampled units. However, the problem can be avoided using a one-step procedure starting with a reasonable initial value which is often as good as the fully iterated estimates (Lee 1991a).

26.3 CONCLUDING REMARKS

In this chapter, I review techniques for dealing with outliers in business surveys. Most techniques are for univariate data even though survey data are often multivariate. Some research done for multivariate outlier detection and a review of this topic were given by Lee et al. (1992). However, more research is needed.

For detection of univariate outliers, the quartile method is most frequently used due to its robustness, simplicity, and nonparametric nature. A testing procedure for outlier detection is not usually available because of the lack of distributional assumptions.

Outliers for the ratio or (multiple) regression estimator are more difficult to detect than univariate outliers. Many methods are available for regression analysis in the case of infinite populations, but most suffer from the masking effect problem or else fail to detect certain type of outliers such as leverage points. To avoid the masking effect problem, the robust regression technique may be used. Quantile regression and regression M- or GM-estimation techniques have good potential. However, application of these methods in sample surveys is a new area, again requiring more research.

Treatment of outliers at the editing stage is usually done by checking the reported values, by follow-ups, or by imputations. For treatment of outliers in estimation, there are basically three approaches: (1) changing the outlier values, (2) reducing the weights of outliers, and (3) using robust estimation techniques such as M-estimation. Winsorization is frequently used for the first approach. The second approach, weight reduction, is often preferred because it is conceptually appealing. However, determination of the weight reduction factor remains a problem requiring more research. The third approach, which is fairly new, has promising potential. An obvious advantage of this approach is that it does not require outlier detection. More research is required to study and implement this approach as well.

Due to lack of models, dealing with outliers in business surveys is more difficult than in the case of infinite populations. Official statistical agencies such as Statistics Canada accumulate a vast amount of information about the business populations in repeated surveys, including census data in some cases. Therefore, it seems quite possible to derive models for the populations using this readily available information. An outlier-robust procedure can be derived more easily and/or an outlier procedure can be optimized using a model assumption.

Treatment of level outliers or robust estimation of levels does not necessarily solve the problem of change outliers and vice versa. Tambay (1988) attempted to tackle both problems in an integrated fashion. Bruce (1991) suggested a time series approach and called for more research.

In general, outlier treatment is always a tradeoff between variance and bias. For small samples, variance is usually the dominating factor in the MSE. On the other hand, the bias dominates when the sample size is large. If the chosen robust estimator is consistent, the bias diminishes as the sample size increases. In general, however, the problem of outliers is less serious when the sample size is large.

As mentioned earlier, I have focused on the outlier problem from the standpoint of the efficiency. However, the validity aspect of the outlier problem is also important. The most crucial issue in this is the provision of an appropriate estimator of MSE of the outlier-treated or robust point estimator. Another pending issue is a way to overcome the skewness of the point estimator when the population is heavily skewed and the sample size is not large enough to invoke asymptotic normality. Bruce (1991) suggested the possibility of using the bootstrap method. However, no concrete result of the bootstrap method has been presented thus far in conjunction with the problem of outliers.

REFERENCES

Andrews, D. F., P. J. Bickel, F. R. Hampel, P. J. Huber, W. H. Rogers, and J. W. Tukey (1972), *Robust Estimates of Location: Survey and Advances*, Princeton, NJ: Princeton University Press.

Barnet, V., and T. Lewis (1984), *Outliers in Statistical Data*, 2nd ed., New York: Wiley.

Basu, D. (1971), "An Essay on the Logical Foundations of Survey Sampling, Part I," in V. P. Godambe and D. A. Sprott (eds.), *Foundations of Statistical Inference*, Toronto: Holt, Rinehart, and Winston, pp. 203–233.

Beckman, R. J., and R. D. Cook (1983), "Outliers," *Technometrics*, **25**, pp. 119–149.

Belsley, D. A., E. Kuh, and R. E. Welsch (1980), *Regression Diagnostics*, New York: Wiley.

Bershad, M. A. (1960), "Some Observations on Outliers," unpublished report, Washington, DC: U.S. Bureau of the Census.

Bruce, A. G. (1991), "Robust Estimation and Diagnostics for Repeated Sample Surveys," *Mathematical Statistics Working Paper* 1991/1, Wellington, Statistics New Zealand.

Chambers, R. L. (1986), "Outlier Robust Finite Population Estimation," *Journal of the American Statistical Association*, **81**, pp. 1063–1069.

Chinnappa, B. N. (1976), "A Preliminary Note on Methods of Dealing with Unusually Large Units in Sampling from Skewed Populations," unpublished report, Ottawa: Statistics Canada.

Cook, R. D. (1979), "Influential Observations in Linear Regression," *Journal of the American Statistical Association*, **74**, pp. 169–174.

Cook, R. D., and S. Weisberg (1982), *Residuals and Influence in Regression*, London: Chapman and Hall.

Cotton, C. M. (1991), "Functional Description of the Generalized Edit and Imputation System," technical report, Ottawa: Statistics Canada.

Dalén, J. (1987), "Practical Estimators of a Population Total Which Reduce the Impact of Large Observations," R & D Report, Stockholm: Statistics Sweden.

Donoho, D. L., and P. J. Huber (1983), "The Notion of Breakdown Point," in P. J. Bickel, K. A. Doksum, and J. L. Hodges, Jr. (eds.), *A Festschrift for Erich Lehmann*, Belmont, CA: Wadsworth.

Earwaker, S. (1987), "Outliers in Administrative Data," paper prepared for the Fifth

Meeting of the Advisory Committee on Statistical Methods, Ottawa: Statistics Canada.

Ernst, L. R. (1980), "Comparison of Estimators of the Mean Which Adjust for Large Observations," *Sankhya*, Series C, **42**, pp. 1–16.

Fuller, W. A. (1991), "Simple Estimators of the Mean of Skewed Populations," *Statistica Sinica*, **1**, pp. 137–158.

Gambino, J. (1987), "Dealing with Outliers: A Look at Some Methods Used at Statistics Canada," paper prepared for the Fifth Meeting of the Advisory Committee on Statistical Methods, Ottawa: Statistics Canada.

Ghangurde, P. D. (1989a), "Outlier Robust Estimation in Finite Population Sampling," unpublished report, Ottawa: Statistics Canada.

Ghangurde, P. D. (1989b), "Outliers in Sample Surveys," *Proceedings of the Survey Research Methods Section, American Statistical Association*, pp. 736–739.

Gwet, J.-P., and L.-P. Rivest (1992), "Outlier Resistant Alternatives to the Ratio Estimator," *Journal of the American Statistical Association*, **87**, pp. 1174–1182.

Hájek, J. (1971), "Comments on 'An Essay on the Logical Foundations of Survey Sampling' by Basu," in V. P. Godambe and D. A. Sprott (eds.), *Foundations of Statistical Inference*, Toronto: Holt, Rinehart, & Winston, p. 236.

Hampel, F. R., E. M. Ronchetti, P. J. Rousseeuw, and W. A. Stahel (1986), *Robust Statistics: The Approach Based on Influence Functions*, New York: Wiley.

Hansen, M. H., W. G. Madow, and B. J. Tepping (1983), "An Evaluation of Model-Dependent and Probability-Sampling Inferences in Sample Surveys," *Journal of the American Statistical Association*, **78**, pp. 776–793.

Hawkins, D. M. (1980), *Identification of Outliers*, London: Chapman and Hall.

Hidiroglou, M. A. (1991), personal communication.

Hidiroglou, M. A., and J.-M. Berthelot (1986), "Statistical Edit and Imputation for Periodic Surveys," *Survey Methodology*, **12**, pp. 73–83.

Hidiroglou, M. A., and K. P. Srinath (1981), "Some Estimators of a Population Total Containing Large Units," *Journal of the American Statistical Association*, **78**, pp. 690–695.

Horvitz, D. G., and D. J. Thompson (1952), "A Generalization of Sampling Without Replacement from a Finite Universe," *Journal of the American Statistical Association*, **47**, pp. 663–685.

Huber, P. J. (1964), "Robust Estimation of a Location Parameter," *Annals of Mathematical Statistics*, **35**, pp. 73–101.

Huber, P. J. (1981), *Robust Statistics*, New York: Wiley.

Hulliger, B. (1993), "Robustified Horvitz–Thompson Estimators," in-house report, Swiss Federal Statistical Office.

Isaki, C. T., and W. A. Fuller (1982), "Survey Design Under the Regression Superpopulation Model," *Journal of the American Statistical Association*, **77**, pp. 89–96.

Koenker, R. W., and G. W. Bassett (1978), "Regression Quantiles," *Econometrica*, **46**, pp. 33–50.

Lee, H. (1990), "Outlier-Resistant Regression Estimators," presented at the Annual Meeting of the Statistical Society of Canada, St. John's, Newfoundland, Canada.

Lee, H. (1991a), "Model-Based Estimators That Are Robust to Outliers," *Proceedings of the Annual Research Conference*, Washington, DC: U.S. Bureau of the Census, pp. 178–202.

Lee, H. (1991b), "Outliers in Survey Sampling," paper prepared for the 14th Meeting of the Advisory Committee on Statistical Methods, Ottawa: Statistics Canada.

Lee, H., P. D. Ghangurde, L. Mach, and W. Yung (1992), "Outliers in Sample Surveys," Methodology Branch Working Paper BSMD-92-008E, Ottawa: Statistics Canada.

Maranda, F. (1989), "Proposal for NFS Estimation," unpublished report, Ottawa: Statistics Canada.

Porter, J. C. (1980), "Censored Estimates for Skewed Distributions—An Application to Automatic Data Inspection," technical report, Chicago: A. C. Nielsen Company.

Rao, C. R. (1971), "Some Aspects of Statistical Inference in Problems of Sampling from Finite Populations," in V. P. Godambe and D. A. Sprott (eds.), *Foundations of Statistical Inference*, Toronto: Holt, Rinehart, and Winston.

Rao, J. N. K. (1978), "Sampling Designs Involving Unequal Probabilities of Selection and Robust Estimation of a Finite Population Total," in H. David (ed.), *Contributions to Survey Sampling and Applied Statistics*, New York: Academic Press, pp. 69–87.

Rivest, L.-P. (1993a), "Statistical Properties of Winsorized Means for Skewed Distributions," unpublished technical paper, Quebec City, Canada: Department of Statistics, Laval University.

Rivest, L.-P. (1993b), "Winsorization of Survey Data," presented at the 49th Session of the International Statistical Institute, Firenze, Italy.

Rivest, L.-P., and D. Hurtubise (1993), "Some Sampling Properties of Winsorized and Truncated Means," unpublished technical paper, Quebec City, Canada: Department of Statistics, Laval University.

Rousseeuw, P. J. (1984), "Least Median of Squares Regression," *Journal of the American Statistical Association*, **79**, pp. 871–880.

Rousseeuw, P. J., and A. M. Leroy (1987), *Robust Regression and Outlier Detection*, New York: Wiley.

Rousseeuw, P. J., and B. C. van Zomeren (1990), "Unmasking Multivariate Outliers and Leverage Points," *Journal of American Statistical Association*, **85**, pp. 633–639.

Särndal, C.E., B. Swensson, and J. Wretman (1992), *Model Assisted Survey Sampling*, New York: Springer-Verlag.

Searls, D. T. (1966), "The Estimator for a Population Mean Which Reduces the Effect of Large True Observations," *Journal of the American Statistical Association*, **61**, pp. 1200–1205.

Smith, T. M. F. (1987), "Influential Observations in Survey Sampling," *Journal of Applied Statistics*, **14**, pp. 143–152.

Srinath, K. P. (1987), "Outliers in Sample Surveys," paper prepared for the Fifth Meeting of the Advisory Committee on Statistical Methods, Ottawa: Statistics Canada.

Tambay, J.-L. (1988), "An Integrated Approach for the Treatment of Outliers in Sub-Annual Surveys," *Proceedings on the Survey Research Methods Section, American Statistical Association*, pp. 229–234.

Thorburn, D. (1993), "The Treatment of Outliers in Economic Statistics," paper presented at the International Conference on Establishment Surveys, Buffalo, NY.

CHAPTER TWENTY-SEVEN

Small Area Estimation: Overview and Empirical Study

J. N. K. Rao[1]
Carleton University

G. Hussain Choudhry
Statistics Canada

Sample surveys are generally designed to provide reliable direct estimates for large areas and major domains (subgroups) of a population. A *direct estimate* uses data only from sample units in the area or domain of interest. For example, farm surveys can provide direct estimates of crop yields of acceptable reliability at the national and state levels and for broad size groups of farms. Similarly, surveys of firms can yield direct estimates with desired coefficients of variation at the level of industry groups. Increasingly, such surveys are also being used to provide estimates for small areas or domains, which are needed in formulating policies and programs, in allocation of government funds, in regional programs, and so on. For example, business statistics for census division by industry group are used to assess local conditions and plan development projects.

Direct survey estimates for a small area are likely to yield unacceptably large standard errors due to an unduly small sample size for the area. This makes it necessary to "borrow strength" from related areas to find indirect estimators that increase the effective sample size and thus decrease the variance. Such estimators are based on either implicit or explicit models that provide a link to related small areas through supplementary data such as admin-

[1]The authors wish to thank Danny Pfeffermann of the Hebrew University for constructive comments. J. N. K. Rao would also like to acknowledge research grant support from the Natural Sciences and Engineering Research Council of Canada.

Business Survey Methods, Edited by Cox, Binder, Chinnappa, Christianson, Colledge, Kott.
ISBN 0-471-59852-6 © 1995 John Wiley & Sons, Inc.

istrative records and recent census counts. Indirect estimators proposed in the literature include synthetic, sample-size-dependent, composite, empirical best linear unbiased prediction, empirical Bayes, and hierarchical Bayes.

In this chapter, we provide an overview of small-area estimators in the context of business surveys. Relative performances of these estimators are studied through simulation using real and synthetic populations. The scope of this chapter is confined to small-area estimation using only cross-sectional data. However, many business surveys are repeated in time with partial replacement of the sampled units. For such repeated surveys further gains in efficiency can be achieved by borrowing strength across both small areas and time. Small-area estimators derived from combining cross-sectional and time-series data have been studied by Pfeffermann and Burck (1990), Choudhry and Rao (1989), and others.

27.1 SMALL-AREA ESTIMATORS

Small-area estimators may be classified into three broad groups: (1) direct estimators, (2) indirect estimators based on implicit models, and (3) model-based estimators. Direct estimators are seldom used unless the sample size is reasonably large for the small area of interest. Nevertheless, we include them mainly for benchmark comparisons in our simulation study. Estimators in group 2 include synthetic, sample-size-dependent, and composite, while those in group 3 include empirical best linear unbiased prediction, empirical Bayes, and hierarchical Bayes. We confine ourselves to simple random sampling from an overall population of size N. The parameters of interest are the small-area totals $Y_i = \sum_{j=1}^{N_i} y_{ij}$, where N_i is the known population size of the ith small area and y_{ij} is the value of a characteristic of interest y for the j-unit in the ith area. In the simulation study, Y_i refers to the total wages and salaries for the ith census division and a particular industry group.

27.1.1 Direct Estimators

A *simple expansion estimator* of Y_i is given by

$$\hat{Y}_i(\exp) = \begin{cases} \dfrac{N}{n} \sum_{j \in s_i} y_{ij} & \text{if } n_i \geq 1, \\ 0 & n_i = 0, \end{cases} \tag{27.1}$$

where s_i is the set of n_i sample units falling in the ith small area and n is the overall sample size. The estimator (27.1) is design-unbiased for the total Y_i; that is, the average value over all possible samples equals Y_i. However, it is conditionally biased if n_i is fixed.

A more efficient estimator is given by the *poststratified estimator*

$$\hat{Y}_i(\text{pst}) = \begin{cases} \dfrac{N_i}{n_i} \displaystyle\sum_{j \in s_i} y_{ij} = N_i \bar{y}_i & \text{if } n_i \geq 1, \\ 0 & \text{if } n_i = 0, \end{cases} \quad (27.2)$$

where $\bar{y}_i$ is the sample mean of the ith area. The poststratified estimator (27.2) is conditionally unbiased for a fixed $n_i \geq 1$, but its variance is also likely to be unacceptably large because the conditional variance is of order n_i^{-1}.

27.1.2 Synthetic Estimators

Unlike direct estimators, *synthetic estimators* "borrow strength" from related small areas, using supplementary x-data (gross business income in our simulation study) with known small-area totals.

A *ratio synthetic estimator* assumes that the population ratios, $R_i = Y_i/X_i$, in a particular industry group are homogeneous; that is, $R_i = R = Y/X$, the population ratio in the industry group, or $Y_i = RX_i$. Estimating R by $\hat{R} = \bar{y}/\bar{x}$ yields

$$\hat{Y}_i(\text{syn}) = (\bar{y}/\bar{x})X_i, \quad (27.3)$$

where $\bar{y}$ and $\bar{x}$ are the overall sample means in the industry group. The ratio synthetic estimator (27.3) has a much smaller variance than the poststratified estimator, since the variance is of order m^{-1} where m is the overall sample size in the industry group. However, it can be heavily biased unless the implicit assumption $R_i = R$ is satisfied.

27.1.3 Composite Estimators

A natural way to balance the potential bias of $\hat{Y}_i(\text{syn})$ against the instability of $\hat{Y}_i(\text{pst})$ is to take a weighted average of the two estimates:

$$\hat{Y}_i(\text{comp}) = a_i \hat{Y}_i(\text{pst}) + (1 - a_i)\hat{Y}_i(\text{syn}), \quad (27.4)$$

where a_i is a suitably chosen weight ($0 \leq a_i \leq 1$). The optimal weight $a_i(\text{opt})$ may be obtained by minimizing the mean square error (MSE) of the *composite estimator* (27.4) with respect to a_i, but the estimated weight $\hat{a}_i(\text{opt})$ can be very unstable. Schaible (1978) proposed an "average" weighting scheme to overcome this difficultly. Simple weights a_i that depend only on the domain counts lead to sample-size dependent estimators considered next.

27.1.4 Sample-Size-Dependent Estimators

Drew et al. (1982) proposed a *sample size-dependent estimator* of Y_i that takes account of the realized relative sample sizes $w_i = n_i/n$. In our context, it is given by

$$\hat{Y}_i(\text{ssd}) = \alpha_i \hat{Y}_i(\text{pst}) + (1 - \alpha_i)\hat{Y}_i(\text{syn}), \qquad (27.5)$$

where

$$\alpha_i = \begin{cases} 1 & \text{if } w_i \geq \delta W_i, \\ (1/\delta)(w_i/W_i) & \text{otherwise,} \end{cases} \qquad (27.6)$$

and $W_i = N_i/N$. The parameter δ in equation (27.6) is subjectively chosen to control the contribution of the synthetic component. The reliance on the synthetic component increases with δ, but $\delta = 1$ is often chosen as a general-purpose value.

Särndal and Hidiroglou (1989) proposed different sample-size-dependent estimators using the supplementary x-data. Their ratio version is given by

$$\hat{Y}_i(\text{ssd*}) = \alpha_i^* \hat{Y}_i(\text{reg}) + (1 - \alpha_i^*)\hat{Y}_i(\text{syn}), \qquad (27.7)$$

where

$$\alpha_i^* = \begin{cases} 1 & \text{if } w_i \geq W_i, \\ (w_i/W_i)^{h-1} & \text{if } w_i < W_i. \end{cases} \qquad (27.8)$$

Here $\hat{Y}_i(\text{reg})$ is an approximately unbiased regression-type estimator which may be written as:

$$\hat{Y}_i(\text{reg}) = N_i[\bar{y}_i + (\bar{y}/\bar{x})(\bar{X}_i - \bar{x}_i)],$$

where $\bar{X}_i = X_i/N_i$. The parameter h in equation (27.8) is subjectively chosen so as to control the contribution of the synthetic component. Särndal and Hidiroglou (1989) suggested $h = 2$ as a general purpose value. Note that $\alpha_i = \alpha_i^*$ if the general-purpose values $\delta = 1$ and $h = 2$ are chosen.

Sample-size-dependent estimators are simple to compute, but they fail to take advantage of the between-area homogeneity when $w_i \geq W_i$ even if n_i is small, because equation (27.5) reduces to $\hat{Y}_i(\text{pst})$ and equation (27.7) reduces to $\hat{Y}_i(\text{reg})$ in this case.

27.1.5 Model-Based Estimators

We assumed $R_i = R$ to obtain the synthetic estimator (27.3). This assumption is similar to the often-used superpopulation model:

$$y_{ij} = \beta x_{ij} + e_{ij}x_{ij}^{1/2}, \quad j = 1, 2, \ldots, N_i, \quad i = 1, 2, \ldots, I \quad (27.9)$$

with independent errors e_{ij} such that $E_m(e_{ij}) = 0$ and $V_m(e_{ij}) = \sigma^2$, where E_m and V_m denote model expectation and model variance respectively, I is the number of small areas, and x_{ij} is the value of the auxiliary variable x for the jth unit in the ith area.

The *best linear unbiased prediction (BLUP) estimator* of Y_i under model (27.9) is given by

$$\hat{Y}_i(\text{blup*}) = \sum_{j \in s_i} y_{ij} + \sum_{j \in \bar{s}_i} \hat{y}_{ij}, \qquad (27.10)$$

where $\hat{y}_{ij} = (\bar{y}/\bar{x})x_{ij}$ is the best predictor of y_{ij} for nonobserved units $\bar{s}_i$ (Brewer 1963, Royall 1970). Estimator (27.10) is approximately equal to $\hat{Y}_i(\text{syn})$ if the sampling fraction n_i/N_i is negligible, irrespective of the size of between-area variation relative to within-area variation. This limitation of model (27.9) can be avoided by using more realistic models that include random area-specific effects. We consider here one such model by introducing random small-area effects a_i into the regression model (27.9). It is given by

$$y_{ij} = \beta x_{ij} + a_i + e_{ij} x_{ij}^{1/2} \qquad j = 1, 2, \ldots, N_i$$

$$i = 1, 2, \ldots, I \qquad (27.11)$$

where a_i and e_{ij} are independent errors with $E_m(a_i) = 0$, $V_m(a_i) = \sigma_a^2$, $E_m(e_{ij}) = 0$, and $V_m(e_{ij}) = \sigma^2$. The ratio $\lambda = \sigma_a^2/\sigma^2$ measures between-small-area variation relative to within-small-area variation. A method for examining the aptness of models of the form (27.11) is given in Section 27.2.1.

Model (27.11) can be extended to the case of multiple auxiliary variables, but it is more difficult to specify the variance of the residuals $y_{ij} - \mathbf{x}_{ij}'\boldsymbol{\beta} - a_i$ where $\mathbf{x}_{ij}$ the vector of values of the auxiliary variables associated with the jth unit in the ith area.

Another type of random effects model arises when only aggregate area-specific auxiliary data, say z_i, assumed to be related to small-area means $\bar{Y}_i$ are available (Fay and Herriot 1979):

$$\bar{y}_i = \beta z_i + a_i + e_i, \qquad i = 1, 2, \ldots, I,$$

where a_i are as before and the e_i's are the sample errors with $E(e_i) = 0$ and $V(e_i) = \sigma_i^2$. It is customary to assume that the sampling variances σ_i^2 are known. This model is not studied here, but the reader is referred to Fay and Herriot (1979), Prasad and Rao (1990), and Ghosh and Rao (1994) for details on small-area estimation under this model.

Empirical Best Linear Unbiased Prediction
The BLUP estimator of Y_i under model (27.11) is given by

$$\hat{Y}_i(\text{blup}) = \sum_{j \in s_i} y_{ij} + \sum_{j \in \bar{s}_i} y_{ij}^*, \qquad (27.12)$$

where y_{ij}^* is the best predictor of unobserved y_{ij}, $j \in \bar{s}_i$. Under model (27.11), y_{ij}^* is an estimator of $E(y_{ij}|a_i) = \beta x_{ij} + a_i$. It is given by

$$y_{ij}^* = \hat{\beta} x_{ij} + \hat{a}_i, \qquad j \in \bar{s}_i,$$

where

$$\hat{\beta} = \left[\sum_i \sum_{j \in s_i} y_{ij} - \sum_i \frac{\gamma_i}{\eta_i} n_i \left(\sum_{j \in s_i} r_{ij} \right) \right] \left[\sum_i \sum_{j \in s_i} x_{ij} - \sum_i \frac{\gamma_i}{\eta_i} n_i^2 \right]^{-1}, \quad (27.13)$$

$$\hat{a}_i = \frac{\gamma_i}{\eta_i} \sum_{j \in s_i} (r_{ij} - \hat{\beta}),$$

$$\eta_i = \sum_{j \in s_i} x_{ij}^{-1}, \qquad (27.14)$$

with $r_{ij} = y_{ij}/x_{ij}$ and

$$\gamma_i = \frac{\sigma_a^2}{\sigma_a^2 + \dfrac{\sigma^2}{\eta_i}} = \frac{\lambda}{\lambda + \eta_i^{-1}}.$$

The estimator y_{ij}^* is obtained from Henderson's (1975) results for general mixed linear models involving fixed and random effects. Note that as $\lambda \to 0$ the predictor y_{ij}^* tends to βx_{ij} and the BLUP estimator tends to equation (27.10), which is approximately equal to the synthetic estimator $\hat{Y}_i(\text{syn})$ if n_i/N_i is negligible.

The BLUP estimator (27.12) depends on the unknown variance ratio $\lambda = \sigma_a^2/\sigma^2$. Unbiased estimators of σ_a^2 and σ^2 can be obtained using the well-known method of fitting constants, also called *Henderson's method* 3. These estimators are given by

$$\hat{\sigma}^2 = (n - I - 1)^{-1} \sum_i \sum_{j \in s_i} \hat{e}_{ij}^2 \qquad (27.15)$$

and

$$\hat{\sigma}_a^2 = \frac{1}{n^*} \left[\sum_i \sum_{j \in s_i} \hat{u}_{ij}^2 - (n - 1)\hat{\sigma}^2 \right], \qquad (27.16)$$

where

$$n^* = \sum_i \eta_i - \frac{\sum_i n_i^2}{\sum_i \sum_j x_{ij}},$$

$\Sigma\Sigma\ \hat{u}_{ij}^2$ is the residual sum of squares obtained from a weighted regression (through the origin) of y_{ij} on x_{ij} with weights x_{ij}^{-1}, and $\Sigma\Sigma\ \hat{e}_{ij}^2$ is the residual sum of squares obtained from a weighted regression (through the origin) of $y_{ij} - \bar{y}_{iw}$ on $x_{ij} - \bar{x}_{iw}$ with weights x_{ij}^{-1}, where $\bar{y}_{iw}$ and $\bar{x}_{iw}$ are the weighted small-area means with weights x_{ij}^{-1} (Stukel 1991). If $\hat{\sigma}_a^2$ turns out to be less than 0, we set it equal to 0.

Substituting $\hat{\lambda} = \hat{\sigma}_a^2/\hat{\sigma}^2$ for λ in equation (27.12), we get a two-step estimator, $\hat{Y}_i$(eblup), called an *empirical best linear unbiased prediction estimator* (EBLUP). Assuming normality of the errors a_i and e_{ij}, Stukel (1991) obtained an accurate approximation to the estimator of mean square error of $\hat{Y}_i$(eblup) under model (27.11), as the number of small areas I increases.

Empirical Bayes

In the empirical Bayes (EB) approach, the posterior distribution of small-area total Y_i, given the data $\mathbf{y}$ and the variance components σ_a^2 and σ^2, is first obtained by assuming that β has a uniform distribution over $(-\infty, \infty)$ to reflect absence of prior information on β. The parameters σ_a^2 and σ^2 are then estimated from the marginal distribution of the data, and inferences are based on the estimated posterior distribution. In particular, the *empirical Bayes estimator* of Y_i is given by the estimated posterior mean $E(Y_i|\ y,\ \hat{\sigma}_a^2,\ \hat{\sigma}^2)$, and its precision is measured by the estimated posterior variance $V(Y_i|\ y,\ \hat{\sigma}_a^2,\ \hat{\sigma}^2)$. The latter, however, can lead to serious underestimation of true posterior variance because it fails to take account of the uncertainty about σ_a^2 and σ^2. Laird and Louis (1987) and Kass and Steffey (1989) have proposed methods of accounting for the underestimation.

Under normal errors a_i and e_{ij}, the estimated posterior mean is identical to the EBLUP estimator $\hat{Y}_i$(eblup), and the estimated posterior variance is identical to the mean square error of the BLUP estimator evaluated at $\hat{\sigma}_a^2$ and $\hat{\sigma}^2$.

Hierarchical Bayes

In the hierarchical Bayes (HB) approach, we specify a prior distribution on the model parameters β, σ_a^2, and σ^2 and then obtain the posterior distribution of Y_i given the data $\mathbf{y}$. Inferences are based on the posterior distribution. In particular, the hierarchical Bayes estimator of Y_i is given by the posterior mean $E(Y_i|\mathbf{y})$; and its precision is measured by the posterior variance $V(Y_i|\ \mathbf{y})$ which takes account of uncertainty about model parameters.

Datta and Ghosh (1991) assumed that β, $(\sigma_a^2)^{-1}$, and $(\sigma^2)^{-1}\lambda^{-1}$ are independently distributed with $\beta \sim$ uniform over $(-\infty,\ \infty)$, $(\sigma_a^2)^{-1} \sim$ Gamma $(\frac{1}{2}a_0,\ \frac{1}{2}g_0)$, and $(\sigma^2)^{-1}\lambda^{-1} \sim$ Gamma $(\frac{1}{2}a_1,\ \frac{1}{2}g_1)$, where $a_0 > 0$, $g_0 \geq 0$, $a_1 > 0$, and $g_1 \geq 0$. Here Gamma $(a,\ b)$ denotes the gamma random variable with density function $f(z) = [\exp(-az)a^b z^{b-1}/\Gamma(b)]$, $z > 0$. Typically, we let $a_0 = a_1 = 0.005$ and $g_0 = g_1 = 0$ to reflect absence of prior information on σ_a^2 and σ^2. Datta and Ghosh (1991) obtained closed form expressions for $E(Y_i|\mathbf{y},\ \lambda^{-1})$ and $V(Y_i|\mathbf{y},\ \lambda^{-1})$ under normality of the errors. They also obtained the posterior distribution of λ^{-1} given $\mathbf{y}$, but it has a complex structure making it necessary to perform one-dimensional numerical integration to get $E(Y_i | \mathbf{y})$ and $V(Y_i | \mathbf{y})$ using the following relationships:

$$E(Y_i | \mathbf{y}) = E_{\lambda^{-1}}[E(Y_i | \mathbf{y}, \lambda^{-1})]$$

and

$$V(Y_i | \mathbf{y}) = E_{\lambda^{-1}}[V(Y_i | \mathbf{y}, \lambda^{-1})] + V_{\lambda^{-1}}[E(Y_i | \mathbf{y}, \lambda^{-1})],$$

where $E_{\lambda^{-1}}$ and $V_{\lambda^{-1}}$, respectively, denote expectation and variance with respect to the posterior distribution of λ^{-1} given the data $\mathbf{y}$.

We did not include the hierarchal Bayes estimate $E(Y_i | \mathbf{y})$ in our simulation study because it was expected to be close to the empirical Bayes or EBLUP estimate, although the hierarchal Bayes variance estimate $V(Y_i | \mathbf{y})$ is always larger than the naive empirical Bayes variance estimate $V(Y_i | \mathbf{y}, \hat{\lambda}^{-1})$. We did not study the performances of variance estimates and associated confidence intervals in the simulation study.

27.2 SIMULATION STUDY

In this section, we examine the relative performances of the previous small-area estimators through simulation, using real and synthetic populations. Comparisons are made under the customary repeated sampling approach as well as under a conditional framework by conditioning on the realized sample sizes in the small areas.

27.2.1 Real Population: Unconditional Comparisons

For the simulation study, we treated a sample of 1678 unincorporated tax filers from Nova Scotia, divided into 18 census divisions, as the overall population. In each census division, units were further classified into four mutually exclusive industry groups; estimates of total for the variable of interest y (wages and salaries) were desired for each nonempty census division by industry group combination (small area of interest). The four industry groups were retail (5.15 units), construction (496 units), accommodation (114 units), and others (515 units). Two of the census divisions were empty in the accommodation group, that is, $N_i = 0$. In the remaining three groups, all the census divisions were nonempty. The average small area population size was 28.6 for retail, 27.5 for construction, 7.1 for accommodation, and 30.7 for others. The smallest census division/industry group combination had 1 unit, while the largest had 130 units. The overall correlation coefficient between y and the auxiliary variable x (gross business income) was 0.42 for retail, 0.64 for construction, 0.78 for accommodation, and 0.61 for others, so that x was not a strong predictor for retail.

To make unconditional comparisons under customary repeated sampling, we selected 500 simple random samples, each of size $n = 149$, from the overall population of $N = 1678$ units, giving an overall sampling fraction of 0.25. From each sample, we calculated the following estimators for each census division by industry group combination: (1) expansion estimator $\hat{Y}_i(\exp)$, (2) poststratified estimator $\hat{Y}_i(\mathrm{pst})$, (3) ratio synthetic estimator $\hat{Y}_i(\mathrm{syn})$, (4) sample-size-dependent estimators $\hat{Y}_i(\mathrm{ssd})$ and $\hat{Y}_i(\mathrm{ssd}^*)$, and (5) EBLUP estimator $\hat{Y}_i(\mathrm{eblup})$ under the nested error regression model (27.11).

For a given estimator $\hat{Y}_i$ and industry group, we computed the average absolute relative bias $\overline{\mathrm{ARB}}$, the average relative efficiency $\overline{\mathrm{EFF}}$ with respect to $\hat{Y}_i(\mathrm{pst})$, and the average absolute relative error $\overline{\mathrm{ARE}}$, defined as follows:

$$\overline{\mathrm{ARB}} = \frac{1}{I} \sum_{i=1}^{I} \left| \frac{\frac{1}{500} \sum_{t=1}^{500} (\hat{Y}_{it} - Y_i)}{Y_i} \right|,$$

$$\overline{\mathrm{EFF}} = \left\{ \frac{\overline{\mathrm{MSE}} \, [\hat{Y}_i(\mathrm{pst})]}{\overline{\mathrm{MSE}} \, (\hat{Y}_i)} \right\}^{1/2}, \qquad (27.17)$$

$$\overline{\mathrm{ARE}} = \frac{1}{I} \sum_{i=1}^{I} \left\{ \frac{\frac{1}{500} \sum_{t=1}^{500} |\hat{Y}_{it} - Y_i|}{Y_i} \right\},$$

where the average is taken over $I = 18$ census divisions for retail, construction, and other groups, and $I = 16$ divisions for accommodation. Here $\hat{Y}_{it}$ is the value of the estimator $\hat{Y}_i$ for the tth Monte Carlo sample ($t = 1, 2, \ldots, 500$), Y_i is the true small-area total, and

$$\overline{\mathrm{MSE}} \, (\hat{Y}_i) = \frac{1}{I} \sum_{i=1}^{I} \left[\frac{1}{500} \sum_{t=1}^{500} (\hat{Y}_{it} - Y_i)^2 \right].$$

To simplify the presentation, we report our results only for the construction industry.

Table 27.1 reports the simulated values of $\overline{\mathrm{ARB}}$, $\overline{\mathrm{EFF}}$, and $\overline{\mathrm{ARE}}$. It is clear from Table 27.1 that the ratio synthetic estimator and the EBLUP estimator perform significantly better than the remaining estimators in terms of $\overline{\mathrm{EFF}}$ and $\overline{\mathrm{ARE}}$, leading to larger $\overline{\mathrm{EFF}}$ values and smaller $\overline{\mathrm{ARE}}$ values. For example, the $\overline{\mathrm{EFF}}$ value for the EBLUP estimator is 261.1, compared to 149.4 for the sample-size-dependent estimator $\hat{Y}_i(\mathrm{ssd}^*)$. The sample-size-dependent estimator $\hat{Y}_i(\mathrm{ssd}^*)$ is slightly better than $\hat{Y}_i(\mathrm{ssd})$, and both are significantly better than the expansion and poststratified estimators. In terms of $\overline{\mathrm{ARB}}$, the synthetic estimator has the largest average absolute relative bias (15.7 percent) followed by the EBLUP estimator with $\overline{\mathrm{ARB}} = 11.3$ percent. The remaining estimators have smaller average absolute relative bias. Overall, the EBLUP estimator is

Table 27.1 Unconditional Comparisons of Small-Area Estimators: Average Absolute Relative Bias $\overline{ARB}$, Average Relative Efficiency $\overline{EFF}$, and Average Absolute Relative Error $\overline{ARE}$; Real Population (Construction)

Quality Measure	Estimator[a]					
	EXP	PST	SYN	SSD	SSD*	EBLUP
$\overline{ARB}$	1.9%	5.4%	15.7%	2.9%	4.1%	11.3%
$\overline{EFF}$	74.8%	100.0%	232.8%	137.6%	149.4%	261.1%
$\overline{ARE}$	44.6%	32.2%	16.5%	24.0%	22.1%	13.5%

[a]EXP = expansion estimator; PST = poststratified estimator; SYN = ratio synthetic estimator; SSD, SSD* = sample-size-dependent estimators; and EBLUP = empirical best linear unbiased prediction under model (27.12).

somewhat better than the synthetic estimator ($\overline{EFF}$ value of 261.1 versus 232.8 and $\overline{ARE}$ value of 13.5 versus 16.5).

We also examined the aptness of model (27.11) by fitting it to the 496 population pairs (y_{ij}, x_{ij}) from the construction group and examining the standardized residuals. We obtained $\hat{\sigma}_a^2 = 1.58$ and $\hat{\sigma}^2 = 1.34$ from equations (27.15) and (27.16). Using these values, we then calculated $\hat{\beta}$ and $\hat{a}_i$ from equations (27.13) and (27.14) and the standardized residuals

$$\frac{\hat{e}_{ij}}{\hat{\sigma}} = \frac{y_{ij} - \hat{\beta}x_{ij} - \hat{a}_i}{\hat{\sigma}x_{ij}^{1/2}},$$

where $\hat{\beta} = 0.21$. A plot of these residuals against x_{ij}-values indicated a reasonable but not good fit. The plot revealed an upward shift in the sense that several values of $\hat{e}_{ij}/\hat{\sigma}$ are larger than 1.0 but none are below -1.0.

We also tried several variations of model (27.11), including a model with an intercept term, but none of them provided a good fit. However, it is gratifying that the EBLUP estimator under model (27.11) performed well despite the so-so fit.

Results for the other industry groups are similar to those reported here for construction (Choudhry and Rao 1988). Furthermore, the $\overline{ARE}$-value for the EBLUP estimator is the largest for accommodation, the smallest industry group, with 114 units, that is, the average absolute relative error tends to increase as the expected small-area sample size decreases. Also, among the three industry groups of roughly equal size, the $\overline{EFF}$-value for the EBLUP estimator is the smallest for retail, the group with the smallest overall correlation coefficient, where $\rho = 0.42$; that is, the average relative efficiency tends to decrease as ρ decreases.

27.2.2 Real Population: Conditional Comparisons

We now turn to conditional comparisons of the estimators by conditioning on the realized sample sizes in the small areas. This is a more realistic approach

Table 27.2 Conditional Comparisons of Small-Area Estimators: Average Absolute Relative Bias $\overline{\text{ARB}}$, Average Relative Efficiency $\overline{\text{EFF}}$, and Average Absolute Relative Error $\overline{\text{ARE}}$; Real Population (Construction)

Quality Measure	Estimators					
	EXP	PST	SYN	SSD	SSD*	EBLUP
$\overline{\text{ARB}}$	25.6%	1.6%	16.8%	4.4%	6.2%	12.5%
$\overline{\text{EFF}}$	87.4%	100.0%	214.9%	125.4%	138.8%	241.2%
$\overline{\text{ARE}}$	38.4%	32.7%	17.4%	25.5%	23.5%	14.5%

because the domain sizes n_i are random with known distributions (i.e., n_i is an ancillary statistic). Särndal and Hidiroglou (1989) also studied conditional performances of small-area estimators.

To make conditional comparisons under repeated sampling, we first selected a simple random sample of size $n = 419$ to determine the sample sizes n_i in the small areas. We then treated n_i as fixed and selected 500 stratified random samples by regarding the small areas as strata, and finally we computed the conditional $\overline{\text{ARB}}$, $\overline{\text{EFF}}$, and $\overline{\text{ARE}}$ using equation (27.17). Table 27.2 reports these values. It is clear from Table 27.2 that the conditional performances are similar to unconditional performances, except for the expansion estimator which leads to substantial conditional $\overline{\text{ARB}}$: 25.6 percent.

27.2.3 Synthetic Populations: Unconditional Comparisons

We now demonstrate that the EBLUP estimator performs better when the assumed model provides a good fit than in the case of an imprecise fit. To this end, we generated a synthetic population of y-values, assuming model (27.11) and using the real x-values associated with the tax filers in the construction group, as follows. Letting $\beta = 0.21$, $\sigma_a^2 = 1.58$, and $\sigma^2 = 1.34$, we generated independent random variates a_i and e_{ij} from N(0, σ_a^2) and N(0, σ^2) and then obtained y_{ij} for each x_{ij} from $y_{ij} = 0.21x_{ij} + a_i + e_{ij}x_{ij}^{1/2}$. Note that the parameter values we chose are identical to the estimates we obtained by fitting the model to the real population of pairs (y_{ij}, x_{ij}). A plot of standardized residuals obtained by fitting model (27.11) to the synthetic population showed an excellent fit as expected.

We selected 500 simple random samples from the synthetic population and then computed $\overline{\text{ARB}}$, $\overline{\text{EFF}}$, and $\overline{\text{ARE}}$ as before. Table 27.3 reports these values. Comparing Tables 27.1 and 27.3, it is clear that $\overline{\text{EFF}}$ increases for the EBLUP estimator and the synthetic estimator, while it remains essentially unchanged for the sample-size-dependent estimators ($\overline{\text{EFF}}$ value of 261.1 increases to 319.1 for the EBLUP estimator and 232.8 increases to 313.3 for the ratio synthetic estimator). Similarly, $\overline{\text{ARE}}$ decreases for the EBLUP estimator and the synthetic estimator, while it remains essentially unchanged for the sample-size-dependent estimators ($\overline{\text{ARE}}$ value of 13.5 decreases to 11.8 for

Table 27.3 Unconditional Comparisons of Small-Area Estimators: Average Absolute Relative Bias $\overline{ARB}$, Average Relative Efficiency $\overline{EFF}$, and Average Absolute Relative Error $\overline{ARE}$; Synthetic Population

Quality Measure	Estimator					
	EXP	PST	SYN	SSD	SSD*	EBLUP
$\overline{ARB}$	2.0%	5.6%	12.5%	2.4%	3.2%	8.4%
$\overline{EFF}$	76.9%	100.0%	313.3%	135.8%	149.9%	319.1%
$\overline{ARE}$	47.4%	35.0%	13.2%	25.9%	23.3%	11.8%

the EBLUP estimator, and 16.5 decreases to 13.2 for the ratio synthetic estimator). The value of $\overline{ARB}$ also decreases for the EBLUP estimator and the synthetic estimator (11.3 versus 8.4 for the EBLUP estimator and 15.7 versus 12.5 for the ratio synthetic estimator).

We next demonstrate that the sample-size-dependent estimators do not take advantage of the between-area homogeneity, unlike the EBLUP estimator. To this end, we generated a series of synthetic populations using the previous parameter values $\beta = 0.21$, $\sigma_a^2 = 1.58$, and $\sigma^2 = 1.34$ and the model

$$y_{ij} = \beta x_{ij} + a_i \theta^{1/2} + e_{ij} x_{ij}^{1/2} \qquad (27.18)$$

by varying the parameter θ from 0.1 to 10 ($\theta = 1$ corresponds to the previous synthetic population). Note that for a given $\lambda = \sigma_a^2/\sigma^2$, the between-area homogeneity increases as θ decreases. Again, we selected 500 simple random samples from each synthetic population and then computed $\overline{EFF}$, $\overline{ARE}$, and $\overline{ARB}$ values. Note that $\hat{\sigma}_a^2$ given by equation (27.16) is now being used to estimate $V(a_i\theta^{1/2}) = \theta \sigma_a^2$. Tables 27.4, 27.5, and 27.6 report the results. It is clear from Tables 27.4 and 27.5 that $\overline{EFF}$ and $\overline{ARE}$ values for the sample-size-dependent estimators remain essentially unchanged as θ increases from 0.1 to 10. On the other hand, $\overline{EFF}$ for the EBLUP estimator and the ratio synthetic estimator is largest when $\theta = 0.1$ (i.e., when the between small-area

Table 27.4 Unconditional Comparison of Small-Area Estimators: Average Relative Efficiency $\overline{EFF}$; Synthetic Population

θ	Estimator					
	EXP	PST	SYN	SSD	SSD*	EBLUP
0.1	76.5%	100.0%	328.3%	136.0%	150.2%	324.3%
0.5	76.7%	100.0%	326.8%	136.0%	150.1%	324.6%
1.0	76.9%	100.0%	313.3%	135.8%	149.9%	319.1%
2.0	77.2%	100.0%	286.4%	135.6%	149.2%	305.0%
5.0	77.9%	100.0%	229.9%	134.7%	147.2%	270.8%
10.0	78.7%	100.0%	181.4%	133.1%	143.7%	239.9%

Table 27.5 Unconditional Comparison of Small-Area Estimators: Average Absolute Relative Error $\overline{\text{ARE}}$; Synthetic Population

θ	Estimator					
	EXP	PST	SYN	SSD	SSD*	EBLUP
0.1	47.0%	34.6%	12.4%	25.6%	23.0%	11.5%
0.5	47.2%	34.8%	12.6%	25.7%	23.1%	11.6%
1.0	47.4%	35.0%	13.2%	25.9%	23.3%	11.8%
2.0	47.7%	35.4%	14.8%	26.3%	23.8%	12.5%
5.0	48.4%	36.2%	19.1%	27.2%	25.0%	14.5%
10.0	49.2%	37.2%	25.3%	28.2%	26.8%	16.7%

Table 27.6 Unconditional Comparison of Small-Area Estimators: Average Absolute Relative Bias $\overline{\text{ARB}}$; Synthetic Population

θ	Estimator					
	EXP	PST	SYN	SSD	SSD*	EBLUP
0.1	2.0%	5.6%	11.5%	2.7%	3.4%	8.1%
0.5	2.0%	5.6%	11.7%	2.5%	3.1%	8.0%
1.0	2.0%	5.6%	12.5%	2.4%	3.2%	8.4%
2.0	2.0%	5.6%	14.1%	2.6%	3.5%	9.3%
5.0	2.0%	5.7%	18.8%	3.4%	4.6%	11.3%
10.0	2.0%	5.7%	25.1%	4.5%	6.2%	13.2%

variation is very small relative to the within-small-area variation) and decreases as θ increases to 10 (i.e., when the between-small-area variation is large relative to the within-small-area variation). Similarly, $\overline{\text{ARE}}$ for the EBLUP estimator and the synthetic estimator is the smallest when $\theta = 0.1$ and increases as θ increases to 10. Note that the EBLUP estimator performs significantly better than the synthetic estimator as θ increases (for example, $\overline{\text{ARE}}$ value of 16.7 versus 25.3 for $\theta = 10$).

Turning to the relative performance of the estimators with respect to $\overline{\text{ARB}}$, Table 27.6 shows that the EBLUP estimator performs significantly better than the synthetic estimator as θ increases (for example, $\overline{\text{ARB}}$ value of 13.2 versus 25.1 for $\theta = 10$). The sample-size-dependent estimators have a smaller $\overline{\text{ARB}}$ compared to the EBLUP estimator for all values of θ.

27.2.4 Synthetic Populations: Conditional Comparisons

We also made conditional comparisons from the synthetic populations, similar to those in Section 27.2.2, for the real population. The conditional comparisons are similar to the unconditional comparisons, except for the expansion estimator which leads to substantial conditional $\overline{\text{ARB}}$; see Table 27.7.

Table 27.7 Conditional Comparisons of Small-Area Estimators: Average Absolute Relative Bias $\overline{ARB}$, Average Relative Efficiency $\overline{EFF}$, and Average Absolute Relative Error $\overline{ARE}$; Synthetic Population ($\theta = 1$)

Quality Measure	Estimator					
	EXP	PST	SYN	SSD	SSD*	EBLUP
$\overline{ARB}$	26.3%	1.8%	12.3%	3.1%	3.9%	8.4%
$\overline{EFF}$	92.1%	100.0%	322.8%	129.9%	154.5%	323.3%
$\overline{ARE}$	42.1%	37.5%	13.0%	28.4%	24.1%	11.9%

Finally, we studied the effect of sample size n_i on the values of $\overline{ARB}$, $\overline{EFF}$, and $\overline{ARE}$. To this end, we calculated two separate values for each measure, obtained by averaging first over areas with $n_i < 6$ only and then over areas with $n_i \geq 6$. Tables 27.8 and 27.9 report these values for $n_i < 6$ and $n_i \geq 6$, respectively, based on the original synthetic population ($\theta = 1$). It is clear from Table 27.8 that $\overline{EFF}$ for the EBLUP estimator is much larger than the values for sample-size-dependent estimators when the domain sample sizes are small (<6); $\overline{EFF} = 384.6$ for the EBLUP estimator compared to $\overline{EFF} = 175.6$ for the sample-size-dependent estimator $\hat{Y}_i(\text{ssd*})$. Similarly, $\overline{ARE}$ for the EBLUP estimator is much smaller than the values for the simple-size-dependent estimators when the domain sample sizes are small (<6); $\overline{ARE} = 14.4$ for the EBLUP estimator compared to $\overline{ARE} = 30.4$ for $\hat{Y}_i(\text{ssd*})$.

Table 27.8 Conditional Comparisons of Small-Area Estimators ($n_i < 6$): Average Absolute Relative Bias $\overline{ARB}$, Average Relative Efficiency $\overline{EFF}$, and Average Absolute Relative Error $\overline{ARE}$; Synthetic Population ($\theta = 1$)

Quality Measure	Estimator					
	EXP	PST	SYN	SSD	SSD*	EBLUP
$\overline{ARB}$	32.0%	2.4%	13.9%	4.7%	6.1%	10.3%
$\overline{EFF}$	102.8%	100.0%	409.9%	139.5%	175.6%	394.6%
$\overline{ARE}$	53.7%	53.5%	14.6%	37.8%	30.4%	14.4%

Table 27.9 Conditional Comparisons of Small-Area Estimators ($n_i \geq 6$): Average Absolute Relative Bias $\overline{ARB}$, Average Relative Efficiency $\overline{EFF}$, and Average Absolute Relative Error $\overline{ARE}$, Synthetic Population ($\theta = 1$)

Quality Measure	Estimator					
	EXP	PST	SYN	SSD	SSD*	EBLUP
$\overline{ARB}$	19.1%	1.0%	10.3%	1.0%	1.2%	6.0%
$\overline{EFF}$	66.3%	100.0%	180.7%	103.6%	107.0%	203.0%
$\overline{ARE}$	27.4%	17.3%	10.9%	16.8%	16.2%	8.8%

Turning to the values for $n_i \geq 6$, we see from Table 27.9 that the poststratified estimator performs as well as the sample-size-dependent estimators, as expected. The average relative efficiency of the EBLUP estimator decreases compared to the case of $n_i < 6$ (203.0 versus 384.6), but the EBLUP estimator maintains its advantage over the sample-size-dependent estimator because its average relative efficiency also decreases [203.0 for the EBLUP estimator versus 107.0 for $\hat{Y}_i(\text{ssd}^*)$]. The average absolute relative error decreased for all the estimators compared to the case of $n_i < 6$, as expected, but the EBLUP estimator maintains its advantage over the sample-size-dependent estimators [8.8 for the EBLUP estimator versus 16.2 for $\hat{Y}_i(\text{ssd}^*)$].

27.3 CONCLUDING REMARKS

In this chapter we compare the relative performance of small-area estimators through simulation, using both real and synthetic populations. Our study suggests that the sample-size-dependent estimators are significantly better than the expansion and poststratified estimators. However, they do not take advantage of the between-area homogeneity, unlike the empirical best linear unbiased prediction estimator. As a result, the EBLUP estimator is significantly better than the sample-size-dependent estimators, especially when the between-small-area variation is not large relative to the within-small-area variation. The ratio synthetic estimator suffers from larger bias, because it gives zero weight to the direct estimator.

The small-area EBLUP estimators do not add up to a reliable direct estimator $\hat{Y}$ of the population total Y in a particular industry group. It is desirable to ensure consistency because the direct estimator is often preferred at the aggregate level. One simple way to achieve consistency is to make a ratio adjustment to $\hat{Y}_i(\text{eblup})$:

$$\tilde{Y}_i(\text{eblup}) = \frac{\hat{Y}_i(\text{eblup})}{\sum_i \hat{Y}_i(\text{eblup})} \hat{Y},$$

where the summation is over the areas i in the industry group. The modified estimator $\tilde{Y}_i(\text{eblup})$ will be somewhat less efficient than $\hat{Y}_i(\text{eblup})$ but avoids possible aggregation bias by ensuring consistency with $\hat{Y}$.

REFERENCES

Brewer, K. R. W. (1963), "Ratio Estimation in Finite Populations: Some Results Deductible from the Assumption of an Underlying Stochastic Process," *Australian Journal of Statistics*, **5**, pp. 93–105.

Choudhry, G. H., and J. N. K. Rao (1988), "Evaluation of Small Area Estimators:

An Empirical Study," paper presented at the International Symposium on Small Area Statistics, New Orleans, LA.

Choudhry, G. H., and J. N. K. Rao (1989), "Small Area Estimation Using Models That Combine Time Series and Cross-Sectional Data," in A. C. Singh and P. Whitridge (eds.), *Proceedings of Statistics Canada Symposium on Analysis of Data in Time*, pp. 67–74.

Datta, G. S., and M. Ghosh (1991), "Bayesian Prediction in Linear Models: Applications to Small Area Estimation," *Annals of Statistics*, **19**, pp. 1748–1770.

Drew, J. D., M. P. Singh, and G. H. Choudhry (1982), "Evaluation of Small Area Techniques for the Canadian Labour Force Survey," *Survey Methodology*, **8**, pp. 17–47.

Fay, R. E., and R. A. Herriot (1979), "Estimates of Income for Small Places: An Application of James–Stein Procedures to Census Data," *Journal of the American Statistical Association*, **74**, pp. 269–277.

Ghosh, M., and J. N. K. Rao (1994), "Small Area Estimation: An Appraisal," *Statistical Science*, **9**, pp. 55–93.

Henderson, C. R. (1975), "Best Linear Unbiased Estimation and Prediction under a Selection Model," *Biometrics*, **31**, pp. 423–447.

Kass, R. E., and D. Steffey (1989), "Approximate Bayesian Inference in Conditionally Independent Hierarchial Models (Parametric Empirical Bayes Models)," *Journal of the American Statistical Association*, **84**, pp. 717–726.

Laird, N. M., and T. A. Louis (1987), "Empirical Bayes Confidence Intervals Based on Bootstrap Samples," *Journal of the American Statistical Association*, **82**, pp. 739–750.

Pfeffermann, D., and L. Burck (1990), "Robust Small Area Estimation Combining Time Series and Cross-Sectional Data," *Survey Methodology*, **16**, pp. 217–237.

Prasad, N. G. N., and J. N. K. Rao (1990), "On the Estimation of Mean Square Error of Small Area Predictors," *Journal of the American Statistical Association*, **85**, pp. 163–171.

Royall R. M. (1970), "On Finite Population Sampling Theory Under Certain Linear Regression Models," *Biometrika*, **57**, pp. 377–387.

Särndal, C.-E. and M. A. Hidiroglou (1989), "Small Domain Estimation: A Conditional Analysis," *Journal of the American Statistical Association*, **84**, pp. 266–275.

Schaible, W. L. (1978), "Choosing Weights for Composite Estimators for Small Area Statistics," *Proceedings of the Survey Research Methods Section, American Statistical Association*, pp. 741–746.

Stukel, D. (1991), *Small Area Estimation Under One and Two-Fold Nested Error Regression Models*, Ph.D. Thesis, Ottawa: Carleton University.

CHAPTER TWENTY-EIGHT

Statistical Problems in Estimating the U.S. Consumer Price Index

Sylvia Leaver and Richard Valliant[1]

U.S. Bureau of Labor Statistics

Price index estimation is, in many ways, much more complicated than estimation of other parameters for business surveys. For example, the number of employees in a universe of business entities at a given point in time is a reasonably straightforward quantity to evaluate. Definitions must be formulated for what constitutes a "business entity" and "employment," but once rules are established, a quantity such as "total employment" can be clearly defined. In price index estimation, however, unique problems exist even at the stage of defining the object of estimation.

The Konüs cost of living index is the economic concept that many government agencies attempt to measure with a price index. Microeconomic theory defines the cost of living in terms of a single consumer faced with price regimens at two time periods. The *cost of living index* is the minimum cost to the consumer of achieving a certain level of utility at Time 2 relative to the minimum cost of achieving the same utility at Time 1. Diewert (1987) reviews economic work done on such indexes. Under certain assumptions, the Laspeyres index functions as an upper bound on a single consumer's true cost of living index. Determining how to aggregate across consumers invokes additional economic arguments. There is no universal agreement among economists or statisticians as to which form of price index should be used in esti-

[1] The authors thank Paul Armknecht and Robert Baskin of the U.S. Bureau of Labor Statistics, David A. Binder and Hyunshik Lee of Statistics Canada, and Phillip S. Kott of the U.S. National Agricultural Statistics Service for their comments and Robert Fay of the U.S. Bureau of the Census for his advice on the use of VPLX software. Opinions expressed are the authors' and do not reflect policy of the U.S. Bureau of Labor Statistics.

Business Survey Methods, Edited by Cox, Binder, Chinnappa, Christianson, Colledge, Kott. ISBN 0-471-59852-6 © 1995 John Wiley & Sons, Inc.

mating the cost of living index for government programs. Ideally, statisticians prefer to take the position that economic theory and reasoning should dictate the form of the *finite population index*—that is, the index that would be calculated if data on all items in the population were available. Given the form, the statistician's task is to estimate the population index based on a sample. Dalén (1992) makes similar points.

For the U.S. Consumer Price Index (CPI), the nominal estimation target for long-term price change from the reference period 0 to a later time *t* is a modified, fixed-base Laspeyres price index. The total CPI is built up from subindexes, many of which are published. The subindexes are referred to in the literature as *elementary aggregates* (Dalén 1992, Turvey et al. 1989). The building blocks of the index are geographical areas crossed with 207 groups of items called *item strata*. These strata are specific and include, for example, fresh whole milk, apples, fuel oil, and sofas.

Suppose that $I_{im}^{t,0}$ is the finite population price index for item stratum *i* and geographical index area *m*. In the case of a constant universe (i.e., population) of items, we would have

$$I_{im}^{t,0} = \frac{\sum\limits_{j \in U_{2im}} \sum\limits_{q \in U_{1ijm}} P_{imjq}^t \, Q_{imjq}^a}{\sum\limits_{j \in U_{2im}} \sum\limits_{q \in U_{1imj}} P_{imjq}^0 \, Q_{imjq}^a}, \tag{28.1}$$

where *j* denotes an establishment or outlet, *q* is an individual item or quote at outlet (*imj*), $P_{imjq}^u (u = 0, t)$ is the price of item (*imjq*) at time *u*, Q_{imjq}^a is the quantity of the item purchased in the expenditure base period *a*, U_{1imj} is the set of all items in outlet *imj*, and U_{2im} is the set of outlets in *im*. A *quote q* at a particular outlet is a very specific item, such as a 6-ounce tube of toothpaste of a particular brand, as distinguished from the item stratum, which is a broad group of items. If $a = 0$, then equation (28.1) defines a standard, *fixed-base Laspeyres index*. Equation (28.1) defines a *long-term price change* from the reference period 0 to time *t*. A *short-term price change* from *s* to *t* (*s* < *t*) is defined by the ratio of the associated long-term changes: $I_{im}^{t,s} = I_{im}^{t,0}/I_{im}^{s,0}$. The index $I_{im}^{t,0}$ can also be written as $I_{im}^{t,0} = C_{im}^t/C_{im}^0$ with $C_{im}^t(C_{im}^0)$ being the numerator (denominator) of the right-hand side of equation (28.1). The C_{im}^u terms are sometimes referred to as *cost weights*, because they are the cost to consumers of a collection of items.

The item-strata/index-area price indexes given by equation (28.1) are aggregated to form higher-level aggregate indexes. Suppose that $\mathcal{I}$ is a set of item strata and *M* is a set of index areas. An *aggregation weight* can then be defined as

$$\tilde{A}_{im}^a = \sum\limits_{j \in U_{2im}} \sum\limits_{q \in U_{1imj}} P_{imjq}^a \, Q_{imjq}^a \tag{28.2}$$

The total across all item strata/index areas in $\mathcal{I}$ and *M* is $\tilde{A}_{\mathcal{I}M}^a = \Sigma_{i \in \mathcal{I}} \, \Sigma_{m \in M}$

$\tilde{A}_{im}^a$. A definition of the *aggregate finite population index* is then

$$I_{\mathcal{G}M}^{t,0} = \sum_{i \in \mathcal{G}} \sum_{m \in M} A_{im}^a \, I_{im}^{t,0}, \tag{28.3}$$

where $A_{im}^a = \tilde{A}_{im}^a / \tilde{A}_{\mathcal{G}M}^a$. We write the aggregate index as equation (28.3) rather than stating it explicitly in terms of prices and quantities because the terms A_{im}^a and $I_{im}^{t,0}$ are estimated separately in the CPI.

If the universe were constant, then the finite population versions of fixed-base Laspeyres indexes in equations (28.1) and (28.3) would be satisfactory. However, the fluctuation of the sets of items and outlets over time must be accounted for in defining the properties of any estimate. Outlets go in and out of business, new items are introduced while others disappear, and other items change radically in quality between times 0 and t. Evaluation of equation (28.1) entails assigning a base-period price to some items that did not even exist in the base period. These problems imply that although useful as an economic concept, a fixed-universe definition of the finite population index is less useful for statistical estimation.

In this chapter we discuss the CPI sample design, index and variance estimation methods, and quality adjustment problems. Empirical results comparing two alternative variance estimation procedures are also presented.

28.1 THE CPI SAMPLE DESIGN

The Consumer Price Index uses a complex multistage sample design implemented by the U.S. Bureau of Labor Statistics (BLS). The general structure of the design is as follows (U.S. Bureau of Labor Statistics 1992, Chapter 19; Leaver et al. 1987). Four stages of sampling are used: geographical areas, outlets, groups of items, and individual items. Eighty-eight geographical primary sampling units (PSUs) are sampled using controlled selection (Dippo and Jacobs 1983, Kish 1965). *Controlled selection* is a restricted random sampling method in which a set of samples is formed, each of which conforms to a specified allowable pattern—for example, a pattern based on geographical distribution (Cochran 1977, pp. 126–127). One of the allowable samples is then selected from the set. For the CPI, 32 PSUs were selected with certainty and the remainder were selected with probability proportional to a measure of size. The geographical index areas referred to earlier are design strata represented by one or more sample PSUs. Each PSU is a county or group of counties and can range in population from several million in large urban areas to about 20,000 in smaller urban areas. Rural areas are excluded from the CPI. A detailed list of the sample PSUs is given by the U.S. Bureau of Labor Statistics (1992, Chapter 19, Appendix 3). Within the sample PSUs, two different outlet and item sample designs are used: one for commodities and services, and one for housing (rent and rental equivalence). In this chapter, we concentrate on commodities and services, paying less attention to housing.

The CPI is unique among Federal government statistics because it relies on information from both household and business surveys. Three separate surveys contribute to data development for the commodities and services portion of the CPI: the Consumer Expenditure (CE) Survey, the Continuing Point of Purchase Survey (CPOPS), and the CPI Pricing Survey. The U.S. Bureau of the Census conducts CE and CPOPS for the Bureau of Labor Statistics in the same PSUs sampled for CPI. BLS conducts the CPI Pricing Survey, which collects price data from retail businesses and households in CPI cities. For the commodities and services component of the CPI, price data are collected from retail establishments and providers of services. For the housing component of the CPI, prices are collected from residents or managers of rental housing units. Figure 28.1 shows the interrelationships of the surveys and their functions for commodities and services.

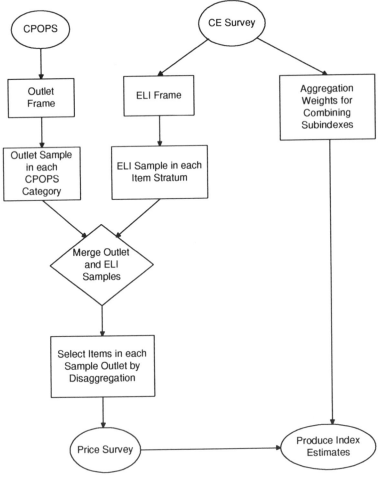

Figure 28.1 Components of the CPI data collection process for commodities and services.

The Consumer Expenditure Survey is a continuing national survey of households that collects detailed expenditure data. During each year, CE conducts approximately 26,000 household interviews, counting multiple interviews at some households (U.S. Bureau of Labor Statistics 1992, p. 174). The normalized aggregation weights A_{im}^a are estimated from annual expenditures derived from CE data accumulated over a 3-year period, currently 1982–1984, to produce the estimated normalized aggregation weights or $\hat{A}_{im}^a$. Thus, the expenditure base period a is not a single time period but rather a time span centered at 1983. CPI indexes are computed so that the average of the December indexes over the 1982–1984 period is forced to be 100. In that sense, period a in the CPI is equal to period 0. An additional feature of the $\hat{A}_{im}^a$ estimates is the use of empirical Bayes methods in which data from area im are combined with data from broader geographical areas. (See Chapter 27 for a general discussion of small-area estimation methods.) Cohen and Sommers (1984) describe the empirical Bayes methods currently in use. Ghosh and Sohn (1990) and Lahiri and Wang (1992) discuss improvements in those methods, which may be implemented in the next CPI revision. Aggregation weights are updated once per decade. In the next revision of the CPI, the $\hat{A}_{im}^a$ will be updated to cover the period 1993–1995.

CE is also used to compile a frame of item groups called *entry level items*. The sample of items is rotated in about 20 percent of the sample PSUs each year. In each PSU scheduled for rotation, entry level items are selected from each item stratum with probabilities proportional to annualized estimates of total consumer expenditures. The measure of size assigned to an entry level item is based on the most recent 2 years of CE surveys conducted in the region, rather than the CE survey for the particular PSU.

The Continuing Point of Purchase Survey is a household survey that collects data on where individuals shop and on how much they spend for various groups of items. For each category of items or services on a checklist, the respondent is asked whether purchases were made during a specified recall period and, if so, the names and locations of all places of purchase and expenditure amounts. The item or service classes, called *CPOPS categories*, are groupings likely to be sold in or by the same type of outlet. CPOPS category definitions are closely related to entry level item definitions but often cover somewhat broader classes of items. Outlets reported in CPOPS provide the sample frames for outlet selection for the CPI price survey. For each CPOPS category containing a sampled entry level item in a PSU, outlets are selected with probabilities proportional to the expenditures consumers reported for base period a in the latest CPOPS done in the PSU. Note that the expenditure base period a actually depends on the PSU containing the outlet, though we have not introduced this complication into our notation. Note also that period a of the CPOPS in a PSU differs from period a in the normalized aggregation weights A_{im}^a estimated from the CE survey. Because the sample of outlets is rotated in 20 percent of the sample PSUs each year, CPOPS is conducted in about one-fifth of the PSUs each year. Consequently, in any given year, the PSU outlet samples come from five different CPOPS years.

In the Pricing Survey, field representatives visit sample outlets, recruit them into the survey, and begin price data collection, which continues periodically thereafter. Sampled items are priced on a monthly, bimonthly, or seasonal basis. For commodities and services, about 19,000 outlets are visited per month, with price collected on about 70,000 items (U.S. Bureau of Labor Statistics 1992, p. 205). Within a sample outlet, an item is selected for pricing by a process called *disaggregation*. The selection probability of an item is designed to be approximately proportional to an item's sales in the outlet as reported by the outlet's respondent. The selection probability of an item within a PSU is thus intended to be closely related to the dollar amount spent by consumers for the item during the expenditure base period. Because of various approximations and operational compromises, however, this objective is not achieved exactly.

28.2 COMBINING TIME-SERIES DATA FROM PANEL SURVEYS

The CPI is a continuing panel survey with a key estimation feature being the combination of data over time. The long-term CPI index estimator is built up over time as a product of one-period price changes as we describe next.

28.2.1 Estimation of a Fixed-Base Index

The sample estimate for item stratum/index area *im* for time t is created in a multiplicative fashion. In the following, *im* subscripts could be added to be more explicit, but to simplify notation, *im* has been omitted. The notation is designed to conform to that of Leaver et al. (1991) and Leaver and Swanson (1992). At time t, the estimated index for *im* is defined as the ratio of the estimated cost of a market basket of goods at time t to the cost at time 0:

$$\hat{I}^{t,0} = \frac{\hat{C}^t}{\hat{C}^0}, \tag{28.4}$$

where the $\hat{C}$'s are estimated costs. Because of the changing universe, the set of goods in a market basket cannot be fixed over time, a point that is key in determining properties of estimators. (See Chapter 4 for further discussion of changing universes.) In the United States a sample estimate at a particular time period u is based on the overlapping sample units (outlets and items) between time periods u and $u - 1$. This ensures that any item used in estimation has a price for both time periods. Define

$$\bar{r}^{u,a}_{u,u-1} = \sum_{j \in s_2^{u,u-1}} \sum_{q \in s_{1j}^{u,u-1}} w^u_{jq} \, r^{u,a}_{jq},$$

where $s_2^{u,u-1}$ is the sample of outlets from im that overlap between times u and $u-1$; $s_{1j}^{u,u-1}$ is the overlap sample of items from outlet imj; w_{jq}^u is a sample weight; and $r_{jq}^{u,a} = P_{jq}^u/P_{jq}^a$ is the long-term, item price relative from a to u. The u, a superscripts denote that prices are from times u and a, while the u, $u-1$ subscripts denote that overlapping sample units between times u and $u-1$ are used in the estimate. The single-period price change from $u-1$ to u is estimated as

$$\hat{R}^{u,u-1} = \frac{\bar{r}_{u,u-1}^{u,a}}{\bar{r}_{u,u-1}^{u-1,a}} ,$$

with the same set of sample units being used in the numerator and denominator. The time t estimated cost is then calculated as $\hat{C}^t = \hat{R}^{t,t-1}\hat{C}^{t-1} = \hat{C}^0 \prod_{u=1}^t \hat{R}^{u,u-1}$. The estimate (28.4) can then be written as

$$\hat{I}^{t,0} = \prod_{u=1}^t \hat{R}^{u,u-1}, \tag{28.5}$$

that is, as the product of one-period price changes. Using the definition of $\hat{R}^{u,u-1}$, equation (28.5) can also be expressed as

$$\hat{I}^{t,0} = \frac{\bar{r}_{t,t-1}^{t,a}}{\bar{r}_{1,0}^{0,a}} \prod_{u=1}^{t-1} \left(\frac{\bar{r}_{u,u-1}^{u,a}}{\bar{r}_{u+1,u}^{u,a}} \right). \tag{28.6}$$

The first factor on the right-hand side of equation (28.6) is a ratio estimate of the change from index reference period 0 to time t. The second is the product of ratios, the numerator and denominator of which are both estimates of change from base period a to time u.

What $\hat{I}^{t,0}$ estimated depends on the form of the sample weight w_{jq}^u. For example, suppose that $w_{jq}^u = P_{jq}^a Q_{jq}^a/\pi_{jq}$ for all u, with π_{jq} being the selection probability of item $imjq$, and that $U_2^{u,u-1}$ and $U_{1j}^{u,u-1}$ are the overlap universes represented by sample outlets $s_2^{u,u-1}$ and $s_{1j}^{u,u-1}$. With E_p denoting expectation with respect to the design, we have

$$E_p(\bar{r}_{u,u-1}^{u,a}) = \sum_{j \in U_2^{u,u-1}} \sum_{q \in U_{1j}^{u,u-1}} P_{jq}^u Q_{jq}^a$$

$$\equiv \mathbf{P}_{u,u-1}^{u'} \mathbf{Q}_{u,u-1}^a ,$$

where $\mathbf{P}_{u,u-1}^u$ is the vector of all time u item prices for all outlets and items in the $(u, u-1)$ overlap universe, and $\mathbf{Q}_{u,u-1}^a$ is the corresponding overlap universe vector containing base period a quantities. Consequently, the design expectation of the weighted mean of price relatives is the cost of items in the $(u, u-1)$ overlap universe evaluated at time u prices and expenditure base period

a quantities. If the sample at each time period is large, then

$$E_p(\hat{I}^{t,0}) \cong \prod_{u=1}^{t} \left(\frac{\mathbf{P}_{u,u-1}^{u'} \mathbf{Q}_{u,u-1}^{a}}{\mathbf{P}_{u,u-1}^{u-1'} \mathbf{Q}_{u,u-1}^{a}} \right)$$

$$= \frac{\mathbf{P}_{t,t-1}^{t'} \mathbf{Q}_{t,t-1}^{a}}{\mathbf{P}_{1,0}^{0'} \mathbf{Q}_{1,0}^{a}} \prod_{u=1}^{t-1} \left(\frac{\mathbf{P}_{u,u-1}^{u'} \mathbf{Q}_{u,u-1}^{a}}{\mathbf{P}_{u+1,u}^{u'} \mathbf{Q}_{u+1,u}^{a}} \right). \qquad (28.7)$$

Thus, with the weights $w_{jq}^{u} = P_{jq}^{a} Q_{jq}^{a} / \pi_{jq}$, the index estimates an unusual type of chained population index in which the universe of items is allowed to change, but in which base period quantities are maintained. If the universe were constant, then the subscripts designating different time periods in equation (28.7) would be unnecessary, the ratios in the parentheses would equal 1, and the expectation would reduce to $\mathbf{P}^{t'}\mathbf{Q}^{a}/\mathbf{P}^{0'}\mathbf{Q}^{a}$; that is, the result is a modified fixed-base Laspeyres index of the type defined in equation (28.1) for item stratum/index area im. Chapter 25 provides additional discussion of weighting for business surveys.

In the United States, individual items or quotes are selected by a process designed to approximate sampling with probability proportional to base period expenditures; that is, $\pi_{jq} = n_j^a P_{jq}^a Q_{jq}^a / \mathbf{P}_a^{a'} \mathbf{Q}_a^a$, where n_j^a is the number of sample quotes selected in sample outlet imj from the base period a universe. The a subscripts on $\mathbf{P}_a^a$ and $\mathbf{Q}_a^a$ denote the universe extant at time a. In practice, the desired values of selection probabilities can only be approximately achieved. Turnover occurs in item samples from period to period due to item substitution, outlet nonresponse, and other factors. Thus, a probability proportional to size (*pps*) sample at time a will not necessarily be a *pps* sample at a later time. Additionally noncertainty PSUs are selected with probability proportional to population rather than expenditures.

The disaggregation process used for item selection within an outlet involves several operational compromises. First, item selection is usually done at a time somewhat later than the expenditure base period a. Second, respondents may not have readily accessible records that allow the desired measures of size to be assigned to each item. In such a case, respondents are allowed to use judgment in assigning the measures. If the measures are distributed in proportion to the base period expenditures $P_{jq}^a Q_{jq}^a$, then we still have $\pi_{jq} = n_j^a P_{jq}^a Q_{jq}^a / \mathbf{P}_a^{a'} \mathbf{Q}_a^a$ initially, in which case $w_{jq}^u = \mathbf{P}_a^{a'} \mathbf{Q}_a^a / n_j^a$. The total base period expenditure $\mathbf{P}_a^{a'} \mathbf{Q}_a^a$ is estimated from the CE survey, conducted separately from the pricing survey. The weight w_{jq}^u actually used in the CPI is built up from several factors (U.S. Bureau of Labor Statistics 1992, p. 190) and reduces to $w_{jq}^u = \mathbf{P}_a^{a'} \mathbf{Q}_a^a / n_j^a$ only in special cases. In those cases, the term $\hat{R}^{u,u-1}$ simplifies to the ratio of unweighted means.

Note that $P_{jq}^a Q_{jq}^a$ will generally not be known for individual items so that this form of π_{jq} allows the weight w_{jq}^u to reduce to something that is quantifiable. In fact, a key reason for using this type of *pps* selection is that data on quantities cannot be easily collected at the item level while data on prices and price relatives can.

Next, consider the estimator of equation (28.3), the aggregate index for a set of item strata/index areas defined by $\hat{I}_{\mathcal{I}M}^{t,0} = \Sigma_{i \in \mathcal{I}} \, \Sigma_{m \in M} \, \hat{A}_{im}^{a} \hat{I}_{im}^{t,0}$. If the item sample used to calculate $\hat{I}_{im}^{t,0}$ is large in each im and the household sample used to calculate $\hat{A}_{im}^{a}$ in the CE survey is large, then $E_{p}(\hat{I}_{\mathcal{I}M}^{t,0}) \cong \Sigma_{i \in \mathcal{I}} \, \Sigma_{m \in M} \, A_{im}^{a}E_{p}(\hat{I}_{im}^{t,0})$ with $E_{p}(\hat{I}_{im}^{t,0})$ being given by equation (28.7); that is, the result is a weighted sum of chained indexes from a changing universe. If the universe were constant, then the approximate expectation would reduce to a modified fixed-base Laspeyres index, but it generally does not.

28.2.2 Problems in Combining Data Below the Basic Aggregation Level

At this point, we should note that the United States has more specific information for combining data to estimate elementary aggregates than many countries have. In some countries, there is no information on what items are sold in a particular outlet or about quantities sold. Items may be purposively selected for pricing and often have no weights of any type associated with them. This lack of information at the item level has led to consideration of a variety of ways of combining item data to estimate an elementary aggregate. Ratios of mean prices, geometric means of price relatives, ratios of harmonic mean prices, and harmonic means of price relatives have all been considered as ways of combining item data in the absence of quantity data (Dalén 1992, Turvey 1989). In the United States, item weights are based on the sample design, and design-based considerations have led to the use of weighted averages of price relatives for estimation of the Laspeyres index.

Economists have criticized the use of averages of price relatives because they fail certain index number tests—time reversibility and circularity, in particular (see Diewert 1987). The desire for the statistical property of design unbiasedness has thus led to an estimator that may have some undesirable economic properties. Use of harmonic or geometric means or weighted means of prices would have no design-based rationale for commodities and services estimation, in the sense of producing a design-unbiased estimator, as long as a Laspeyres index is the target of estimation. A different sample design is used for the rental housing component of the CPI, on the other hand, and ratios of weighted mean prices are used to estimate elementary aggregates (U.S. Bureau of Labor Statistics 1992, p. 181).

When no weight data are available for calculating microindexes, the concerns expressed earlier about what is to be estimated remain relevant. In cases where probability sampling is not used at the microlevel, Valliant and Miller (1989) proposed use of statistical models as an obvious approach to analyzing properties associated with different ways of combining microdata. Using this approach the desired form of the population index is defined first. Though this step will often be difficult, it is necessary. Models fitted for prices or price relatives can then be used to study properties of different estimators of elementary aggregates to determine whether or not they estimate the desired population index. As Szulc (1989) notes, elementary aggregates may be composed of somewhat heterogeneous items, so the model-based approach may not be free of problems.

Another proposed approach is to use various economic index tests (e.g., monotonicity, proportionality, price/time reversibility) to choose among the different methods for combining data, as in Dalén (1992). Without considering whether a candidate estimator actually estimates the desired population index, the test approach seems less defensible statistically than the model-based approach. Note that if a form other than a Laspeyres price index is chosen as the estimation target on the grounds of its being a better approximation of true cost of living, then one may be led to geometric means or other nonstandard ways of combining microdata (Moulton 1993).

28.3 QUALITY ADJUSTMENT AND SPECIAL ESTIMATION PROBLEMS

The CPI uses a variety of imputation and special estimation procedures that may not be present in other types of surveys. Questions of statistical interest are how (and whether) the different procedures should be accounted for in calculating quantities such as bias and variance. This section describes the procedures used for commodities and services and the rationale for each. Other applications of imputation are described in Chapter 22.

Price data collection in the CPI has two distinct phases: initiation and pricing. Nonresponse can occur at either stage. At the initiation stage, sample outlets are contacted by a BLS field representative, and their eligibility for CPI is determined. At the pricing stage, current period prices are collected on sample items in the eligible outlets that participate in the survey. For estimation, the relevant response rate is the percentage of eligible units that are used in an index estimate. In 1991, 93.2 percent of eligible outlets and 86.3 percent of eligible price quotes could be used in estimation (Longacre 1992). Usable data may not be obtained from an outlet because the outlet cannot be contacted or refuses to participate. Seasonality is one of the primary reasons that usable quotes may not be collected from a cooperating outlet. The category most affected by seasonality is "apparel and upkeep," in which 57.4 percent of eligible quotes were used in estimation in 1991. Because out-of-season items are eligible for pricing should they be available, seasonally unavailable items are counted as nonresponses.

Although the fixed-based Laspeyres concept requires that the same set of items be measured through time, the reality is that products frequently disappear, products are replaced with new versions, and new products emerge. The U.S. Bureau of Labor Statistics (1992, p. 191) defines three situations in which substitutions are made to account for quality change or for changes in item specifications. The situations are categorized as (1) directly comparable, (2) comparable after direct quality adjustment, and (3) noncomparable. In each case, a new base-period price P^{a*} is imputed for an item either implicitly or explicitly in order to use the new item specification in future periods. Price relatives for subsequent times u needed for evaluating estimate (28.5) are then computed as $r^{u,a} = P^{u*}/P^{a*}$, with P^{u*} being the quality-adjusted period u price.

If old and new item specifications are considered to be *directly comparable*, the new item is simply considered as a continuation of the old. Consequently, the implicitly estimated base price of the new item is the same as that of the old.

Direct quality adjustment is most easily explained by illustration using the CPI automobile and apparel components, where this adjustment is most often made. At the time of annual model changeover, price adjustments are made to new cars and trucks to account for quality differences from the prior year's models. Adjustments are made in the new car index to account for structural and engineering changes that affect safety, performance, durability, carrying capacity, and other factors associated with manufacturer's costs. Apparel presents special problems because of frequent style changes (Liegey 1993). When an item is discontinued in an outlet, considerable effort is expended to find a comparable substitute for the item. Regression modeling is used to identify the most important characteristics of an item that go into determining its price; this is usually referred to as *hedonic regression*. In a woman's suit, for example, the presence of a jacket, skirt, or pants and the fiber content of the material all contribute to the suit's value in the eyes of the consumer. A substitute is sought that most nearly matches the old item's price-determining characteristics. These important characteristics are listed on data collection documents so that field agents can attempt to hold those characteristics constant when finding substitutes.

In each case of direct quality adjustment, a synthetic and previous period price for an item is created as $P^{t-1,*} = P^{t-1} + Q$, where P^{t-1} is the observed price for the substitute and Q is a quality adjustment. The adjustment Q may come from analyst judgment, hedonic regression, or other sources. The base-period price of the substitute is then imputed as $P^{a*} = P^a P^{t-1,*}/P^{t-1}$ where P^a is the base price of the previous item.

Quality adjustments are not made when a comparable substitute cannot be found. Two types of imputation are made in such *noncomparable* cases. The first type is common for food and service items. The rate of price change between the old item and the noncomparable new item is assumed to be the same as for the overall item stratum/index area *im* containing the new item. The base price of the new item is imputed as $P^{a*} = P^t/\hat{I}^{t,a}$ where $\hat{I}^{t,a}$ is an index of change from the expenditure base period, computed similarly to equation (28.4), for the group of items in the area. For other kinds of items, particularly new cars and apparel, price changes for comparable substitutions are used to estimate changes for noncomparable substitutions. The base-period price for a noncomparable substitution is imputed as $P^{a*} = P^a(P^{t-1}\hat{R}_c^{t,t-1}/$

P^t), where $\hat{R}_c^{t,t-1}$ is the estimated change defined in Section 28.2.1 for the item stratum/index area based only on comparable and quality-adjusted substitutions.

In the variance calculations described in Section 28.5, only variation due to the noncomparable substitution procedures is reflected. This is done by a separate imputation of a base-period price within replicate subsamples. The imputations for substitutions that are directly comparable or comparable after direct quality adjustment are treated as deterministic and not subject to sampling variation. Any variation associated with these two types of quality adjustment is the result of judgment differences by analysts on item comparability or the variance in hedonic regression adjustments. At the present time, CPI has no methods in place to reflect such variation. We do not consider this to be an important deficiency, but the topic deserves additional research.

28.4 VARIANCE ESTIMATION ALTERNATIVES

Though the need for precision estimates for price indexes published by national governments was recognized some years ago (McCarthy 1961), variance estimation is not routine for many programs. The CPI has a variety of sources of variation that might be accounted for in variance estimation. These sources include estimation of aggregation weights, sampling of items for pricing, and imputation for missing prices. In this section, we address the question of whether to account for variance in estimated aggregation weights and then discuss two alternative methods of variance estimation.

28.4.1 Conditional Versus Unconditional Variances

When calculating the estimator of an aggregate price index defined by equation (28.3), estimated aggregation weights $\hat{A}_{im}^a$ are used together with estimated indexes $\hat{I}_{im}^{t,0}$. The aggregation weights come from a household survey of consumer expenditures. The indexes $\hat{I}_{im}^{t,0}$ come from the Pricing Survey of establishments conducted in the same set of PSUs used for the Consumer Expenditure Survey. As described in Section 28.2.1, the item weights w_{jq}^u used in $\hat{I}_{im}^{t,0}$ are also based, in part, on data from the Consumer Expenditure Survey.

In estimating the variance of an aggregate index estimator

$$\hat{I}_{\mathcal{G}M}^{t,0} = \sum_{i \in \mathcal{G}} \sum_{m \in M} \hat{A}_{im}^a \, \hat{I}_{im}^{t,0}, \qquad (28.8)$$

a choice must be made as to whether to calculate variances conditional on the set of weights $\hat{A}_{im}^a$ or unconditionally, accounting for variation in the weights. Related work has been done by Balk and Kersten (1986) and Biggeri and Giommi (1987), who considered estimation of the aggregation weights a potentially important source of variation. Various arguments can be made in favor of calculating either conditional or unconditional variances. From a purely design-based point of view, the fact that $\hat{A}^a_{im}$ and $\hat{I}^{t,0}_{im}$ are sample estimates implies that they should be treated as random unless some convincing conditionality argument can be marshaled to the contrary. On the other hand, the aggregation weights are held constant for approximately 10 years in the United States while the values of the $\hat{I}^{t,0}_{im}$ are re-estimated every time period. Thus, there is some logic in conditioning on the $\hat{A}^a_{im}$'s when calculating a variance.

The degree to which conditional and unconditional variances differ from each other depends on whether long-term or short-term change is being considered. A brief, informal analysis illustrates this. First, write the aggregate index as $\hat{I}^{t,0}_{9M} = \hat{A}^{a'}_{9M}\hat{I}^{t,0}_{9M}$ where $\hat{A}^a_{9M}$ and $\hat{I}^{t,0}_{9M}$ are vectors of estimated weights and indexes with the definitions based on equation (28.8). A short-term change from time s to t ($s < t$) is then defined as $\hat{I}^{t,s}_{9M} = \hat{A}^{a'}_{9M}\hat{I}^{t,0}_{9M}/\hat{A}^{a'}_{9M}\hat{I}^{s,0}_{9M}$. Expanding the short-term change around the finite population values of the weight and index vectors and doing some algebra produces

$$\hat{I}^{t,s}_{9M} - I^{t,s}_{9M} \cong \frac{\mathbf{A}^{a'}_{9M}\,\hat{\mathbf{d}}^{t,s} + \hat{\mathbf{A}}^{a'}_{9M}\,\mathbf{d}^{t,s}}{I^{s,0}_{9M}}, \tag{28.9}$$

where $\hat{\mathbf{d}}^{t,s} = \hat{\mathbf{I}}^{t,0}_{9M} - I^{t,s}_{9M}\hat{\mathbf{I}}^{s,0}_{9M}$, $\mathbf{d}^{t,s} = \mathbf{I}^{t,0}_{9M} - I^{t,s}_{9M}\mathbf{I}^{s,0}_{9M}$, and $\mathbf{I}^{u,0}_{9M}$ ($u = t$ or s) is the vector of population indexes with elements defined by equation (28.7). The unconditional design variance of the estimated index is then approximately

$$\text{var}_p(\hat{I}^{t,s}_{9M}) \cong \frac{\mathbf{A}^{a'}_{9M}\,\text{var}_p(\hat{\mathbf{d}}^{t,s})\,\mathbf{A}^a_{9M} + \mathbf{d}^{t,s'}\,\text{var}_p(\hat{\mathbf{A}}^a_{9M})\,\mathbf{d}^{t,s}}{(I^{s,0}_{9M})^2}. \tag{28.10}$$

Treating the estimated aggregation weights as constants in equation (28.9) yields the approximate conditional variance as

$$\text{var}_p(\hat{I}^{t,s}_{9M}\,|\,\hat{\mathbf{A}}^{a'}_{9M}) \cong \frac{\mathbf{A}^{a'}_{9M}\,\text{var}_p(\hat{\mathbf{d}}^{t,s})\,\mathbf{A}^a_{9M}}{(I^{s,0}_{9M})^2}. \tag{28.11}$$

In large samples from the Consumer Expenditure Survey and the Pricing Survey, the ratio of the unconditional variance to the conditional variance is then approximately

$$1 + \frac{\mathbf{d}^{t,s'}\,\text{var}_p(\hat{\mathbf{A}}^a_{9M})\,\mathbf{d}^{t,s}}{\mathbf{A}^{a'}_{9M}\,\text{var}_p(\hat{\mathbf{d}}^{t,s})\,\mathbf{A}^a_{9M}}.$$

The numerator of the second term above is a linear combination of the variances and covariances of the aggregation weights, while the denominator corresponds to the conditional variance of the short-term index. Roughly speaking, the more variable the short-term index is, the less important the variance of the aggregation weights will be. We examine the possibility empirically in Section 28.5.

28.4.2 Methods of Variance Estimation

CPI variance estimation is difficult for many reasons. Controlled selection of PSUs implies that a strictly design-based estimator of variance is difficult, if not impossible, to construct; the methods in this section ignore this feature of the design. Two standard methods of variance estimation in complex surveys are Taylor series linearization and replication. The form of the long-term estimator defined by equations (28.5) and (28.6) is complicated; for short-term change found by taking ratios of long-term indexes, the complications are greater. This complexity makes implementation of a full linearization estimator difficult. For example, the number of covariances to be directly estimated grows over time, as illustrated by Valliant (1991). In addition, covariances between index components must be estimated for each time period between the base and the current period. The sheer size of the first-stage sample, considering establishments as the first-stage units in certainty PSUs, also leads to compromises in implementing either linearization or replication variance estimators.

Two methods employing versions of linearization and replication are described in this section and are compared empirically in Section 28.5. The first is a hybrid method that combines random-group estimation of the cost-weight covariance matrix with linearization for price-change variance estimation. We contrast national- and regional-level variance estimates produced by this hybrid methodology with those produced by VPLX software (Fay 1993) using stratified, random-group replication. A major difference in the two methods is that the VPLX estimator is implemented in such a way that its effective degrees of freedom are substantially larger than those of the hybrid estimator.

Before describing these methods, sample design features relevant for variance estimation need to be noted. Samples from all index areas (sampling strata) are divided into two or more disjoint subsets, used as replicate panels, which historically have been called *half-samples* because in most areas the number of subsets is two. For each certainty PSU, these replicates are subsamples of the PSU's outlet sample. For noncertainty strata, a replicate consists of the entire sample for one or two sample PSUs in the stratum. One-half of the replicates in a stratum are designated *odd* and the other half *even*. CPI item and outlet selection is performed independently for each PSU replicate. In certainty PSUs, independent sampling is done in seven major item groups: food, shelter, apparel, transportation, medical care, entertainment, and other commodities and services. The item groups serve as substrata in the certainty PSUs; the samples from each item group are divided among the PSU replicates. This feature was used in the VPLX estimator to create additional replicates. Cost-weight calculations done separately by replicate are the key ingredients in the two variance estimation methods.

As a notational convenience for variance calculation, let $\hat{C}_{im}^t = \hat{A}_{im}^a \hat{I}_{im}^{t,0}$. We concentrate on short-term price changes for item-strata/index-area aggregates defined as

$$\hat{I}^{t,s} = \frac{\sum\limits_{9,M} \hat{C}^t_{im}}{\sum\limits_{9,M} \hat{C}^s_{im}} = \frac{\hat{C}^t}{\hat{C}^s},$$

where the cost weights $\hat{C}^t$ and $\hat{C}^s (s < t)$ are defined by the last equality. Subscripts 9 and M could be added to the index and cost-weight estimators for more specificity but have been omitted to simplify the notation.

A Hybrid Linearization-Replication Estimator

The first variance estimate we consider is a hybrid between the linearization and random group methods. The variance of $\hat{I}^{t,s}$ obtained using the standard linearization approximation method is

$$\text{var}_p(\hat{I}^{t,s}) = (\hat{I}^{t,s})^2 \ [\text{relvar}_p(\hat{C}^s) + \text{relvar}_p(\hat{C}^t) - 2 \ \text{relcov}_p(\hat{C}^t, \hat{C}^s)],$$

$$(28.12)$$

where, for random variables X and Y, the *relative covariance* is defined as $\text{relcov}_p(X, Y) = \text{cov}_p(X, Y)/[E_p(X)E_p(Y)]$ and the *relative variance* is defined as $\text{relvar}_p(X) = \text{relcov}_p(X, X)$. The hybrid method is based on approximation (28.12) and uses random group estimates of these components. We take the case of a short-term index for a particular major area, defined by region and population size. There are eight such major areas in the United States. Indexes that are aggregations of major area indexes have variances calculated by adding variances across the major areas.

Suppose there are n index areas in a major area. Consider the $2n \times 1$ vector $\hat{\mathbf{C}}$ of full sample cost weights for an item aggregate 9, whose elements are the cost weights for each of the n index areas in the major area in months t and s:

$$\hat{\mathbf{C}} = (\hat{C}^t_{9m_1}, \hat{C}^t_{9m_2}, \ldots, \hat{C}^t_{9m_n}, \hat{C}^s_{9m_1}, \hat{C}^s_{9m_2}, \ldots, \hat{C}^s_{9m_n})',$$

where $m_1, m_2, \ldots, m_n$ are the index areas in M and $\hat{C}^u_{9mj} = \Sigma_{i \in 9} \hat{C}^u_{imj}$ and $u = t$ or s. Similarly, denote by $\hat{C}^u_{19mj}$ the cost weight based only on replicate samples denoted as *odd* and by $\hat{C}^u_{29mj}$ the estimates from *even* replicates. The two $2n \times 1$ vectors of estimated cost weights from the *odd* and *even* replicates are $\hat{\mathbf{C}}_1$ and $\hat{\mathbf{C}}_2$, respectively. Next, define a $1 \times n$ row vector $\delta = (\delta_1, \ldots, \delta_n)'$ of 0's and 1's to designate index areas to be included in the estimator of (28.12). Each component of the vector is 1 if index area j is included in the estimator and 0 if not. A $2 \times 2n$ aggregation matrix based on δ is

$$\mathbf{\Delta} = \begin{bmatrix} \delta & \mathbf{0} \\ \mathbf{0} & \delta \end{bmatrix},$$

with **0** being a $1 \times n$ vector of 0's. The 2×2 matrix of variance and covariance estimators needed to evaluate approximation (28.12) is

$$\mathbf{W} = \tfrac{1}{2} \mathbf{\Delta} \left[\sum_{r=1}^{2} \mathbf{D}_r \, \mathbf{D}_r' \right] \mathbf{\Delta}',$$

where $\mathbf{D}_r = \hat{\mathbf{C}}_r - \hat{\mathbf{C}}$ is the difference between one of the replicate sample cost weight vectors and the full sample vector with $r = 1$ or 2 denoting *odd* or *even*. The hybrid variance estimator is then

$$v_h(\hat{I}^{t,s}) = \mathbf{L}' \mathbf{W} \mathbf{L},$$

where $\mathbf{L}' = [1/\hat{C}^s, -\hat{C}^t/(\hat{C}^s)^2]$. For each geographical aggregate larger than an index area, estimates of between-index-area covariances for each pair of different index areas in the same major area in the aggregate are included in the hybrid variance estimate.

When $\mathcal{I}M$ denotes all items and all index areas, $n = 42$ and the hybrid estimator has 42 degrees of freedom, roughly speaking. From the certainty PSUs, 30 degrees of freedom were obtained after collapsing smaller certainties. The remaining noncertainty PSUs were grouped into 12 sets based on region and population size, and two replicates formed within each set. In some cases, grouping in the original design replicates was done to reduce the number of replicates to two within each of the 12 sets. Each of the $30 + 12$ groups contributes roughly 1 degree of freedom to the hybrid estimator, bringing the total to 42.

The hybrid method was adopted some years ago as being computationally expedient, though it was realized that more stable methods of variance estimation would be preferred. The hybrid estimator uses data poorly in two ways. First, few degrees of freedom are obtained in the certainty PSUs. In the pricing survey, the first-stage units in the certainty PSUs are establishments. Establishment sample sizes per certainty PSU range from about 200 to 1800, but the hybrid method picks up only a single degree of freedom in each. Second, strata containing noncertainty PSUs are grouped together to form the region/size sets, thereby losing more potential degrees of freedom.

VPLX Estimation

An alternative method of variance estimation that makes better use of available data was implemented using the VPLX software. In contrast to the hybrid methodology, VPLX produced direct estimates of short-term price change using stratified, random-group replication. The variance estimate for a given index was then based on the variance among the replicate index estimates.

For item-stratum/index-area aggregate $\mathcal{I}M$, VPLX constructed replicate cost weights $\hat{C}^u_{imr}$ ($u = t$ or s) for each of the $r = 1, 2, \ldots, G_m$ replicates in index area (sampling stratum) m. The same number of replicates was used for each item group in an index area so that the number of replicates G_m does not depend on i. This was done by retaining replicate imr, deleting the other $G_m - 1$ replicates in the index area, and weighting the units in replicate imr to produce an estimate for the full index area. The cost weight for aggregate $\Im M$ corresponding to replicate imr was then

$$\hat{C}^u_{(imr)} = \hat{C}^u_{imr} + \sum_{i',m' \in \{\Im M - im\}} \hat{C}^u_{i'm'}, \tag{28.13}$$

where $\{\Im M - im\}$ denotes the set of item strata/index areas omitting im. Replicate estimates of short-term price change were derived by taking ratios of replicate cost weights:

$$\hat{I}^{t,s}_{(imr)} = \frac{\hat{C}^t_{(imr)}}{\hat{C}^s_{(imr)}}.$$

The stratified, random group estimator of $\hat{I}^{t,s}$ is given by

$$\nu_{\mathrm{VPLX}}(\hat{I}^{t,s}) = \sum_{\Im,M} \frac{1}{G_m(G_m - 1)} \sum_{r=1}^{G_m} [\hat{I}^{t,s}_{(imr)} - \hat{I}^{t,s}]^2,$$

where the first sum is over all im in the item-strata/index-area aggregate $\Im M$.

The use of the seven major item groups as strata in each certainty PSU had important advantages in producing a more stable variance estimate. Two hundred twenty-four degrees of freedom were obtained in the certainty PSUs, and 19 were obtained in the noncertainty PSUs. The resulting degrees of freedom associated with the VPLX variance estimate at the all-items/all-cities level was 243, which was much greater than the 42 degrees of freedom for the hybrid variance estimator. Note that even this approach does not pick up all potential degrees of freedom in the certainty PSUs. The algebraic degrees of freedom associated with VPLX estimates at the major item-group/all-cities level, 51 (32 from certainty and 19 from noncertainty PSUs), was larger than that for the hybrid estimator, though not as dramatically greater as at the all-items level. Though the VPLX implementation was an improvement over that of the hybrid, VPLX nonetheless does not take full advantage of the hundreds of establishments that are the first-stage units in the certainty PSUs.

28.5 EMPIRICAL RESULTS FOR CPI

Unconditional variances were estimated using both methods with data from January 1987 through December 1991. Before presenting these detailed results, we note some differences from earlier published CPI empirical work. The first published estimates of conditional price-change variance of CPI were given by Leaver (1990) for 1978 to 1986. The following year unconditional variance estimates for the same period were published (Leaver et al. 1991).

Variance estimates for the five years following the 1987 CPI revision were published later (Leaver and Swanson 1992).

The 1987–1991 estimates differ from those computed for earlier years in two important ways. The first difference is that the revision expenditure estimates $\hat{A}_{im}^a$ for December 1986, which were updated from the 1982–1984 CE, were independently estimated for each replicate in each index area. Thus, the variances computed for the 1987 revision index series *directly* incorporate the sampling variance attributable to the estimation of expenditure weights from the 1982–1984 CE *plus* variability due to projecting these weights forward to December 1986. The second difference between the estimates is that the 1987 revised CPI variances, estimated using the hybrid method, incorporate between-index-area covariances for higher-level geographical areas, such as regions, city-size classes, and all cities combined. This incorporation affects some estimates because imputations for missing prices in a given area are sometimes made using data from a different geographical area. Imputations for apparel are frequently made in this way.

Figure 28.2 contrasts VPLX unconditional, standard-error estimates for 1-month price change at the all-cities level with those produced by the hybrid method for all items and five of the seven major item groups. Figure 28.3 provides the same contrast for 12-month price-change standard errors. The jagged solid line in each figure is VPLX, while the circles are the hybrid method. VPLX estimates are generally less variable across time, owing to VPLX's larger number of degrees of freedom. Using the Lowess method (Cleveland 1979), nonparametric smooths of VPLX and hybrid estimates are shown in each figure panel. Despite substantial variation over time, the smoothed curves imply that the VPLX and hybrid methods do generally estimate the same thing. The level of standard error is nearly constant across time for a 1-month change. Some fluctuation over time is evident for a 12-month change, with medical care as the most obvious case where VPLX and hybrid methods differ.

Table 28.1 gives the median price change and median, unconditional, VPLX-estimated standard errors for 1-, 6-, and 12-month intervals for all items and seven major groups for January 1987 through December 1991. It also gives ratios comparing VPLX-estimated 6- and 12-month price-change standard errors to 1-month price-change standard errors, along with ratios labeled "relative stability VPLX/hybrid." The relative stability ratios compare the standard deviations of VPLX price-change standard errors over the 60-month period to those of hybrid standard errors over the same time period. Note that this is an indirect measure of stability. The variance of the VPLX and hybrid variance estimators at each time point would be more direct measures, but those are more difficult to estimate.

Median VPLX-estimated price-change standard errors in Table 28.1 for all items and major groups are hardly different from those estimated earlier by the hybrid methodology (Leaver and Swanson 1992). With the exception of apparel, VPLX standard error estimates are more modulated, exhibiting fewer and less extreme fluctuations than their corresponding hybrid estimates in Fig-

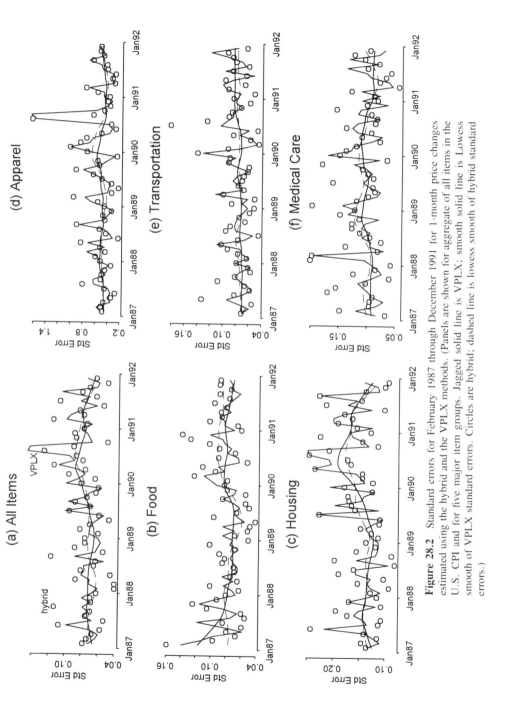

Figure 28.2 Standard errors for February 1987 through December 1991 for 1-month price changes estimated using the hybrid and the VPLX methods. (Panels are shown for aggregate of all items in the U.S. CPI and for five major item groups. Jagged solid line is VPLX; smooth solid line is Lowess smooth of VPLX standard errors. Circles are hybrid; dashed line is lowess smooth of hybrid standard errors.)

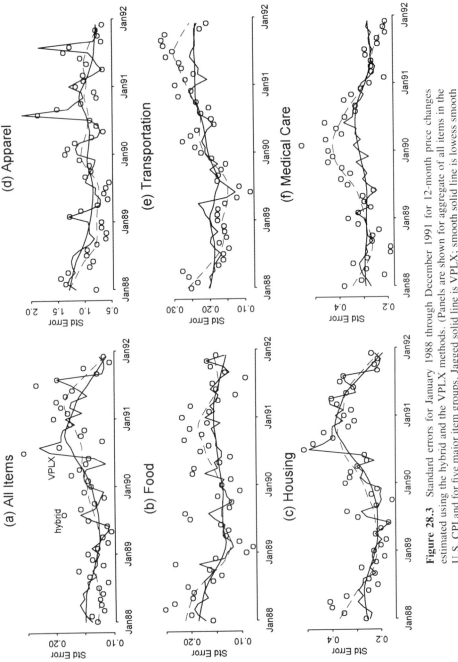

Figure 28.3 Standard errors for January 1988 through December 1991 for 12-month price changes estimated using the hybrid and the VPLX methods. (Panels are shown for aggregate of all items in the U. S. CPI and for five major item groups. Jagged solid line is VPLX; smooth solid line is lowess smooth of VPLX standard errors. Circles are hybrid; dashed line is lowess smooth of hybrid standard errors.)

Table 28.1 Median Price Change, Unconditional VPLX Standard Errors by Major Group for 1-, 6-, and 12-Month Intervals During the period January 1987 Through December 1991[a]

Major Group	One-Month Median PC	One-Month Median SE (VPLX)	Relative Stability VPLX/ Hybrid	Six-Month Median PC	Six-Month Median SE (VPLX)	Ratio 6-Month/ 1-Month SE	Relative Stability VPLX/ Hybrid	Twelve-Month Median PC	Twelve-Month Median SE (VPLX)	Ratio 12-Month/ 1-Month SE	Relative Stability VPLX/ Hybrid
All items	0.356	0.071	0.94	2.296	0.128	1.81	0.80	4.641	0.144	2.01	0.81
Food	0.295	0.075	0.69	2.418	0.123	1.75	0.59	5.110	0.149	2.11	0.47
Housing	0.272	0.141	0.93	1.913	0.238	1.70	0.90	3.875	0.273	2.04	0.91
Apparel	−0.035	0.478	1.07	1.664	1.009	2.16	0.84	4.279	0.928	1.88	0.98
Transportation	0.318	0.076	0.70	2.100	0.168	2.16	0.84	4.279	0.205	1.88	0.98
Medical Care	0.592	0.097	0.71	3.822	0.225	2.30	0.55	8.112	0.290	3.15	0.34
Entertainment	0.382	0.184	0.69	2.303	0.353	1.99	0.60	4.571	0.442	2.39	0.61
Other Commodities and Services	0.428	0.110	0.91	3.821	0.264	2.55	0.76	7.636	0.393	3.58	0.56

[a]PC, price change; SE, standard error. Relative stability VPLX/Hybrid is the ratio of the SE of the VPLX SE's across time to the SE of the hybrid SE's across time.

Source: U.S. Consumer Price Index, Bureau of Labor Statistics.

Table 28.2 Ratios of Median Unconditional to Median Conditional Standard Errors Across Time for Major Item Groups[a]

Major Group	One-Month	Two-Month	Six-Month	Twelve-Month
All Items	1.03	1.05	1.07	1.08
Food	1.02	1.04	1.05	1.04
Housing	1.04	1.01	1.03	1.08
Apparel	1.18	1.19	1.24	1.14
Transportation	1.03	1.03	1.04	1.02
Medical Care	0.99	0.99	1.00	1.00
Entertainment	1.05	1.03	1.04	1.07
Other Commodities and Services	1.07	1.06	1.00	1.10

[a]Standard errors estimated by VPLX.

Source: U.S. Consumer Price Index, Bureau of Labor Statistics.

ures 28.2 and 28.3. It is also interesting that within any item group in Table 28.1, the relative degree of modulation of standard errors tends to increase as the interval lengths. Ratios in Table 28.1 of standard deviations of price-change standard errors over the 60-month study interval range from 0.34 for the 12-month price change for medical care to 1.07 for the 1-month price change for apparel. These ratios generally show VPLX to be much more stable than the hybrid method.

Stratified, random group replication by means of VPLX was also employed to compute estimates of short-term price change standard errors that are conditional on the values of the December 1986 expenditure estimates. Recall that $\hat{C}^t_{im} = \hat{A}^a_{im} \hat{I}^{t,0}_{im}$. For the unconditional variances, this cost weight was recomputed for each replicate so that the aggregation weight $\hat{A}^a_{im}$ was computed separately in each replicate. Conditional cost weights $\hat{C}^t_{imj}$ were computed so that the same set of values $\hat{A}^a_{im}$ was used in each replicate. VPLX formed replicates as described above, replacing unconditional random-group-level cost weights with their conditional values in equation (28.13). Replicate index values were computed as ratios of conditional replicate cost weights. Table 28.2 presents the ratio of the median over time of the unconditional, estimated standard error to the median over time of the conditional, estimated standard error for major item groups and 1-, 2-, 6-, and 12-month change. Median unconditional standard error is at most 8 percent larger than median conditional standard error, with the notable exception of apparel.

28.6 CONCLUDING REMARKS

CPI variance estimation is an important concern, and U.S. methods are still being refined. The VPLX approach described here is preferred to the previously studied hybrid method because VPLX produces a more stable estimator.

A number of other statistical issues should be studied in more depth, however. Apparel is subject to missing data problems due to seasonality, and apparel estimates behave substantially differently from other estimates. The effect of imputations on index variances has not been examined in detail. How and whether to account for quality adjustments in variance estimation is an open question. Finally, a change in the method of estimating elementary aggregates may have implications for sample design and weighting that must be considered carefully.

REFERENCES

Balk, B., and H. M. P. Kersten (1986), "On the Precision of Consumer Price Indices Caused by the Sampling Variability of Budget Surveys," *Journal of Economic and Social Measurement*, **14**, pp. 19–35.

Biggeri, L., and A. Giommi (1987), "On the Accuracy and Precision of the Consumer Price Indices: Methods and Applications to Evaluate the Influence of the Sampling of Households," *Bulletin of the International Statistical Institute*, **LII**, Book 3, pp. 137–154.

Cleveland, W. S. (1979), "Robust Locally Weighted Regression and Smoothing Scatterplots," *Journal of the American Statistical Association*, **74**, pp. 829–836.

Cochran, W. G. (1977), *Sampling Techniques*, New York: Wiley.

Cohen, M., and J. Sommers (1984), "Evaluation of Methods of Composite Estimation of Cost Weights for the CPI," *Proceedings of the Survey Research Methods Section, American Statistical Association*, pp. 466–471.

Dalén, J. (1992), "Computing Elementary Aggregates in the Swedish Consumer Price Index," *Journal of Official Statistics*, **8**, pp. 129–147.

Diewert, W. E. (1987), "Index Numbers," in J. Eatwell, M. Milgate, and P. Newman (eds.), *The New Palgrave: A Dictionary of Economics*, Vol. 2, London: Macmillan, pp. 767–780.

Dippo, C., and C. Jacobs (1983), "Area Sample Redesign for the Consumer Price Index," *Proceedings of the Survey Research Methods Section, American Statistical Association*, pp. 118–123.

Fay, R. (1993), VPLX: *Variance Estimates for Complex Samples*, Washington DC: U.S. Bureau of the Census.

Ghosh, M., and S. Y. Sohn (1990), "An Empirical Bayes Approach Towards Composite Estimation of Consumer Expenditure," unpublished report, Washington DC: U.S. Bureau of Labor Statistics.

Kish, L. (1965), *Survey Sampling*, New York: Wiley.

Lahiri, P., and W. Wang (1992), "A Multivariate Procedure Towards Composite Estimation of Consumer Expenditure for the U.S. Consumer Price Index Numbers," *Survey Methodology*, **18**, pp. 279–292.

Leaver, S. G. (1990), "Estimating Variances for the U.S. Consumer Price Index for 1978–1986," *Proceedings of the Survey Research Methods Section, American Statistical Association*, pp. 290–295.

Leaver, S. G., J. E. Johnstone, and K. P. Archer (1991), "Estimating Unconditional Variances for the U.S. Consumer Price Index for 1978–1986," *Proceedings of* the *Survey Research Methods Section, American Statistical Association,* pp. 614–619.

Leaver, S. G., and D. Swanson (1992), "Estimating Variances for the U.S. Consumer Price Index for 1987–1991," *Proceedings of the Survey Research Methods Section, American Statistical Association,* pp. 740–745.

Leaver, S. G., W. L. Weber, M. P. Cohen, and K. P. Archer (1987), "Item-Outlet Sample Redesign for the 1987 U.S. Consumer Price Index Revision," *Proceedings of the 46th Session, International Statistical Institute,* **LII,** Book 3, pp. 173–185.

Liegey, P. (1993), "Adjusting Apparel Indexes in the CPI for Quality Differences," in M. Foss, M. Manser, and A. Young (eds.), *Price Measurements and their Uses,* Chicago: NBER–University of Chicago Press, pp. 209–226.

Longacre, J. (1992), "Calculating Response Rates in the Consumer Price Index Program," *Monthly Labor Review,* **115,** pp. 37–39.

McCarthy, P. (1961), "Sampling Considerations in the Construction of Price Indexes with Particular Reference to the United States Consumer Price Index," in G. Stigler (ed.), *The Price Statistics of the Federal Government,* Washington DC: National Bureau of Economic Research, pp. 197–232.

Moulton, B. R. (1993), "Basic Components of the CPI: Estimation of Price Changes," *Monthly Labor Review,* **116,** pp. 13–24.

Szulc, B. (1989), "Price Indexes Below the Basic Aggregation Level," in R. Turvey, D. J. Sellwood, B. J. Szulc, H. W. J. Donkers, M. A. Marret, L. C. Clemments, T. J. Woodhouse, and K. M. Hanson (eds.), *Consumer Price Indices. An ILO Manual,* Geneva: International Labor Office, pp. 167–178.

Turvey, R. (1989), *Consumer Price Indices. An ILO Manual,* Geneva: International Labor Office.

U.S. Bureau of Labor Statistics (1992), *BLS Handbook of Methods,* Bulletin 2285, Washington: U. S. Government Printing Office.

Valliant, R. (1991), "Variance Estimation for Price Indexes from a Two-Stage Sample with Rotating Panels," *Journal of Business and Economic Statistics,* **9,** pp. 409–422.

Valliant, R., and S. M. Miller (1989), "A Class of Multiplicative Estimators of Laspeyres Price Indexes," *Journal of Business and Economic Statistics,* **7,** pp. 387–394.

CHAPTER TWENTY-NINE

Probabilistic Detection of Turning Points in Monthly Business Activity

Danny Pfeffermann[1]
Hebrew University

Tzen-Ping Liu
Statistics Canada

François Ben-Zur
Bank of Israel

Many business and economics analysts regard the detection of turning points as one of the most challenging aspects of their work. Existing procedures for prediction of turning points can be classified into three broad groups: (1) procedures based on signal detection from leading indicators, (2) procedures employing time-series models for probabilistic forecasting of turning points, and (3) procedures combining methods (1) and (2) by incorporating the leading indicators variables in the time-series models.

The concept of a turning point is not uniquely defined in the literature. In this study, we use the following definitions for downturns and upturns. Denote by $\{Z_t, t = 1, 2, \ldots\}$ the time series of interest. The series Z_t is said to have a *downturn* (DT) of order (k, m) at time t if

$$Z_{t-k} \leq \ldots \leq Z_{t-1} > Z_t \geq Z_{t+1} \geq \ldots \geq Z_{t+m}. \qquad (29.1)$$

The series Z_t is said to have an *upturn* (UT) of order (k, m) at time t if

$$Z_{t-k} \geq \ldots \geq Z_{t-1} < Z_t \leq Z_{t+1} \leq \ldots \leq Z_{t+m}. \qquad (29.2)$$

A common choice for (k, m) used in empirical illustrations is $k = 3, m = 0$.

[1]Most work on this study was carried out while I served as a research fellow at Statistics Canada.

Business Survey Methods, Edited by Cox, Binder, Chinnappa, Christianson, Colledge, Kott.
ISBN 0-471-59852-6 © 1995 John Wiley & Sons, Inc.

We use the term *leading indicator* in a broad sense to refer to any economic or business time-series whose movements tend to signal changes in the behavior of the series of interest. Thus, the leading indicator series may consist of early, preliminary estimates of the same series or another related series, or it may consist of a series whose movements are known to precede the movements of the series of interest.

Recent studies on the prediction of turning points emphasize the use of time-series models, as classified under procedures (2) and (3) above. See, for example, Kling (1987), LeSage (1991, 1992), and Zellner et al. (1990, 1991). The use of time-series models is advantageous because it permits prediction of the points in time at which changes are expected to occur, combined with explicit probability statements about the likelihood of the occurrence of such changes.

In the studies cited above, the time series of interest consist of annual census data so that they are unaffected by seasonal variations or survey errors. When analyzing monthly time series collected from surveys, prediction of turning points in the raw estimates is of little substantive interest because the estimates are subject to survey errors. One way to overcome this problem is to try to predict turning points in the corresponding population (census) values. However, this method is of limited use for seasonal series where the goal is to detect turning points in the seasonally adjusted figures or trend levels. To illustrate this point, suppose that in the population from which the sample is taken, the monthly values Y_t $(t = 1, 2, \ldots, N)$ can be decomposed as

$$Y_t = L_t + S_t + \varepsilon_t, \tag{29.3}$$

where Y_t represents the value for month t, L_t represents the trend level, S_t represents the seasonal effect, and ε_t is a white noise with variance σ_ε^2. The series Y_t is not itself observed, but is estimated using the survey estimates $y_t = Y_t + e_t$, where e_t is the survey error. Thus, the survey estimators can be decomposed as

$$y_t = L_t + S_t + \varepsilon_t + e_t. \tag{29.4}$$

The decomposition equation (29.4) with $I_t = \varepsilon_t + e_t$, referred to as the *irregular component*, is commonly used by statistical bureaus for seasonal adjustment. Turning points unaffected by survey errors are associated with the series $Y_t = y_t - e_t$. If the goal of the analysis is to detect turning points in the trend, attention is focused on the series L_t. In some situations, the main interest is in the series $SA_t = Y_t - S_t = L_t + \varepsilon_t$, the seasonally adjusted values in the population. The analysis proposed in this chapter applies to either one of these unobservable components.

In this chapter, we illustrate the computation of probabilities of turning points in the trend levels of time series obeying decomposition (29.4), by considering time series consisting of monthly estimates measuring business activity. The time-series models fit to the data incorporate preliminary estimates as well as a quarterly series that is believed to precede the movements of the trend levels. We begin the discussion by describing the data sources used for the study and the models fit to these data. We then outline the technical details regarding the estimation of the model and the computation of turning points probabilities. We conclude by presenting empirical results illustrating the procedure.

29.1 DATA SOURCES AND STUDY OBJECTIVES

The data we use to illustrate the method were collected as part of the Canadian Survey of Manufacturing (CSOM). The CSOM produces monthly estimates of the values of Canadian manufacturing shipments, inventories, and orders. The surveyed population consists of the manufacturing establishments for which yearly census data are available plus large births identified after the census. The CSOM sample is stratified, with the strata defined by type of industry, province, and three size levels with total shipments as the size variable. Except for the large-size strata which are certainty strata, the stratum samples are selected by systematic sampling. The CSOM does not involve any rotation pattern, so the same sample is surveyed every year except for the omission of deaths and the inclusion of new births. For variance estimation purposes, the systematic sample selection within strata is treated as a simple random sample. The current CSOM sample consists of about 9000 establishments.

CSOM monthly shipments estimates are ratio estimators that update the last census' values by the corresponding ratio of the sample estimate in the particular month to the estimate derived from the same sample units using the previous census data. (The census data are produced with a time delay of almost 3 years.) All nonrespondents and dead records are imputed using five different imputation methods. A particular imputation method is selected from a decision table derived from analysis of historical data.

The preliminary estimates derived from the first wave of data collection are revised in the subsequent 3 months as data from more establishments become available. Three more benchmark revisions are made based on the results of new censuses of manufacturers, yielding a total of six revisions to each preliminary estimate. Several benchmark revisions are made because of the time delay in the production of the census data. For example, the preliminary estimate for August 1992 was first released in October 1992, and it was derived by updating the census data for 1989. This preliminary estimate was revised in November 1992, December 1992, and January 1993. The first benchmark revision took place in March 1993 when the 1990 census data became available. The second benchmark estimate for August 1992 was published in March 1994, with the final estimate expected to be published in March 1995. This final estimate is the only estimate that uses 1992 census data.

To obtain early information on possible shifts in trends of the CSOM series,

Statistics Canada conducts the Business Conditions Survey. BCS is a voluntary quarterly mail survey which uses the same initial sample as the CSOM, with response rates of about 50 percent. Survey questions refer to (1) management's expectations as to production volume in the next 3 months and (2) the current business situation in terms of orders received to date, the backlog of unfilled orders, and the inventory of finished goods. Questions are answered as either "higher than normal," "normal," or "lower than normal." Questionnaire responses are weighted based on census shipment values. Missing data are not imputed, and missing responses are not compensated for by nonresponse adjustments. Because of low response for some industries, Statistics Canada only publishes the aggregate proportions for manufacturing industries as a whole.

Based on recommendations of subject-matter experts at Statistics Canada, we chose to analyze three series: (1) total shipments of all manufacturing industries, (2) shipments of furniture and fixtures, and (3) shipments of transportation equipment. Because of storage limitations, every time an estimator is revised, the old estimate is discarded. From hard-copy publications we recaptured the preliminary and second set of estimates for January 1981 to June 1992 and the final benchmark estimates for January 1981 to December 1989. (The benchmark estimates for 1990 became known to us much later and were used for model diagnostics. See Section 29.4.2 for details.) For these three series, the differences between the first and second set of estimates turned out to be negligible, so we restricted the analysis to preliminary and final estimates only. Because of space limitations, results presented in subsequent sections pertain only to the total shipments of all manufacturing industries (TSM) series. The analysis of the other two series followed the same stages and yielded similar results.

29.2 MODELS FITTED TO THE SERIES

Our study addresses two main questions associated with the total shipments of all manufacturing industries series:

- Can simple models be constructed that predict the final benchmark estimates and their unobservable components (like the trend and the seasonal components) from knowledge of the preliminary estimates?
- Can Business Conditions Survey information be used as leading indicators to predict population values and their unobservable components and hence predict possible turning points in the level of the series?

With regard to the first question, it is important to mention that the time gap between the preliminary estimates (published with a lag of about 2 months) and the final estimates is between 2 and 3 years. Any method that predicts the benchmark estimates more accurately than the preliminary estimates should, in principle, also improve the prediction of the corresponding unknown population values. The use of Business Conditions Survey data is intended to improve prediction of future trends in the time series as compared to the use of preliminary estimates only.

29.2.1 Models Relating the Preliminary and Final Estimators

Figure 29.1 displays the preliminary and the final TSM estimates. For the time period considered, the preliminary estimates are systematically lower than the final estimates and the differences between the two estimates tend to increase with increasing values of the final estimates.

A simple model accounting for the observed relationship between the preliminary estimators Y_t^P and the final estimators Y_t^F is

$$Y_t^P = \beta_0 + \beta_1 Y_t^F + \xi_t, \tag{29.5}$$

where ξ_t are random errors. The errors ξ_t are serially correlated, but a first-order autoregressive model accounts for these correlations. (A second-order model was required for the other two series.) The model is

$$\xi_t = \lambda_1 \xi_{t-1} + \nu_t; \qquad E(\nu_t) = 0, \qquad E(\nu_t \nu_{t-k}) = \delta_k \sigma_\nu^2,$$

$$\delta_0 = 1, \qquad \delta_k \equiv 0, \qquad k \geq 1. \tag{29.6}$$

Table 29.1 shows the results obtained when fitting the model defined by equations (29.5) and (29.6) using the AUTOREG procedure of the Statistical Analysis System (SAS). The values in parentheses show the corresponding standard errors. The column entitled "REG RSQ" shows the ordinary R-square correlation after transforming the data, so that the transformed observations are independent. The column entitled "TOT RSQ" shows the correlation between the observations Y_t^P and the predictors $\hat{Y}_t^P = \hat{\beta}_0 + \hat{\beta}_1 Y_t^F + \hat{\lambda}_1 \hat{\xi}_{t-1}$. As expected from Figure 29.1, the results shown in the table indicate a very close fit.

29.2.2 Models Relating Trend Levels of Shipments to BCS Data

The Business Conditions Survey questionnaire requests management's opinions, "allowing for normal seasonal conditions." Thus, the survey responses are natural candidates to serve as early leading indicators for trend movements in the corresponding shipments series. To identify relationships between the two series, we regressed the trend levels of the final benchmark estimates, as obtained from X-11 ARIMA method (Dagum 1988), against lagged values of the trend levels and regressor variables representing the Business Conditions Survey information. The most suitable model has the general form

$$L_{3t+k} = \theta L_{3t+k-1} + \gamma_{k0} + \gamma_{k1} \text{RSL}_{3t+1} + \eta_{3t+k}; \qquad t = 0, 1, \ldots;$$

$$k = 1, 2, 3, \tag{29.7}$$

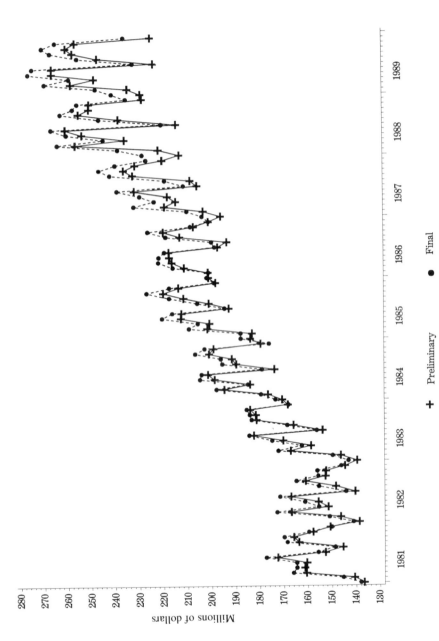

Figure 29.1 Preliminary and final estimates for the total shipments of all manufacturing industries series, 1981–1989.

Preliminary   +   Final   •

572

Table 29.1 Empirical Results When Regressing Y_t^P Against Y_t^F

Series	Regression and Autoregression Coefficients			REG RSQ	TOT RSQ
	$\hat{\beta}_0$	$\hat{\beta}_1$	$\hat{\lambda}_1$		
TSM	648	0.94	0.46	0.98	0.99
	(215)	(0.010)	(0.10)		

where L_{3t+k} is the trend level and η_{3t+k} is white noise. The regressor RSL is the ratio $[(\Sigma_i w_i h_i)_{SA}/(\Sigma_i w_i l_i)_{SA}]$, where $(\Sigma_i w_i h_i)_{SA}$ is the seasonally adjusted (SA) weighted percentage of respondents reporting a "higher backlog of un-filled orders than normal" and $(\Sigma_i w_i l_i)_{SA}$ is the seasonally adjusted weighted percentage of respondents reporting a "lower-than-normal backlog." (The weights are proportional to census total shipments; see Section 29.1.1.) A similar model was identified when TSM trend levels were estimated by fitting the basic structural model (Harvey 1989) to the final benchmark estimates. The trend-level estimates obtained by these two procedures were used for model identification purposes only. The trend levels, regression coefficients, and the noise variance were re-estimated as part of the estimation of the overall model (see Section 29.3).

The relationship defined in equation (29.7) employs separate coefficients γ_{k0}, γ_{k1} depending on whether the monthly trend levels refer to the first month of the quarter (January, April, July, October), the second month of the quarter, or the third month. The reason for this is that the Business Conditions Survey is quarterly, so that the relationships between the monthly trend levels of the shipment series and the Business Conditions Survey data depend on the position of the months in the quarter. Interestingly, Statistics Canada uses the differences $[(\Sigma_i w_i h_i)_{SA} - (\Sigma_i w_i l_i)_{SA}]$ as an indicator of the movements of the gross domestic product (GDP) of manufacturing industries. As explained in Statistics Canada releases, unfilled orders are the stock of orders generating future shipments. However, we found the ratio rather than the difference to be more powerful in predicting trend movements.

29.2.3 Time Series Models for Population Values and Final Estimates

The time plot in Figure 29.1 reveals two important features: (1) the series has an upward trend overall but with changing slopes, and (2) the series is very seasonal, with seasonal effects of approximately constant amplitude. Having observed these features, we assume that the population (census) values Y_t follow the additive decomposition of equation (29.3). The seasonal effects are modeled as

$$S_t = \sum_{j=1}^{11} S_{t-j} + \eta_{st}; \qquad E(\eta_{st}\eta_{s(t-k)}) = \delta_k \sigma_s^2, \qquad (29.8)$$

which implies that the sum of every 12 successive effects fluctuates around zero. When $\sigma_s^2 = 0$, the model imposes fixed seasonal effects.

We consider two different models for the trend levels.

Model A:

$$L_t = L_{t-1} + R_t, \qquad R_t = R_{t-1} + \eta_{rt};$$

$$E(\eta_{rt}) = 0, \qquad E(\eta_{rt}\eta_{r(t-k)}) = \delta_k\sigma_r^2. \tag{29.9}$$

Model B: The model defined by equation (29.7).

Model A does not include the Business Conditions Survey data. It can be written alternatively as

$$(L_t - L_{t-1}) - (L_{t-1} - L_{t-2}) = \eta_{rt}, \tag{29.10}$$

which can be viewed as a local approximation to a linear trend. (The special case $\sigma_r^2 = 0$ defines an exact linear trend.)

Comparing the relationships defined in equations (29.7) and (29.9) reveals that for θ close to 1 (θ was estimated as 0.98 in the empirical study), the major difference between the two models in the short run is in the stochastic structure postulated for the slopes. Thus, while in equation (29.9) the slope R_{3t+k} is assumed to evolve as a random walk without the use of external information, in equation (29.7) the slope $\Delta_{3t+k} = \gamma_{k1}\mathrm{RSL}_{3t+1}$ is modeled as a function of the survey data. In times of abrupt shifts in the trend, the slope Δ_{3t+k} can be expected to respond more rapidly than the slope R_{3t+k}, assuming of course that the trend shifts are reflected in the survey responses. It is in this respect that the survey responses are expected to act as leading indicators.

Next we consider the final benchmark estimates. These estimates can be assumed to be unbiased, but since essentially the same establishments are surveyed each month (see Section 29.1), we expect the survey errors $e_t = (Y_t^F - Y_t)$ to be serially correlated. A simple model accounting for the serial correlations of the survey errors is the autoregressive model

$$e_t = \Phi_1 e_{t-1} + \Phi_2 e_{t-2} + \cdots + \Phi_p e_{t-p} + w_t, \tag{29.11}$$

where $E(w_t) = 0$ and $E(w_t w_{t-k}) = \delta_k\sigma_w^2$. In our empirical study we found that the order $p = 3$ yields an adequate fit.

29.3 MODEL ESTIMATION AND COMPUTATION OF TURNING POINTS PROBABILITIES

The separate models defined by Equations (29.4), (29.5), (29.6), (29.8), (29.9), and (29.11) (henceforth Model A) or equations (29.4) to (29.8) and

(29.11) (henceforth Model B) can be combined and written compactly in the *state-space form*

$$\mathbf{Y}_t = \mathbf{X}\boldsymbol{\alpha}_t, \quad \text{where} \quad \boldsymbol{\alpha}_t = T_t\boldsymbol{\alpha}_{t-1} + \boldsymbol{\eta}_t; \quad E(\boldsymbol{\eta}_t\boldsymbol{\eta}'_{t-k}) = \delta_k\mathbf{Q}. \quad (29.12)$$

The left-hand equations are the *observations equations* and the right-hand equations are the *system equations*, with $\boldsymbol{\alpha}_t$ defining the *state vector* and T_t the *transition matrix*. Writing the model in state-space form has the advantage that it permits the prediction of the state vectors and hence the population valu ς and other components in a rather simple way by means of the Kalman filter. In our case, $\mathbf{Y}'_t = (Y^P_t, Y^F_t)$ and for Model A we have

$$X = \begin{bmatrix} 1, & \beta_1, & 0, & 0, & 0, & \beta_1, & \mathbf{0}'_{10}, & \beta_1, & \beta_1, & \mathbf{0}'_{(p-1)}, & 1 \\ 0, & 1, & 0, & 0, & 0, & 1, & \mathbf{0}'_{10}, & 1, & 1, & \mathbf{0}'_{(p-1)}, & 0 \end{bmatrix},$$

a matrix of order $2 \times (18 + p)$ where $\mathbf{0}'_k$ defines a row vector of zeroes of order k,

$$\boldsymbol{\alpha}'_t = [\beta_0, \quad L_t, \quad L_{t-1}, \quad L_{t-2}, \quad L_{t-3}, \quad S_t, \quad S_{t-1}, \quad \ldots ,$$

$$S_{t-10}, \quad \varepsilon_t, \quad e_t, \quad e_{t-1}, \quad \ldots , \quad e_{t-p}, \quad \xi_t]$$

$$\boldsymbol{\eta}'_t = [0, \quad \eta_{rt}, \quad 0, \quad 0, \quad 0, \quad \eta_{st}, \quad \mathbf{0}'_{10}, \quad \varepsilon_t, \quad w_t, \quad \mathbf{0}'_{(p-1)}, \quad \nu_t]$$

and $\mathbf{Q} = \text{DIAG}[0, \quad \sigma^2_r, \quad 0, \quad 0, \quad \sigma^2_s, \quad \mathbf{0}'_{10}, \quad \sigma^2_\varepsilon, \quad \sigma^2_w, \quad \mathbf{0}'_{(p-1)}, \quad \sigma^2_\nu, \quad 0]$, a diagonal matrix of order $(18 + p)$.

Notice that by multiplying the second row of X by $\boldsymbol{\alpha}_t$ we obtain $Y^F_t = (L_t + S_t + \varepsilon_t) + e_t = Y_t + e_t$, whereas multiplying the first row of X by $\boldsymbol{\alpha}_t$ yields $Y^P_t = \beta_0 + \beta_1(L_t + S_t + \varepsilon_t + e_t) + \xi_t = \beta_0 + \beta_1 Y^F_t + \xi_t$, which is the same as equation (29.5). For Model B, the state vector includes also the γ-coefficients of the model holding for the trend with the matrix X and the vector $\boldsymbol{\eta}'_t$ extended accordingly. Because of space limitations, we do not present the corresponding transition matrices, but these can be identified very easily from the relationship $\boldsymbol{\alpha}_t = T\boldsymbol{\alpha}_{t-1} + \boldsymbol{\eta}_t$ and the model equations. The lagged values L_{t-3} (under Model A) and L_{t-2} and L_{t-3} (under Model B) have been added to the state vector as a convenient way to obtain the covariances between the corresponding trend-level estimators, needed for the computation of the turning point probabilities.

The *Kalman filter* consists of a set of recursive equations that can be used to update the estimators of current state vectors, smooth the estimators of past state vectors, and predict the values of future state vectors, as new data become available. The filter also yields the variance-covariance ($\mathbf{V}$-$\mathbf{C}$) matrices of the prediction errors. The filter equations are presented in numerous publications; see, for example, Anderson and Moore (1979) and Harvey (1989). Harvey also discusses the basic structural model as defined by equations (29.4), (29.8), and (29.9) of this chapter.

The actual application of the Kalman filter requires estimation of the unknown model parameters and initialization of the filter. The set of model parameters consists of the following: the slope coefficient β_1; the autoregression coefficients λ_1 and $\phi' = (\phi_1, \ldots, \phi_p)$; the variances σ_ν^2, σ_r^2 (or σ_η^2), σ_s^2, and σ_w^2; and, for the trend model defined by equation (29.7), the coefficient θ. We omit the technical details of the estimation, but the procedures used are similar to the procedures employed by Pfeffermann (1991). [Note that the intercept β_0 of equation (29.5) and the γ-coefficients of equation (29.7) are included as fixed coefficients in the state vector.]

29.3.1 Prediction of Population Values, Final Estimates, and Trend Levels

Under model (29.12), prediction of the population values and their components is a simple exercise. Let $\hat{\mathbf{a}}_{89}$ and $\hat{\mathbf{P}}_{89}$ denote the estimated state vector and V-C matrix at month $t = 89$ (the last month for which the final estimate is known), when fitting the model to the series of preliminary and final estimates (Y_t^P, Y_t^F), $t = 1, 2, \ldots, 89$. Having estimated the model parameters, we estimate the state vectors for months $t > 89$ by applying the Kalman filter to the state-space model holding for the preliminary estimates. This model is the same as model (29.12) but with X containing only the first row. (The system equations remain unchanged.) The Kalman filter can be initialized at time $t = 89$ with $\hat{\mathbf{a}}_0 = \hat{\mathbf{a}}_{89}$ and $\hat{\mathbf{P}}_0 = \hat{\mathbf{P}}_{89} = \hat{E}[(\hat{\mathbf{a}}_{89} - \mathbf{a}_{89})(\hat{\mathbf{a}}_{89} - \mathbf{a}_{89})']$.

Let $\hat{\mathbf{a}}_t$ define the estimate of the state vector for month $t \geq 89$ and let $\hat{\mathbf{P}}_t = \hat{E}[(\hat{\mathbf{a}}_t - \mathbf{a}_t)(\hat{\mathbf{a}}_t - \mathbf{a}_t)']$. The predictor of the population value $Y_t = L_t + S_t + \varepsilon_t$ and the estimated prediction variance are obtained as

$$\hat{Y}_t = \hat{L}_t + \hat{S}_t + \hat{\varepsilon}_t = \mathbf{c}'\hat{\mathbf{a}}_t, \qquad \hat{V}(\hat{Y}_t) = \hat{E}(\hat{Y}_t - Y_t)^2 = \mathbf{c}'\hat{\mathbf{P}}_t\,\mathbf{c}, \quad (29.13)$$

where $\mathbf{c}'$ is a vector of zeros and ones defined appropriately.

The predictor of the corresponding final estimate and the estimated prediction variance are obtained as

$$\hat{Y}_t^F = \hat{L}_t + \hat{S}_t + \hat{\varepsilon}_t + \hat{e}_t = \mathbf{d}'\hat{\mathbf{a}}_t; \qquad \hat{V}(\hat{Y}_t^F) = \hat{E}(\hat{Y}_t^F - Y_t^F)^2 = \mathbf{d}'\hat{\mathbf{P}}_t\,\mathbf{d},$$

$$(29.14)$$

with $\mathbf{d}'$ appropriately defined. Estimators of the trend levels L_t or the population seasonally adjusted values $Y_t - S_t = L_t + \varepsilon_t$ are obtained similarly to equation (29.13).

Under the model, the only rationale for predicting the final estimates is "model diagnostics." In fact, $\text{Var}(\hat{Y}_t - Y_t) \leq \text{Var}(\hat{Y}_t^F - Y_t^F)$, but by comparing the predictors $\hat{Y}_t^F$ with the actual final estimates as they become available, one can assess model performance. Recall, however, that the final estimates are produced with a time lag of 2–3 years so that the current model performance can be better assessed by comparing the preliminary estimates Y_t^P with their predictors from the preceding month $t - 1$. See Section 29.4 for both kinds of empirical comparisons.

By using the smoothed estimates of the state vectors, we can predict the population value and the final estimate for a past month $t - k$ by using all preliminary estimates known by month t. These predictors are more accurate under the model than the predictors based on data available at time $t - k$, although the differences are generally small when t is large and k is small. The smoothed predictors have the same structure as in equations (29.13) and (29.14), but with $\hat{\boldsymbol{\alpha}}_{t-k}$ replaced by the smooth estimator $\hat{\boldsymbol{\alpha}}_{(t-k)|t}$.

The **V-C** matrices $\hat{\boldsymbol{P}}_t$ produced by the Kalman filter assume that model parameters are known. In practice, these parameters have to be estimated, (see Section 29.3.1), which adds another component of variance to the prediction errors. A simple method to account for this extra source of variation is described in Section 29.3.2.

Comment: An alternative, straightforward, and appealing way to predict the final estimates (and hence the population values) is by reversing the regression model in equation (29.5) so that the final estimates form the regression's dependent variable and the preliminary estimates define the independent, regressor variable. Another advantage of this procedure is that it does not rely on the decomposition of equation (29.3) and the added model assumptions. The disadvantages of this procedure are that:

- It does not permit the estimation of the level of the series as defined by the trend levels or the population seasonally adjusted values.
- There is no effective way to account for serial correlations between the regression residuals as in equation (29.6) because at a current month t, the final estimates for at least the previous 27 months are unknown.
- The use of the time-series model permits the estimation of survey errors $e_t = Y_t^F - Y_t$ for any given month t and hence yields better predictors for the population values under the model than does the use of the final estimates alone.

The use of the reversed regression relationship is not considered further in this chapter.

29.3.2 Posterior Distribution of the State Vectors

In what follows, we adopt a Bayesian formulation for the Kalman filter as proposed by Harrison and Stevens (1976) and by Meinhold and Singpurwalla (1983). By this formation, the recursive equations used for prediction, updating, or smoothing of the state vectors can be thought of as recursive algorithms for updating the corresponding posterior distributions of the state vectors. Specifically, consider the smoothed predictors of the state vector and suppose first that the vector of model parameters denoted by λ is known. Then, under the assumption of normality of the model error terms, the posterior distribution of $\boldsymbol{\alpha}_t$ given the data $\boldsymbol{D}_n = [\boldsymbol{Y}_1 \ldots \boldsymbol{Y}_n]$ until some month $n \geq t$ is normal, or

$$\boldsymbol{\alpha}_t \mid \boldsymbol{D}_n \sim \mathrm{N}[\hat{\boldsymbol{\alpha}}_{t|n}(\lambda_0), \hat{\boldsymbol{P}}_{t|n}(\lambda_0)], \qquad (29.15)$$

where $\hat{\boldsymbol{\alpha}}_{t|n}(\lambda_0) = E(\boldsymbol{\alpha}_t \mid \boldsymbol{D}_n, \lambda = \lambda_0)$ is the smoothed predictor of the state vector for time t as obtained by the smoothing algorithm with $\lambda = \lambda_0$ and $\hat{\boldsymbol{P}}_{t|n}(\lambda_0) = E\{[\hat{\boldsymbol{\alpha}}_t - \hat{\boldsymbol{\alpha}}_{t|n}(\lambda_0)][\boldsymbol{\alpha}_t - \hat{\boldsymbol{\alpha}}_{t|n}(\lambda_0)]' \mid \boldsymbol{D}_n, \lambda = \lambda_0\}$ is the posterior V-C matrix.

In practice, the vector λ is unknown and thus we have

$$f(\boldsymbol{\alpha}_t \mid \boldsymbol{D}_n) = \int f(\boldsymbol{\alpha}_t, \lambda \mid \boldsymbol{D}_n) \, d\lambda = \int f(\boldsymbol{\alpha}_t \mid \lambda, \boldsymbol{D}_n) g(\lambda \mid \boldsymbol{D}_n) \, d\lambda, \quad (29.16)$$

where $g(\lambda \mid \boldsymbol{D}_n)$ denotes the posterior density of λ. By a well-known result, if n is sufficiently large, then under some regularity conditions, $g(\lambda \mid \boldsymbol{D}_n)$ approaches a normal distribution with mean λ_n [the maximum likelihood estimator (MLE) of λ] and V-C matrix $\hat{\boldsymbol{\Lambda}}_n$, (the inverse of the information matrix evaluated at $\lambda = \hat{\lambda}_n$). Notice that both $\hat{\lambda}_n$ and $\hat{\boldsymbol{\Lambda}}_n$ are computable by the method of scoring used in this study.

To compute the mean and V-C matrix of the posterior distribution $f(\boldsymbol{\alpha}_t \mid \boldsymbol{D}_n)$, we apply a technique used previously by Hamilton (1986). By this technique, realizations $\lambda_{(k)}$, $(k = 1, 2, \ldots, K)$ are generated from the asymptotic normal posterior distribution of λ. Next, the smoothing algorithm is applied with each $\lambda_{(k)}$, yielding realizations $[\hat{\boldsymbol{\alpha}}_{t|n}(\lambda_{(k)}), \hat{\boldsymbol{P}}_{t|n}(\lambda_{(k)})]$ for the posterior mean and V-C matrix of $\boldsymbol{\alpha}_t$. The posterior mean $E(\boldsymbol{\alpha}_t \mid \boldsymbol{D}_n)$ is computed as

$$E(\boldsymbol{\alpha}_t \mid \boldsymbol{D}_n) = E_\lambda[E(\boldsymbol{\alpha}_t \mid \boldsymbol{D}_n, \lambda)] \triangleq \frac{1}{k} \sum_{k=1}^{K} \hat{\boldsymbol{\alpha}}_{t|n}(\lambda_{(k)}) = \overline{\boldsymbol{\alpha}}_{t|n}. \qquad (29.17)$$

The posterior variance is computed as

$$V(\boldsymbol{\alpha}_t \mid \boldsymbol{D}_n) = E_\lambda[V(\boldsymbol{\alpha}_t \mid \boldsymbol{D}_n, \lambda)] + V_\lambda[E(\boldsymbol{\alpha}_t \mid \boldsymbol{D}_n, \lambda)]$$

$$\triangleq \frac{1}{k} \sum_{k=1}^{K} \hat{\boldsymbol{P}}_{t|n}(\lambda_{(k)}) + \frac{1}{k} \sum_{k=1}^{K} \{[\hat{\boldsymbol{\alpha}}_{t|n}(\lambda_{(k)}) - \overline{\boldsymbol{\alpha}}_{t|n}][\hat{\boldsymbol{\alpha}}_{t|n}(\lambda_{(k)}) - \overline{\boldsymbol{\alpha}}_{t|n}]'\}.$$

$$(29.18)$$

29.3.3 Computation of Turning Point Probabilities

We illustrate the computation of the turning point (TP) probabilities by considering TPs in trend levels. (The same procedure can be applied for the computation of TPs in other component series such as the population seasonally adjusted values.) Let $\boldsymbol{L}_t' = (L_{t-3}, L_{t-2}, L_{t-1}, L_t)$ represent the trend levels at months $(t-3, t-2, t-1, t)$. Following the definitions given in the introduction with $k = 3$ and $m = 0$, a downturn (DT) occurs at month t if $L_{t-3} \leq L_{t-2} \leq L_{t-1} > L_t$ whereas an upturn (UT) occurs at month t if $L_{t-3} \geq L_{t-2} \geq L_{t-1} < L_t$. The specification of $k = 3$, $m = 0$ [see equations (29.1) and (29.2)] corresponds to the definitions employed in other related studies.

Applying the estimation procedures described in the previous sections yields for every month t the posterior mean $\hat{\mathbf{L}}_t = E(\mathbf{L}_t \mid D_n)$ and V-C matrix $V(\mathbf{L}_t \mid D_n)$ $= E[(\mathbf{L}_t - \hat{\mathbf{L}}_t)(\mathbf{L}_t' - \hat{\mathbf{L}}_t') \mid D_n]$, where D_n represents the data available until month n. For $n < t$ the vector $\hat{\mathbf{L}}_t$ refers to future months, whereas when $n > t$ the vector $\hat{\mathbf{L}}_t$ refers to past months. Assuming that $\mathbf{L}_t \mid D_n$ is multivariate normal, the computation of the TP probabilities can be carried out very easily by Monte Carlo simulations.

For example, in the empirical study of Section 29.4, the probabilities $P_D(t)$ $= \Pr[L_{t-3} \leq L_{t-2} \leq L_{t-1} > L_t \mid D_n]$ are computed as

$$P_D(t) \triangleq n(L_{t-3} \leq L_{t-2} \leq L_{t-1} > L_t)/10^4, \tag{29.19}$$

where $n(L_{t-3} \leq L_{t-2} \leq L_{t-1} > L_t)$ is the number of times that the levels $\mathbf{L}_t$ satisfy the inequalities in the brackets when simulating 10,000 observations from the posterior normal distribution of $\mathbf{L}_t$. Another set of probabilities of interest are the conditional probabilities for given patterns of the trend levels. For example, the forecast probability $FP_{DC}(t) = \Pr(L_{t+1} < L_t \mid L_t \geq L_{t-1} \geq L_{t-2}, D_n)$ is computed as

$$FP_{DC}(t) \triangleq [n(L_{t+1} < L_t \geq L_{t-1} \geq L_{t-2})/10^4]/[n(L_t \geq L_{t-1} \geq L_{t-2})/10^4].$$

$$\tag{29.20}$$

The probabilities defined by $FP_{DC}(t)$, but with L_t replaced by the original series realizations Y_t, correspond to the probabilities commonly computed when analyzing raw annual data. The condition $[Y_t \geq Y_{t-1} \geq Y_{t-2}, D_t]$, for example, represents in this case the observed pattern in the last 3 years (and the years before), whereas Y_{t+1} is the yet unobserved value. In our case, the trend levels are never observed, giving rise to the computation of unconditional probabilities like $P_D(t)$.

29.4 EMPIRICAL RESULTS

We fit the two models defined in Section 29.2 (referred to there as Model A and Model B) using the procedures described in Section 29.3. Originally, the data being analyzed contained the preliminary estimates for January 1981 to June 1992 and the final estimates for January 1982 to December 1987; later on we augmented the data set with the final estimates for 1988 to 1989 and very recently with the final estimates for 1990. Unless stated otherwise, the results presented here refer to the augmented data set for January 1981 to December 1989.

29.4.1 Analysis of Prediction Errors

It is common to assess the goodness of fit of time series models by analyzing the behavior of the one-month-ahead prediction errors $e_t = Y_t - \hat{Y}_{t|t-1}$, where $\hat{Y}_{t|t-1}$ are the model based predictions at time $t-1$. Plotting the standardized prediction errors $\tilde{e}_t^P = [(Y_t^P - \hat{Y}_{t|t-1}^P)/\hat{SD}(e_t^P)]$ and $\tilde{e}_t^F = [(Y_t^F - \hat{Y}_{t|t-1}^F)/\hat{SD}(e_t^F)]$ against time (not shown here) gives no indication of any model failure. Testing for normality of the standardized residuals yields P-values of 0.31 and 0.32 for $\tilde{e}_t^P$ and $\tilde{e}_t^F$, respectively, under Model A and 0.74 and 0.44 under Model B. These P-values give no reason to reject the hypothesis that the prediction errors are normally distributed.

In Table 29.2 we present summary statistics that permit comparison of the closeness of fit of the two models. The statistics pertaining to the preliminary estimates employ the observed prediction errors for July 1983 to June 1992 ($n = 108$), whereas the statistics pertaining to the series of final estimates employ the prediction errors for July 1983 to December 1989 ($n = 78$).

The results in Table 29.2 indicate a good fit under both models; in particular, note the very small relative absolute prediction errors. Model B, which incorporates the Business Conditions Survey information, is clearly the better model. We also computed the same statistics separately for the first month of each quarter (i.e., for months $3t + 1$, $t = 10, 11, \ldots$), the second month, and the third month [see equation (29.7)]. Under both models and for both preliminary and final estimates, the prediction errors pertaining to the first month of the quarter were higher than those for the other two months. This phenomenon could be explained under Model B as an indication that the survey data are more informative for the second and third month shipments than for the first month shipments; but we are unable to explain why this happens under Model A.

29.4.2 Prediction of the Final Estimates

The data we originally analyzed contained the final estimates for 1981 to 1987 only. Having estimated the two models from these data, we next used the models to predict the final estimates' values in the following three years, using the known values of the preliminary estimates and the Business Conditions Survey data for January 1981 through June 1992. Table 29.3 shows summary statistics of the empirical prediction errors $D_t^F = (\hat{Y}_t^F - Y_t^F)$ as obtained for 1988 to 1990 ($n = 36$) under the two models and when predicting the final estimates by the corresponding preliminary estimates Y_t^P. The notation $SD[|D_t^F|/Y_t^F]$ signifies the standard deviation of the relative absolute errors.

The results shown in Table 29.3 indicate that both models yield much better predictors than use of the preliminary estimates alone, the latter being always below the final estimates. As expected, the averages of the absolute relative errors, as obtained under the two models by use of the smoothed estimators based on the extended data, are lower than the corresponding averages of the one-month-ahead absolute relative prediction errors shown in Table 29.2. The

Table 29.2 One-Month-Ahead Prediction Errors

Series	Preliminary Estimates				Final Estimates			
Statistic	$\frac{1}{n}\Sigma\, e_t^{\mathrm{P}}$	$\frac{1}{n}\Sigma\, |e_t^{\mathrm{P}}|$	$\left[\frac{1}{n}\Sigma\,(e_t^{\mathrm{P}})^2\right]^{1/2}$	$\frac{1}{n}\Sigma\,\dfrac{|e_t^{\mathrm{P}}|}{Y_t^{\mathrm{P}}}$	$\frac{1}{n}\Sigma\, e_t^{\mathrm{F}}$	$\frac{1}{n}\Sigma\, |e_t^{\mathrm{F}}|$	$\left[\frac{1}{n}\Sigma\,(e_t^{\mathrm{F}})^2\right]^{1/2}$	$\frac{1}{n}\Sigma\,\dfrac{|e_t^{\mathrm{F}}|}{Y_t^{\mathrm{F}}}$
Model A	−0.07	5.61	7.06	0.025	−0.53	4.73	5.89	0.022
Model B	−0.10	4.90	6.24	0.022	−0.11	4.00	5.12	0.018

Table 29.3 Prediction of Final Estimates for the Years 1988–1990

Statistic	$\frac{1}{n}\,\Sigma\,D_t^F$	$\frac{1}{n}\,\Sigma\,|D_t^F|$	$\frac{1}{n}\,\Sigma\,\left|\dfrac{D_t^F}{Y_t^F}\right|$	$SD\left(\dfrac{|D_t^F|}{Y_t^F}\right)$
Model A	1.26	3.13	0.013	0.010
Model B	0.67	3.00	0.012	0.010
Y_t^P	−7.83	7.83	0.031	0.015

two models seem to perform equally well in terms of the average absolute prediction errors, but the predictions obtained under Model B are somewhat less biased. The absolute prediction errors, under Model B are smaller than the corresponding absolute prediction errors under Model A in 24 out of the 36 months.

We re-estimated the two models based on the preliminary and final estimates through December 1989 and then used them to predict the final estimates for 1990. Table 29.4 contains the values of the final estimates and the corresponding absolute relative prediction errors as obtained under the two models and when predicting the final estimates by the corresponding preliminary estimates.

Table 29.4 reveals similar results to those of Table 29.3 except that the preliminary estimates are very close to the final estimates in the last 9 months of the year, which is very different from the picture revealed in 1988 and 1989. It is still to be seen if this phenomenon repeats in 1991. The general conclusion from Tables 29.3 and 29.4 is that both models produce very good predictors for the final estimates, which leads us to believe that the same is true when predicting the corresponding population values.

Unlike the results in Table 29.2 where Model B is shown to be superior to Model A, the results in Tables 29.3 and 29.4 suggest that both models perform equally well. This seeming contradiction has a simple explanation. When predicting future values of the final estimates, with the corresponding concurrent and consecutive preliminary estimates yet unknown, the early information available from the Business Conditions Survey data incorporated in Model B is valuable in improving the predictions obtained under Model A. However, when the preliminary estimates are known for many months ahead as in Tables 29.3 and 29.4, they indicate possible changes in the behavior of the final estimates, and the survey data no longer convey extra information.

29.4.3 Computation of Trend-Level Turning Point Probabilities

Figure 29.2 shows the trend levels obtained from the smoothed state vectors under the two models (see Section 29.3.1), and for comparison it also shows the trend levels produced by X-11 ARIMA. As can be seen, all three trend curves evolve similarly over time, but the trend levels obtained under Model B are more erratic. Which of the three trends is the correct one? The answer to this question depends on one's definition and understanding of the trend

Table 29.4 Final Estimates and Absolute Relative Prediction Errors for 1990 (Estimates Divided by 100)

	Jan	Feb	Mar	Apr	May	Jun	Jul	Aug	Sep	Oct	Nov	Dec
Final estimates	230.70	238.80	267.90	248.30	272.30	267.90	230.40	245.60	251.00	270.30	249.70	219.10
$|D_t^F|/Y_t^F$, Model A	0.02	0.01	0.01	0.02	0.00	0.01	0.00	0.01	0.02	0.01	0.01	0.02
$|D_t^F|/Y_t^F$, Model B	0.01	0.00	0.01	0.02	0.01	0.01	0.02	0.02	0.02	0.01	0.02	0.02
$|D_t^F|/Y_t^F$, Y_t^P	0.04	0.05	0.02	0.02	0.02	0.01	0.01	0.01	0.01	0.01	0.01	0.02

584

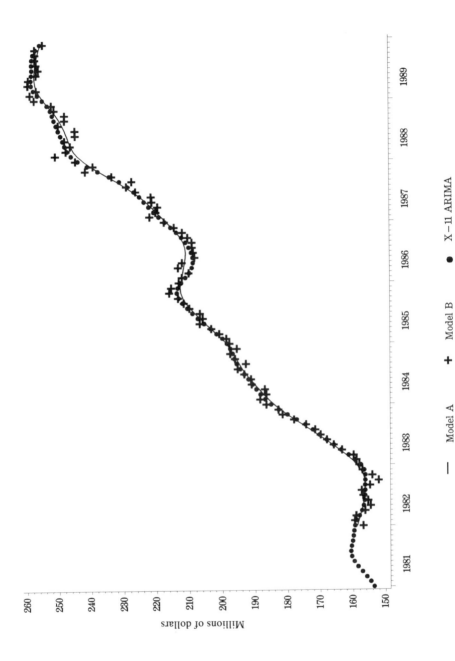

Figure 29.2 Smoothed trend levels as produced by X-11 ARIMA and under the models.

— Model A   + Model B   ● X – 11 ARIMA

notion. Clearly, the trend levels obtained under Model B are too erratic and could be smoothed, but we chose to leave them without further smoothing so as not to suppress short terms trends that are real.

We computed TP probabilities in the trend levels under both models using the procedures discussed in Section 29.3.3. As it turned out, under Model A the downturn probabilities defined by equation (29.19) and the upturn probabilities using an analogue definition never exceed 0.29 for DT events and 0.21 for UT events. Moreover, by restricting to sequences $(L_{t-1}, L_{t-2}, L_{t-3})$ such that $\Pr(L_{t-1} \geq L_{t-2} \geq L_{t-3}) \geq 0.5$ or $\Pr(L_{t-1} \leq L_{t-2} \leq L_{t-3}) \geq 0.5$, even the conditional DT and UT probabilities $P_{DC}(t) = \Pr(L_t < L_{t-1} | L_{t-1} \geq L_{t-2} \geq L_{t-3})$ and $P_{UC}(t) = \Pr(L_t > L_{t-1} | L_{t-1} \leq L_{t-2} \leq L_{t-3})$ are always below 0.35 except for one month for which $P_{DC}(t) = 0.47$. This phenomenon is explained by the fact that the trend levels produced under Model A are very smooth functions of time so that changes in direction become a rare event. The estimated variances and covariances of the trend level estimators are much larger under Model A than under Model B, resulting in unstable V-C matrices for the vectors $\mathbf{L}'_t = (L_{t-3}, L_{t-2}, L_{t-1}, L_t)$.

Figure 29.3 shows the trend levels obtained under Model B for the time period of January 1982 to July 1992, along with the corresponding TP probabilities computed using equation (29.19) (with an analogue equation for UT events). The trend levels for January 1990 to July 1992 were estimated based on the preliminary estimates and Business Conditions Survey data only. For presentation sake, we excluded probabilities lower than 0.4 from Figure 29.3.

When computing TP probabilities for an observable series like the raw estimates, the validity of these probabilities can be assessed by studying their performance in predicting actual TP events. Because the trend is never observed, this is impossible in our case. Moreover, it is clear that the TP probabilities computed under the model depend on the estimated trend levels. Nonetheless, some interesting observations can be made.

Consider UT events first. Five months that exhibit what seems to be real UTs are November 1982, July 1986, May 1988, April 1991, and February 1992. We focus on these months because around the first three months UTs are observed in the X-11 ARIMA trend and for February 1992 a UT is found in the trend levels estimated under Model A. Notice that except in July 1986 where the change in level is very mild, the UTs observed for the other four months are well predicted by the corresponding UT probabilities. A relatively high probability for a UT is also computed for April 1986, $[P_D(t) = 0.52]$, but this probability can be classified as a false alarm. On the other hand, the other sporadic UT events observed in the graph in May 1987 and August 1989 are assigned a low UT probability.

Next we consider DT events. Three months that exhibit real DTs are December 1985, March 1988, and June 1991, and to some extent also March 1989. These DT events are indeed assigned relatively high DT probabilities. On the other hand, for seven other months with relatively high DT probabilities, the decline in the trend is only for one or two months. These probabilities

586

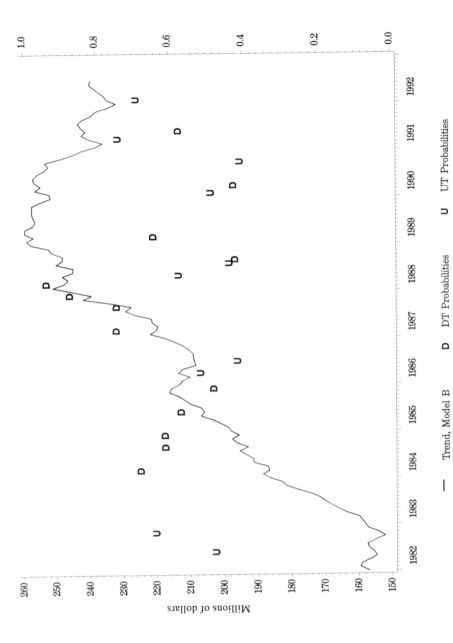

Figure 29.3 Smoothed trend levels under Model B and turning points probabilities.

should be classified as false alarms. As in the case of UT events, several other sporadic DTs are assigned very low DT probabilities.

The fact that sporadic TPs are occasionally assigned relatively high TP probabilities is a direct consequence of the TP definitions used in this study. Clearly, by considering larger values for m in definitions (29.1) and (29.2), the probabilities of false alarms would be decreased although this would also decrease the probabilities of real TP events.

We also computed TP forecast probabilities as defined by equation (29.20) with an analogue expression for UT events. As it turns out, what seem to be real TP events are in most cases forecasted with relatively high probabilities; on the other hand there are almost the same number of false alarms. In this respect, notice that while the trend-level estimates use all available data for all years, the computation of TP forecast probabilities is based on only the past and current data.

29.5 CONCLUDING REMARKS

In Section 29.1.2, we raise two questions which for the most part are successfully addressed. Our study indeed shows that simple models can be constructed that permit better predictions of the final estimates than the use of preliminary estimates alone. Also, by using the Business Conditions Survey data, future values of the final estimates can be predicted reliably. The results are somewhat less satisfactory regarding the computation of TP probabilities because the definitions of TP events used in this study (requiring only one month of change in direction) give rise to a relatively large number of false alarms. While this problem could be handled by redefining TP events to include more months after the first change occurs, another, perhaps more promising, direction is to attempt to identify closer relationships between the shipment series and Business Conditions Survey data. In our study, the answers to only one question are included in the model, but it would seem that by using the answers to other questions, a better fit could be obtained. The inclusion of additional variables in the model fitted to the trend may also result in a smoother trend curve.

REFERENCES

Anderson, B. D. O., and J. B. Moore (1979), *Optimal Filtering*, Englewood Cliffs, NJ: Prentice–Hall.

Dagum, E. B. (1988), *The X-11 ARIMA Seasonal Adjustment Method—Foundations and User's Manual*, Ottawa: Statistics Canada.

Hamilton, J. D. (1986), "A Standard Error for the Estimated State Vector of a State-Space Model," *Journal of Econometrics*, **33**, pp. 387–397.

Harrison, P. J., and C. F. Stevens (1976), "Bayesian Forecasting (with discussion)," *Journal of the Royal Statistical Society, Series B*, **38**, pp. 205–247.

Harvey, A. C. (1989), *Forecasting Structural Time Series Models and the Kalman Filter*, Cambridge University Press.

Kling, J. L. (1987), "Predicting the Turning Points of Business and Economic Time Series," *Journal of Business*, **60**, pp. 201–238.

LeSage, J. P. (1991), "Analysis and Development of Leading Indicators Using a Bayesian Turning-Points Approach," *Journal of Business and Economic Statistics*, **4**, pp. 305–316.

LeSage, J. P. (1992), "Scoring the Composite Leading Indicators: A Bayesian Turning Points Approach," *Journal of Forecasting*, **11**, pp. 35–46.

Meinhold, R. J., and N. D. Singpurwalla (1983), "Understanding the Kalman Filter," *The American Statistician*, **37**, pp. 123–127.

Pfeffermann, D. (1991), "Estimation and Seasonal Adjustment of Population Means Using Data from Repeated Surveys," *Journal of Business and Economic Statistics*, **9**, pp. 163–175.

Zellner, A., C. Hong, and G. M. Gulati (1990), "Turning Points in Economic Time Series, Loss Structures and Bayesian Forecasting," in S. Geisser, J. Hodges, S. J. Press, and A. Zellner (eds.), *Bayesian Likelihood Methods in Statistics and Econometrics: Essays in Honor of George A. Barnard*, Amsterdam: North Holland, pp. 371–393.

Zellner, A., C. Hong, and C. Min (1991), "Forecasting Turning Points in International Output Growth Rates Using Bayesian Exponentially Weighted Autoregression, Time-Varying Parameter, and Pooling Techniques," *Journal of Econometrics*, **48**, pp. 275–304.

CHAPTER THIRTY

Combining Design-Based and Model-Based Inference

K. R. W. Brewer

Australian National University

Survey sampling (also known as finite population sampling) is unique within statistics because there are two distinct probability distributions that can serve as sources of inference. One, peculiar to survey sampling, describes the way the sample is selected. This distribution is exactly known because the designer imposes it on the population. The other distribution, shared with all other areas of statistics, is a structure innate to the population itself and is unknown but capable of being modeled.

In design-based inference, the bias, variance, and mean squared error (MSE) are defined in terms of the expectation E_p over all possible samples under the sample design p. In the model-based or prediction theory approach, the relevant expectations E_ξ are over all possible realizations of a stochastic model ξ (usually a linear regression model), which connects a variable of interest y with a set of auxiliary or benchmark variables $\mathbf{x}$. In this chapter the design-based concepts will be referred to as p-bias, p-variance, and p-MSE, and the model-based ones will be referred to as ξ-bias, ξ-variance, and ξ-MSE. Using radically different sources of inference, researchers are likely to adopt radically different approaches to survey design and analysis. Despite these differences, the two approaches have been converging in their notions of best practice (Smith 1984, 1994).

On one hand, many design-based statisticians now acknowledge that the explicit use of modeling can aid them in making more useful inferences. Särndal et al. (1992) caters to the expressed needs of this group. In fact, competent design-based sampling statisticians have implicitly used population models

Business Survey Methods, Edited by Cox, Binder, Chinnappa, Christianson, Colledge, Kott. ISBN 0-471-59852-6 © 1995 John Wiley & Sons, Inc.

from the beginning. For evidence of this, see Basu's (1971) entertaining elephant fable and the lively discussion concerning it.

On the other hand, the model-based school has long recognized that randomization can play a useful, if secondary, role. Royall (1976) likened this to the use of a coin toss at the start of a ball game. Model-based statisticians have also recognized that even the purposive selection of a balanced sample cannot provide robustness against ξ-bias caused by an unknown regressor and that "random sampling can help to provide protection against bias like this" (Royall 1992, p. 184). Yet another useful role played by randomization is in sample control, which randomization permits through the use of permanent random numbers (PRNs) within sample frames (see Chapter 9). It would be difficult, if not impossible, to duplicate this role using balanced sampling or indeed any other kind of purposive sampling intended to ensure useful statistical properties for the estimator.

Moreover, there are at least two arguments to suggest that both forms of inference are useful in their own right. When analyzing data from a sample that has already been selected, it is plausible to argue that inferences depending on samples that might have been, but were not in fact, selected are beside the point. But when designing a future survey, the set of all samples that could be selected is a legitimate focus of attention. If the population model is later found to have broken down in some unanticipated fashion, design-based estimates and measures of precision remain unaffected by this discovery. Conversely, if it is discovered that the sample is atypical, recourse to an appropriate model can often do much to compensate for the problem. Additionally, whenever small domain estimation (see Chapter 27) requires the use of estimators not supported by design-based inference but the estimators for certain aggregations of those domains are so supported, it would be useful if the former automatically summed to the latter.

Thus, finding a combined estimation strategy—one which is not only tolerant of either interpretation but also close to the currently accepted best practices of both—would seem a worthwhile activity. In this chapter, I present two examples of current business surveys, one using only traditional probability sampling inference and the other using a model-dependent estimation strategy. I then describe what kind of survey design could accommodate both sets of requirements and show that the changes required from both sides are relatively simple.

30.1 A DESIGN-BASED RETAIL SURVEY

In 1950, the Australian Bureau of Statistics (ABS) designed its first retail survey. During the ensuing decade, the Quarterly Survey of Retail Establishments underwent considerable development and eventually became a remarkably reliable source of retail statistics for its time. ABS has since redesigned the survey several times, carrying out a fundamental restructuring in 1982 and a somewhat less far-reaching redesign in 1988. Despite the changes made in these redesigns, however, many basic features have not changed since the 1950s. In particular, the survey remains design-based in its estimation approach. [The following description has been loosely based on Boal (1991), but the factual material has been severely condensed and many comments have been added.]

The present Quarterly Retail Survey is based on management units (loosely speaking, enterprises), a change from the previous use of single-location establishments. The survey emphasizes the estimation by industry of a quantity called "turnover" (the precise definition of which is unimportant). The selection procedure is list-based and uses a form of stratified, equal-probability sampling without replacement. Management units are stratified by state, industry, and size. Dalenius–Hodges (1959) optimal size boundaries are calculated on the basis of the most recent retail census figures for "takings." (As with "turnover," the precise definition of "takings" is unimportant, but it is almost identical with turnover.)

The separate stratum ratio estimator $t_R(\mathbf{y}_h)$ of $T(\mathbf{y}_h)$, the hth stratum total of the estimand vector $\mathbf{y}_h$, is given by

$$t_R(\mathbf{y}_h) = r_h T(\mathbf{x}_h), \tag{30.1}$$

where $T(\mathbf{x}_h)$ is the hth stratum total of the benchmark or supplementary variable $\mathbf{x}_h$ and where

$$r_h = \frac{\sum\limits_{j=1}^{n_h} y_{hj}}{\sum\limits_{j=1}^{n_h} x_{hj}}.$$

Here, y_{hj} is the turnover for the jth enterprise in the hth stratum, and x_{hj} is the corresponding retail census "takings." The sample units are labeled from 1 to n_h, and the remaining population units are labeled from $n_h + 1$ to N_h. The p-variance of $t_R(y_h)$ is given approximately by

$$V_p[t_R(\mathbf{y}_h)] \cong \frac{N_h(N_h - n_h)}{n_h} S_{Rh}^2,$$

where

$$S_{Rh}^2 = \frac{\sum\limits_{j=1}^{N_h} (y_{hj} - R_h x_{hj})^2}{N_h - 1}$$

and

$$R_h = \frac{\sum\limits_{j=1}^{N_h} y_{hj}}{\sum\limits_{j=1}^{N_h} x_{hj}}.$$

The estimator of p-variance used for this survey is

$$\nu_p[t_R(\mathbf{y}_h)] = \frac{N_h(N_h - n_h)}{n_h} s_{Rh}^2,$$

where

$$s_{Rh}^2 = \frac{\sum\limits_{j=1}^{n_h} (y_{hj} - r_h x_{hj})^2}{n_h - 1}.$$

Each state by industry sample is allocated across size strata using the variant of optimal or Neyman allocation appropriate for separate stratum ratio allocation. Ideally, the sample fraction is proportional to S_{Rh}; more realistically, it is proportional to an estimate of S_{Rh} based on related data, such as a previous retail census. The sample is selected using a permanent random number technique known as synchronized sampling (Hinde and Young 1984). For estimation purposes, this technique is conditionally equivalent to stratified simple random sampling without replacement, but it can be used to rotate the sample in a controlled fashion and to minimize overlap with other samples selected from the same frame (see Chapter 9).

Using the ratio estimator $t_R(\mathbf{y}_h)$, each estimand variable will have its own sample weight if different x variables are used for each estimand variable. This is not a problem when there are comparatively few estimand variables and the survey's primary emphasis is on estimating totals, means, and ratios. It can represent a severe problem, however, when the number of estimand variables is large and the survey's emphasis is on the relationships between them. In this chapter, such a survey is described as *multipurpose*, while the former type is referred to as *limited-purpose*. A model-based multipurpose survey is the subject of the next section.

30.2 A MODEL-BASED FARM SURVEY

The Australian Bureau of Agricultural and Resource Economics (ABARE) conducted its first farm surveys in the late 1940s. Several of these surveys were consolidated into the Australian Agricultural and Grazing Industry Survey in

1978. Until the late 1980s when it obtained access to an ABS sample frame, ABARE was unable to use probability sampling for its farm surveys. So, originally by necessity, ABARE adopted a model-based approach to survey estimation. It has continued with this approach by choice even after the introduction of probability samples. Bardsley and Chambers (1984) describe ABARE's method for constructing a single set of sample weights for all estimand variables. (The following is a condensed version of ABARE's approach, again with my comments.)

Model-based sampling statisticians often work in terms of two linear regression models simultaneously. The first, ξ_1, is a very simple model that is used for estimation or, as they often prefer to say, prediction. Their estimator or predictor is at least ξ_1-unbiased and ideally would be the best linear unbiased predictor/estimator (BLUP/BLUE) under ξ_1. The second, ξ_2, is a generalization of ξ_1 that includes all additional regressor variables assumed to be possibly relevant. In the simplest case, where the first model is linear and homogeneous in a single variable and the regression estimator is a ratio estimator, the additional regressors in the second model might include an intercept term, polynomial terms in the original regressor variable, or completely unrelated regressors. The assumption is that if the first model is underspecified, this is most likely because one or more of the additional regressors in the second model is needed to complete the specification.

Rather than regress on all these additional variables, which would almost certainly involve overspecification and hence instability in the estimates of the regression coefficients, many model-based statisticians prefer to ensure that the predictor/estimator from their first model is effectively unbiased under the second model by choosing a sample that is balanced on all the regressors in the second model (Royall and Herson 1973). A sample is said to be *perfectly* (*approximately*) *balanced* on a variable x when the sample mean of x is exactly (nearly) equal to its population mean. When the sample is balanced on all regressors in the second model, each bias term in the first model's BLUP/ BLUE becomes identically zero under the second model. Usually, the bias term for any given additional regressor can be made equal to zero by balancing the sample on that regressor. However, zero bias for the intercept term is achieved by balancing the sample on the (single) regressor in the first model (Royall and Herson 1973).

For Bardsley and Chambers (1984), balanced sampling was not an option. They faced a serious dilemma; because their survey was multipurpose in character, the same weights were required for all survey variables. Consequently, the same large set of regressor variables was needed for each survey variable. So if they based their estimation or prediction procedure on their second model, their estimated regression coefficients would have had high ξ_2-variance, and some would probably have carried the wrong sign. But if they used the small first model instead, which had only an intercept term as a benchmark and was therefore severely underspecified, the corresponding estimates would have had high ξ_2-bias. Bardsley and Chambers chose to trade off the ξ_2-variances against the ξ_2-biases by minimizing a combined quadratic form; the result turned out to be in the form of a generalized ridge regression. The mathematical description given below has been condensed from their published article but also uses notation and text from Cowling et al. (1993).

Consider (1) a population made up of N distinguishable elements and (2) a sample that is a subset of n distinct elements from it. The vector-valued benchmark regressor is represented by the $N \times m$ matrix $X = \{x_1 \ldots x_m\}$. The population totals $T(x_1), \ldots, T(x_m)$ are known. Let $\mathbf{1}$ and $\mathbf{h}$ be N-vectors such that $\mathbf{1}$ takes the value unity for each element, while $\mathbf{h}$ takes the value unity for each sample element and zero otherwise. The restriction of any vector or matrix to the sample is indicated by the subscript s.

Under the model, the realizations y_j of the random variable Y_j corresponding to the survey variable $\mathbf{y}$ have the population total $T(\mathbf{y})$, which is estimated from the sample. In practice, there would be many survey variables, and this fact has an important bearing on the estimation procedure, but it is not necessary to distinguish between them in the notation. A vector of any one of those variables will be denoted by $\mathbf{y}$ and assumed to obey the (large) model

$$\xi_2: \quad \mathbf{y} = X\boldsymbol{\beta} + \mathbf{v}, \quad E_\xi(\mathbf{v}) = \mathbf{0}_N, \quad E_\xi(\mathbf{vv'}) = \sigma^2 A^2 \quad (30.2)$$

where $\boldsymbol{\beta}$ is an m-vector of unknown parameters, $\mathbf{v}$ is an N-vector of random variables, σ is an unknown scale parameter, and A is an $N \times N$ diagonal matrix with elements a_j. For any particular $\mathbf{y}$, some variables in X are irrelevant, but each such variable is required for at least one $\mathbf{y}$. Hence no X variable can be ignored, but regression of any particular $\mathbf{y}$ on all the variables in X involves overspecification. The a_j^2 are the variance functions of the y_j. Because a moderate misspecification of the variance functions does not affect the accuracy of the estimates seriously, the a_j^2 are commonly assumed to be simply related to a known measure of size variable and, therefore, known up to a constant. Bardsley and Chambers used sheep equivalents [number of sheep + (8 × number of beef cattle) + (12 × number of dairy cattle) + (12 × hectares of wheat)] as their measure of size and as their assumed variance function.

For the typical population total $T(\mathbf{y})$, we seek a linear estimator

$$t(\mathbf{y}) = \mathbf{w'y} = \mathbf{w}_s'\mathbf{y}_s,$$

where $\mathbf{w}$ is the N-vector containing the sample weights w_j for the sample units and zeros elsewhere. The required properties of $t(\mathbf{y})$ and w_j are as follows: (1) the prediction error $t(\mathbf{y}) - T(\mathbf{y})$ should be small for all $\mathbf{y}$, (2) the weights w_j should be positive and preferably not less than unity, and (3) the weights w_j should depend only on X not on $\mathbf{y}$. The prediction error is

$$t(\mathbf{y}) - T(\mathbf{y}) = \mathbf{w}_s'\mathbf{y}_s - \mathbf{1'y} = (\mathbf{w} - \mathbf{1})'(X\boldsymbol{\beta} + \mathbf{v}) = \mathbf{f'\boldsymbol{\beta}} + (\mathbf{w} - \mathbf{1})'\mathbf{v},$$

where $\mathbf{f} = X'(\mathbf{w} - \mathbf{1})$ can be interpreted as the bias with which the vector of benchmarks is estimated. The prediction error has ξ_2-expectation $\mathbf{f'\boldsymbol{\beta}}$ and ξ_2-

variance

$$\sigma^2(\mathbf{w} - \mathbf{1})'A^2(\mathbf{w} - \mathbf{1}) = \sigma^2[(\mathbf{w}_s - \mathbf{1}_s)'A_s^2(\mathbf{w}_s - \mathbf{1}_s) + (\mathbf{1} - \mathbf{h})'A^2(\mathbf{1} - \mathbf{h})].$$

Only the first expression within brackets depends on the w_j's.

Bardsley and Chambers minimized the following expression with respect to $\mathbf{w}_s$:

$$Q = (\mathbf{w}_s - \mathbf{1}_s)'A_s^2(\mathbf{w}_s - \mathbf{1}_s) + \lambda^{-1}\mathbf{f}'C\,\mathbf{f},$$

where C is an $m \times m$ diagonal matrix of nonnegative costs associated with the benchmark biases and λ is an arbitrary scalar. The cost associated with the single variable in the small model was chosen to be effectively infinite. The other costs varied in size depending on the relative importances of the corresponding regressors. After considerable algebra, they obtained

$$\mathbf{w}_s = \mathbf{1}_s + A_s^{-2}X_s(\lambda C^{-1} + X_s'A_s^{-2}X_s)^{-1}(X'\mathbf{1} - X_s'\mathbf{1}_s).$$

The resulting estimator of $T(\mathbf{y})$ was

$$t_{BC}(\mathbf{y}) = \mathbf{w}'\mathbf{y} = \mathbf{1}_s'\mathbf{y}_s + (X'\mathbf{1} - X_s'\mathbf{1}_s)'\mathbf{b}_{BC}, \qquad (30.3)$$

where

$$\mathbf{b}_{BC} = (\lambda C^{-1} + X_s'A_s^{-2}X_s)^{-1} X_s'A_s^{-2}\mathbf{y}_s$$

was the generalized ridge estimator. Hence, λ was described as the ridge parameter.

When λ is very small, $\mathbf{b}_{BC}$ is effectively the BLUE under the large model ξ_2. When λ is very large, however, every diagonal element in the expression $\lambda C^{-1} + X_s' A_s^{-2}X_s$ is dominated by the corresponding element of λC^{-1} except where that element is effectively zero. So $(\lambda C^{-1} + X_s' A_s^{-2}X_s)^{-1}$ has very small values in all diagonal elements except where the corresponding element of λC^{-1} is effectively zero (i.e., for the regressors of the small model). So when λ is very large, $\mathbf{b}_{BC}$ is effectively the BLUE under ξ_1. Intermediate values of λ provide compromises between these two BLUEs. The diagonals of C are chosen to ensure that the more important a regressor is, the longer it retains its influence as λ increases.

Bardsley and Chambers arrived at an appropriate choice of λ by examining two ridge-trace plots. The first was a plot of the change in the n components of the weight vector $\mathbf{w}_s$ as λ changed. The second was a plot of the changes in the m components of the bias vector $\mathbf{f}$ as λ changed. The best value of λ was chosen subjectively. It had to be large enough to ensure that all sample weights were at least unity and that none was excessively large. At the same time, it had to be small enough to ensure that the relative biases for the additional benchmark variables were acceptable. These two requirements resulted, in practice, in the restriction of λ to quite a narrow range. Over the past 10 years, ABARE has found that plausible estimates are regularly obtained for all variables investigated with this strategy. The most subjective of the inputs is not the choice of λ but the set of costs in C.

Because the biases in the estimates of the benchmarks were known, Dunstan and Chambers (1986) were able to estimate not merely the ξ_2-variance but the actual ξ_2-MSE of the estimator $t_{BC}(\mathbf{y})$. Their mean squared error estimator is straightforward to use but too complex to develop in this chapter.

30.3 BRIDGING THE GULF

It is easiest to begin spanning the rather daunting gulf between the survey designs described in Sections 30.1 and 30.2 from the design-based side.

30.3.1 Model-Assisted Survey Sampling

The ratio estimator used in the ABS Retail Survey, dropping the stratum subscript, I write as $t_R(\mathbf{y})$, is identified as a design-based estimator, but has quite important model-based properties. Indeed, its design-based credentials are limited. For example, it is not p-unbiased. As n increases, however, its squared bias decreases more rapidly than its variance, a property readily formalized as design or p-consistency (DC-ness) combined with asymptotic p-unbiasedness (ADU-ness) using the asymptotics of Brewer (1979), Isaki and Fuller (1982), or Särndal and Wright (1984).

Nevertheless, even DC-ness and ADU-ness combined provide little basis for confidence in an estimator, if not supported by model-based properties. It is easy to construct populations in which the y_j and x_j are all positive, but the y_j tend to decrease as the x_j increase. In this case, the variance of $t_R(\mathbf{y})$ would be larger than the variances of estimators such as the Horvitz–Thompson (1952) estimator:

$$t_{HT}(\mathbf{y}) = \sum_{j=1}^{n} y_j \pi_j^{-1},$$

where π_j is the inclusion (selection) probability of unit j. This estimator has impeccable design-based properties (Hanurav 1968).

The other design-based property commonly adduced in favor of $t_R(\mathbf{y})$, its Cochran-consistency, is even less impressive. Cochran (1953, p. 13) states that "... a method of estimation is called *consistent* if the estimate becomes exactly equal to the population value when $n = N$, that is, when the sample consists of the whole population." While this property is obviously one that we would prefer to have, it tells us nothing about the behavior of the estimate when even one population unit is missing from the sample.

The real attraction of $t_R(\mathbf{y})$ for design-based statisticians is that it behaves well in practice for populations commonly encountered. This, in turn, is largely because these populations characteristically look as if they had been generated by a special case of the ξ_2 of equation (30.2), namely,

$$\xi: \qquad \mathbf{y} = \beta\mathbf{x} + \mathbf{v}, \qquad E_\xi(\mathbf{v}) = \mathbf{0}_N, \qquad E_\xi(\mathbf{vv}') = \sigma^2\mathbf{A}^2. \quad (30.4)$$

The estimator $t_R(\mathbf{y})$ is ξ-unbiased under this model and is the BLUP for $T(\mathbf{y})$ when $a_j^2 \propto x_j$. It is this combination of model-based and design-based properties that makes $t_R(\mathbf{y})$ a successful and popular estimator.

Nevertheless, it is possible to find a more efficient estimator without sacrificing DC-ness or ADU-ness if the variance function is specified more accurately. The BLUP of $T(\mathbf{y})$ under (30.4) is actually

$$t_{BLU}(\mathbf{y}) = \sum_{j \in s} y_j + \frac{\sum\limits_{j \in s} y_j x_j a_j^{-2}}{\sum\limits_{j \in s} x_j^2 a_j^{-2}} \left(\sum_{j=1}^{N} x_j - \sum_{j \in s} x_j \right),$$

which reduces to $t_R(\mathbf{y})$ when $a_j^2 \propto x_j$. In practice, however, it is usually found that the variance function is better described by $a_j^2 \propto x_j^{2\gamma}$, where $0.5 < \gamma \le 1$, in which case $t_{BLU}(\mathbf{y})$ does not reduce to $t_R(\mathbf{y})$ and no longer possesses DC-ness or ADU-ness when the π_j are equal.

For $t_{BLU}(\mathbf{y})$ to possess DC-ness and ADU-ness, the π_j must be approximately proportional to $a_j^2 x_j^{-1}$ or $x_j^{2\gamma-1}$; the proportionality would be exact if the finite population correction could be ignored (Scott et al. 1978). For $0.5 < \gamma \le 1$, this would allocate inclusion probabilities that increased with size; but for values of γ close to 0.5, larger units would not be favored very much. The question arises as to whether another set of weights might not provide inclusion probabilities that favored the large units to a greater extent and more than compensated for departure from the BLUP.

Fortunately, when there is only a single regressor, it is easy enough to find a condition on the predictor weights that ensures DC-ness and ADU-ness; one can then optimize the weights and the inclusion probabilities simultaneously subject to this condition. To this end, we write a more general ξ-unbiased predictor of $T(\mathbf{y})$ under equation (30.4) as

$$t_Q(\mathbf{y}) = \sum_{j \in s} y_j + \frac{\sum\limits_{j \in s} y_j q_j}{\sum\limits_{j \in s} x_j q_j} \left(\sum_{j=1}^{N} x_j - \sum_{j \in s} x_j \right). \quad (30.5)$$

It is convenient to scale the weights q_j by requiring that the p-expectation of $\sum_{j \in s} q_j$ — that is, $\sum^N q_j \pi_j$ — be equal to $N - n$. Brewer (1979) showed that the required condition for ADU-ness was then $q_j = \pi_j^{-1} - 1$ and that the $p\xi$-variance of the resulting estimator,

$$t_B(\mathbf{y}) = \sum_{j \in s} y_j + \frac{\sum\limits_{j \in s} y_j(\pi_j^{-1} - 1)}{\sum\limits_{j \in s} x_j(\pi_j^{-1} - 1)} \left(\sum_{j=1}^{N} x_j - \sum_{j \in s} x_j \right), \quad (30.6)$$

was minimized when $\pi_j \propto a_j$. When this estimator is rewritten as

$$t_B(\mathbf{y}) = \sum_{j \in s} y_j \pi_j^{-1} + \frac{\sum_{j \in s} y_j (\pi_j^{-1} - 1)}{\sum_{j \in s} x_j (\pi_j^{-1} - 1)} \left(\sum_{j=1}^{N} x_j - \sum_{j \in s} x_j \pi_j^{-1} \right),$$

it is recognizable as a special case of the Generalized Regression Estimator (GREG) of Cassel et al. (1976) (see Chapter 25).

In matrix notation, the GREG is

$$t_{\text{GREG}}(\mathbf{y}) = \mathbf{1}_s' \mathbf{\Pi}_s^{-1} \mathbf{y}_s + (\mathbf{1}'X - \mathbf{1}_s' \mathbf{\Pi}_s^{-1} X_s) \mathbf{b}_{\text{GREG}}, \qquad (30.7)$$

where $\mathbf{\Pi}_s = \text{diag}_n(\pi_j)$. Important properties of $t_{\text{GREG}}(\mathbf{y})$ are that it is ξ-unbiased under the model ξ_2 of equation (30.2), that it tends to $t_{\text{HT}}(\mathbf{y})$ as the sample size increases, and that its asymptotic variance is at the Godambe–Joshi lower bound for a p-unbiased estimator (Godambe and Joshi 1965). The optimal values of the π_j that minimize this lower bound are proportional to the square roots of the variance function; that is, $\pi_j \propto a_j$ (Särndal et al. 1992). As indicated in Section 30.2, the true variance function is never known exactly. Because the optimum is flat, an approximation such as an overall measure of size raised to the 1.5 power can often be regarded as sufficiently close. Remarkably, all these properties of the GREG hold for a wide range of choices of $\mathbf{b}_{\text{GREG}}$. It is the small sample performance of $t_{\text{GREG}}(\mathbf{y})$ that really determines the optimal form of $\mathbf{b}_{\text{GREG}}$, and this makes it rather difficult to discover. A plausible suggestion is given in Section 30.3.2.

The $t_{\text{GREG}}(\mathbf{y})$ of equation (30.7) may be described as the archetypal estimator of model-assisted survey sampling. Although ξ-unbiased, it is essentially a design-based estimator. Suppose, for instance, that we choose $\mathbf{b}_{\text{GREG}}$ to be the BLUE of $\boldsymbol{\beta}$, that is,

$$\mathbf{b}_{\text{BLU}} = (X_s' A_s^{-2} X_s)^{-1} X_s' A_s^{-2} \mathbf{y}_s, \qquad (30.8)$$

and we rewrite $t_{\text{GREG}}(\mathbf{y})$ in the model-based form

$$t_{\text{GREG}}(\mathbf{y}) = \mathbf{1}_s' \mathbf{y}_s + (\mathbf{1}'X - \mathbf{1}_s' X_s) \mathbf{b}_{\text{MGREG}},$$

where $\mathbf{b}_{\text{MGREG}}$ is the implicit estimator of $\boldsymbol{\beta}$. The resulting expression for $\mathbf{b}_{\text{MGREG}}$ is then too complicated to be readily interpretable. Model-assisted survey sampling—essentially, optimization within a class of sampling strategies under the assumption that the population was generated by a model such as the ξ_2 of equation (30.2)—is like an incomplete bridge. The traveler can get closer to the other side but cannot actually step foot on it.

30.3.2 Combined Design-Based and Model-Based Inference

Isaki and Fuller (1982) and Särndal and Wright (1984) suggested that a model-based or prediction format for the GREG could be obtained directly by requiring that $\mathbf{b}_{\mathrm{GREG}}$ be chosen such that

$$
t_{\mathrm{GREG}}(\mathbf{y}) = \mathbf{1}_s' \boldsymbol{\Pi}_s^{-1} \mathbf{y}_s + (\mathbf{1}'X - \mathbf{1}_s' \boldsymbol{\Pi}_s^{-1} X_s) \mathbf{b}_{\mathrm{GREG}}
$$

$$
= \mathbf{1}_s' \mathbf{y}_s + (\mathbf{1}'X - \mathbf{1}_s' X_s) \mathbf{b}_{\mathrm{GREG}}.
$$

This yields the condition

$$
\mathbf{1}_s' (\boldsymbol{\Pi}_s^{-1} - I_n)(\mathbf{y}_s - X_s \mathbf{b}_{\mathrm{GREG}}) = 0. \tag{30.9}
$$

To solve equation (30.9), we can impose the weak requirement that $\mathbf{b}_{\mathrm{GREG}}$ be of the projection form, $\mathbf{b}_{\mathrm{GREG}} = (Q_s' X_s)^{-1} Q_s' \mathbf{y}_s$, where Q_s is any $n \times m$ matrix of rank m such that $Q_s' X_s$ is positive definite, and can also impose the stronger requirement that $(\boldsymbol{\Pi}_s^{-1} - I_n)\mathbf{1}_s$ be spanned by the column space of Q_s, that is, that there is some m-vector $\boldsymbol{\alpha}_s$ such that $Q_s \boldsymbol{\alpha}_s = (\boldsymbol{\Pi}_s^{-1} - I_n)\mathbf{1}_s$. Then,

$$
\mathbf{1}_s' (\boldsymbol{\Pi}_s^{-1} - I_n)(\mathbf{y}_s - X_s \mathbf{b}_{\mathrm{GREG}}) = \boldsymbol{\alpha}_s' Q_s' [\mathbf{y}_s - X_s (Q_s' X_s)^{-1} Q_s' \mathbf{y}_s]
$$

$$
= \boldsymbol{\alpha}_s' [Q_s' - Q_s' X_s (Q_s' X_s)^{-1} Q_s'] \mathbf{y}_s
$$

$$
= 0,
$$

satisfying equation (30.9). However, if we choose $\mathbf{b}_{\mathrm{GREG}}$ to be the $\mathbf{b}_{\mathrm{BLU}}$ of equation (30.8), then $Q_s = A_s^{-2} X_s$, and the column space of $A_s^{-2} X_s$ does not, in general, span $(\boldsymbol{\Pi}_s^{-1} - I_n)\mathbf{1}_s$. The requirement that $(\boldsymbol{\Pi}_s^{-1} - I_n)\mathbf{1}_s$ be in the column space of Q_s may be satisfied by modifying either X or A. In what follows, I explore modifying X. Brewer (1994) discusses modifying A.

Isaki and Fuller (1982) and Särndal and Wright (1984) achieved the desired property by adding one or two columns to X and then using the redefined BLUE of $\boldsymbol{\beta}$. Adding a single column with $a_j^2 (\pi_j^{-1} - 1)$ in the jth position is all that is necessary in this case, but there are two reasons for hesitating before adopting this solution. The first is that all estimators of $T(\mathbf{y})$ effectively work by making predictions for the nonsample values. The addition of spurious explanatory variables, however much they might improve the observed fit for sample values, makes predictions based on small samples distinctly less efficient. This is particularly the case because the $a_j^2 (\pi_j^{-1} - 1)$ values are likely to be fairly well correlated with some linear combination of the original columns of X and especially if the optimum selection probabilities are used, in which case $a_j^2 (\pi_j^{-1} - 1) \propto \pi_j (\pi_j^{-1} - 1)$. The other reason is that the optimal ratio estimator, the $t_{\mathrm{B}}(\mathbf{y})$ of equation (30.6), is not a special case of this $t_{\mathrm{GREG}}(\mathbf{y})$.

Both concerns can be met by the use of an instrumental variables (IV) estimator (Brewer et al. 1988). This may be written

$$
\mathbf{b}_{\mathrm{IV}} = (X_s^{*\prime} A_s^{-2} X_s)^{-1} X_s^{*\prime} A_s^{-2} \mathbf{y}_s,
$$

where X^* differs from X only in that a single regressor variable column is replaced by the vector $\mathbf{z}$, of which the typical element is $z_j = a_j^2 (\pi_j^{-1} - 1)$.

The optimal choice for the column of X to be replaced is the one most nearly correlated with $\mathbf{z}$ (if there is an intercept term) or the one most nearly proportional to $\mathbf{z}$ (if there is not).

For a limited purpose survey such as that described in Section 30.1, the choice of $\mathbf{b}_{IV}$ for $\mathbf{b}_{GREG}$ effectively completes the specification of the optimal design, but a few comments are relevant.

If the optimum inclusion probabilities are used, $\mathbf{b}_{IV}$ can also be written as

$$\mathbf{b}_{IV} = (X_s^{*\prime} \mathbf{\Pi}_s^{-2} X_s)^{-1} X_s^{*\prime} \mathbf{\Pi}_s^{-2} \mathbf{y}_s. \tag{30.10}$$

This alternative form for $\mathbf{b}_{IV}$ can be used even if the inclusion probabilities are not optimal. The estimator $t_{GREG}(\mathbf{y})$ retains its DC-ness and ADU-ness in these circumstances, and $z_j = \pi_j(1 - \pi_j)$. The large-sample loss of efficiency that it suffers is almost entirely because of the nonoptimality of the inclusion probabilities. The consequent suboptimality of the formula for $t_{GREG}(\mathbf{y})$ itself is asymptotically negligible.

The use of $t_{GREG}(\mathbf{y})$ with $\mathbf{b}_{IV}$ seems to be particularly appropriate for surveys that require much small domain estimation (see Chapter 27). If the synthetic estimates for these small domains use values $\mathbf{b}_{IV}$ obtained from larger domains where $t_{GREG}(\mathbf{y})$ is appropriate, they automatically sum to the large domain totals.

Sample control (exercised in the present ABS Retail Survey through synchronized sampling) could be continued through the use of some form of unequal probability PRN sampling (see Chapter 9). My choice would be PRN collocated sampling because the variability in sample size is smaller than for Poisson sampling and because the second-order inclusion probabilities π_{jk} are still simple in form and sufficiently close to $\pi_j \pi_k$ for the original Horvitz–Thompson (1952) variance estimator to be efficient (Brewer et al. 1984). This estimator can be used in place of the more usual Sen–Yates–Grundy variance estimator (Sen 1953, Yates and Grundy 1953) to build up one for the GREG along the lines suggested by Särndal et al. (1992, Chapter 6) or by Kott (1990).

Because the design described in Section 30.1 already uses an implicit model and optimal allocation within each size stratum, little direct improvement in efficiency will result. However, as with the GREG, the use of the variance function to determine optimal inclusion probabilities means that size strata are no longer necessary. So without increasing the total number of strata, a finer industry stratification can be used, usually improving the efficiency of the estimates. For example, confectioners are likely to behave more differently from health food stores than small other-food stores are from large other-food stores!

Three possible variances can be estimated for $t_{GREG}(\mathbf{y})$, and all three are fully meaningful if $\mathbf{b}_{IV}$ of equation (30.10) is used as the estimator of $\boldsymbol{\beta}$. These are its p-variance, its ξ-variance, and its $p\xi$-variance. Estimating the p-variance, however, involves the second-order inclusion probabilities. For samples of more than 2 units per stratum drawn with unequal probabilities without replacement, this involves a choice between evils. The most convenient selection procedures are systematic, but their π_{jk} values are tedious to calculate and give rise to highly unstable or even biased variance estimators. Most other available methods are complicated in execution, their π_{jk} values are again tedious to calculate, and the resulting variance estimators are still markedly less efficient than the model-based alternatives. In my opinion, the least of the evils are those associated with collocated sampling. However, the sample size is a random variable which may be considered a disadvantage. Estimators of ξ-variance and $p\xi$-variance do not involve the π_{jk} and hence are more convenient as well as more efficient.

30.3.3　Combined Inference For Multipurpose Surveys

For multipurpose surveys, each sample unit must carry a single weight, regardless of the variable being estimated. The same problem therefore arises with the GREG as arose in Section 30.2 for the BLUP—namely, that the number of regressors is excessive and the estimator of β unstable (see Chapter 25). The procedure required for modifying the ordinary regression estimator or predictor into one that uses ridge regression is not very different from the original Bardsley–Chambers procedure described in Section 30.2.

Without loss of generality, we can assume that it is the first column of X^* that contains z. We then require that the first diagonal element in C^{-1} be zero, ensuring that the regression on whatever column of X has been replaced in X^* by $z = A^2(\Pi^{-1} - I_N)\mathbf{1}$ retains its full influence, regardless of the value chosen for the ridge parameter λ.

The required estimator of β may be written

$$\mathbf{b}_{\mathrm{BCIV}} = (\lambda C^{-1} + X_s^{*\prime} A_s^{-2} X_s)^{-1} X_s^{*\prime} A_s^{-2} \mathbf{y}_s,$$

and we require of $t_{\mathrm{BCIV}}(\mathbf{y})$, the estimator or predictor that uses it, that

$$t_{\mathrm{BCIV}}(\mathbf{y}) = \mathbf{1}_s^{\prime} \Pi_s^{-1} \mathbf{y}_s + (\mathbf{1}^{\prime} X - \mathbf{1}_s^{\prime} \Pi_s^{-1} X_s) \mathbf{b}_{\mathrm{BCIV}}$$

$$= \mathbf{1}_s^{\prime} \mathbf{y}_s + (\mathbf{1}^{\prime} X - \mathbf{1}_s^{\prime} X_s) \mathbf{b}_{\mathrm{BCIV}},$$

or equivalently,

$$\mathbf{1}_s^{\prime}(\Pi_s^{-1} - I_n)(\mathbf{y}_s - X_s \mathbf{b}_{\mathrm{BCIV}}) = 0. \tag{30.11}$$

Let δ denote the row m-vector $(1, 0, \ldots, 0)$ so that $\delta C^{-1} = \mathbf{0}_m^{\prime}$. Then, writing

$$D = C^{-1}(X_s^{*\prime} A_s^{-2} X_s)^{-1} X_s^{*\prime} A_s^{-2},$$

we have $DX_s = C^{-1}$ and $\delta D = \mathbf{0}_n^{\prime}$, so that

$$\delta(\lambda D + X_s^{*\prime} A_s^{-2}) = \delta X_s^{*\prime} A_s^{-2} = \mathbf{1}_s^{\prime}(\Pi_s^{-1} - I_n)$$

for all λ. Hence,

$$\mathbf{1}_s'(\mathbf{\Pi}_s^{-1} - \mathbf{I}_n)(\mathbf{y}_s - \mathbf{X}_s \mathbf{b}_{\text{BCIV}})$$

$$= \delta(\lambda D + X_s^{*\prime} A_s^{-2})[\mathbf{y}_s - X_s(\lambda C^{-1} + X_s^{*\prime} A_s^{-2} X_s)^{-1} X_s^{*\prime} A_s^{-2} \mathbf{y}_s]$$

$$= \delta(\lambda D + X_s^{*\prime} A_s^{-2}) \mathbf{y}_s - \delta(\lambda C^{-1} + X_s^{*\prime} A_s^{-2} X_s)$$

$$\cdot (\lambda C^{-1} + X_s^{*\prime} A_s^{-2} X_s)^{-1} X_s^{*\prime} A_s^{-2} \mathbf{y}_s$$

$$= \lambda \delta D \mathbf{y}_s$$

$$= 0,$$

satisfying condition (30.11). Alternatively, the definition of $\mathbf{z}$ may be simplified to $\mathbf{\Pi}(\mathbf{I}_N - \mathbf{\Pi})\mathbf{1}$, in which case the formula for $\mathbf{b}_{\text{BCIV}}$ and the derivation of the conditions under which equation (30.11) is satisfied require $\mathbf{\Pi}_s$ to be written for A_s throughout.

Selection procedures often have values of π_{jk} that ensure that their design-based variance estimators are inherently unstable. The Dunstan and Chambers (1986) estimator of ξ_2-MSE, though complex in form, does not suffer this disadvantage and can be easily modified to apply to the prediction form of the estimator recommended here. Finally, it would again seem possible to control the sample using PRN collocated sampling or some other unequal probability variant of PRN sampling.

30.4 CROSSING FROM THE OTHER SIDE?

It has been comparatively easy to build the bridge beginning from the design-based side of the gulf because many estimators used by design-based statisticians already possess quite obvious model-based properties. The converse is not true. Model-based statisticians typically insist that randomization "is neither necessary nor sufficient for rigorous statistical inference" (Royall 1976, p. 473). When it comes to protection against model breakdown, they tend to follow one of two routes. The first is to stay with the BLUP and rely on the selection of balanced samples to provide protection against model breakdown (Royall and Herson 1973). The other is to accept that samples will never be more approximately balanced and to adjust the estimator so as to make the best possible use of the existing sample (Bardsley and Chambers 1984).

The Royall and Herson strategy involves stratification by size x_j and allocation of the sample to stratum h according to the formula $n_h \propto N_h \text{RMS}(a_{hj})$, where $\text{RMS}(a_{hj})$ is the root mean square of the a_j within stratum h. (Royall and Herson assumed that $a_j^2 \propto x_j$, but the generalization to arbitrary a_j^2 is straightforward.) They then required that the sample within each stratum be balanced on the x_j^k, $k = 1, 2, \ldots, K$. The classical ratio estimator $t_R(\mathbf{y}_h)$ of equation (30.1) was used to estimate the stratum total $T(\mathbf{y}_h)$. This estimator is the BLUP when $a_j^2 \propto x_j$ and is ξ-unbiased for all variance functions a_j^2. The special feature of the Royall and Herson scheme is that because the sample is balanced, $t_R(\mathbf{y}_h)$ is still ξ-unbiased when the model includes an intercept term and polynomial terms in $x_j^2, x_j^3, \ldots, x_j^K$.

It can be shown that if the stratum sample size is unity for each stratum, the implicit overall Royall and Herson estimator is virtually identical with the $t_B(\mathbf{y})$ of equation (30.6), and the size distribution of their balanced sample is close to that of the typical sample selected with inclusion probabilities $\pi_j \propto a_j$. This leaves the use of balanced sampling for the Royall and Herson strategy and of probability sampling for the combined inference strategy as the only difference of substance between them. It might be supposed that balanced sampling would perform better for model breakdowns involving an intercept and/or polynomials in $x_j^2, x_j^3, \ldots, x_j^k$ and that probability sampling would be better for breakdowns involving unknown regressors, but Tam and Chan (1984) have proven that the tendency towards approximate balance on unknown regressors provided by strict randomization is at least retained, and almost always improved, if the balanced samples are selected using restricted randomization.

It might well be concluded from this that the combined inference strategy has absolutely nothing to offer to someone whose philosophical predilection is for model-based inference. But this would be to ignore the need to control sample rotation and overlap in the context of business surveys. All sample selection procedures that permit this kind of control (see Chapter 9) work essentially in terms of probability sampling and do not seem to lend themselves to the picking and choosing necessary for restricted randomization. Balanced sampling designs appear to be practicable mainly for one-time-only surveys.

For repeated surveys, therefore, the modeler must often find it necessary to use the Bardsley and Chambers approach, which accepts whatever sample is available and tailors the estimator accordingly. A recent paper by Royall (1992) is relevant here. The author generalizes the concept of balance to samples for which $t_R(\mathbf{y}_h)$ is not the BLUP. He distinguishes between *simple balance*, which is the balance defined in Section 30.2 above, and *balance* in a more general sense, which is a property of the interrelation between the sample and the estimator. For instance, the Royall and Herson samples are simply balanced within individual strata. The overall Royall and Herson sample, which does not possess moments in common with the population and therefore is not simply balanced, is nevertheless balanced in relation to the implicit estimator for the total of the entire population.

Balance in this sense is equivalent to calibration as described by Deville and Särndal (1992). An estimator $t(\mathbf{y})$ of $T(\mathbf{y})$ is *calibrated* on the set of m explanatory variables $\mathbf{X}$ when the sample weights w_j are such that $\mathbf{w}_s' \mathbf{X}_s = T(\mathbf{X})$. Calibration puts only m conditions on the w_j, so Deville and Särndal completed the specification by minimizing the departures from the π_j^{-1} weights of $t_{\mathrm{HT}}(\mathbf{y})$, using various distance measures. The resulting estimators approach the Horvitz–Thompson expansion estimator asymptotically and are therefore design-consistent and asymptotically design-unbiased.

It is possible to choose the distance measure to minimize the departures of the q_j in the $t_Q(\mathbf{y})$ of equation (30.5) from the $\pi_j^{-1} - 1$ of equation (30.6) in such a way that the w_j will seldom be less than unity. This may obviate the need for ridge regression entirely. However, a personal communication from J. N. K. Rao indicates that ridge regression could be introduced without prejudice to the design-based properties of $t_Q(\mathbf{y})$.

Despite the formal equivalence of the relationships between sample and estimator, calibration of the estimator is *not* a perfect substitute for the selection of a balanced sample. The ideal is to be able both to choose the estimator and to specify the characteristics of the sample to suit it. Accepting a less-than-optimal sample, and then adjusting the estimator so as to compensate for its deficiencies, can only be a second-best option; removal of potential ξ-bias almost inevitably incurs an increase in ξ-variance. However, because the calibration strategy achieves the same relationship between estimator and sample as is found in balanced sampling and in the end yields an estimator having design-based properties, the calibration strategy reinforces the bridge with materials from the model-based side of the chasm. For further development of these ideas, see Brewer (1994).

Finally, it must be conceded that the arguments in favor of a combined inference strategy carry considerably less weight for small samples than they do for large ones. I would be prepared to use a pure modeling strategy in circumstances where I was confident that the model would be reasonably adequate, that the required sample size was small, and that sample control was not an issue. These three conditions do sometimes coexist in business surveys.

REFERENCES

Bardsley, P., and R. L. Chambers (1984), "Multipurpose Estimation from Unbalanced Samples," *Applied Statistics*, **33**, pp. 290–299.

Basu, D. (1971), "An Essay on the Logical Foundations of Survey Sampling, Part One" (with Discussion), in V. P. Godambe and D. A. Sprott (eds.), *Foundations of Statistical Inference*, Toronto: Holt, Rinehart, and Winston, pp. 203–242.

Boal, P. (1991), "Retail Survey," in *Current Sample Surveys*, unpublished internal working document, Canberra: Australian Bureau of Statistics.

Brewer, K. R. W. (1979), "A Class of Robust Sample Designs for Large Scale Surveys," *Journal of the American Statistical Association*, **74**, pp. 911–915.

Brewer, K. R. W. (1994), "Survey Sampling Inference: Some Past Perspectives and Present Prospects," *Pakistan Journal of Statistics*, **10**, pp. 213–233.

Brewer, K. R. W., L. J. Early, and M. Hanif (1984), "Poisson, Modified Poisson and Collocated Sampling," *Journal of Statistical Planning and Inference*, **10**, pp. 15–30.

Brewer, K. R. W., M. Hanif, and S.-M. Tam (1988), "How Nearly Can Model-Based Prediction and Design-Based Estimation Be Reconciled?" *Journal of the American Statistical Association*, **83**, pp. 128–132.

Cassel, C.-M., C.-E. Särndal, and J. H. Wretman (1976), "Some Results on Generalized Difference Estimation and Generalized Regression Estimation for Finite Populations," *Biometrika*, **63,** pp. 615–620.

Cochran, W. G. (1953, 1963, 1977), *Sampling Techniques*, 1st, 2nd, and 3rd eds., New York: Wiley.

Cowling, A., R. L. Chambers, R. Lindsay, and B. Parameswaran (1993), "Mapping Survey Data," *Proceedings of the International Conference on Establishment Surveys*, Alexandria, VA: American Statistical Association, pp. 150–157.

Dalenius, T., and J. L. Hodges (1959), "Minimum Variance Stratification," *Journal of the American Statistical Association*, **54,** pp. 88–101.

Deville, J.-C. and C.-E. Särndal (1992), "Calibration Estimators in Survey Sampling," *Journal of the American Statistical Association*, **87,** pp. 376–382.

Dunstan, R., and R. L. Chambers (1986), "Model-Based Confidence Intervals in Multipurpose Surveys," *Applied Statistics*, **35,** pp. 276–280.

Godambe, V. P., and V. M. Joshi (1965), "Admissibility and Bayes Estimation in Sampling Finite Populations, I, II and III," *Annals of Mathematical Statistics*, **36,** pp. 1707–1742.

Hanurav, T. V. (1968), "Hyperadmissibility and Optimum Estimators for Sampling Finite Populations," *Annals of Mathematical Statistics*, **39,** pp. 621–642.

Hinde, R., and D. Young (1984), "Synchronised Sampling and Overlap Control Manual," unpublished internal working document, Canberra: Australian Bureau of Statistics.

Horvitz, D. G., and D. J. Thompson (1952), "A Generalization of Sampling Without Replacement from a Finite Universe," *Journal of the American Statistical Association*, **47,** pp. 663–685.

Isaki, C. T., and W. A. Fuller (1982), "Survey of Design Under the Regression Superpopulation Model," *Journal of American Statistical Association*, **77,** pp. 89–96.

Kott, P. S. (1990), "Estimating the Conditional Variance of a Design-Consistent Regression Estimator," *Journal of Statistical Planning and Inference*, **24,** pp. 287–296.

Royall, R. M. (1976), "Current Advances in Sampling Theory: Implications for Human Observational Studies," *American Journal of Epidemiology*, **104,** pp. 463–477.

Royall, R. M. (1992), "Robustness and Optimal Design under Prediction Models for Finite Populations," *Survey Methodology*, **18,** pp. 179–185.

Royall, R. M. and J. Herson (1973), "Robust Estimation in Finite Populations I," and "Robust Estimation in Finite Populations II: Stratification on a Size Variable," *Journal of the American Statistical Association*, **68,** pp. 880–889 and 890–893.

Särndal, C.-E., B. Swensson, and J. H. Wretman (1992), *Model Assisted Survey Sampling*, New York: Springer.

Särndal, C.-E., and R. L. Wright (1984), "Cosmetic Form of Estimators in Survey Sampling," *Scandinavian Journal of Statistics*, **11,** pp. 146–156.

Scott, A. J., K. R. W. Brewer, and E. W. H. Ho (1978), "Finite Population Sampling and Robust Estimation," *Journal of the American Statistical Association*, **73,** pp. 359–361.

Sen, A. R. (1953), "On the Estimate of the Variance in Sampling with Varying Probabilities," *Journal of the Indian Society for Agricultural Statistics*, **18,** pp. 52–56.

Smith, T. M. F. (1984), "Present Position and Potential Developments: Some Personal Views on Sample Surveys," *Journal of the Royal Statistical Society*, Series A, **147,** pp. 208–221.

Smith, T. M. F. (1994), "Sample Surveys 1975–1990: An Age of Reconciliation?" (with discussion), *International Statistical Review*, **62,** pp. 5–34.

Tam, S.-M. and N.-N. Chan (1984), "Screening of Probability Samples," *International Statistical Review*, **52,** pp. 301–308.

Yates, F., and P. M. Grundy (1953), "Selection without Replacement from within Strata with Probability Proportional to Size," *Journal of the Royal Statistical Society*, Series B, **15,** pp. 253–261.

Past, Present, and Future Directions

CHAPTER THIRTY-ONE

The Evolution of Agricultural Data Collection in the United States

Rich Allen[1]
U.S. National Agricultural Statistics Service

Vicki J. Huggins and Ruth Ann Killion
U.S. Bureau of the Census

Farms represent a special subset of businesses. More than any other type of operation, they combine business and family considerations. Many farms are operated by a single person, but others involve quite complicated ownership and operation arrangements. In addition, agriculture includes many different products and production practices that change over time.

In this chapter we trace major issues and developments in agricultural data collection in the United States from 1790 to today. We also describe major changes in methodologies with explanations of factors that encouraged the changes. Methodology for both agriculture censuses and the current agricultural survey programs changed extensively as the United States evolved from an overwhelmingly rural and agricultural nation to a highly urbanized society with only a small fraction of the population engaged in agriculture. For more

[1]The authors benefited extensively from reviews by the following individuals: Doug Miller and Jim Liefer of the U.S. Bureau of the Census; Kenn Inskeep of the Evangelical Lutheran Church in America; B. Nanjamma Chinnappa and George Andrusiak of Statistics Canada; and Bill Arends, Jerry Clampet, Bob Schooley, Paul Bascom, and Lue Yang of the U.S. National Agricultural Statistics Service. The authors also thank Mary Ann Higgs for her administrative support and thank Priscilla Simms, Marsha Milburn, and Hazel Beaton for typing and graphics assistance. The views expressed in this chapter are attributable to the authors and do not necessarily reflect those of the U.S. Bureau of the Census or the U.S. National Agricultural Statistics Service.

Business Survey Methods, Edited by Cox, Binder, Chinnappa, Christianson, Colledge, Kott.
ISBN 0-471-59852-6 © 1995 John Wiley & Sons, Inc.

detailed discussion of historical developments, see Bidwell and Falconer (1925), Brooks (1977), National Agricultural Statistics Service (1989a), Statistical Reporting Service (1969), Taylor and Taylor (1952), and Wright and Hunt (1900).

31.1 AGRICULTURAL STATISTICS IN THE UNITED STATES

Early requirements of agricultural statistics in the United States stemmed from subsistence farmers' need to know about current and improved farming practices. Farmers were primarily concerned with increasing production to improve their own family's food and fiber supplies and to enable the purchase of goods that they could not produce. Local and state farming societies were interested in farming methods because farming products and practices were geographically similar. Gradually, the need for national agricultural statistics increased as better marketing channels opened, production expanded westward into new farming territories, production levels and commercialization increased, and pricing arrangements evolved.

For many years, published agricultural statistics were designed largely to provide government with production information, and this remains a major concern of government surveys. Such data are essential for administering farm income support and disaster relief programs and to develop new legislation. Because much agricultural production is exported, current statistics enable orderly marketing in both domestic and foreign trade channels. Additionally, many national and state programs use combinations of census and current statistics for allocating funds for extension services, research, and soil conservation projects. Private industry also uses production data for construction and marketing decisions. Purchasers of farm products, suppliers of farm inputs and services, producers, and producer organizations need good information.

31.1.1 Present Structure of Official Agriculture Data Collection

The agriculture census program of the U.S. Bureau of the Census and the current statistics program of the U.S. National Agricultural Statistics Service (NASS)[2] complement each other quite well. For example, data collected by NASS on prices, production levels, marketing patterns, inventories, and production expenses are used in farm income estimates that are part of the U.S. national accounts. When census data are available, estimated farm income components are revised and benchmarked to current census levels.

The Census Bureau is located within the U.S. Department of Commerce, whereas NASS is an agency in the U.S. Department of Agriculture (USDA). Because the two sets of statistics are created by different organizations, there

[2]NASS was previously called the Statistical Reporting Service. NASS is used throughout this chapter for consistency.

is no internal pressure that the census must support what the current statistics have been indicating. There is a good working arrangement and cooperation between NASS and the Census Bureau, however. NASS aids in census mail list creation, and its area frame probability sample data are important indicators of list coverage. NASS staff assist in reviewing state and county census data during the editing phase. The two organizations also participate in joint research projects.

31.1.2 Current Program of the U.S. Bureau of the Census

Conducted by the Census Bureau every 5 years, the agriculture census represents the only effort to collect a broad set of data from all farms in the United States. Farm operation and family characteristics are collected along with data on acreage, land use, irrigation, crops, vegetables, fruits and nuts, nursery and greenhouse products, livestock and poultry, value of sales, type of organization, farm workers, and characteristics of operators. The census provides the only set of small area data covering farm characteristics and minor production items.

The agriculture census is currently a mailout/mailback enumeration of all farms and ranches. Response is mandatory, and the Census Bureau implements extensive follow-up procedures using mail, certified mail, and telephone calls to improve response. For about a 30 percent sample, additional data are collected on items such as (1) use of commercial fertilizers, (2) use of chemicals, machinery, and equipment, (3) expenditures, (4) market value of land and buildings, and (5) farm-related income. This survey is based on a stratified, systematic sample design within counties to ensure reliable statistics at the county level.

Following each agriculture census, the Census Bureau generally conducts one or more follow-on surveys to collect detailed data on relatively narrow areas of interest. Past surveys collected information about farm characteristics such as land ownership, expenditures, finances, irrigation practices, and energy use. There also have been censuses of horticulture specialties.

31.1.3 Current Program of the U.S. National Agricultural Statistics Service

The U.S. agricultural estimation system for current statistics uses sample surveys to measure production and inventories of major crop and livestock items along with information on prices, labor usage, and disposition of livestock. Expected production levels of major crops are forecast monthly, starting 3 or 4 months before harvest. Because many statistics can impact on volumes and prices for commodity futures and cash market trading, strict procedures and laws govern the preparation and release of these statistics to the public.

Many survey vehicles are used, with some programs conducted weekly, monthly, and quarterly in addition to surveys conducted on an annual or semiannual basis. National and state estimates are routinely produced; county estimates are created for items needed to administer federal and state farm programs. Most surveys are based on probability samples, but some use designated panels of reporters and others are based on contacting individuals who can answer for an entire segment of a specific population. When data can be reported more efficiently, a few surveys approach agribusinesses such as grain elevators, food processors, seed companies, livestock packing houses, and feed companies. Administrative data such as exports, imports, and marketings are used either directly or as a check on estimation levels.

Nearly 400 statistical reports are issued by NASS each year, and more than 9000 reports are issued by the state statistical offices. Six of the current statistics series are included in the principal economic indicators series of the United States (Statistical Reporting Service 1983b).

31.1.4 Definition of Farm

The definition of a *farm* has changed over time. The Secretary of Commerce sets the definition after collaboration and agreement with all relevant agencies and advisory groups. The definition is not codified in U.S. law. Since 1975, the definition has been "any place which sells, or normally would sell, $1000 or more of agricultural products in the reference year." Because of the low threshold, individuals who keep just a few head of cattle or sell fruits and vegetables qualify even if they do not regard themselves as farmers. As a result, the greatest difficulty in conducting a complete census or representative survey is in accounting for the vast number of small farms. For example, in the 1987 census, approximately 50 percent of all farms had less than $10,000 in total value of agricultural products sold, 60 percent had fewer than 180 acres of land, and 30 percent had fewer than 50 acres (U.S. Bureau of the Census 1989).

Table 31.1 provides the farm definitions used during different periods of U.S. history. Although defining a farm appears to be a simple task, it requires a tremendous amount of discussion, analysis, and communication to satisfy federal, congressional, and private concerns. For example, for the 1974 census, USDA and the Commerce Department both supported raising the threshold to $1000 because of increased prices and changes in the structure of agricultural operations. However, the U.S. Congress disagreed. Thus, the 1974 all-farm preliminary reports were published with the same definition of "farm" used as in the previous 1959–1969 censuses. Only the final reports were based on the new definition with a threshold of $1000 (U.S. Bureau of the Census 1979).

31.1.5 Definition of Counties and Townships

The major geographic administrative subdivision within a state is the *county*. There are over 3300 counties within the 50 states of the United States. Their

Table 31.1   The U.S. Farm Definition Throughout History

Period(s)	Definition
1850–1860	No acreage requirement, but a minimum of $100 of total value of agricultural products sold (TVP).
1870–1890	A minimum of 3 acres or $500 TVP.
1900	Acreage and minimum sales requirements removed.
1910–1920	A minimum of 3 acres, with $250 or more TVP or full-time services required for at least one person.
1925–1945	A minimum of 3 acres, with $250 or more TVP.
1950–1954	A minimum of 3 acres or $150 or more TVP. If a place had sharecroppers or other tenants, land assigned to each was treated as a separate farm. Land retained and worked by the landlord was considered a separate farm.
1959–1974	Any place with 10 acres or more, and with $50 or more TVP, or any place with less than 10 acres but at least $250 TVP. If sales were not reported, average prices were applied to reported estimates of harvested crops and livestock produced to arrive at estimated sales values.
1978–Present	Any place that had, or normally would have had, $1000 or more TVP.

size varies depending upon the physical size, geography, and population density of the state. Many counties are divided into *townships* of about 36 square miles each. County and township agricultural reports were important at various times in history; counties remain important units for publication of both census and current agricultural statistics.

31.2 AGRICULTURAL STATISTICS BEFORE 1880

In 1790, the United States was essentially an area along the Atlantic Ocean with an average width of only 255 miles from the coast and a population of less than four million people, over 90 percent of whom were involved with agriculture. Tobacco was the principal agricultural export and accounted for about 44 percent of total exports, or $4.4 million. Farm operations were centered around villages in the northeastern states; were often large, multiple family plantations in southern states; and were isolated farmsteads in the middle states (Economic Research Service 1993).

31.2.1 Origins of Information

Before 1840, agricultural information was derived by asking knowledgeable individuals to assess current conditions and supplies of crops and livestock. An interesting sidelight is that President George Washington is perhaps responsible for the first agricultural survey. In 1791 Washington wrote to friends in several states listing agricultural questions that he wanted them to answer. This is the first known occasion when an actual survey form was used to collect expert opinions. Collection of expert opinions was the major source of agricultural statistics for the next 100 years, except for the agriculture censuses (Statistical Reporting Service 1969).

Few advances in the collection of agricultural statistics occurred during the early 1800s because agriculture was mostly for subsistence. Broad-based collection of agricultural statistics was unnecessary. An exception was found in the South, which was more commercially oriented and needed marketing information. Some states and agricultural societies collected and provided agricultural information on a small scale to local farms and businesses.

31.2.2 Westward Expansion

By the mid-1800s, the country was growing rapidly and expanding westward, agricultural surpluses existed in some areas, and state efforts to collect agricultural information were uncoordinated. The need for national data became more obvious.

Archibald Russell, a young scholar who prepared a book of proposals for collecting U.S. resource data with the 1840 population census, wrote that an economic understanding of the United States could be greatly assisted by agricultural data because agriculture was the leading business. In 1838, President Martin Van Buren recommended to Congress that the 1840 Census of Population be expanded to collect information "in relation to mines, agriculture, commerce, manufacturers, and schools." As a result, 37 questions about agriculture were included in the 1840 population census (Taylor and Taylor 1952, pp. 258–259). At that time, slightly over 70 percent of the U.S. population reported that they engaged in agriculture. All information was collected by face-to-face interviews. The questions concentrated on production without recording acres harvested or yields, limiting comparisons or linking of other information to the census (Wright and Hunt 1900). Table 31.2 shows the range of questions and the increase in the number of questions relating to agriculture in the decennial years of 1840 to 1890.

In 1839, the U.S. Congress appropriated $1000 for the Patent Office to collect agricultural statistics on new seed varieties and other agricultural information. It was hoped that providing annual statistics would guard against monopolies or exorbitant prices. By linking back to the 1840 census, the Patent Office issued its first report for 1841. Because census estimates of crop areas harvested were unavailable, estimates of population change in each state and territory were used as indicators of production changes. Other information came from agricultural societies, agricultural papers, and knowledgeable individuals.

The Patent Office attempted to track crop failures or other unusual conditions in making its annual reports. However, in 1849 the Patent Office came under new leadership that did not support agricultural statistics except for

Table 31.2 Number of Agriculture Census Questions, 1840 to 1890

Items of Inquiry	1840	1850	1860	1870	1880	1890
Name of Person Conducting Farm		1	1	1	1	1
Color of Person Conducting Farm						1
Tenure					3	3
Acres of Land		2	2	3	4	5
Acres Irrigated						1
Value of Implements, Machinery, and Livestock		3	3	3	3	3
Cost of Building and Repairing Fences					1	1
Cost of Fertilizer Purchased					1	1
Wages Paid for Farm Labor				1	1	1
Weeks of Hired Labor upon Farm					2	2
Estimated Value of All Farm Productions				1	1	1
Forest Products	7			1	2	2
Grass Lands and Forage Crops	1	3	3	3	5	22
Sugar	1	3	3	3	8	17
Cereals: Barley, Buckwheat, Corn, Oats, etc.	6	6	6	7	12	27
Rice	1	1	1	1	2	3
Tobacco	1	1	1	1	2	4
Peas and Beans		1	1	1	2	4
Peanuts						3
Hops	1	1	1	1	2	4
Fiber: Cotton, Flax, Hemp, Broomcorn, Wool	3	6	7	5	10	19
Horses, Mules, Asses, Sheep, Goats, Dogs	2	3	3	3	12	22
Neat Cattle[a] and Their Products	1	3	3	3	8	10
Dairy Products	1	2	2	3	3	11
Swine	1	1	1	1	1	4
Poultry and Eggs	1				3	8
Beeswax and Honey	1	1	2	2	2	4
Onions						4
Potatoes	1	2	2	2	4	6
Nurseries, Orchards, and Vineyards	5	2	2	2	12	42
Other	3	4	4	4	1	19
Total	37	46	48	52	108[b]	255[b]

[a]Neat cattle includes working oxen, beef cattle, and dairy cows.

[b]Number of inquiries or details called for in the general schedule of agriculture only; additional inquiries on special schedules of agriculture, not common to the general schedule or other special schedules, are not included.

Source: Wright and Hunt (1900).

prices, and the first continual series of production estimates ended. In spite of an outcry to reinstate the crop reports, the Patent Office made no further attempt to estimate annual production.

Because of complaints about many errors in the 1840 census, the 1850 census included major changes in organization and data content. Legislation was passed to define the duties of census enumerators and the consequences of neglecting their duties. The organizational structure initiated in the 1850 census has continued through present-day censuses. The Census Bureau established the farm definition as any place that had $100 or more in value of sales of agricultural products. A temporary census board within the Department of the Interior was established to oversee the conduct of the 1850 census, a procedure that was followed until 1900 (Wright and Hunt 1900).

By 1860, commercial corn and wheat production was well underway, with the west developing these crops rapidly. The northern states began developing other agriculture interests such as dairy and feed crops. Cotton surpassed tobacco as the major agricultural export crop, with its approximately $100 million average value accounting for half of all export value (Economic Research Service 1993).

31.2.3 Immigration and Agriculture

Foreign migration increased into the United States as a result of Ireland's potato famine and the German Revolution of 1848. Many immigrants were lured into agriculture. Scientific invention was applied to create and improve farm machinery. The nation became increasingly dependent on national and international markets as it moved from subsistence farming toward commercialization (U.S. Census Office 1902).

The average farm size in 1850 was just over 200 acres. However, the Homestead Act, which made 160 acres of new agricultural land available to settlers, and the breaking up of southern plantations during the 1860s caused the average farm size to decline. Not until 1950 did average farm size again exceed 200 acres (see Figure 31.1).

Between 1849 and 1862, when there were no federally collected annual statistics, agricultural societies continued to publish their "interpretations" or estimates. A new Commissioner of Patents in 1856 encouraged governors and other prominent individuals to create estimates for their areas. However, none of these efforts resulted in a consistent collection of statistics.

The present-day program of current agricultural statistics traces its development to events occurring in the 1860s. In 1862, Orange Judd, editor of the *American Agriculturalist*, (a popular farm magazine of the time), asked readers to designate a person in every town to fill out a monthly form on crop area and crop prospects. Respondents were asked to evaluate the current month's data relative to a base of 10 as an average crop and to reply in whole numbers, with each unit above or below 10 indicating a 10 percent departure. Judd received between 1000 and 1500 responses for various months in 1862. This response established the pattern of monthly information used by the USDA to the present day.

Judd's groundbreaking survey was discontinued in 1863 when Congress established the U.S. Department of Agriculture, which began its own collection of statistics. Congress acted in response to a grass-roots clamor throughout the nation, swelled by the agriculture press, for better data than that produced by

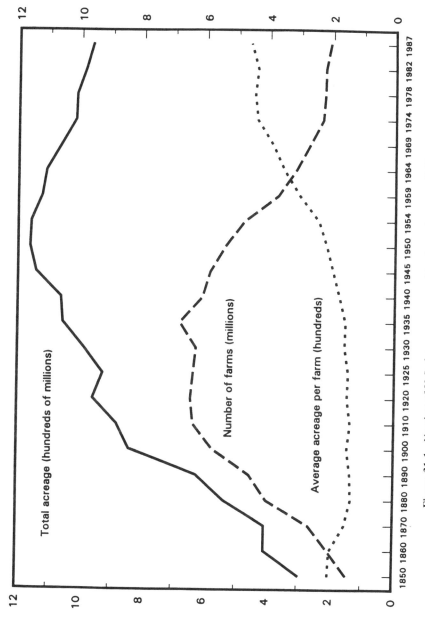

Figure 31.1 Number of U.S. farms, average and total area in acres: 1850 to 1987.

the Patent Office and the decennial census. Using the methods of Orange Judd, the USDA's first report covered estimates for 1859 and 1862 based on 1860 census figures (crop year 1859) and data received on questionnaires sent to every county during the winter of 1862–1863.

31.2.4 USDA Initiatives

The Office of the Statistician in USDA was created in 1863, and monthly reports of crop conditions were initiated immediately. One reporter and five assistants per average size county were chosen to provide local agriculture assessments. Reports were sent by mail to Washington, DC on a designated day each month. The purpose of monthly crop reports was explained by the USDA in 1863 as follows: "Ignorance of the state of our crops invariably leads to speculation, in which oftentimes, the farmer does not obtain just prices, and by which the consumer is not benefitted . . . the true condition of these crops should be made known. Such knowledge, while it tends to discourage speculation, gives to commerce a more uniform and consequently, a more healthy action" (Taylor and Taylor 1952, pp. 180–181).

Beginning on January 1, 1867, cattle, hog, and sheep numbers and values were estimated, along with wheat and corn prices for 1866. Price data were estimated annually until 1908, when monthly prices were instituted (National Agricultural Statistics Service 1989b).

There was little change in the agricultural data collected in the 1860 and 1870 population censuses. There was increased call for information on the acreage of crops such as wheat, barley, and oats, but this information was first available in the 1880 census. Inquiries in the 1860 census relating to "what crops are short" and "usual average crop" may have been useful in crop prediction for intercensal years, but these questions were dropped in the 1870 census. The first statistical atlas was published based on the 1870 results, which showed the geographic distribution of crop production (see Table 31.2).

31.3   AGRICULTURAL STATISTICS FROM 1880 TO 1915

By 1880, large cattle operations and wheat production were established on the Great Plains. While only 50 percent of the population was employed in agriculture, agriculture still accounted for over 75 percent of U.S. exports. Improved tools for production and harvesting had been developed and were used widely (Economic Research Service 1993). A census would need to reflect the growing diversity of the agricultural economy. The census of 1880 marked a turning point in agricultural statistics. Cooperation between the census and current statistics programs peaked as J. R. Dodge, Chief of the Division of Statistics in USDA, served as Chief Statistician for development of the 1880 census. He was given considerable freedom in revising the census questionnaire; he responded with added questions on area harvested and yield. These data were important as a census benchmark as well as assisting with intercensal predictions. The basic questionnaire had 108 questions about agriculture, while special schedules contained 1572 questions (Statistical Reporting Service 1969).

An innovation for 1880 was the presentation of results in cartographic and tabular forms. Comprehensive questions on renting arrangements and mechanization were added to the 1880 census as well. One other significant feature, which aided interpretation of current statistics surveys, was collection of information on average and largest cereal crop yields by region or locality (Wright and Hunt 1900).

Periodic reports issued by USDA were aided by the 1880 census information on acreage and yields; estimation of actual yield was shifted to the end of the growing season. In 1884, reporters were asked to consider 100 as the condition of a full crop, not that of an average crop. Another important improvement in procedures was the 1888 initiation of weighted averages based on county acreage for state estimates, because counties varied greatly in area and production potential.

The 1890 census also saw significant changes. Bronsen C. Keller, an economic and social scholar, in association with two other men from St. Louis, sent a letter to people around the country in 1889 urging them to insist that indebtedness information be collected in the 1890 census. Keller believed that indebtedness was leading to loss of farms through foreclosure and hence decreasing ownership. Through this grass-roots effort, data on five questions on indebtedness were collected in the 1890 population census and the results were classified by farm and nonfarm homes.

As agricultural statistics drew more attention and the number of census users increased, the accuracy of monthly crop reports began to be questioned. In 1895, the National Board of Trade passed a resolution stating the following (Taylor and Taylor 1952, p. 204):

> Whereas, The monthly and yearly crop reports of the U.S. Department of Agriculture have in recent years been confusing, misleading, and manifestly erroneous in important particulars; . . . if the crop reporting service . . . is to be continued, . . . every needful effort be made for ensuring the fullest degree of efficiency . . . completeness and accuracy of data, ...

In response, the National Board of Trade established the Committee on Crop Reports, which made several proposals for improving agricultural statistics. Based on its recommendations, Congress passed a law in 1909 making it a crime to divulge any statistical information ahead of its scheduled release. Official township reporting was recommended; by 1896 there were 28,000 township reporters, 9000 to 10,000 county reporters, and 6000 to 7000 assistants to state statistical agents. Reports were also received from 15,000 grain dealers, millers, and elevator operators and from 123,000 farmers.

Interpreting this vast amount of information was a difficult task. In 1905, the Crop Reporting Board was established to improve interpretation. This Board, consisting of state statisticians and experienced headquarters statisticians, assisted the Chief Statistician in reviewing indications and setting estimates, and it continues in this role today for major reports.

The 1900 agriculture census was similar to that of 1890 but added questions on tenure, total value of farm buildings, and ownership of rented farms. In addition, Congress provided publication guidelines that required that data be tabulated by race and gender of farm operators. Several processing innovations were introduced for this census as well. Punch cards and electronic tabulating machines were adopted following their successful use in the 1890 population census. Because of the large number of crop cards, a sorting machine was developed to facilitate their use in the 1990 agricultural census. Notably, the 10-key keypunch machine was used for farm census cards 20 years before its use for the population census.

Large differences occurred between USDA estimates and the 1900 census results for crop acreage and production. In almost all instances, USDA estimates were significantly lower than census results. Statistics for the number of acres for two of the nation's largest crops, corn and wheat, differed by 16 percent and 18 percent, respectively. A committee of inquiry investigated this and found that the census provided a low base in 1890 and that the USDA estimates for subsequent years included accumulated error from faulty yearly percent change ratios applied to the census base. The committee recommended improvements in training of census enumerators and clerks, improved editing procedures, and more verification during data processing and tabulation. They also recommended that a census of population be conducted every 5 years, especially for the collection of agricultural information (Statistical Reporting Service 1969).

31.3.1 U.S. Bureau of the Census Established

In 1902, a permanent Bureau of the Census was established in the Department of the Interior. It was transferred to the new Department of Commerce and Labor in 1903. When that Department split in 1913, the Census Bureau was placed in the Department of Commerce. Establishment of a permanent bureau created a more stable environment for the census program which promoted better planning, greater comparability between censuses, further evaluation opportunities, increased time for systems development, and an extended basis for producing additional statistics upon demand.

Specialized censuses on irrigation and on drained land were added to the agriculture program in 1910. These two censuses remained part of the program through 1950. Table 31.3 lists special censuses and surveys related to agriculture that have been conducted by the Census Bureau since 1890 (U.S. Bureau of the Census 1987).

31.4 AGRICULTURAL STATISTICS BETWEEN WORLD WARS I AND II

By 1915, gasoline-powered tractors and combines were developed for extensively farmed areas. New varieties and disease-resistant strains of plants were

Table 31.3 Census Bureau Special Censuses and Surveys by Reference Year

Year	Title(s)	Year	Title(s)
1890	Census of Horticulture	1956	Farm Mortgage Indebtedness
1905	Cotton Ginnings[a]	1959	Census of Horticulture,
1910	Census of Irrigation		Census of Irrigation,
1920	Census of Irrigation,		Irrigation in Humid Areas,
	Census of Drained Land		Census of Drained Land
1930	Census of Horticulture,	1964	Survey of Farm Workers,
	Census of Irrigation,		Survey of Hired Farm Workers,
	Census of Drained Land,		Survey of Farm Indebtedness
	Farm Mortgage Indebtedness[b]	1965	Survey of Nonfarm Income and
1935	Survey of Part-Time		Source
	Farming,	1969	Census of Horticulture,
	Farm Mortgage Indebtedness		Census of Irrigation,
1940	Census of Irrigation,		Census of Drained Land
	Census of Drained Land,	1970	Agriculture Finance Survey
	Farm Mortgage Indebtedness	1979	Census of Horticulture,
1945	Farm Mortgage Indebtedness		Farm Finance Survey,
1950	Census of Irrigation,		Survey of Farm Energy Use,
	Census of Drained Land,		Farm and Ranch Irrigation[b]
	Census of Horticulture,	1984	Farm and Ranch Irrigation
	Farm Mortgage Indebtedness	1988	Farm and Ranch Irrigation,
1955	Irrigation in Humid Areas[b]		Census of Horticulture,
			Agricultural Economics and Land
			Ownership Survey

[a]The Cotton Ginnings Survey collected data twice a month through the ginnings season every year from 1905 up to the present. Conducted by the Census Bureau until 1991, the survey is now conducted by NASS.
[b]The titles of the Survey of Farm Mortgage Indebtedness, the Survey of Irrigation in Humid Areas, and the Farm and Ranch Irrigation Survey have been shortened to better fit the table format.

developed. The average value of U.S. agricultural exports was approaching $2 billion a year, about 45 percent of all exports. Only about 30 percent of the population was now engaged in agriculture, but the need for commercial agricultural information was growing. There was also increased awareness of the importance of marketing and a pressing need for reliable information on supplies of food and fiber (Economic Research Service 1993).

Several important improvements to current agricultural statistics were made between 1910 and 1920. Official annual estimates of the numbers of farms by state began in 1910. Starting in 1911, monthly condition information was converted to yield forecasts for crops like wheat, oats, corn, and tobacco. Once the validity of the monthly forecasts was proven, more market-volatile crops such as cotton were added. The *par procedure* was developed to interpret yield by adjusting the 10-year average yield per acre by the ratio of current condition to average condition. As agricultural statisticians searched for more mathematically based yield-forecasting procedures, regression models based on year-to-year relationships were developed. By 1929, regression models were in common use for monthly yield forecasts (Statistical Reporting Service 1969).

An important factor in heightening interest in agricultural statistics was the onset of World War I in 1914. Information was needed on available food and feed supplies at county, as well as state, levels. By 1920, state and national estimates on 29 crops were being produced, compared to 13 crops 10 years earlier, and condition reports were being issued on 44 crops—about double that reported in 1910 (Statistical Reporting Service 1969).

In 1919 the chief of the Bureau of Statistics (USDA) reported many requests for agricultural data during World War I:

> A vast amount of information was compiled and furnished in response to inquiries received by telephone, telegraph, letter, or personal call of representatives of the Food Administration, the War Trade Board, the War Industries Board, the Military Intelligence Office of the War Department, the Tariff Commission, the Federal Trade Commission, the Council of National Defense, other departments of federal and state governments, congress, and private individuals. . . .

Following the armistice of November 11, 1918, the demand for special agriculture information declined, but the demand for food shipments to war-torn Europe continued. There was a general reluctance to discontinue the products and services provided by the USDA during the war, so many of them continued (Statistical Reporting Services 1969, p. 62).

Many states developed statistics programs, which meant that different state and federal estimates might be published for the same commodity, and respondents might be contacted by multiple organizations for the same information. Because of the emphasis on increased agricultural statistics during World War I and associated respondent burden issues, state and federal organizations in Wisconsin agreed in 1917 to share expenses for data collection and publication. Many states quickly followed suit, and today agreements exist in every state involving state departments of agriculture or state agricultural universities or both. These agreements are truly unique within the United States. State funding provides for special publications, surveys, or other services not covered by the federal program. All employees (state or federal) work under the NASS-appointed state statistician, who usually has a state department of agriculture title and state government duties.

To improve data reliability, sampling by 1918 shifted largely from panels reporting for their locality to panels reporting for individual farms. This provided a clearly defined basis for comparison of production levels. This sampling of farmers in every township became the monthly Farm Report Survey. Special livestock reporter lists were also established. In 1922, looking for methods to improve livestock statistics, a proposal was adopted to use free delivery of mail in rural areas to provide a broader base of reporters for major reports. Rural mail carriers were given a supply of card-type questionnaires to drop off at a sample of farms along mail delivery routes. Completed responses were forwarded to the state statistician by the postal service.

Rural carrier cards collected data for only the current year. Before that time, information was requested for both the current and the past year and a current compared to historic interpretation was made. Interpretation of rural carrier acreage surveys was based on calculation of ratios of acreage of specific crops to total cropland or to land in farms. However, ratios were usually biased upward because progressive farms tended to be sampled and farmers without row crops often did not report. The approach used to adjust for bias was the *ratio relative calculation* in which the current year ratio of a crop to total land is divided by the previous year's ratio to estimate true change. By 1928 an effort was made to look at reports from the same farm on subsequent surveys. Matching provided valuable information on year-to-year changes, but was extremely time-consuming and difficult.

From the beginning, complaints were made that an agriculture census every 10 years was inadequate for an industry with such large fluctuations. For its yearly estimates for crop and livestock items, USDA relied on the agriculture censuses for a new base or benchmark each 10 years. Occasionally, due to methodologies for projecting yearly changes and initial bias in census figures, annual estimates differed widely from the same year's data available later in farm census findings (for example, the previously described differences observed in 1900). Data users engaged in agricultural research, farm management, and business investments needed more current census information on agriculture activities (Statistical Reporting Service 1969).

Proposals were therefore floated for annual agriculture censuses or for a census every 5 years at the very least. In 1909, Congress mandated that the Department of Commerce conduct a mid-decade agriculture census. Because of World War I preparations, the first 5-year Census of Agriculture was not conducted until 1925. The 5-year census greatly assisted annual crop and livestock estimation and provided improved information for decision-making.

In addition to providing measures of production, the census of agriculture provides data on the effects of technological changes on agriculture and on social and economic characteristics of farm operators. Many questions were added in 1920. Farm operators were asked if they had gas or electric lighting in their homes and if they owned tractors, automobiles, or trucks. Additional questions in 1930 and 1940 queried farmers about the kinds of roads adjoining their farms, whether telephones were available, and the presence of new equipment such as combines and milk machines. The presentation of data by type of farm in 1930 was a valuable contribution to analysis of agriculture production. It provided a basis for much discussion and planning concerning the needs of farms in the early 1930s.

Socioeconomic questions on topics such as hired farm labor, farm versus nonfarm employment, income, race, and tenure of farm operators were asked in all 20th-century censuses. In 1920, for example, the census found that 61 percent of the rural population and 30 percent of the total population were engaged in farming (U.S. Bureau of the Census 1983).

Other expansions of statistics were made around 1925. The number of crops included on monthly farm reports nearly doubled. Questions on milk cows and milk production, hens and layers, numbers of eggs produced, and farm labor were added to the monthly farm report. The percentage of U.S. farms having various livestock species is shown in Table 31.4, with comparisons at intervals of about 10 years. As percentages of farms with milk cows and chickens declined, those questions were removed from the monthly farm report in the 1970s and specific surveys were developed for these data (Bureau of Agricultural Economics 1933).

The 1930s brought the next major changes in agricultural estimates. During the period of extremely dry weather, floods in the South, and the critical economic conditions of the depression, some agricultural products, particularly hogs, were overproduced, resulting in very low prices. Federal farm relief was needed to pull the country out of the agricultural slump that began after World War I. As a part of that effort, many government emergency programs were established to provide financial support to farmers. One 1933 program called for controlling the supply of hogs by selective destruction of a portion of supply. Thus, good information on supply was essential. In less than 2 years, 90 additional professional staff members were hired for state statistical offices specifically to develop hog estimates county by county (Brooks 1977).

The 1933 Agricultural Adjustment Act and its successor, the Soil Conservation and Domestic Allotment Act of 1936, were critical milestones in the government's approach to agricultural policy and to the statistical work to support it. The USDA was given unprecedented authority and funds to alleviate distress situations in agriculture (Statistical Reporting Service 1969).

One shortcoming of procedures used throughout the 1930s was that up-to-date lists of farms were not available and no other probability sampling frame existed. In 1938, research to divide the entire land area of the United States into sampling units began. The area sampling approach showed promise, and in 1943 the USDA and the Bureau of the Census jointly funded work at Iowa State College (now Iowa State University) to create the master sample of agriculture. This master sample created segments of land with definite boundaries that contained an average of four farms per segment. The sample was first used to measure coverage in the 1945 agriculture census.

World War II stimulated the use of sampling in the agriculture census in an effort to reduce cost and meet time limitations for special statistics needed during that period. In the 1940 census of agriculture, data were tabulated sep-

Table 31.4   Percent of Farms with Livestock, Censuses of 1910 to 1987

Item	1910	1920	1930	1940	1950	1959	1969	1978	1987
All Cattle	83.1	83.1	76.4	79.4	75.5	72.1	63.0	59.6	56.3
Milk Cows	80.8	69.2	70.8	76.2	67.8	48.3	20.8	13.8	9.7
Hogs and Pigs	68.4	75.2	56.2	61.8	56.0	49.8	25.1	19.7	11.7
Chickens	87.7	90.5	85.4	84.5	78.3	58.5	17.3	10.7	6.9

arately for large and small farms to identify their contribution to production levels and to assist in food supply decisions. Sampling as an enumeration methodology was introduced in the 1945 census, when county-level data were collected through a conventional all-farms canvas, while selected data at various geographic levels were obtained by sampling (U.S. Bureau of the Census 1979).

31.5 AGRICULTURAL STATISTICS FROM WORLD WAR II TO THE PRESENT

The years surrounding World War II saw some of the largest productivity changes in U.S. agriculture. The sad condition of agriculture in the early part of the century began improving. The surplus food problem began to vanish, and programs to increase production ensued. Farm production reached a high during World War II, despite labor loss and difficulty in obtaining machinery.

Commercial fertilizer use tripled in a 20-year period, and production was further augmented by continued improvements in hybrid seeds. Use of irrigation also increased as the country set war-time production goals. The structure of farms changed as *vertical integration*—the ownership of multiple stages of the production, marketing, and distribution functions by one organization— started in the poultry industry, and new marketing techniques such as frozen foods shifted production patterns. In 1950, about one-eighth of the labor force was farmers; by 1990, this was down to 2.6 percent. Agricultural exports are currently about 15 percent of all U.S. export value (Economic Research Service 1993).

After the turn of the 20th century, data users began requesting information in addition to production quantities and sales by product. In determining census of agriculture content, designers had to balance two contradictory issues: (1) demand by data users for more detailed data and (2) the need to keep respondent burden to a minimum. Experiments to tailor report forms to reflect different characteristics of farm operations in various regions were introduced during the 1940s and 1950s. From the 1945 to the 1959 censuses, questions were added to identify emerging farm operational patterns such as landlord–tenant operations or multiple operations owned by corporations. Technical improvements in processing also continued. Mechanical editing of data captured on punch cards began in 1940, followed by the development of modern computer technology, which improved publication timing and controlled the enormous processing responsibilities. The world's first general-purpose electronic computer, the UNIVAC system, developed to the Census Bureau's specifications and installed in 1951, was used for part of the 1950 population census and later to process 1954 agriculture census data (U.S. Bureau of the Census 1983).

Until 1950, agriculture censuses used personal enumeration—that is, farm-to-farm canvassing. Drawbacks were delays because of bad weather, smaller pools of census enumerators over time, and difficulty in locating absentee farm operators. For the 1950 census, the Bureau introduced mail questionnaires with questions phrased as if they were being asked by an interviewer. Questionnaires were delivered to rural route box holders, who were asked to complete the report forms and hold them until an enumerator came. This moderately successful system was used through the 1964 census.

Throughout history, the agriculture census had been taken in conjunction with the decennial census. But in 1950 the agriculture program split off; by mid-decade, agriculture censuses were independent of the decennial census. In the 1970s, the decision was made to change the timing of the census of agriculture to coincide with all other U.S. economic censuses in years ending in 2 and 7. This change was accomplished between 1974 and 1982.

The use of sampling techniques in the census of agriculture program expanded with the introduction of random samples for follow-on surveys of farms with specific characteristics. Following the 1954 agriculture census, a mail sample survey of farm expenditures was conducted and follow-on surveys such as irrigation and horticultural specialties have been included in every subsequent census of agriculture.

As the use of agricultural statistics grew after World War II, data users requested more current information. Several states developed cooperative arrangements with the Weather Bureau and with the Federal–State Extension Service to obtain informed opinions on crop progress and fieldwork operations each week to supplement the monthly *Crop Production* reports. By 1958, this popular "weekly weather crop" approach was expanded to all states with submission of state summaries to NASS headquarters for national release.

Other than weekly weather crop, agricultural statistics have emphasized developing probability-based methodology to improve the quality and stability of estimates and forecasts. In 1957, a long-range plan for improving USDA agricultural statistics was presented which called for development of a scientifically distributed area frame sample of farms to strengthen state and national crop and livestock estimates. Since 1964, a June agricultural survey has been conducted in all states except Alaska. This survey, which yields sampling errors of 1 percent or less for estimates of major crops, has become the backbone of improved crop and livestock estimating procedures.

Area frame sampling was extremely successful for crop estimates but did not provide the same efficiency for livestock totals because they can vary tremendously (from zero to many thousands) in relatively small land holdings. One means of stabilizing estimates and sampling errors was the creation of lists of large livestock producers to be surveyed with certainty. Because of the high costs of face-to-face interviewing, the area frame sample is fully enumerated only once a year in June.

Another livestock survey approach, the probability mail survey, was tried in the mid-1960s. All available information for a livestock species such as hogs was used to create a list sampling frame for that species. Samples were drawn by strata which improved the stability of estimates, but because lists were incomplete, data expansions did not cover total production, and this method was abandoned.

Ongoing internal and external research provided an improved solution in the early 1970s to livestock estimation difficulties. H. O. Hartley at Texas A&M University developed multiple-frame sampling, which combined the relatively low cost of list sampling with the complete universe coverage of the area frame survey. (Area frame and multiple frame sampling are explored in more detail in Chapter 11.) The June agricultural survey was a natural vehicle for determining completeness of the list frame. Once the base area frame survey had been conducted each year, subsequent mail or telephone surveys would include samples from the list frame, with supplements derived from the area frame. In addition to livestock, the multiple-frame approach was tried for grain stocks, farm labor, and production of specialty crops and was adapted in the mid-1970s to economic surveys of farm operators. Multiple-frame surveys were successful in improving the consistency of estimates.

Meanwhile, the Census Bureau was introducing a mailout/mailback enumeration procedure in the 1969 agriculture census. This method of enumeration was more cost-effective and allowed farmers to complete questionnaires at their convenience, permitted unhurried access to records, and gave respondents a chance to review and correct forms before returning them to the Bureau. To ensure good response rates, six or seven mail follow-ups, as well as telephone enumeration of large farms, were conducted (U.S. Bureau of the Census 1992).

This approach has several problems, including the need for development of a complete mailing list and for ensuring complete and timely response. Identifying small farm operators is a special problem because they constantly enter and exit the universe and are not adequately covered by administrative lists. No single list source identifies all farms. Sources such as government farm program records, farm tax forms, livestock inspection lists, and so forth, include farm operator names but also contain individuals such as landlords who are not farm operators. Some operators do not appear on any list.

Budget efficiencies, as well as the convenience of mailout/mailback, outweighed these drawbacks. Since 1945, the Census Bureau has evaluated coverage for each census of agriculture. Net coverage for number of farms has generally ranged from 85 to 93 percent. Coverage of agricultural production has consistently been above 95 percent (Clark and Vacca 1993). Despite problems with a mail census, overall coverage is only marginally lower than face-to-face enumeration conducted prior to 1969.

Major list frame development for the agriculture census program began prior to the 1969 mailout/mailback census. In 1976, NASS also began a major mail list frame development effort to support a mailout/mailback mode of data collection for their current surveys. The primary difference between the two mail list programs is that NASS built their mail list once with a capability of routine updates and maintenance whereas the Census Bureau develops a current mail list of farm operators for each census. Both mail list programs include (1) computer software routines to convert and standardize name forms from multiple lists, (2) matching all portions of names, address, and identifiers across records, (3) prediction of the probability of farm or nonfarm status based on combinations of data sources, and (4) creation of outputs for sampling and list maintenance purposes. The NASS mail list is a source for the census mail list as are Internal Revenue Service farm tax records. Both agencies strive to improve their mail list by using mathematical modeling to improve match success rates and reduce duplication (see Chapter 20). About 20 percent of active name records on a state's list frame change in some way each year, demonstrating the high volatility in the farm universe.

Emphasis on probability survey techniques had a significant effect on data collection methods. Funds were not available for extensive face-to-face interview follow-up of nonrespondents, so both agencies began using telephone follow-up in the late 1970s. To improve the quality of telephone interviewing, the agencies began research on the use of computer-assisted telephone interviewing (CATI) with interactive editing about 1980.

Another probability methodology improvement introduced by the USDA was the development of procedures to determine crop yield and production by in-field visits, counts, and observations. Since the 1960s, these objective yield surveys have been conducted for corn, wheat, soybeans, and cotton, and procedures have been developed for a wide range of tree and field crops. These surveys result in forecasts of the number of "fruit" (ears of corn, bolls of cotton, number of hazelnuts, etc.) present at harvest plus a forecast of weight per fruit. Forecast models use historic information for the same time period and maturity stage. Objective yield surveys have been extremely successful, but they are expensive because monthly on-site visits are required. They are only used in major producing states, usually covering 75 to 80 percent of U.S. production for a given crop.

Since 1972, NASS has used aerospace remote sensing as a data source. NASS became a leader in the automatic classification of full satellite scenes of digital data involving many million pieces of information. The June agricultural survey area frame segments are an ideal sample for training computer discrimination models and for judging the precision of classifications. If cloud-free imagery can be obtained, classification of the satellite data after training usually yields sampling errors equivalent to increasing the ground-based sample by three to five times. However, the satellite data cannot provide information for acreage determination earlier than conventional means. Thus, the value to the NASS is for review of season ending estimates of planted and harvested acreage (Statistical Reporting Service 1983b).

31.6 AGRICULTURAL STATISTICS FOR THE 1990s AND BEYOND

Over the years, the development of agricultural statistics has provided many innovations in the field of statistics and data collection. These include (1) obtaining reports as variances from a norm, (2) design of keypunch and sorting equipment, (3) matching cases for developing change estimates, (4) the seminal work in area samples at Iowa State University, (5) continuing research into list frame development, (6) using questionnaire design techniques to improve data quality, (7) using multiple frame samples and estimation, and (8) developing advances in techniques such as interactive editing.

The benefits to basic statistical theory and data processing techniques will continue as agriculture data collectors address today's issues. These issues can be divided into three basic areas: management concerns, technological developments, and societal changes. Both the Census Bureau and NASS, as well as other groups collecting agriculture data, face these challenges.

Management concerns include such matters as controlling costs, ensuring appropriate coverage levels, increasing data quality, and ensuring respondent confidentiality while providing maximum data to the users. As U.S. federal budgets become more restrictive, agencies have to determine cost-effective and cost-reducing measures while providing increasingly convincing arguments of the need for data collection in the agriculture sector. As more data collection efforts depend on complete list frames, both U.S. agencies need to keep abreast of constantly changing operating and marketing arrangements while maintaining substantial evaluation programs to determine list completeness. The application of many federal and state laws depends on accurate agricultural sector data. Both agencies need to keep a vigilant eye on developments to ensure data quality, timeliness, and comparability. Efforts are needed to ensure that customer needs are met.

The confidentiality of respondent data is of utmost importance. The goodwill of data reporters depends on the agencies ensuring that individual data not be publicly disclosed. The Census Bureau, in particular, is developing an extensive body of theory related to cell suppression theory which protects respondent data in tabular data presentation (see Chapter 24). Application of the theory can be difficult, but it is necessary to ensure respondent confidence and cooperation. The counterbalance to protecting respondent data is providing useful data to users. The more data are suppressed to ensure respondent confidentiality, the less data are available for users. This poses a constant dilemma, because the users pay for the data collection. The future holds many such policy development and statistical innovation challenges.

Agricultural technological development is most challenging because it is hard to predict future issues. As technological advances are made in the agricultural community (such as sustainable agricultural practices, more direct marketing of commodities like fruits and vegetables, or growth in the number of farmers producing new crops for specific markets), it will be important to gather data about the changes. This task is complicated by the need to reach consensus about definitions before attempting measurement.

Technology is advancing rapidly in the fields of data collection and statistical estimation. How will data collection activities change as more respondents use computers and computer-driven communications and technology? Data collectors have an obligation to keep up with—and preferably ahead of—such trends.

Societal issues run the gamut from maintaining acceptable response rates in a diverse society to changes in questionnaire characteristics as farms become larger and small farmers cease operations. Both agencies are currently devoting effort to the areas of questionnaire design, respondent burden, and customer needs. For example, NASS holds yearly data user meetings around the country, focusing on different types of statistics each year, and the Census Bureau has increased its presence at agriculture related meetings. Again, the issue of customer expectations is difficult; meeting the needs of data users while not requiring too much of an ever-shrinking community of data providers is a challenge that will require perseverance and creativity on the part of both agencies. The issues of respondent burden may force innovative solutions involving special contacts and wider use of database techniques.

The goal of both U.S. agencies is to be quality-oriented in the future. The needs for data on the agricultural sector will not diminish as the federal, state, local, and private sectors strive to meet their mandates. Planning, research, marketing, and management of farm and rural programs in this country will continue to depend on quality data collected through innovative techniques. Both agencies are determined to meet the challenges of the future as they have in the past.

REFERENCES

Bidwell, P. W., and J. I. Falconer (1925), *History of Agriculture in the Northern U.S.*, Carnegie Institution of Washington.

Brooks, E. M. (1977), *As We Recall: The Growth of Agricultural Estimates, 1933–1961*, Washington, DC: Statistical Reporting Service, U.S. Department of Agriculture.

Bureau of Agricultural Economics (1933), *The Crop and Livestock Reporting Service of the United States*, Miscellaneous Publication No. 171, Washington, DC: U.S. Department of Agriculture.

Clark, C. Z. F., and E. A. Vacca (1993), "Ensuring Quality in U.S. Agricultural List Frames," *Proceedings of the International Conference on Establishment Surveys*, Alexandria, VA: American Statistical Association, pp. 352–361.

Economic Research Service (1993), *A History of American Agriculture, 1776–1990*, Washington, DC: U.S. Department of Agriculture.

National Agricultural Statistics Service (1989a), *The History of Survey Methods in Agriculture (1863–1989)*, Washington, DC: U.S. Department of Agriculture.

National Agricultural Statistics Service (1989b), *Agricultural Production and Prices—125 Years, A Historical Review*, Washington, DC: U.S. Department of Agriculture.

Statistical Reporting Service (1969), *The Story of the U.S. Agricultural Estimates*, Miscellaneous Publication No. 1088, Washington, DC: U.S. Department of Agriculture.

Statistical Reporting Service (1983a), *Framework for the Future*, Washington, DC: U.S. Department of Agriculture.

Statistical Reporting Service (1983b), *Scope and Methods of the Statistical Reporting Service*, Miscellaneous Publication No. 1308, Washington, DC: U.S. Department of Agriculture.

Taylor, H. C., and A. D. Taylor (1952), *The Story of Agricultural Economics in the United States, 1840–1932*, Ames, IA: The Iowa State College Press.

U.S. Bureau of the Census (1979), *1974 Census of Agriculture Procedural History*, Vol. 4, Part 4, Washington, DC: U.S. Department of Commerce.

U.S. Bureau of the Census (1983), *1978 Census of Agriculture Procedural History*, Vol. 5, Part 4, Washington, DC: U.S. Department of Commerce.

U.S. Bureau of the Census (1987), *1982 Census of Agriculture—History*, Vol. 2, Part 4, Washington, DC: U.S. Department of Commerce.

U.S. Bureau of the Census (1992), *1987 Census of Agriculture—History*, Vol. 2, Part 4, Washington, DC: U.S. Department of Commerce.

U.S. Bureau of the Census (1989), *1987 Census of Agriculture U.S. Summary and State Data*, Vol. 1, Part 51, Washington, DC: U.S. Department of Commerce.

U.S. Census Office (1902), *Census Reports: Twelfth Census of the U.S.: Farms, Livestock, and Animal Products*, Vol. V, Part 1, Washington, DC: U.S. Government Printing Office.

Wright, C. D., and W. C. Hunt (1900), *History and Growth of the U.S. Census: 1790–1890*, Washington, DC: U.S. Government Printing Office.

CHAPTER THIRTY-TWO

A History of Business Surveys at Statistics Canada: From the Era of the Gifted Amateur to That of Scientific Methodology

David A. Worton and Richard Platek[1]

Statistics Canada (retired)

This chapter traces the history of business surveys at Statistics Canada from the middle of the 19th century, when the first surveys were taken, to the last decade of the 20th century. It describes the evolution of business survey methodology from the early collection of business data as part of the decennial population censuses, through the development of ad hoc business surveys and periodic censuses with the gradual adoption of random sampling techniques and more standardized concepts and procedures, to current methodologies driven by analytical users' needs and aided by the capabilities of computer technology.

During the 19th century, many countries developed systems of official statistics. At the same time, statistics flourished as a science. Yet theoreticians had little overall impact on practical developments. There were exceptions, of course, such as Adolphe Quetelet and Ernst Engel, who served as official statisticians in Belgium and Germany, respectively. In vital and health statistics,

[1]The chapter draws on a history of official statistics in Canada, 1851–1971, now in preparation by David A. Worton, and on an unpublished history to 1975 of survey methodology in Statistics Canada by Richard Platek. The authors gratefully acknowledge the advice given by Michael J. Colledge of the Australian Bureau of Statistics (formerly of Statistics Canada) on developments during the most recent years covered by the chapter. The views expressed should be attributed to the authors only and not to Statistics Canada.

Business Survey Methods, Edited by Cox, Binder, Chinnappa, Christianson, Colledge, Kott.
ISBN 0-471-59852-6 © 1995 John Wiley & Sons, Inc.

theory and practice developed hand in hand under such luminaries as William Farr in England and Edward Jarvis in the United States. But the great bodies of census data and statistics of trade, transportation, finance, agriculture, and labor developed in the English-speaking world until almost the mid-20th century were the work of "gifted amateurs," men with no formal statistical training who earned their credentials on the job.

North America's first census was the Census of New France (then a colony of France, later to become the province of Quebec in Canada), conducted by Jean Talon in 1666. This census not only counted people, but also measured the wealth of industry and agriculture, the value of local timber and mineral resources, and the number of domestic animals, seigneuries, government buildings and churches.

Statistics of agricultural and industrial activities in North America were collected during the 19th century in conjunction with the decennial population censuses. This early practice of piggybacking supplementary inquiries (which were not feasible otherwise) on to the prestige and bankroll of constitutionally prescribed demographic censuses went further and endured longer in Canada than in the United States. Section 32.1 recalls this practice in Canada as well as the gifted amateurs who conducted these surveys. As the 20th century dawned and Canada developed into a major trading nation, pressures emerged for improved and extended statistics. These led to the establishment in 1918 of a central statistical office, namely, the Dominion Bureau of Statistics.[2] (Legislation: The Statistics Act, 1918) Section 32.1 also traces the first quarter century of the bureau's history before the advent of probability sampling.

Section 32.2 covers the period after World War II during which the demand for economic statistics exploded and a quantum increase in user sophistication occurred. Major changes in the survey process were introduced along with probability sampling. Section 32.2 explores these changes as well as the organizational changes put in place up until the present day.

32.1 THE ERA OF THE GIFTED AMATEUR: 1851 TO 1945

The province of Canada was formed in 1841 from Upper and Lower Canada, which became the provinces of Ontario and Quebec after Confederation in 1867. Canada's economy was in transition from the dominance of the fur trade to a more broadly based resource economy emphasizing agriculture. Agriculture meant land settlement which required immigrants. Thus, there had to be quarantine facilities at ports of entry and, in times of epidemics, centralized control of public health. This spectrum of interrelated responsibilities fell to the Department of Agriculture, which even registered patents because the majority of new inventions related to farming methods.

---

[2]In 1971, the Bureau was renamed Statistics Canada.

32.1.1 The Heyday of the Decennial Census

Continuity in modern Canadian statistics began with the province's first census in 1851. Censuses were subsequently conducted every 10 years; these were the principal source of statistics of agricultural and industrial activity until 1911.

The Censuses of 1851 and 1861

The legislative basis for the 1851 census was the Act of 1847, which required the collection of vital and criminal statistics and "all such information relative to the Trade, Manufactures, Agriculture and population of the Province as it may be able to obtain." The entire executive responsibility for the survey fell on the Secretary of the Board of Registration and Statistics. The 1851 census had both a personal and an agricultural component, the former covering a great deal more than purely demographic data. It also required a listing of "every Shop, Store, Mill and Manufactory, with their return of capital, produce, rent, number of hands employed, etc." (Lovell 1853, p. vi).

Barely a month before the date of the census, the census commissioners were appointed and, through them, enumerators. No training was provided; the census schedules were thought to be self-explanatory. Completed returns were to be sent back through the municipalities, which were optimistically required to examine them "and to cause any defect or inaccuracy that may be discovered therein to be supplied or corrected as far as may be possible." Difficulties naturally ensued from these casual arrangements. In general, the census of Lower Canada was thought to have been taken with greater care than that of Upper Canada, "where unfortunately many of the Enumerators proved themselves wholly unfit for the duties assigned to them." It was clear, however, that most respondents thought that "the Census had some direct or indirect reference to taxation—and in this belief the Enumerators were frequently received most ungraciously."

The 1861 census, an undertaking of similar scope, was conducted in much the same way, but in the haste of compiling its overdue findings, a promised administrative report failed to appear.

The Appointment of Joseph Charles Taché

Political embarrassment at the quality of the 1851 and 1861 censuses and the failure to achieve the general objectives of the statistical legislation of 1847 led in 1864 to the appointment of Joseph Charles Taché as Deputy Minister of Agriculture with special responsibility for statistics. Educated as a physician, he also had legislative, journalistic, and academic experience. His immediately preceding position was that of Chairman of the Province's Board of Prison and Asylum Inspectors. His appointment was literally an act of nepotism; his uncle, Sir Etienne Paschall Taché, had twice served as joint Premier of the Province. Nevertheless his patrons chose wisely. Taché served with distinction through Confederation in 1867 until his retirement in 1888, and he is recognized as the father of modern Canadian statistics.

Upon appointment, he conducted a searching analysis of the procedures and findings of the 1851 and 1861 censuses and condemned their reports outright as "a priori nearly worthless for they give as facts figures which express absolute impossibilities" (Taché 1866). These latter were for the most part the consequence of clumsy efforts to balance tables which had been improperly compiled in the first place: "The addition of the columns do not always agree; but they do sometimes agree in totals, while they quite disagree in the details. . . . I have learned, by consulting the traditions of the office, that such a wonderful result was obtained by a high-handling of figures—called at the time— to make them correspond." Examples cited included the following:

- "The shipyards of Upper Canada are represented as having built no ship, but that 0 of ship is valued at \$74,700."
- "Twelve mills in the County of Norfolk are said to have manufactured only 5,100 barrels of flour out of 139,000 bushels of grain; but, on the other side, 15 mills in the County of Middlesex have manufactured 23,775 barrels of flour out of only 35,000 bushels of grain."

Taché offered a number of recommendations for correcting these and other shortcomings. Most importantly, there had to be "a regular, permanent, well chosen and properly paid staff of statistical clerks." He asked for, and received, two such clerks. To put this seemingly modest request into perspective, the department's headquarters staff at the time, covering many responsibilities other than statistical, amounted to only 17.

The Censuses of 1871 to 1911

Following Confederation in 1867, Taché, now the Deputy Minister of the new Dominion Department of Agriculture, was faced with the conduct of the 1871 census. This undertaking was of much greater scope than its predecessors of 1851 and 1861, covering four provinces (Ontario, Quebec, New Brunswick, and Nova Scotia).

One lesson from the earlier experience was the need for early planning and careful training of commissioners and enumerators. An instruction manual was prepared that disassociated the census from any connection with taxation and emphasized the confidentiality of all information. The schedules were distributed ahead of time so that respondents knew what was required and could avoid the errors of hasty recall.

Most problems with the 1851 and 1861 censuses stemmed from the use of unskilled enumerators to collect both straightforward demographic characteristics and more complex economic information. So it was a gamble in 1871 to attempt a more detailed census, comprising 211 questions organized in nine schedules. Two schedules made up the personal census, while the remaining seven covered the returns of: (1) public institutions, real and personal estate; (2) cultivated land and products; (3) livestock, animal products, home-made fabrics, and furs; (4) industrial establishments; (5) products of the forest; (6)

shipping and fisheries; and (7) mineral products. There are no records of public criticism of the census findings, nor of personal dissatisfaction on Taché's part.

The 1881 census covered seven provinces (Ontario, Quebec, New Brunswick, Nova Scotia, Prince Edward Island, Manitoba, and British Columbia), together with the Northwest territories, as then defined.[3] It was leaner in content, comprising eight schedules and 172 questions, but this time Taché acknowledged some criticism of the agricultural and industrial material. Anticipating future problems of industrial classification, he noted that "Many errors of the critics have their source in the fact that several known industries are mixed in various establishments which must be recorded under one title, because it is impossible to discriminate between them and make up accounts for each separate element of the joint undertaking" (Dominion Department of Agriculture 1885). At this time, no classification systems were in place to ensure consistency between the industrial, occupational, and commodity detail of successive censuses.

Meanwhile, in the United States, dissatisfaction was beginning to surface with the "piggyback" approach. Writing in the *Quarterly Journal of Economics* in 1888, Francis A. Walker, Superintendent of the U.S. censuses of 1870 and 1880, condemned industrial statistics produced in conjunction with demographic information as intrinsically inadequate and too infrequent. The census, he argued, should be confined to the enumeration of population and the collection of agricultural information, with other statistical inquiries being conducted intercensally. Walker, a sometime lawyer, Civil War general, and president of the Massachusetts Institute of Technology, was probably the archetype of the gifted amateur. He was later described as "king among Census takers, and the greatest all-round master of the science of statistics" (North 1918, p. 25).

But Walker's advice went unheeded in Canada, where the requisite administrative continuity and funding for intercensal work was not put in place for two more decades. Taché retired in 1888 and was succeeded in his statistical duties by George Johnson, who was given the title of Dominion Statistician. Johnson, a journalist rather than an administrator, was appointed to launch an annual statistical abstract authorized in the new statistical legislation of 1879 (Legislation, Census and Statistics Act, 1879). He had made his name as editor of a special handbook prepared for the Colonial and Indian Exhibition held in London in 1886.

The prescription for the census of 1891, conducted by Johnson, was for more of the same, with nine schedules and 216 questions. The Minister's Report for 1890 stated that "great care has been exercised to preserve the essential features of previous census-taking, while at the same time introducing new features in order to bring the statistics obtained up to the most modern view of what a census should be as a national stocktaking" (Minister of Agriculture, Canada, 1891). An interesting methodological feature of the 1891 census was

---

[3]In 1905, these became the provinces of Saskatchewan and Alberta.

the experimental use of automatic tabulating equipment, which had been successfully used in the 1890 census of the United States.

Economic growth faltered during the early 1890s, but began to pick up as the end of the decade approached. Consequently, the newly appointed Special Census Commissioner, Archibald Blue, took the piggyback approach much further. The census of 1901 was an unprecedented statistical undertaking with 11 schedules and 561 questions. On the schedule for manufacturing alone, the number of questions more than tripled. The Minister of Agriculture, in his report for 1901, said "It was felt important that a good deal of information which had not been before gathered in Canadian censuses should be obtained." (Minister of Agriculture, Canada, 1901). Referring to the appointment of Archibald Blue as Special Census Commissioner, the Minister noted that "the elaboration of the schedules for this purpose required the attention of an expert who could devote his whole time and attention to this particular labour." Blue had earlier served as Ontario's Deputy Minister of Mines and had the distinction of having a mineral named after him—blueite. He came to Ottawa at a higher salary than the Deputy Minister, and he effectively sidelined Johnson, who spent the remaining 4 or 5 years of his career working with the Yearbook.

Blue's mandate extended further than the decennial census, but it was noteworthy that he conducted the 1911 census along much the same lines as that of a decade earlier, with 13 schedules and 549 questions. Blue's Censuses of 1901 and 1911 were later assessed by M. C. Urquhart as better that those of his predecessors. Indeed, Urquhart (1987) noted that "They suffer little in comparison with those of later dates."

32.1.2   1901 to 1918: Years of Change

While continuing to support the concept of the omnibus census, Blue recognized the need for intercensal information; and, largely at his initiative, legislation was enacted in 1905 to establish the permanent Census and Statistics Office with a mandate to conduct intercensal inquiries (Legislation: Census and Statistics Act, 1905). With an increase in staff to 21, a census of manufactures was conducted in 1906, a census of dairy production in 1907, and an agricultural census of Ontario, Quebec, and the Maritime provinces in the same year. This new material was collected, not by direct enumeration, but through the mails from Ottawa. In 1905, a Dominion-wide (i.e., national) system of agricultural statistics was instituted, which yielded annual estimates of many variables in the decennial census of agriculture, in addition to providing monthly crop reports. The emphasis on agriculture arose partly from the need to honor Canada's reporting obligations to the newly established International Institute of Agriculture. Unhappily, these initiatives were taken independently of those providing annual statistics of agriculture in the majority of the provinces. As Coats later pointed out, the various sets of figures were "at all but complete cross purposes, coinciding neither as to time, definitions, nor general methods" (Coats 1946).

The Department of Agriculture never enjoyed a monopoly in the production of statistics. Other departments produced economic statistics as a by-product of administration in such areas as trade, banking, insurance, railways, and canals. These statistics were becoming increasingly important as the economy matured.

A completely new arrival on the statistical scene at the turn of the century was the Department of Labour, which was set up in 1900 to collect and publish labor market information in a new monthly publication, the *Labour Gazette*. The greatest of the gifted amateurs, Robert Hamilton Coats, made his statistical debut as associate editor of the *Gazette*, under W. L. Mackenzie King, the Deputy Minister of Labour. Coats, a graduate of the University of Toronto in classics and economics, had worked as a journalist before coming to Ottawa.

From the start, the *Gazette* provided information on wage rates, hours of work of representative trades, and the cost of living, but it was fragmentary and irregular. At first, data were collected through the *Gazette*'s network of local correspondents, but gradually a more rigorous approach was taken through the direct mailing of printed schedules to potential respondents.

Coats repeatedly urged on King the creation of a separate Wages and Cost of Living Branch with a more systematic and extended program of regular data collection and publication, including the development of weighted cost of living indexes. He advocated the reduction of the information collected to ''a system of index numbers so that comparisons might be made on a mathematical basis both as between wages and the cost of living in the several cities and provinces and also as to the relation between living expenses and the remuneration of labour at different points'' (Coats 1904). Later, he addressed the technical issue of weighting the various components of an aggregated index number so that their relative importance would be appropriately reflected, and he suggested the construction of a budget for a family of five with an income of two dollars per day.

By the end of the decade, the cost of living had become a vexing social and political issue. King, now the Minister of Labour, outlined in the February 1910 issue of the *Labour Gazette* plans for addressing the problem through a more comprehensive and systematic program of statistics (Department of Labour, Canada, 1910). The first element was a monthly table in the Gazette showing ''the retail prices of 34 commodities which enter largely into the cost of living in the more important centers of population across Canada.'' The second element was a monthly summary of fluctuations in the wholesale prices of 235 commodities ''principally accounting for the trade of Canada, and representing the more important phases of its industrial activity.'' This new series was buttressed with a special benchmark study on the course of wholesale prices between 1890 and 1909 (Coats 1910). The study was hailed in the July 1910 issue of the *Labour Gazette* as ''undoubtedly the most comprehensive on the subject of prices ever published in Canada.'' (Department of Labour, Canada, 1910). It attracted favorable attention the world over and established Coats as an authority in the field.

The statistical work of the Department of Trade and Commerce also assumed growing importance during this period. As noted by Sir John A. Macdonald in the House of Commons Debates on June 10, 1887, the department had been set up with the objective of "developing and maintaining everything connected with our trade and commerce, whether it be home or foreign trade." The statistics with which the department worked were those of imports and exports, generated by the Department of Customs, which had long published them with little or no adornment. However, the Department of Trade and Commerce soon began to repackage them with interpretative analyses. The department's first annual report in 1892 showed a sophisticated appreciation of the problems of making valid international comparisons of imports and exports, citing differences in valuation, the lack of a common statistical period, inconsistent bases for showing the countries of origin and destination, and so on.

With its global outlook, the Department of Trade and Commerce portfolio became the natural focus for the imperatives towards statistical reform that emerged in the new century. The formation of the Census and Statistics Office had been such a reform, but was seen as insufficient. The agent of the eventual changes was the Honorable George Eulas Foster, appointed Minister of Trade and Commerce in October 1911 by a new Conservative government.

First, the Census and Statistics Office was transferred from the Department of Agriculture to the Department of Trade and Commerce. Soon after, on Foster's recommendation, a Departmental Commission on Official Statistics, of which Coats was a member, was appointed:

> . . . to inquire into the statistical work now being carried on in the various Departments, as to its scope, methods, reliability, whether and to what extent duplication occurs; and to report to the Minister of Trade and Commerce a comprehensive system of general statistics adequate to the necessities of the country and in keeping with the demands of the time.[4]

When the commission reported 6 months later, it confirmed Foster's view of the fragmentary and poorly coordinated nature of official statistics in Canada and recommended "that there be created a Central Statistical Office to organize, in cooperation with the several departments concerned, the strictly statistical work undertaken by the Dominion Government." (Report of the Departmental Commission on the Official Statistics of Canada, 1913). As regards the decennial census, the report recommended that it would be desirable to limit its scope and to make separate arrangements for industrial statistics. In this connection, it proposed an annual census of production "embracing the chief products of agriculture, forestry, fisheries, mining and manufactures."

Foster liked the report and, in June 1915, he appointed R. H. Coats as Dominion Statistician and Controller of the Census. Coats took a twofold approach to his new responsibilities. First, a detailed plan was needed to implement the commission's recommendations, in particular to determine the pro-

---

[4]Order-in-Council No. 1485, 1912.

cedures for cooperating with other Dominion departments and the provinces. Second, there was an immediate requirement to put in place the annual census of production.

This latter concern led to a repeat of the 1906 postal census of manufactures, covering calendar year 1915. Although the results were well received at the time, they did not find favor with Urquhart and Buckley (1965, p. 456), who later compiled the first edition of *Historical Statistics of Canada*. Only establishments with outputs of $2500 or more were enumerated, and the data could not be adjusted to full coverage. Accordingly, they were omitted from the historical time series for manufacturing. This was ironic, because it was probably the first attempt to enforce the mandatory reporting requirements of the 1905 Census and Statistics Act. Very persistent follow-up efforts were made, and four cases were eventually brought to court.

Coats' thinking on the broader issues was formally submitted to Foster in a summary paper entitled "A National System of Statistics for Canada—Centralization, Reorganization and Enlargement of Canadian Statistics." (Coats 1916). This provided the blueprint for the establishment of a centralized statistical office, built around the nucleus of the existing Census and Statistics Office, and a field-by-field outline of the programs it should carry out. The decennial census was to be a census of population and agriculture, relieved of the albatross of industrial and other complex inquiries but combining the two household-based topics of inquiry for administrative convenience. Elsewhere, Coats argued that a proper scheme of industrial statistics, covering fishing, mining, lumbering, and manufacturing, involved "(a) a comprehensive inquiry covering all phases of industry once every ten years; and (b) an annual postal inquiry in intercensal years, limited in the main to production and designed to keep the figures of the industrial census up to date."

32.1.3 1918 to 1945: The Last Years of the Amateurs

Following enactment of a new Statistics Act on July 1, 1918, the Dominion Bureau of Statistics was set up (Legislation, The Statistics Act, 1918). Its early years were ones of high achievement. Building on the nucleus inherited from the Census and Statistics Office, the organizational framework outlined in 1916 was put in place through transfers from other departments and through the creation of new branches. As the framework took shape, existing programs were strengthened and elaborated and new ones undertaken. This resulted in a doubling of the Bureau's staff to about 450 people between 1918 and 1923.

Coats' earliest priorities were (1) the establishment of a national scheme of vital statistics and (2) the conduct of the reorganized decennial census in 1921. But great strides were also made in economic statistics. Coats' annual report for 1922 described the progress with respect to: (1) monthly and annual statistics of agriculture; (2) annual statistics of fisheries, mines, forestry, dairying, central power and general manufactures; (3) statistics of foreign trade, transportation, and communication; (4) the establishment of an internal trade branch and of a program of price statistics; and (5) an agreement with the Department of Labour outlining the respective responsibilities of the two bodies in the area of labor statistics (Dominion Statistician, 1923).

Sedley A. Cudmore and Herbert Marshall, two able lieutenants recruited from the University of Toronto, assisted Coats in much of this work. Cudmore succeeded Coats as Dominion Statistician in 1942, followed by Marshall in 1945. Thus both were important transitional figures between the early and modern methodological approaches.

The arrangements for an annual census of industry ran counter to Coats' 1916 plans for comprehensive censuses every 5 years, interspersed with annual updates. This was largely because the areas of statistical collection in which the Bureau had been working towards cooperative arrangements with the responsible dominion and provincial departments—fisheries, mines, forestry, dairying and electric power—were already on an annual basis. These arrangements were summarized in a document prepared for the 1935 Commonwealth Statistical Conference held in Ottawa:

> (1) A uniform method and technique is arrived at in conference between the Bureau and the Dominion and Provincial Departments concerned; (2) the Bureau . . . prints and provides the standard forms and schedules as agreed upon; (3) the Provincial Government Departments in most cases undertake the collection and visaing[5] of the data; (4) the Bureau compiles the schedules according to an agreed plan; (5) the publication of the data is made on a Dominion-wide basis by the Bureau, the provinces being given their own data for use in any way desired; (6) the Dominion Departments use the Bureau as their statistical agency and obtain from its appropriate branches such statistical services as they require (The Dominion Bureau of Statistics, 1935).

The lack of an unambiguous and consistently applied product classification was a potential Achilles' heel for the new industrial census. In May 1918, Coats set up a committee under Professor W. C. Clark of Queen's University to develop workable guidelines. The committee recommended that products be classified according to chief component material but also allowed the use of purpose or source of origin categories for separate and supplementary analyses. This *tripartite classification*, as it came to be called, was energetically promoted by Coats during the 1920s among his international colleagues,[6] but they were generally unconvinced, preferring a single classification based on mixed principles.

The Bureau took no further action for a quarter of a century until the increasing variety and sophistication of Canadian industrial processes made the principle of chief component material progressively more difficult to apply. What emerged soon after the World War II, under the leadership of N. L.

---

[5]An archaic word meaning "checking."

[6]Notably at the British Empire Statistical Conference of 1920 and successive Conferences of Labour Statisticians, called by the International Labour Organization.

McKellar, was a system which augmented the criterion of chief component material with that of purpose. McKellar later played a major role in developing the International Standard Classification of all Economic Activities, adopted by the United Nations Statistical Commission in April 1948. His work in developing the Canadian Standard Industrial Classifications of 1948 and 1960 became a *sine qua non* for the postwar task of integrating economic statistics.

The policy of diligently pursuing and, if necessary, prosecuting delinquent respondents, begun with the 1915 Census of Manufactures, was continued after World War I for the Census of Industry. The Assistant Chief of Industrial Statistics, J. A. Schryburt, spent much of his time preparing cases and assisting counsel in presenting them in court. In October 1920, 43 cases were disposed of in a single day in Toronto, and a few weeks later, 94 actions were launched in Montreal. In later years, delinquents continued to be followed up, but fewer and fewer of them were taken to court. After one 1923 trip through Ontario, covering 29 working days, Schryburt reported that he had "cleaned up" almost 400 delinquents.

The success experienced with industrial statistics pointed the way towards a similar census of distribution activities. Accordingly, in the 1921 census, a list of the establishments engaged in wholesale and retail trade was collected on a special schedule. With some refinements, this served as the basis for a special postal inquiry covering 1923. Returns were obtained from some two-thirds of the target population. Budgetary stringency made it impossible to follow up, but many nonrespondents were known to be small. It was therefore possible to assert in the report, which did not appear until 1928, that "while the data is [sic] incomplete, it is considered to be indicative of real conditions" (Dominion Bureau of Statistics 1928, p. 5).

Continued growth in the relative importance of distribution and related activities led to a second attempt to measure them in 1931. A Census of Distribution was being planned in the United States, and it was decided to make the Canadian Census as comparable as possible. One major difference, however, was that the latter also covered service establishments. Lists of establishments were developed in the same way as before, but vigorous follow-up was now possible. It was estimated, for instance, that only a little more than 5 percent of the retail stores were missed. Apart from its considerable intrinsic value, this census also laid the foundation for an annual survey of retail and wholesale trade.

The rapid expansion of the Bureau's staff and programs during its first five years was interrupted by government-wide staff cuts in the mid-1920s. MacGregor (1939) later expressed the view that little of value was accomplished in the decade and a half that followed: "Henceforth, the work of the Bureau was to consist of routine interrupted, modified and extended at many points by a multitude of forces other than the guiding principles of the first six years."

This state of affairs was attributed largely to the Bureau's status as "the tail end of the Department of Trade and Commerce'' under the thraldom of generally unsympathetic—or indifferent—Ministers and Deputy Ministers, who were vulnerable to shortsighted demands from special private sector interests. Such pressures were thought to have a pernicious effect on the way the Bureau worked:

> New sources of material are added before old ones have been tested, audited, revised and written up in a proper and scientific manner. The system rarely gets beyond the stage of extensive inquiry, rough editing and "general purpose" material. In the effort to serve everybody, no one is served well; long-term problems of major national importance are given less attention than the work of providing school boys and the daily press with statistics without tears.

This criticism was harsh and overstated, but not without elements of truth. Years later, Herbert Marshall (1953) was to concede that "During this period, quantitative considerations had sometimes to be given priority over considerations of quality in order to get things started.''

Yet there was in fact much progress after 1924. In spite of another round of staff reductions in the depression years, Coats increased the staff to almost 600 by 1939, and further elements of the 1916 blueprint were put in place. The real problem was a lack of external guidance. Coats repeatedly argued the case for an expert advisory body, but he was ignored. The Trade and Commerce Department, to which the Bureau was administratively subordinate, had no genuine understanding of what the Bureau was trying to do.

The situation began to change in the late 1930s, following the formation of the Bank of Canada and the later creation of the Rowell–Sirois Commission, which carried out the most far-reaching examination ever conducted of the Canadian economy and its fiscal policy. With the outbreak of World War II in 1939, the management of the war economy became almost an industry in its own right. The Bureau made a significant contribution to the war effort, both through the assumption of ad hoc tasks for which its clerical staff and facilities were well-suited and through the extension and elaboration of its statistical programs. Its staff increased from approximately 600 in 1939 to 1200 in 1945. But it could not satisfy many demands made by the new users. In particular, they found the Bureau's work on national income statistics inadequate. This led to a completely new start, which eventually gave rise to the present system of national accounts.

In 1942, R. B. Bryce of the Department of Finance advised his Deputy Minister, W. C. Clark, that: "the time has come, or is rapidly approaching, when the whole system should be reviewed and overhauled to fit it for the far different world of the 1940s—both war and aftermath." (Bryce, 1942). The turnaround began under Cudmore; his tenure as Dominion Statistician (1942–1945) brought the first steps to transform Coats' legacy into a statistical system capable of meeting more rigorous needs. The application of scientific methodology was an indispensable part of the transformation.

32.2 THE ERA OF THE DEVELOPMENT AND APPLICATION OF SCIENTIFIC METHODOLOGY: 1945 TO DATE

The term *scientific methodology* as used in the rest of this chapter means the application of probability sampling and related statistical disciplines to survey design and implementation, from the specification of survey objectives and quality requirements of the end product through questionnaire design, sample design, data collection and processing, editing and imputation to quality assessment, analysis, and the provision of access to users. It had its beginnings in the Bureau in the development in 1945 of the Canadian Labour Force Survey, a sample survey of households to provide current information on the characteristics of the population and labor force for the departments that were planning and monitoring postwar reconstruction. (Keyfitz and Robinson 1949.)

Probability sampling in the collection of official statistics had only come into use in North America when the Works Progress Administration of the United States instituted a national monthly sample survey of households in 1940.[7] This had required the development of a new methodology to deal with a universe of unprecedented size and uneven geographic distribution. The Canadian Labour Force Survey adopted a similar questionnaire, and, Nathan Keyfitz and his colleagues drew heavily upon the advice of their U.S. counterparts, William Hurwitz and Morris Hansen. This was the beginning of a process of United States–Canadian scientific cooperation that continues today.

In subsequent years, the gradual application of scientific methodology in social statistics proceeded uneventfully. Notable examples were census quality checks and postcensal surveys. In the field of economic statistics, however, comparable progress came more slowly. The field was extremely compartmentalized, and integration of its various parts was complicated by lack of common elements such as frame information and classification systems. Traditional subject-matter managers, after decades of virtual autonomy, were skeptical about the value of new approaches. A number of factors, operating over more than three decades, gradually brought about change. One was the requirement of the national accounts for integration of economic statistics at the micro-level, rather than in terms of aggregates. After working with individual surveys, it also came to be realized that ad hoc approaches would have to be replaced by regular surveys and Bureau-wide standards.

32.2.1 1945 to 1965

During the first postwar decade, some agricultural information was collected by quota surveys which simulated random sampling by requiring so-called crop correspondents to furnish information from their own and neighboring farms according to specified criteria. Otherwise, the only notable application of sampling in business surveys was in a pilot survey of the road transport industry

---

[7]The work was taken over by the U.S. Bureau of the Census in 1942.

in the province of Manitoba in 1954. This subsequently gave rise to a 10 percent national survey, conducted quarterly, which provided the first statistics of motor transport traffic. In 1948, the Statistics Act was amended to authorize the collection of statistics by sampling, thereby legalizing a proven mathematical technique that increased the Bureau's flexibility, and made survey-taking more economical and reliable.

Simon Goldberg, who had earlier been responsible for the National Accounts, was appointed Assistant Dominion Statistician for Integration in 1954. This was an important psychological step in promoting awareness among the subject-matter divisions of the requirements of the system of national accounts (SNA) for integrated statistics. He did valuable work in breaking down divisional barriers and smoothed the way for later inputs of scientific methodology.

Initially, however, the ability to provide such inputs in the area of business statistics was as weak as the demand for them. The methodological functions outlined above were slow to develop and had few experienced practitioners, not only in the intrinsic skills required but also in the ability to work effectively with potential clients. A nucleus of methodology staff was first developed in the division responsible for conducting sample surveys. Eventually, it was recognized that progress in methodology required separate organizational arrangements. In the early 1960s a Sampling and Survey Research Staff was set up under Ivan Fellegi. The staff reported to Simon Goldberg, who thus had another instrument of integration at his disposal.

While the new staff had a clear and accepted mandate for household surveys, this was not the case for business surveys, which were rapidly increasing in number, mostly in isolation from one another. Computerization was also in its infancy and perhaps added to the problem. The concept of project management as a coordinated multidisciplinary approach to the organization and conduct of statistical operations was an idea whose time had not yet arrived. Even so, there were some notable accomplishments during the 1955–1965 period. Methodologists became involved in such projects as the design of a post-census sample survey of areal units to provide estimates of coverage and data quality of the 1956 Census of Agriculture, the redesign of the Retail Trade Monthly Survey, initial work on the development of a central list of businesses, and quality control methods for data capture operations which were introduced in the Retail Trade Merchandising Survey.

32.2.2 1965 to 1975

During this period, the now fully articulated system of national accounts began to have a major impact. It provided a conceptual framework for business surveys, but demanded more accurate and timely data and identified gaps and discrepancies in the statistics at the micro level. Again, the use of administrative data, including those from taxation sources, was made possible by changes in the 1971 Statistics Act, which renamed the Bureau as Statistics Canada (Legislation, The Statistics Act, 1971). This provided compelling incentives for the reduction of survey-taking costs and respondent burden—a trend that has continued to the present time.

All these considerations pointed to the desirability of focusing on the common elements of survey processing as distinct from subject-matter content and thus posed a formidable challenge to Statistics Canada's methodological capability. The growth of the latter was reinforced by major organizational changes, notably the establishment in 1972 of a new Statistical Services Field, as part of an agency-wide reorganization put into effect by the recently appointed Chief Statistician of Canada, Sylvia Ostry. Under Ivan Fellegi as Assistant Chief Statistician, the new Field covered Systems and Data Processing, Regional Operations, and a group of methodology divisions, two of which were concerned with business surveys. One benefit of the centralization of methodology and systems was the emergence of a central research capability which made it possible to replace ad hoc approaches to particular surveys with generalized goals and agency-wide standards.

One important application of methodology in business-based surveys during 1965 to 1975 related to the monthly Employment and Payrolls Survey. For businesses with 20 or more employees, attention focused on the development of an automatic edit and imputation module with a diagnostic output of the edit failure and imputation. This was designed to direct the editing staff to records with a significant impact on the estimate measured in terms of response rates, edit failure rates, and the amount by which the value of the estimate was changed to pass the edit. This was the first example of fully automated and selective editing at Statistics Canada. But the system had limitations and caused operational headaches because of its inflexibility.

The concurrent sample survey of businesses with fewer than 20 employees was revised to permit the collection of the data as part of the Job Vacancy Survey. This was a major undertaking which, with the financial support of the Department of Manpower and Immigration, attempted to parallel the Labour Force Survey as a source of information on the demand side of the labor market. There were thus high expectations for its successful outcome. From the beginning, however, awkward definitional problems had to be faced, such as the notion of a job vacancy and a job vacancy reporting unit.[8]

The survey, conducted twice monthly, was primarily a mail survey that reinterviewed a subsample of business units. As Platek (1987) described it:

> The survey is basically a two phase design in which a first phase population sample is further subsampled prior to the interview phase. The population is stratified by industry, location, and size of reporting units prior to the selection of the mail sample. The mail sample is then stratified by location and within location by response class (respondents with vacancies, respondents without vacancies, nonrespondents) prior to the interview phase. The design is thus a complex one and a

---

[8]Statistics Canada's counterparts in the U.S. Bureau of Labor Statistics were always skeptical about the feasibility of such a survey and never attempted one themselves.

challenging task has been to produce systems for maintaining such a complex design that are operationally feasible.

Data from the survey were published from June 1970, but their practical usefulness was questionable particularly because unemployment rates steadily increased during the 1970s. The survey was discontinued later in the decade. From a purely methodological standpoint, however, the experience furnished useful, if somewhat negative, lessons.

The 1971 Census of Merchandising provided a similar lesson, because it was planned on the basis of an imperfectly understood and improperly controlled frame. Compiled during the June census enumeration of households and farms, the frame caused difficulties and delays in analyzing and tabulating data and led to the cancellation of the Census of Merchandising planned for 1976. However, this experience gave considerable impetus to work on a computerized central register of business units.

Another instance of misplaced optimism was the automation of the Industrial Selling Price Index. Progress turned out to be slower than hoped because of confusion and conflict between the relative roles of the methodologists and the computer systems analysts. Subject-matter clients eventually took charge, with the methodological and computer systems staffs serving in consultative roles. Although it took some time, it was eventually possible to report that the responsible division, Prices, had "revised the industrial selling price indexes with enlarged coverage, computerized processing, new weights and a 1971 time base." (Statistics Canada, Annual Report 1974–1975, p. 5).

However, the 1965–1975 period was not without some success stories in the area of business surveys. One of the earliest was the Survey of Current Shipments, Inventories and Orders, the design of which incorporated an automated imputation procedure for nonresponse. Later, planning for a system for the Census of Manufactures was begun, and in the 1973–1974 Annual Report, it was noted that: "Among the benefits of those phases of the Division's automation program already completed is the availability of all publishable data in machine-readable form. The next, and most complex, phase is editing, imputation and multipurpose retrieval which will be implemented progressively." (Statistics Canada 1975).

In the area of merchandising and services statistics, experience with current surveys was more successful than with the 1971 Census of Merchandising, although it is clear that many challenging problems related to frames. For example, the new design of the Retail Sales Survey, an early example of a multiple-frame survey (see Chapter 11), was described as follows (Dominion Bureau of Statistics, 1970):

> . . . a sample drawn from a master list that has been created and is to be updated on a continuing basis from lists presently available within the Bureau (and) a supplementary set of locations listed in a sample of areas that will estimate the relevant characteristics of locations either missing from the master list or recently come into existence.

In the area of transportation statistics, the longstanding Motor Transport Traffic Survey, which had been one of the earliest to use sampling, was replaced uneventfully by a For-Hire Trucking Origin and Destination Survey. The survey was kept under continuing review and analysis. From 1973 to 1974, modifications were made to both the stratification variables and the survey's scope, and from 1974 to 1975, more efficient sampling procedures permitted a 40 percent reduction in sample size and a substantial decrease in field costs. Taxation data were used to replace direct data collection for the Census of Construction, and the new system lasted for 15 years.

32.2.3 1975 to Date

If this section were to have a subtitle, it would surely be "The darkest hour is that before the dawn." In 1978, the Statistical Services Field was formally dismantled by the then Chief Statistician, Peter Kirkham, with the various specialized methodological divisions being allocated to the corresponding subject-matter fields. The rationale for the decentralization of methodological services was that direct integration into the management structures of their former clients would increase responsiveness. This argument perhaps had some public relations appeal because Statistics Canada was suffering from the worst press in its history, and any decisive step on its part was bound to make a good impression. The move did not receive unanimous support. The methodologists argued that it was a blow against professional collegiality, that there could no longer be an overview of the function, that central research and training would suffer, and so on.

In retrospect, neither the hopes nor the fears were fully justified. Mutual understanding and responsiveness were a function of more profound considerations than of organizational arrangements, and the slow trend of improvement hardly faltered. Again, professional collegiality was sufficiently robust not to be devastated by decentralization. This was also a time when Statistics Canada began to suffer significant resource cuts after a decade or more of continuous growth to a strength of more than 4000. At the time, some argued that Statistics Canada was now able to cure itself of a severe case of indigestion: Too many resources had come too easily and had not always been allocated wisely. The Job Vacancy Survey was perhaps a case in point. These cuts initiated a trend of continually striving to do more or better with the same or less. Thus, it may have been necessity that eventually made bedfellows of the methodologists and subject-matter managers.

In 1980, after a short interregnum under Lawrence Fry, Deputy Minister of Health and Welfare, Kirkham was succeeded by Martin Wilk. Wilk was the first Chief Statistician to be formally qualified in mathematical statistics, although his skills had hitherto been exercised in scientific research rather than official statistics. He restored the centralization of statistical methodology, supporting and actively fostering the process of reconciliation by encouraging subject-matter staffs to concentrate on analysis of products and client needs rather than on monitoring data collection and processing. To this end, he integrated the collection and processing operations of many surveys for efficiency reasons. This process was also facilitated by the gradual emergence of a new breed of subject-matter professionals who had user-friendly computing facilities at their disposal and to whom technology was no longer a threat.

When Wilk retired in 1985, Ivan Fellegi, who for some years had served as Assistant Chief Statistician, Social Statistics, was appointed to succeed him. This was another kind of first—the first Chief Statistician qualified in mathematical statistics to have come up through the ranks.

A significant indication of Statistics Canada's more recent flexibility and eclecticism was the major methodological and operational undertaking, the Business Survey Redesign Project (BSRP), managed by the Assistant Chief Statistician for Business and Trade Statistics, Jacob Ryten, during the latter half of the 1980s. This merits attention as an integrated approach to what were superficially different problems in various subject-matter areas, but which had in fact numerous common elements.

From the mid-1970s, there were growing concerns about the quality of primary economic statistics. As noted earlier, computerization of the national accounts data required the confrontation of survey outputs from various sources at the microdata level, and the process revealed substantial conceptual and operational inconsistencies. This was a reflection of the longstanding piecemeal approach to survey development. A central business register was developed during the 1970s, but was not being fully used by the surveys as a common starting point for defining the target units. At the same time, there were growing concerns about response burden. The number and complexity of requests for data had grown steadily, and it was becoming clear that many questions being asked could not be readily answered from conventional accounting records. In any case, there was a widespread perception in the business community that the paper burden was becoming excessive.

These concerns about weaknesses in Statistics Canada's infrastructure led to the Business Survey Redesign Project. Its first phase, the Infrastructure Project, turned out to be a massive undertaking, ultimately spanning the 6 years from 1985 to 1991, during which it was the focal point for the vast majority of all development work in economic statistics. The strategic objectives were (1) to provide facilities for the integration of economic statistics at the microlevel, (2) to ensure comprehensive and nonduplicated coverage of all economic activities, and (3) to foster the replacement of business units reporting through direct surveys by comparable information gathered from administrative sources.

These goals required the redesign and introduction of a new central register for business surveys,[9] a new income tax data acquisition program covering both payroll deduction data and income tax returns, and the adaptation of existing business surveys so that they could make full use of the centralized

---

[9]See Chapter 2 for an overview of current work on frames and business registers.

services. As a test case for the latter goal, it was decided to redesign the program of wholesale and retail trade surveys.

There now exists a business register with many novel features, which is realizing benefits not achieved by its predecessor of the 1970s. Income tax data are being used systematically and efficiently. Frame information for large businesses can be recorded in the form provided by the businesses as well as in terms of the statistical units defined by Statistics Canada. The register has the capacity to provide an image of its current or past contents and to track businesses through time. It also has automated scheduling of clerical tasks. Using the new business register as a frame, monthly wholesale and retail trade surveys have been redesigned. The Survey of Employment, Payrolls and Hours also uses the register as a frame and is being redeveloped to take advantage of additional data items now available from administrative sources (Anderson and Vincent 1993, Dolson 1993). Not all objectives were fully achieved. Some remain elusive. But the project's most enduring and significant achievement was to break down the barriers between separate survey operations and to emphasize the importance of integration.

A sequel to the project has been the initiation of the Large Enterprise Statistics Project, which includes among its objectives the derivation and recording of legal, operating, and statistical reporting units for very large businesses and the consequent modification of data collection procedures. It also aims at facilitating the integration of economic data through the collection of data concerning intercorporate financial transfers within large businesses, as well as through the construction of a multipurpose output statistical database.

The future of business surveys is not likely to be a straight-line continuum from present practices. The environment within which business operates may change quite radically and, with it, information needs. In Chapter 35, for instance, Ryten suggests that the balance of priority between small, current inquiries and large, detailed, but slow surveys and censuses may need to be redrawn. The gifted amateurs of the past have been replaced by dedicated professionals with a variety of complementary skills who work in a constructive mode of interdisciplinary cooperation to analyze user needs, develop appropriate collection and processing procedures, provide relevant analysis, and facilitate user access. It augurs well for the future that there now exists a capability to anticipate issues and plan for their resolution.

REFERENCES

Anderson, B., and R. Vincent (1993), "Strategies for the Redesign of a Major Business Survey," *Proceedings of the International Conference on Establishment Surveys*, Alexandria, VA: American Statistical Association, pp. 787–792.

Bryce, R. B. (1942), "Memorandum for Dr. Clark, March 7," Public Archives of Canada, Record Group 19, **445,** File 111-IR.

Coats, R. H. (1904), "Memorandum to W. L. M. King," *Collection and Publication of Statistics Policy, 1904-09*, File 4, **48,** Department of Labour Records, AG27.

Coats, R. H. (1910), *Wholesale Prices in Canada, 1890-1909*, Ottawa: Government Printing Bureau.

Coats, R. H. (1916), "A National System of Statistics for Canada—Centralization, Reorganization and Enlargement of Canadian Statistics," unpublished paper, Ottawa: Census and Statistics Office.

Coats, R. H. (1946), "Beginnings in Canadian Statistics," *Canadian Historical Review*, **XXVII,** pp. 109-130.

Department of Labour, Canada (1910), *Labour Gazette*, Issues of February and July.

Dolson, D. (1993), "On Redesigning Canada's Establishment Based Employment Survey," *Proceedings of the International Conference on Establishment Surveys*, Alexandria, VA: American Statistical Association, pp. 793-798.

Dominion Bureau of Statistics (1928), *Census of Trading Establishments*, Ottawa: King's Printer.

Dominion Bureau of Statistics (1970), *Annual Report of the Dominion Bureau of Statistics, 1968-69*, Ottawa.

Dominion Department of Agriculture (1885), *General Report of the Census of Canada, 1880-81*, **IV,** Ottawa.

Dominion Statistician (1923), *Annual Report for the Year Ended March 31, 1922*, Ottawa: King's Printer.

Keyfitz, N., and H. L. Robinson (1949), "The Canadian Sample for Labour Force and Other Population Data," *Population Studies, II, 1948-49*, Cambridge: University Press.

Legislation:
- An act respecting the Board of Registration, and the Census and Statistical Information, Consolidated Statutes of Canada, 1847, 22 Victoria, Cap. 33.
- Census and Statistics Act, 1879, 42 Victoria, Cap. 21.
- Census and Statistics Act, 1905, 4-5 Edward VII, Cap. 5.
- The Statistics Act, 1918, 8-9 George V, Cap. 43.
- The Statistics Act, 1971, 19-20 Elizabeth II, Cap. 15.

Lovell, J. (1853), "First Report of the Secretary of the Board of Registration and Statistics on the Census of Canada for 1851-52," Quebec.

MacGregor, D. H. (1939), "External Forces Governing the Development of the Bureau," Public Archives of Canada, Record Group 31, **1418,** File: Miscellaneous, Part 1.

Marshall, H. (1953), "The Role of the Dominion Bureau of Statistics in the Post-War World," Presidential Address to the joint meeting of the Canadian Historical Association and the Canadian Political Science Association.

Minister of Agriculture, Canada (1891), *Annual Report for the Year 1890*, Ottawa.

Minister of Agriculture, Canada (1901), *Annual Report for the Year 1901*, Ottawa.

North, S. N. D. (1918), "Seventy Five Years of Progress in Statistics: The Outlook for the Future," in J. Koren (ed.), *Memorial Volume, 75th Anniversary of the American Statistical Association*, New York: Macmillan.

Platek, R. (1987), *The History of Survey Methodology in Statistics Canada up to 1985*, unpublished manuscript, Ottawa: Statistics Canada.

Report of the Departmental Commission on the Official Statistics of Canada (1913), Ottawa: King's Printer.

Statistics Canada (1975), *Annual Report of Statistics Canada, 1973–74*, Ottawa.

Statistics Canada (1976), *Annual Report of Statistics Canada, 1974–75*, Ottawa.

Taché, J. C. (1866), "Memorial to the Board of Registration and Statistics," Report of the Minister of Agriculture for the Province of Canada for the Year 1865, Ottawa.

The Dominion Bureau of Statistics: Its Origin, Constitution and Organization (1935), Ottawa: King's Printer.

Urquhart, M. C. (1987), "Three Builders of Canada's Statistical System," *Canadian Historical Review*, **LXVIII**, pp. 414–430.

Urquhart, M. C., and Buckley (1965), *Historical Statistics of Canada*, Cambridge University Press, Toronto, The MacMillan Company of Canada, Ltd.

Walker, F. A. (1888), "The Eleventh Census of the United States," *Quarterly Journal of Economics*, **2**, pp. 138–156.

Worton, D. A. (1993), *History of Official Statistics in Canada, 1851–1971*, unpublished manuscript, Ottawa: Statistics Canada.

CHAPTER THIRTY-THREE

The Role of National Accounts and Their Impact on Business Surveys

Rodney J. Lewington[1]
Statistics New Zealand

In a fully articulated statistical system, the national accounts are but a part of an information set that records the nation's economic performance. This performance is initially recorded in accounting and other business records. National accounts summarize the major transactions conducted in the economy. Like business accounts, they show the values of goods and services produced and the costs involved in production. But national accounts go further, recording how the product is used for consumption and exports and what portion is used for investment and to increase inventories.

Basic economics texts provide a description of national accounts. Reference to these and to the relevant national statistical agency's publications are a necessity for those involved in conducting business surveys. The United Nations System of National Accounts (UNSNA) is the definitive text on national accounts. The 1968 version is currently followed by most statistical agencies, but will be replaced in the mid-1990s by the 1993 revision.

The main end-user of national accounts and business survey information is government, which uses it to manage the economy. Economic commentators, academic theorists, and international agencies need similar information. To be worthwhile, the information has to be summarized to give it coherence and to relate it to economic policy. The national accounts provide the conceptual frame

[1]The author acknowledges input from speakers and other participants at the International Conference on Establishment Surveys in developing his chapter. Special thanks are due to Jeff Cope and David Archer of Statistics New Zealand for contributing useful comments. The author is responsible for the final version.

Business Survey Methods, Edited by Cox, Binder, Chinnappa, Christianson, Colledge, Kott.
ISBN 0-471-59852-6 © 1995 John Wiley & Sons, Inc.

for such a model. Essentially Keynesian in philosophy, the national accounts also have a place for transactions relevant to other views of economic activity.

In developed statistical systems, the business survey is the main conduit by which the initial record of activity is obtained from business records and fed into the national accounts. In some countries, such surveys are based on the local production unit or establishment. In others, the surveys obtain information primarily about the legal unit or enterprise (see Chapter 3). The national accounts add to this information system, providing definitions of the economy's boundary and concepts for defining and classifying statistical units and their transactions.

The national accounts require that business surveys conform to its definitions and provide quality information that is relatable over time and between surveys. In the ideal information system, the national accountant and business survey staff cooperate to provide linkage between the micro- and macro-statistics. That is, they negotiate between the business' records of production and financial activity on the one hand and the national accounts' production, income and outlay, financial accounts, and balance sheets on the other hand.

This chapter describes how national accounts provide guidance on what information should be collected in business surveys and on the strategy that should be adopted in collecting business and financial statistics. Business surveys are only one source of information on the production and distribution of income. Government activities generate administrative records that can be used to supplement or replace information obtained from business surveys. The role of national accounts in ensuring consistency and accuracy across these data sources is also discussed in this chapter. Finally, the importance of having an information system that provides full and unduplicated coverage and that consistently defines transactors and transactions is emphasized. Well-designed business surveys that collect information from the relevant level within businesses are an integral part of this information system.

33.1 NATIONAL ACCOUNTS: THEIR PURPOSE

National accounts describe a nation's economic activity so that the government, its advisors, and its critics can manage and monitor the economy. When fully used, national accounts provide data for modeling past and future economic activity, for examining options, and for providing feedback to government and commentators. The accounts provide international comparisons, and, for smaller debtor nations, have a very real impact on the international credit rating of the country and its major companies. Thus, national accounts are an important tool in describing and managing a nation's economy and provide internationally accepted, quantitative measures of the economy's performance at the aggregate level. This performance may be based on production but also includes the distribution and redistribution of income, investment, consumption, savings, wealth, and overseas borrowing.

Measures such as gross domestic product and national income are the main aggregates provided by the national accounts. These accounts identify the relative and absolute contribution to the economy by industries and institutions and provide a geographic analysis of economic activity. Inventories and balance sheet items such as capital investment and overseas debt are also important national accounts measures.

The UNSNA provides a conceptual basis for defining the statistical units that make up the economy and the transactions between them. Together with international standard classifications, the national accounts provide definitions and classifications for industries and institutional sectors.

Compiling national accounts requires information from business surveys and other statistical and administrative sources. These sources include tax records, migration counts, merchandise trade from export and import records, banking transactions, and household surveys. The national accounts are a prime force behind the integration of economic statistics and have a major influence on social and demographic statistics as well.

The national accountant's information needs range over the whole economy and frequently require details not readily available from business accounting records. The classification of businesses demanded by the national accountants sometimes appears unreasonable to survey statisticians. In all of this, it must be recognized that national accounts are only a means to an end; it is the uses of national accounts that dictates their information requirements. For statistically advanced countries, these demands are accepted as legitimate and indeed form the core of their statistical agencies' strategies for economic statistics.

33.2 THE BUSINESS SURVEY AS INTERMEDIARY BETWEEN THE COMMERCIAL WORLD AND NATIONAL ACCOUNTS

National and commercial accountants each record the activity of the real world as they see it, aiming for the form most useful to their audience. Commercial accounts meet the needs of the business' managers, owners, the banks financing their operations, and taxation authorities. National accounts meet the needs of those who formulate government economic policy and those who provide commentary on this policy. Surveys must approach businesses in a way that allows respondents to complete the questionnaire in spite of all their variations in recording data. At the same time, the business survey must provide information that is consistent across surveys and can be readily formulated into national accounts.

The information need for national accounts covers a wide range of transactions. They include the value of production, the kinds of goods and services produced, and uses of the product. The information used by national accounts covers the costs of production, the income generated in this process, and the distribution and use of income. Fully articulated, the national accounts also record transactions in, and holdings of, financial assets. To collect this variety of information requires approaching several levels in the hierarchy of business organizations. For transactions such as interest, dividends, savings, and capital investment, the enterprise with its balance sheet and profit and loss accounts is the relevant entity to interview. For production-related transactions, the establishment with its records of profit and loss account and (possibly) manufacturing account, wage books, and records of quantities is best able to provide information. For subnational regions, information may only be available from records kept at locations.

National accountants are typically required to use all information sources to minimize government's data collection cost. Consequently, a statistical agency's processing of administrative data is also influenced by national accounting requirements. Administrative sources generally relate to the enterprise or enterprise group and include information on income, profits, tax, and wages paid. Somehow, all these sources have to be integrated into a single, coherent information set.

There are other users of business surveys besides the national accountant. Many users are not as specific in their needs as the national accountant but will often find greater utility in business statistics that are integrated with and follow national accounting precepts. Thus, the survey statistician will often benefit from the use of national accounting guidelines in designing business surveys.

33.2.1 Transactors

The UNSNA defines the extent of the national economy, distinguishing between those transactors to be included in it and those considered not a part of the economy. It goes further by providing guidelines on what production to measure and what activities to exclude. It views the economy as made up of several kinds of economic actors. These actors may be producers or consumers; they may be financial intermediaries or providers of government services. The definition of roles defines the units for national accounts and influences the scope of business surveys and their definition and classification of statistical units.

Defining the National Economy

The national accounts attempt to measure the business conducted within the country. To do this, they define the geographic boundary and the population of economic actors to be included in business surveys. In national accounting terms, the latter are the *resident economic agents*. With insignificant exceptions, both are consistent with the boundaries prescribed by the International Monetary Fund for balance of payments statistics.

The starting point for national accounts is the definition of the physical boundary of the economy in geographic terms. This boundary is normally the political frontier of the country plus any economic zone in the surrounding sea that the country controls for fishing or mineral exploitation purposes. The boundary may or may not include territories, protectorates, and/or other geographic areas outside the home country. This variant depends on the information users. National accounts include special customs or export zones within a territory, although statistical surveys may need to identify these separately.

Refinements to the boundary definition center on the principle that activities included in the national accounts should be under the potential control of the national government. Activities taking place within the geographic boundary but outside the control of the government are excluded. Thus, embassies, consulates, and military bases of foreign countries have special treatments. In addition, enterprises that have establishments (factories, farms, commercial offices, etc.) that are active within the country's geographic boundary must be included in the country's national accounts and hence in its business surveys. This includes establishments owned in whole or in part by foreigners. Similarly, foreign-based establishments of home-based enterprises are excluded from the country's national accounts. It follows, then, that business surveys carried out by a national statistical agency should aim to provide complete and unduplicated coverage of the activities of resident enterprises and establishments that operate within the geographic boundary of the nation.

Defining the Production Boundary

Practical difficulties prevent business surveys from covering all activities defined by the national accounts. These difficulties include problems in identifying enterprises and establishments, the lack of adequate accounting records for some businesses, and the cost of collecting data from smaller businesses.

Data on unmarketed transactions are particularly difficult to collect. In developed countries, a clear line separates the market economy from the subsistence economy, the latter being minimal or restricted to agriculture. Less-developed economies have more difficulty in collecting data on their subsistence economies and may have to accept less-than-optimal coverage for their business surveys.

Even in developed countries, small family businesses and businesses with low turnover are a problem. To control costs, business surveys often have a size cutoff below which they do not survey. As a result, the national accountant must estimate the contribution that unsurveyed businesses make to gross domestic product. In industries where economies of scale are important and larger units dominate, this is not a problem. It can, however, cause significant undercoverage in industries such as construction and retailing where small units are common.

In agriculture, there can be significant subsistence activity as well as many family-operated farms. The national accountant has to accept the difficulties faced by business surveys in obtaining information for these cases. Data on quantity may be hard to obtain, and the value of self-produced outputs consumed by farm households is not readily determined by reference to market prices. The quantum and the value of labor of the farm household is also difficult to measure, and surrogate values can only be estimated.

It is useful to remember that the business survey is a vehicle for collecting information from one set of records and presenting this in a different format. If there are no records maintained by businesses, then a business survey may not be the appropriate vehicle for obtaining information. The national statistical agency and the national accountant will then have to look elsewhere. In the case of subsistence industries, the statistics may be obtained from household income and expenditure surveys, consumption surveys, and time-use surveys. Where units and transactions are excluded from business surveys, the scope and boundary need to be clearly defined. This makes it apparent to users what economic activity has been excluded.

Defining the Statistical Unit

In Chapter 3, Nijhowne discusses the statistical units to be incorporated in business registers. The proposed units are consistent with the UNSNA concepts, although the conceptual view of the business world may be too simplistic to be readily operationalized. In general, national accounts conceive of two types of economic players. In UNSNA, there is distinction between those that produce and those that own and consume:

- *Producers* take resources and create a new product or service. In doing so, they add value and contribute to the gross domestic product.
- *Owners* sell resources to producers. These resources consist of land, labor, and capital. These same owners may also own financial assets or incur financial liabilities. They use their income to purchase goods and services for final consumption or to save for investment.

These distinctions are not between actors but between roles. An individual person, company, or government agency often plays more than one role. This duality of roles also leads to a requirement that statistical units be defined and classified in different ways depending on the transactions being recorded.

Larger commercial organizations are based on units such as cost and profit centers, accounting groups, and locations that are owned by the business. The accounting system arranges these components in a hierarchy that allows the owners (or their representatives) to monitor and influence the smallest branch. In the small business or family farm, there may be only a single unit.

The national accountant's needs for information at each level in this hierarchy are very similar to those of the business manager and owners. It is no coincidence, therefore, that national accounting information is obtained by business surveys that identify statistical units that are the same or very similar to commercial units. Where commercial practice and national accounts concepts differ, the survey statistician has to make judgments in defining the statistical unit and the information to be obtained. Where there are differing practices among commercial accounts, the statistician's decision-making is simplified by referring to national accounts' guidelines.

The Classification of the Transaction Unit

Making sense of the information collected from the many units that are included in a business survey requires that information be aggregated. Conventionally, this aggregation is by geographic region, industry, and the owner's institutional sector. These three classifications provide the basic structural information that planners use in monitoring the economy at national and regional levels. The technology is now available to present survey information for different groupings. The only restrictions are the requirement to maintain confidentiality (see Chapter 24) and the need for an adequate number of businesses in groupings so as to meet sampling error requirements (see Chapter 8).

Geographic classifications, by necessity, are particular to each country. Industry and sector classifications are prescribed in internationally accepted standard classifications (see Chapters 2 and 3). All countries have obligations to report economic activity to international agencies who require that international standard classifications be applied. The two international classifications of interest are the International Standard Industrial Classification of all Economic Activities (ISIC) Revision 3 (U.N. Statistical Office 1990) and the classification of Institutional Sectors in the UNSNA.

On the other hand, the needs of planners and policy makers within the country are reflected in the demands they make on the national accountant. Thus, the national requirements for the business sector classifications may differ from the international requirements. The degree to which the national accounts dictate classifications for business surveys depends on the extent to which the government intervenes in economic activity and the way in which it wishes to monitor the result. A free enterprise philosophy requires no specific industry analysis and thus provides little guidance to national accountants or survey statisticians in defining classifications.

In a centrally planned economy, the groupings of establishments in the presentation of production statistics must follow the planning mechanism. This need was apparent in New Zealand in the 1970s when the government adopted an indicative planning policy and set up joint government and private councils for some 15 industrial groups. Production and financial statistics had to be classified by industry groups that fitted each council's purview. In addition, an institutional sector classification was used to distinguish those enterprises that were under the direction of government from those that were privately owned and could only be advised and encouraged in making investment and production decisions.

33.2.2 Transactions

In many statistically advanced countries, the strategy for collecting economic statistics assumes that a business' commercial accounts at the establishment-level contain all needed production information such as the costs of materials, fuel, services, labor, indirect taxes, and the depreciation of equipment and buildings. Alternatively, the establishment may be defined as the lowest-level unit able to provide this information. On the income side, the accounts list sales and (possibly) subsidies and support payments. Information on inventories is also assumed to be available.

Profit is the residual on the commercial establishment account. This may be trading profit, gross profit, or something similar, depending on the industry and the business' size and complexity. Purchases and sales of capital (investment) items are usually available at the establishment level. The national accounts' production account covers the same general flows of resources and the value of output as an establishment's operating account. In the national accounts, the residual is *operating surplus*. The accounting records for a business at the enterprise level go further and include the cost of financing and the appropriation of surpluses to dividends, interest, direct taxes, and so on.

The definition of transactions in the national accounts are generally unsuitable for direct inclusion in a questionnaire to be completed from commercial accounting records. What can be done, however, is to ask a series of questions that together can be used to calculate the national accounting transactions. For example, the national accounts' "compensation of employees" is aggregated from questions on salaries and wages paid in a period, supplemented by information on directors' fees, shareholders' and proprietors' drawings, employers' contributions to superannuation funds and health insurance, and the value of housing and vehicles provided to employees.

A mapping of national accounting transactions against commercial accounting entries as shown in Table 33.1 can be adapted to an individual country's commercial accounting practice. What the national accountant asks is that questions in business surveys not group business account items together that go into different national account transactions.

The transactions mentioned are all relevant to the production process and would generally be available from records maintained at the establishment level. Some transactions are also relevant at the enterprise level. These include the operating surplus and capital formation. In addition, there are items required by national accounts on the distribution of the operating surplus and business' borrowing and lending. Such transactions include direct taxes, interest receipts and payments, donations, dividends, the incurrence of financial liabilities, and the acquisition of financial assets. For these transactions, the national accounts' definition is very close to that of commercial accounts.

The information mentioned above is used to compile the current price national accounts. Additional information is required to compile the price deflators required to calculate constant price estimates. This includes detailed analysis of goods and services produced and purchased. Inter-industry input–output studies also require detailed commodity analysis within the national accounts framework. Productivity studies require information on labor input, including the number employed and hours worked. All need to be related to the relevant national accounting transactions.

In addition, three national accounting transactions need special mention. These are the costs of banking services embedded in interest flows, the service element of insurance premiums, and the valuation of stock change. Estimating these transactions requires more detailed information than is readily available from commercial accounts. However, such information is often available from other records maintained by firms and can be collected from the business sur-

Table 33.1   Mapping of National Accounts Transactions to Commercial Accounting Transactions

National Accounting Transactions	Commercial Accounting Transactions
Intermediate Consumption	Purchase of materials, fuels, and nonlabor services less increase in inventories of these items. Finance charges but not interest.
Compensation of Employees	Salaries and wages. Directors' fees and honoraria. Employer contributions to superannuation funds and insurance schemes. Vehicles, dwellings, food, etc., provided to employees.
Consumption of Fixed Assets	Depreciation. Value of investment goods written off as obsolete.
Indirect Taxes	Value-added taxes, excise duties, and sales tax. Local authority and state rates and other taxes on land. (Excludes direct taxes such as income tax.)
Subsidies	Grants by government to producers on current account. The losses of government enterprises also need to be considered.
Gross Output	Sales of goods and services plus increase in stocks of work in progress and finished goods. Excludes the sale of goods previously forming part of the establishment's capital.
Operating Surplus	A residual in the national accounts. It is approximately equivalent to profit before direct tax is deducted and without the inclusion of interest paid or received.
Increases in Stock	The value of the change in inventories. The national accountant also needs to know the basis of valuation and to distinguish the change in stocks of purchases from those of work in progress and finished goods. Growing timber and livestock is treated differently in the national accounts of different countries. The treatment will determine what information is required from business surveys.
Gross Fixed Capital Formation	The value of purchases less the value of sales of plant and machinery, vehicles, buildings, improvement to land, roads, railways, etc. Includes also the value of items produced within the establishment for its own use.
Land	Land and its embodied improvements and subsoil deposits when purchased or sold with the land.

vey. Alternatively, if statistical information from different sources is well-integrated, then the national accountant can draw on these sources.

33.2.3 The Valuation of Transactions

The UNSNA has conventions on the valuation to be applied in recording transactions. These conventions follow the principle that transactions should be valued at *market price*, defined as what a willing, arm's length buyer would pay to a willing seller. This is the norm in commercial accounting records, but problems do arise. Where transactions occur between establishments under common ownership, the market value may not be recorded in commercial accounting records; the value may be at cost or may be understated. Alternatively, taxation and other regulations may make it beneficial for a multinational business to overstate values.

   The valuation of the output of government services presents particular problems. As a deliberate policy, some government enterprises price their output below cost. Urban transport, health services, and education services commonly fall into this category. Even where there is no deliberate policy, the accounting systems of government need to be examined to ensure that costs are consistent with those recorded in the market economy. Inconsistencies include cash rather than accrual accounting and different treatment of capital purchases and depreciation. Often costs are not charged to the activity that uses the inputs. For example, government agencies are not commonly charged for office space, printing services, and payroll administration.

   When such practices occur, it may be impossible for a business survey to collect market value. What can be done, however, is to identify the deviant practices and collect information that allows revaluation to market value or, in the case of government, an estimate of the full cost of activities in ways consistent with national accounting valuation principles.

33.3 LINKING INFORMATION FROM DIFFERENT SOURCES: THE PLACE OF NATIONAL ACCOUNTS

In the ideal and fully integrated statistical system, every statistical collection identifies transactors according to the same set of rules and defines transactions in the same way, and every statistical unit is classified by industry and institutional sector according to standard classifications applied in the same way. Furthermore, the system's business surveys provide full and unduplicated coverage of market and nonmarket activities of national, state, and local governments.

   In such a system, the consistency and complete coverage of the annual business surveys is complemented by surveys using the same principles to provide short-term statistical indicator series. For example, the monthly retail trade survey is selected from the same business register and asks questions that are consistent with those asked by the annual retail survey. Similar approaches are taken for surveys of employment, manufacturing, construction, and agriculture. In New Zealand, the principle is extended to the population census where place of work is classified using the business register's industrial classification. The national accounts concepts provide a guide for designing subannual surveys and the supporting statistical infrastructure of registers and classifications.

Where government records form part of the statistical information system, administrative records should be brought to the same level of consistency. This can be done, for example, by identifying the individual legal entities in the administrative tax record and providing the industry, institutional sector, and location classifications from the business register. Unfortunately, for the national accountant in many countries this is not possible because of legal constraints or privacy considerations.

Such an ideal statistical system makes statistical information from a variety of surveys readily relatable. For example, labor input used in productivity measures matches the volume of product, and lending by financial institutions is comparable with investment in, and production by, industry. As most survey statisticians know, this ideal is seldom met. Some countries have standard classifications, but they are applied by different survey statisticians to units identified uniquely for each survey, often with only partial knowledge about the business unit's activity. As a result, units differ between surveys and classifications are applied differently.

There are tools that can minimize these consistency problems. A major tool is a *business register*, a centralized list of business entities that is used by every business survey. Some countries have these; others, including the United States, have not yet developed this statistical infrastructure. In Chapter 5, Archer discusses the concepts and practices necessary to build and use a business register. A centralized list of businesses provides a common classification system to aid in data integration. It ensures that each survey's coverage is known and, more importantly, provides quantitative indications of uncovered activities.

Other advantages accrue when the business surveys for every industry use the same transactions, the same definitions, and even the same questions. This type of standardization needs a conceptual framework, and the national accounts are convenient and internationally accepted.

33.4 THE AUDITING FUNCTION OF THE NATIONAL ACCOUNTS

The national accounts provide principles that guide the design and content of business surveys. In recent years, many countries have demonstrated that the national accounts also have an important role in monitoring the accuracy and coverage of statistical series derived from business surveys, in deciding what information needs should be given priority, and in identifying gaps in the information needed to monitor an evolving economy's activities.

The auditing function comes about because the national accounts bring together information in a way that checks one set of statistics against another. Differences between measures of the value of a nation's production and its expenditure on that production show up in comparisons of the national accounts measures of gross domestic product and expenditure on gross domestic product. In compiling inter-industry input–output studies, every country spends considerable effort reconciling the measure of the source of each commodity with its apparent disposal.

In trying to bring consistency to such flows, the national accountant examines source statistics in great detail and identifies potential and actual reasons for discrepancies. Often these differences are because of inconsistent definitions or differing industrial or commodity classifications. Other reasons for discrepancies include differences in valuation and in timing of the observations. These discrepancies can occur because of the business survey's design or simply because information about a business reaches one survey before another. Sometimes the errors are caused by tardiness in updating business registers, inconsistent sampling and imputation methods, faults in design or implementation, and errors in aggregation.

Irrespective of how such errors occur and whether in fact they are errors, the national accountant can point to inconsistencies at the macro-level and hence advise on shortcomings of survey designs and operations. By using this advice, statistical agencies can improve the quality of survey output.

Because the national accounts cover all activities, the national accountant is also in a position to identify gaps in the statistical coverage of the economy. These gaps can result from industries or institutional sectors not being covered by statistical surveys. This may take the form of whole industry or sector classes that are not surveyed. It may also take the form of omissions from surveys of entities such as small businesses, rural businesses and households, institutional dwellings such as barracks and hospitals, and off-shore activity such as mineral exploration and fishing.

Where administrative sources are used for statistical purposes, their coverage may not be consistent with other surveys, leading to gaps or duplication in coverage. Collection of statistics by several agencies increases the chance of such errors. The national accountant will often be able to identify these gaps and overlaps; it is then the responsibility of the survey statistician to remedy or at least quantify them.

Statistical agencies can also improve their effectiveness by examining the end use of their outputs and adjusting resources to best meet identified needs. The national accounts are one of the main series used by government to manage the economy. A weakness in the source data may have a disproportionate effect on policy implementation.

Studies of measurement errors and their effects are relatively new (see Chapter 15). However, these studies provide a formalized method of establishing what is important for national accounts and what this should mean in terms of accuracy of source data such as that derived from business surveys. Thus, the national accounts provide feedback on data quality to those concerned with the policy, design, and operation of business surveys. This in turn can increase the robustness, relevance, and efficiency of business and financial statistics.

33.5 PLACE AND PERIODICITY: BUSINESS SURVEYS IN THE STATISTICAL INFORMATION SYSTEM

The users of national accounts and their supporting business statistics operate in a changing world. They require information on the structure of the economy, past performances, and the current situation. No single survey can provide all this. In many countries, statistical practice has developed that provides detailed information every few years on the economy's structure and tracks activity levels annually, quarterly, and monthly in various parts of the economy. The coverage of industry and the institutional sector and the periodicity of business surveys are dependent on the demands of government policy formulation and monitoring.

A common pattern in many countries is to have an inter-industry input–output study every 5 years drawing on full coverage of the economy by business surveys. This may be supplemented by inter-industry studies at more frequent intervals based on less statistical data. The inter-industry study provides information on the flow of goods and services between industries. It forces a reconciliation between sales and purchases and between income and expenditure and thus provides a definitive, balanced statement on which other macroeconomic series can be based.

Next, countries have annual national accounts firmly based on business surveys that collect details about the economy's important industries. Each country identifies those industries that are key for development and growth or critical for sustaining existing activity.

Finally, there are quarterly and monthly estimates of the main national account aggregates based on summary information collected from businesses. This provides government and their advisors with information for monitoring policy results and for fine-tuning the instruments that guide or control economic activity. The activities covered in every country will normally be indicators of domestic consumption, capital formation, and export demand. Such information may come from business surveys or from other sources. In addition, each country will monitor the critical sectors using quarterly or monthly surveys to provide broad aggregates such as manufacturing turnover, production of key agriculture commodities, or receipts from tourism.

The extent to which a national statistical agency can meet this prescription is dependent on the demands of government and the resources allocated. It is also very much dependent on the way in which the official statistical system is organized. Where a single agency is responsible for the collection of all business surveys, there is no excuse for surveys that are not integrated in terms of coverage and content. In countries where individual agencies are responsible for the statistical coverage of specific industries, there must be coordination and subordination of agency needs to reach national consistency.

33.6 IS IT APPROPRIATE TO BASE STATISTICAL STRATEGY ON NATIONAL ACCOUNTS?

In Chapter 35, Ryten challenges the current statistical strategy and asks if the information now being produced by Canada, the United States, the United Kingdom, Australia, New Zealand, and others is meeting current and potential needs. The concern addressed in this section is the extent to which the national accounts and business surveys force data into a form that is either unavailable or not useful. While national accounts are a coherent system providing information for economists, its concepts have to be balanced by a recognition of the way in which the information is used.

The basic information for national accounts comes from commercial accounts via business surveys and administrative records. National accounting information is then used to make judgments on economic policy, which are then implemented. These policy prescriptions need to relate to real-world decision units and transactions. Whether the government uses the "stick" of direction, controls, and taxation or the "carrot" of subsidies and support payments, these policies have to be written into laws or bureaucratic guidelines. These laws or guidelines, in turn, need to be related to the commercial records of the affected businesses. This suggests that national accountants and survey statisticians should not move too far from the realities of commercial accounts in the way they collect and present information.

The problem is exacerbated when commercial accounts do not record all transactions at a particular establishment or location. For many large, multi-location businesses, the local unit does not keep records of all production costs. In the service, finance, insurance, and information industries, the location is almost immaterial. Even for the manufacturing industry, administration and marketing may contribute a high proportion of the value added off-site. These and other reasons often leave the balancing item from the firm's record of plant performance as contribution to gross profit or contribution to sales.

Thus, smaller business units are not always able to provide the full information needed to complete the national accounts prescription. One has to question the wisdom of having the survey statistician estimate measures that are missing. Creativity to remedy lack of information in the basic records and also to impute information for nonrespondents brings problems to the final data user. The need to minimize costs leads to an increasing tendency to minimize sample size, and survey methodologists are developing more refined methods for imputation and editing. There is a danger that these artificially completed data sets will give users a false impression of data quality.

At a minimum, the survey statistician has to declare the methods used for missing data treatment and flag those items that have been imputed (see Chapter 22 and David and Czajka 1993). An argument can also be made for putting resources into collecting data rather than into post-collection adjustment (see Chapters 21 and 34).

Industries engaged in finance, business, and personal services are able to switch resources from one product to another. For example, insurance providers can become brokers overnight, and bankers can and have moved out of merchant banking into financial advice. Even in manufacturing, there is an increased ability to change outputs. This suggests that there are some activities where a fine industrial disaggregation is irrelevant.

The ability to quickly change the output of an establishment, together with the lack of full cost information, provides a strong argument for adopting a statistical unit more akin to the enterprise than the local activity unit for production statistics (see Chapter 35). The French statistical system tends in this direction. The European Communities (1993) definitions of statistical units has moved toward accepting the idea that unit definitions have to consider information availability. The kind of activity unit (KAU) definition requires the availability of ". . . at least the value of production, intermediate consumption, manpower costs, the operating surplus and employment and gross fixed capital formation." The UNSNA definition also recognizes this need.

Developments in the use of microdata sets for economic modeling add pressure to ensure that the end product of the economic information system can be related back to commercial records. In social policy formulation, micro-simulation models using individual household records representing all households in the economy has become commonplace. Modeling the economy by fitting numerical data to a theoretical concept of how the economy works has increased over the second half of this century. It is technically feasible to extend economic modeling to the micro-level. Postner (1986, p. 217) concluded that putting macroeconomic accounts on a microdata foundation ". . . is at least partly capable of practical implementation." His paper provides a historical perspective to the development of the idea of linking microdata (business) and macrodata and discusses the consistency issues.

Consequently, the survey statistician should now look ahead to the time when economic modeling will require *unit data* information. Those are data sets that provide information about each individual business and its economic activities. McGuckin (1993, p. 275) argues that "A broad range of [economic] issues cannot be addressed without microdata on establishments and the firms that own them." Such uses impose a further need for integrating business statistics and national accounts. These models, one would anticipate, will stay within the Keynesian framework and therefore use the national accounts concepts and numbers. In this scenario, unit data incorporated in the model needs to account for all transactions that contribute to the national accounts. That is, the sum of transactions of the individual units needs to add to the industry and sector totals within the national accounts. This can only occur if the national accounts transactors and transactions remain closely related to business records.

It should be apparent that the choice of the appropriate statistical unit will be increasingly important. This unit must reflect real-world management and organization and not be imposed on the enterprise. The cost will be that value added by industry, and location will become increasingly vague. This will reflect the real world's lack of homogeneity of economic activity at the establishment level.

This is not to say that national accounts are an unsuitable framework. Business and financial statistics still require a conceptual framework, but the national accounts framework may need to be adjusted to recognize the data available in business records and the technical capabilities offered by information technology. For its part, national accounts may have to sacrifice some conceptual purity to provide better links between the (micro) source data and the macroeconomic statistics.

33.7 CONCLUDING REMARKS

Surveys of enterprises and their constituent establishments should be part of a complete information system about an economy. Business surveys are an important component covering the structure, the levels of activity, and the interactions between businesses within the economy. They are concerned with production and the assets used in production. They also cover the financial transactions and the distribution of income. Other statistics cover the details of consumption and relations with the rest of the world.

The national accounts are a part of this information system. They provide the framework and concepts that integrate the whole. By observing national accounts concepts and prescriptions, business surveys make their greatest contribution to statistical information. Where practical constraints make it difficult to meet national accounting concepts, the survey statistician contributes best to the information system by knowing, revealing, and, where possible, quantifying the deviation from national accounts and standard classifications.

The economic statistics now being collected may not meet all needs. Increased influence of multinationals, environmental degradation, resource usage, and social issues are not fully recognized by standard national accounts. National accounts are not meeting all perceived needs, but there is a continuing demand from the needs they are meeting. The solution is not to drop national accounts but to extend the information system of which they form a part.

With or without such changes to meet changing user demand, there is a need for information. This need can only be met efficiently by national statistical agencies having comprehensive strategies for economic statistics that use the national accounts to determine the nature and content of business surveys.

REFERENCES

David, M. H., and J. L. Czajka (1993), "Metadata for Establishment Surveys: Corporation SOI," *Proceedings of the International Conference on Establishment Surveys*, Alexandria, VA: American Statistical Association, pp. 594–599.

McGuckin, R. H. (1993), "The Importance of Establishment Data in Economic Research," *Proceedings of the International Conference on Establishment Surveys*, Alexandria, VA: American Statistical Association, pp. 275–282.

European Communities (1993), *Official Journal of the European Communities*, **36**, L76, 4 ff.

Postner, H. H. (1986), "Micro Business Accounting and Macroeconomic Accounting: The Limits to Consistency," *The Review of Income and Wealth*, Series 32, No. 3, pp. 217ff.

U.N. Statistical Office (1990), *International Standard Industrial Classification of all Economic Activities—Revision 3 (ISIC Rev 3)*, Series M, No. 4, Rev. 3, New York: United Nations.

U.N. System of National Accounts (1968), *Studies in Methods*, Series F, No. 2, Rev. 3, New York: United Nations.

U.N. System of National Accounts (1993), *Studies in Methods*, ST/ESA/STAT/SerF/2/Rev 4, New York: United Nations.

CHAPTER THIRTY-FOUR

Quality Assurance for Business Surveys

Greg Griffiths and Susan Linacre[1]
Australian Bureau of Statistics

Users' increasing capacity to incorporate statistical information in their decision-making and research processes challenges statistical agencies to improve
the quality and breadth of their information base. This increasing demand is
supported by the rapidly changing technological environment and the increasing skills of the workforce. As a consequence, survey data are applied to a
multiplicity of purposes, increasingly exposing them to tests of validity and
usability. Users are demanding an integrated view that brings together families
of statistics. Because data from a variety of sources are increasingly combined,
the design of individual data collections must facilitate integration. In Chapter
33, Lewington describes the effect that one such integrated view, the system
of national accounts, should have on the strategy statistical agencies use in
collecting economic statistics.

While demand for an integrated view is increasing, data collection from
businesses is also increasing in complexity as management structures change
within and across businesses more frequently. Technology is changing the way
businesses set up their information systems, adding further complexity to data
collection, as well as providing opportunities for innovative data collection and
processing. Within this demanding environment, statistical agencies remain
responsible for continuing cost-effective conduct of data-gathering activities.

This chapter examines the processes involved in conducting a business survey and discusses quality issues relevant to each step, using examples drawn

---

[1]The views expressed in this chapter are the authors' and do not necessarily reflect those of the
Australian Bureau of Statistics. Where quoted or used, they should be attributed to the authors.

Business Survey Methods, Edited by Cox, Binder, Chinnappa, Christianson, Colledge, Kott.
ISBN 0-471-59852-6 © 1995 John Wiley & Sons, Inc.

primarily from the Australian Bureau of Statistics (ABS). Sample design and estimation are not covered in this chapter, which deals with the nonsampling error aspects of quality assurance. While the chapter looks at the individual survey components, it is recognized that quality output is the result of viewing these components as part of a larger process and balancing resources across them to achieve the best overall outcome.

34.1 ESTABLISHING USER REQUIREMENTS

A business survey provides users with quality outputs when its outputs are timely, easy to interpret, relevant to issues of concern, of known and acceptable reliability, and a good value for money spent. To achieve quality outputs, it is important to understand how the outputs are used in decision-making or research. Ideally, the survey planner should maintain a strong cooperative relationship with users, discussing design issues from a client orientation using client-based language. Both parties can then work together through survey testing and development, balancing timeliness and cost against accuracy and comprehensiveness of topic coverage.

Most surveys, however, have many users so that maintaining this cooperative relationship can be infeasible. Even if this were feasible, a few major survey users are likely to account for a significant portion of the total value added through use of survey output by all users. A strong relationship with these major users invariably benefits most other users.

Once users' needs are known, consideration must also be given to the form the output should take. Besides being interested in descriptive statistics, users often want the answers to questions such as: Is the economy growing? Is consumption rising? Is production falling? and so on. Many business surveys provide economic indicators that attempt to address these questions. In these cases, the object of interest is a derived variable, smoothed to remove the effects of seasonality and irregularities such as transient economic effects and volatility due to sampling and nonsampling errors. Thus, understanding users' needs allows surveys to be designed not simply in terms of descriptive statistics but in terms of its final, usable outputs.

Often, users are interested not in a single survey but in integrated statistics drawn from several surveys. In this case, survey quality requires compatibility of the survey's output with the output from other surveys. It may be that the main statistic of interest is derived from a number of survey sources. Then the quality objectives of the design for output from any one source must consider the role it plays in deriving the integrated statistic.

In addition to involving users in the design, planners should develop an ongoing relationship with key users to obtain feedback on the actual value to them of survey output. Where surveys are user-funded, continued funding is one measure of how well user needs are being met. Close monitoring of how survey outputs impact on decision-making and research provides another measure. In this context, it is useful to monitor the adjustments analysts make to survey outputs prior to use and monitor their judgments of the data's fitness for use. Conducting market research of the general user community to determine their view of the relevance, timeliness, and production of usable outputs can also provide useful feedback to the survey agency, as well as quantitative measures over time of how well user needs are being met.

34.2 CONCEPTS, ITEM DEFINITIONS, AND QUESTIONNAIRE DESIGN

User needs from a survey will generally be expressed in terms of concepts for which some quantitative measure is required. These concepts must be translated into appropriate data items and classifications. The use of common data items and classifications is a major requirement for ensuring relatability of statistical information across data collections. At the same time, it is imperative that the requested data and the classifications used are unambiguously understood by respondents and that the relevant information is readily available. ABS recently amended its classification of business units into a hierarchy of enterprise group, management unit, establishment, and location to ensure that the information sought from these units is based on how activity and financial data are maintained within the businesses.

To allow users to identify gaps between their concepts and those measured by individual surveys, classifications and item definitions need to be documented, and that documentation must be made available. ABS is developing a meta-data strategy to support the use of standard concepts and to provide a repository for survey documentation of interest to data users. Besides the definition of concepts and classifications, ABS aims to store information on the questions used to obtain the information from respondents as well as other relevant information such as data derivation and transformations. This strategy has largely been prompted by the work of Rosen and Sundgren (1991), who provide insight into the need for a common statistical language across statistical collections.

Fundamental to survey quality is the ability to embody concepts and data item definitions effectively in usable survey instruments. Traditionally, the assumption has been made that business survey data are provided by people with a good understanding of economic concepts, that they derive the data directly from business records, and hence that provision of the data is relatively straightforward. Unfortunately, this is not always the case.

The economic concepts traditionally used in business surveys are not easily understood by respondents. Vital to developing appropriate procedures for designing and testing business questionnaires is the recognition that the quality of statistical estimates is largely influenced by the respondent's ability to understand questions and to have the data available. Questionnaire design must emphasize:

- creating a clear and uncluttered form that provides appropriate and easily understood explanations;
- evaluating the questionnaire in observational studies and pilot tests to ensure that it is workable and that requested data are available, and
- analyzing feedback to improve the questionnaire for subsequent surveys.

In addition, a questionnaire history file for each continuing survey is useful to record all questionnaire changes and their reasons, whether the change is due to user requirements or adaptation to respondent characteristics. Data from studies supporting these changes should also be kept. ABS's use of this approach has resulted in substantial improvements in its business survey questionnaires.

For information not normally kept, the business must be given adequate warning of the need to set up suitable systems to record information. In this case, the survey agency can conduct seminars for businesses that explain the survey's purpose and the concepts behind the collection. Surveyors may even undertake field visits to complex businesses to identify relevant contact points and assist with setting up appropriate recording systems (see Chapter 6).

An ABS example illustrates the process of developing a workable survey instrument. For the Economic Activity Survey (EAS), an analysis of error was undertaken for a pilot data collection of 255 businesses using a questionnaire containing 58 questions. Errors were detected in 3 percent of the answers, with nearly half of the questionnaires free of error. Figure 34.1 shows the number of errors for the 12 most error-prone questions. Four questions (28a, 2a, 2b, 10) accounted for one-third of all errors detected. Question 28a—a data check question of little relevance to most respondents—was often left blank. It was dropped from subsequent surveys.

Questions 2a and 2b asked for the number of employees and the number of working proprietors and partners, respectively. Edits of total employment against total wages and salaries indicated that working proprietors and partners were occasionally omitted or were reported as employees. In the full survey, respondents were asked for the total employment figure as well as its components. A post-enumeration survey later uncovered no subsequent problems with these items.

Question 10 asked for other income. The high error rate for this question in the pilot test reflected respondents' problems in interpreting a previous question on sales of goods and services. In particular, respondents in financial industries had difficulty with the sales of services concept. For the full survey, slight changes were made to the inclusions/exclusions list to decrease the level of error. However, post-enumeration surveys still indicated that respondents had difficulty with the sales of goods and services question and were attributing more income than appropriate to other income. As a result, subsequent surveys have adopted the strategy of using questionnaires customized for each industry and delineating the "sales of goods and services" concept more precisely for the banking, insurance, and community service industries.

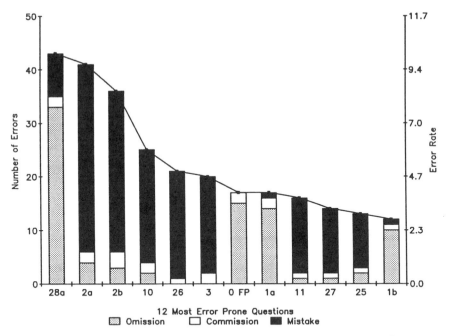

Figure 34.1 Analysis of errors: 1988–1989 economic activity survey. (From Australian Bureau of Statistics.)

34.3 FRAME CREATION, MAINTENANCE, AND MONITORING

The frame for a business survey is usually drawn from a business register, a continuously maintained list of businesses. The register provides the basis for an integrated view of the economy across surveys. The extent to which the register is accessible, comprehensive, and up-to-date determines a survey's ability to mirror real activity in the target population.

Unlike units in household surveys, business survey frame units vary greatly in their contribution to estimates. The largest units contribute a substantial portion of the total estimate. Their structure can be extremely complex; their operations may span many industries and locations. Takeovers, mergers, and internal reorganizations can have significant impact on survey reporting units and data availability. Considerable error can be introduced into estimates through omission or double counting of any of these large organizations, or by failure to identify appropriate contacts for different types of statistical data. Given their potential for error in survey estimates, particular emphasis should be placed on getting correct reporting arrangements for these largest units. One way to achieve this is to set aside specific staff with the assigned responsibility for ongoing liaison with the largest enterprise groups. These assigned individuals ensure that the structure as recorded on the business register remains up-to-date, that reporting arrangements are appropriate, and that these key organizations know who to contact for queries and feedback (see Chapter 6).

For smaller businesses, the problems are different and revolve around identifying "births" and "deaths" in this volatile population. Frequently, several coverage sources are used to improve coverage of small units, introducing the possibility of substantial duplication. Other frame imperfections of concern include (1) biases introduced by the undercoverage of in-scope units, and (2) inefficiencies induced by the inclusion of out-of-scope units, and (3) the use of poor quality auxiliary information. Wright and Tsao (1983) provide an annotated bibliography of issues relating to the conduct of surveys from imperfect frames. The papers presented at the annual Round Table on Business Frames present an excellent overview of international activity to improve register quality.

Even with a single good source that covers small units, time lags between the birth of a business and its capture on the register (most commonly through tax, employment, or other administrative records) can lead to undercoverage. These effects can be compounded by the choice of formal definitions used to decide the birth and death of a business (see Chapter 4). For example, should the sale of a business to new owners be interpreted as the death of one business and the birth of a new one? Is the sale merely a continuing business undergoing a change in the value assigned to its ownership variable? Deciding the formal definition for what constitutes a new business must agree with how register sources record these changes.

These formal definitions have natural ramifications for birth rates and death rates within the register's population. A survey using the register as a frame and mirroring the register's definitions of new and deceased businesses will identify defunct units as they are surveyed. Any corresponding birth will take time to be fed through the register; the resulting undercoverage will be reflected in estimates.

For continuing surveys, strategies can be used to minimize the effects of undercoverage. For example, when administrative records are used for frame updates, new deaths detected in the sample might be assumed to offset as yet unreported births. If the assumed model is correct, the treatment of deaths as "nonrespondents" rather than as zero reports could lead to more accurate estimates. Alternatively, a new unit located at the address of a selected defunct unit can be included in a survey as a proxy until the new unit enters the register naturally. Care needs to be taken with the implementation of approaches such as these. The use of different approaches across surveys can lead to disparate movements in conceptually similar estimates.

For business surveys monitoring indicators of economic performance, irregular frame updating can cause frame-induced changes in the economic indicators and mask real changes in economic performance. They can affect the survey's ability to provide clear signals of turning points. Given the important role of the register in enabling business surveys to reflect real changes in economic activity, effective monitoring tools of register performance are needed.

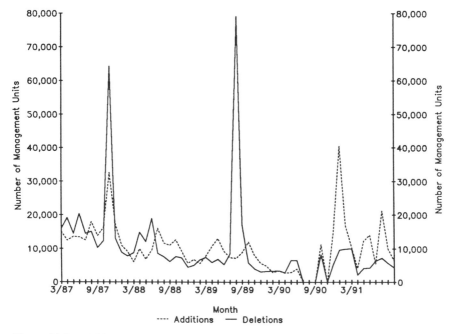

Figure 34.2 Additions and deletions: ABS business register. (From Australian Bureau of Statistics.)

Figure 34.2 shows birth and death patterns in the ABS business register from March 1987 to September 1991. Note that the flow of units on and off the register was not stable. Underlying economic processes as they affect business life cycles are filtered through the register's administrative and maintenance processes. Clearly, the potential exists for movement estimates from continuing surveys to be driven more by frame maintenance practices than by underlying economic processes. Disentangling these effects provides an as yet unresolved challenge for survey organizations.

Figure 34.3 plots key percentile points in the distribution of time for ABS register units to move from each of the register's data sources through its maintenance system, resulting in either a death, birth, or amendment. This period has been called the *gestation period*. Note that the gestation period may be long-tailed and differ considerably by source. If run regularly, this graph can identify problems with particular updating sources.

Figure 34.4 shows the distribution of time to move through the register's maintenance system as a function of the unit's complexity. Complexity is measured here in terms of the number of associated units potentially requiring amendment as a result of changes to the initial unit. In this particular case, following the decision of the ABS to allocate resources to maintain very large units, the category of most complex units (those with over 100 other register

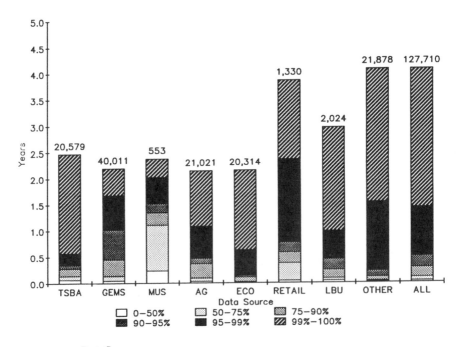

Data Sources:

TSBA	Tax Stamp Book Applicants
GEMS	Group Employer Management System
MUS	Multi-Establishment Enterprise Unit Survey
AG	Agricultural Census
ECO	Economic Censuses and Surveys
RETAIL	Retail Census
LBU	Large Business Unit Maintenance

Figure 34.3 Gestation period: births from March 1992 to September 1992. (From Australian Bureau of Statistics.)

records in potential need of consequential amendment) is processed relatively quickly.

Figure 34.5 compares the estimates of employment from a register-based, economy-wide survey of employers (the Survey of Employee Earnings) with those from an area-based household survey (the Monthly Labor Force Survey). It charts the change in the ratio of employment estimates from the two surveys. The increasing ratio over recent years is indicative of increasing undercoverage of employment on the register. Investigations have not yet indicated the exact cause of this phenomenon.

Duplicates can be a problem for registers based on overlapping sources. To measure register duplication, a sample of register units can be matched against the full register with follow-up evaluation of near matches. A simpler duplicate check involves sampling localities and then searching the register for duplicate

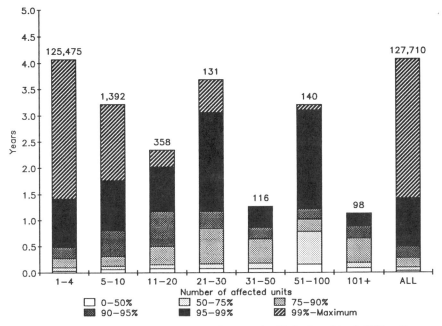

Figure 34.4 Gestation period by number of affected units: births from March 1992 to September 1992. (From Australian Bureau of Statistics.)

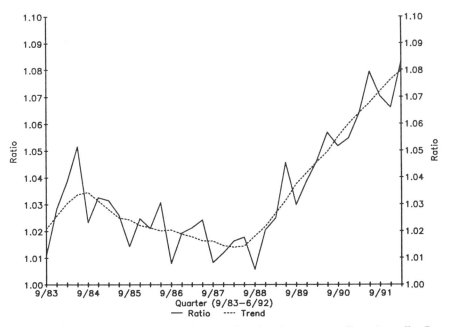

Figure 34.5 Employment estimates ratio: labor force/employer survey. (From Australian Bureau of Statistics.)

locations within each locality. The within-locality duplication rate for business locations is consistently below 1 percent for the ABS register. No difference has been found in duplication rates between industries or between the central business district and other localities. ABS uses only one substantial updating source, employers registered through the Australian Taxation Office, and hence a low duplication rate is to be expected.

The quality of the register's auxiliary information can be evaluated through surveys that collect this information. For example, the correlation between the register's employment benchmarks and those collected in employment surveys can be monitored over time as an indicator of benchmark data quality. The register's industry information can similarly be evaluated using surveys where activity information is collected.

34.4 DATA COLLECTION AND PROCESSING

Collection control for a survey involves classifying all selected units as either (1) live, responding, and in scope; (2) live, nonresponding, and in scope; (3) live and out of scope; or (4) defunct. Nonresponding units can be difficult to classify. For volatile populations, this is an important issue. Different approaches for follow-up and classification of nonresponding units as live or defunct can lead to discrepancies between estimates from different surveys.

Data collection aims to achieve as high a response rate as possible in as short a period as possible. This can be facilitated by targeting the initial collection and follow-up strategies to respondent needs or characteristics. As Werking and Clayton note in Chapter 18, data collection methodologies such as computer-assisted telephone interviewing (CATI) make it possible to pre-schedule requests for information. Then, survey requests can be made at a time convenient for the respondent, and soon after the required information becomes available. CATI can assist substantially in targeting response follow-up. Using control system counts of stratum response levels or an "approximate percentage of the final stratum estimate reported to date" (based on data reported in previous periods) can also identify priority areas for follow-up.

Targeting response follow-up can itself lead to quality problems, however. Resource constraints often lead to large nonrespondents being targeted ahead of small nonrespondents, where the measure of size is based on the previous period's data. Often, imputation for nonresponse is also based on the previous period's data. If this selective approach to follow-up and imputation is ignored, bias can result. This may be significant when the level of imputation is relatively high, as with preliminary estimates. For example, in ABS's Survey of Capital Expenditure, nonresponse follow-up for preliminary estimates is targeted at units with high levels of expected capital expenditure as reported in the previous period. Imputation is based on the stratum ratio of current actual

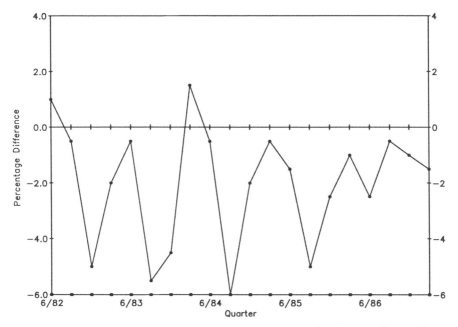

Figure 34.6 Percent difference: final minus preliminary estimates of capital expenditures. (From Linacre 1987.)

to previous expected expenditure, which is then applied to the previous expected expenditure value for the unit. Figure 34.6 shows the percentage difference between the preliminary and final estimates of capital expenditures over several years. Note the inherent underestimation in preliminary estimates introduced by the follow-up targeting approach. In this case, the final estimates of capital expenditure contain a contribution from units that in the preceding period reported expectations of low levels of capital expenditure but whose actual expenditure markedly exceeded their previous expectations.

Problems associated with targeting nonresponse follow-up to large units within strata can be overcome by predefining a random preliminary sample for intensive follow-up prior to collection. Another compromise is to follow up all large units, but only allow a predefined random sample of units to contribute to imputation for nonrespondents. Nonetheless, whatever method is used for continuing surveys, it is valuable to routinely compare preliminary estimates based on lower response levels with later estimates to monitor the extent of nonresponse bias.

An effective monitoring system facilitates building quality during data processing. Such a system monitors the progress of returns through the processing cycle, the resources used during each processing phase, and the effect of changes made during processing.

34.5 EDITING

Evaluation studies of editing in business surveys, undertaken in a number of agencies (e.g., Latouche and Berthelot 1992, McDavitt 1992), have resulted in similar conclusions:

- Quality should be built into collections through the questionnaire development process, not edited into the collection during processing.
- Continuing surveys should be analyzed on an ongoing basis to provide feedback from editors to questionnaire designers and data collection managers.
- Editing effort should be concentrated on those data problems that significantly affect survey output; that is, it should not be dispersed through all data problems, large and small.

Analyzing error rates by question and size and industry of the respondent can improve questionnaire design and data collection strategies for subsequent cycles. Survey designs should include an edit evaluation system, with evaluation information captured as part of the survey's management information system during the collection process.

When data must be consistent across surveys, across-survey editing or data confrontation at the macro-level or micro-level can provide important feedback about the conceptual and operational frameworks of the different data collections. For example, through reconciling production and financial data provided by very large businesses, ABS discovered inconsistencies at the unit level. When possible, very large businesses should assume responsibility for assuring data consistency across the different collections. Coordination of data collection and special support to large businesses to encourage this can be cost effective.

Targeted editing has the potential to increase quality, because trained resources can be directed to areas of greatest effect (see Chapter 22). Such editing can reduce costs and respondent load and improve timeliness. For censuses of businesses, efficiency demands that resources be targeted at the returns of larger, more complex businesses, while considering the impact that smaller units have on small area or fine commodity data. For sample surveys, weighting should also be taken into account during the targeting process.

ABS has tested a significance editing approach for its quarterly collection of labor earnings statistics (McDavitt 1992). Statistics Canada investigated a similar approach (Latouche and Berthelot 1992). The targeting technique uses historical information to assess the likely impact on survey estimates from editing of the unit's current value. In McDavitt's study quarterly movement in the average weekly earnings estimate (AWE) is the estimate of interest. Only those units reporting values with the highest likely impact on movement are followed up during editing; all other units reporting questionable values are imputed. An early ABS study of this approach indicated that resources spent on editing could be nearly halved, while maintaining quality and substantially reducing respondent burden. Figure 34.7 shows that after ranking potential edits by their impact on movement, the marginal benefit in the estimate falls off rapidly after 10 percent of the previously edited returns are edited; no gain at all is made after 40 percent of the previously edited forms are edited.

Where historical information does not exist at the unit level, an analytic approach that sets edit bounds based on logical relationships between collected variables can also target effort to reduce costs and increase quality. Where a clear relationship exists between two variables—for example, wages and salaries versus number of employees—past data can be used to model the relationship and set acceptance regions in line with resource availability and the required accuracy of output. For models with lower levels of fit, the relationship provides less discriminatory power and the acceptance region is larger. Management information, collected during editing for units falling on either side of the acceptance region, allows revision of edits during processing if necessary.

For surveys with many output cells (e.g., business censuses producing small area estimates by industry or commodity level), an analytic approach can target cells with different relationship patterns from the majority of cells. This approach could be in terms of concurrent relationship between two variables or relationship between current and past estimates. Graphical analysis can

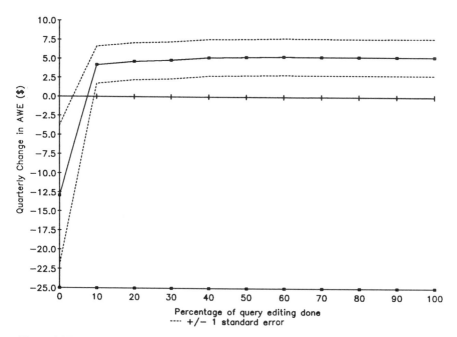

Figure 34.7 Change in average weekly earnings (AWE) estimates: June to September 1991. (From McDavitt 1992.)

identify outlying estimates; it can also provide insights that aid the user in interpreting estimates.

Many statistical agencies are developing generalized systems for survey processing to ensure that development and support effort focuses on a few multi-use products, with training and operating costs kept low and "best practice" approaches embedded in the systems. As each collection moves from its current system to a generalized system, the full processing strategy should be reviewed to ensure that the added capabilities of the generalized system are used for quality improvements as well as for efficiency gains.

34.6 OUTLIER DETECTION AND TREATMENT

Editing concerns itself with the accuracy of a survey's reported data. A related issue is the identification of outliers in a collection. As Lee describes in Chapter 26, *outliers* are units whose reported values are considered accurate, but whose weighted responses distort the sample estimate. Such units are often handled by reducing the weight of the outlier along with increasing the weight of the remaining units in the stratum. The alternative is to leave the weight fixed, but reduce the unit's value from the reported level.

Either of the above approaches reduces the standard error of the estimate but increases the potential for bias. If sufficient information is available to estimate the population distribution, strategies can be developed to minimize the mean squared error. Lee discusses many such strategies in Chapter 26. However, it is the extreme end of the distribution that must be estimated based on past data, and this will always be difficult to do. There must be some means of monitoring the use of outlier approaches in a business collection. While the bias at the stratum level may be small relative to the variance, it will invariably be unidirectional and possibly substantial at the aggregate level. Management information should include information by stratum on the number of outliers treated in the survey and, where possible, should also include an estimate of the resultant bias.

34.7 ANALYSIS AND DISSEMINATION

Users are often interested in adjusted statistics rather than raw survey estimates. Such adjustments include constant price estimation, seasonal adjustment, and trend estimation. These adjustment processes are themselves subject to error. Where series have high innate irregularity or where seasonality is changing and requires estimation based on a small number of observations, the errors in seasonal adjustments can be quite significant. These errors will generally result in revisions to the estimates as subsequent years' data become available. As well as taking steps to minimize error in adjusted figures, users often benefit when the collection agency provides a measure of the likely revision level for use in sensitivity analyses. The accuracy of estimation of seasonal factors and the extent of revision depends on the variability of the irregular component in the series. Hence, the standard deviation of the irregular component can be used to provide analysts with the likely revision level to seasonally adjusted statistics.

The seasonally adjusted series includes the irregular/residual component of the series as well as the trend or signal which interests many users. Moreover, the residual component is frequently the major contributor to period-to-period movement in economic series (Zarb 1992). In the ABS monthly series on retail trade, irregular factors accounted for over half of the seasonally adjusted gross movement in 55 percent of monthly movements during the last 10 years. Therefore, users who assume that period-to-period movements in seasonally adjusted series represent underlying movements in the economy will frequently be led astray. Smoothed, seasonally adjusted series can reduce the problem.

However, the current end of such smooth series may suffer from substantial revision. As for the seasonally adjusted series itself, the extent of revision of the smoothed, seasonally adjusted series depends on the standard deviation of the residuals. This relationship can be used to provide a measure of the likely revision to the smoothed series based on the history of each series. For a number of business surveys, Table 34.1 gives the estimated percent revision to seasonally adjusted data after 1 year and after 5 years and also gives the estimated percent revision to smoothed, seasonally adjusted or trend data after the one, two, and three periods following (generally months or quarters). Note that the smoothed series is generally robust to the revisions in seasonally adjusted series.

Effective use of raw or analyzed data from a business survey requires that decision-makers and researchers understand the concepts behind the data and its quality aspects. This understanding can be helped through such mechanisms as regularly scheduled user seminars on complex statistics (for example, with the release of each set of national accounts data) and through providing training to users on the nature and implications of seasonal adjustment and trend analysis processes used to transform the series.

Survey outputs should also describe the conceptual basis of the statistics, the methodology used, and data quality issues such as coverage, sampling error, response rates, response biases, imputation rates, nonresponse bias, benchmarking and revision effects, and the comparability of statistics over time and with other data sources. Basic information should be provided in each relevant publication with details in separate explanatory publications.

To encourage the effective use of its outputs, the ABS commissioned communications consultants to research the uses made of its publications, including their readability. This research included observation studies of decision-makers and analysts using publications to answer questions, and it highlighted their difficulties and misinterpretations. The study results are being used in designing more effective publications.

Table 34.1 Expected Revisions to Seasonally Adjusted and Trend Estimates (as a Percent of Estimate)

Indicator	Average Absolute Percent Change in Irregular	Seasonally Adjusted Expected Revision After:		Trend Expected Revision After:		
		1 Year	5 Years	1 Period	2 Periods	3 Periods
Residential Building Approvals	4.3%	2.2%	0.9%	2.2%	1.1%	0.4%
Retail Turnover	0.7%	0.3%	0.1%	0.3%	0.2%	0.1%
Basic Iron Production	4.2%	2.1%	0.8%	2.1%	1.1%	0.4%
Housing Finance	3.5%	1.7%	0.7%	1.7%	0.9%	0.4%
Gross Operating Surplus Trading Enterprises	2.5%	1.2%	0.5%	1.2%	0.6%	0.3%
Manufacturing Companies	2.2%	1.1%	0.5%	1.1%	0.6%	0.2%

Source: Australian Bureau of Statistics.

34.8 MANAGEMENT OF THE SURVEY PROCESS

Thus far, the stages of a statistical collection have been treated as separate and almost independent. The challenge for the survey designer is to ensure process quality for each stage of collection, but at the same time to see the collection as a whole and as part of the larger set of integrated statistics. Resources must be allocated to provide the highest overall quality to users, shifting them to the area where they are most effective. Similarly, feedback from each process must be used to improve other processes: for example from editing to questionnaire design, from analysis to estimation, and from discussion with users after data release to specification of future survey data items and survey design. The survey designer needs to understand potential sources of error or poor quality for statistical outputs (e.g., lack of timeliness) and attempt to conceptualize the resultant error as a function of resources applied across error-reduction tasks (Linacre and Trewin 1989).

Not only are there many error sources for each collection, but the error structure arising from these sources is modified by the data transformations that occur during survey processing and analysis. For example, the error structure of repeated surveys is modified in developing smoothed estimates of the series or in integrating the survey's outputs with outputs from other sources.

The following example illustrates this. Within the ABS, the main use of the Survey of Employment and Earnings is in the compilation of the national accounts. However, the survey outputs are not used directly but are combined with other survey data as part of a model to derive gross wages and salaries. By modeling the sampling variance of this estimate as a function of the means, variances, and correlations of the different survey components, it was discovered that halving the employment survey sample from 21,000 to 10,000 would increase the standard error of the national accounts estimate by only 14 percent (raising it from 0.43 percent to 0.50 percent of its level). Similarly, transformations of error structures in seasonal adjustment and trend estimation processes affect the optimal survey design if the design is specified in terms of these outputs.

To date, problems faced by the survey manager in modeling the total transformed error as a function of allocated resources are largely unresolved. Instead, the process for achieving survey quality relies on the manager deriving a complete picture of (1) error sources and their magnitude and (2) resource usage options and their relationship to error reduction, together with the effects of transformations inherent in the proposed use of survey outputs. Armed with this picture, the manager can rank alternative resource allocations according to their likely levels of effectiveness. For a repeated survey, a planned series of evaluations will build such a picture. This approach leads to continuous quality improvement within processes over time, along with significant advances as major imbalances in resource allocation and new opportunities to improve quality are identified.

34.9 CONCLUDING REMARKS

Achieving quality in business surveys requires understanding of users' needs for survey output and an understanding of the various survey processes' impact on quality. Gaining this understanding may mean discarding conventional wisdom, questioning the way things have always been done, and seeking information that describes process performance and identifies problem areas. Once the process is understood, efforts can be directed at continuous quality improvement, monitored by ongoing performance measures. The emphasis should be on designing survey processes that are ''right first time'' and that can evolve to ''stay right'' rather than relying on post hoc processes (e.g., large-scale editing and respondent recontacts) to ensure quality.

Achieving quality in a cost-effective way also requires an understanding of the contributions that survey processes have on overall quality, along with an appreciation of the interrelationships between survey processes. It will frequently require an understanding of the survey as but one part of an integrated family of statistics.

REFERENCES

Latouche, M., and J. M. Berthelot (1992), ''Use of a Score Function to Prioritize and Limit Recontacts in Editing Business Surveys,'' *Journal of Official Statistics*, **8**, pp. 389–400.

Linacre, S. (1987), ''Investigating the Accuracy of Preliminary Capital Expenditure Estimates,'' *Bulletin of the International Statistical Institute, 46th Session*, pp. 261–262.

Linacre, S. (1991), ''Approaches to Quality Assurance in ABS Business Surveys,'' *Bulletin of the International Statistical Institute, 48th Session*, pp. 487–511.

Linacre, S. J., and D. J. Trewin (1989), ''Evaluation of Errors and Appropriate Resource Allocation in Economic Collections,'' *Proceedings of the Annual Research Conference*, Washington, DC: U.S. Bureau of the Census.

McDavitt, C. (1992), ''A Study of Significance Editing on the Survey of Average Weekly Earnings,'' unpublished report, Canberra: Australian Bureau of Statistics.

Rosen, B., and B. Sundgren (1991), ''Documentation for Reuse of Microdata from the Surveys Carried Out by Statistics Sweden,'' Internal Report 1991-0628, Stockholm: Statistics Sweden.

Wright, T., and H. J. Tsao (1983), ''A Frame on Frames: An Annotated Bibliography,'' in T. Wright (ed.), *Statistical Methods and the Improvement of Data Quality*, New York: Academic Press.

Zarb, J. (1992), ''Smarter Data Use,'' *Australian Economic Indicators*, ABS Cat No. 1350.0, Belconnen: Australian Bureau of Statistics.

CHAPTER THIRTY-FIVE

Business Surveys in Ten Years' Time

Jacob Ryten
Statistics Canada

Past attempts to peer into the future of economic statistics have often resulted in incorrect forecasts, as do most long-term forecasts, because they had no grounding in explanatory economic mechanisms (e.g., Moser 1977, USSR Central Statistical Board 1977). Many predictions assume that past trends will continue in the future, in particular that:

- the System of National Accounts will continue to serve as the framework of governmental economic policy, resulting in ongoing interest in the economic aggregates embodied in the accounts; and
- computing costs will decrease while computer equipment and software capabilities increase, leading to expanding use of computer technology.

This discussion of future directions, while just as likely to prove wrong, builds upon an additional perspective: that the compilation of economic statistics will be heavily influenced by the relation between businesses' internal costs and external prices as well as by the nature of their contractual obligations.

Economic literature emphasizes the importance of prices and contracts on the boundaries of firms and their internal behavior. In 1937, Ronald Coase, subsequently a Nobel Prize winner, attempted to explain why the market mechanism stops at the firm's boundaries. Generally, there is little dispute about the overall value of the market in signaling surpluses and scarcities and in allocating resources. However, those features are not fully used within a firm. The firm must incur costs to collect and use price information and to establish contracts governing internal transactions. When these costs offset the advantages that accrue from the unfettered use of the market mechanism, up goes the

Business Survey Methods, Edited by Cox, Binder, Chinnappa, Christianson, Colledge, Kott. ISBN 0-471-59852-6 © 1995 John Wiley & Sons, Inc.

boundary of the firm. Operations within a firm are not so much governed by the coordinating forces of the market as by a coordinator who plans, organizes, and directs activities with the goal of maximizing the firm's profits.

Coase did not go on to estimate the costs of collecting price information and to study how they explain the size distribution of firms. I do not propose to do so either. Instead, I draw up a balance of sheet of factors that help or impede data collection from businesses and describe their implications for existing and prospective surveys. Factors that are likely to impact on the willingness of businesses to collect and use various types of information are also discussed. I also examine the circumstances affecting the continuing viability of the establishment as a unit for the collection of economic data; review old and new demands for information and the rising demand for longitudinal data; consider tradeoffs that future circumstances may engender between frequency, detail, administrative records, surveys, infrastructure, and current operations; and, finally, suggest an approach that government statistical offices can adopt to accommodate the changes that are likely to affect businesses over the next decade.

35.1 DATA COLLECTION: INCENTIVES AND DISINCENTIVES

There can be no progress in statistical compilation if businesses do not wish to provide the data or, even worse, do not compile the basic data for their own use. Some of the following trends may serve as incentives, whereas others may serve as disincentives to businesses to compile these records and to provide the data sought by statistical agencies. To meet the demands of the next decade, the statistical agency must understand the impact of each of these trends.

35.1.1 Computing Costs and Accessibility

Over the last decade, there has been a dramatic improvement in the prices and accessibility of computing which should continue well into the next decade. Three features dominate this effect, all of which contribute to easier and better data collection: (1) cheaper prices per unit of memory and faster retrieval times, (2) easier access for lay persons, and (3) greater integration between computing and telecommunications.

Because of easier access and reduced costs, computing will continue being diffused throughout the business world and eventually reach businesses of all sizes in all industries. The range of computing applications is also bound to increase beyond basic accounting functions such as payroll, inventory control, and tax accounting. New computing applications may produce information that is of interest to business management as well as to the statistical data collector.

Integration between computing and telecommunications should result in standardization of messages being communicated and the way they are accessed and filed. Projecting from current trends, it can be safely predicted that electronic data interchange (EDI) will become a major element in data collection in the last half of the 1990s, simplifying communications within the business, communications among businesses, and business communications with the government (see Chapter 19).

Accelerating these trends can benefit the data collector substantially. As an example, statistical agencies, tax authorities, and tax and cost accountants could come to an understanding that standardizes variables, concepts, and conventions and specifies the technology required to access basic records. If this agreement reduces the cost of the business' interactions with the government demonstrably, then the gains may offset most other impediments. In part, such developments are predicated on the inventiveness, proactivity, and flexibility of statistical agencies.

35.1.2 Increased Competition and Internationalization

The moves toward the promotion of freer trade, the formation of large trading blocs, and the geographic diversification of production create a more competitive world economic environment. Competition will become fiercer as more and more markets are opened up and as protectionist or regulatory barriers come down. The growing internationalization of large enterprises also brings with it added competition even when these enterprises form strategic alliances and coalitions among themselves.

Increased competition reduces the life expectancy of the average company. Businesses that can reallocate resources and adapt rapidly to market changes will have a competitive advantage. The flexibility to change and to optimize operations usually requires tighter management, which, in turn, presupposes superior information and the capability to use it intelligently. Such use demands information internal to the firm as well as for the industry as a whole. Hence, the information needed by the business is likely to overlap with the data sought by statistical agencies.

Internal information includes information on rates of return of alternative resource allocations open to the firm's management. Cost information must be tracked so that adjustments can be made in response to changes in relative prices, and processes can be switched to more efficient lines of business. Generally, better internal information also means greater integration of production and financing accounts.

The demand for better external information is just as important for data collection in the future. Survivability requires better information on performance in different markets. Performance means gains in market share. To the extent that official data provides businesses with timely and detailed information on markets and labor conditions, statistical agencies will have the necessary quid pro quo to justify their demand for prompt and accurate information from businesses.

35.1.3 Indirect Taxes

Another environmental factor likely to change collection conditions is the trend toward greater government dependence on indirect taxes such as Customs tariffs, excise duties, value-added taxes, and sales taxes. Although many industrialized countries gain greater revenues from income taxes, in most countries indirect taxation has grown enormously in absolute terms in spite of steady reductions in customs tariffs.

Whatever the yield, indirect taxation systems require records that reflect current activity. Customs tariffs—the oldest of indirect tax systems—support the most detailed records which, in turn, support the most detailed collection of economic data, namely, merchandise imports and exports. More recently, value-added taxes have given rise to administrative records that are proving to be a rich source from which to compile current, detailed, economic statistics.

If government's dependence on indirect taxes continues to increase, then business firms may systematically collect information on purchases and sales in even greater detail than needed to comply with government demands. The obligation to provide information to support the collection of indirect taxes is also likely to improve the conditions affecting data collection. It may offset the declining government demand for data related to direct grants and subsidies to industry, as well as the declining extent to which government regulations go hand in hand with requirements for detailed information.

35.1.4 The Role of Government

Increased competition and greater government reliance on indirect taxes are not the only changes in the business environment with an effect on data collection. There are others that are likely to make collection costlier, more cumbersome, and sometimes outright impossible.

Among these trends is the universal move toward rethinking the role of government vis-à-vis business, resulting in initiatives designed to present a new image of government to businesses:

- as a provider of services,
- as a facilitator,
- as a remover of impediments to the free functioning of markets, and
- as a guarantor of businesses' capability to operate in internal and external markets, untrammeled by regulatory obstacles.

Statisticians are particularly affected by one result of this trend—the move to reduce government-imposed paperwork. Ways to achieve this admittedly desirable objective include:

- rationalizing forms by making them compatible,
- standardizing forms where possible,

- simplifying the contents of forms, or
- doing away with forms outright.

It is the last of these, outright abolition of paperwork, which can hinder statistical data collection.

There is a motto in many customs offices throughout the world: When enforcement meets simplification, the former had better stand aside. Thus, when the requirement for more and better information collides with the need to show greater understanding toward the public and prevent excessive zeal in enforcement, the latter takes precedence. Trends in North America and other parts of the Anglo-Saxon world justify this notion, although it is less marked for continental members of the European Union. It is obvious, then, that data collectors must position themselves alongside taxation departments and influence the movement toward simplification so that it does not sweep away all accounting records—the useful with the superfluous.

Another step that usually accompanies a drive to simplify paperwork is more reliance on public honesty. The reason invoked is that compliance derives from intrinsic respect for the government rather than from the deterrent effect of the government's powers of compulsion and sanction. In neither case does filing of documents increase the rate of compliance. The reasonable measure to take is to warn the public that their records may be audited and that only then do they need to produce them as evidence. However, such measures often have a detrimental impact on data collection.

Governments usually surround simplification drives with bureaucratic controls of paper burden such as

- agencies that estimate and control burden,
- agencies that ration quotas for putative increases in burden,
- procedures to request authorization to increase or change burden, and
- targets for mandatory reductions in burden.

Although these controls tend to make collection more difficult in the short run, they may have the salutary effect of bringing needed revisions in what is asked from businesses and how these data are obtained.

35.1.5 The Transnational Enterprise

The worldwide trend toward internationalization of large enterprises may also impede data collection. International or transnational enterprises operate across national boundaries, with bonds between the transnational parent and the national subsidiary that allow the former to provide strategic services to the latter. Increasingly, transnational enterprises are centralizing the accounting of current operations (other than accounting for tax purposes).

Centralized records can cause serious problems for the data collector. As

Lewington notes in Chapter 33, data on domestic activities occurring within the country's geographic borders are required to estimate gross domestic product, a key macroeconomic measure. Domestic subsidiaries may not have access to the accounts kept for it by the transnational parent. Domestic enterprises may be unable to net out foreign subsidiaries from the enterprise's accounts. In addition, serious ambiguities may exist in the reporting of intercorporate transactions that could affect the quality and usefulness of the data.

35.1.6 The Net Balance of Incentives

Which of these trends dominate in the future is at the heart of predicting the future of business surveys. The issues boil down to the following questions:

- How will businesses' need to allocate resources efficiently and their desire to get better intelligence on markets affect the availability of accounting records for industry statistics?
- As governments are driven to demonstrate the quality of their stewardship to their electorates and the outsiders whose attention they seek, how strong and sustained will their requirement be for objective and reliable statistical information?
- What will be the response problems caused by the greater interdependence between local units and their transnational parents and the strong move toward simplification of government administrative requirements?

Only crystal ball gazing can answer these questions.

35.2 WILL THE ESTABLISHMENT CONTINUE AS A SURVEYABLE UNIT?

If the future of the establishment as a unit for which useful data can be compiled is dubious, there is little point in discussing impediments to data collection from establishments. In the following, the reasons that led to the decision to survey a unit—whatever its name—below the level of the enterprise and how the establishment fits in with these reasons are examined, and the evidence for and against its continuing viability are weighed.

35.2.1 Traditional Assumptions

In the next decade, governments will continue to organize their budgets, develop economic plans, and view the international economic environment in accordance with the structure of the system of national accounts. Therefore, the statistical infrastructure that makes possible the estimation of national production will continue to be required. As long as there are national accounts that include the accounts of the industries in which production originates, information will be needed for the unit to which the production account relates.

However, it is not obvious how such a production unit should be defined. France and countries influenced by French statisticians build up their production accounts from statistics about enterprises. Other countries such as Canada and the United States compile this information from sub-enterprise units. Neither option causes detectable weaknesses in the results, nor is there pressure to modify the target unit to improve the quality of the production aggregates.

Besides their use for national production accounts, statistical agencies survey establishments for at least two reasons: (1) to link production characteristics (what is produced, how it is produced, and how much is produced) with the workforce responsible for production and with the geographic location where production occurs and (2) to match the unit where production takes place with the finest possible industrial classification. The first reason is an expression of the need to know the economic utilization of the nation's physical space and in particular how that space is used by the workforce. The second reason relates to the twofold quest to provide (1) data for more narrowly defined markets which entails ever-finer descriptions of commodities produced and (2) data for more narrowly defined techniques of production (also known as production functions) adopted by producers to serve those markets.

Both reasons are deeply rooted in the Industrial Revolution. (They owe as much to Charles Dickens as to Alfred Marshall!) For production purposes, workers congregate in a physical space where there are equipment and other assets that are used to produce goods. These means of production are under a single stewardship (Coase's "coordinator" or Marshall's "entrepreneur") responsible for day-to-day decisions affecting those plant operations which do not require the replacement of capital assets or use of sources of financing (Coase 1937, Marshall 1961). Should the plant close, jobs will be lost from the local area. Should the plant expand, jobs will be created and new sites will be established to store, manage, produce, package, or distribute products. The notion of the use of space underlies instances where the establishment is defined as a location at which production—broadly defined to include economic activities such as shopping, storing goods, or banking—takes place.

35.2.2 The Breakdown of Traditional Assumptions

The Marshallian roots of the notion of production functions are clear. In the Victorian world, capital equipment was specialized and, once installed, dictated the nature of outputs, the location where production took place, the material inputs required, and the size of the workforce required to operate it efficiently.

The notion of the use of space breaks down when the place of production becomes irrelevant. Emerging industries for which computing, telecommunications, and fast transportation are essential props may not require concentration in space of buyers, sellers, and goods. For instance, a segment of the modern travel industry requires no more than an operator, a computer, and a telephone line to conduct business. Automated teller machines have eliminated the need to link deposits and withdrawals of money with manned bank counters. In these instances, the precise location of the operation is irrelevant. For activities where face-to-face contact is unnecessary, the notion of the establishment as a local unit where production takes place is only applicable by unduly stretching the original concept.

Providing details for narrowly defined markets and production techniques may no longer be as valid as it once was. Competition together with innovation creates new markets and destroys old ones. To survive, firms must be in a permanent state of readiness to adjust to changes in consumer preferences—changes that are often caused by innovations. Therefore, the firm's workforce and capital equipment must be versatile and able to turn out new commodities in response to changes in demand. When this is true, the *production function*—a well-defined, stable relationship between a precise list of inputs and an equally precise list of outputs—may be so transient as not to be observable or of little consequence in describing how industry operates.

Naturally, this point can only be made within limits. Versatility does not imply that steel mills can become providers of financial services overnight. But within a broadly defined sector of activities, firms need not stay attached for long to any specific activity or finely defined industry.

Accordingly, the notion of production functions may no longer be applicable. Of course, there are traditional manufacturing and mining industries to which the concept still applies, although even there modern industries are quite different. However, it cannot be assumed that a basis exists for applying fine distinctions derived from detailed industry classifications for today's (and even less for tomorrow's) distribution and finance industries, for providers of technical advice, and for the "dynamic duo" of computing and telecommunications.

35.2.3 Definitions of the Sub-Enterprise Unit

For the time being, most countries show no weakening in their approach to compiling industry statistics for a sub-enterprise unit. The definitions of such a unit have converged, and now only two major variants survive. One definition is firmly anchored in geography and is used successfully for traditional manufacturing industries, agriculture, mining, and energy. The definition centers on the location where production takes place and on the identification of a capital structure that houses the resources required for production.

The second definition which is less successful has to do with the locus of autonomous decisions. For example, the establishment is where decisions affecting the production account are made: whether the workforce should work two or three shifts; whether product A should be manufactured in the summer and product B during the winter; whether grade A or grade B inputs should be used for current production; and so on. By implication, there are decisions that are beyond the scope of the establishment's coordinator. Examples are whether a new activity should be financed through the regular budget or a loan, whether liquid assets should be held in gilt-edged securities or in high-yield earners, and whether the company's housing estate should be leased to a third party.

The first definition points in the direction of the space where resources are congregated; the second relates variables to the unit whose management governs their behavior. There are other definitions, but they are not intellectually satisfying. The Canadian variant brings in the capacity to report and the notion of a unit limited by provincial boundaries that can provide the components needed to calculate value added. Such a definition is no more than a useful convention for data collection and hardly qualifies as the theoretical underpinning of a target unit for industry statistics.

As the production of goods becomes less important to the gross domestic product of developed countries, the first of the two standard definitions—based on geography—may increasingly apply to fewer cases. As for the second—scope of autonomous decisions—it is difficult to apply in practice. Depending on the type of management imposed by the head office, it can lead to a heterogeneous assortment of units differing in size, scope, and industrial purity. Moreover, as management techniques change even more rapidly than in the past, authority and decision-making may bounce from head office to local plants and back at great speed.

35.2.4 The Need to Keep Detailed Business Records

Against this background, two factors provide incentives to collect data for parts of an enterprise. The first factor is the expected availability of records. The second relates to the nature of increased competition. In searching for alternative ways of employing capital, investors—particularly international ones—are likely to require more refined measures of profitability. As complex operations are restructured with greater frequency, decisions about whether particular operations should be closed down will have profound implications on local workforces. In turn, contracts with those workforces will make it necessary to demand better information on the profitability of local activities.

To these factors must be added the argument of relevance to respondents. Businesses will need incentives to ensure compliance. For those respondents who approach data from a small-area perspective, the only relevant data are those that refer to locally describable operations. For example, environmental perspectives relate to locally defined operations and are likely to reinforce the pressure for compilation of information on local units.

But there are impediments. For data on the establishments to be useful, they must account for the outgoings required to finance the gross value of production. When the latter are the direct expenses required by production plus depreciation of capital assets, little ambiguity exists in estimating value added. As capital-labor ratios increase and centrally provided services assume an increasing share of production costs, the ability to relate costs to production decreases. This ability need not vanish entirely, but to be retained the analyst or reporting business must allocate centrally incurred costs to local production processes. Such allocations seldom make good economic sense. Rather, the purpose they serve is to minimize the effect of taxes, immunize business from fluctuations in exchange rates, and accommodate terms stipulated in labor contracts.

Versatility of production implies that productive units potentially span a broader range of industrial activities than is apparent from a mere inspection of their inputs and outputs at a particular time. For manufacturing, agriculture, mining, and energy, this is not strictly true because fixed assets and the workforce's abilities keep the range of possibilities within narrow confines. It is, however, the case for service industries, where conversions from one activity to another can and do occur fairly frequently. For example, companies operating within (1) insurance and banking, (2) architecture and real estate development, (3) accounting and management consulting, (4) airlines operation and reservation services, and (5) computing and telecommunications can move from one end of their industry to the other with comparative ease. Neither their human nor their fixed capital are so specialized as to impede rapid adjustments. In all these cases, fixed capital can be leased, moved, or discarded with comparative ease.

35.2.5 Changing Business Boundaries

The notions spelled out above imply a redrawing of the boundaries of the firm. This is a continuation of a trend that began in the past decade. Large firms began shedding internal service functions such as security, maintenance, computing, and legal and accounting advice and instead contracted with specialized providers for the services. To use Coase's framework, the boundaries of the firm were redrawn as it replaced internal coordination with specific contracts subject to the market mechanism. Accordingly, the average number of establishments and supporting units per enterprise probably decreased while the number of enterprises increased. If this conjecture is correct, what truly increased was the number of activities subject to the discipline of the market.

Where such shedding has not taken place, the meaningfulness of available data may be limited by the growing number of cases where the provision of services (e.g., design, access to a central base of information, communications, advertising) is centralized and significant. The local unit will be of interest only in those cases where geography plays a role. Finally, where versatility and adaptability are key to success, describing commodities and workforce skills will be more important than the firm's current activity.

Thus, while there is still a broad range of activities for which the geographic distribution of employment is of consequence, the number of industries for which the location of the workforce is of little consequence is growing.

35.2.6 Net Balance

These conditions suggest that over the next decade, there will be less cause to identify units with the characteristics of today's establishment and to apply the results to all industries. Rather, there will be a miscellany of purposes, each of which will have different requirements. Thus, for data on employment and occupational distributions and for environmental and other small-area purposes, the standard will be a local unit whether that local unit is engaged in a pure activity or in mixed operations. For more traditional industry statistics, location will continue to be an important variable for many major activities, whereas for other activities location will be little more than a postal address used for contact. For the assignment of industrial classification codes, in some cases the unit to be classified must be finely defined and meet tests such as stability, specialization of production, and so forth. Those are the tests required to satisfy the need to estimate industry production functions. In other cases, the industrial classification will be interesting only at a broad level of aggregation. Much more important will be the enumeration of the commodities produced and the skills of the workforce employed to produce them.

35.3 NEW DEMANDS FOR DATA

Both traditional demands for data (that show no signs of abating) and new demands require (1) statistics that result from accounting for expenses incurred during production and (2) all receipts accruing to the company as it disposes of its production. The completeness of these two accounts guarantees the presence of elements necessary to compute value added. Completeness also requires the availability of information on services purchased—internally or externally—which, in turn, presupposes major changes in internal accounting.

There are cases where the firm is not prepared to shed the provision of internal services to outside bodies because it fears that it can only guarantee continuity and quality of supply from inside. In such cases, if the firm wishes to prevent the growth of internal services from swamping its operations and the coordinator's contractual powers are not sufficient to keep internal expenses under control, the firm may choose to simulate an internal market complete with shadow prices. If that is the outcome, the data collector's objectives are not threatened.

35.3.1 Specific Requirements

The following issues are apt to color demand over the balance of the decade.

Comparability and Confidentiality
International comparability of industry statistics can no longer be left to superficial measures such as approximate concordances. Such comparability will

soon be the object of negotiation—within trade blocs or within the General Agreement of Tariffs and Trade (GATT)—and will therefore need official sanction. As the required resource investment in securing international comparisons gets underway, new challenges are bound to turn up. For example, stricter comparability may not be achievable unless confidential data can be shared among interested parties, not all from the same country. No legislation governing statistical institutions defines the role of honest broker where such problems arise.

Rates of Return on Invested Capital
Strict international comparability will also be a basic condition for inter-country comparisons of rates of return on invested capital. This assumes that rates of return can be estimated, which should be possible at the enterprise level but may not be for lesser units. In a Marshallian world, rates of return could be made to apply to establishments. In today's world, too much capital may be invested in the undifferentiated provision of services to the operating parts of the enterprise. Therefore, no computable rate of return will satisfy strict comparison tests. Industrially well-classified rates derived from establishments will fail the criterion of completeness; well-estimated rates of return will apply to cases too heterogeneous to be fully comparable.

Financing
In spite of increased difficulties in accounting for the expenses of the establishment, surging demand can be expected for information on how establishment operations are financed. Pressure will come from circumstances related to foreign operations of national subsidiaries (establishment trade) and wider concerns about rates of return and the complexity of intercorporate and intracorporate arrangements.

Whatever the reasons, collecting financing data for establishments is not simple. Establishment surveys deal with financing episodically and even then only in part. Enterprise surveys tend to deal with financing more systematically. Even enterprise surveys which request balance sheet information are insufficient at tracing financing flows and at distinguishing flows that are intra-enterprise from those that come from outside sources. Whereas much emphasis has been placed on relating head office expenses to establishment activities, future concerns might switch the emphasis to relating enterprise head offices to the enterprise group and might also switch the emphasis from direct financing of production to the maze of financing flows that move resources from one enterprise group to another.

Environmental Spinoffs
Additional pressures for data related to the location of production will come from environmental watchers and policy analysts. Enterprise-level statistics are of little interest for the environment, unlike location-specific information about smoke stacks, smelters, nuclear power reactors, and oil refineries. The

question posed will be whether statistics on resource management and environmental contamination can be made to relate to only those branches of industry for which location statistics are meaningful notions.

Occupational Skills and Training
In the absence of internal accounting systems that include shadow prices, the value of internal services can be measured through the occupational distribution of the labor force at head offices. The systematic collection of occupational distributions should complement information on centrally located capital equipment that provides services to all internal units.

Establishment Trade
Foreign operations of national subsidiaries, also referred to as *establishment trade*, can be expected to become a key item of contention at the next GATT round of international trade negotiations. The activities of foreign subsidiaries can no longer be described in terms of purely financial operations—that is, in terms of the foreign direct investment required to own them and the foreign assets owned or controlled as a result of accumulated investment and its returns.

Complementary information on current operations of its foreign subsidiaries is likely to become important in reviewing the vulnerabilities of a country's balance of payments. For example, the operations of manufacturing subsidiaries may change the balance between imports and domestic production, create new demands for the export of strategic services, compete with true domestic exports in third country markets, and so on.

Getting such information requires that a country's trading partners cooperate in assembling information on these operations. Such cooperation may lead to a series of important steps in the process of harmonizing industry statistics: the variables collected, classifications and methods of assignment used, and the units targeted.

35.4 LONGITUDINALITY: NEW DEMAND

Longitudinality, or the ability to track the same entity over time (be that entity a business in the legal sense of the word or an establishment in the sense of a physical unit where production takes place), which has been a concern, is now much closer to becoming a requirement. The ability to track people with their endowment of human capital, jobs, and businesses, however, is still in its formative stage. Statistical activities carried out experimentally in many countries include employment dynamics, business demography, life cycles of new entrants into an industry, transitional probabilities for newly created businesses to reach middle age, and so on. The experimental nature of the activities results from lack of agreement on conventions and interpretations of basic results.

35.4.1 Favorable Conditions

Conditions favorable to the development of longitudinal files of businesses or of their employees include (1) adequate computing capacity, (2) the availability of business registers with timestamps of the important events in the life of a business, and (3) the administrative drive towards standardization of records with unique identifiers. The important timestamps in the life of a business include its date of birth, death, changes in activity, and structural transformations that change its ownership such as takeovers, mergers, and so on. These events only have operational definitions that need a conceptual foundation to ensure that their application does not run counter to economic theory.

Unique identifiers may introduce new perspectives on what data can be gathered and on future demands for such data. For example, unique identifiers required to track commodity shipments from the factory gate to the distributor may provide the missing tie-in between production, warehousing, transportation, and distribution. Unique identifiers are already providing a way to relate business structural characteristics (e.g., size, employment, activity) to current sales and purchases.

In general, record linkage techniques have made it possible to put together the information contained in separate administrative files (see Chapter 20). At Statistics Canada, for instance, employee records can be linked to the companies in which they worked so that an employed individual's odyssey can be traced through different industries, locations, sizes, and so forth.

Though interpretation of these statistics is controversial, their publication generates enormous interest. Many reasons explain the success of these statistics. For example, no alternative sources exist for banking institutions to estimate the survival probabilities of newly created businesses or to gauge those probabilities in the life of a business year by year. There is no reasonable alternative source of information to assess the impact on the labor force of the adjustments to which manufacturing has been subject over the past decade.

35.4.2 Deterrents to Longitudinal Data Compilation

There are deterrents to the development of these longitudinal databases. The ability to link administrative files is widely perceived as a dangerous tool in the hands of less-than-responsible governments—one that could lead to political or personal abuse. Without accessible administrative files or in the presence of public opposition to linking them, response burden from surveys designed to achieve similar results are likely to be resisted especially if the survey does not produce relevant results for respondents.

Conventions necessary to compile longitudinal profiles have yet to be established. Examples include the convention on how to date a business birth—even the notion of birth is not clear. Unlike a physical establishment, an enterprise can be born as a result of the transformation of other businesses or out of nothing. Conventions are needed to define how to date business deaths and how to establish that a business has remained the same entity over time rather than being the juxtaposition of one death and one birth. Such conventions must be decided before there is progress on longitudinality (see Chapter 4).

For the rest of the decade, progress can be expected on business dynamics or business demography. By the end of the decade, there may be a body of definitions sufficiently well articulated to create a conceptual framework for data compilation and analysis. In the face of the primitive condition of the underlying concepts, it is unlikely that these files will be developed before the 21st century.

35.5 OTHER MATTERS: SELECTED TRADE-OFFS

Shifting priorities will force considerations of tradeoffs. In virtually every industrialized country, the system of industry statistics consists of a series of annual surveys or quasi-censuses in which questions about outputs and inputs are tied to questions about employment and fixed assets. The situation for short-term indicators is less standard. As the expense of conducting censuses increases and tends to offset the value of the information gathered and the timeliness of its availability to the public, the net value of such censuses is apt to be questioned.

Few countries have adopted the approach of investing sufficiently in their survey infrastructure to support quick turnaround business surveys to answer specific questions. In Canada, this facility is still in its infancy, but there are two or three notable examples of such a capacity. The trick consists in relying on a business register (generally a complete list of businesses classified by main economic activity) to mount a regular survey and then to use that survey as a derived frame for an ad hoc supplementary survey (see Chapter 5). This approach was used in Canada to study the contribution of advertising to survivability, the speed with which high technology is disseminated throughout industries, and the shifts between wholesaling and manufacturing as the latter tends to be contracted out.

The tradeoffs between administrative and survey data are not resolved. It is not possible to assess the extent to which statistical offices will be able, in the remaining years of this decade, to influence other government agencies in terms of the administrative records the latter generate, their contents, and conditions of access. There are success stories in Australia. In France, the proximity of the statistical office to the government's taxation unit creates a unique situation in which the chief statistician has leverage on his colleague in charge of taxation. In Canada, disappointments have been tempered by successes. In time there should be more standardization and greater awareness on the part of government departments of the virtues of collecting administrative information that can be fitted into a statistical framework.

35.6 CONCLUDING REMARKS

On balance the following predictions can be made about the demand for and supply of economic statistics likely to prevail at the end of the decade:

- The future collection of industry statistics will be mostly through electronic data interchange (EDI) using existing records—part of a business' information system.
- The establishment as a single integrating unit of economic statistics will fade away to be replaced by more flexible units: the enterprise for financial and performance information and the local unit for employment.
- Industry classifications will preserve their detail for certain major activities but become global for service industries where interest will focus on occupational distributions of the workforce and the range of commodities produced.
- The ability to link units longitudinally will continue to be developed albeit subject to political control. This must be helped by conceptual developments in describing the life cycle of businesses.
- The future will see more ad hoc surveys, but their success will require substantial investments in infrastructure.
- Statistical offices will find that they must work out a close cooperative relationship with their counterparts who compile administrative files.
- The days of strictly national development of business surveys may be over. Instead, we may be entering an era of joint ventures involving groups of countries organized within supranational entities or motivated by cultural similarities and trading relations. This could result in exciting changes in law, custom, methods, techniques, and the scope of our activities.

To respond to these challenges, national statistical agencies must adjust today's structure of surveys, censuses, and administrative registers as well as today's capabilities. The difficulty is that to produce the right information at the right time, these adjustments must take place even before these trends become pronounced. This will need a willingness to court risk.

If these predictions come about, to stay relevant statistical agencies will have to alter the range of outputs with no expectation of infusion of resources to make the transition. A thorough review of resource allocation within agencies may reveal that they are not well balanced, with some areas surveyed with unnecessary frequency or excessive detail. Too many resources may be locked into regular surveys or censuses with large invested capital that makes even marginal change difficult to envisage. But the price of relevance is rapid adaptation, which can only prevail if we start disengaging now from the costlier structural surveys and formulate principles to guide the reallocation of resources from areas that consume relatively too many resources to areas where measurement needs are emerging.

Within the next decade, we should expect new questions that owe their roots to technology, internationalization of business, liberalization of markets, the effects of these trends on employment, and so on. Many of the new questions are unique, for example:

- Is there evidence that transfer pricing between members of a transnational enterprise is systematically out of line with market prices?
- What is the value of the services that subsidiaries receive from the head office?
- At what rate is manufacturing subcontracting its more basic activities abroad?
- To what extent is national production out of line with demand in international markets?
- What is the role that domestic prices play in market gains or losses abroad?
- At what rate is the environment becoming contaminated through traditional economic activities?
- What effect will continuing production at the present rate have on natural resources?

Many questions cross industry boundaries and are addressed to more than one level within the enterprise. It is doubtful that such questions can be answered by traditional surveys, which were forged to provide regular and consistent information on standard variables. For this reason, new capabilities must be created.

It is no longer safe to rely on economy wide censuses, except for development and quality improvement of sampling frames. There the level of questioning should be comparatively light. It is advisable to concentrate resources on the maintenance of a business register as the sampling frame. The statistical agency should regard the register not only as a list of industrially classified business names, addresses, and size measures but also as a repository of everything known about the structural features of each listed business (i.e., its profile). A centrally maintained register is the key device to support small and topical surveys, the workbench to relate different items of information to each other, and the institutional memory of what is known about the business world and its behavior and transformation over time.

More than 60 years ago, Maynard Keynes testified before the Macmillan Committee on the causes of unemployment in Great Britain.[1] Keynes argued

---

[1]In the same session, Keynes enumerated as basic requirements for a future system of economic statistics, virtually all the short-term indicators that countries in the Western world use today. Minutes of the Macmillan Committee, December 5, 1930.

that in conditions of rapid change in the economic environment, it is counter-productive to invest heavily in censuses of production and wiser to spend time and talent developing new types of current statistics designed to track rapid change. The wisdom of the comment has been forgotten. Ahead of us there still lies a choice between small, current inquiries and detailed, slow, large-scale surveys or censuses; the speed of change would appear to favor the former.

Investing heavily in developing a classification of the commodities produced by service industries is advisable and more worthwhile than investing excessively in their detailed industrial classification. Statistical agencies must vouch for the accuracy of their estimates of real product, which cannot be done until the measurement apparatus for the new service industries is firmly in place.

Many goods and services originate in head offices and are then transferred within complex enterprises at nominal prices. The cost of those services cannot be estimated until more is known about what head offices do and who they employ by occupation. It is worth investing slowly but incrementally in surveys to describe the activity and employment of head offices. This will require a much closer relationship between statistical compilers and business executives. Statistical agencies ought to remind themselves periodically that measurement with no theory is pernicious and that measurement without direct contact with what is measured is error prone.

A clear basis has to be established for questioning today's detail and frequency of measurement. It is not wise to attribute it entirely to the needs of national accounting because these needs do not constitute an obvious boundary. Until we define what is the best balance of error, timeliness, and detail in statistics for macroeconomic analysis, little can be done to define limits for the gathering of microeconomic statistics.

New forms of data gathering are bound to be created as the result of

- the drive toward a better understanding of business records,
- the need to increase contact with business over the nature of information considered strategic, the wording, and the conceptual framework used by business for decision-making,
- the differences between that framework and the microeconomic scaffolding of the national accounts, and
- the pressure to adopt modern methods of data transmission.

We might be surprised to discover how comparatively little information business has to make key decisions and how much that information varies from industry to industry and case to case. We may be appalled to discover that the business can only provide the seemingly routine information we request by making extreme assumptions. This should cause us to revise many unchallenged assumptions and to view data previously assumed to be hard as based on rough estimates. Drawing closer to businesses and delving into what underlies business decisions can only result in more interesting, meaningful, and reliable data for compilations.

REFERENCES

Coase, R. (1937), "The Nature of the Firm," *Economica*, new series, **IV,** No. 13–16, pp. 386–405.

Marshall, A. (1961), *Principles of Economics*, 9th [variorum] ed., London: Royal Economic Society, cf. Book IV, Chapters XI & XII.

Moser, C. (1977), "Environment in Which National Statistical Services Will Work in Ten Years' Time," Paper CES/SEM. 8/3, presented at the Seminar on Statistical Services in Ten Years' Time, Conference of European Statisticians, Washington, DC.

USSR Central Statistical Board (1977), "Environment in Which National Statistical Services Will Work in Ten Years' Time," Paper CES/SEM. 8/2, presented at the Seminar on Statistical Services in Ten Years' Time, Conference of European Statisticians, Washington, DC.

Index

Accounting records, 25, 29, 50–52, 55,
     240–243, 260–261, 269–270
Accredited Standards Committee, 345
Accuracy, 38, 99, 118, 227–229, 246, 250–252,
     255, 267, 372–376
   of coding, 38, 99, 118
   evaluation of, 250, 267
   issues, 246
   quality of lists, 372–376
Activity, 33–34, 49–63, 67, 74
   ancillary, 50, 53–54, 67
   business, 49–50
   changes in, 67
   defined, 50
   economic, 50–54, 56–62
   industrial, 52
   principal, 52–54, 57, 59, 62–63
   secondary, 52–54, 57, 62–63, 74
Actual quality, 253–255
Address parsing software, 379
Adjusted jackknife variance estimator, 416
Administrative record, 13, 36–38, 41, 43–44,
     80, 91–96, 101, 240–243, 408–411,
     413–415, 612
   data sources, 36–38, 41, 43–44, 80, 91–97
   international use of, 240–243
   lists, 36
   quality control of, 99
   secondary use of, 242
   use for frames, 96, 101
   use in estimation, 13
   use in imputation, 408–411, 413–415
Administrative units, 30–31
AERO, 412
Aerospace remote sensing, 628
Agency for Health Care Policy and
     Research, U.S., 219, 224, 225, 228, 231
Aggregate change, 65–67
Aggregate finite population index, 544–545
Aggregate method, 393
Aggregation weight, 544–547, 564
   contribution to index variance, 554–556
   defined, 544

   replicate estimates of, 564
Agribusiness, 612
Agricultural Adjustment Act, 624
Agricultural Census of Ontario, Quebec and
     the Maritimes, 638
Agricultural Classification Error Survey,
     U.S., 311
Agricultural and Grazing Industry Survey,
     Australia, 592
Agricultural survey, 248, 396, 592, 609–632
Agriculture Census, U.S., 380, 609–611, 615, 620,
     623, 625, 627
Agriculture survey units, 94
Agriculture Tax Data Program, Canada, 435
Agriculture Whole Farm Database Project, 435
All commodity volume, 211–214
Allocation of samples, 135–139
   defined, 135
   model-assisted, 136–137
   multipurpose, 139
   multivariate, 138–139
   Neyman, 135–136, 592
   optimal, 135
   proportional, 135
   proximal, 136–138
   x-optimal, 137
American Agriculturist, 616
American National Standards Institute, 345, 349
Analytical unit, 57
Ancillary activities, 50, 53–54, 67
Ancillary unit, 53–55, 61
   defined, 53–54
   and geographical classification, 54–55
   and industrial classification, 54, 61
Annual Motor Carrier Freight Survey,
     Canada, 435
Annual Refiling Survey, 115
Annual Retail Survey, Canada, 397
Annual Retail Trade Survey, U.S., 311
Annual Survey of Financial Accounts,
     Sweden, 389
Annual Survey of General Purpose
     Government Finances, U.S., 343

711

British Empire Statistical Conference, 642
Bryce, R. B., 644
Burden indicator, 88
Bureau of the Census, U.S., 38, 141, 143–144, 149, 190, 200, 238, 268, 271, 294, 304, 306, 310–315, 339–352, 380, 413, 434, 436, 609–611, 616, 620, 621, 624–627, 629–630
Bureau of Labor Statistics, U.S., 38, 99, 115–116, 119–123, 125–127, 284–297, 317–336, 394, 426, 429, 436, 438–439, 545
Bureau of Statistics, U.S., 622
Business:
  activity, 49–51
  approaching in surveys, 243–244
  characteristics, 56, 248
  defined, 2–3, 5–6, 23, 49–50
  historic dimension, 86
  inter-business relationships, 95
  large, 21, 29–30, 36, 49–53, 55, 59–61, 64, 89, 101–114
  legal structure, 29, 50–51, 55, 102
  multiple establishment units, 95, 133
  operational structure, 29, 50–51, 55, 102
  organizational structure, 25, 27, 29, 30–32, 40, 50–52
  profiling, 30, 39, 55, 95, 101–114, 677, 679
  skewed distribution, 4, 21, 29–30, 133, 387, 677
  small, 6, 21, 29–30, 36, 49–51, 61, 64
  statistical structure, 28, 32, 39, 55
  transactions, 250, 339
  universe of, 49, 51
  volatile nature, 4, 133
Business Conditions Survey, Canada, 570
Business demography, 41–42, 77–80, 88–115, 127, 705
Business entity, 3
Business frame, 4–7, 21–44, 49–50, 677–682
Business transactions, 250, 339
Business register, 4–7, 21–44, 51, 54–55, 61–62, 85–86, 96–97, 101–115, 119, 125–127, 651, 664–666, 677, 704–705, 707
  commercial uses, 41
  creation, 35–40, 358, 380
  data sources, 35–40, 96–98
  defined, 22, 36, 665
  maintenance of, 35–40, 54–55, 61, 85–115, 127
  proving surveys, 39, 94, 97, 115, 119, 122
  unique features, 4–7
  uses, 36, 40–42, 86–89, 115, 125–126
Business register direct survey, see Proving survey
Business Survey Redesign Project, Canada, 650
Business surveys:
  and national accounts, 657–658
  contrasted to social surveys, 1–2, 25–26, 237–238, 260, 305

defined, 2–3, 23
international survey of, 237–256
unique features of, 1–16, 25, 133

Calibration, 486–489, 603
Callback, 325
CALMAR program, 488
CAMLIS system, 380
CANEDIT system, 412
Canada:
  Confederation of, 634–636
  Dominion of, 634, 636–638, 640
  Province of, 634
Capacity constraint, 457
CAPI, see Computer assisted personal interviewing
Capital expenditure survey, 92
Capital formation, 663
CATI, see Computer-assisted telephone interviewing
Causal modeling, 269
Cell suppression, 448, 450–453, 462, 629
  complementary, 452
  defined, 450
  oversuppression, 462
  primary, 450
  theory, 629
Censored estimator, 511
Census, 144
  of population, 60, 63
Census of Agriculture, Canada, 434, 435–436, 610–611, 625–626
Census of Agriculture, U.S., 311, 358–359, 375, 376, 380, 623–624, 646
  frame, 358–359
  matching, 358–359, 380
Census of Construction, Canada, 413, 435, 649
Census of Construction Industries, U.S., 446, 448
Census of Dairy Production, Canada, 638
Census of Drained Land, U.S., 620–621
Census of Distribution, U.S., 643
Censuses, Canada, 635–638, 643
Census of Horticultural Specialties, U.S., 311
Census of Industry, Canada, 642–643
Census of Irrigation, 620–621
Census of Manufactures, Canada, 638, 641, 643, 648
Census of Manufactures, U.S., 144, 262, 389, 436, 648
Census of Merchandising, Canada, 648
Census of Population, U.S., 614, 618, 625
Census of Retail Trade, U.S., 262, 344, 348, 448
Census and Statistics Act, Canada, 641
Census and Statistics Office, Canada, 638, 640–641
Census of Wholesale Trade, U.S., 448
Central Statistical Office, United Kingdom, 38, 93, 99

WILEY SERIES IN PROBABILITY
AND MATHEMATICAL STATISTICS

ESTABLISHED BY WALTER A. SHEWHART AND SAMUEL S. WILKS
Editors
Vic Barnett, Ralph A. Bradley, Nicholas I. Fisher, J. Stuart Hunter, J. B. Kadane, David G. Kendall, David W. Scott, Adrian F. M. Smith, Jozef L. Teugels, Geoffrey S. Watson

*Now available in a lower priced paperback edition in the Wiley Classics Library.

*Now available in a lower priced paperback edition in the Wiley Classics Library.

*Now available in a lower priced paperback edition in the Wiley Classics Library.